THE ART OF PROBLEM SOLVING IN ORGANIC CHEMISTRY

Third Edition

MIGUEL E. ALONSO-AMELOT
University of the Andes
Department of Chemistry
Merida, Venezuela

WILEY

This edition first published 2023

Edition History
First edition © 1987 by John Wiley & Sons, Inc.
Second edition © 2014 by John Wiley & Sons, Inc.

Registered Office
John Wiley & Sons, Inc., 111 River Street, Hoboken, NJ 07030, USA

For details of our global editorial offices, customer services, and more information about Wiley products visit us at www.wiley.com.

Wiley also publishes its books in a variety of electronic formats and by print-on-demand. Some content that appears in standard print versions of this book may not be available in other formats.

A catalogue record for this book is available from the Library of Congress

Paperback ISBN: 9781119900665; ePub ISBN: 9781119900689; ePDF ISBN: 9781119900672

Cover Image: © Garsya/Shutterstock
Cover Design: Wiley

Set in 10/12pt Times LT Std by Integra Software Services Pvt. Ltd., Pondicherry, India

C9781119900665_071223

Printed and bound by CPI Group (UK) Ltd, Croydon, CR0 4YY

To Adela Tarnawiecki, best wife and friend
Our son Gabriel, a tireless problem solver in troubled times
To the memory of Professor Ernest Wenkert, friend and mentor

CONTENTS

PREFACE

A chemical reaction necessarily implies the transformation of a body of substance into something different. From the household experience of boiling an egg, the natural spoilage of cabbage, the conversion of solar energy and carbon dioxide into biomass and chemical energy and tens of thousands of organic phytochemicals, to the manufacture of sophisticated organic materials, the *mechanism* by which the transformation of matter takes place remained an open question for most of humankind's history. People, even knowledgeable scientists of the Renaissance and the late eighteenth century, were not aware of the permanent change of complex organic substances around them by natural and human-made forces. When science finally discovered this everlasting change, the next questions were how it happened and whether there were laws governing change or they were expressions of the whims of chance.

For centuries, these fundamental questions have driven people to develop endless hypotheses, theoretical proposals, and schemes based on religious beliefs when no rational explanation was available. Others walked the rough trails of rationalizations supported by solid theoretical considerations and experimental demonstration. In the end, elucidating the mechanisms of change has been one of the most taxing challenges to understand the forces of nature.

In all its complexity, the enormous gamut of organic reactions and their mechanisms of transformation constitutes a fundamental part of this change as well as being *a fantastic training ground for the mind of the chemist, a route to the inner workings of molecules, and a most rewarding and amusing learning experience*.

This book intends to take any reader in possession of intermediate to advanced organic chemistry knowledge to this realm of molecular change through a series of time-proven techniques for solving problems posed by reaction mechanisms. The strategies are collected herein as an expansion of the previous two successful editions of *The Art of Problem Solving in Organic Chemistry* published by Wiley in 1986 and 2014. Multiple new sections, one additional chapter compiled in the "Toolbox", Part I, and an updated set of 50 fully discussed reaction-mechanism problems extracted from the current research literature gathered in "The Problem Chest", Part II have been compiled this new edition.

Earth is the only place in the known universe where there is evidence of complex organic substances – but see Reference [1]: natural alkaloids, polyphenolics, porphyrins and isoprenoids, RNA, DNA, functional proteins, and a wide assortment of complex molecular aggregates of organic matter constitute our living materials. Many of these substances are susceptible to light; oxidizing, reducing, and hydrolytic agents in the environment; acids and bases; as well as catalysts in mineral surfaces and soluble media. Digestive juices from all kinds of animal consumers, microbial and cytoplasmic enzymes, and swarms of natural reactive species incessantly transform organic compounds into different substances. All known living forms live within this organic space to occupy an exceedingly thin crust and seas of our planet.

Furthermore, the range of temperatures that organic materials resist is very narrow at a planetary scale: few non-catalytic reactions occur below −100 °C, most organic compounds undergo profound structural change at 450–600 °C and char at higher temperatures [2].

Because this is a book about putting the reader's mind into a potent thinking mood from the start, let me bring an example in the form of an accessible mechanism problem: what organic molecules do at very high temperature without the aid of catalysis, solvents, or reagents, just heat, this time in a controlled manner (Scheme P.1) [3]. Be my guest to try a mechanism.

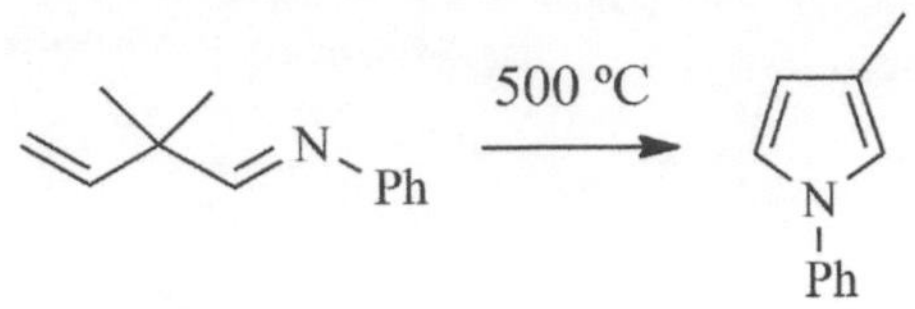

SCHEME P.1 Example of controlled pyrolysis of a simple organic compound. The heating method, known as vacuum flash pyrolysis, allowed exposure of the imine to 500 °C for only 20 seconds. [3] / Royal Society of Chemistry.

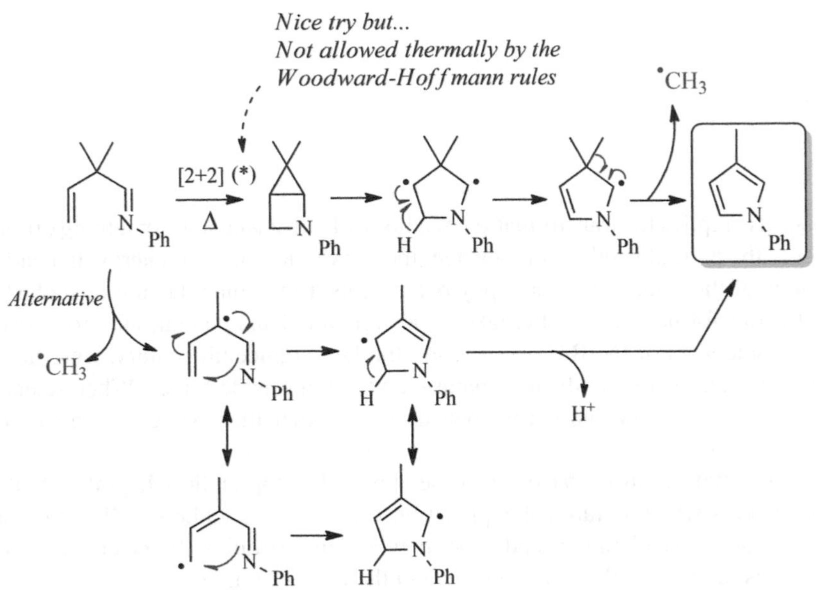

SCHEME P.2 Solutions to Scheme P1. The radical pathway is preferred under the indicated conditions.

(Hint: counting carbon atoms in starting material and product brings about a surprise.) A couple of feasible (and unfeasible) routes are shown at the end of this preface (Scheme P.2).

Interatomic bond stretching and energy transfer from heat (thermolysis, pyrolysis) and electromagnetic radiation (photons) overcome bonding forces and lead to rearrangement or fragmentation to smaller molecules.

Time, pressure, and heat also convert large masses of natural organics from past forests and animal residues into mineral tars, bituminous charcoal, and petroleum containing thousands of organic compounds, and fossilized organic matter resembling the carbon skeleton of the original molecules. As a matter of fact, it was from the distillation and patient separation of coal tars that the chemistry of aromatic compounds came of age in the middle of the nineteenth century, launching organic chemistry to the sophisticated branch of science it is today.

And yet, parts-per-billion quantities of alkanes, alkenes, aromatics, alcohols, aldehydes, ketones, carboxylic acids, halogenated hydrocarbons, and others have been detected in fumaroles that emanate in the course of volcanic eruptions [4–6]. The origin of these compounds has been traced to two possible contributors: i) heating of soil or sedimentary layers containing ancient organisms [4] and ii) transport by atmospheric air entering the porous volcanic ash and rocks to incorporate as fumarole steam discharges [6]. No one has been able to show that hot magma contains any organic substances at all due to the high temperature and it continues to be a topic of debate [6].

Conversely, it is firmly established is that *problem solving is one of the best means currently available to educate future specialists in the sciences and maintain a competitive professional profile*. Advanced undergraduates, graduate students, lecturers, and professionals in the areas of organic chemistry are bound to improve their performance substantially through problem solving. This book pursues to improve these skills by describing in practical terms the principles, methods, thinking tools, and problem analysis techniques, as a hopeful improvement on earlier editions.

Miguel E. Alonso-Amelot. PhD
2023

REFERENCES AND NOTES

1. Recovered meteorites falling to Earth have been shown to contain "substantial quantities" of organic matter, primarily carbon-rich carbonaceous chondrites from the primitive solar system nebula. The amino acid glycine, dipeptides, anthracene, and several other polyaromatic hydrocarbons with alkyl substituents, nitrogen heterocycles such as purines, pyrimidines, as well as sugars, alcohols, amines aldehydes, ketones, carboxylic acids, amides, and so forth have been identified, and the list is continually being expanded. Various amounts and compositions have been observed in other celestial bodies including asteroids, comets, and dust particles of which several thousand tons fall on Earth every day. The origin of these compounds is not yet clear, but one thing is certain: These are not contaminants from Earth substances. This is a hot subject of current space research. See, for example: Martins Z. *Handbook of Astrobiology*, pp. 177–194, V.M. Kolb (Ed.) CRC Press, Boca Raton, 2019. Martins Z, Chan QHS, Bonal L, King A, Yabuta H. *Space Sci. Rev.* 2020;216: art no. 54, DOI: 10.1007/s11214-020-00679-6, references cited therein and several others.
2. Sewage sludge is one of the human-made chemical pools that contain a vast variety of waste organic chemicals with environmental impact. The following paper accounts for the fate of sludge when heated between 350 and 550 °C. Hu Y, Lu Y, Ma W, Wang L, Wibowo H, Huang Z, Yu F. *Energy* 2019;23(12):2258. DOI:10.3390/en12122258.
3. Grimme W, Härter MW, Sklorz CA. *J. Chem. Soc. Perkin Trans.* 1999;2:1959. DOI:10.1039/A901770D.
4. Stoiber RE, Legglett DC, Jenkins TF, Murrmann RP, Rose Jr WI. *GSA Bull.* 1971;82(8):2299–2302. DOI:10.1130/0016-7606(1971)82[2299:OCIVGF]2.0.CO;2.
5. Frische M, Garofalo K, Hansteen TH, Borchers R, Harnish J. *Environ. Sci. Pollut. Res.* 2006;13:406. DOI:10.1065/espr2006.01.291.
6. Schwandner FM, Seward TM, Giże AP, Hall K, Dietrich VJ. *Geochim. Cosmochim. Acta* 2013;101:191–221. DOI:10.1016/j.gca.2012.10.004.

ACKNOWLEDGEMENTS

A third edition may seem a reviewed and updated version of a previous one. But this is not the present case. I decided to rewrite the manuscript, if not from scratch, from a wider conceptual background, to be able to include additional and updated aspects of the substratum of discovering and understanding organic reaction mechanisms: the rules of electron existence and flow, their fight for stereochemical space and the manner through which the unknown may be pierced through by orderly reasoning. While keeping the previous editions scheme of the buttressing of problem solving skills, this expanded approach turned into a four year-long endeavor that required the contribution of a number of people I would like to thank. In addition to students, readers of the previous editions and colleagues with whom I discussed and tested sections of the manuscript and problem analysis methods, I wish to thank professors Tatsuhiko Nakano and Ernest Wenkert, both mentors in their own style, and specially my wife Adela Tarnawiecki for her support and encouragement to stay on track all along. This edition also profited tremendously from the professional work of people in John Wiley & Sons, Skyler Van Valkenburg, Kubra Ameen, Richa John, Sabeen Aziz, and other people who worked hard behind the scenes to improve the language and composition. My special thanks to senior editor Jonathan Rose who invited me to write this third edition and managed with encouragement the process all along. Without them the publication of this book would have never been possible.

WHERE DOES THIS BOOK LEVEL START, HOW FAR DOES IT TAKE YOU?

1 OVERVIEW

You may wish to learn that the second edition of this series earned a high position in 2020 among the "The 90 Best Organic Chemistry Books of All Times in the World" (twenty-ninth echelon), according to BookAuthority [1].

This third edition embraces the tradition of the previous two editions (1986 and 2014) but offers additional concepts, techniques, and updated examples. These concepts expand the previous editions to new limits of problem solving as a tool for advanced organic chemistry. The approach continues to focus on the multiple purpose of developing problem-solving skills to fasten the rules of nature and reactivity of organic compounds in your knowledge baggage and create a state of mind to face the challenges of finding solutions to the wide scope of problems found in student and professional life.

The Art of Problem Solving in Organic Chemistry, now in its third edition was designed to be a useful participant of the chemical space constituted by *organic reactions, reactive intermediates, and mechanisms of molecular transformations*; that is, a complimentary source for courses aimed at covering these subjects, a demanding workbook for individual or group study to gain expertise and competitive skills, and a self-contained textbook for courses on organic reactions and problem analysis at intermediate and advanced levels. It can also be used as text for a one semester course covering advanced electron redeployment in organic reactions and stereochemistry to be applied to specific problem-solving strategies presented exclusively in this workbook.

Problems have existed for people to solve since the epoch of early humans living in bivouacs in trees. Eons later, organic chemistry began creating its own sort of problems since the middle of the nineteenth century when it faced the classic question: What in the world is benzene? When top scientists of the 1850s jumped on the bandwagon, they had learned just a few years earlier that carbon atoms had four valences and could be bonded to other carbon atoms to form compounds. They also knew that the composition of the tar fraction at the 60 °C boiling point, called benzene, had a C_6H_6 composition but nothing else. Heated debate around benzene's molecular structure continued until Kekulé produced his famous hexagon (1865), which remains a continuing research subject.

If you happened to be a research chemist in the 1860s and wished to participate in this brawl with a brilliant display of imagination, you might have ended up with *217 different perfectly valid* C_6H_6 *combinations*, as Professor Nagendrappa collected in 2001 [2]. Obviously, our chemical reasoning has changed enormously along with problem complexity.

2 THE TWO WAYS TO PROBLEM SOLVING OF ADVANCED ORGANIC REACTION MECHANISMS

Approaches to problem solving have been a widespread discussion among theoreticians of chemical education for a long time. A central aspect is this:

When facing a given chemical reaction, **the brain apparently switches from one state to another depending on whether the target product is shown or not** [3]. This switch completely changes the student's perception of the problem.

If no structural details of the product are provided (see Scheme I.3 of Chapter 1), the subject applies the chemical principles they are familiar with, which depend on the accumulated knowledge, to devise a **pushing-forward attitude toward possible products**. Chemical education researchers admit that the applied reasoning may lead to misconceptions and mistakes, but the problem solvers at least attempted to apply what they knew in search of a solution. Therefore, an educational end is satisfied [3, 4].

In comparison, **when the target product is shown the student "dramatically" changes the solving strategy to a "domain-general" type.** This means focusing attention on the target product and then working backward to the reagents in search of the one mechanism that will justify the product, which these researchers contend is a tunnel-like approach [3, 4]. Accordingly, the students do not shake up their knowledge basket to examine other possibilities. Hence, this strategy "may not be the most effective at assessing mechanistic reasoning, because it can limit the student's thought processes" [4]. I disagree and show this point of view throughout this book referencing non-elementary reaction-mechanism design. Science is, after all, propelled by disagreement and rational discussion.

3 THE VASTNESS OF ORGANIC CHEMISTRY: THE FIRST CHALLENGE

The scope of organic chemistry is so immense that defining its limits is hardly possible. To cope with this, the central question is cognitive memory retrieval versus intelligent use of correlative lines of thought. **This book is intended to provide the skills necessary to turn this branch of science into a more accessible field through correlation judgment** by way of **specific problem-analysis strategies as a framework for applying the concepts that support every reaction-mechanism proposal**.

This view is applicable from the simple problems of molecular transformation to very complex riddles. The learning process involved when coming across intelligent problem solving from accessible to challenging arenas inevitably advances the reader to an enhanced proficiency and response capacity to the variety of problems posed by advanced studies and career practice.

Because **this book is meant to be read and get deeply involved in, by way of the many examples and questions,** its basic design fits the model of a **workbook for the individual reader and discussion group alike**: A reading instrument in company with paper/pencil or chalk/blackboard and a great deal of **organized thinking**.

It may also become a cherished addition to a personal collection, along with the first and second editions; a valuable source for the lecturer to retrieve **definitions and examples of principles, reactions, mechanisms, reagents**, and so forth, described in the detailed indexes at the end of the book.

The present section seeks to shed light on the **bottom line**. Because this book is based on a large collection of reactions to illustrate principles entrenched in increasingly advanced concepts, the following case (Scheme 0.1) illustrates the starting level of difficulty. *Difficulty is inversely proportional to the study level and the fraction of knowledge readers retain, in addition to practice, perception, and association thinking*. Therefore, advanced masters' and PhD students may find this first problem somewhat elementary, whereas college seniors and non-major chemistry students may feel a bit challenged. It is up to you to find your place in this scale when you try it.

4 ASSESSING YOUR COGNITIVE LEVEL: A FIRST TEST

Imagine that one morning you are walking down the hall at your favorite university and a couple of students in your class, Jack and Helen, approach you as the more advanced student (or dependable lecturer) you are, with a question. Jack wants to know whether there is a mistake in product **3** of Scheme 0.1 or not. The reaction appeared in the chemical literature, but Jack is not sure he copied it correctly from the abstract, the internet service is under maintenance, the library is closed for repairs, and he needs a quick answer. Would you be able to help Jack?

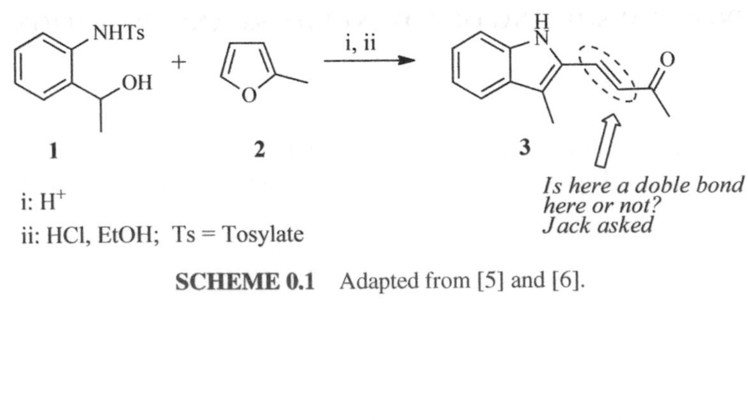

SCHEME 0.1 Adapted from [5] and [6].

A memory-gifted chemist may have recognized remnants of the Reissert indole synthesis (Scheme 0.2) [7]. If you did, you might probably proceed to scan your memory bank for a pre-established mechanism stored previously in the "Reissert module" of your gray matter. The core (dashed block of Scheme 0.2) emerges as a quite elementary and unchallenging mechanism.

This manner of solving problems (association with a reaction stored fresh in your memory) is not rational, artful reasoning but simple memory search and rescue very much like computers do effortlessly in a hum; a time-efficient ice-cold response, *but devoid of the pizzazz and learning power of creative chemistry.* The latter is the result of a balanced combination of educated intuition, imagination, and solid chemistry principles. This is what this book intends to develop or enhance in the reader. By the way, the basic Reissert mechanism will not satisfy Jack's question. The only way is to sort out the reaction mechanism anew.

If you did not pick up the Reissert reaction at first, you are bound to walk along a far more interesting mind adventure: Creativity. "*Imagination is more important than knowledge*" Albert Einstein was once quoted. Your chalk and blackboard analysis may go like this:

Firstly, after a birds-eye look at reactants and product in Scheme 0.1:

1. You recognize the furan carbons on the right appendage of indole **3.**
2. Therefore, the reaction involves a well-known acid-catalyzed coupling of furan and the secondary carbinol **1**. The product of this typical electrophilic addition may be stable enough to be isolated.
3. Next, ring opening of the furan elicited by strong acid and heat.
4. Further closing to the indole with previous, acid-catalyzed removal of the tosylate protecting group on N.
5. Some final touches… finish the sequence (Scheme 0.3).

Because the end product **10** is not like the α,β-unsaturated ketone suggested by Jack, structure **3** was not correctly copied. Sorry Jack.

Helen, who watched thoughtfully every part of your sequence, said she was a bit baffled because a minor modification of the starting compounds yielded the double bond in question, and she proceeded to show the equation scribbled in her notebook (Scheme 0.4). In addition, she claimed this was the correct product. How could it be? Now is your turn, as the chemistry expert

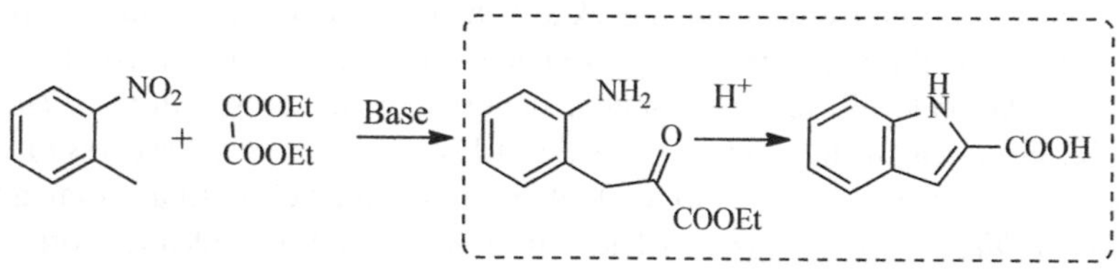

SCHEME 0.2 The basic Reissert indole synthesis discovered in the late nineteenth century

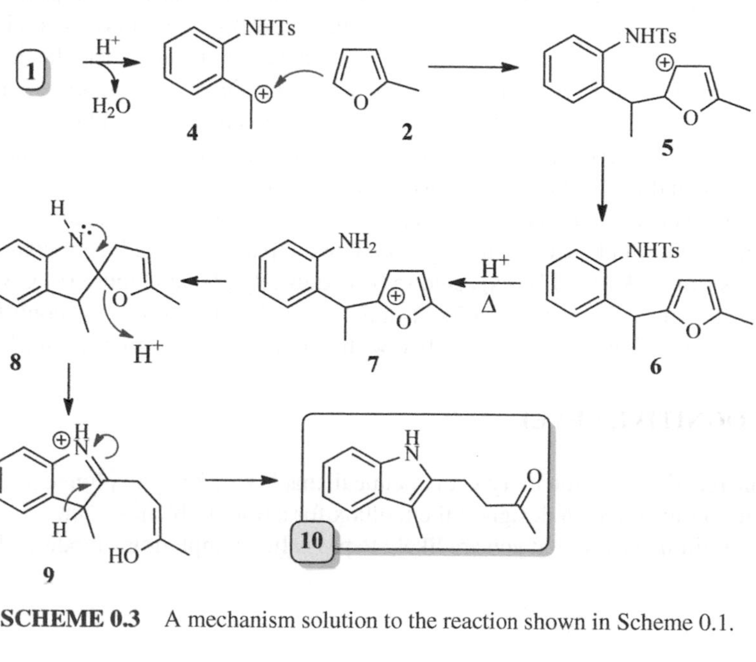

SCHEME 0.3 A mechanism solution to the reaction shown in Scheme 0.1.

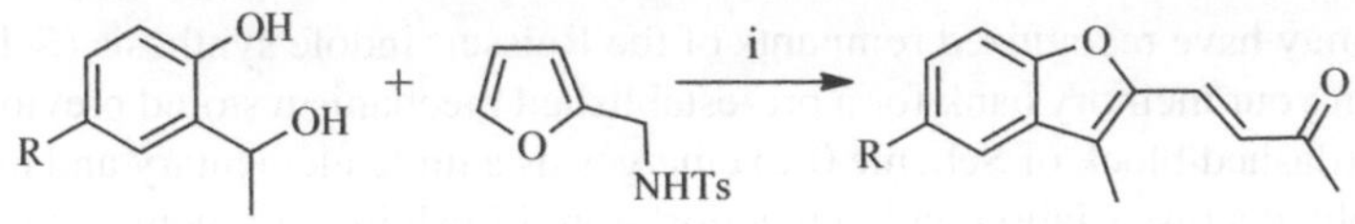

11 12 13

i: H_3PO_4, HOAc
R = OMe

SCHEME 0.4 Adapted from Reference [8]. © 2015, Elsevier Ltd, used with permission license no. 5393560381056.

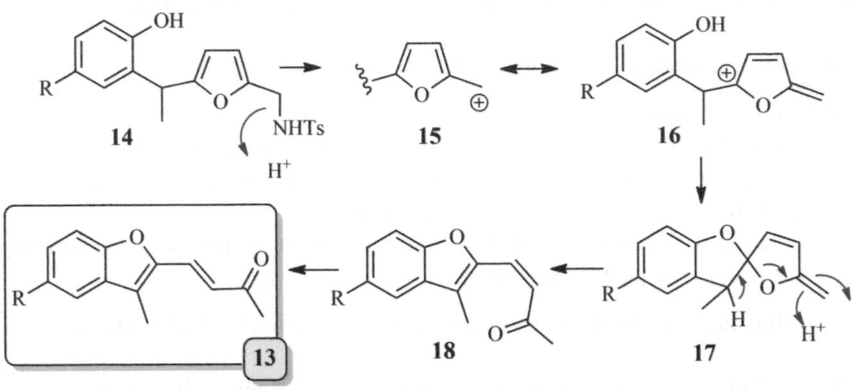

SCHEME 0.5 A mechanism modification introduced by a tosylate substituent at the methyl appendage.

of the group, to take the lead and ask yourself as a potential reader of this book if you can tackle this Jack–Helen dilemma within the next few minutes. Please refrain from having a peek at Scheme 0.5 until after you have tried your own.

At first glance you may have noticed that placing the *N*-tosylate (*N*-Ts) unit on the 2-methyl furan plays a decisive role. If you are familiar with relative oxidation levels, reviewed in great detail in the first two editions of *The Art of Problem Solving in Organic Chemistry*, and in this edition as well, you will know that **12** *is a more oxidized species than* **2** in Scheme 0.1 because of the *N*-Ts substituent. This feature may be the key to the higher oxidation level of **13** as compared with **10,** (an extra C=C bond). (*The oxidation level is highlighted in Chapter* 4.) The point now is to put this idea to work in your reply by modifying the mechanism.

Let us suppose that you assumed the first furan-phenol electrophilic addition in the acidic medium had already occurred and start from **14**, which is a homolog of **6** in Scheme 0.3. It is tempting (and rather boring) to repeat the **7** → **8** → **9** sequence and play it from there. *Paying attention to the* N-*Ts end may be more interesting*, though. This is a good leaving group in acid, better yet if the $C^{(+)}$ left behind is well stabilized by π-conjugation (resonance) as in this particular instance. The scheme evolves smoothly from this starting point onwards (Scheme 0.5). In the end, the side chain is adorned with a C=C bond. Helen feels enthralled, except that the primary product is *cis* **18**. This is of no consequence, you explain, because reaction conditions are suitable to promote acid-catalyzed isomerization to *trans* **13** owing to greater thermodynamic stabilization.

Besides, if you were a proficient drawer of three-dimensional (3D) renderings, it would be easy for you to convince Jack and Helen that methyls in ketone and benzofuran sections exert electrostatic (steric) repulsion against each other, enough to force the side chain out of the molecular plane and hence, decrease *p* orbital overlap of the overall π conjugation, thus reducing resonance energy (as in **18-3D**), whereas *trans* **13** would be expected to be exquisitely planar (Figure 0.1). A quick molecular mechanics calculation, in fact, reveals a cis/trans difference of $18.5\,kcal\,mol^{-1}$, enough to trim down the contribution of **13** to a mere speck in a putative equilibrium. Jack and Helen walk away satisfied with your *intelligently creative analysis* of this problem.

5 ASSESSING YOUR COGNITIVE LEVEL

Here are two more accessible reactions for you to try your chemical knack at before you plunge into the actual body of this book and become a proficient reaction-mechanism designer: the toolbox for a true problem solver.

Take these reactions as a preliminary test that you are likely to pass after completing a moderately advanced course of undergraduate organic chemistry.

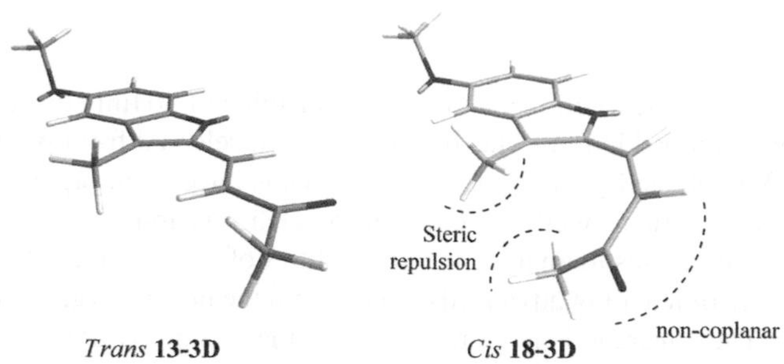

FIGURE 0.1 Perspective rendering of cis and trans compounds **18** and **13** of Scheme 0.5 showing the out-of-plane distortion of the enone owing to steric repulsion.

TEST No. 1:

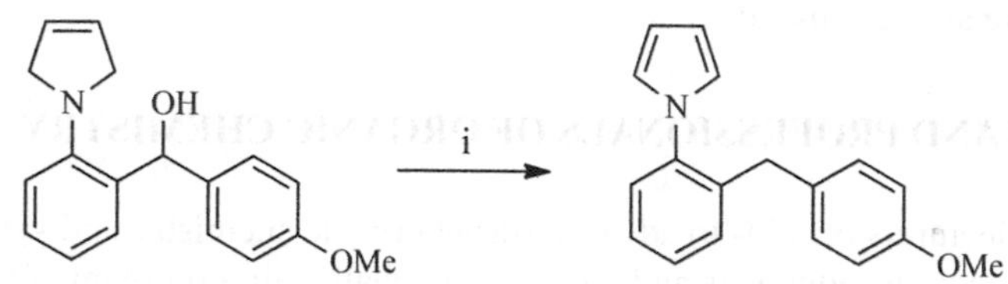

19 **20** 81% yield

i: *p*-toluenesulfonic acid (*p*-TSA); Dichloroethane (DCE), 60 °C

SCHEME 0.6 [8] / Elsevier.

Level of difficulty for this book: 1/10.
Hint: involves one curious step; oxidation status shifts from one part of the molecule to another section.

TEST No. 2:

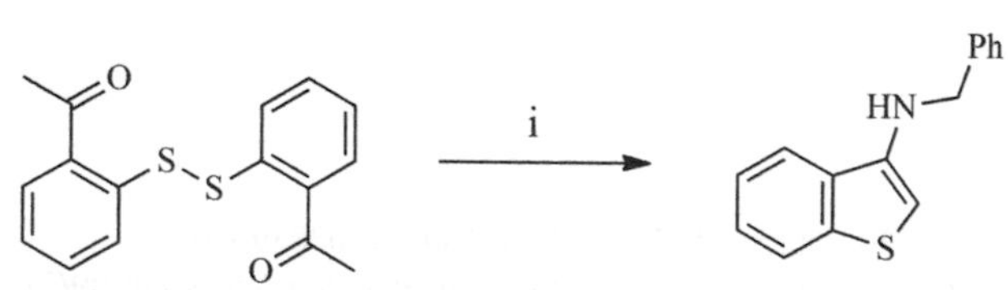

21 **22** 48% yield

i: benzylamine (2 molar excess); acetonitrile, reflux

SCHEME 0.7 [9] / Elsevier.

Level of difficulty for this book: 2/10.

6 HOW FAR WILL THIS BOOK TAKE YOU?

Shuffle the book pages and your question will be satisfied. Take a look at the content, move half way through and check the chemistry of a particular subject or problem on that page, and finally, allow your fingers to select any of the last 10 and most thorny problems in this collection. Round it all up and this will be the scope of this book.

7 TEXT ORGANIZATION

The book is divided in two main blocks: i) the toolbox and ii) a new collection of 50 fully discussed organic reaction-mechanism problems, in addition to several more embedded in "The Toolbox", Part I, selected from published research in recent years as well as valuable and seminal work published years ago. These problems have been organized by increasing difficulty, even though this term is contingent on my criteria as well as the readers level, a very imprecise feature.

At the heart of Part I is a series of strategies to analyze the possibilities of reaction mechanisms between reactants and products that I have tested over many years of teaching advanced courses from the late undergraduate up to advanced graduate levels of organic chemistry and research. The coverage is not a re-staging of the second edition of this book but an expanded pursuit of electron properties, function and redeployment, the inner and complex question of the origin of steric effects that influence mechanism, the tools designed for problem analysis of chemical transformations and their mechanism possibilities, and applications in contemporary organic chemistry.

Part II, "The Problem Chest" compiles an inspiring and personal selection of mechanism problems extracted from published results from several countries, organized by increasing difficulty. The mechanisms are fully discussed with alternative pathways of my own, followed by comparisons to related reactions from other sources as warranted. Whenever possible, data from quantum chemical calculations are also discussed.

8 A NOTE TO LECTURERS AND PROFESSIONALS OF ORGANIC CHEMISTRY

In this third edition of the series, lecturers of mid- to advanced-level organic chemistry and synthetic method courses may find inspiration for bringing modern reaction sequences and increasingly challenging problems for students to take home and discuss. On the other hand, **solving these problems individually as a mind training tool** is one of the reasons some of my colleagues appreciated the previous editions.

This book should also serve as a source of cases of certain sophistication from the current literature applicable to various course levels. Furthermore, mechanism elucidation and hypothesis proposal are much needed subjects for the advanced chemistry syllabus. A teacher's guide to this book is planned for in the near future.

In closing:

Societies of the world continue to face increasingly demanding challenges that need bright minds with the capacity for problem analysis to produce viable solutions. Serious problems to our survival are legion. Solving reaction mechanisms of organic chemistry, I believe, is an excellent training ground for developing these capacities, which also apply to many fields of scientific and technological endeavors, the design of development policies, and exploring the unknown ahead.

REFERENCES

1. https://bookauthority.org/award.
2. Nagendrappa G. *Resonance- J. Sci. Educ.* 2001;6(5):74–78. DOI: 10.1007/BF02839086.
3. Bodner GM, Domin DS. *Univ. Chem. Educ.* 2000;4(1):24–30. https://rsc/3Cht26g Accessed: 09/21/2021.
4. DeCocq V, Bhattacharyya G. *Chem. Educ. Res. Pract.* 2019;20:213–228. DOI:10.1039/C8RP00214B.
5. Butin AV, Stroganova TA, Lodina IV, Krapivin GD. *Tetrahedron Lett.* 2001;42:2031–2033. DOI:10.1016/S0040-4039(01)00066-1.
6. Butin AV, Smirnov SK. *Tetrahedron Lett.* 2005;46:8443–8445. DOI:10.1016/j.tetlet.2005.09.057.
7. Gribble GW. Reissert indole synthesis. In: *Indole ring synthesis: from natural products to drug discovery*, Ed: Gribble GW, Chapter 40, pp 332–337. Wiley, New York, 2016.
8. Zhen L, Lin C, Du H-J, Dai L, et al. *Tetrahedron* 2015;71:2839–2843. DOI:10.1016/j.tet.2015.03.077.
9. Pöllnitz A, Silvestru A. *Tetrahedron* 2015;71:2914–2921. DOI:10.1016/j.tet.2015.03.058.

PART I
THE TOOLBOX

1

INTRODUCTION TO PROBLEM ANALYSIS IN ADVANCED ORGANIC REACTION MECHANISM

CONTENTS

1.1 OVERVIEW

This chapter describes the first steps to take during problem analysis when seeking to devise a reasonable mechanism for an unfamiliar organic reaction.

Organic compounds adorned with two or more functional groups may follow more than one route to become something else, often turning mechanism proposals into a heated debate, for example, in class interactions or research group seminars.

Given the reaction A+B... → C+D... two general impressions and their consequences crop up: i) It may be explained readily if this is an easily recognizable reaction – hence not a problem, just an exercise and a few unknowns, ii) it may not have any solution in sight and requires further study – a true problem.

Everyone knows that most reaction mechanism problems, and others, are better solved in steps; better yet, when there are practical techniques especially designed to enhance the understanding of the whole and from there, devise feasible processes. These techniques have been organized in the first four chapters: i) the general approach; ii) sources of error and basic ways to solutions; iii) electron deployment and flow required for reactions to go; iv) stereochemistry; and v) a set of especially designed strategies to extract the most learning from the application of the first three chapters' content.

Problems are not there for causing trouble but for learning something new. Pondering options, examining alternative routes of action, making decisions, and drawing a successful plan constitute a most rewarding and enjoyable experience; it is a game with complex rules.

This first chapter focuses on the basic *problem analysis* strategies. In the end, choosing one or more of these strategies for each problem enables you to extract the most profitable substance of each mechanistic riddle. For a practical understanding,

The Art of Problem Solving in Organic Chemistry, Third Edition. Miguel E. Alonso-Amelot.

each description is dotted with embedded problems for you to try, then compare your solution with the ones provided here. Following the tradition set by previous editions of this book, more than one mechanism pathway is often presented, sometimes in disagreement with the proposal of original research authors. The design is what you might expect of a text and a workbook, containing the theoretical basis and refresher sections to enhance the mindset you need to solve problems.

1.2 THE FIRST THREE STEPS IN PROBLEM ANALYSIS

At the risk of stating the obvious, a good start would be to:

1. Admit that the reaction may be a bit of a challenge and therefore a true mechanism problem.
2. Ensure that you understand all involved structures and reagents/solvent/conditions.
3. Find out whether sufficient information is provided for drawing a feasible solution.

1.2.1 A Bird's Eye Overview

When faced with any given reaction showing reagents and products, the first question is whether it is accessible to you, nothing different from a straightforward application of well-known concepts, or on the contrary, a true problem: *a situation such that the solution is not in sight after a first assessment of the elements in it*. This has nothing to do with the common meaning: "a situation regarded as unwelcome and harmful." No!

To get this ball rolling, please take a look at the reaction in Scheme I.1.

Now, after glancing at Scheme I.1 for a few seconds, go back to the three basic questions posed above. If your answers were yes, no, and yes, respectively, this reaction takes the stature of *a problem*, hence, you have some interesting work ahead.

1.2.2 Change the Molecular Rendering to a Familiar Framework

Trick no. 1: Look at three-dimensional (3D) renderings from a different perspective. Once you do, the mechanism appears as a simple operation (Scheme I.2)

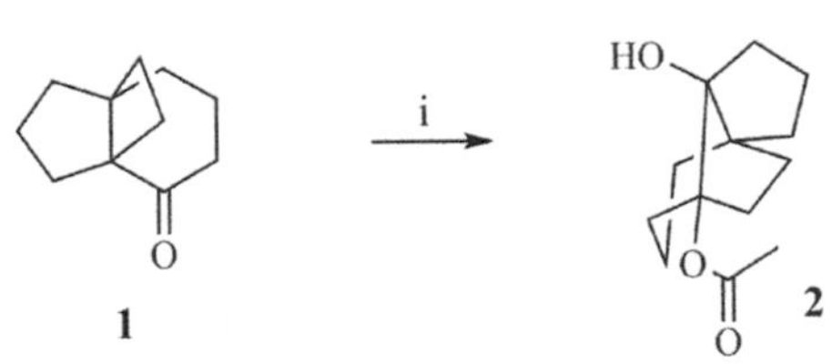

i: AcOH, pTSOH (cat), 50 °C, 144 h, 68% yield

SCHEME I.1 Adapted from Reference [1]. ©1984, the Royal Society of Chemistry, used with permission.

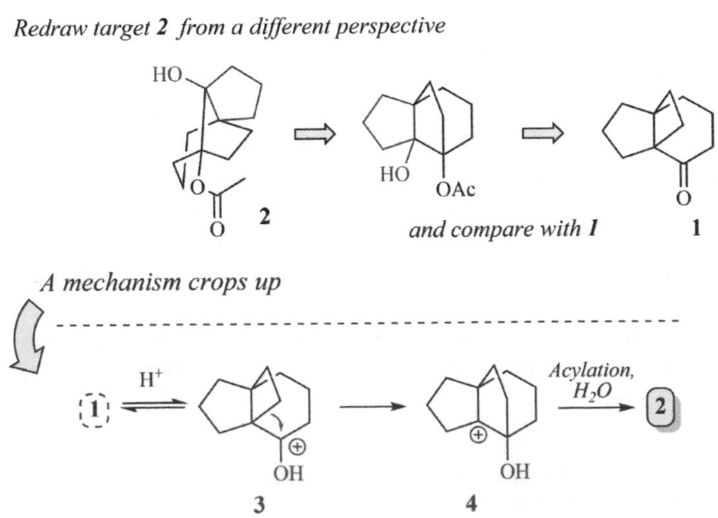

SCHEME I.2 Example of looking at three-dimensional (3D) molecular structures from different perspectives to better envision a reaction mechanism.

1.2.3 Go for the Relevant, Skip the Superficial Information

Research is customarily published with great experimental detail for others to confirm the results. In reaction mechanisms, these details are of great importance since a change of solvent/cosolvent, temperature, a particular ligand in a transition metal catalyst, and so forth determines great changes in yield or even products. Yet, there are other pieces of information with a blurring effect that take you away from the marrow of the mechanisms involved. In such cases apply:

Trick no. 2: Get the most ***relevant*** *information you can.*

Difficulty in problem solving is also an inverse function of cognitive load. In addition to the accumulated knowledge, the success of the first assessment depends on the information provided. Scheme I.3 illustrates this by proposing a mechanism problem a different way; a trivial exercise or perhaps a more complex, learning proposal [2].

With this much information, and a second impression, the reaction of Scheme I.3 is not an exercise but a kind of riddle. All one can say at the start is:

1. If anything occurred at all, that was possibly an isomerization because the molecular formula of **5** and **6** are the same.
2. It did not involve the D_3-methoxy group, since it appears in **6**.

Now, let us add some more real-world information: the ^{1}H nuclear magnetic resonance (NMR) spectra of both compounds (Figure I.1, Scheme I.4). This becomes an exercise in basic NMR spectroscopy that will shed light on this reaction, a daily occurrence in most organic chemistry laboratories around the world. Can you make the structure of **6** using this level of data?

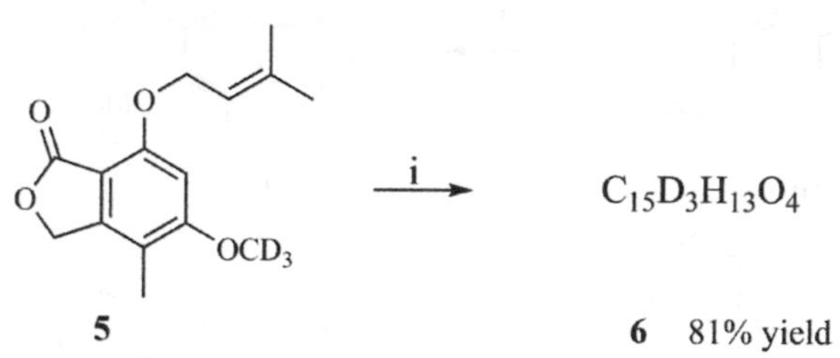

i: dimethylaniline, MHDS, 200 °C,

SCHEME I.3 MHDS = hexamethyldisilazane. [2] / Elsevier.

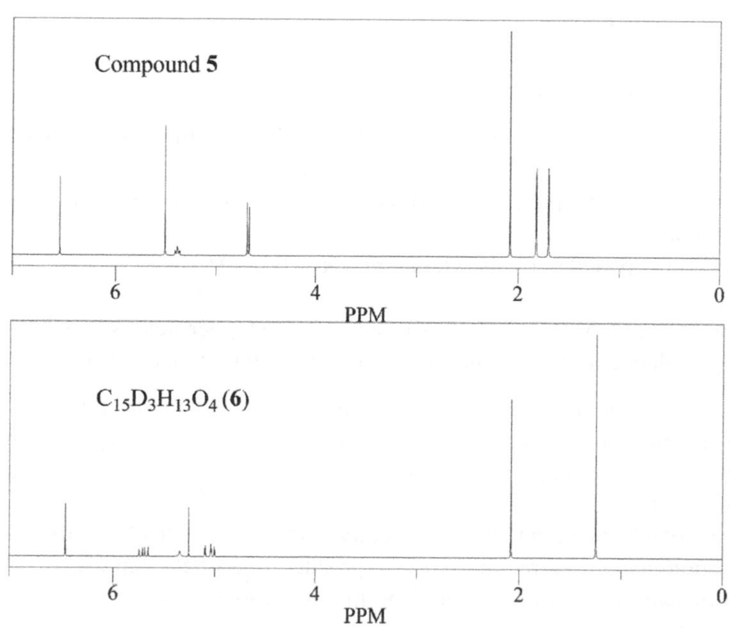

FIGURE I.1 Simulated ^{1}H NMR spectra of compounds **5** and **6**. Simulation by the author.

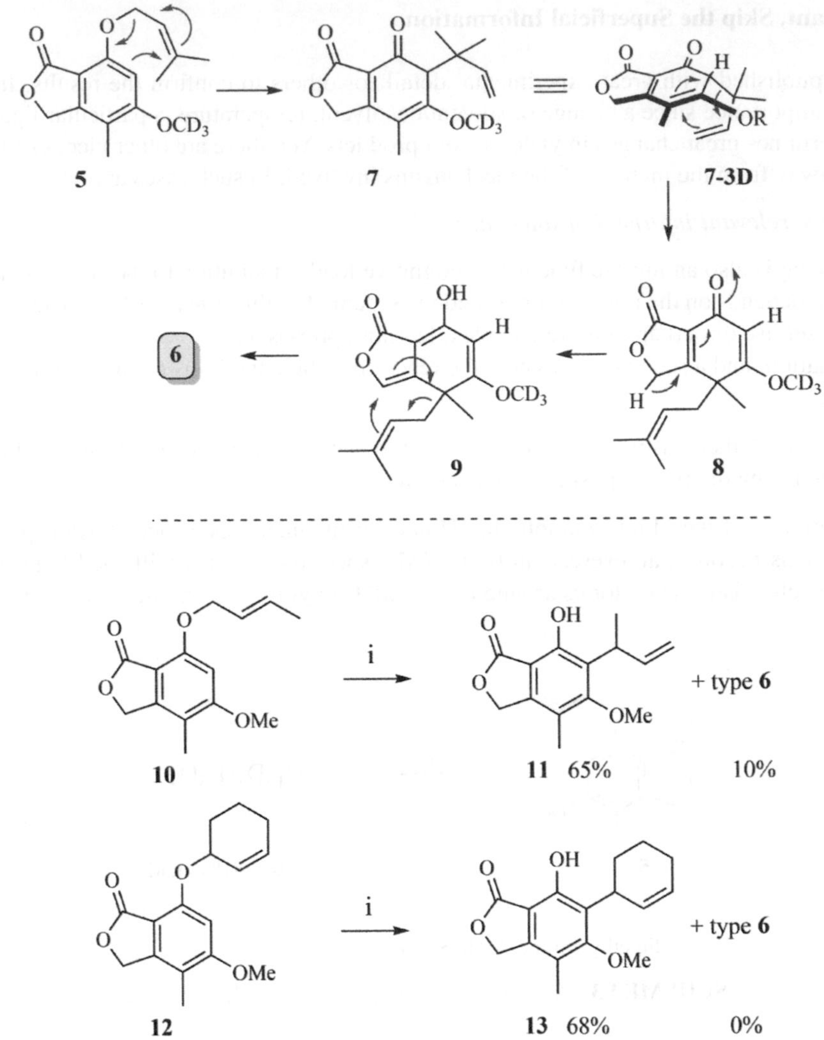

SCHEME I.4 Top: A cascade of allowed Claisen rearrangements leading to observed product **6**. Bottom: influence of the appendix structure on the Claisen rearrangement outcome. [2] / Elsevier.

A few hints from the ^{1}H NMR spectra, in case you needed them:

1. The methyl substituent on the aromatic ring remains in place in **6** and is the only aromatic proton at the low field end, hence chemical activity does not involve either of these carbons.
2. Two vinylogous methyls at around 1.8 ppm appear at a quaternary alkyl carbon, meaning that the butenyl appendage is bonded differently in **5** and **6**.
3. A monosubstituted vinyl unit is revealed by the vicinal proton ABX system at low field, likely stemming from the oxo-butenyl chain.
4. The lactone methylene protons at 5.51 ppm in **5** are shifted a bit to a higher field, suggesting increased alkyl substitution.
5. Additionally, the 100% prevalence of the CD_3 unit in **6** suggests an intramolecular transfer of the vinyl side chain.

If your inferences were correct, you may have come up with the structure (Figure I.2)

At this point, the **5** → **6** reaction may appear much more accessible: a tandem or walking Claisen rearrangement, (Scheme I.4, top section). Claisen discovered the first example of [3.3] sigmatropic rearrangement in 1912 that carries his name. Thus, the problem becomes a reassuring exercise.

The story behind the discovery of **6** from **5** is an intriguing oddity. Authors submitted four other closely related lactones (**10, 12**, and two others) to similar conditions and recovered the expected products of a single step Claisen migration of the side chain [2]. (Two of these are depicted in Scheme I.4, lower section). You may wish to advance an explanation by focusing on the space requirements of the putative transition states.

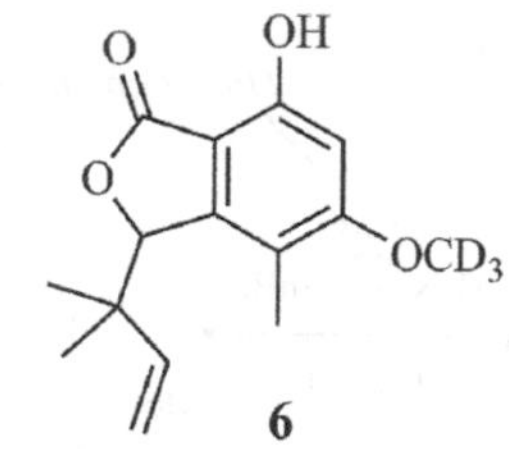

6

FIGURE I.2 A likely structure for **6** from ^{1}H-NMR data.

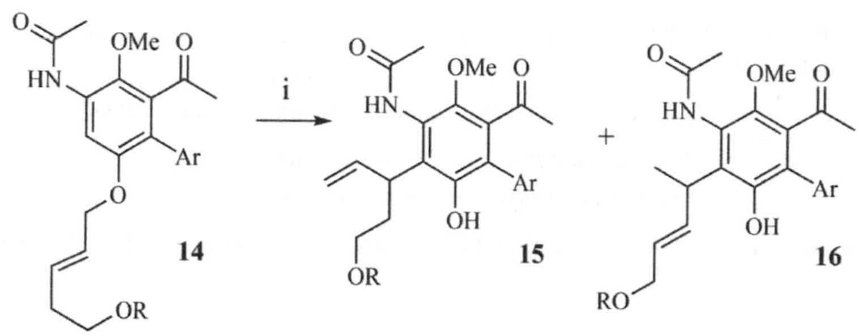

i: Et_2NPh, 210 °C
Ar: *p*-MeO-Ph; R: *p*-MeO-Bz
15:16 = 1:1

SCHEME I.5 A standard thermolysis that illustrates the appearance of a true problem. [3] / Elsevier.

Stereochemical considerations are only relevant in transient product **7-3D,** so depicted to demonstrate that the intervening C=C bonds in the electrocyclic reaction can be positioned at close distance, whereas compound **5** or reagents do not offer restrictions to the conformation of intermediates to furnish any enantioselectivity, and **6** is obtained as a 1 : 1 enantiomeric mixture.

1.3 MOVING BEYOND THE PRIMARY ANSWERS

Let us see if such a well-known process may turn into a *true problem*. Prominent Professor Tohru Fukuyama and coworkers found an unexpected product, **16,** in a 1 : 1 ratio relative to the desired compound **15** during a standard thermolysis in diethyl aniline (Scheme I.5) [3]. Does the **14** → **15** + **16** conversion constitute another exercise for you or a real problem? The mechanism is discussed at the end of this chapter (Scheme I.9). Sulfur and nitrogen equivalents have been applied many times to the synthesis of complex structures (e.g. [4, 5]).

1.4 DRAWING A PRELIMINARY OUTLINE FOR GUIDANCE

A perspective view of the reaction is always helpful. This first approach allows finding reactive spots and bonding, organize hypotheses and bring about possibilities of action. Scheme I.6 illustrates this first level analysis with an apparently accessible reaction [6].

Scribbling notes around the structures helps to clarify some key features of Scheme I.6:

1. There are two new C–C bonds in **18**.
2. There is one extra carbon atom in **18** relative to **17**, which proceeds from the amino-tin reagent (there is no other source) and becomes the centerpiece for the cyclization leading to the five-membered ring of **18**.
3. Therefore, you should start by linking this reagent to **17** and work it out from there.
4. Then, the cycloaddition step anticipated by the quick sketch must be sorted out from a still unknown molecular setup.

As you check out Scheme I.7, which follows the mechanism suggestions of the reference authors [6], one runs into the fancy chemistry of iminium ylides and bromo(trialkyl)stannane that not everyone is familiar with. In addition, the all-cis configuration of substituents in **18** demands careful consideration of possible transition states. Authors [6] developed an instructive

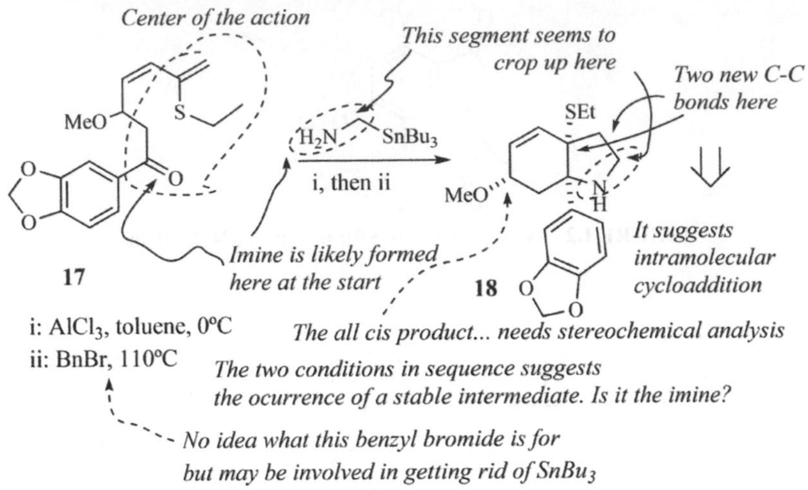

SCHEME I.6 A first assessment and unanswered questions to approach the **17** → **18** conversion. [6] / American Chemical Society.

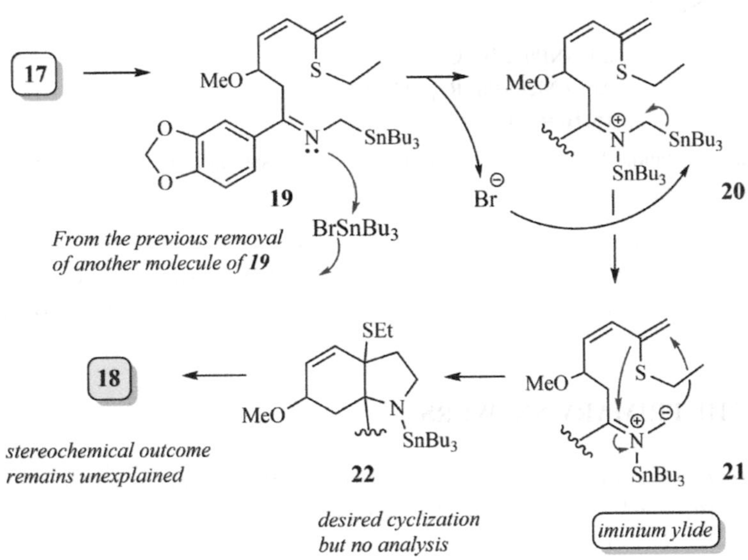

SCHEME I.7 A mechanism sequence explaining the formation of target **18**. [6] / American Chemical Society.

retro-synthetic plan. At this stage of the workbook though, it is perhaps wiser to keep this reaction in mind for once the other chapters in "The Toolbox", Part I, have been read.

This example illustrates that while one can extract a considerable amount of information and guidance from a preliminary outline, it is probably not sufficient to plunge headfirst into a feasible mechanism. Other tools, such as those comprising the following chapters, will be necessary.

1.5 INTUITION AND PROBLEM SOLVING

According to contemporary psychology, people observe and interpret reality through two parallel channels: the intuitive (System I) and the rational (System II). System I is of a more animal nature, it comes from the Latin word *tuere*, meaning "to look at or watch over"; in other words, there is no conscious deliberation. System II appears to be exclusively human, although some animals show signs of inferential reasoning. Chemists thread this sort of rationalization while devising solutions to reaction mechanism.

Scientists, like anybody else, use both Systems I and II every conscious day of their lives with mixed emphasis and results. Short-lived daydreaming and micro-intuition, which escape conscious processing, mark everyone's life, even awarded

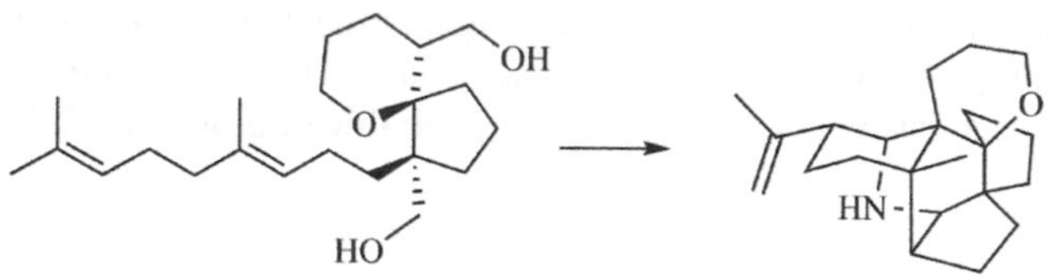

22 23

SCHEME I.8 Some reactions like this one may require more advanced assessment techniques to find feasible solutions. Such an analysis is developed in *Problem 32* of the "Problem Chest" chapter. [8] / American Chemical Society.

scientists. It is common wisdom that the scientific endeavor and intuition are not independent of each other, and in fact, intuition makes scientific knowledge possible, to the point that this is an increasingly important and still unresolved issue in the development of artificial intelligence today. Additionally, a frequent conflict between intuitive and rational processing of information has persisted throughout history [7].

However, when it comes to general problem solving of any kind, System I gets the lead because it works much faster and provides a trainload of instinctively adapted answers, whereas System II may take several seconds or minutes to produce the first feasible solution.

The latter is particularly true while attempting to solve organic reaction mechanisms of some complexity or *with insufficient information*. However, not much learning is derived from intuition alone.

To shed some light on this hypothetical division of the mind, please have a look at the reaction of Scheme I.8 [8]. Before you perform a preliminary sketch, turn off System II on purpose, explore what System I has to say about compounds **22** and **23**, and how to proceed to connect them. Intuition alone should take only a few seconds. Mind that I am not asking you to solve the problem but envision a way *intuitively*.

The organo-chemical *instinct* may have driven you to a few conclusions:

a) The conversion is not straightforward.
b) The tethered isoprenyl group plays a role in making so many bonds.
c) A cyclic amine appears in **23** whereas **22** shows none. There is a printing error or a missing key reagent (indeed, I skipped this compound on purpose to place the problem within a setting of incomplete information and thus, promote the intervention of System I).
d) The mechanism is momentarily out of reach of the problem-solving strategies described so far. Do not worry, you will find this reaction and all its components with a revealing discussion in Problem 32 of "The Problem Chest", Part II.

1.6 SUMMING UP

1. Solving organic reaction mechanism problems puts all your capacities to work: accumulated knowledge, mind responsiveness, imagination, association thinking, and mind concentration along with an inevitable bit of intuition, logic reasoning, and a shed of courage to dare thinking out of the box (occasionally).
2. All this high-power cerebral commotion is aimed at a single highly focused objective: elucidating the reaction mechanism reasonably well, which demands well-organized and constructive thinking aided by the toolkit of strategies offered in this and following chapters.
3. Planning is much more productive. Problems need to be identified, first as true challenges, and analyzed carefully. Proposed solutions need to be explored and assessed against good chemistry grounds. A single sketch encompassing preliminary ideas brings an integrated view for fresh options to show up. The amount of information taken in by skimming a plot, figure, or sketch is enormous. Exploit it!
4. Overwhelm, distraction, and stumbling into dead ends are common and frustrating by-products of problem solving that can be avoided effectively by following an orderly plan based on the application of basic concepts that most advanced students and practitioners dropped long ago for being too elementary. Additionally, a hefty measure of practice and focused perseverance always pays off.
5. Although intuition is a valuable tool in interpreting our world and facing many daily situations, it is usually of limited use in hard-science problem analysis but useful when evidence is scant. Solutions to mechanistic problems should never be left entirely to intuition, as bad chemistry will rear its ugly head.
6. Ultimately, problem solving as part of a profession is a game, no matter how challenging, for which one develops an irresistible taste over time.

It is all about orderly thinking in the brain. This organ, weighing no more than 2% of your body weight, uses up 20% of the total oxygen you inhale while burning 25% of your daily glucose storage. There has got to be a jolly good evolutionary reason for this, and **the brain's problem analysis capacity stands as a probable and powerful driving force**. It is never a bad idea to put it to work for good reasons.

1.7 SOLUTION TO THE 14→16 CONVERSION IN SCHEME I.5

Because **14**→**15** rolls over the well-known Claisen rearrangement track, **16** is an unanticipated product, and hence, the more interesting route. Scheme I.9 is an application of the well-known thermal *Conia-ene reaction* [9]. Arrows in **25** portray the formal electron redistribution of the ensuing step but bypass interesting aspects of cyclopropane (cp) breakage prompted by the double effect of dienyl and carbonyl groups on the quaternary cp carbon. Similar species are not uncommon and provide chemical frameworks of substance.

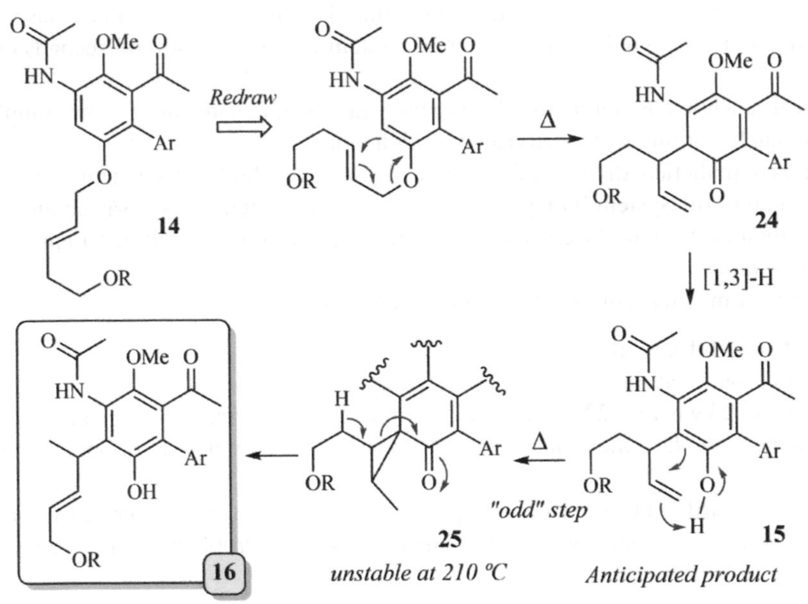

SCHEME I.9 Solution to the **14** → **16** reaction depicted in Scheme 1.5. [3] / Elsevier.

REFERENCES AND NOTES

1. Tobe Y, Yamashita S, Kakiuchi K, Odaira Y. *J. Chem. Soc. Chem. Commun.* 1984;1259–1260. DOI:10.1039/C39840001259.
2. Smith DB, Elworthy TR, Morgans Jr, DJ, et al. *Tetrahedron Lett.* 1996;37:21–24. DOI:10.1016/0040-4039(95)02101-9.
3. Fukuyama T, Li F, Peng G. *Tetrahedron Lett.* 1994;35:2145–2148. DOI:10.1016/S0040-4039(00)767081-5. Accessed: 08/20/2022.
4. Hui Z, Jiang J, Qi X, et al. *Tetrahedron Lett.* 2020;61(24):article 151995. DOI:10.1016/j.tetlet.2020.151995.
5. Majumdar KC, Nandi RK. *Tetrahedron* 2013;69(34):6921–6957.
6. Pearson WH, Kropf JE, Choy AL, Lee IY, Kampf JW. *J. Org. Chem.* 2007;72:4135–4148. DOI:10.1021/jo0703799.
7. Greek philosophers of the fifth and fourth centuries BCE dealt with the intuitive versus the rational. Aristotle said there were not two but five types of disposition to arrive to the truth of things: [*Episteme*] or scientific knowledge, obtained through induction and deduction; [*Techne*] or technical skills applied rationally to make objects designed to unveil the real world; [*Phronesis*], the practical wisdom, a rational capacity for politics and to secure a good life; [*Nous*], the true intuition to apprehend the irrational, the unexplained principles, and the dispersed, unarticulated events of reality from which scientific knowledge is built; and [*Sophia*], a term closely associated to modern philosophy, a combination of intuition and scientific knowledge; that is *Episteme* plus *Nous* work together to grasp the highest of Nature.
8. Heathcock CH, Kath JC, Ruggeri RB. *J. Org. Chem.* 1995;60:1120–1130. DOI:10.1021/jo00110a013.
9. Hack D, Blümel M, Chauhan P, Philipps AR, Enders D. *Chem. Soc. Rev.* 2015;44:6059–6093. DOI:10.1039/C5CS00097A.

2

ELECTRON FLOW IN ORGANIC REACTIONS

CONTENTS

2.1 OVERVIEW

There are three essential properties of organic molecules as we know them: resist media conditions (temperature, radiation, electromagnetic fields), interact with the chemical space to create macromolecular aggregates, and undergo conversion to other molecules under relatively mild conditions. In turn, this is a function of two basic characteristics: i) the electron distribution and flow to create and cleave interatomic bonds, build molecular scaffolds, and intermolecular associations; and ii) the spatial distribution of the composing elements, we call as stereochemistry. Chapters 2 and 3 focus on these two foremost subjects of chemical performance, without which reaction-mechanism design and problem solving would hardly be possible.

Molecules are a positively charged core, contained in a negatively charged cloud, organized in molecular orbitals, and comprising a flimsy proportion of molecular mass (Note 1). Electrons come and go within this envelope of molecular orbitals in

The Art of Problem Solving in Organic Chemistry, Third Edition. Miguel E. Alonso-Amelot.

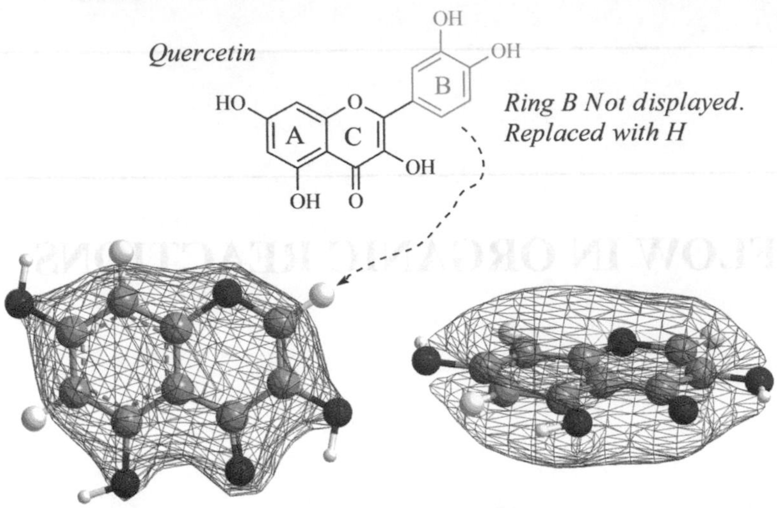

FIGURE II.1 Top and side views of Hückel's molecular orbital surface of a low-energy bonding π molecular orbital (MO) of the core structure quercetin, a flavone (rings **A** and **C** shown; the outer ring **B** was excluded for simplicity). H atoms protrude as they have no π component to share in the cloud. Other bonding-occupied molecular orbitals of higher energy (not shown) envelop the valence electrons of more than two contiguous atoms, which furnishes a combined electronic shroud around the nuclei frame.

accordance with the rules of quantum physics (Figure II.1). Nuclei are endowed with vibrational modes around a given average point in space, which is the reference coordinate for the estimation of interatomic distances.

During molecular transformations selected nuclei are dragged to different positions within the molecular framework, added to, reorganized within, or separated from it by the forces of electron redistribution in σ and π bonds, which leads to changes in molecular architecture.

Electrons also flow from one molecular package to another during intermolecular rapports *at very close range*, made possible by orbital interaction. This gap is about twice the sum of the van der Waals radii of interacting atoms, that is 2.5–4 Å, depending on atom size.

The energy to move the eight required eye muscles as you read this line and process information in your brain is provided by highly controlled electron transfers and associated fields in inner-membrane protein complexes of the muscle cell mitochondria during electronic respiration.

It is a bit surprising to realize that a general theory fully explaining the intimate physics of interatomic bonding by pairs of electrons between nuclei, beyond mathematical models emanating from solutions of the Schrödinger equation and *ab initio* methods, is not yet with us.

Above all, electron sharing within the interatomic space defined by atomic-orbital overlap and conjugation continues to give us a comfortable and intuitive construction of the covalent bond. Decent molecular renderings and a concise use of arrows showing electron traffic is helpful for you, and others watching over your shoulder, to understand your thought process. As a complement, computer quantum-mechanical calculations provide a feasible framework to further refine reaction mechanisms.

Specialists in these tools welcome mechanistic proposals by well-prepared organic chemists to work out detailed routes, energy profiles, and shapes of transition states governed by electronic deployments, favorable molecular conformations, and an assortment of molecular properties. From these data, the quantification of coupling–decoupling energy and kinetics can be obtained.

Undoubtedly, **handling electrons skillfully is essential for the proficient problem solver of organic reaction mechanisms**. A set of 12 issues and embedded practical rules that govern electrons within organic molecules, in the context of reaction mechanisms, are condensed in this chapter.

2.2 INTRODUCTION

Problem analysis consists of a series of connected steps. The rules that govern electron flow within and between molecules is essential in this analysis but is not sufficient for an imaginary mechanism to reflect the feasibility of molecular conversion.

Electrons form part of a molecular context that may allow or forbid electron transit. Before beginning to move electrons around by curly arrows, consideration must be given to other factors associated with:

✓ Molecular architecture.
✓ Electronic density and character of intervening functional groups.
✓ Physical distance between them and stereochemical configuration.
✓ Conjugation.
✓ Strain of intermediates or transition states and associated $\Delta H^{\ddagger}$.
✓ Energy type employed (heat, microwave, visible to ultraviolet (UV) radiation).
✓ Contribution of acids, bases, catalysts, and the solvent medium as activators.

2.2.1 A Word or Two on Notation

2.2.1.1 Lewis Notation and Line Renderings Molecular depictions need to be clear, precise, and informative, no doubt. The **Lewis notation** is popular among lecturers of organic chemistry principles. Professor Gilbert Lewis (1875–1946) was first to propose the covalent bond (1916) based on a universally accepted notion that is still valid today: the covalent bond results when one pair–or more–of electrons located between two linked atoms are shared and satisfy the valence shells of both atoms.

Lewis also introduced a manner notating molecules that was immediately adopted by chemists of his time and used up to the present. All valence electrons, bonding and non-bonding, and elements including hydrogens are shown (e.g. Scheme II.1 top left).

However, as molecules developed into more and more complex scaffolds, the Lewis structures became impractical. Although they persist in several organic chemistry textbooks today, they are rarely, if at all, used in research reports. *Line drawings* are preferred wherein only heteroatoms are shown and C and H are skipped, except when hydrogens take part in a mechanism, Cs in cummulenes (C=C=C…), for underlining these atoms for particular reasons.

Specific non-bonding electrons that participate in resonance structures, the stabilization of intermediate species, or in reaction processes are explicitly portrayed in addition to those that require emphasis for other reasons (e.g. radicals, carbenes, nitrenes, and similar). We will use this contemporary line notation throughout, as you have been able to appreciate in previous chapters' schemes.

Why do I take this approach? See the example of Scheme II.1, top part, and compare Lewis' versus line drawings. If not convinced, have a look at the simple reaction sequence shown in the lower section of Scheme II.1 [1], and think what it would be like to draw the full Lewis structures and whether they would be of any advantage to your understanding of the reaction course.

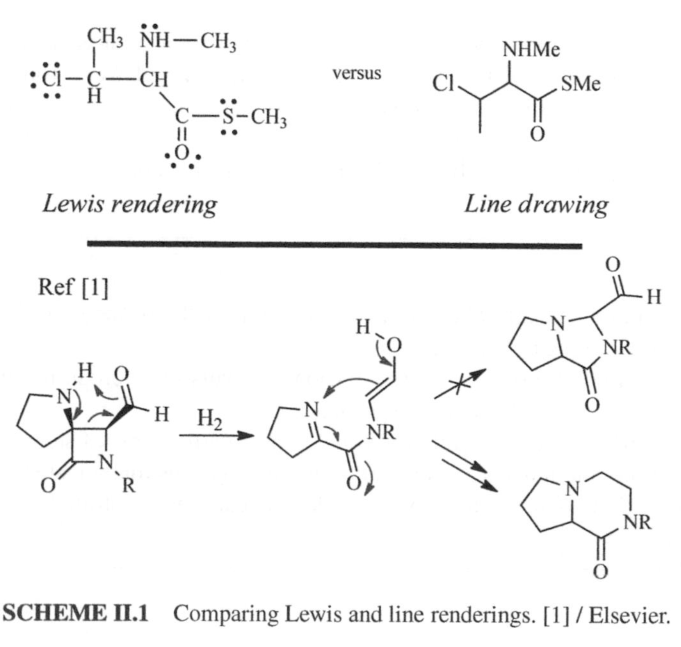

SCHEME II.1 Comparing Lewis and line renderings. [1] / Elsevier.

2.2.1.2 Curved or "Curly" Arrows These arrows are commonplace to indicate the origin, direction, and target of the electrons deployed to a lower electron density or vacant orbital. They are usually indicated in red throughout this book but occasionally occur in green to distinguish between forked options.

Arrows save a lot of word-based explanations but continue to be a superficial rendering of vastly more complex concepts. In your mechanism designs, use arrows sparingly to avoid blurring molecular renderings.

2.3 ELECTRONS IN COVALENT BONDING AND REDEPLOYMENT

2.3.1 A Preliminary Review of the Essentials

Germane to reaction mechanisms is the intimate acquaintance with the way electrons perform in bond formation and rupture.

2.3.1.1 Electrons and Covalent Bonds: The Still Unsolved Fundamental Questions The following issues around electrons and bonding are still not fully understood:

a) Uncertainty of what the chemical bond is.
b) Bonding electrons as negative charge shields.
c) Electron responses to electromagnetic forces only.
d) Enthalpy and entropy as echoes that govern the reaction course.

2.3.1.1.1 Uncertainty The intimate physical explanation of atoms being held together in molecules through covalent bonds "*has remained obscure to most scientists*" to this date, according to eminent Professor G. Whitesides (Note 2). "*Most general chemistry books avoid this question or advance incorrect explanations*" [2]. Scholars describe the covalent bond as the result of static force-field energies, based on the notion that lowering bonding energy is due to a combination of the electrostatic potential energy developed between the two nuclei at bonding distance and the accumulated electronic charge in the bond region by mechanically positioned electrons.

However, *the nature of the physical forces* that maintain the bundle of electrically charged particles at such a short range escapes this contention. Recent advanced calculations are forcing us to get acquainted with novel paradigms in regard to mass, charge, and covalent bonding [2]. While these pending issues generally do not interfere with reaction mechanistic proposals, they become a matter of concern for theoreticians applying computer calculations.

2.3.1.1.2 Electrons as Negative Charge Shields Electrons in covalent bonds further act as a negative charge shield that prevents positively charged nuclei to repel each other and fly apart. This question will be expanded in Section 3.1.2.

While this concept allows most of us to remain within our chemical comfort zone, electron shielding is just one contributor to bonding.

2.3.1.1.3 Electromagnetism as the Leading Force Particles with electronic charge *respond to electromagnetism only*, whereas the other three fundamental forces of the universe, gravity, strong, and weak forces, have no influence on them. Similarly, electromagnetic fields are created by charged particle movement. Because molecules contain several electrons and protons in motion according to the degrees of freedom, *one can pronounce electromagnetism the foremost force governing non-nuclear chemistry*. This force is expressed in a handful of electronic effects, the cornerstone of reaction mechanism (see box).

A reminder in a nutshell

✓ Inductive effect: (±*I*) is transmitted through σ bonds; it wanes rapidly with distance: $-C^{\delta\delta\delta+}-C^{\delta\delta+}-C^{\delta+}-X^{\delta-}$.
✓ Resonance: transmitted through conjugated π bonds.
✓ Mesomeric effects: (±*M*) a substituent property where non-bonding pairs in heteroatoms enter in conjugation with π bonds in aliphatic and aromatic systems. A manner of resonance.
✓ Field effects: Through-space influence of sterically proximal bonds; participants of stereoelectronic effects.
✓ Neighboring group participation (anchimeric assistance): Functional groups that are sterically positioned to affect reaction centers in the same molecule two, three, four, or more skeletal carbons (or atoms) away; π bonds and heteroatoms typically act in this manner.

2.3.1.1.4 Enthalpy, Entropy Enthalpy and entropy echo molecular structure and thus, are consequential to electromagnetic forces contained within the molecules themselves and the particle conglomerate where they reside.

Therefore, as leading determinants, reaction mechanisms recognize attraction and repulsion forces according to charge sign, full or partial, localized or delocalized, and the energy involved. The small mass and charge of electrons concentrate the capacity of migration within molecules attendant on a series of constraints. Thus, σ and π bonding changes result primarily from electronic relocation to available atomic/molecular orbitals once limitations have been overcome.

In essence, we feel comfortable in mechanism design as we move across the following landscape:

1. As atoms approach bonding distances, the electrostatic potential interactions in the mechanically accumulated charge in the bond (by electron placement in the interatomic space) are not bonding but just the opposite: *antibonding*. And so, electron spins of opposite sign alone do not offer a physical explanation for the bonding force.
   However:
2. Interatomic electron delocalization in the bonding zone creates a critical contribution to bond strength by lowering the kinetic energy component of the bonding energy.
3. As bonds form, atomic orbitals undergo intra-atomic orbital *contraction*, which do not drive bonding per se but create side effects. These contractions further enhance the electron delocalization.
4. As bonds begin to form and atomic-orbital overlap increases, the electronic wave function is extended from one atom to two or more, which in turn contributes to lowering further the kinetic energy component of bonding (more negative, hence more stabilizing effect).
5. Therefore, as Schmidt, Ivanic, and Ruedenberg categorically put it [2]: "*There remains no room for doubt that covalent-bond formation is driven by the quantum-mechanical attenuation of kinetic energy due to thè interatomic delocalization of electrons, i.e. by the quantum-mechanical kinetic drive of electron waves toward expansion.*"

2.4 PRACTICAL RULES GOVERNING ELECTRON REDEPLOYMENT

So much said, we move on to a set of manageable concepts for electron redeployment in reaction mechanism. These directives are clustered in **12 issues**:

1. Electrons reside within orbitals.
2. The electron shield concept contributes to covalent bonds.
3. Atomic and molecular orbital (AO/MO) overlap, a requirement for σ and π bond formation.
4. Electron traffic and stereochemistry.
5. Electron energy level and accessibility.
6. Electron flow and molecular active sectors.
7. Electron flow and compatible AO types.
8. Electron flow: delocalization of bonding and non-bonding electrons, resonance.
9. Electron traffic and electron-density differences: high and low electron-density zones (HEDZ and LEDZ).
10. Electron traffic on account of LEDZ alone.
11. Inverting the natural electron flow: umpolung.
12. One-electron flow: radicals and carbenes as special cases.

Concepts supporting each issue are now discussed with numerous illustrative examples and problems along the way, a training ground for more challenging ones to come in "The Problem Chest", Part II. Some of these topics may sound elementary, others less so, but compiling them in one chapter is a practical way of retrieving this information whenever necessary.

2.4.1 Issue 1: Electrons Reside Within Orbitals

☑ *Electrons in molecules are always associated with atomic and molecular orbitals and do not run free in the reaction medium.*

2.4.1.1 The Question of Orbital Restrictions and Electron Deployment It is firmly established that electrons are mostly distributed in organic molecules within quantum probability spaces defined by wave functions from solutions of the Schrödinger equation. These are the familiar AOs and MOs conceived by valence-bond theory (VBT) and expanded by several increasingly sophisticated quantum mechanics-based MO models.

This principle spans:

- ✓ Carbon and heteroatom molecular frameworks and their functional groups, be they neutral, anionic, cationic, radical, carbenic, or a combination of these.
- ✓ Molecules with hydrogen and dative bonds.
- ✓ π-Complexes.
- ✓ One-electron transfers.
- ✓ Electron transfers in pairs.
- ✓ Reductions involving alkaline metals in liquid ammonia or other electron sources in water, alcohols, and dimethyl sulfoxide as well as electrons on metal oxide surfaces.

The latter species have been regarded classically as "electrons in solution" and "solvated electrons". Such species result from ultra-fast interactions of photo-excited or dissolving metal elements (e.g. Li, Na) electrons with solvent molecules and their polarization. This concept is of great importance in chemistry and biological systems. It is currently being explored in depth [3].

Recent trendy discoveries of electron transfer between nanocrystals capable of trapping photons and redox enzymes require the two species to be in close proximity [3]. Others have observed conjugated organic compounds serving as electron bridges between the electrode surface and receptor substances in the solution body. The very close distance between the involved species strongly suggests the passage of electrons through overlapping MOs, not across space.

Therefore, managing intra- or intermolecular electron traffic requires a good handling of AOs and MOs. The σ-MO sets of a given compound form a bonding continuum across the molecular framework, which maintains all atoms together and influences other σ bonds through inductive and field effects. The π bonds form such continuums only if certain conditions are met: electronic conjugation and configuration allow a parallel *p*-AO arrangement. Resonance leads to the joint involvement of conjugated systems to redistribute electron density and create predictable reaction patterns.

Figure II.2 offers some examples of π MOs of representative isolated and conjugated double bonds.

Figure II. 2 *highlights*:

1. Non-conjugated dienes I and II display non-overlapping π MOs and therefore perform independently of the other. However, separate (non-conjugated) unsaturations in a single molecule can interact with each other in cycloadditions as the many examples of intramolecular Diels–Alder reactions demonstrate. These cycloadditions are employed extensively in the construction of molecular complexity as Scheme II.2 illustrates [4]. Once starting material (**1**) is redrawn conveniently as **1a** by 180° rotation of σ-bonds C^a–C^b and C^c–C^d, the positioning of diene and MO-independent dienophile becomes ideal for the ensuing [2+4] cycloaddition.

   The ease of this distant interaction is underscored by the low temperature (−78 °C) and high yield obtained (85%) when zinc chloride was used as the Lewis-acid promoter acting on C=O in **1b**. The gripping action of zinc in this bidentate complex drives the C=C–R^2 alkenyl group above the molecular plane forcing the final trans ring fusion in target **2**. The exquisite stereoselectivity of the reaction thus attained is lost (cis : trans ratio = 1) when the temperature is raised to 200 °C and no Lewis-acid promoter is necessary.
2. Structure **III** in Figure II.2 portrays the case of π-bond extension through a "bridge" of non-bonding pair (NBP) of electrons on N between two C=C units. A low-energy extended π MO engulfs the NBP, which allows transit of electron density from one end of the molecule to the other. Enamines R_2N–C_α=C_β and enolate/vinyl ethers or esters $^{(-)}O$–C_α=C_β/RO–C_α=C_β develop similar π MOs and express nucleophilic character at the β carbons that arise from the neighboring anion or NBP.
3. Interruptions caused by inserting a saturated carbon in **IV** create a partition of the extended π MO in two sections separated by a zero-probability trough. Through this space, π-electron flow is not expected or reduced to a minimal expression. The enamine unit to the left in **IV** encloses C=C and nitrogen's NBP in a single π MO as in **III**.

   Interestingly though, this extended Hückel estimation does not limit the π MO to the isolated C=C but also includes the allyl CH_2, which can be seen on the right side π MO of **IV**. This can be explained by the coplanarity of the allyl C–H bonds and the C=C *2p* orbitals, which leads to a certain measure of σ–π interaction. A similar σ–π MO envelopment of R–CH_2–C=C–R is found in **V**. However, the strength of this interaction should not be overemphasized and cannot be translated into resonance or delocalization, only to a limited enhancement of allyl proton or **H–C–C=O** reactivity. Once the carbanion is formed after proton withdrawal, AO interaction coalesces into one π MO providing stability.
4. This enhancement is nonetheless particularly strong when other stabilization factors come into play: e.g. **aromaticity**. This seemingly simple and yet complex concept is stated by the well-known Hückel's aromaticity rule: a cyclic, planar, and conjugated structure (be it neutral or anionic) displaying $(4n+2)$ π electrons (n = integer > 0) is aromatic and possesses a more negative enthalpy relative to the non-aromatic structure.

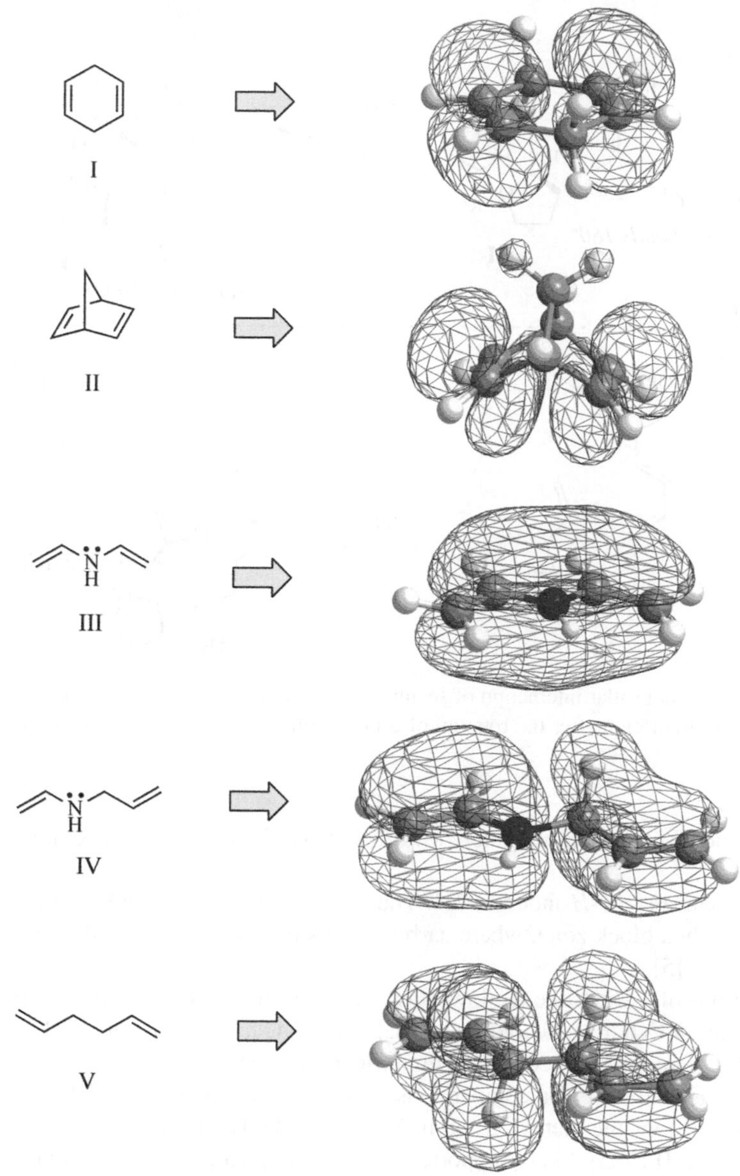

FIGURE II.2 Bonding π MOs of selected dienes of independent and conjugated bonds. Note the effect of the non-bonding pair in N on the shape of the MO that represents active versus inexistent conjugation.

This feature is portrayed in Figure II.3, where it is worth noting that:

4a. As the methylene unit of cyclopentadiene (**VI**) is engulfed in the π MO of the diene, electron density is redistributed *unevenly* across the ring. The "bulge" of the calculated π MO in the CH_2 zone of **VI** reveals a decrease in π-electron density since the volume of the MO expression in this zone is larger for the same number of electrons.
In addition, the methylene protons in **VI** exhibit the most pronounced positive character (calculated Mulliken charge: +0.189 au [atomic units]) compared to cyclopentane (+0.156 au) or C^4–H of cyclopentene: +0.159 au for the pseudoequatorial H and +0.165 au for the pseudoaxial H at the homoallylic position upon which the σ inductive effect of the C=C bond is minor [5].

Consequently, the cyclopentadiene's CH_2 is a more potent attractor of bases than the mentioned homologs, all other things being equal. As this proton is removed, the C–H σ bond increases its *p* character while the transition state (TS) progresses toward **VIIa**, thus modifying the geometry of the π MO to the symmetric electronic configuration of the cyclopentadiene anion, **VIIb**. With six electrons in the cyclic, delocalized system, **VIIb** complies with Hückel's

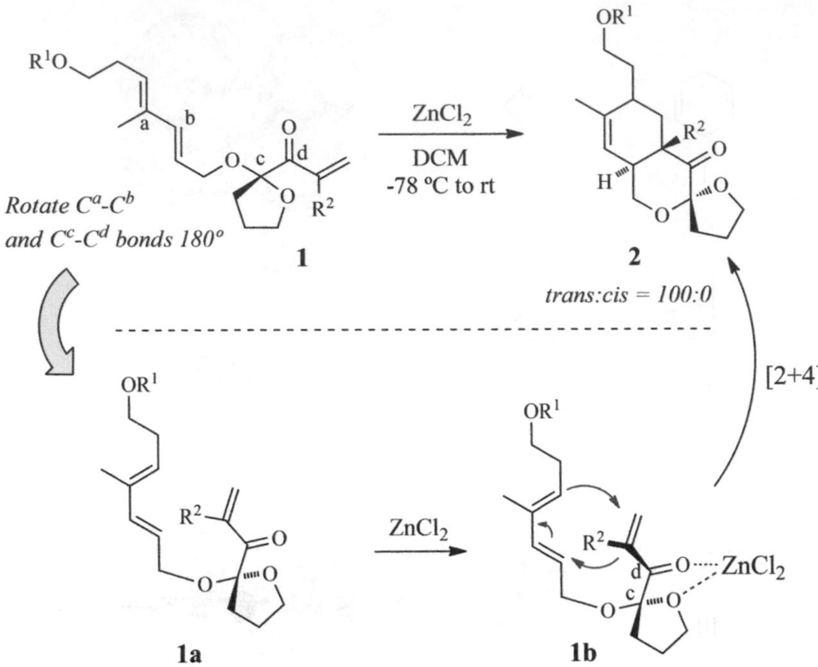

SCHEME II.2 An example of intramolecular interaction of seemingly distant π MOs that come sufficiently close for this intramolecular cyclization owing to a favorable configuration after the rotation of σ bonds linking the two unsaturated moieties. [4] / American Chemical Society.

aromaticity rule. The end result is an inordinately low pKa(15) of **VI** for an organic compound, only one log unit below that of water.

4b. The case of **VI** is reproduced in 1-*H*-indene **VIII**. The asymmetry of the π MO is marked by a relative reduction of orbital volume in the dashed block zone, where carbon atoms have high electron density (HED) (calculated Mulliken charge: −0.166 to −0.208) [5].

The methylene protons of **VIII** evoke the lowest electron density of all H atoms (+0.193). The π MO adds 10 electrons in its passage to anion **IXb**, thus becoming aromatic as well, although its $pKa = 21$ is considerably lower than that of **VI**. Again, the π MO of anion **IXb** becomes symmetric as expected.

4c. CH_2=CH–CH_3 (**X**) is inserted here for comparisons with a non-aromatic case. The strongly asymmetric π MO of propylene, which also envelops the methyl group in **X**, mirrors the Hückel charges of C atoms. The highest electronic charge materializes in C^1 (−0.1785), likely supported by hyperconjugation of the methyl C–H bonds rendering electron density (+0.056 H charge in average).

Passage to anion **XIa** and delocalization in **XIb** also furnish a symmetric π MO, without distinction of the $C^{(-)}$ position in the molecular ends, as predicted. Allyl carbon radicals are distributed similarly along conjugated unsaturations.

2.4.2 Issue 2: The Electron Shield Concept Contributes to Covalent Bonds

☑ *Electrons in covalent bonds serve as negatively charged shields against positively charged nuclei.*

The early model portending that electrons create electromagnetic fields that shield the positive charge of nuclei sharing a covalent bond is well accepted and experimentally observable in the *chemical shift* of nuclear magnetic resonance (NMR) spectroscopy. Supporting evidence, useful for reaction-mechanism design, is now examined in two scenarios: carbocations and carbon dications.

2.4.2.1 ***Carbocations*** When one of the covalently bonded atoms bears a positive charge, as in carbocations, one expects the electron-density distribution in bonds around this atom to be shifted toward the positive nucleus by electromagnetic attraction and provide stabilization.

This shift is relayed to bonds beyond the first link, which is the basis of the positive inductive effect (+*I*) of vicinal substituents and hyperconjugation of neighboring C–H bonds. The positive charge is partially delocalized. The combined effects create a partial electronic charge deficit in the surrounding molecular zone, dispersing charge, and hence, stabilizing the cation.

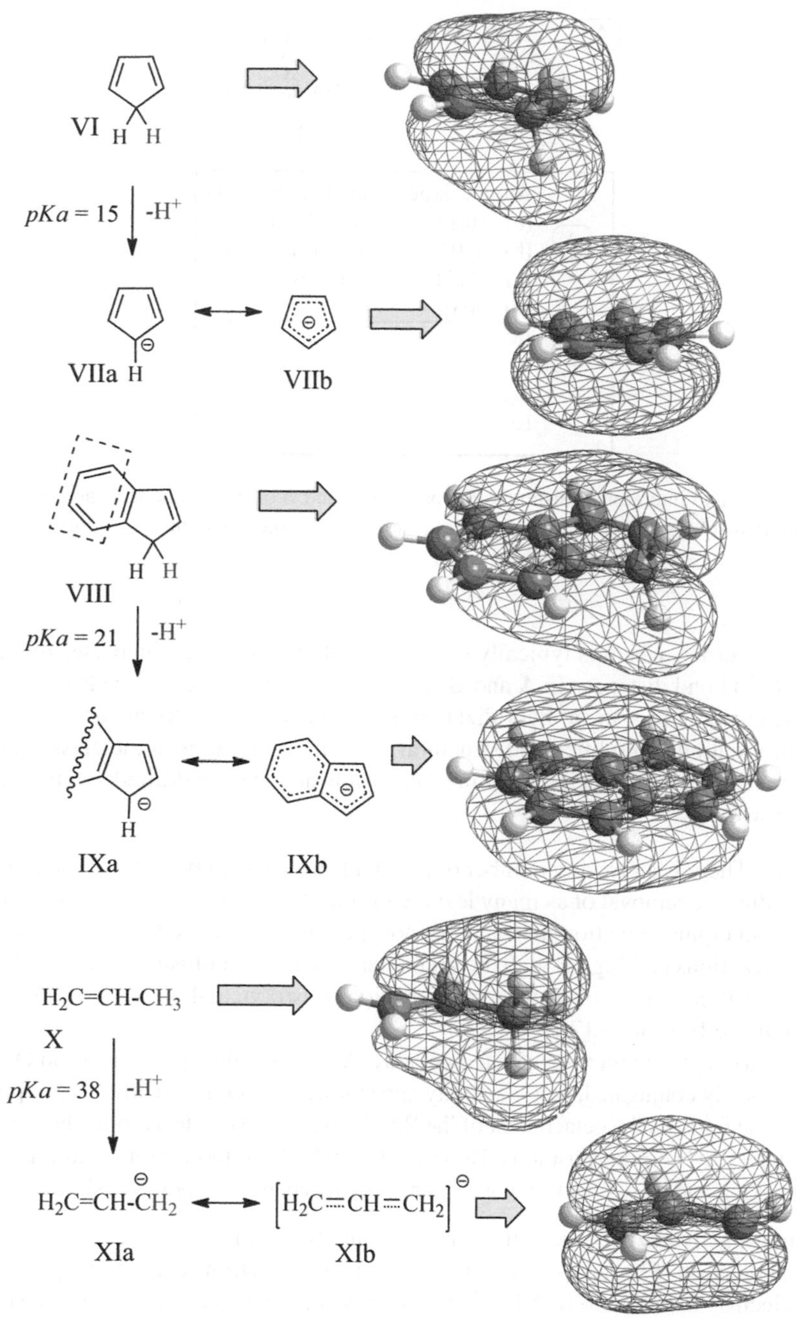

FIGURE II.3 Bonding π MOs of selected cyclic alkenes with methylene interruption of conjugation and their anions, and influence of π MOs on the *pKa* of alkyl protons (red). Propylene is included for comparison. See text for comments.

Many stable hydrocarbon carbocations have been studied by combining extensive calculations with NMR studies of organic compounds bearing nucleofuges in super-acids at low temperature [6]. Electronic shifts prompted by $C^{(+)}$ are observed by a decrease of electron density in α and β carbons as well as a sensitive shortening of the neighboring C–C bonds.

The NMR spectra of specific compound such as the *tert*-pentyl cation indicate the occurrence of structures **A** and **B** (see Figure II.4, top section) [6]. The end methyl group (a) in **A** is out of the plane of the four atoms composing the carbocation, whereas all carbon atoms are coplanar in **B.** The energy difference of **A** and **B** rotamers is only 0.5 kcal mol^{-1}, so they contribute almost equally to the rotational equilibrium in solution. The structural difference has a strong impact in the electronic distribution of the entire molecule: in **B,** all carbons show a deficit of electronic charge, whereas vicinal carbon C(b) retains a residue of electronically negative character in **A**. (See table in Figure II.4.)

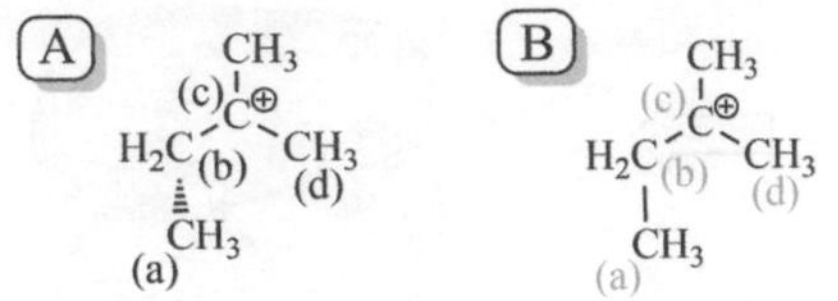

	Atom charge	Bond distance (Å)
A	(a) +0.15	(a) - (b) 1.587
	(b) -0.03	(b) - (c) 1.439
	(c) +0.74	(c) - (d) 1.461
	(d) +0.07	
B	(a) +0.07	(a) - (b) 1.457
	(b) +0.01	(b) - (c) 1.455
	(c) +0.75	(c) - (d) 1.457
	(d) +0.08	

FIGURE II.4 Electronic charges and interatomic distances of rotamers **A** and **B** of the *tert*-pentyl carbocation, which underlines the substantial influence on electronic density distribution and bond lengths arising from the apparently uneventful rotation of the C(b)–C(c) bond. Adapted from [6].

As a result the C–$C^{(+)}$ bond order, which is typically one in neutral hydrocarbons, increases to 1.5 as the C–$C^{(+)}$ bond gets shorter [compare the C(a)–C(b) bond distances in **A** and **B**, 1.587 versus 1.457 Å, respectively].

When combined with the change of sp^3 to sp^2 hybridization of $C^{(+)}$ and the geometrical variation from tetrahedral to trigonal (planar), the molecular conformation is modified and structural strain may appear as in the classical case of bridgehead $C^{(+)}$ in bicyclic scaffolds. These are high-energy structures but not impossible and may be drafted safely in hypothetical mechanisms. Very many examples are known.

2.4.2.2 Carbon Dications There is a growing number of transient species with two positively charged atoms. These include formal two carbenium ions after the removal of as many leaving groups, $R^{(+)}$–$CH_4^{(+)}$ molecules with a pentavalent carbon atom, and others that rarely appear in organic reaction mechanisms not specifically designed to produce such species.

However, there are some reactions passing through dication intermediates that illustrate the shielding effect of covalent-bond electrons mentioned in this section and the overpowering repulsion of two vicinal $X^{(+)}$–$X^{(+)}$, as the conceptually outstanding molecular commotion of Scheme II.3 shows [7].

The triggering factor is just a couple of protons at low temperature. All non-bonding pairs on N and O, and the π electrons totaling 14 electrons per side are exquisitely conjugated, hence stability appears secured. Or is it really? Try to put forward a mechanism.

After a preliminary view, one foresees the detachment of the RN–NR units and create a σ bond between furan carbons C^4 and $C^{4'}$ while the amino groups occupy the two *para* positions. However, C^4 and $C^{4'}$ are too far removed for a simultaneous or concerted occurrence of N–N bond breaking and C^4–$C^{4'}$ forming operations. Our mechanism hypothesis shown in Scheme II.4 goes like this:

- ✓ Double protonation of the hydrazine core at the start; it should be the driving force.
- ✓ The resulting vicinal dication (**5**) exerts strong repulsion, which σ electrons cannot overpower.
- ✓ Stabilization by π-electron donation to R–$NH_2^{(+)}$–N… in **5** is not allowed as the N valence shells are complete. No resonance arrow pushing, please.

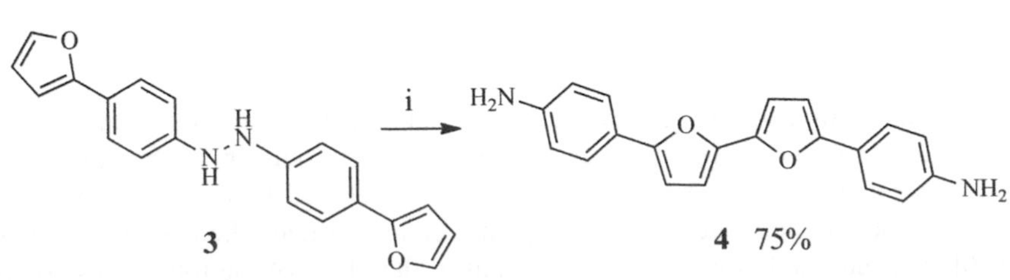

3 **4** 75%

i: HCl 0.1N, 85% EtOH, 0 °C

SCHEME II.3 A case of extreme isomerization through a forced stereochemical arrangement. [7] / American Chemical Society.

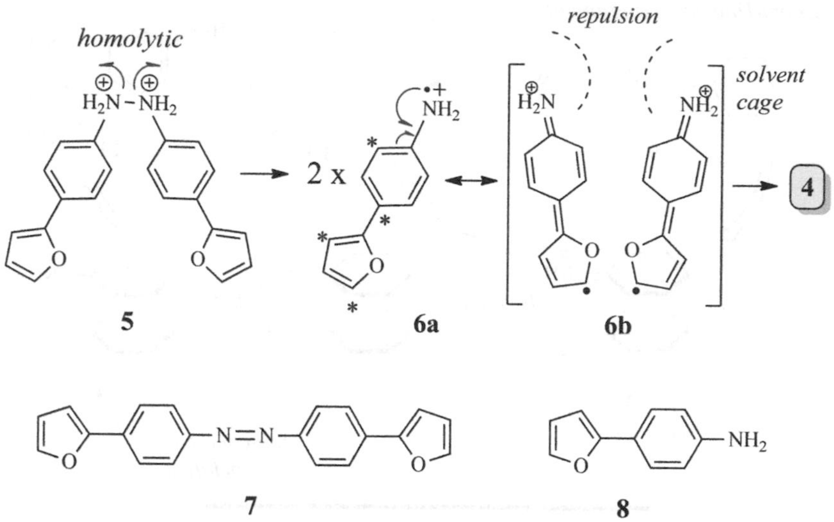

SCHEME II.4 Non-concerted pathway explaining the 3→4 isomerization of Scheme II.3

- ✓ The $N^{(+)}$–$N^{(+)}$ bond homolytic fracture ensues, since this is a highly symmetric structure. Two *N*-radicals (**6a**) are produced.
- ✓ The N-radicals are delocalized throughout the π cloud, with C• expression marked with asterisks and extending to the furan rings (**6b**). This furnishes radical character to furan rings carbons C^4 and $C^{4'}$.
- ✓ Coupling of C^4 and $C^{4'}$ radicals before the two separate units escape the solvent cage created by the polar medium ($EtOH/H_2O$) yields target **4.**

This hypothesis is compatible with the isolation of compounds **7** and **8** in 10% yield each, (the latter, an escapee of the solvent cage) corresponding to disproportionation and termination, respectively, two typical outcomes of free radicals.

Professors Park and Kang who discovered this reaction in South Korea had a different idea, though [7]: the mechanism had to go in a *concerted* fashion instead of our *stepwise* proposition, because the reaction was a furanoid homolog of the benzidine rearrangement (BR).

To refresh your memory, the BR is depicted in Scheme II.5, top part. Two major compounds, **10** and **11** are produced from dibenzohydrazine (**9**) and protic acid. A great deal of work has been devoted to this rearrangement over several decades and the N–N bond breaking is established as taking a part in the TS of the rate determining step. Detailed isotope effects supported the conclusion that **10** is the product of a *concerted* N–N *bond breaking and* C–C *bond-forming process* passing through an **A**-type TS (Scheme II.5). This is formally recognized as a thermally allowed suprafacial [5,5] sigmatropic rearrangement. On the other hand, justification of **11** by way of a concerted reaction would require a thermally non-allowed suprafacial [3,5] process. Therefore, the cation radicals **B** and **C** would be better attuned to the evidence.

Application of the concerted pathway **A** to the extended conjugated system **8** involves a face-to-face TS **D** (Scheme II.5) stabilized by Dewar's π–π cloud interaction or complex. After counting the number of participating atoms in formal conjugation, this amounts to *an unprecedented [9,9] sigmatropic rearrangement!* Ever since this discovery was reported in 1997, only the thiophene homolog of **5** is believed to follow the same concerted route. The debate around the concerted versus stepwise mechanisms and the species involved continues [8, 9].

There is no shortage of mechanism pathways open to these diarylhydrazines in protic acid, especially when asymmetric electronic effects are introduced in the aromatic rings (see, for example, Scheme II.6).

2.4.3 Issue 3: AO/MO Overlap is a Requirement for σ and π Bond Formation

☑ *For electrons belonging in separate atoms to form covalent bonds, suitable overlap between the intervening orbitals must occur first.*

This assertion amounts to asking:

a) How near the pertinent atoms get to make the hypothetical mechanism practicable? (*R*: *< 2.5–2.8Å in second-row elements to reach a TS.)*

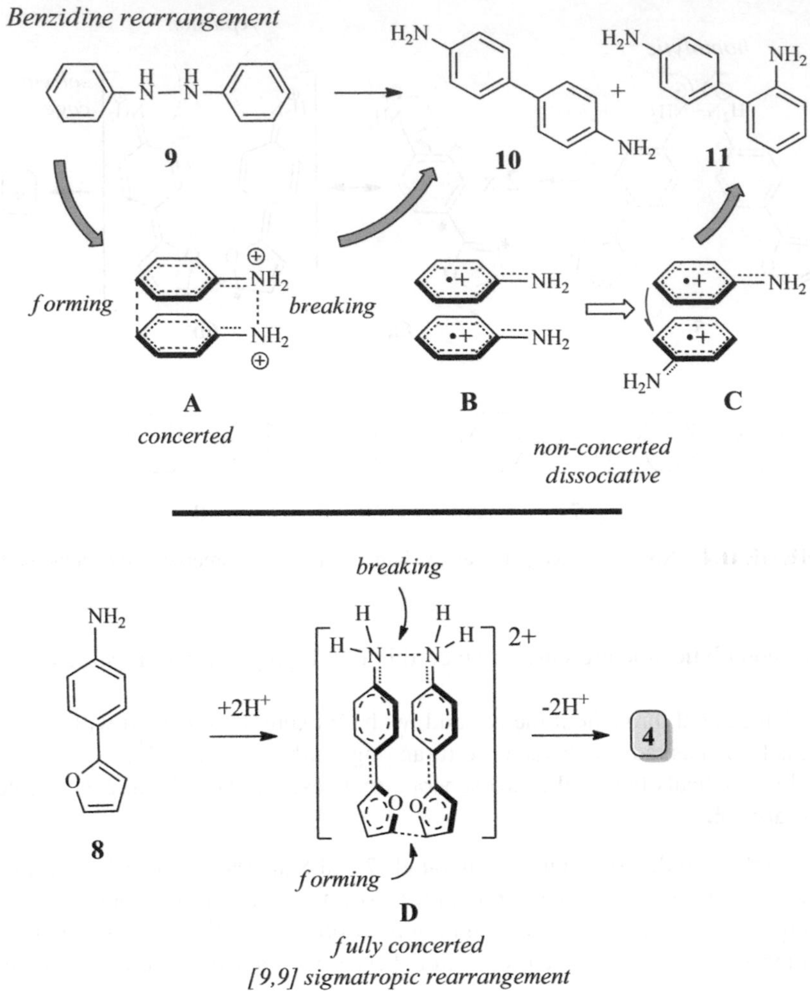

SCHEME II.5 A concerted route for the 3 → 4 isomerization based on the benzidine rearrangement, as proposed by authors [7].

b) Are involved orbitals coaxial, parallel, or orthogonal, independently of their proximity? *R: Coaxial (e.g.* $sp^3 + sp^3$*) for* σ *bonding or parallel (p + p) for* π *bonding. Orthogonal: no bonding.*
c) Are these orbitals compatible? (*R: of closest energy level and according to the highest occupied MO [HOMO] to the lowest unoccupied MO [LUMO] status, see below*).

This is relevant in the following cases:

- A filled AO (anion, NBP, singlet carbene) + a vacant AO (cation) or LUMO.
- A filled AO (anion, NBP, carbene) + *p* AO in a π system (as in addition reactions).
- A filled AO (anion, NBP) + developing *p* AO (as in bimolecular nucleophilic substitution [S_N2] reactions).
- A one electron-occupied AO (radical) + a filled AO (in propagation, radical addition reactions, oxidations, reductions) considering that only two electrons can occupy any AO/MO.
- Molecular associations such as:
  - Hydrogen bonding,
  - Bonding to transition metal catalysts,
  - Electrocyclic reactions,
  - Sigmatropic rearrangements, and
  - Dipolar, stepwise, and concerted cycloadditions.
  - The list comprises all bond-forming reaction types. You do not have to etch this into your memory, just acknowledge their existence.

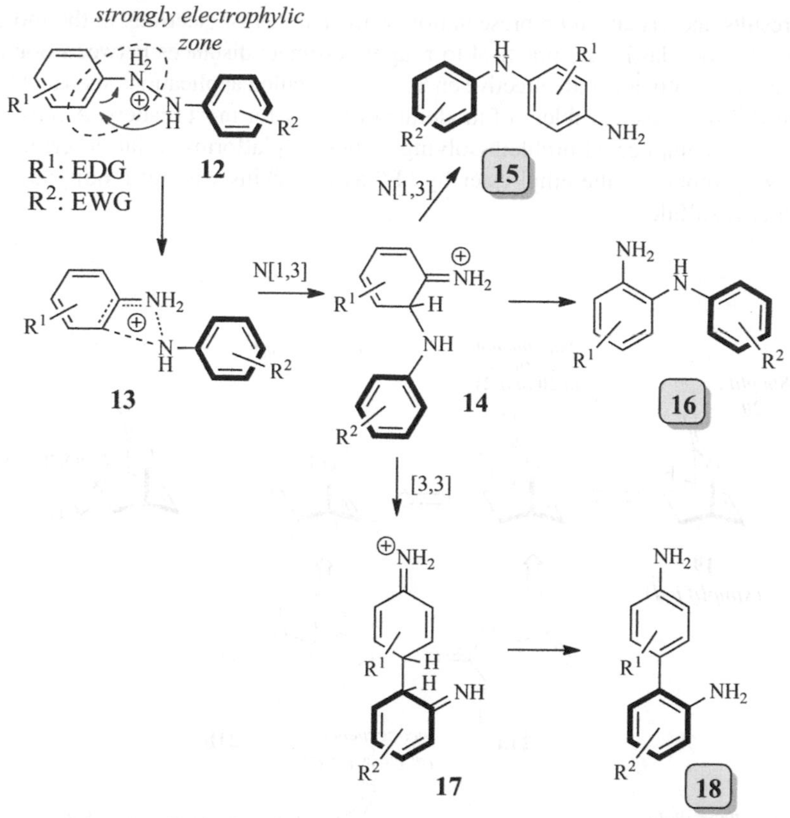

SCHEME II.6 One of the aromatic rings has been emphasized for the sake of clarity. Framed numbers indicate final observed products. EDG = electron donating group. EWG = electron withdrawing group (notation used throughout this book).

The research literature offers a quarry of examples in which atomic separations are critical to allow or exclude suitable AO overlap. Please have a look at Scheme II.7 [10] and design a feasible mechanism after a preliminary analysis in terms of what has been discussed so far.

Before moving on to solve the mechanism of Scheme II.7, let us point out that when proposing transient structures in the way to a target, the MO renderings described in the previous section (Figures II.2 and II.3) are not practical due to the increasing

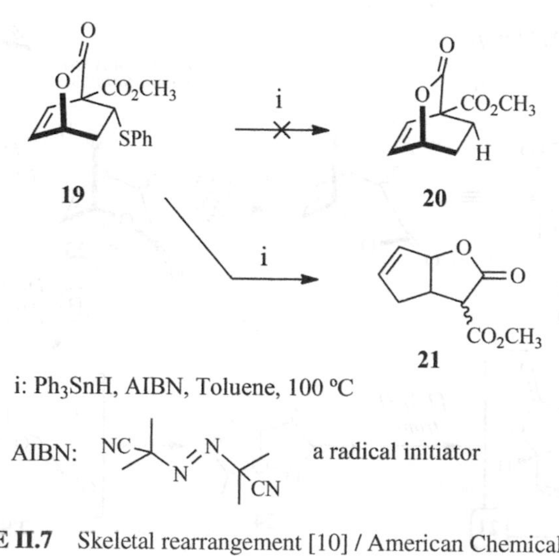

SCHEME II.7 Skeletal rearrangement [10] / American Chemical Society.

uncertainty of calculation results and visual MO representation as the number of electrons in the molecule increases. Molecular models, either Dreiding, or of good plastic, are practical to roughly estimate distances between bonding elements, better yet if computational molecular rendering software with effective energy minimization applications such as MMFF94s, at least is at hand.

A quick look at Scheme II.7 suggests a problem of moderate difficulty for most readers. A reasoning sequence is shown in Scheme II.8 based on the still uncomplicated problem-solving technique platforms so far described herein. We begin by simplifying a bit compound **19** by replacing the ethyl ester by "E" as it remains without modification in the final product and ignoring temporarily the phenyl sulfide.

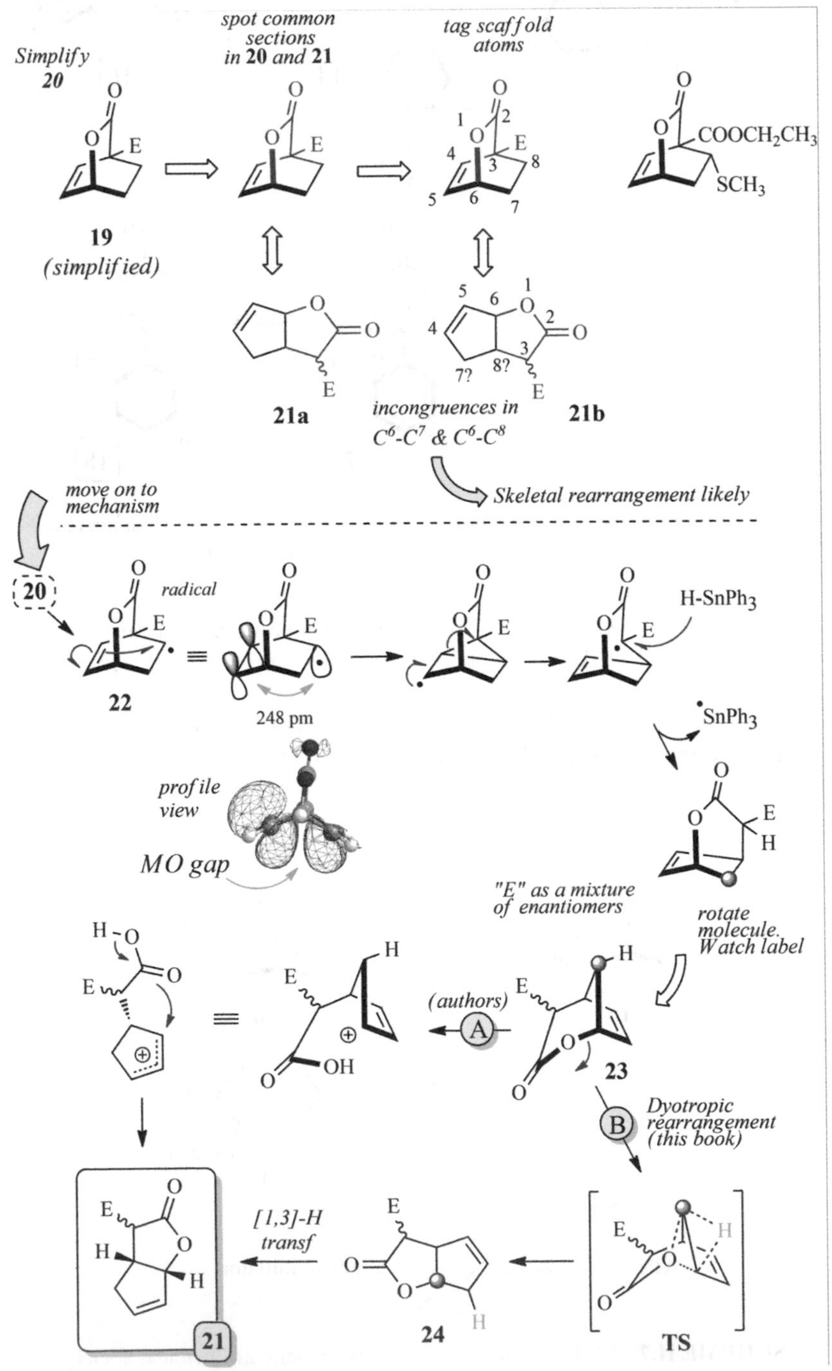

SCHEME II.8 Mechanism analysis and sequence of the reaction of Scheme II.7.

After realizing that the lactone moiety appears in both starting material and target, one can use it to examine the rest of the carbon scaffold relative to the lactone and determine whether changes have occurred there. Because the [2.2.2] bicyclic structure rips open during **19** → **21**, it is difficult to foresee the way it does so beyond knowing that some C–C dislocation must have occurred. Tagging atoms in the backbone of **19** and identify them in product **21**, a technique we will use repeatedly in this workbook, is one helpful instrument for the task of revealing bond formation and cleavage.

Reasoning sequence highlights:

- After adding pertinent atom labels in **19** (simplified version; Scheme II.8, top section) and translating to **21**, one sees that connectivity is maintained except for an incongruence around C^6 to carbons C^7 and C^8. This is sufficient evidence to propose that the scaffold was cleaved and underwent rearrangement.
- Since we have no other clues, we will have to *push forward* compound **19** toward the target while keeping an eye on **21**. Please, follow the yellow arrow to a feasible mechanism and carry on.
- The mixture of triphenyl tin hydride and azobisisobutyronitrile (AIBN) at high temperatures–and at much lower temperatures–is an excellent radical promoter, and the best candidate is the removal of the phenyl-sulfide substituent (set aside temporarily). Indeed, this was the original objective to synthesize **20,** but **21** got in the way. Radical **22** arises as the first active species in the sequence.
- *The central question arises*: could $C^{(\bullet)}$ interact with any radical-sensitive part of the rest of the molecule or is it too far removed from the radical-reactive C=C bond? Computer models place the $C^4 \leftrightarrow C^8$ distance between the atomic centers at 2.48 Å (248 picometers or pm). According to our previous discussion, it is still within AO reach for electron transfer. However, a MO depiction (color inset, profile view) shows a gap between the C=C π MO and the radical AO (as *endo*).
- In this inter-orbital gap, the probability of electrons occupying this zone is very low, hence electron flow should not be favored, and no reaction beyond this point should ever take place. How would you explain continuation to the target? There are two ways:
  1) Molecules are not rigid entities, but are endowed with degrees of freedom, translational, rotational, torsional, and vibrational, as you know. Only rotation, torsion, and above all, vibrations spanning symmetrical and unsymmetrical stretching, bending, scissoring, waging, and twisting, do change them. Thus, the position of atoms and distance to other atoms varies constantly around an average value, which is the one provided by theoretical calculations based on a set of rigid atomic coordinates. Advanced calculations in consonance with infrared spectroscopy have established vibrational amplitudes in C–C bond distances of simple alkanes on the order of ±0.079 Å [11]. AOs and MOs accompany these motions and orbital overlap will reduce the gap in question to possibly reach momentary overlap for electron transfer. (See following box; check Note 5 also).
  2) The discussion of point 2.4.3.1 in reference to the quantum tunneling effect (below).
- Once this roadblock is by-passed, the actual scaffold rearrangement proceeds until the stabilized radical reaches termination stage in **23** by H transfer from excess triphenyl tin hydride. However, **23** is not the isolated product, so further conversions lay ahead.
- Two routes ending at target product **21** are conceivable. Pathway **A** (proposed by Markó and coworkers [10]) moves along the short, self-explained heterolytic route although there is no elicitor, be it ring strain, highly polar solvent (which toluene is not) or protic acid in the medium, save perhaps for high temperature (100 °C). Route **B** in turn (my own design), is somewhat feasible all the same, by making use of a dyotropic rearrangement (see Scheme II.15 and accompanying text) based on the concomitant [1,2] shift of sterically opposed H and O atoms passing through the depicted TS. Pay attention to the labeled carbon (red ball) originally located in the bicyclic bridge, which ends up in the angular position of **24**. Final isomerization of the C=C bond furnishes target **21**.

a) Dreiding and computer models show the distances between atom nuclei, and not between atomic orbitals. In fact, *p* and sp^3 AO lobes extend away from nuclei centers. Depending on their spatial direction, the sp^3 lobe defining the space of the radical may be much closer to the other radical counterpart.
b) Multinuclear molecules are not stone-rigid entities but endowed with vibrational modes that constantly affect bond distances and dihedral angles. The computed interatomic distances are *average values* only and do not estimate the shorter distances stemming from these bond vibrations, whose dimension in Å is not known.
c) The familiar pear-shaped or elongated AOs and MOs are drawn with a sharp edge at the terminus as if ending abruptly in a zero-probability space. These orbital images only illustrate the volume where the probability of finding electrons is 90%. The remaining 10% diffuses away as the wavefunction defining the orbital wanes into space. The edge-to-edge contact

(*continued*)

between orbitals does not have a concrete meaning in mechanism design owing to the diffuse nature of this boundary. Therefore, orbital interaction may be extended (cautiously) beyond the interatomic distance usually taken as reference.

d) In mechanism design, we can only presume that molecular conformations offering the best possible orbital overlap are more likely to participate in a TS favorable to electron traffic in anticipation of products than other TSs that do not.

2.4.3.1 AO/MO Limits and the Quantum Tunneling Effect A few pages back it was stated that as electrons populate molecules in the quantum atomic and MOs, electron redeployment and migration occur only after two atoms or molecules undergoing reaction are close enough in the TS to attain some degree of orbital overlap. This picture leads to the notion that electrons are strictly excluded from abandoning AOs and MOs during reactions and released into the reaction medium.

True enough, but not 100% accurate.

MO shapes and location define the probabilities of spatial π electron distribution and redeployment in inter and intramolecular reactions that proceed with >99% probability. However:

1. *AOs and MOs are nothing but brilliant mathematical solutions to the Schrödinger equation*: wave functions ψ that define probability spaces with favorable energy for quantum particles, nucleons, or electrons, to exist *somewhere* in atoms considering the Heisenberg's uncertainty principle. Yet, AOs/MOs have no "edges or borders" that distinctly separate the *yes* and *no* spaces for electrons to cuddle up, as the classical balloon-shaped renderings lend us to believe.

   The way we all draw AO and MO as solid objects is just an instructive, practical, artistic exercise to provide a layperson's perception of quantum things. But is it reasonable to sketch electrons as single dots in radicals or a couple of tiny blotches to show NBP at exact places in a chalk/board mechanism the Lewis way? Well, yes if you realize that this is just a rather crude model, a manner of expressing such a fuzzy and elusive concept as a quantum particle in humanly macroscopic terms, as well as sticks representing bonds.

   That said, we get some comfort from learning that squaring the wave function ψ^2, which defines the shape of an occupied AO, is a measure of the probability of finding electrons within this space. This probability is about 0.9, that is 90%. This number drops rapidly as one gets away from the drawn AO or MO "border", meaning that there is still a 10% probability for electrons to exist beyond this edge. *Remote intramolecular migration* occurs by quantum particles in the 10% probability space.

2. *The question of energy barriers and the tunneling effect* (QTE)

   Chemical reactions and the mechanisms assisting them occur entirely in the quantum realm. Imagine now that a quantum particle – an electron in this chapter's context – with kinetic energy *Ek* approaches a barrier hill of *Ho* potential energy and *L* wide. If $Ek > Ho$, the electron surpasses the energy barrier seamlessly and carries on to the other side with 100% probability of success. In classical terms of the transition-state theory (TST), we assume that there is sufficient energy in the system (e.g. heat, radiation) to overcome the activation energy $\Delta E^{\ddagger}$ and reactants proceed to products.

   Conversely, when $Ek < Ho$, two things can happen at the quantum level:

   a. Unable to surmount the barrier, the particle and its associated wave package bounces back with 100% probability, as intuitively expected, in consonance with the classical TST. No reaction occurs unless more heat or radiation energy is applied.
   b. The particle portion cannot go further up the hill and goes back, but part of the wave portion continues its forward course *inside and across the barrier*. If the barrier's width *L* is infinite, this wave will decay inside the barrier and eventually vanish. But if *L* is finite, the wave portion will have a chance to get across the energy hill and emerge at the other side, albeit attenuated, depending on the barrier length, not its height.

Because the wave portion equally represents the quantum particle portion, there will be a certain probability to the passing particle-wave electron for appearing on the other side of the barrier, despite $Ek < Ho$, that is, in regions of space where electrons were not supposed to go. One concludes that *the electron tunneled through the barrier*. This phenomenon is the quantum tunneling of electrons (QTE), also known as quantum-mechanical tunneling (QMT, see box).

Quantum Tunneling, Electrons, and Lab Bench Reactions

It is well established that *quantum particles* (QP) are not subatomic chunks (dots and arrows drive us to believe in our sketches) in motion but objects possessing the dual wave–particle characteristic, first proposed by Louis de Broglie in his 1924 PhD thesis (Nobel Prize, 1927). His hypothesis has been experimentally demonstrated several times including in recently studied organic reactions.

The wave package associated with a moving QP that shows the two basic characteristics of transversal sinusoidal waves: wavelength and amplitude. The former is known as the *De Broglie wavelength* λ. This parameter is inversely proportional to the QP mass and velocity according to De Broglie Equation (2.1) for particles moving at non-relativistic speeds (10^{-2} at most of the speed of light [$2.999 \times 10^8 \, m\, s^{-1}$]):

$$\lambda = h / mv \tag{2.1}$$

h = Plank's constant $6.6262 \times 10^{-34} [m^2 kg\, s^{-1}]$
m = mass of the particle [kg]
v = particle velocity [$m\, s^{-1}$]

Given that electrons are the main character of this chapter, let us calculate λ of an electron of 1 eV of kinetic energy *Ek* in an AO. *Ek* (in *joules*) must be translated first into "v" to apply Equation 2.1. For non-relativistic velocities, *Ek* and "v" are associated by Equation 2.2

$$Ek = ½mv^2 \tag{2.2}$$

Solving for v:

$$v = (2Ek / m)^{0.5} \tag{2.3}$$

Given that:

$$m_{(electron)} = 9.11 \times 10^{-31} \text{ kg}$$
$$1 eV = 1.602 \times 10^{-19} J$$

Then:

$$v = (5.66 \times 10^{-10}) / (9.54 \times 10^{-16}) = 5.93 \times 10^5 ms^{-1}$$

Applying Equation 2.1:

$$\lambda = h / mv = (6.6262 \times 10^{-34}) / [9.11 \times 10^{-31} \times 5.93 \times 10^5] = 1.23 \times 10^{-9} \text{ m}$$

This is 12.3 Å, a much longer distance than from two remote positions in large organic molecules and the expected width *L* of electron-born fields like those associated with steric hindrance. Indeed, a number of intra- and intermolecular reactions involving electron, hydrogen atom, proton, hydride, and carbon atom shifts occur by quantum tunneling competing against the classical thermally activated reaction course and have been observed in very low temperature settings (a few Kelvin) in which thermal reactions are exceedingly low or inexistent [12–16].

From the reaction-mechanism perspective, the quantum potential energy barrier *Ho* may be an electromagnetic field offered by other electron clouds existing in another part of the same molecule that hinder intra- or intermolecular electron (proton or hydride) migration, or a separate compound attempting to form a TS with the first.

QPs that tunnel physically across the *Ei* < *Ho* hill do not emerge instantly on the forbidden side but after a measurable time lapse. The success of this passage depends on the QP wavelength λ: the longer λ (and shorter distance *L*) the higher probability for successful tunneling.

However, these increasingly sophisticated findings, no matter how superb, are beyond the scope of this book, but can be called upon in many reactions in which AOs/MOs are farther than bond distances (e.g. 3–4 Å). Indeed, there is an increasing number of organic reactions that may be accompanied as well by a rate enhancing QTE.

2.4.4 Issue 4. Electron Traffic and Stereochemistry

☑ *Electron flow is strongly contingent on molecular configuration and conformation. Stereochemical restrictions must be overcome to allow for electron traffic between molecular orbitals involved in bond-forming operations.*

This is a logical offshoot of *Issue 2.4.3* and a governing feature of regio- and stereoselective processes, and reactions confronting differential steric configuration and encumbrance.

The steric configuration is an intimate molecular property associated with several characteristics and reactivity:

✓ Molecular and local group polarity and interaction between them.
✓ Steric strain, hindrance to σ-bond rotation.
✓ Limits to degrees of freedom and entropy changes.
✓ Enthalpy of formation.
✓ Electron flow between sections of the structure through resonance or impediment to it.
✓ Intramolecular associations from van der Waals and London forces.
✓ Thermal or radiation stability.
✓ Accessibility to solvents and solubility.
✓ Susceptibility to external reagents–including catalysts–that govern the reaction possibilities.

Scheme II.9 provides an introductory perception of the impact on product outcome of a seemingly irrelevant detail regarding the configuration of just one chiral carbon in a 13-carbon scaffold (C*). For a solution of the reaction mechanism see Problem 12 in "The Problem Chest", Part II.

Stereochemical influences on mechanism are legion and deserve a separate chapter (see Chapter 3).

2.4.5 Issue 5: Electron Energy Level and Accessibility

☑ *Electrons at higher energy levels are more accessible for reaction than at those at lower energy levels.*

The roots of this rule are found in the AO model: In multiple covalent bonds, electrons in π orbitals are more exposed to outer reagents than electrons hidden in the coaxial casket of σ bonds between atoms linked by multiple bonds. In the molecular orbital model, electrons in π bonding MOs lay higher in the energy scale than those in σ orbitals, so you will never see the σ orbital reacting first then π orbitals in a C=X bond (X: C, heteroatom). You should, nevertheless, care about the σ-bonding capacity of NBP in C=X (N, O), which occupy sp^2 AOs.

The topology (spatial arrangement) of these MOs reflects this feature. This is illustrated by acrolein MOs drawn in Figure II.5 at the same scale. The energy difference between bonding σ and π MOs is about 30 eV. As a result, the reactivity pattern of acrolein is governed by the π MO rather than the σ C–C or C–H bonds. Electromagnetic effects are transmitted through the σ and π architecture in response to the electronegative difference between C and O, giving rise to a significant dipole moment ($\mu = 3.11$ D)–higher than that of water ($\mu = 1.85$ D).

The reactive dominance of multiple bonds does not rule out σ-bond activity in compounds where π bonds are few or absent. Substitution, elimination, and structural rearrangements are typical reactions involving σ bonds, some of which are seemingly inactive. Scheme II.10 depicts one such case that involves a deep-rooted transformation of the molecular scaffold elicited by a strong base (t-BuO$^-$) and heat [18]. The only C=O present is a suspect of triggering this turmoil.

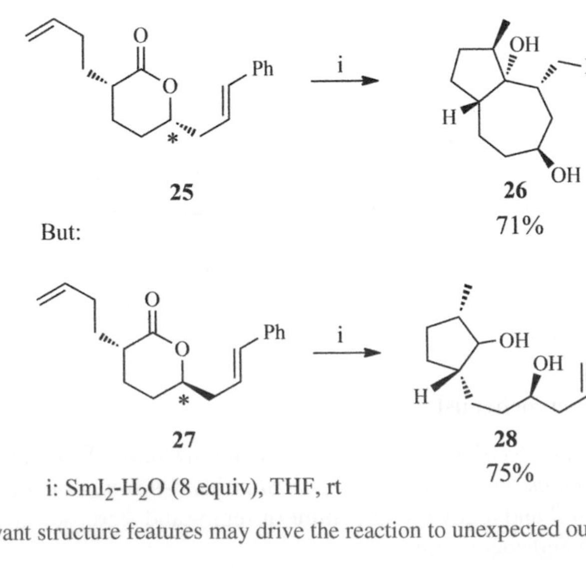

SCHEME II.9 Seemingly irrelevant structure features may drive the reaction to unexpected outcomes. [17] / American Chemical Society.

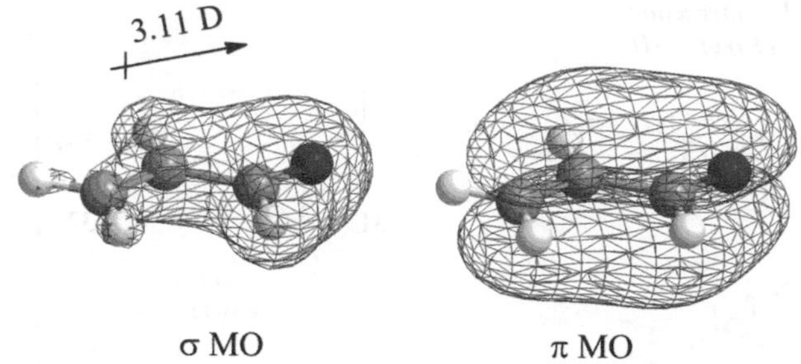

FIGURE II.5 Hückel's molecular orbital surfaces of acrolein drawn at the same scale. Left: σ_1 MO (E=−51.17 eV) showing polarization of the EW effect of carbonyl and the bulged contribution of the NBPs. Right: π_1 MO (E=−21.16 eV) wrapping the entire α,β-unsaturated carbonyl moiety. Arrow indicates the direction of the dipole moment vector.

Carbonyls in strong alkali often do this through the acid character of the α–CH, as you know. The difficulty in **29** is that both α-carbons are blocked (the bridgehead CH is orthogonal to the C=O π bond and hence much less acidic). Any ideas to bypass this roadblock?

A preliminary analysis of what may be happening is shown in the lower part of Scheme II.10. Recognizable molecular sections in **29** and **30** are earmarked to track the bicyclic ring opening at C(α)–C=O. Following these hints, you can perhaps go ahead and envisage a feasible mechanism. Once you do, check out a likely answer in Scheme II.11.

Because there is a strong non-nucleophilic base (t-BuO$^-$) we are limited to developing anions arising from proton abstraction. The question is which proton. Three additional questions should be solved:

a) Find a way to dislodge [CH_3O^-] and keep it available for a later attack on C=O as this cannot occur simultaneously *because of distance*.
b) Alternatively determine whether the unimolecular elimination (E1)cb of CH_3OH might be performed by t-BuO$^-$ abstraction of the vicinal H (β-relative to C=O) on the ring if appropriately activated by C=O.
c) Cleave the C(α)–C=O bond with stabilization of the presumed anion thus formed.

These ideas are worked out on a simplified version of **29** in Scheme II.11by removing the molecule's front section, which screens the intervening elements on the back side. It will be put back in place later. The 3D energy-minimized (MMFF94s) model (**29-3D**) I ran unveils the proximity of the C=O p and the C(β)–H sp^3 AOs, providing activation of this H to become sufficiently attractive to the strong base clutches. Once the C(β) anion is formed, the sequence moves forward with relative ease.

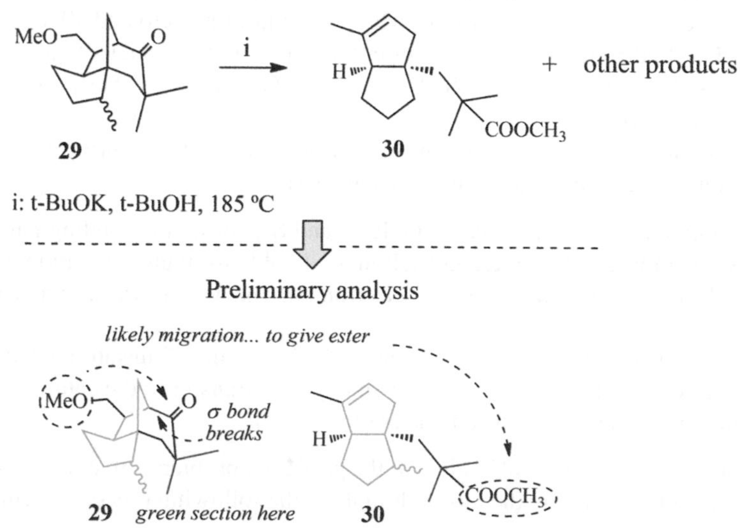

***39* and *30* both contain 15 carbons, no loss of fragments**

SCHEME II.10 The reaction **29**→**30** was adapted from [18] / Canadian Science Publishing.

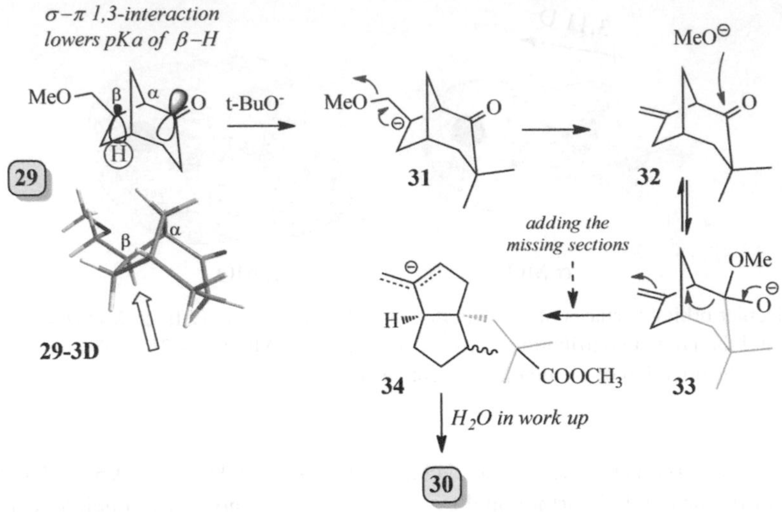

SCHEME II.11 A likely mechanism for the reaction of Scheme II.10.

Target **30** was recorded as the carboxylic acid (we kept the methyl ester as a clue), which was formed during work up [18].

A great deal of mechanism problems, from accessible to extremely difficult ones develop from σ bonds almost exclusively, no matter how active multiple bonds may be.

2.4.6 Issue 6: Electron Flow and Molecular Active Sectors

☑ *Electrons are redeployed more favorably between electronically active sectors, but not always.*

Useful to the organic reaction problem solver is the quick identification of possible electron-activity sites in starting materials and intermediates in domino or cascade reactions. Choices are easily picked from an ever-growing menu of functional groups.

One comes across compounds loaded with functionalities to play with, as in many trendy *multi-component reactions* (MCR) (Scheme II.12, top part) [19]. By contrast, others appear almost denuded of functional groups although impressive molecular complexity can be built from them with the use of very few reagents and a smart synthetic design (Scheme II.12, bottom) [20] (see also Problem 41 of "The Problem Chest" in this edition and Problem 7 [idem], of the second edition). Of note, potential functional carbinols in **40** are blocked by protective groups, which leave just one keto and allene functions to operate.

Functional groups can be classified following a variety of criteria such as σ-X (nucleofuge), nucleophile, or their electrophilic quality, such as π bonds and heteroatoms with NBPs (chiefly N, O, halides). Other properties include conjugation and polarity that may reinforce their properties.

For mechanistic purposes, finding the natural direction of electron redeployment during functional groups interactions is handy to organize them according to their qualitative electron density (ED):

High (HED): generally good nucleophiles, anions of O, P, N, and S ylides, non-bonding pairs in N, O, and R_x**CH**-EWG units (as in ketones, esters, imines, sulfoxides, and sulfones) suitable for tautomerization, proton abstraction, and carbanion generation as well as groups with accessible electrons in alkenes, alkynes, and aromatics with electron donor (EDo) substituents.

Low (LED): generally good electrophiles, from $C–X^{\delta+}$ and $C=C–X^{\delta+}$ as in α,β-unsaturated aldehydes, ketones, and esters or other EW groups, and Lewis-acid active groups that undergo ED transfer to the catalyst and a consequential decrease of their own ED enhancing their electrophilic character (Table II.1).

Besides the abridged collection of functionalities of Table II.1, the proficient problem solver of reaction mechanisms should pay attention to several molecular attributes. Among the most relevant are the following (check Scheme II.13 as you read)

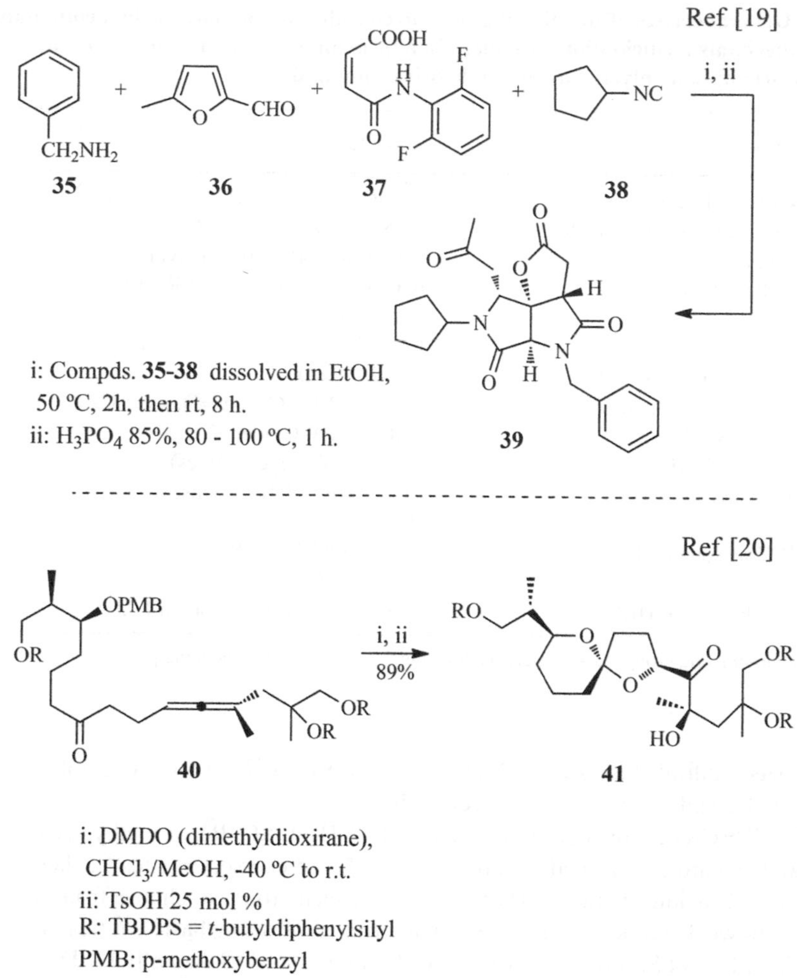

SCHEME II.12 [19] / American Chemical Society; and [20] / American Chemical Society. For a wholesome discussion, see Problem no. 39 and 43, respectively in the second edition of this book.

1. **Ring strain**: This is observed in many alicyclic compounds. The simplest examples arise in cyclopropyl and cyclobutyl units in starting materials but is no longer present in the products. Epoxides and aziridines are particularly susceptible (see Problem 7 in "The Problem Chest", Part II) as well as small carbocycles with EWG (Scheme II.13, section **A**), EDGs, or a combination of both. *Ring contraction* is often observed (Scheme II.13, sections B and D). [x.y.z] multiple cyclic models offer opportunities for ring cleavage and rearrangement (see, for example, Scheme I.1 and Problem 49).
2. **Cyclization comfort**: whenever alicyclic scaffolds are built in intermediates or targets, five-, six-, or seven-membered rings, either all carbon, heterocyclic, or during transient hydrogen bonding are more energetically accessible than larger or smaller rings. This tenet does not rule out cyclizations to larger rings, smaller carbocycles such as cyclopropyl groups, which often show up as intermediates, or end products.
3. **In acidic media**, Brønsted or Lewis, look for potential $C^{(+)}$ sites in carbon-nucleofuge groups: diazonium salts ($N{\equiv}N^{(+)}$–CR_3), halides, C–OR (R=H, alkyl), epoxides and thioepoxides, dihydrofurans, acetals, and dihydropyrans, previously O– protonated or bonded to Lewis acids, as well as sulfonic acid derivatives R–SO_2–O– (tosylates, mesylates, triflates). These processes also occur within the same molecule (Scheme II.13, section **C**).
   a. In potential $C^{(+)}$ sites, look for stabilization through resonance or *+I* effects. Polar solvents and solvent intervention–high dielectric constant–(water, dimethyl sulfoxide [DMSO], hexamethylphosphoric triamide [HMPA], dimethyl formamide [DMF], water-dioxane, methanol, or tetrahydrofuran [THF] mixtures). Solvents with low dielectric constant (pentane, hexane, benzene, chloroform, dichloromethane, dichloroethane) do not contribute to the stabilization of $C^{(+)}$, yet cations are often proposed in the latter three solvents. Thus, pay attention to the solvent in your problems.
   b. Consider $C^{(+)}$ intermediates in: unimolecular nucleophilic (S_N1) substitutions, E1 eliminations, and as starters of skeletal rearrangements.

TABLE II.1 General classes of functional groups to consider first in most carbon compounds at the early stage of mechanism elucidation, organized in approximate decreasing order of reactivity. The list is designed for orientation only, not as comprehensive compilation

High electron density	Low electron density
Anions of C, N, O, S, $H^{(-)}$; halides	$C^{(+)}$ /C=C–$C^{(+)}$/ Aryl–$C^{(+)}$/EDG–$C^{(+)}$
Free radicals of C, O, S, H, halogen atoms	C=$N^{(+)}R_2$, X–C=O (X=Cl, Br)
Strong basic reagents	R(C=O)–O–(C=O)R (anhydrides)
Photo-excited (p-triplet) states	R(C=O)–OR, R(C=O)–NR_2, RO–(C=O)–OR
(−)CN	Epoxides
Carbenes, nitrenes	R–C≡N
P and S ylides: $RnX^{(+)}$–$C^{(-)}$; X=P, S	C=NR / C=O
C=C–$X^{(-)}$ / C≡C–$X^{(-)}$ (X=O, N, C)	C=C–C=NR / C=C–C=O, idem C≡C···
C=C–XR / C≡C–X–R (X=O, N, S)	C=C–C–EWG, idem C≡C···
C=C–X / C≡C–X (X=C, H)	C–C–Y (Y = nucleofuges)
C–C–$X^{(-)}$ (X=O, N, S)	RO–CH(R')–OR (acetals)
NBP of N, O	R_2C=C=O
EWG–CH_n–EWG as source of $C^{(-)}$	RN=C=O; RN=C=S
EWG–CH_3 idem	R_2C=C=CR_2'
Lewis bases (as inductors of HED)	Lewis acids (as inductors of LED)

EWG = electron withdrawing group; EDG = electron donating group; NBP = non-bonding pair

c. Search for polarized multiple bonds $C^{(\delta+)}$=X (X = C, O, N, S) or $C^{(\delta+)}$≡C–EWG combinations, which have excellent functionalities for 1,2 and 1,4 addition and cycloadditions
d. Allylic (R–CH_2–CH=CR_2), propargylic methylenes (R–CH_2–C≡C–R), or methynes are also sources of *cations*.
e. **In basic media**, disregard cations and identify potential $C^{(-)}$ precursors at sites flanked by at least one strong EWG that will enhance C–H acidic character. There are several such groups to choose from for preferential abstraction by HOMO bases, from weak amines and $HO^{(-)}$ to an assortment of strong basic entities including metal hydrides, alkyl derivatives of C, N, O, and Si possessing Li, Na, or K counter ions (Scheme II.13, **D**).
f. Allylic (R–CH_2–CH=CR_2), propargylic methylenes (R–CH_2–C≡C–R), or methynes equally serve as sources of *anions and radicals* as well as cations depending on reaction conditions and solvents.

4. If present, evaluate the **base power**: there is a plethora of anionic and neutral bases, from moderate to exceedingly strong (superbases), and anionic or neutral with carbon, nitrogen, or phosphorous centers, developed in the past decades that you may wish to get familiar with [24].
5. **In neutral media**, nucleophiles and electrophiles can be spotted among the moderate C=C, C≡C functions in polarized systems by conjugated EDG (as in RO–C=C, R_2N–C=C), or EDG (as in C=C–C=O, C≡C–C=O) and charged species arising from catalyst binding.
   a. C=X (X=C, N, O, S) and C≡C are susceptible to [n+m] cycloadditions, such as the Diels–Alder reaction, which act as dienophiles upon a breadth of dienes. Other electrocyclic reactions are also accessible to apparently inactive C=C units, such as the Cope rearrangement, ene-, and Conia-ene reactions (e.g. Scheme I.13, part **D**).
   b. *Retro-electrocyclic reactions* should be suspected during simplification of bicyclic carbon scaffolds or odd structural changes.
6. Consider radical intermediates under conditions of heat, UV radiation, or specific reagents for radical generation. The classical reagent to induce radical formation has been Bu_3SnH (TBTH) created *in situ* by the reduction of the alkyl-tin halide by lithium-aluminum hydride, as in Schemes II.7 and II.8. Novel radical inducers and more environmentally friendly options are replacing TBTH (for a review see Reference [25]).

2.4.7 Issue 7: Electron Flow and Compatible AO Types

☑ *Electron bonding interaction occurs in σ–σ, σ–π, and π–π combinations*

Additional help for the problem solver is considering the three general forms of AO interactions that command the construction of covalent bonding in organic compounds: σ–σ, σ–π, and π–π. Each type defines specific mechanisms ahead of starting

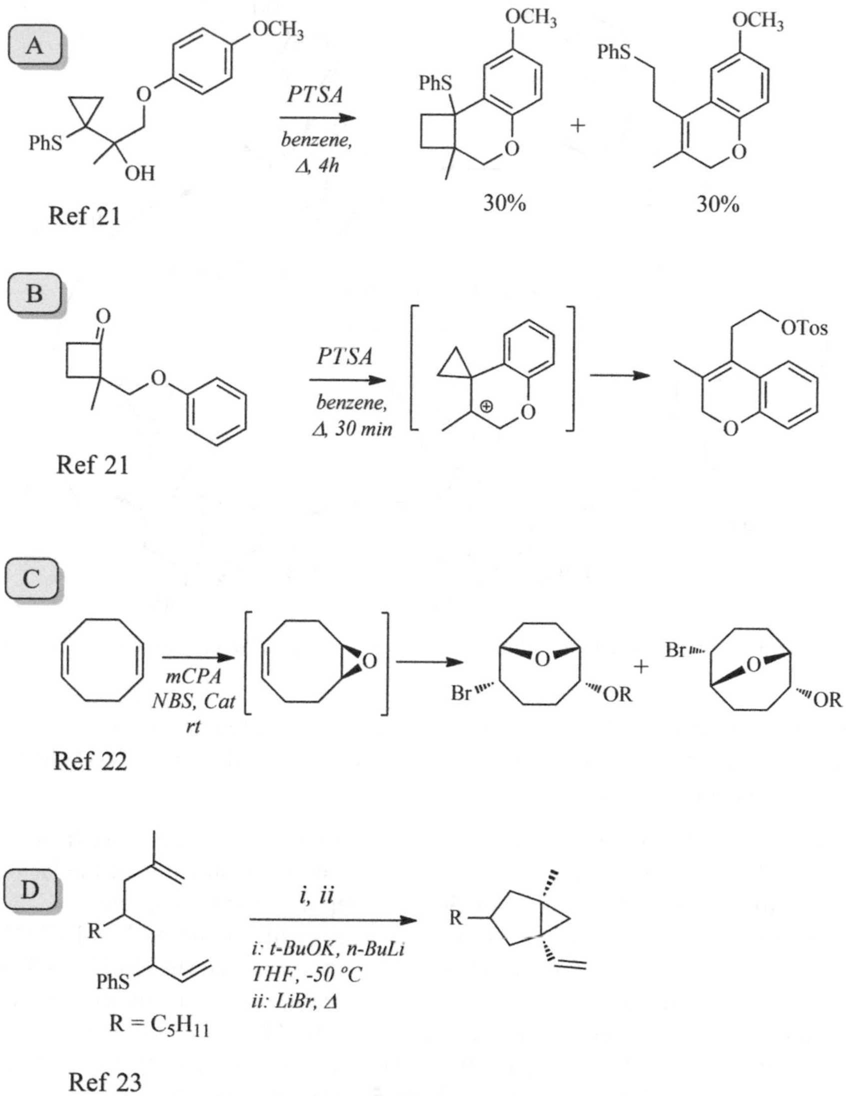

SCHEME II.13 Adapted from: A and B, [21] / American Chemical Society; C, [22] / American Chemical Society; D, [23] / American Chemical Society. Feasible mechanisms to these reactions are shown in Scheme II.43 at the end of this chapter.

materials, transition states, stereochemical features, and energy profiles. While this sounds quite elementary, reviewing the ensuing points will prove useful.

2.4.7.1 σ–σ Interactions In general, NBPs in N, O, and S that trigger σ bonding locate the reactive site in an sp^3 hybrid orbital. This is a common occurrence in σ–σ rapports during proton capture, S_N1 and S_N2, and Wagner–Meerwein skeletal rearrangements. Up to this chapter, several interactions involving electron redeployment between two σ AOs in reference to other processes have cropped up, for example sections of reactions A, B, and C of Scheme II.13. Abundant cases also appear in cascade reactions through a series of σ–σ interactions accompanied by molecular rearrangements, which may be confused with remote C–H activation of apparently dormant molecular sections (Scheme II.14) [26]. We need an activation center in **42** to awaken dormancy, the exocyclic C=C in this case by means of the electrophilic addition of BF_3. Subsequently, a σ–σ cascade does the rest to reach target **43**.

Of note, the [1,2]-group transfers take place in a suprafacial manner, (on the same plane of the molecule they are initially located), hence the original configuration of **42** is preserved.

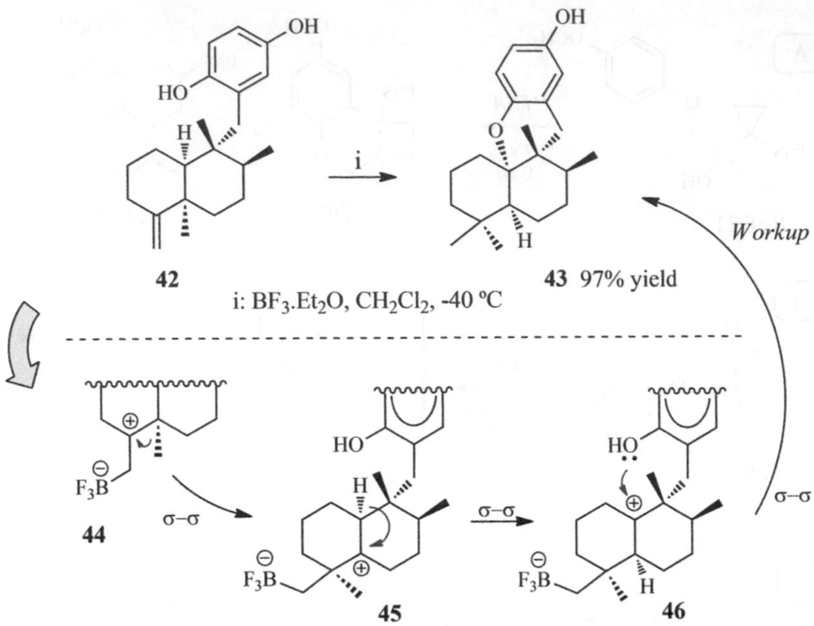

SCHEME II.14 Activation of a remote proton through a cascade of Wagner Meerwein rearrangements. [26] / Elsevier.

2.4.7.1.1 The Dyotropic Rearrangement Other σ–σ interactions take a different course; for example, observe the ring contraction of lactone **47** induced by 2% of trimethylsilyl triflate (TMSOTf) in toluene at 0 °C (Scheme II.15) [27].

One can rationalize this conversion through **three** pathways:

Route **A**: A straightforward route based on the loss of the angular OH aided by TMSOTf or freed as such in the medium. A tertiary $C^{(+)}$ (**49**) is produced, triggering the [1,2]-*O* shift **49** → **50** ring contraction of the lactone and the return of the loose $HO^{(-)}$ before it diffuses away and traps cation **50**. Both steps are σ–σ interactions.

Route **B**: A very close version of the preceding interpretation starts by the known S_N2 of TMSOTf + alkyl alcohols. The angular triflate **52**, among the best nucleofuges available, would then dissociate to a full $C^{(+)}$ **49** as above. Both sequences face trouble since an epimeric mixture, rather than the diasteromer **48,** should be the end product. Besides, toluene is a poor supporter of carbenium ions. A carbenium ion-free scheme would be more suitable.

Route **C:** The non-ionic sequence would accompany the triflate cleavage with a concomitant [1,2]-*O* shift of the lactone oxygen atom via two conceivable transition states, **53** and **54**. Stereospecifically defined *transoid* triflate-spirolactone **55** would remain unchanged after hydrolytic work up to the observed carbinol target **48**.

Authors of this work [27] opted to cut short the reaction time seeking to trap a **48**-like product before water elimination to give the exocyclic alkene (they actually worked with a more complex carbon scaffold, but it would bring visual confusion in the complete renderings). Their strategy succeeded to validate framed structure **53** in the form of the carbinol–not the triflate–in which OH and lactone oxygen atoms would undergo concerted [1,2] migration. This is formally a *dyotropic rearrangement* (DR) [28, 29].

The term is coined from the classical Greek *dyo* means "two", in reference to the intramolecular transfer of two σ-bonded atoms or functional groups located on neighboring carbons, or separated by three or more atoms, as long as the migrating groups are situated at close range.

Two types of DR have been recognized:

Type-1 DR: concerted, but asynchronous, [1,2] shift of two neighboring atoms, be they H, alkyl, aryl, or heteroatom-bearing groups; formally, a σ–σ exchange (Scheme II.16).

Type-2 DR: two σ-bonded vicinal atoms or substituents migrate either [1,2] or longer over to a C=C bond; formally in the σ–π domain.

In compliance with the Woodward–Hoffmann rules, both thermal DR migrations are *suprafacial*, and do not involve ionic or radical intermediates. This condition offers opportunities for sterically defined molecular construction.

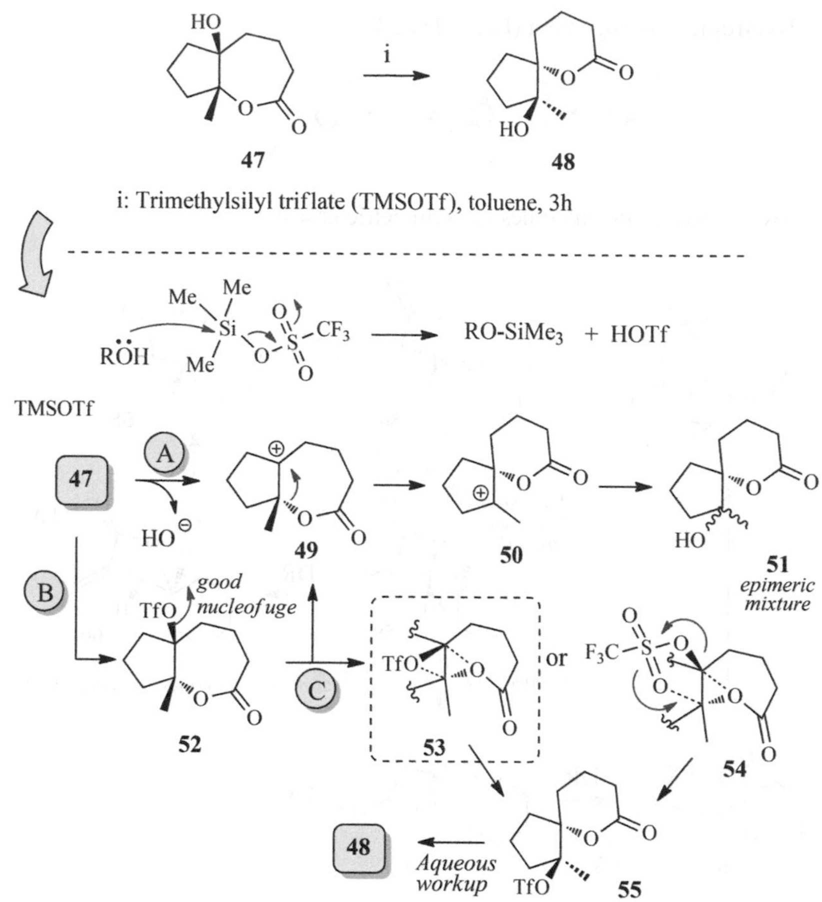

SCHEME II.15 The **47** → **48** conversion can be explained by three different routes A, B and C, which involves the *dyotropic rearrangement* shown in intermediate **54**. [27] / American Chemical Society.

Scheme II.16 illustrates one model example in the σ–σ domain (Type 1). However, the DR also occurs in the σ–π combinations (Type 2). DR is not a superfluous oddity but a powerful synthetic tool since it is regioselective and stereospecific, high yielding, and can be applied to a wide range of molecular structural types [27–30]. Note that both migrations are suprafacial, one (the methyl) occurs above the exocyclic C–C and the other under it. This is equivalent to a double S_N2 process *with inversion of configuration in both vicinal carbons* involved in the migration. Therefore, the shifting units preserve their relative 3D position in the molecule.

> *Takeaway*: whenever you have a carbocation chain sequence and get tripped up by an uncomfortable $C^{(+)}$, try a type-1 dyotropic TS while taking care of the stereochemistry.

2.4.7.2 ***σ–π Interactions*** Abundant examples of this type of bond construction and cleavage include the rich universe of nucleophilic addition–elimination of esters and amides (e.g. hydrolysis, alkylation); 1,2-additions to polar C=X bonds (X: O, N, S) by elements bearing NBPs (see Section 2.4.7.2.1) or anions; 1,4 additions to α,β-unsaturated C=X synthons (X: O, N); [1+2] and [3+2] cycloadditions of ylides, carbenes, and metal carbenoids; and several other types.

In the area of neutral organic functionalities with higher energy demands, there is also a collection of hydrogen-alkene or alkyne concerted shifts belonging to the ene- rearrangements. These are combinations of an enophile section and C=C bond bearing an allyl H atom, which are conceptually related to the classical diene + dienophile pericyclic cycloaddition commanded by the celebrated Diels–Alder reaction (Scheme II.17).

Hydrogen transfers of the σ–π class can occur between two different molecules, ene and enophile, which include five general types (**B** to **F** of Scheme II.17), as well as within a compound having the ene and enophile components embedded in the structure. These are known as **sigmatropic rearrangements (SRs)**. Formally speaking, the extended ene reactions **C** and **F** are SRs.

Dyotropic rearrangement (DR) - Type 1

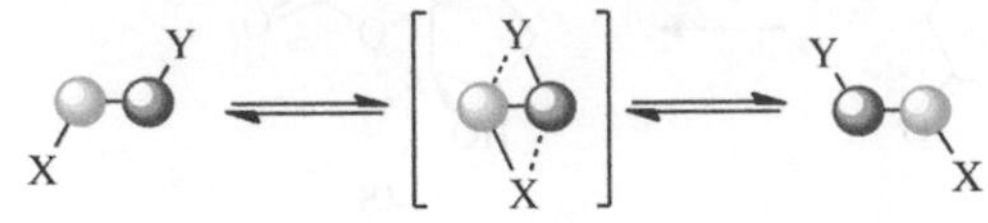

Two possible DR outcomes in asymmetric cases:

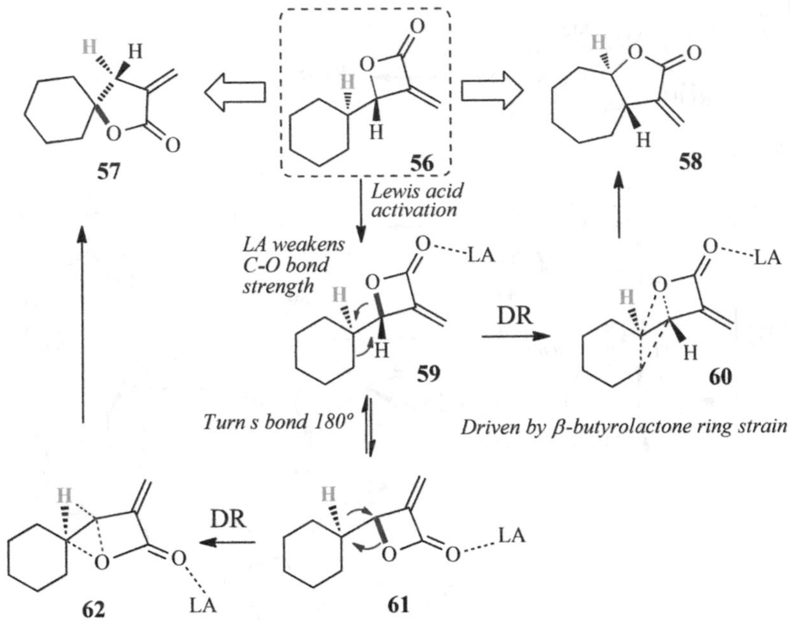

Dyotropic rearrangement - Type 2

SCHEME II.16 Types of DR and a simplified model of DR driven by α-methylene-γ-butyrolactone ring strain.

Reminder 1: for the *numbering system of SRs* the carbon atom (or other) linked to the migrating atom is labeled as no. 1, and the receiving carbon is numbered according to the chain length of *n* atoms comprising the involved π system and where the new *s* bond will be formed (check cases **C** and **H** in Scheme II.17). SRs boil down to [1,*n*] atom shifts. Side loops and cyclic sections are ignored for numbering purposes, as in case **F**, the extended Conia-ene reaction, another instance of [1,5] SR. Thus, one can expect thermal [1,3], [1,5], and [1,7] SRs to occur smoothly.

Reminder 2: According to the Woodward–Hoffmann rules (WHR), and application of the frontier molecular orbital theory (FMO) that govern these conversions, the course of the migrating H, metal element, or alkyl group relative to the plane of symmetry of the molecule to form a new σ bond with the target vinyl carbon can occur on the same side of the molecular plane where it is located at the start (a *suprafacial* course) or by cutting across this plane and binding to the target from the opposite side (an *antarafacial* course). A heat-induced SR may be allowed or forbidden by the WHR depending on:

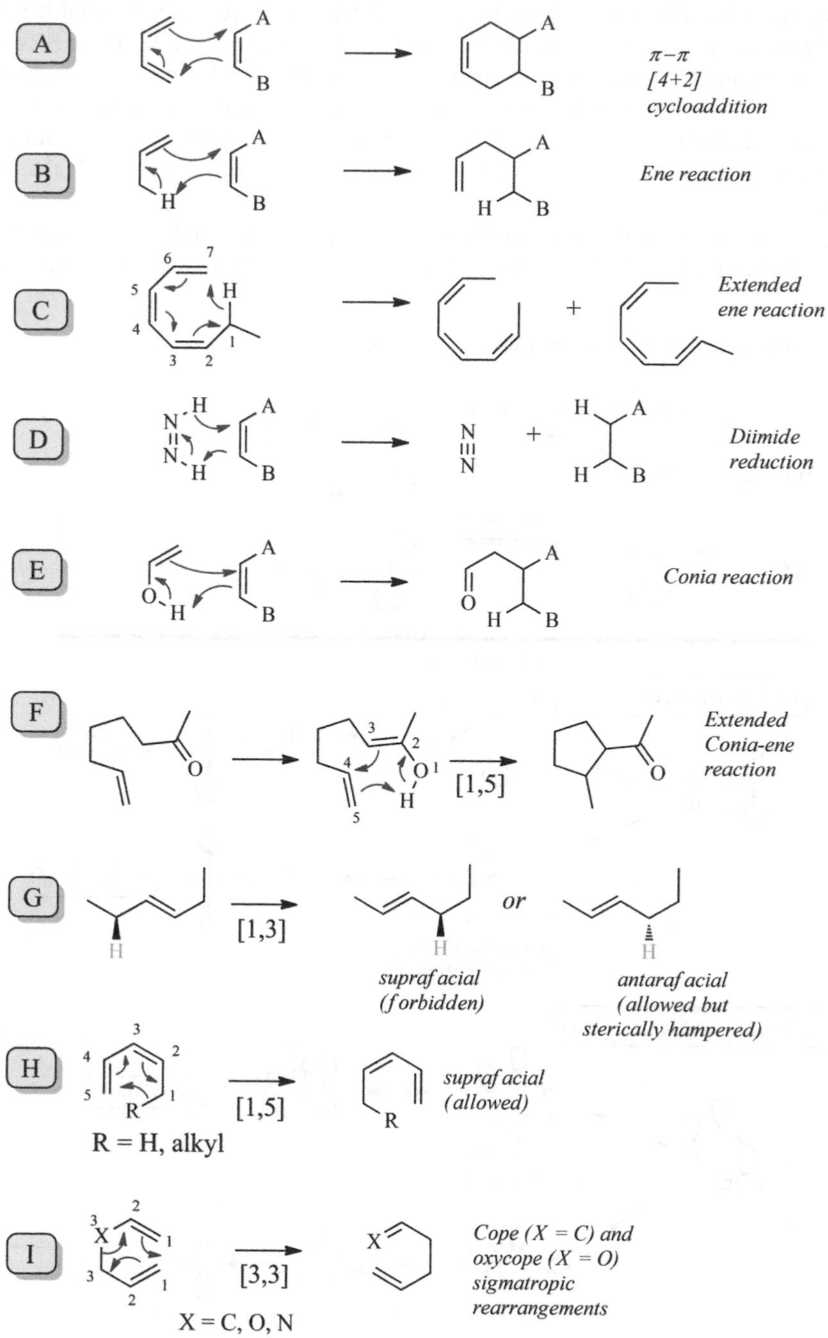

SCHEME II.17 Some common examples of SRs and the [2+4] cycloaddition (reaction **A**). The list is not a comprehensive SR collection but a reminder.

a. *The phase compatibility (±) of the involved orbitals* of migrating and receiving atoms, in the Hückel AO and/or the FMO frameworks.
b. *The number of electrons involved in the TS* assumed for the atom transfer, fulfilling the requisites of Hückel's aromaticity rule ($4n+2$ π electrons, n = integer) in the cyclic TS.
c. *The topology* or 3D shape of the molecular frame.

During problem solving it is advisable not to appeal to special circumstances to include forbidden transitions.

Figure II.6 illustrates the application of these concepts to selected cases to show whether these SRs transitions are forbidden or allowed in suprafacial and antarafacial situations, defined in the top section for 1,3-pentadiene and 1,3,5-heptatriene. While the suprafacial transfer, shown in **63** does not demand special conditions, the antarafacial approach requires that the π bond must be forced out of the natural coplanarity imposed by the conjugated system to allow the migration of the shifting group to the opposite side of the plane of the molecule. Loss of π-bond character implies a higher $\Delta H^{\ddagger}$ than the suprafacial transfer, hence it is less competitive.

The antarafacial [1,3]-X shift is physically unattainable if one accepts a concerted σ–π transfer. Orbital-phase incompatibility (dark versus light color) between H s^1 AO and the p AO it pursues to bind also highlights this difficulty.

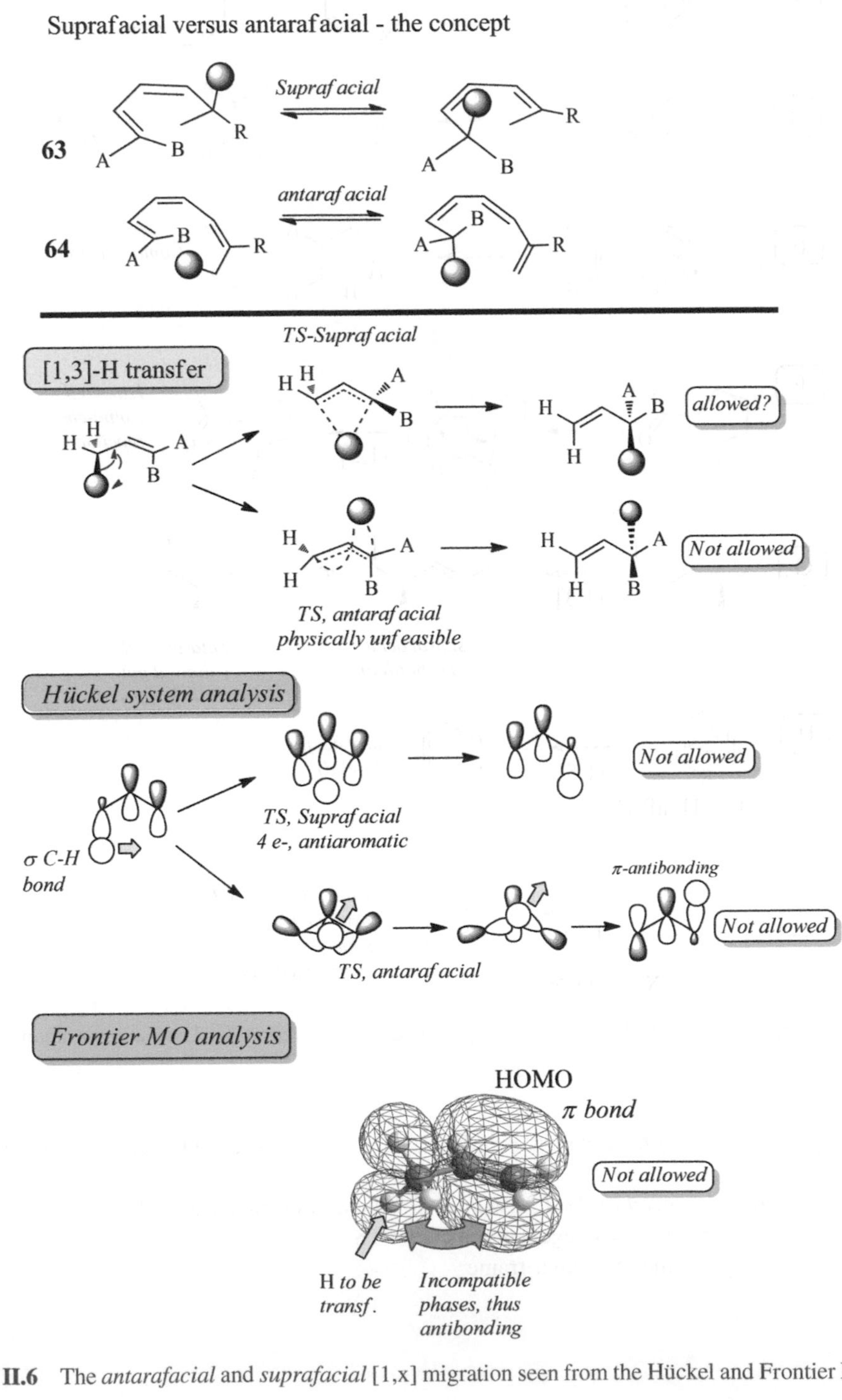

FIGURE II.6 The *antarafacial* and *suprafacial* [1,x] migration seen from the Hückel and Frontier MO analyses.

In the Hückel system, the analysis of the number of electrons involved in the TS ought to be consistent with the aromatic ($4n+2$ π electrons) or antiaromatic ($4n$ π electrons) character in this TS, as it is assumed to be cyclic, hence a closed orbital loop is electronically harmonious all around.

In the FMO treatment, the HOMO is examined to determine phase compatibility of the interacting σ and π orbitals. In propene, for example, the phases do not match, and therefore, the thermal suprafacial [1,3]-H is not allowed.

A similar analysis for the three treatments to predict whether the thermal [1,5]-H transfer is allowed or forbidden is shown for a pentadiene system in Figure II.7. All three conclude that the suprafacial transfer is thermally permitted [31].

Yet, when electromagnetic radiation (e.g. UV) is the energy source, *supra-* versus *antarafacial* symmetry H transfer permission leads to the opposite predictions (Table II.2). This can be rationalized by the promotion of electrons in the π HOMO to a higher energy, photo-excited state π* that furnishes a complementary orbital-phase matching in the TS.

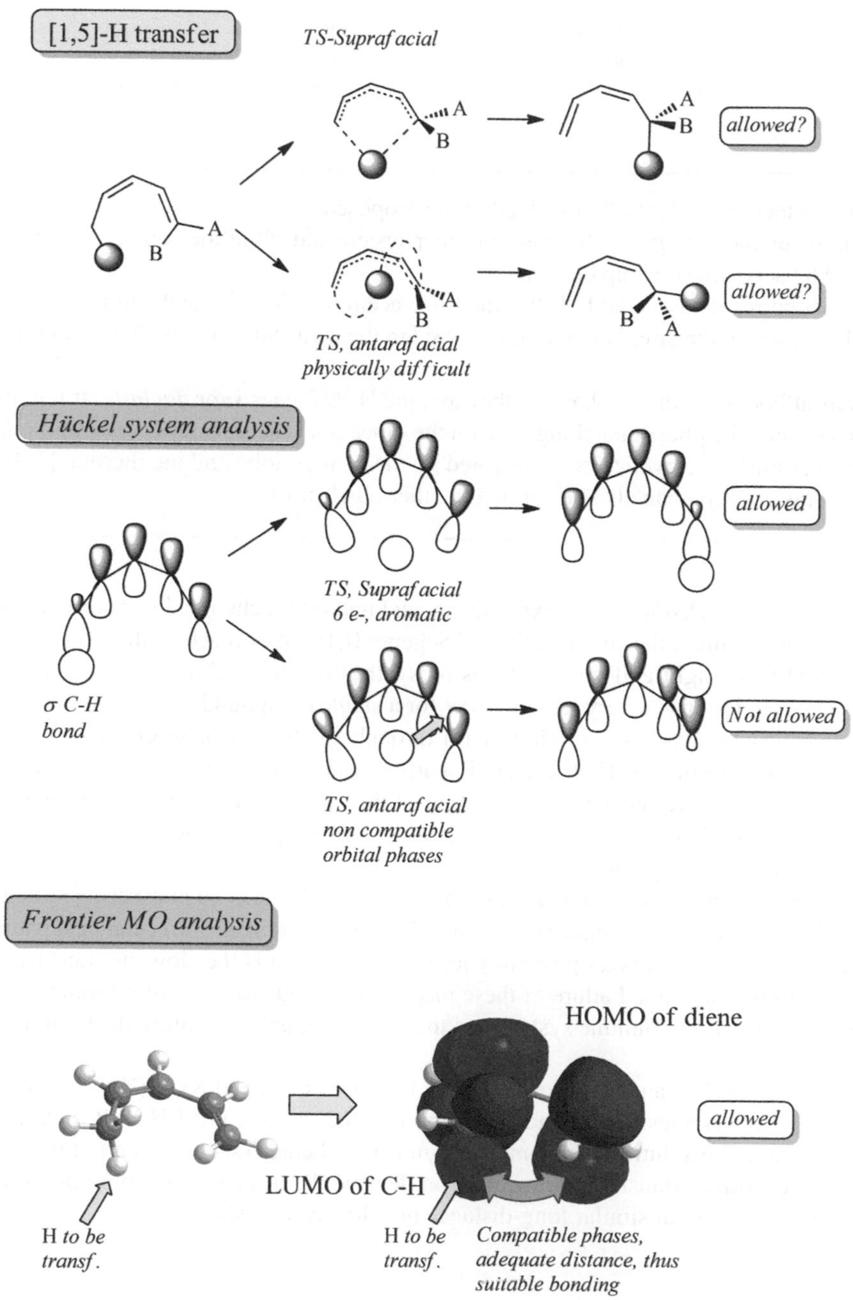

FIGURE II.7 The *antarafacial* and *suprafacial* [1,x] migration seen from the Hückel and Frontier MO analyses.

TABLE II.2 Prediction of the allowed sigmatropic transition topology in [1, *m*]-H transfers according to the number of electrons in the TS and energy input to the reaction

No. electrons in the TS	Energy source	
Hydrogen shifts	**Thermal**	**Light**
4*n*	antarafacial	suprafacial
[1,3]-**H**, [1,7]-**H**		
4*n* + 2	suprafacial	antarafacial
[1,5]-**H**		(rare)
Alkyl *(sp^3)* **shifts**		
4*n*	Suprafacial with inversion of configuration	
[1,3], [1,7]		
4*n* + 2	Suprafacial, with retention of configuration. Complicated by competing reactions.	
[1,5]		

1. Count the number of π electrons potentially involved in the proposed TS.
2. Draw the corresponding elementary *p*-AO diagram for the π system and fill in the counted electrons, one pair per AO. HOMOs and the LUMO sets will show up clearly.
3. Select the top *p*-set that corresponds to the LUMO (the first vacant *p collection*) and add the C–H σ bond to be transferred on the left (do not forget the small counter lobe). Assign the σ HOMO to it, with 1*s* (circle) of H and *sp*^{3} AO of carbon in phase (+/+).
4. See if the two orbitals at both ends are in phase. If they are, the H shift goes *suprafacially*; if not, the hydrogen (or migrating group) needs to seek the phase-matching lobe on the other side; the transfer is *antarafacial*.
5. In alkyl transfers, ring strain in the cyclic TS is mitigated by the rear *sp*^{3} lobe and the thermal [1,3]-alkyl suprafacial transfer is plausible. This criterion can also be applied to other alkyl shifts

2.4.7.2.1 A Touch of Experimental Reality Lab experiments and natural biochemical transformations offer an ample menu of electrocyclic reactions. For example, the first reaction of Scheme II.18 takes place in the skin of all vertebrates, including yours when you are exposed to sunlight and vitamin D_3 is biosynthesized from cholesterol. As much as 25% of skin fat in human adults is composed of cholesterol, a very large reserve for a single compound.

This sequence is preceded by the oxidation of cholesterol through a dehydrogenase enzyme. When the Sun's UV-B light penetrates the skin, dehydrocholesterol (**65**) (Scheme II.18, part A) undergoes a first sigmatropic cleavage (Cope rearrangement) to the heptatriene partial structure **66**. The configuration of this intermediate, known as previtamin D_3, appears ideally suited for a [1,7]-H shift (green H). This H transfer is *antarafacial*. The compound backbone is sufficiently flexible to put the pieces in a suitable position for this to happen.

The mechanism essence is demonstrated by placing an angular CD_3 at C^{10} (steroid numbering system), and the deuterium is transferred to the underside of the molecular plane (the α side). Hence, antarafacial shifts can be a natural occurrence.

The conversion of **67** to the vitamin D_3 takes place in your blood stream. It is the slow cis-trans isomerization of the dienyl function to bridge the two carbocycle units. Failure of these mechanisms leads to the severe limitation of calcium metabolism, bone mineral accrual, weakening of the immune system and muscle tissue, and premature death. It all stems from an unlikely *antarafacial* H transfer.

[1,5]-H shifts are not limited to C=C and C–H combinations but spans C=X and X=X (X: heteroatoms) bonds as well.

The lower part of Scheme II.18 is a testing ground for you to find the site of a [1,5]-H shift. Indium triflate works as a new generation Lewis-acid promoter. The solution to this mechanism is in Scheme II.44 at the end of this chapter.

In addition to SRs, remote hydride transfers to low electron-density alkenes also constitute σ–π transfers. Replacing H by some metallic elements also brings about similar long-distance metal-to-carbon shifts.

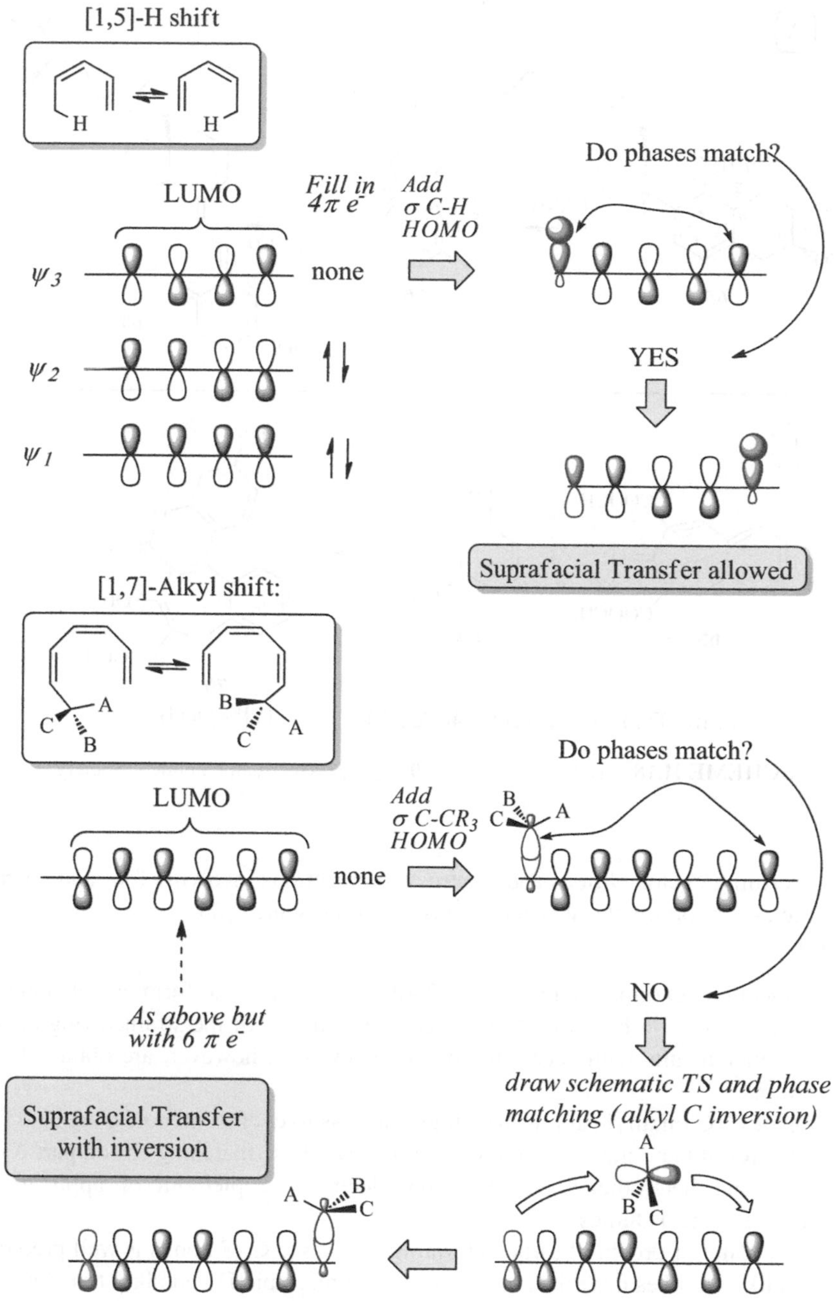

FIGURE II.8 A practical method to determine the thermally allowed H and alkyl sigmatropic shifts. Read box description.

2.4.7.3 π–π Interactions Compounds exhibiting two or more C=C or C≡C bonds at appropriate distances are prone to react through π–π interaction in predictable ways. Considerable complexity can be built, and rather astonishing cascade sequences down the line are not infrequent.

Electrophilic additions ([2+*n*]; *n*=2, 3, 4, 5) between two separate alkenes, alkynes, allenes, or multiple C=X (X = N, O) bonds, which can also occur within the same molecule in pericyclic reactions with appropriate stereochemistry, belong in this type of interaction. Multiple bond activation is provided by heat, UV light, transition metal, Brønsted- or Lewis-acid catalysis, and radical initiators, which enhance sp^2–sp^2- rapports.

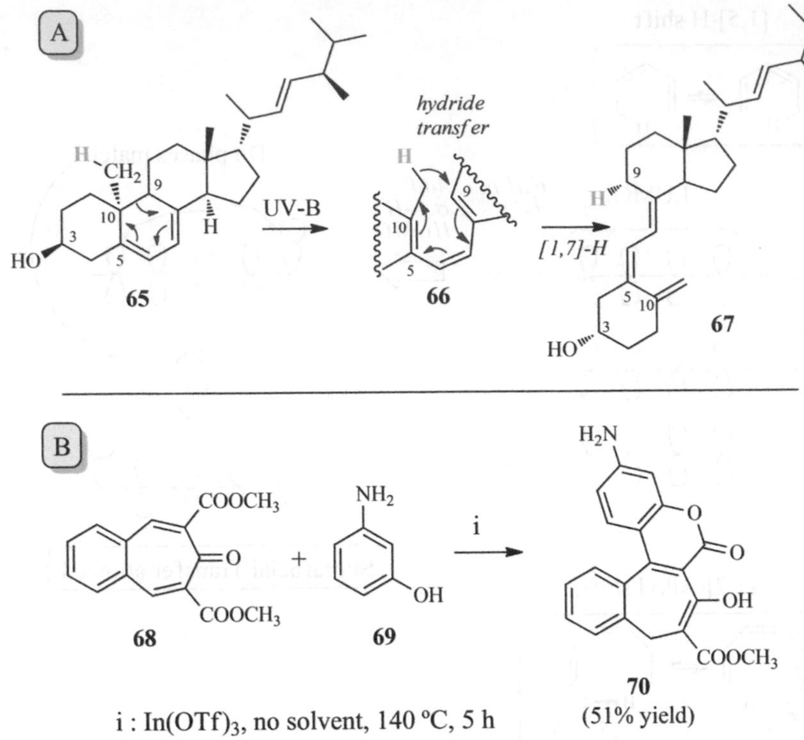

SCHEME II.18 The reaction **68**→**70** is [32] / American Chemical Society.

Examples of C–C bond-forming cycloadditions and electrocyclic reactions involving four and six π electrons are shown in Figure II.9, among which the all-important Nazarov reaction is well represented [33].

Highlights of Figure II.9:

✓ Ethylene derivatives (section **A**) dimerize by way of [2+2] cycloaddition to form cyclobutanes of well-defined geometry, but the reaction is ineffective because of the reversible cleavage of the strained ring back to ethylene under the reaction conditions. Better results with access to "box carbocycles", however, are obtained in intramolecular [2+2] photolytic cycloadditions.

✓ [2+3] cycloadditions of C=C on allyl cations, which give access to cyclopentenoids, are also feasible. Orbital interaction between the C=C HOMO and the allyl cation LUMO show phase matching (inset, part A). As for [2+4] cycloadditions between C=C and heterodienes involving 6π electrons, a plethora of applications inundate published organic-synthesis and theoretical studies.

✓ The stereochemical outcome of prochiral carbons forming the new σ C–C bonds is well predicted by the WHRs using AO and MO schemes like the ones described herein for the sigmatropic H and alkyl transfers. Interacting π–π orbitals are scrutinized according to orbital-phase and HOMO/LUMO concepts.

✓ In any event, this issue raises the question of the rotation of the two C=C units undergoing bonding. Figure II.9, section **B** shows the two possible modes, conrotatory (in the same direction) or disrotatory (in the opposite direction), in *cis* 2,4,6-octatriene. It all depends on the orbital-phase combination of the HOMO in this compound. The *thermal coupling* of the extreme $\Delta^{2,6}$ C=C bonds passes through a TS fashioned by a *disrotatory* twist, as shown to the left of the central structure in section **B**. If the end methyls are trans, relative to the carbon chain, it is easy to see them ending cis in the cyclohexadiene product. By contrast, the photolytic coupling occurs by a *conrotatory* spin of the extreme orbitals C=C to achieve orbital-phase matching. This is due to the photoexcitation of one electron to a higher energy MO in which the orbital phase is the opposite in the Hückel's ground model, very much like the analysis of [1,5]-H shift of Figure II.7. After a quick review, you should be able to figure out the present case by yourself. The outcome of the photo-induced coupling should now be *trans*.

✓ The C^1–C^5 coupling of two sp^2 carbons (or heteroatoms) cross-conjugated with a C=O unit, as in section **C** of Figure II.9, is prompted by protic catalysis. This process, known as the Nazarov reaction, has been applied many times to organic

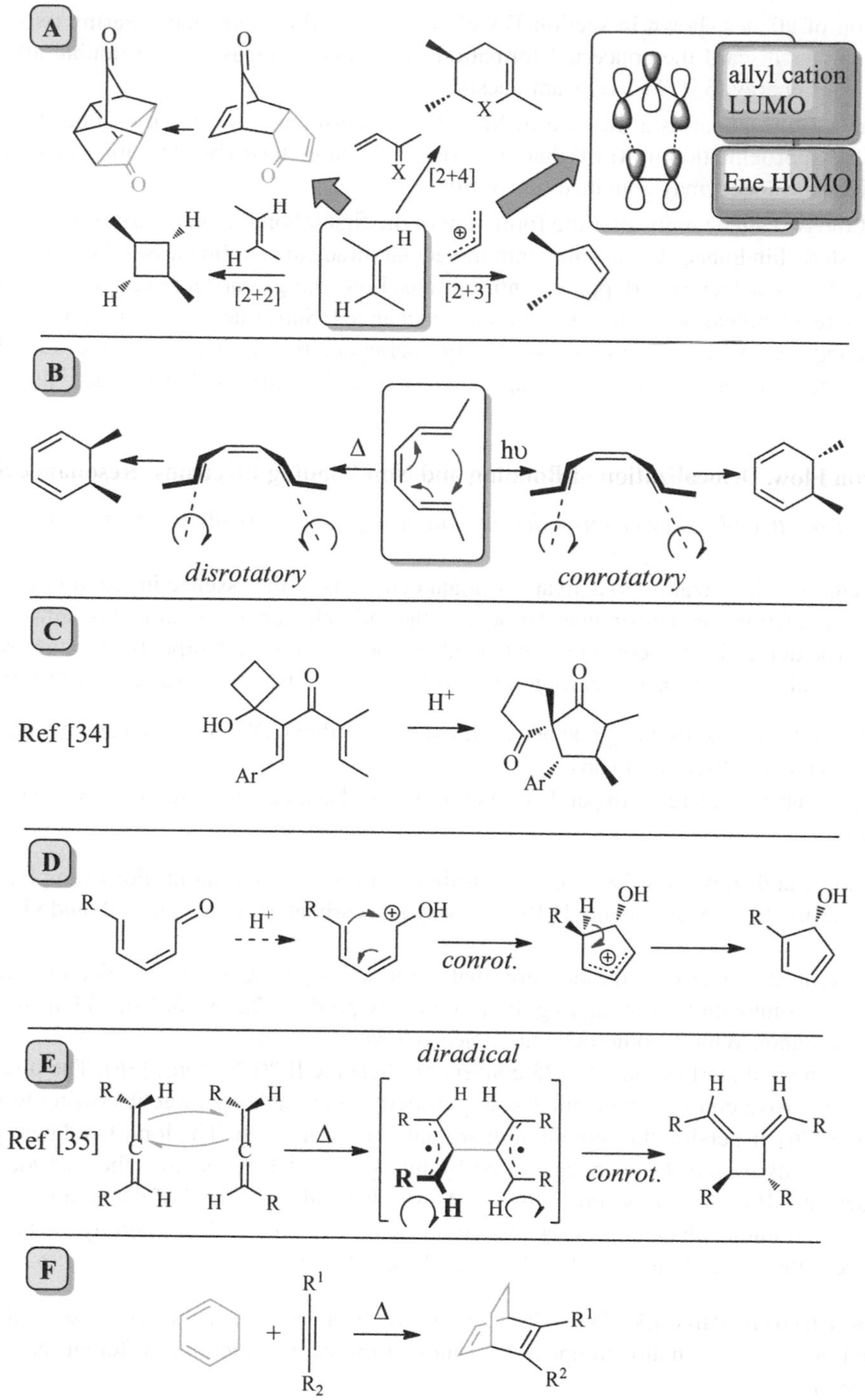

FIGURE II.9 Various modes of cyclization discussed in the text. Sections C and E were adapted from refs [34] and [35] /American Chemical Society, respectively, with permission.

synthesis. The conversion of section **C** is an accessible problem that you may wish to solve, paying attention to the observed stereochemistry [34]. A likely solution is shown in Scheme II.45 at the end of this chapter.

✓ The intramolecular bonding interaction C=C to C=O in its $C^{(+)}$–O–$LA^{(-)}$ manifestation (section **D**) was developed as an electrocyclic reaction under the WHR. It proceeds through a non-aromatic, 4π–electron TS, and conrotatory closing. Whereas the cationic intermediate is potentially one enantiomer, the identity is lost during the elimination step to the cyclopentadiene carbinol.

✓ The cycloaddition of allenes shown in section **E** yield ring-strained cyclobutanes bearing two exocyclic C=C bonds. Computational studies discard the concerted formation of the two σ bonds since a double allyl radical intermediate shows a lower Gibbs energy [35]. Two relevant questions are:
- Given that the starting allene is a mixture of M and P stereoisomers (see Note 4 about the M/P nomenclature of allenes), it is the approximation of M+P that gives rise to the more favorable TS rather than M+M or P+P on steric grounds. Molecular models may convince you of this.
- The planes of the two allene units after the formation of the first σ bond rotate nearly 90° and become almost orthogonal to reduce steric hindrance. The ensuing intramolecular diradical coupling arises from the conrotatory twisting for the orthogonal AOs to achieve overlapping confirmations. Low-energy barriers to coplanarity and methylene rotation maintain the stereochemical position of substituents unchanged. Substituents end up trans in the carbocycle.

✓ Finally, alkynes and heteroalkyne relatives also undergo pericyclic reactions whereby the triple bond plays the dienophile role. Therefore, electron withdrawing groups and Lewis-acid catalysis enhance reactivity (section **F**).

2.4.8 Issue 8: Electron Flow, Delocalization of Bonding and Non-bonding Electrons, Resonance Stabilization

☑ *Hot reactivity spots of relative high/low electron density can be defined accessibly by electron delocalization according to conjugation rules.*

Electron redeployment by delocalization to activate dormant centers is of the essence in reaction-mechanism design. Their impact on the definition of reaction mechanism goes far beyond the early elementary treatment of substituted benzene and *pKa*/*pKb* of acids and bases. The delocalization concept is extended to other alicyclic and aliphatic unsaturated substituents, bearing cations or not, anions, radicals and carbenes in carbon, and NBPs in N and O atoms in keeping with two conditions:

a) Conjugation with unsaturations, meaning one bond away from a π orbital. (You will probably be impressed by Scheme II.19 about what resonance-delocalization can do).
b) Suitable orbital geometry, in reference to parallel versus orthogonal and intermediate intervening angles in rigid or sterically hindered scaffolds.

Electron delocalization can deeply alter the electron distribution of a particular atom. For example, follow the progress of the carbon atom in the starting form **A** of Scheme II.19 in its way to a carbene **D**, vinyl anion **B**, and vinyl cation **C** through the formalities of resonance.

Conjugation effects frequently facilitate the decision-making tasks of problem solving. The top section of Scheme II.20 depicts the convenience of conjugation in predicting the more likely product, **72** or **73,** from **71** in strong acid (HCl [hydrochloric acid] 0.1 N) in methanol. Which would be your choice and why? (See Note 3).

A more exciting mechanism underlies the **74**→**75** conversion (Scheme II.20, bottom) [36]. The direction of electron redeployment is marked by extensive conjugation from 2,4-dihydropyrrol and methoxy moieties owing to their respective NBPs. They contribute to cation (**76**) dispersion throughout both aromatic rings in **77a/b**. The long-lived benzylic carbocation favors hydride transfer from the dihydropyrrol appendage aided by nitrogen's NBP. Mind that the hydride-cation distance in the energy-minimized structure (MMFF94s, this author) is 4.06Å, well beyond the "edge" of the acting 2*s* and 2*pz* orbitals.

Worth noting also: the intramolecular electron transfer amounts to an internal redox whereby the heterocycle is oxidized to a pyrrole synthon, whereas the *sec*-carbinol is reduced to a methylene bridge.

2.4.8.1 Non-bonding Electron Pairs (NBPs) Up to now we have come across NBPs several times. Ubiquitous in heteroatoms, active and necessary for many compounds to move forward to something different, NBPs are an indispensable tool for mechanism design.

As a brief reminder, valence-shell electrons are organized around atoms in molecules in three categories: inner shell, bonding, and non-bonding. Regarding the latter two:

1. ***Bonding pairs***: positioned in pairs between two atoms filling *s* and sp^x (x=1,2,3) and sp^xd^y hybrid orbitals.
2. ***Non-bonding pairs***:
   a. Neutral heteroatoms, N, O, P, S, and halogens possess non-bonding electrons in pairs, which complete the stable octet in the second-row elements, or 18 in the third-row elements (generally not shown in P, S, Cl, only exceptionally).
3. Specific highly reactive transient intermediates may have atoms with:
   a. one unpaired electron or lone electron (seven electrons in second-row elements; radicals in C, N, O).
   b. one unstable NBP in an octet (carbanions, anions of O, N, halides).
   c. one NBP pair in a sextet (singlet carbenes, nitrenes).
4. Devoid of NBP in an electron sextet (carbocations, oxonium, nitrenium, sulfonium ions).

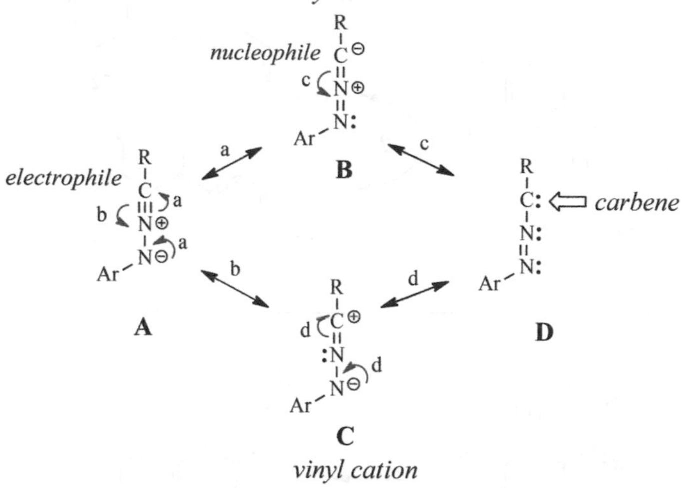

SCHEME II.19 For the sake of clarity, only NBPs of neutral N atoms are shown.

NBPs operate either as Lewis bases, or nucleophilic entities, and electron pumping for redistribution to lower electron-density sectors of the molecule, enhance nucleophilic character as well as provide stabilization to carbocations across conjugated systems (e.g. Scheme II.20). Conjugation of NBP AOs and multiple bonds or vacant AOs as in $C^{(+)}$ is of the essence.

Germaine to organic reactivity and biochemistry, NBPs are located on O and N atoms in uncharged functional groups. Our life chemical works depend on this (namely, no NBPs = no organic life, period).

NBPs in elements farther removed from carbon in the periodic table such as halogens heavier than chlorine and third-row elements play much less relevant roles. These electrons are in *d* rather than *p* AOs. The *d*-AO volume is much larger than that of *p* or sp^3 AO, hence lower ED and less effective *p-d* orbital overlap for delocalization and H bonding. In these cases, inductive effects acquire more relevance. For this reason, sulfoxides are often portrayed as $R_2S^{(+)} \rightarrow O^{(-)}$ rather than $R_2S{=}O$. Sulfones and higher oxidation-level sulfides are nevertheless portrayed with S=O bonds, for conventional reasons.

In addition, NBPs generally operate as nucleophiles *directly* or *indirectly* on susceptible low electron density (LED) centers. There is a large collection of N, O, and S nucleophiles other than their anions to choose from, in which their NBPs are the main bonding characters. Scheme II.21 illustrates the *direct* involvement of NBPs in an S_N2 reaction [37]. Here we have two instructive issues:

a) The attack by the hydrosulfide **80** on aziridine **79** occurs at the secondary carbon C^2 rather than the primary end C^1, contrary to the primary ≫ secondary C reactivity pattern in bimolecular reactions.
b) There is retention of configuration of C^2 in product **51** when inversion would be expected.

Given that the yield of **51** is quite satisfactory, it reflects the dominant mechanism.

Highlights and takeaways of Scheme II.21

✓ While two amine units are available for $BF_3{\cdot}Et_2O$ complexation, the aziridine N is less sterically hindered, therefore driving the electron-density decrease.
✓ The aziridine (as aziridinium ion complex **82**) may release ring strain by route branches A or B, the latter furnishing a more stable developing δ^+ at secondary C^2 by virtue of alkyl *+I* effects. Route A and its outcome (**85**) is thus discarded.
✓ Through fork **B**, **86** is sensitive to hydrosulfide S_N2 attack on C^2 from the backside thus following branch **C** to diaminosulfide **87**, the anticipated S_N2 product with inversion of configuration. Yet, **87** was not observed either, hence route **C** is discarded as well.
✓ Then, how can C^2 afford retention of configuration in final product **81**? By the old expedient of *neighboring group participation*, this time due to the sterically well-defined dibenzylamino functionality, leading to a second aziridinium ion **88**.
✓ Once formed, **88** can progress to **89** and the observed product **81** via a hydrogen-bonded hydrosulfide, which facilitates substitution on C^2, in a pseudo five-membered ring TS **90**.
✓ Neighboring group participation can persist on account of the incorporated sulfide as in **91** to furnish a second assisted S_N2 in excess alkyl hydrosulfide. Indeed, product **92** was reported by authors of this work [37].
✓ It is noteworthy that this series of activation and substitution jobs occurred with total absence of π bonds, conjugation, or any related phenomena; *all this commotion occurred on* ***account of NBPs in N and S alone***.

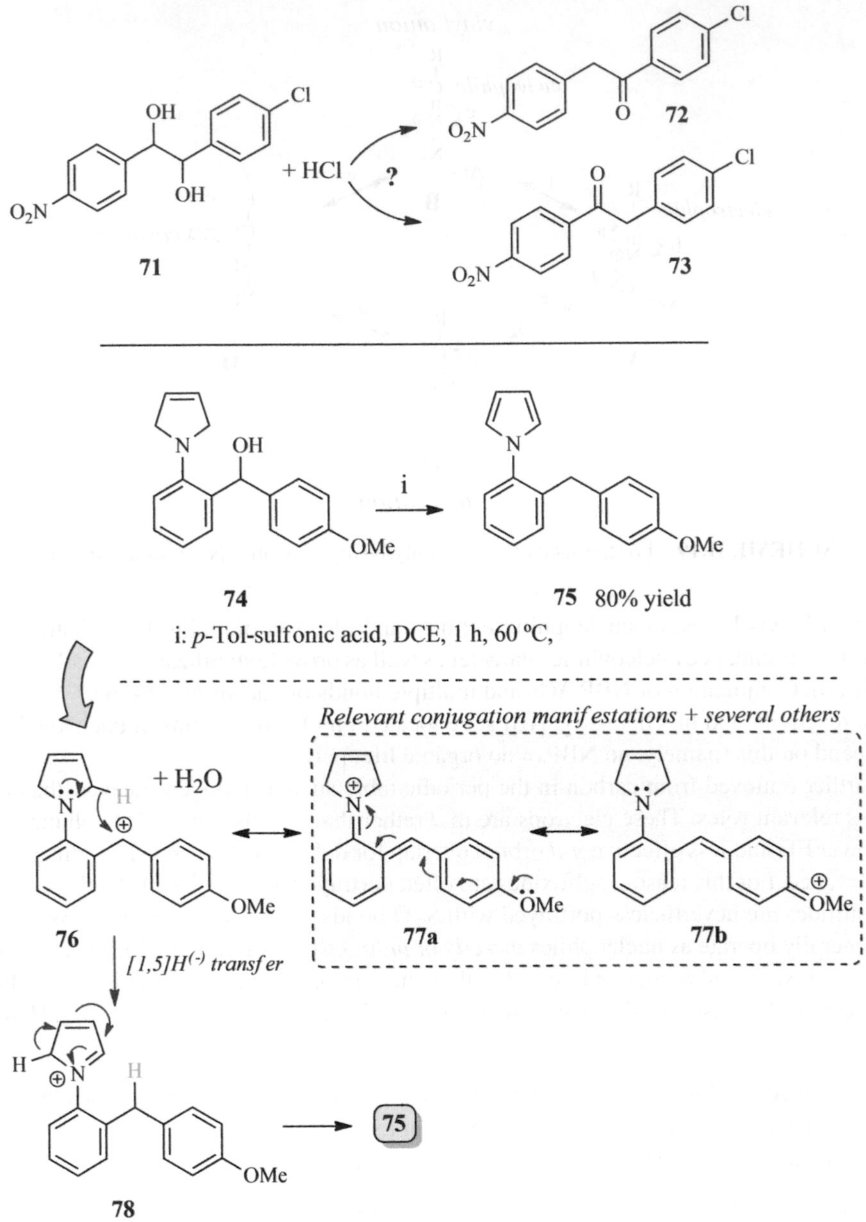

SCHEME II.20 The **74**→**75** conversion was adapted from [36] / Elsevier.

The *indirect* involvement of NBP is also a potent and frequent tool to consider in reaction mechanism. The reaction of Scheme II.22 [38] shows direct and indirect NBP participation in a multiple-component synthesis, which did not work according to plans because of the intrusion of unwanted NBPs.

At first sight, product **96** seems unrelated to the starting materials. But a second look unveils molecular patterns in **96** that resemble the components of the mixture with minimal changes. The middle section of Scheme II.22 shows this, accounting for the 10 skeletal carbons of the two sections, green and purple.

Our mechanism analysis would go like this:

a) Green and purple sections are linked through two *methylamine* bridges (in black).
b) *Methylamine* is the neat solvent.
c) A third black bond (C–C) links these two sections. Being at the β carbon of a C=C–C=O substructure, this black bond should come from a Michael 1,4-addition.

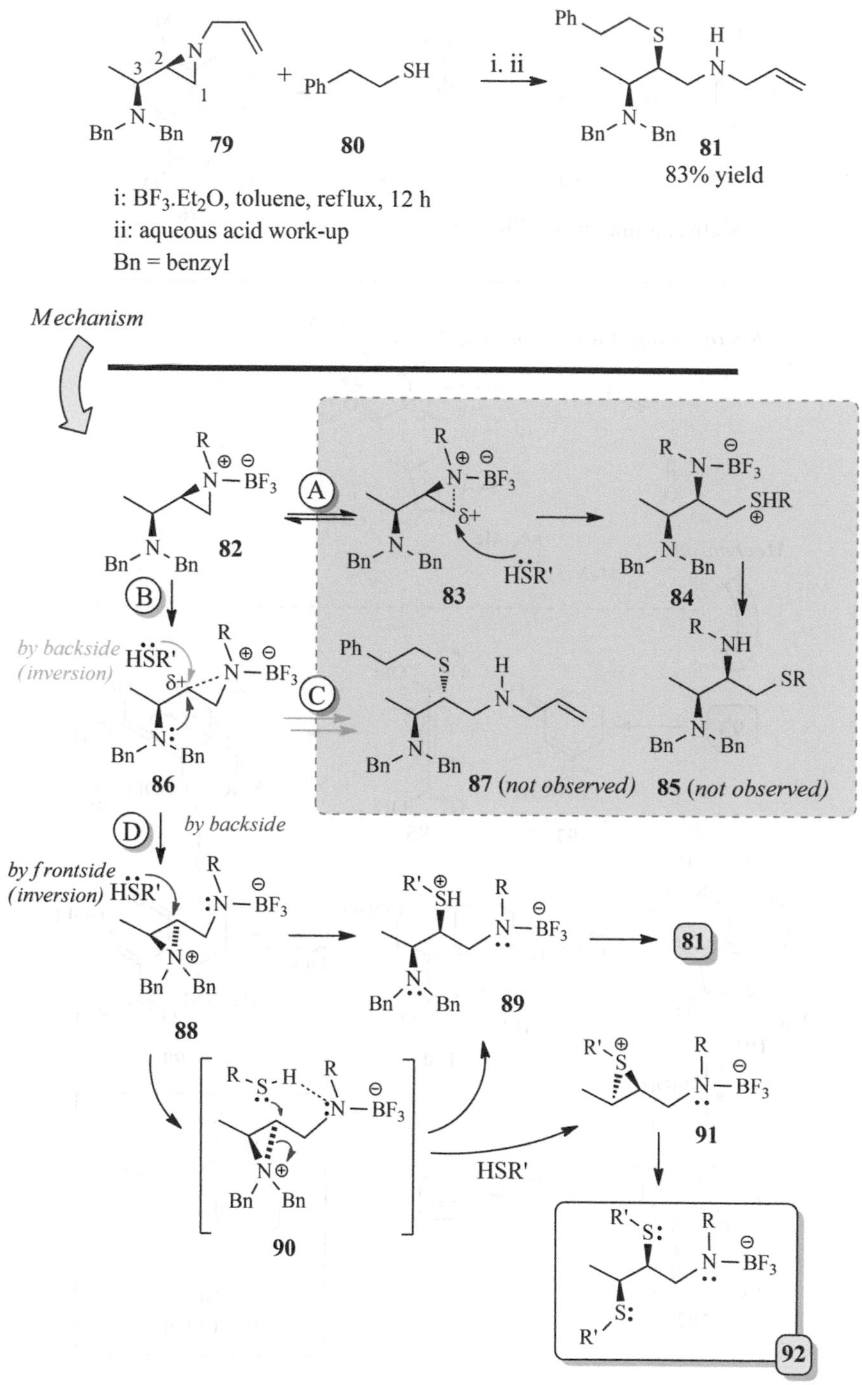

SCHEME II.21 The prevalence of non-bonding electron pairs impinges strongly on the reaction course providing several alternative pathways and end products. [37] / American Chemical Society.

d) Methylamine can not only justify the replacement of C=O by an N function but could promote the 1,3-diketone (**93**) alkylation to create this black C–C bond via enamine **97** in slightly basic medium.
e) Meanwhile, ethyl bromoacetate **94** sat unaltered, while the other three components plotted links between them.

The bottom section of Scheme II.22 shows a feasible transit to target **96** in agreement with this line of reasoning.

Highlights and takeaways of Scheme II.22:

✓ Without the participation of NBP, the reaction design would be nonsense.
✓ NBPs in the amines operated as *direct* nucleophiles or indirect elicitors of carbon nucleophilicity (enamine).

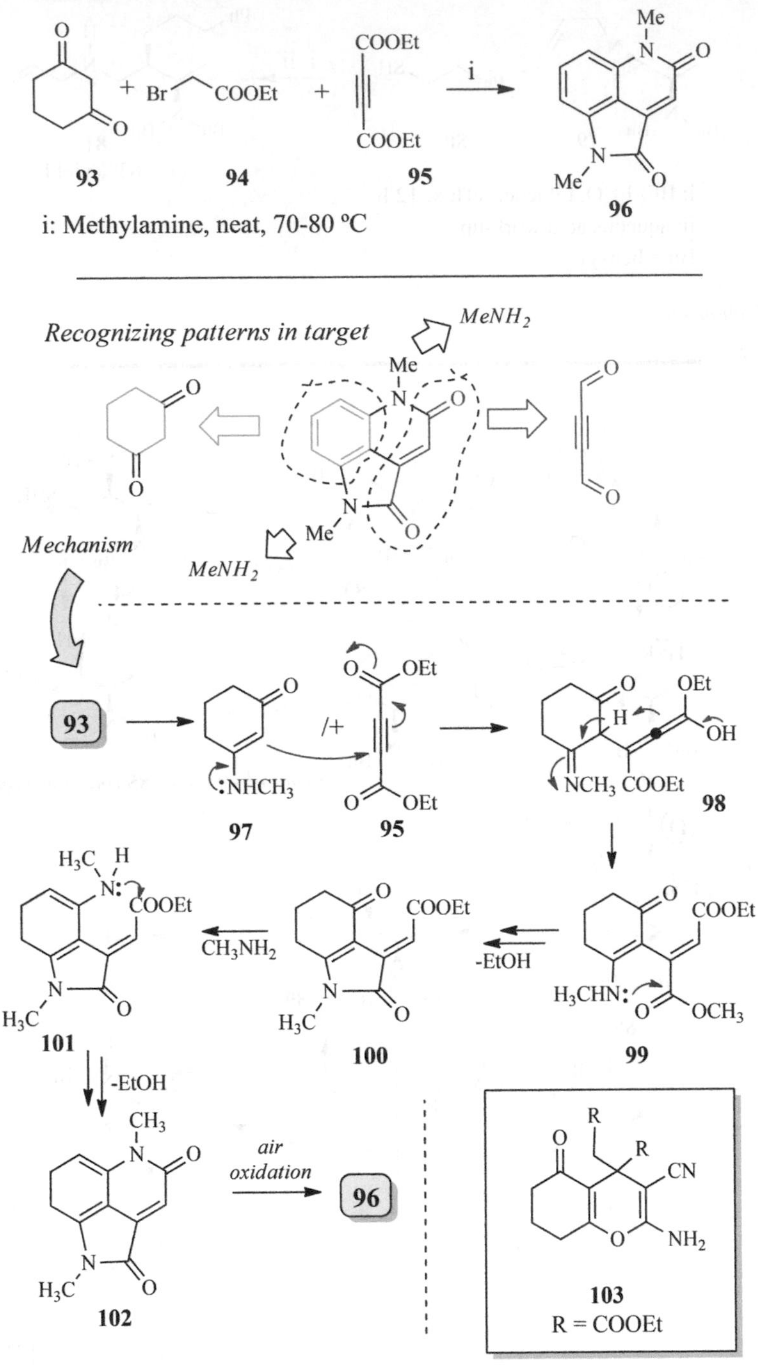

SCHEME II.22 Top: a first simple case of multiple component reaction (MCR) for building structural complexity. The analysis begins with recognizing the starting materials in the target (middle section), and turning their connection in the form of a mechanism cascade (bottom). [38] / Elsevier.

- ✓ Ethyl bromoacetate was expected to form an anion on the α-ester carbon after proton subtraction by the methylamine solvent, but it was overrun by faster enamine formation and alkylation on *bis*-carboxyethyl-acetylene (**95**).
- ✓ Take note of the allene-enolate abstraction of the allylic proton in **98**. If you bother to figure out the stereochemistry of the π orbital of the allene performing this role, you will discover a satisfactory overlap with the C–H σ bond.
- ✓ However, there are a few steps in the sequence that require protons (**97** → **98**; **98** → **99**; **101** → **102**) but the medium was formally water-free. Two moles of ethanol are produced along the way, so two moles of a protic source would be required.
- ✓ Authors [38] were after chromenes of type **103** (see inset, Scheme II.22) by a practical one-pot procedure. To their good fortunes, the reaction worked in the desired direction with a good yield using more effective nucleophiles than ethyl bromoacetate, such as ethyl 2-cyanoacetate and dicyanomethane. By now, you should be able to draw the corresponding mechanisms.

2.4.9 Issue 9: Electron Traffic and Electronic Density Differences

☑ *Electrons flow from regions of high to low electron density.*

– A story of HEDZ and LEDZ

Although this tenet is straightforward, some people may occasionally be tempted to violate this elementary rule when despairing to find an answer to a difficult problem.

In this section, we will develop the meaning and origin of the acronyms HEDZ and LEDZ – high and low electron-density *zones*, respectively – in molecules frequently used in this workbook. These units affect other nearby atoms so their influence is *zonal* at the molecular level, rather than affecting individual atoms only.

The high–low electron-density spectrum is a fundamental electronic characteristic of functional groups appended or integrated in much less electronically functional carbon scaffolds.

HEDZ (high electron-density molecular zones, electron donors – EDG) and LEDZ (low electron-density molecular zones, electron acceptors or electron withdrawers – EWG) are easily recognized during problem analysis by looking at specific traits in the starting material(s).

HEDZ or electron donors

- ✓ Neutral nucleophiles carrying NBPs: R_nX (X = O, N, S, P, halogens).
- ✓ Anionic nucleophiles: e.g. alkyl lithium, Grignard reagents, R_2CuLi, etc.
- ✓ Reducing agents, hydride transferors.
- ✓ Potential anion precursors:
  - R_nXH (X = O, N, S) and EWG–CH_n-EWG, (n = 1,2)
  - Allyl and propargyl methylenes
  - RCH_n(C=O), RCH_n(sulfones), (n = 1,2)
- ✓ C=C–X and C≡C–X (X = O, N with their NBPs), typically enolates, enamines.
- ✓ C=C and C≡C *per se*.
- ✓ Arenes with EDG substituents (e.g. alkoxy, amino, alkyl).
- ✓ Zerovalent metals.

LEDZ or electron acceptors

- ✓ Potential carbocation precursors: sp^3 carbon atoms bearing nucleofuges suitable for S_N1 or E1 reactions.
- ✓ X–C=C–Y and X–C≡C–Y (X = EWG, alkyl, aryl, or additional EWG; Y = EWG, CN, NO_2, and others).
- ✓ Polar $C^{(\delta+)}=X^{(\delta-)}$ functions (X = N, O, S) amenable to nucleophilic addition or addition–elimination reactions: carbonyls in carboxylic esters and acids, amides, anhydrides, susceptible to Lewis- or Brønsted-acid catalysis.
- ✓ Conjugated polar C=C–$C^{(\delta+)}=X^{(\delta-)}$ and C≡C–$C^{(\delta+)}=X^{(\delta-)}$ functions (X = N, O, S).
- ✓ Arenes with EWG substituents (NO_2, CN, esters, ketones, aldehydes, sulfonates esters, ammonium ions).

These traits may change from HEDZ to LEDZ by conjugation with EWGs, electron transfer to and from external agents, bond cleavage, and the inversion of polarization (umpolung).

Once formed, radicals (see Section 2.4.12.1) are electron-deficient atoms and do not properly rate as HEDZ or LEDZ. Carbenes are more difficult to predict in this regard since they can be nucleophilic or electrophilic and require separate treatment (see Section 2.4.12.2).

2.4.9.1 Detecting Potential Donors and Acceptors During the initial steps of problem analysis, the identification of one HEDZ and/or LEDZ in the reactants – if any present – helps in approaching solutions. Ideally, an HEDZ may have a LEDZ counterpart that meets the compatibility principle (see box).

This task is eased by reviewing functional groups organized in Table II.1 (see Section 2.4.6). On the left column, nucleophiles, Lewis bases, elements more electronegative than carbon (not many, N, O, S, halides), all operate as HEDZ inducers. On the right column are LEDZ candidates.

HEDZ/LEDZ compatibility principle

- ✓ HEDZ + LEDZ ➨ bonding most likely.
- ✓ HEDZ + HEDZ ➨ bonding unlikely.
- ✓ LEDZ + LEDZ ➨ bonding unlikely.

(*continued*)

✓ HEDZ → LEDZ conversion or the opposite may take place through *umpolung* or polar inversion (e.g. **R–CH$_2$–Br** [a LEDZ] + 2Li → **R–CH$_2$–Li** [a HEDZ] + LiBr).

✓ A mild LEDZ may be converted to a potent LEDZ intermediate by protic acid, catalysis, or high energy (heat or light) under appropriate conditions:

$$R_3C-OH \rightarrow R_3C^{(+)} \text{ or } R_3C(\bullet)$$

✓ Similarly, a modestly reactive HEDZ may be transformed into an active HEDZ intermediate by catalysis or base.

HEDZ/LEDZ associations rule over a large proportion of inter and intramolecular reactions, but not exclusively. So, after deciding to take up a reaction mechanism as a problem, it is a good idea to identify HEDZ and LEDZ among the molecular zones, moieties, or reagents in the mixture to build a bonding design. A relatively straightforward (for advanced students) reaction is depicted at the top of Scheme II.23 [39].

To add some more glamour, I have concealed the primary product **107** for you to figure it out from the convergence of two ways of reasoning:

1. from starting materials in the forward direction while keeping an eye on final product **108**, and
2. from **108** reasoning things in reverse mode to reveal who **107** might be. By application of such a procedure you start getting some practice with the deconstruction strategy described in detail in Chapter 4.

Forward reasoning: we have three compounds to consider for HEDZ/LEDZ identification (see the second section of Scheme II.23). Notice that dicyanomethane (**106**) would sit inoperative were it not for triethylamine awakening its latent HEDZ nature. In fact, the CN groups in **106** are not only EWGs but LEDZs sensitive to nucleophilic addition, as we shall see momentarily. Having our first HEDZ in hand, we need its counterpart, an LEDZ. The obvious choice is *p*-chlorobenzaldehyde (**105**); these two pieces are likely to form covalent bonds.

Reverse reasoning: after a first look at target **108,** one easily recognizes the pieces of all the starting materials and the construction of two C–C and two C–N bonds. The aryl aldehyde is trapped by the two HEDZs precisely at the more negatively charged carbon atoms, whereas the southern heterocycle develops binding rapports with the chloro-arene and cyano moieties.

With this information it is possible to propose a mechanism, adding the fact that in the second step a stronger base (potassium carbonate), higher temperature, and an aprotic, anion stabilizing solvent (dimethyl formamide) provide conditions for the nucleophilic addition–elimination of N on the western arene portion. With these elements of judgment to operate, you should be able to work out the mechanism and unveil the structure of compound **107**. A feasible sequence is shown at the end of this chapter (Scheme II.46).

Occasionally, all components in the mixture require a jump-start for their interaction to proceed in the HEDZ/LEDZ manner. In Scheme II.24 [40], β-ketoester **110** is a potential HEDZ in alkali similar to the previous Knoevenagel condensation of Scheme II.23, except for the neutral medium.

On its side, *ortho*-disubstituted benzene (**109**) includes a triflate (OTf) – a good nucleofuge – and a seemingly inactive trimethylsilyl substituent next to OTf. It seems an uneventful compound until one heats it. There is also excess (2.5 equivalents [eq]) cesium fluoride, whose role is unknown to you (perhaps). But as an alkali metal of the Li, Na, and K group and the most strongly reactive of these, it might as well operate as a Lewis acid. In fact, you do not need a moderate-to-strong base to promote HEDZs in β-ketoesters and β-diketones if you think of keto-enol equilibria.

Additionally, the Cs counterion is fluoride, which would have a role to play by attacking the silane the S_N2 way. Thus, the bimolecular elimination (E2) of TMS and OTf prompted by $F^{(-)}$ would yield a triple bond embedded in benzene, that is *benzyne*, a truly reactive LEDZ. This mild access to benzyne was developed by Professor Kobayashi in Japan in 1983 [41], sparking a stream of synthetic applications thereafter [42].

Having activated HEDZ and LEDZ sections, you still must explain the cleavage of the β-ketoester in product **111**. Try your mechanism before having a peek of the solution at the end of this chapter (Scheme II.47).

2.4.9.2 *What Makes a Given Functional Group a Natural HEDZ or LEDZ?*

Memorizing lists of functional groups according to their electron donor or acceptor character like those shown in Table II.1 may be helpful when a fast response to a reaction-mechanism problem is necessary. Yet, it is even better to rationalize the HED/LED property for functionalities we are not familiar with. A few helpful concepts drive your mechanism proposals in the correct direction.

The ED or EW ability of functional groups is primarily defined by Pauling's electronegativity scale of its elements and an array of electronic effects:

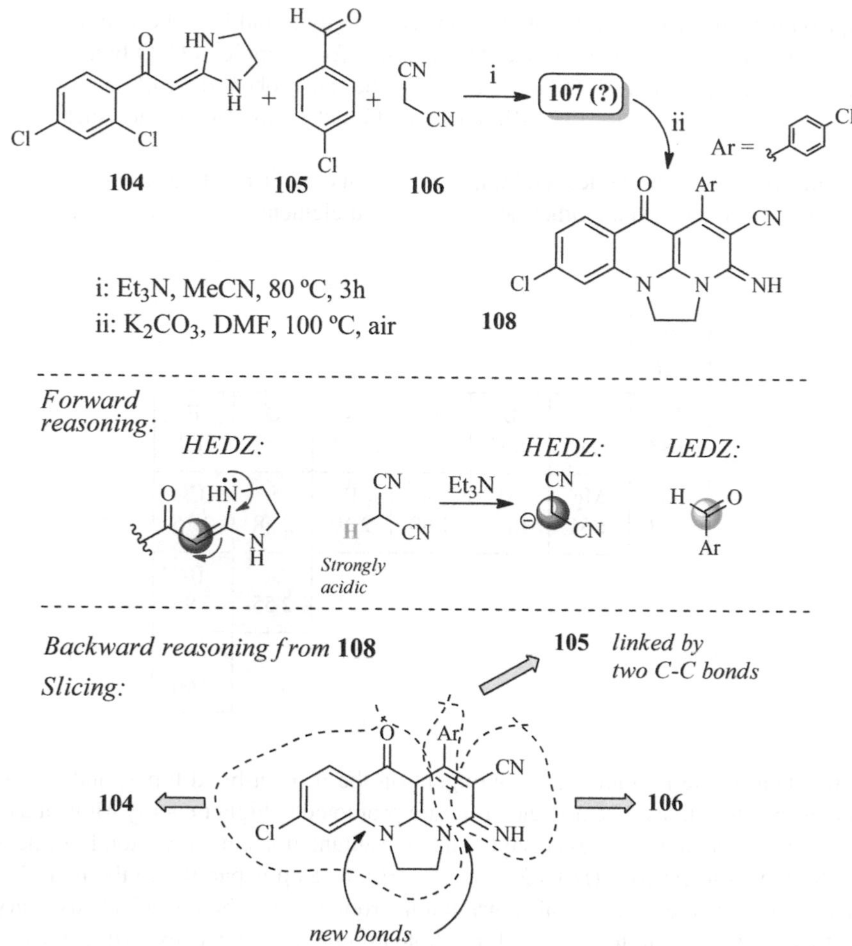

SCHEME II.23 Approaching a mechanism by (1) identifying zones of high and low electron density as likely centers of activity, and (2) recognizing backbone sections of starting compounds in the target. [39] / Elsevier.

- ✓ Bond polar moment and polarization.
- ✓ HOMO/LUMO accessibility of bonding electrons and steric restrictions.
- ✓ Ground or excited state status.
- ✓ Capacity for stabilization of partial or full charges developing in the transition states with or without solvent intervention.
- ✓ Field effects of neighboring functionalities.
- ✓ Activation by catalysts.
- ✓ Conjugation-delocalization.

The several decades-long systematic study of *pKa* of series of substituted carboxylic acids, rate of hydrolysis of their esters, and several other reactions of substituted arene influence on the reaction center brought a most welcome classification of functional groups in regards to this question. Subsequent quantum-chemistry calculation methods vastly expanded the experimental observation, so we can safely assign high and low electron-density moieties.

2.4.9.3 A Reminder Takeaway

2.4.9.3.1 Dipole Moment, Polarizability, Hückel Atom Charges, Quantum Atomic Electron Density Elements differ in their affinity for electrons in their valence shells, the well-known electronegativity (EN) concept. In a covalently bonded pair of atoms A–B of dissimilar EN, charge separation will occur as bond electrons are more attracted to the atom of the pair with greater EN. In such an asymmetric array, a $(\delta^+) \rightarrow (\delta^-)$ dipole necessarily arises, its dimension or dipole moment μ is a function of the EN_A–EN_B difference and the average distance between the nuclei, the bond length. The μ vector becomes a commanding property for reactivity of that bond and relays the charge differences through bond and field effects to the neighboring sections of the molecule. This effect decreases rapidly with distance.

By combining the approximate electronegativity of elements commonly found in carbon compounds (table below) one can obtain the relative μ of individually bonded pairs in qualitative terms: for example, a C–O bond should be more polar than a C–N, since O is more electronegative than N. Both bear a $C^{(\delta+)}$. By the same token, the carbon–boron bond is less polar than a C–H linkage and create a $C^{(\delta-)}$. Likewise, the EN differences of C and Li in R_3C–Li are such that C turns into a virtual carbanion.

Selected electronegativites (Pauling) for elements of major interest in organic reactions are shown below. Colors indicate the elements with lower or higher EN than carbon. Other heavier, unlisted elements are continuously being added to the organic-synthesis chest.

H 2.20						
Li 0.98	Be 1.57	B 2.04	C 2.55	N 3.04	O 3.44	F 4.0
Na 0.93	Mg 1.31	Al 1.61	Si 1.90	P 2.19	S 2.58	Cl 3.16
					Se 2.55	Br 2.96
						I 2.66

Electron shifts and attendant dipole moments are contingent on the σ- or π-bond types, and whether electrons are in the energy ground state or photo-excited states (because electrons are promoted to higher energy MOs of a different topology). The spatial position of atoms in the molecule impinges strongly on the resultant molecular μ vector. For linear molecules (e.g. HCN, hydrogen cyanide) or built along a linear axis ($H_2C{=}X$, where X = O, N, S) μ is parallel to the main-bond axis, but in angular molecules such as $CH_3CH_2NH_2$ there are topological contributions from diagonal bonds, which also vary with each low-energy conformer contributing to the average molecular conformation observed in the μ experimental determination. Thus, the comparison of μ values between different molecules should be observed with care.

Families of closely related compounds shed light on the effect of heteroatoms and bond types on carbon functionalities. Compare, for example, μ of simple C–O, C–N, and C–S compounds in the ground state shown in Table II.3.

Highlights and takeaways from Table II.3:

- ✓ Data in columns are organized by heteroatom in increasing order of electronegativity.
- ✓ The dipole moments of methyl and ethyl N, S, and O compounds (rows 1 and 2) and C=X (X = N; S; O) of row 3 respond to the expected electron drift as EN increases (N < S < O).
- ✓ The high μ value of ketones (2.7 D) and aldehydes (> 3.0) brings a 40–50% ionic character $C^{(+)}$–$O^{(-)}$ contribution to the resonance hybrid, which explains the marked LEDZ property of these moieties in the direction $C^{(\delta+)}{=}O^{(\delta-)}$. The LED at

TABLE II.3 Selected dipole moments μ (Debyes) of carbon-parent compounds σ and π bonded to the most common hetero-elements. Adapted from [36]

Compound (μ-D)		
CH_3NH_2 (1.32)	CH_3SH (1.52)	CH_3OH (1.69)
$EtNH_2$ (0.99)	EtSH (1.39)	EtOH (1.67)
Et_2NH (0.90)	Et_2S	Et_2O
	(1.58)	(1.17)
$H_2C{=}NH$	$H_2C{=}S$	$H_2C{=}O$
(2.14)	(2.57)	(3.04)
	Acetone (2.69)	DMSO 4.09
HCN (2.98)	CH_3CN (3.92)	NH_2CN (4.52)
CH_3NH_2 (1.32)	CH_3–N=O (2.05)	CH_3–NO_2 (3.46)

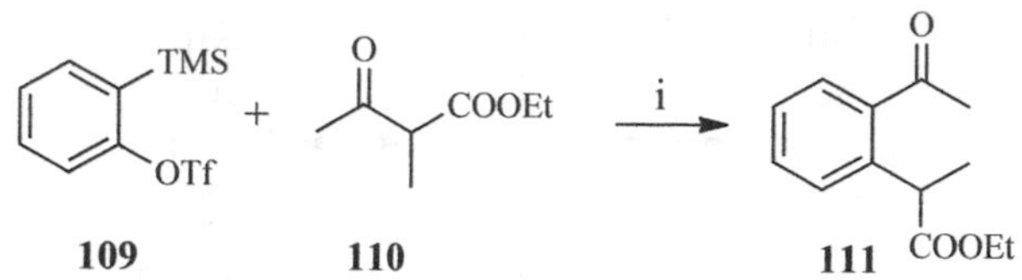

109 110 111

i: CsF (2.5 eq.), MeCN, 80 °C

SCHEME II.24 A straightforward interaction of **109** + **110** seems unlikely unless either compound is previously activated. For a solution of this problem see Scheme II.47 at the end of this chapter. [40] / American Chemical Society.

C claims ED from the neighboring methylene in alkyl ketones, reducing the C–H bond order and increasing its acidic character for abstraction by bases.

- ✓ Along the same lines, the major ionic character of dimethyl sulfoxide (DMSO) shown by the $(CH_3)_2S^{(+)} \rightarrow O^{(-)}$ over the formal $(CH_3)_2S{=}O$ canonical structures explains the proton removal from DMSO by strong bases furnishing a powerful carbon base.
- ✓ In the cyanide series, there is a sharp increment in μ as the electron donor capacity of the substituent increases the +*I* effect of methyl in acetonitrile and the +*R* effect of the amino group by way of its NBP resonance: $H_2N^{(+)}{=}C{=}N^{(-)}$ enriching the ED at the *sp* N, hence its contribution to μ.
- ✓ In the methyl amine/methyl nitrite/nitrosomethane at the bottom of the table, the marked increase in μ obeys to the stronger electron pull by the NO and NO_2 substituents relative to NH_2.

The dipole moment is not a cause by itself, but the consequence of electronic phenomena of participant-bonded atoms. At the beginning of this book the flexibility of the electron cloud in molecules was emphasized as a fundamental characteristic of the electron cloud. As charged particles with a very small mass, electrons are driven by electromagnetic forces to preferred molecular zones over others across the MOs they occupy singly or in pairs. As a result, the electron cloud is not uniform, forming zones of higher and lower *ED* around nuclei according to their electronegativity, electron redeployment by way of conjugation, and volume of the nuclei's orbitals. Local partial atom charges arise according to these electron drifts.

Aniline and nitromethane illustrate this feature in the classical resonance redistribution of π ED throughout the benzene ring. Partial charges are distributed unevenly and are responsible for enhancing the electrophilic addition at the ortho and para carbons in aniline; so much so that bromination with molecular bromide goes non-stop to 2,4,6-tribromo aniline and a rate several orders of magnitude greater than the bromination of toluene.

Along with the old faithful (+*R*) model of aniline, Figure II.10 also compiles the partial atomic charges (Mulliken) I obtained from applying a B3LYP (Becke, 3-parameter, Lee–Yang–Parr) hybrid function at the 6-31G level calculation. Mind that these charges have nothing to do with the formal charges estimated from electron counting with pencil and paper (Note 6), but are the result of the *Mulliken population analysis*, which is based on the quantum distribution of electrons in all AOs of each element and their combination of all intervening atoms in the molecule.

These electron bundles build partial charges expressed in positive or negative scalars. Considering all atoms, their algebraic sum should be zero if the molecule is neutral. The sign is taken as the relative charge density that may be interpreted as the $\delta^{(+)}$ or $\delta^{(-)}$ character but not in absolute terms, as do bare cations and anions (+1.0 and −1.0).

Partial charges in aniline reveal a strongly electron-deficient *epi* C^1 atom (+0.390) emerging from two factors:

a) the C/N electronegativity difference feeding the −*I* inductive effect of the NH_2 group relative to C, and
b) the strong dipole moment μ reinforced by the $C^1{=}N^{(+)}H_2$ canonical structures after dispersing the nitrogen's NBP into the ring.

Mulliken charges duly reflect this qualitative view as they are noticeably larger in ortho and para carbons compared with the *meta* carbon.

Replacing the amino group (EDG) with nitro (EWG) reverses the electron drift from the ring to the substituent. Mulliken partial charges pinpoint the deactivation of C^2 and C^4 toward electrophilic reagents relative to C^3. As a result, the rate of electrophilic aromatic substitution decreases sharply; nitration relative to benzene ($K_{rel}=1$) goes from phenol ($K_{rel}=10^3$), to toluene ($K_{rel}=25$), to chlorobenzene ($K_{rel}=5\times10^{-3}$), to nitrobenzene ($K_{rel}=10^{-7}$).

2.4.9.4 Quantum-mechanical Computations and Mechanism Due to the importance of partial atomic charges (PAC) to understand chemical reactivity and mechanisms, a variety of computational methods have been developed, such as Hückel, Gaussian Hartree–Fock (HF), Gamess density functional theory (DFT) tools, and several others [43]. These methods rarely coincide in PAC and electron-density numbers, in part due to the number of functional bases employed, the occurrence and

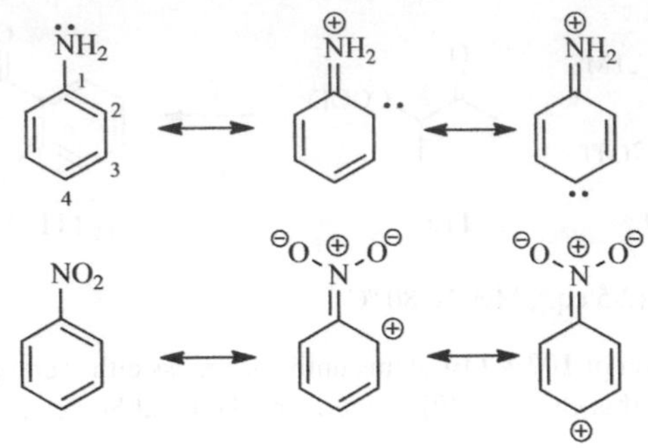

Mulliken Charges maps

Aniline	
C^1	+0.390
C^2	-0.247
C^3	-0.175
C^4	-0.236
N	-0.011

Nitrobenzene	
C^1	+0.208
C^2	-0.122
C^3	-0.227
C^4	-0.157
N	+0.223
O	-0.405

FIGURE II.10 Examples of uneven distribution of Mulliken atom charges (au) in aniline and nitrobenzene and the traditional explanation by resonance canonical structures. Calculation method: B3LYP at the 6-31G level by M.E. Alonso-Amelot (this book).

choice of more than one energy minimum, contingent on conformational possibilities and limitations of computer power as the molecular size, and the number of electrons, increases alongside the complexity of calculations.

For you to have a better grasp on the meaning of PACs, I have run Gamess-DFT and Gaussian restricted Hartree–Fock (RHF) jobs for several, small, closely related molecules to get the various PACs these methods yield and compared them in Table II.4. The difference (Δ) relative to methane was also added to normalize the resulting PACs from each method.

Highlights and takeaways from Table II.4:

- ✓ While PACs differ for each compound, they remain loosely comparable (rows). All three calculation sets provide a similar decreasing trend (columns) (Figure II.11) from negative to positive PACs and yield roughly analogous Δ values (row comparisons for the same compounds).
- ✓ Methane (entry **1**) possesses the carbon atom with the highest electron population among this compound collection. Reason: as C is more electronegative than H and there are four hydrogen atoms in methane, the C atom attracts C–H σ-bond ED increasing its PAC. This is confirmed by replacing one H by a methyl (ethane, entry **2**) or NH_2 (entry **5**) causing a substantial decrease in the carbon electron population and the corresponding PAC.
- ✓ On the other hand, replacing H with Li further increases the ED on carbon as expected, yielding a PAC of −0.698 au (Mulliken, Gamess/6-31G) (not shown). Such electron charge concentration furnishes a potent base/carbanion character governing methyl lithium performance at large.
- ✓ π Carbon-to-carbon bonding (see entries **2, 3,** and **4**) modulates the PAC of carbon atoms. In the absence of EWG, alkenes and alkynes maintain their EDG capacity in addition to reactions and complexation with transition metals.
- ✓ Heteroatoms pull down carbon PACs in proportion to their electronegativity (entries **5** and **6**). Carbon atoms in methyl amine and methanol jump over the barrier of EDG to EWG in regard to the carbon atom and show increased electrophilic character, which is fostered by π bonding (compare entries **6** and **7**).
- ✓ The central *sp* carbons in allene, ketene, and carbon dioxide (entries **8, 9,** and **10**) reach minimum PACs in consonance with their high susceptibility to nucleophilic agents.

2.4.10 Issue 10: Electron Traffic on Account of LEDZ Alone

☑ *The electron cloud may be put in motion by the development of an LED center causing deep-rooted rearrangements, hydride transfer from unsuspected sites, and other processes.*

This assertion is easily understood by considering that the electron cloud in molecules acts as a shroud of negative charge in electromagnetic equilibrium with the collection of positive atom nuclei. This condition is deeply perturbed by the loss of electromagnetic balance. This imbalance may be due to the departure or arrival of a specific substituent, a single electron to a

TABLE II.4 Partial atomic charges (au) of Lowdin and Mulliken obtained by Gamess-B3LYP and Gaussian restricted Hartree–Fock (RHF) at the 6-31G level quantum methods for a series of low molecular-weight organic compounds, representing common hydrocarbon sections and moieties bearing heteroatoms. Calculations by M. E. Alonso-Amelot (this book)

		GAMESS-DFT				GAUSSIAN RHF	
Entry	COMPOUND	Lowdin	Δ	Mulliken	Δ	Mulliken	Δ
1	CH_4	−0.461	0	−0.577	0	−0.622	0
2	CH_3–CH_3	−0.316	0.145	−0.400	0.177	−0.452	0.170
3	CH_2=CH_2	−0.225	0.236	−0.256	0.321	−0.328	0.294
4	Acetylene	−0.158	0.303	−0.217	0.360	−0.324	0.298
5	CH_3–NH_2	−0.200	0.261	−0.214	0.363	−0.195	0.427
6	CH_3–OH	−0.138	0.323	−0.166	0.411	−0.131	0.491
7	H_2C=O	+0.055	0.516	+0.094	0.671	+0.161	0.783
8	CH_2=C*=CH_2	+0.009	0.470	+0.218	0.795	+0.054	0.676
9	CH_2=C*=O	+0.214	0.675	+0.332	+0.909	+0.486	1.108
10	O=C=O	+0.293	0.754	+0.588	1.165	+0.894	1.516

(*) Carbon atom of shown partial atomic charge

LUMO at the start of a reduction reaction, the subtraction of electrons during oxidation, the interaction of a Lewis acid or base, and an individual proton from the medium. The cloud is subsequently distorted, and electron flow activated, seeking to re-establish the electromagnetic equilibrium. A reaction process is thus triggered.

Some of these electron redeployments elicited by a humble proton proceed through one or two steps taking place in a corner with a large molecular backbone, e.g. **112**→**115** [44] (Scheme II.25, reaction **A**), while others involve several more atoms.

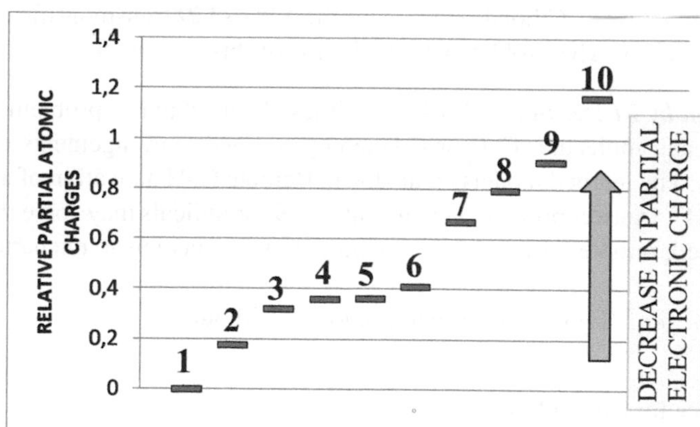

Ploted compounds

1	CH_4	**6**	H_3C-OH
2	H_3C-CH_3	**7**	H_2C=O
3	H_2C=CH_2	**8**	H_2C=C^*=CH_2
4	HC≡CH	**9**	H_2C=C^*=O
5	H_3C-NH_2	**10**	O=C=O

FIGURE II.11 Comparative Mulliken partial atomic charges relative to methane (Δ [PAC]) of carbon atoms associated with a selection of descriptive functional groups. Base of data: Mulliken PACs obtained by the B3LYP method at the 6-31G level (see Table II.4, column 6 for detailed data). Calculations by M. E. Alonso-Amelot (this book).

In this context, the case of *trans*-oleic acid (**116**) (reaction **B**) is of interest in that acylation under standard Friedel–Crafts arylation under $AlCl_3$ catalysis, the incorporation of 1 mol of benzene occurs singly at carbons C^7 to C^{17} [45]. How is this possible if there were nothing but inactive methylenes save for just one C=C bond? Could you propose an answer? Think for a while and then read on.

The $\Delta^{9,10}$ of **116** is the only functionality in the hydrocarbon chain of any interest to the Lewis acid, leading to a $C^{(+)}$ at C^9 or C^{10}. Before these carbocations are attacked by benzene, used here in large excess, they undergo E1 elimination furnishing either $\Delta^{8,9}$ or $\Delta^{10,11}$ unsaturations, respectively. These monoalkenes will either repeat the $C^{(+)} \rightarrow$E1 running cycle down to the methylene tail or be attacked by benzene. Mind that the benzene trapping kills the $C^{(+)}$ sprint, hence mono aryl-stearic acids, 11 isomeric compounds in actual fact, are produced [45].

2.4.10.1 Hidden LEDZs Triggering Deep-rooted Skeletal Rearrangements One or more occult LEDZs may crop up after the removal of electrons from the molecular backbone leaving behind a radical or carbenium ion. A substantial electron reorganization ensues until a stable structure is attained.

Appealing exhibitions of electronic distortions prompt deep-rooted skeletal rearrangements like **118**→**119** (Scheme II.25, reaction **C**). A pen-and-paper mechanism (via carbocation **120**→**124**) shown after the customary yellow arrow was examined by the magnifying lens of density functional calculations. Author Paul Rablen [30] found this to be an exothermic reaction ($\Delta G^o = -22.5\,kcal\,mol^{-1}$), due to the liberation of ring strain, and confirmed most of the proposed steps (numbers under each intermediate are Gibson energies relative to cation **122**). He also added that the **120**→**121**→**122** conversion could be condensed in a single step through a *DR* (see Scheme II.16) whereby the two C–C bonds involved would migrate through a single TS $12.7\,kcal\,mol^{-1}$ above the preceding **120** cation. The migration would be asynchronous, however.

Even more impressive, electron traffic reaching epic proportions occurs in the stereospecific bromonium ion-promoted cascade cyclization of squalene triepoxide carbonate **125** (Scheme II.26, reaction **A**) to **127** in a one-pot biomimetic reaction [46]. It involves the redeployment of 12 electrons across six moieties from head to tail of the chain. *N*-bromo succinimide serves as the source of the $Br^{(+)}$ species whereas HFIP (hexafluoroisopropanol) offers a highly polar medium to support the several $C^{(+)}$ arising along the reaction progress.

As you can surmise, the mechanism unfolds easily once you redraw the straight chain rendering of **125** as the partially folded **125-B** resembling the cyclized product **127**. Other cation-initiated cascades go through complex Wagner–Meerwein rearrangements mechanisms (see, for example, Problem 60, in the second edition of this book).

Pumping electrons into strained molecules through the attack of nucleophiles or reductive reactions creates HEDZs, which also cause deep perturbations initiated by C–C bond cleavage, as the **128**→**129** transmogrification shows (Scheme II.26, reaction **B**) [47, 48] (see also Problem 41 in "The Problem Chest", Part II of this workbook).

2.4.10.2 Remote C–H Activation by LEDZ and Radicals At times, the mechanism problem solver does not find satisfactory HEDZ/LEDZ reactivity patterns and molecular C–C or C-heterocycle bond rearrangements elicited by EWGs and EDGs to explain a given chemical reaction. The learned models break down. Remote C–H activation of an otherwise dormant C–H bond undergoing attendant hydride or $H^{(\bullet)}$ transfer prompted by distant LEDZ or radicals may solve these riddles. While the latter are not uncommon and easily reasoned, remote intramolecular hydride transfers take us to uncharted territory.

2.4.10.2.1 Remote Hydride Transfer Two requisites one cannot do without:

a) Potent LEDZ.
b) Steric adequacy and distance for a feasible TS.

Scheme II.27 provides an instructive case from Professor Akiyama and coworkers in Japan [49]. Try your own answer before looking at Scheme II.28 and then compare.

A quick comparison of **130** and **131** suggests that a C–C bond is created between the labeled C* and that the C=C linkage of the northern part of **130** is reduced to a σ linkage. The sparking reagent is scandium triflate, a potent Lewis acid that enhances the LED activity of **130** by withdrawing ED from the C=C–(C=O)$_2$ π system. This moiety resembles well-known Meldrum's acid (**132**) (Scheme II.28, top section). The anomalously high acidity of the α proton (pK_a 7.3 in DMSO at 25 °C) in **132** has been rationalized in terms of an in-phase orbital interaction between σ (CH) and adjacent π (C=O) orbitals (as in **133-AO**) rigidly positioned in the six-membered ring leading to optimal conjugation and anion stabilization.

On that account, isoelectronic canonical structures **134a/b** should be the prominent contributors to the resonance hybrid of starting compound **130** and furthered by scandium triflate. This intermediate is portrayed in Scheme II.28 as a preferred conformation properly structured for hydride transfer to the cation. A standard intramolecular Friedel–Crafts addition–elimination follows. Alternatively, **136** could have cyclized to spirocycloheptane **139** competitively (green curly arrows) owing to the HED/LED proximity. This adduct was not observed among the products by the Japanese researchers, however [49].

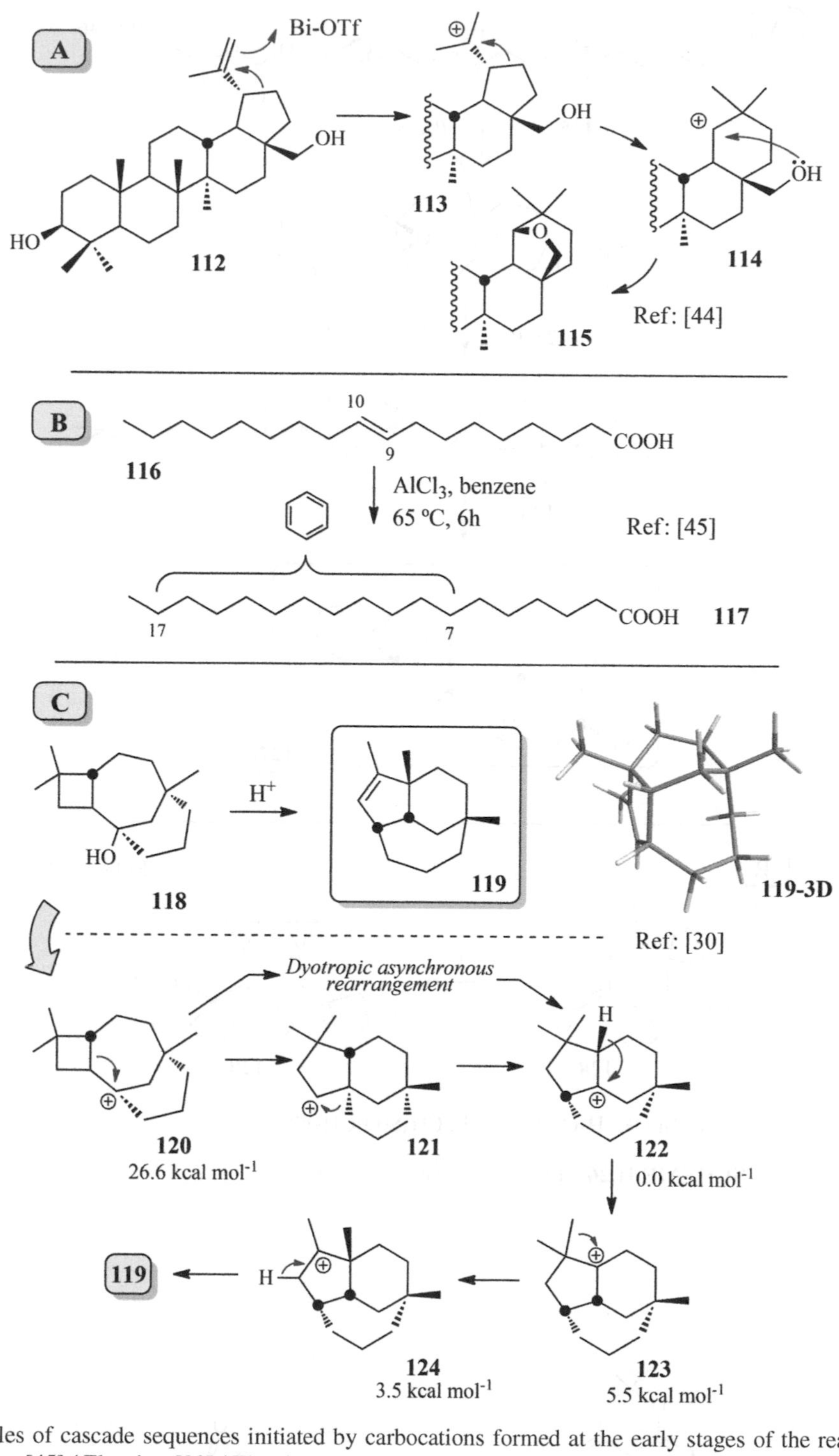

SCHEME II.25 Examples of cascade sequences initiated by carbocations formed at the early stages of the reaction progression. [44] / Royal Chemistry of Society; [45] / Elsevier; [30] / Elsevier.

Before analyzing the feasibility of **139,** observe Scheme II.29, top part, and analyze it for a moment to check whether this is a mechanism problem at your reach. Then, please come back to this line and continue reading.

A previous report by an independent research group in Canada [50] proposed a homologous spirocyclic product (framed **144,** Scheme II.29) that was obtained in nearly quantitative yield by using the same scandium catalyst *at room temperature*, so the Friedel–Crafts addition–elimination would not proceed. You can appreciate the similarities of Schemes II.28 and II.29 as far

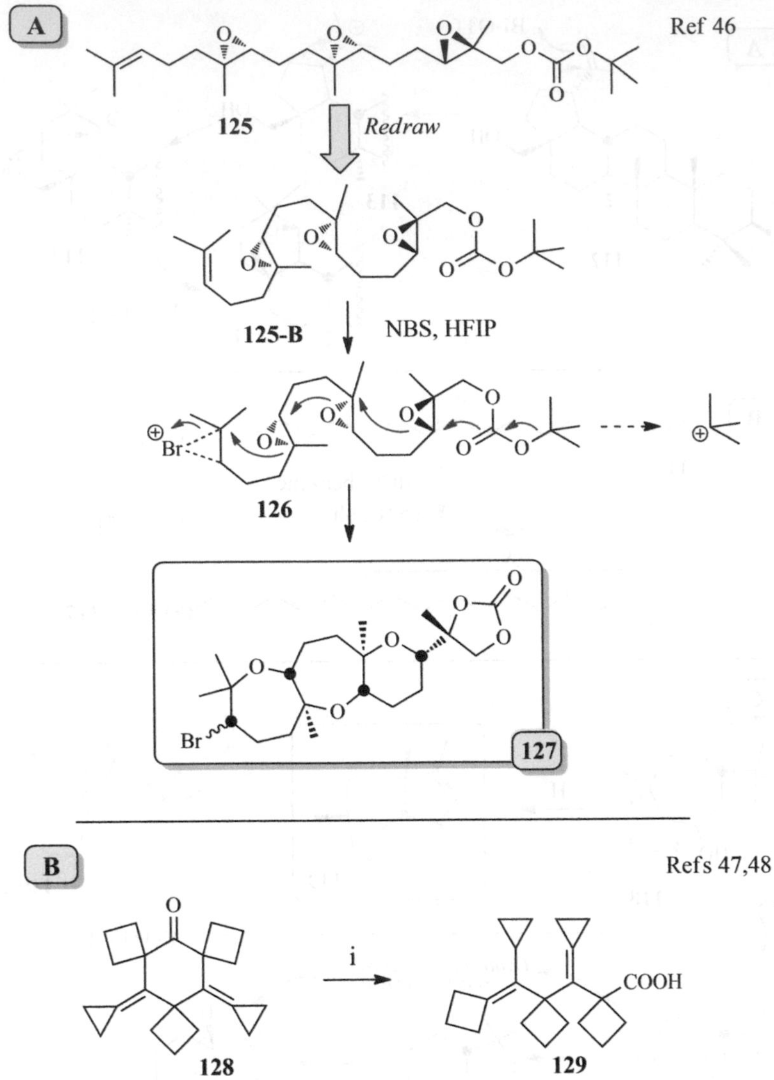

i: tBuOK, H_2O (20 mmol), $CH_3S(O)CH_2]^{(-)}$, 40 °C

SCHEME II.26 [46] / American Chemical Society; [47] /Elsevier.

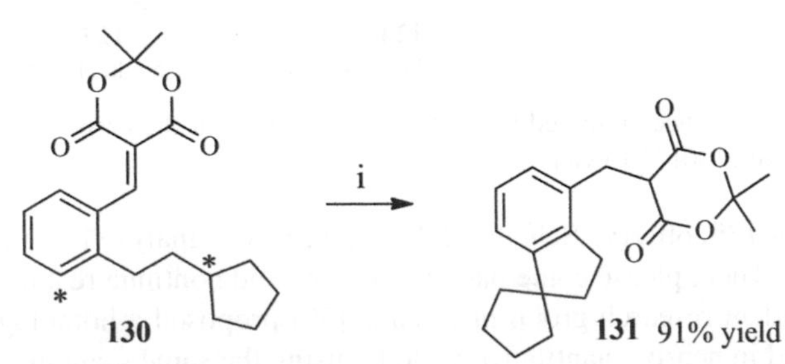

i: $Sc(OTf)_3$ (A powerful Lewis acid), $ClCH_2CH_2Cl$, Δ, 24 h

SCHEME II.27 An example of remote C-H bond activation owing to a distant low electron density moiety. [49] / American Chemical Society.

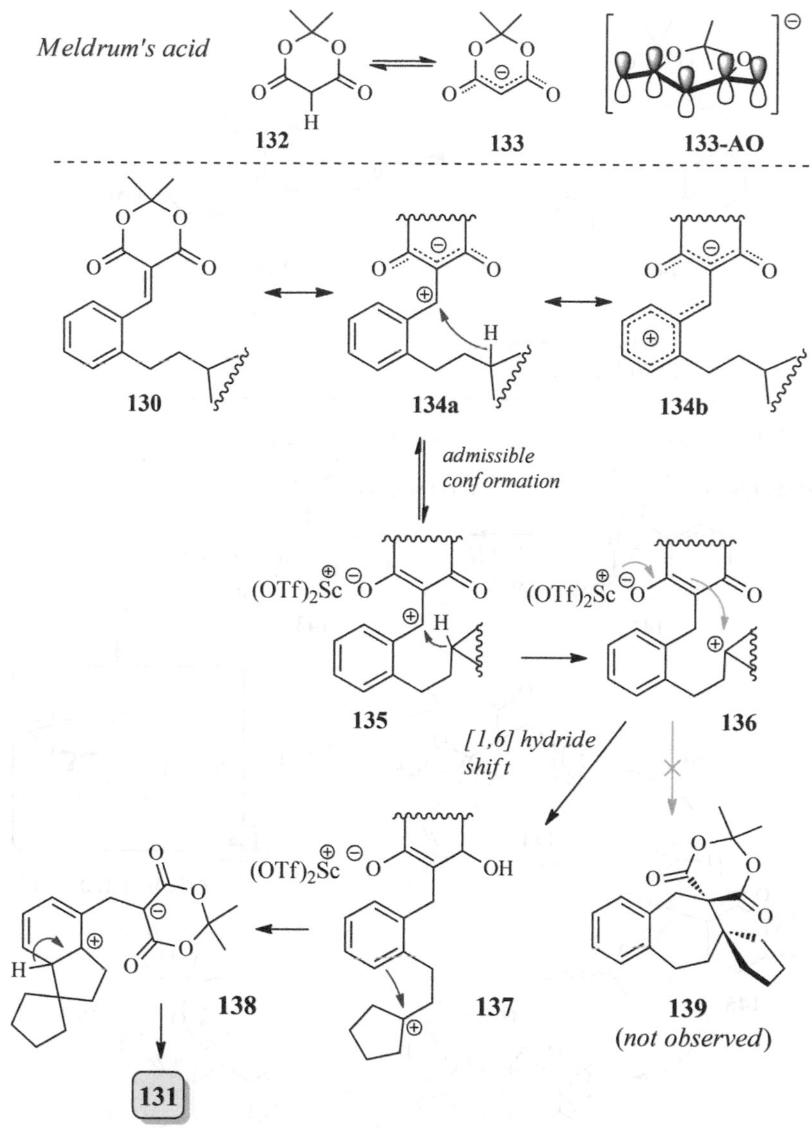

SCHEME II.28 One mechanism solution to Scheme II.27 by exploiting the features of Meldrum's acid. [49] / American Chemical Society.

as products **139** and **144**. In their hands, the latter was further converted to the tetracyclic ketone **141** by increasing the temperature to 70 °C.

The overall, domino-type sequence of Scheme II.29 spanned five reaction types in a row: [1,5] hydride shift → intramolecular alkylation → Friedel–Crafts addition–elimination → hemiketal fragmentation → and β-ketoester (**147**) decarboxylation to your desired target **141**. Authors [50] did not comment on the mechanism, so any blunders in Scheme II.29, if any, are exclusively mine. Did your mechanism design resemble Scheme II.29?

2.4.11 Issue 11: Inverting the Natural Electron Flow, Umpolung

☑ *HEDZ to LEDZ electron traffic can be reversed by changing the natural polarization of functional groups by proper manipulation.*

Up to this point, we have treated HEDZ and LEDZ as they come out of the bottle. Occasionally though, manipulating the core of some functional groups may convert natural LEDZs to HEDZs and vice versa. This abrupt change is known as *umpolung*, a German word meaning "turn-over."

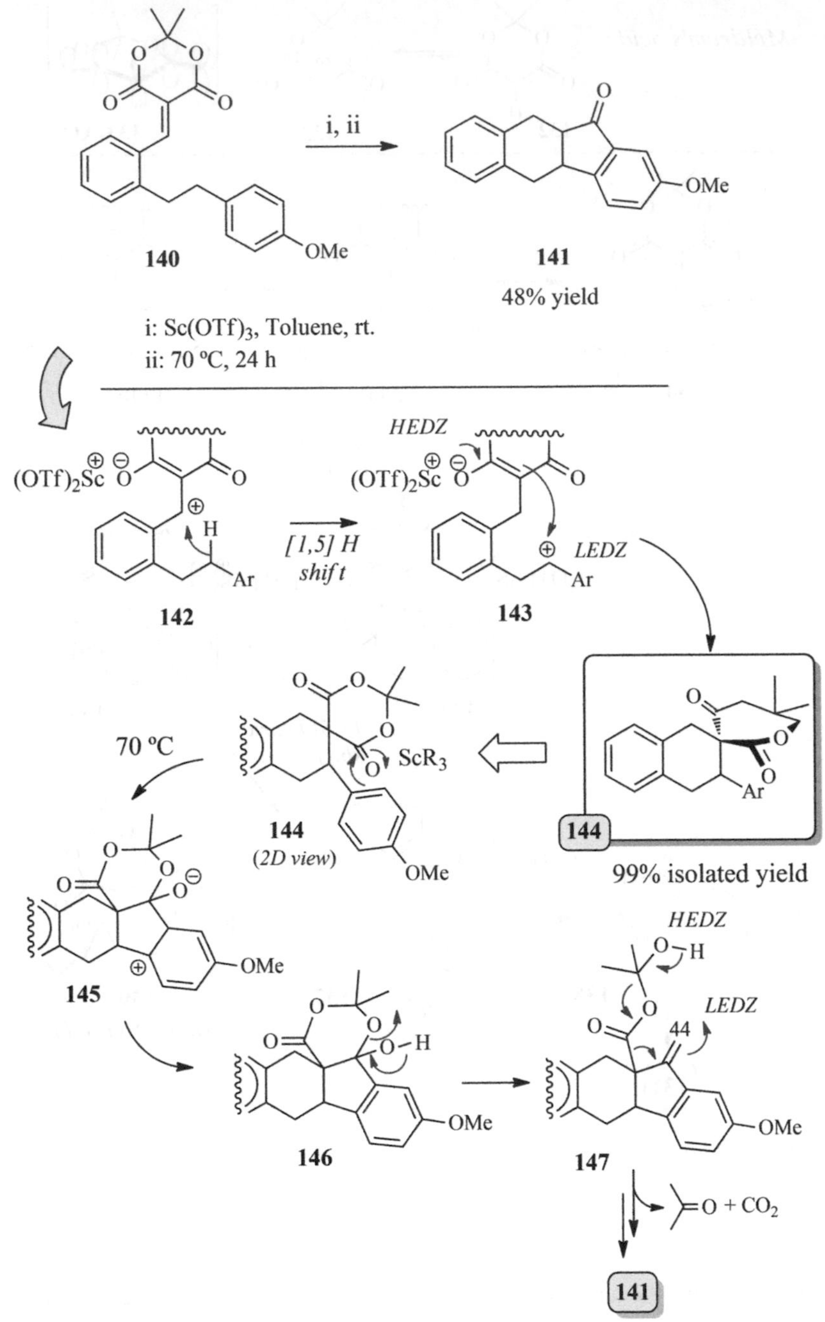

SCHEME II.29 Similar reaction of Scheme II.28 discovered years earlier [50] and a feasible mechanism. [50] / Elsevier.

Chemically speaking, umpolung is understood as *inversion of polarity*. For example, the natural $C^{(\delta+)}{=}O^{(\delta-)}$ polarization of an aldehyde **A** can be inverted to an acyl anion equivalent **B** (Figure II.12) in order to create a σ bond with electrophiles and regenerate the initial C=O during workup.

2.4.11.1 Umpolung Successful Accomplishments Umpolung has been in the mind of organic chemists for a long time. Umpolung acquired revolutionary dimensions in the 1970s when carbanions were created from aldehydes by the previous conversion to thioacetals followed by strong base to remove the ex-aldehyde proton [51] (Scheme II.30, bottom section). However, a formal polarity inversion was introduced more than seven decades before by Professor Françoise Victor Grignard (1871–1935) in 1900 when he discovered, during his doctorate thesis, that mixing Mg° filings and an alkyl (vinyl or aryl) halide in anhydrous diethyl ether, adding a crystal of iodine to accelerate the reaction, a powerful alkylating reagent (RMgX, where

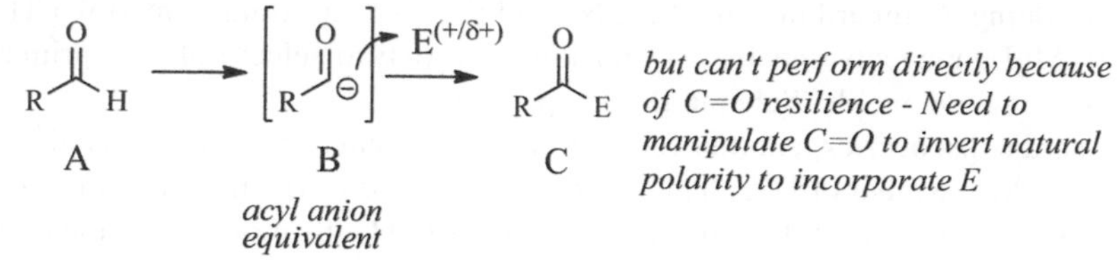

FIGURE II.12 The umpolung principle.

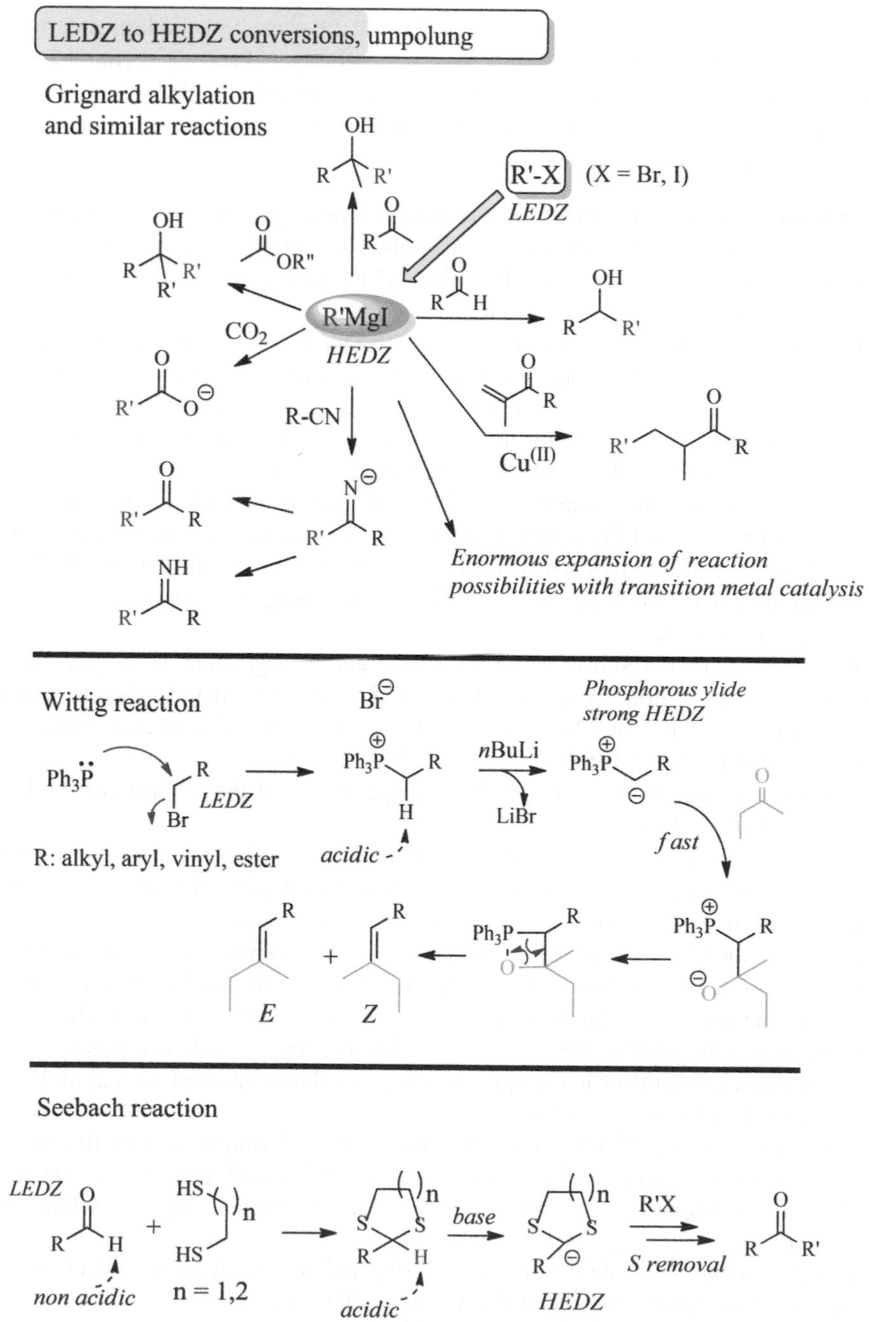

SCHEME II.30 Selected seminal methodology in polarity inversion (LED → HED) of widespread use today. Other methods continue to appear in the literature.

X = Br, I) was produced. In so doing, Grignard (awarded the Nobel Prize in 1920) transformed R_2CH–Br (or I), a typical LED moiety, into robust HED R–MgBr reagents capable of attacking a variety of electrophiles, primarily C=O derivatives, in predictable alkylative additions in good yields (Scheme II.30, top section).

Half a century after Grignard's discoveries, another form of umpolung came about in the hands of Professor Georg Wittig (1897–1987) [52], also a Nobel Prize winner (1979, shared with Herbert C. Brown) when he managed to convert primary alkyl, allyl, and benzyl halides to *phosphorous ylides*. By this procedure, the LED C–Br moiety became a relatively stable carbanion after strong base treatment (*n*-BuLi). The reaction that bears Wittig's name is a choice method to build di- and trisubstituted alkenes regiospecifically from aldehydes and ketones (Scheme II.30). There are several variants that allow E or Z stereoselectivity depending on the stability of the ylide by the R group or phosphine used. A detailed review is available for interested readers [53].

Grignard conditions can also deliver interesting unanticipated results. For example, submitting cyclopropene sulfone carboxylate ester **148** to *n*-BuMgCl in THF at low temperature was supposed to yield tertiary carbinol **149** (Scheme II.31, section **A**). Authors Liu and Ma found, instead, a quite different product, which moderately yielded allene **150** after exposure to propargyl bromide and a catalytic amount of copper cyanide [54]. Take this as your next problem assignment. The answer is provided in Scheme II.48 at the end of this chapter.

2.4.11.2 The Nitrogen Heterocycle Carbenes (NHC) in Carbonyl Umpolung Grignard's conceptual breakthrough opened a broad avenue to C–C bond construction, later expanded to other alkaline and transition metals to replace Mg (Zn, Li, Cu, Al, Sn, Ti) as well as the organoboron compounds, and others that offer catalytic opportunities (e.g. Pd, Ni, Co, Fe, Au) and modulate excess reactivity of the Grignard reagents.

One class of these, alkyl or aryl zinc reagents, are much less reactive than alkyl magnesium equivalents. Their reaction with aldehydes and ketones are slow, even sluggish whereas other functional groups are impervious to them. Moderate reactivity means better control, added regioselectivity, and protecting some functionalities is unnecessary. In addition, transmetallation or catalysis with copper complexes enhances the R_2Zn reactivity, converting it into a useful alkylation agent [56–58].

The design of *ad hoc* ligands in these complexes addresses the question of regio and enantioselectivity, which was achieved after a great deal of systematic work. A commendable example worth reading and following is the work of Professor Kiyoshi Tomioka and collaborators in Kyoto, Japan [55] who pursued the regio- and enantioselective β-alkylation of α,β-unsaturated ketones through Grignard reagents catalyzed by $CuBF_4$ and substituted chiral imidazole cations (SICs) (Scheme II.31, section **B**). SICs can also be configured as *N*-heterocyclic *carbenes* (NHCs), an emerging class of synthons. They are bonded to the transition metal available, copper in this case.

From 3-methylcyclohexenone as model substrate for increased steric congestion of the β-carbon and thus testing the 1,4-addition to its limits, two competing primary adducts **152** and **153** were anticipated, whereas **154** would be produced by elimination of **153** during acid workup. The central objective was to assess the induction of enantioselectivity of the β-alkylation by the asymmetry motif of the chiral catalyst **155**.

Alkylation would not proceed in the absence of $Cu(OTf)_2$ independently of the addition of the imidazole catalyst **155a**, hence the *copper carbenoid* seemed the best option.

Authors [55] did not elaborate on the mechanism. Meanwhile, several hypotheses have been suggested but most of them have been debunked and a consensus has not been reached as of this edition [59]. A few feasible proposals are depicted in Scheme II.32, section **A** as an informative problem solution in the umpolung context.

At the start, the NHC gets bonded to the carbonyl substrate before the Grignard β-alkylation step to incorporate a chiral space and exert enantiomeric control. The question turns into the sort of NHC-carbonyl bond formed. In the absence of a proton source there are basically two bonding options for NHC on C=O with diverging outcomes (**a** and **b** in Scheme II.32, section **A**).

Option **a:** NHC **155-Cu** receives the NBP of the oxygen atom, thus the copper NHC behaves as an LEDZ.

Option **b**: commences with the electron flow in the opposite direction, thus NHC performs as an HEDZ on the electrophilic carbonyl carbon; a double-faceted electron flow terms.

Once route **a** begins, the primary adduct **156** may progress toward the β-alkylation through the copper-alkene complex **157** in case the substrate ketone can acquire such a configuration, or take in the Grignard reagent by virtue, in both cases, of the LED created at the β-carbon. The organo-magnesium intermediate **159** blocks any further alkylation while catalyst **155-Cu** is recovered for another catalytic cycle.

Route **b** may eventually produce an unstable four-membered ring **162** whose cleavage collects the electron flow from the Grignard C–C bond formation and regenerates catalyst **155-Cu** in its original form.

There is experimental evidence supporting either route. For example, rhodium carbenes derived from α-diazocarbonyl compounds (esters, ketones) like **164** (Scheme II.32, section **B**) yield transient *carbonyl ylides* such as (**166**). These species can be trapped in various ways including [3 + 2] dipolar cycloaddition on suitable dipolarophiles like fumaric ester **167** [60].

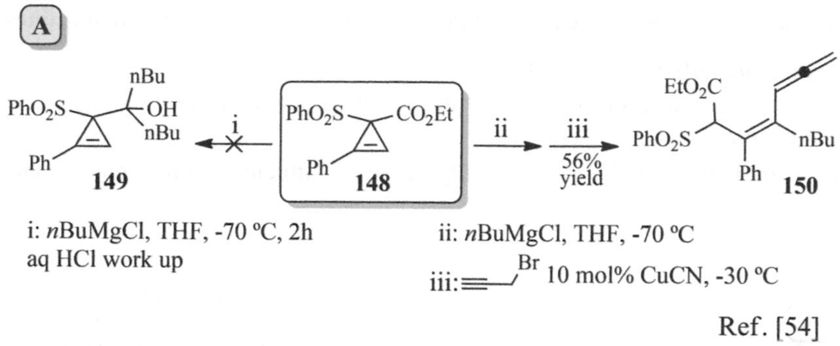

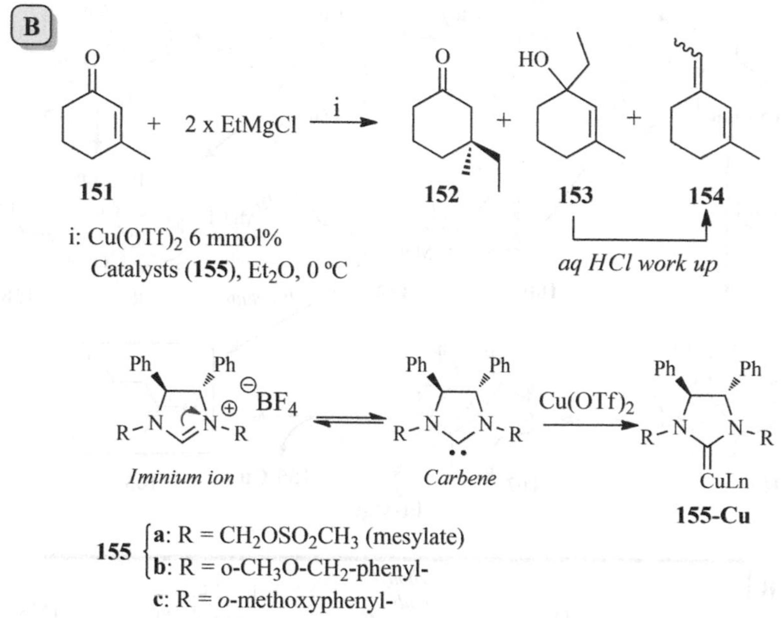

selected results:

Catalyst	% yield	152/154	ee%
155a	34	42:58	21
155b	40	54:46	49
155c	99	>99:1	77

[ref. 55]

SCHEME II.31 Reaction A [54] / Royal Society of Chemistry. **Reaction B** [55] / American Chemical Society.

That the cyclization goes through a concerted TS is suggested by the conservation of the fumaric ester configuration in adduct **168**.

The indirect interaction of NHC in Michael additions to α,β-unsaturated ketones has been exploited in various ways such as the β-amination shown in Scheme II.32, section **C** [60]. Here, the NHC species intervenes on the carbonyl substrate only after hydrogen transfers from the amine owing to the basic character of the metal carbenoid. This transfer is almost irreversible but provides an imidazolium cation that conveniently places the ketone and amine moieties through H bonding and electrostatic attraction, a sort of molecular pincer, to expedite the β-amination of the LEDZ substrate.

2.4.11.3 Umpolung of C=O through Imine Derivatives The imine group $R_nC{=}NR'$ ($n = 1, 2$), or Schiff base, is the nitrogen homolog of $R_nC{=}O$ with some similarities and differences. It is often found in compounds of biological activity. A takeaway reminder is provided herein.

Of special interest to this section, imines show a sort of double personality in the sense that **imines primarily perform as:**

- ✓ N-nucleophiles owing to the NBP in a sp^2 AO.
- ✓ Mild C electrophiles, which require strong nucleophiles or Lewis-acid activation.
- ✓ Discreet dipoles that enable access to heterocycles.
  - Their low polarity can be reversed by electronically active substituents and become a manner of carbonyl umpolung equivalent.

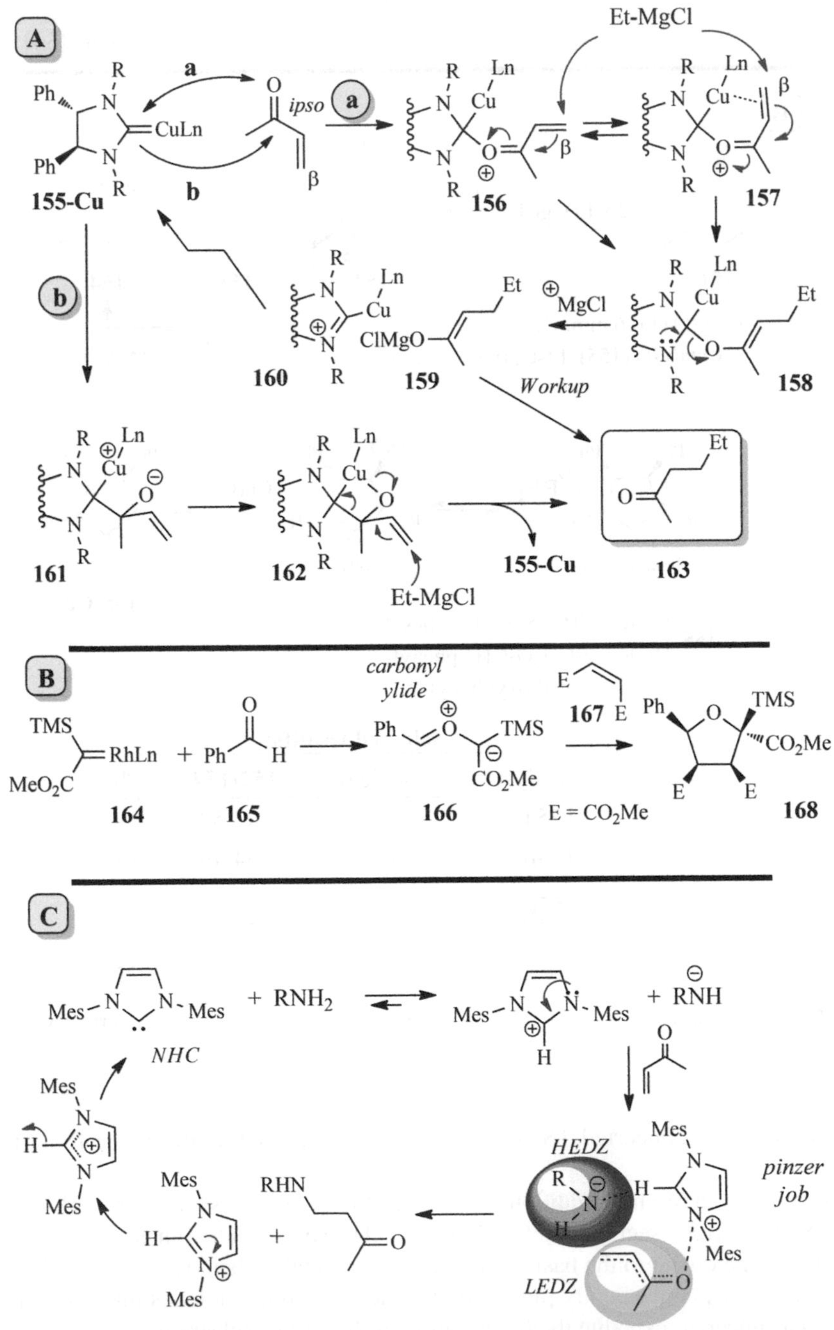

SCHEME II.32 Section **A:** two feasible mechanisms of the interaction of NHCs studied by [54] / Royal Society of Chemistry in their catalysis of Grignard reagent alkylation of α,β-unsaturated ketones. Section **B:** An example of a similar rhodium carbene acting on aldehydes, known to form carbonyl ylide intermediates (**166**), which operate as 1,3 dipoles over LED alkenes like **167**. Section **C:** adapted from [60] / Royal Society of Chemistry.

The peculiar properties of imines are rooted in the smaller difference in electronegativity between C and N (3.0–Pauling) relative to C and O (3.5–Pauling). Electron populations governing dipole moment, local atomic charge, and reactivity obey this electronegativity pattern in the imine moiety itself and the extended α,β-conjugation. Figure II.13 collects calculated Mulliken electron charges of common imine moieties with H atoms summed up in heavy elements.

Noteworthy features of Figure II.13:

- ✓ At the parent moiety level, both μ and atomic charges (ACs) converge in suggesting a far more electrophilic C=O of aldehydes than C=NH of aldimines.
- ✓ For this reason, aldimines and ketimines (from ketones) do not react with mild nucleophiles such as dialkyl zinc and copper agents without prior activation. This difference in reactivity enables the use of regioselective nucleophilic additions on C=O in the presence of C=NR without using protective groups in the latter.
- ✓ While α,β-unsaturated synthons (bottom block) distribute the positive ACs in C^1 and β carbons in C=O derivatives allowing β-addition reactions, the effect is poorer in imines. Yet β-addition can be accomplished by proper manipulation of substituents (see below).
- ✓ EWG appendages on N (top right block) enhance the reactivity of C=N toward nucleophiles, as shown by the significant increase in μ and the enhanced deficit of AC of the azomethine carbon atom. A thermodynamic effect is also at play since

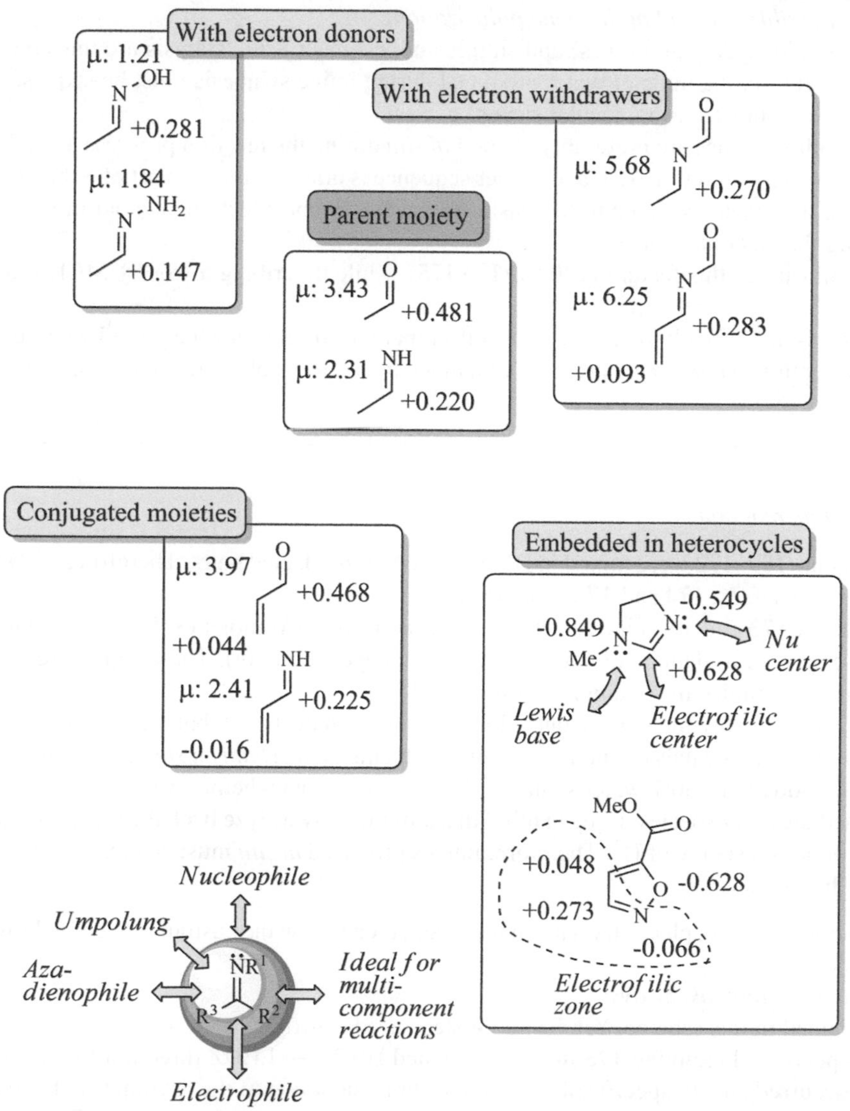

FIGURE II.13 Mulliken atom charges (atomic units) of the sp^2 carbon atom with H atoms summed up in heavy elements and dipole moments (μ) (Debyes) in aldimines and α,β-conjugated synthons in comparison with similar aldehydes. Calculation base: RHF/6-31G (Alonso-Amelot, this book).

the $N^{(-)}$ of the addition intermediate is stabilized by the neighboring EWG. Several EWGs were tested successfully in C^1 and β additions.

✓ As a part of heterocycles, imines take up the influence of accompanying heteroatoms and functionalities at both ends, which alter AC accordingly, thus impinging on the electrophilic–nucleophilic character of the C=N atoms (bottom right block).

In a nutshell, the reactivity profile of imines (Figure II.13 bottom left) parallels that of ketones and aldehydes except for a much reduced reactivity compass that can still be encouraged by suitable substituents and catalysis. Resourceful imines are a true asset for the problem solver.

Two defining examples of imine performance with normal polarization:

Case 1 C=N as Moderate Electrophile

The smaller electronegativity difference and μ of the C=N bond create difficulties to alkylation, yet they open opportunities for the modulation and *possible reversal of the bond polarization*.

Imines produced from aldehydes (aldimines) and simple amines under acid catalysis are unstable and easily hydrolyze back to the carbonyl. Therefore, the mole of water produced during imine synthesis must be sequestered from the reaction medium by a suitable absorbent such as molecular sieves.

Because of this instability, imines are preferably formed *in situ* during the reaction progress as in the well-known, century-old Mannich reaction (discovered in 1912) (a reminder sequence is drawn at the top part of Scheme II.33). You cannot see the imine moiety among the reactants added to the flask. But, as this chapter section is about imines, you may assume that imines do appear during the reaction course.

The second section illustrates this feature (**170** + **171** → **175**), while describing a worthy MCR growing on the Mannich design [61].

While the mechanism is not difficult to discern, it is conceptually instructive for the point I want to make here: the modulation of the imine electrophilic nature in the presence of a much more electrophilic C=O, two of them in fact, in this one-pot reaction.

Keynotes and hints of this reaction:

✓ The four components **169–172** were mixed in DMSO, and not in selected pairs. Therefore, LEDZs ketone and aldehyde compete for the two HDEZs, **170** and **172**, concurrently.

✓ Stoichiometry: target **173** contains nine skeletal C atoms and two *N*-Ar moieties. Relative to 1 mmol of cyclohexanone, there is a molar excess of arylamine (3 mmol) and formaldehyde (6 mmol). These figures result from authors Wei and coworkers' efforts to optimize the reaction conditions [61].

✓ L-Proline (**172**), added in 20% molar ratio only, does not appear in the target, but the reaction would not proceed without it. Therefore, proline acts as catalyst–ancillary component, and *the mechanism should include catalyst recovery step(s)*.

✓ Formaldehyde was added in a 36% water solution. We must figure out whether water participates as a reagent or simply stabilizes intermediate polar species. Keep in mind that aldimines hydrolyze back the aldehyde easily in water and protic catalysis, and proline is a source of H^+. Therefore, imines produced *in situ* must be consumed faster by the electrophilic addition than by hydrolysis.

With these considerations in mind, please try a mechanism and then follow the customary bent yellow arrow in Scheme II.33 to check up a feasible mechanism

Highlights of Scheme II.33 *and takeaways*:

The sequence is explained in this scheme. Yet the major steps need clarification.

Enamine **174** intercepts *N*-aryl aldimine **175** in the anticipated HEDZ → LEDZ direction to afford iminium ion **176**. This first addition probably occurred enantiospecifically due to the steric induction of the asymmetric L-proline ancillary, although this will never be known as **176** evolves toward the thermodynamically more stable enamine **179** destroying all evidence of a chiral α carbon, in preparation for a second addition step via **180**. Up to this point two moles of formaldehyde have been consumed.

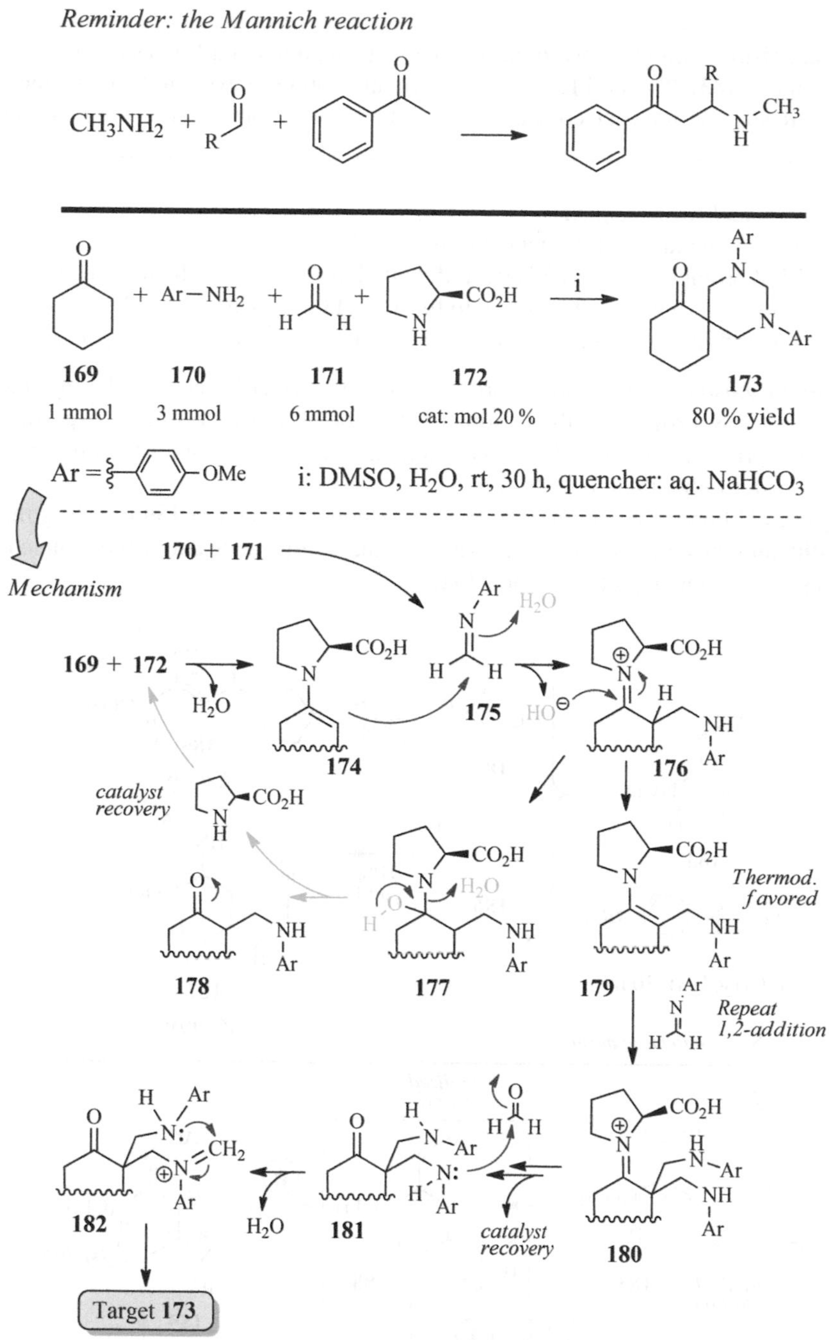

SCHEME II.33 A multi-component reaction involving occult imine chemistry and a likely mechanism. [61] / American Chemical Society.

Diaminoketone **181** is ideally suited to trap a third mole of formaldehyde in a pincer-like bite via strongly LED iminium ion **182**. The molar excess indicated above is thus justified.

> ***Case 2 The C=C–C=NR Moiety Acting as an LEDZ at the β-carbon.***
>
> The Michael-type amination of α,β-unsaturated imines, albeit uncommon, is perfectly feasible in satisfactory yields and some degree of stereocontrol. Take the example of Scheme II.34 as your next problem assignment by putting aside the mechanism depicted under the dotted line.

(continued)

At first glace, the mechanism is straightforward thanks to the visible traits of a Michael-type 1,4-addition of amines **184** and **186** on the starting imine (**183**). The problem begins when attempting to explain the difference in the stereochemical outcome. Both reactions are strictly stereoselective affording either *E* or *Z* isomers, contingent on the nucleophile.

Keynotes and evolution of this mechanism sequence

Keep an eye on Scheme II.34 as you read the following discussion, please.

First, imine **183** is our LEDZ prompted by the EW effect of diethyl phosphonate, which drives ED away from the conjugated C=C–C=N moiety in addition to offering steric hindrance to the 1,2-addition. This leaves the 1,4-addition of the amine free to proceed to enamine zwitterion **188** (dashed frame), which may evolve along two competing pathways:

(**A**) *N–N H transfer to the Z-isomer.* H transfer is probably intermolecular because of the intramolecular remoteness and rigid cisoid structure of **188**. Progress to the E-isomer is thus blocked. Contrastingly, experiments show that this is the preferred pathway for propargylamine (**186**), allyl amine, and aniline for somewhat obscure stereoelectronic reasons that favor a fast H transfer rather than route **B** [62].

(**B**) The enamine rigid C=C bond can also be expressed as canonical structure **189** enabling the 180° turn of this bond (green arrow) to **190**, and thus, mitigate the steric strain of the phosphonate and newly bonded amine. The more stable *E*-enamine emerges after quenching of zwitterion **191**.

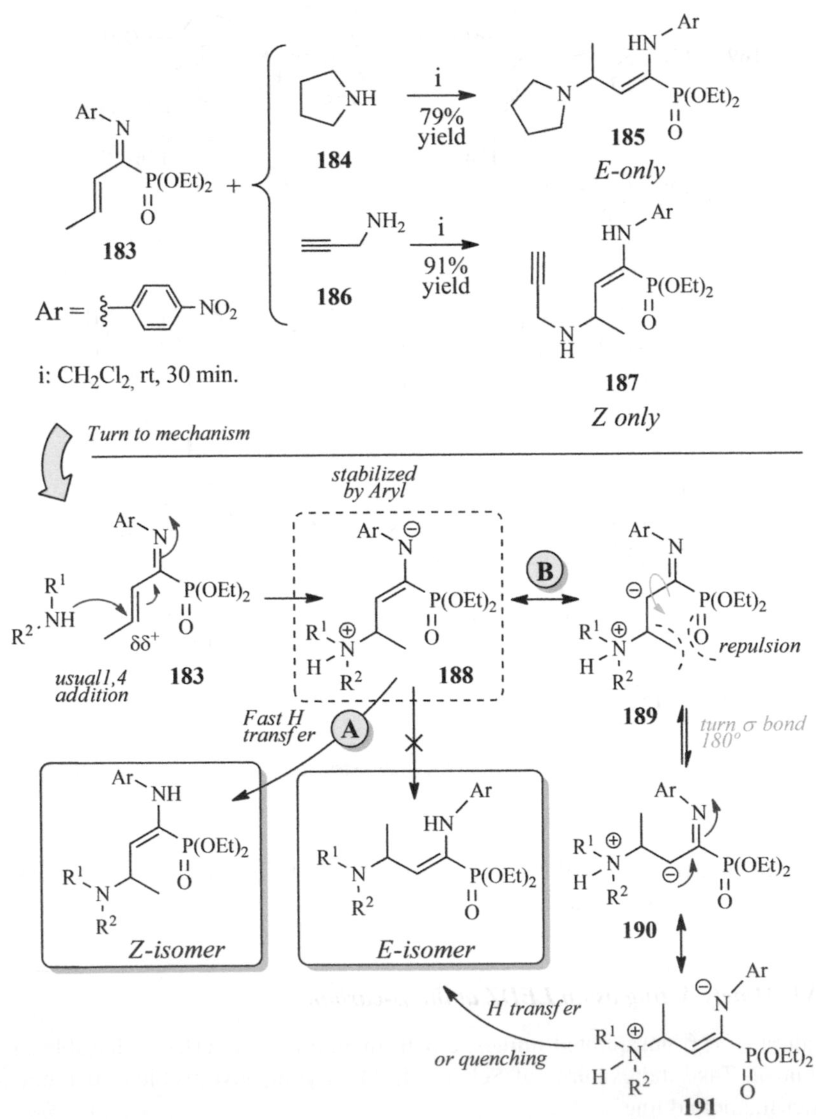

SCHEME II.34 Contrast between stereochemical outcomes in very similar reactions caused by apparently irrelevant changes in the amine nucleophile. [62] / American Chemical Society.

This pathway seems faster than route (**A**) because of the increased steric repulsion depicted in **188** with sterically more demanding amines such as pyrrolidine (**184**) and various anilines tested (not shown). Yet, petite methylamine with no special steric requirements also gives 100% of the E-isomer (88% yield). In cases like this, hitherto unreported computational *ab initio* calculations may be ideal to further explore this curious outcome.

2.4.11.4 The Hydrazone Way to C=O Umpolung The calculated Mulliken ACs of a variety of imines shown in Figure II.13 (upper left block) suggested that electron donors would decrease the imine electrophilicity (+0.220 in non-substituted imine versus +0.147 in the C=N–NH_2 moiety). In principle, such ACs on the carbon atom are insufficient to procure a true nucleophilic character.

A way around is to operate indirectly to build an anion at the hydrazone carbon and create conditions for it to trap electrophiles: the true umpolung. This was accomplished by the Shapiro reaction shown in Scheme II.35. In the original design, a tosyl hydrazone (**192**) and a strong base such as methyl or butyl lithium were necessary to remove the two mildly acidic protons (blue H) for the reaction to go in the way of *polarity inversion equivalent*.

In forming a vinylogous anion **197,** the Shapiro model furnishes an extra C=C created during the umpolung process as a result of the removal of the imine β-proton. This bond is not always desirable for specific synthesis purposes and must be removed by reduction (hydrogenation, for example).

While effective, the Shapiro reaction suffers from experimental complications that have been solved with use of dialkyl magnesium or cesium carbonate instead of the alkyl lithium base followed by the addition of an electrophile [63, 64]. Given the higher acidity of the N–H proton of the hydrazone, a milder base would, in principle, pick up this proton, selectively leaving the β-CH inactive. The N anion so created would be delocalized down to the α-carbon (**194b**) to provide the desired charge inversion and the ensuing nucleophilic activity at this site.

Alternatives to the Shapiro reaction have been developed more recently (Scheme II.36). For reviews, see References [65, 66].

Highlights and takeaways of Scheme II.36:

Suppose one gets a monoalkyl or aryl hydrazone (**200**) from an aldehyde and alkyl or (aryl) hydrazine under mineral-acid catalysis as usual and purify it so you start with a clean material. Exposing it to LEDZs like the ones in Table II.1 (right column) without a catalyst in neutral medium (Scheme II.36, route **A**) will come to nothing. Only some powerful electrophiles like the dimethyl iminium **201** (known as the Vilsmeier–Haak reagent/reaction) [67] will be sensitive to hydrazone (**200**) but chiefly delivers the *N*-chloroalkylamine **202**. The ensuing elimination/hydrolysis yields the *N*-acylated **204**, but *not the umpolung product*.

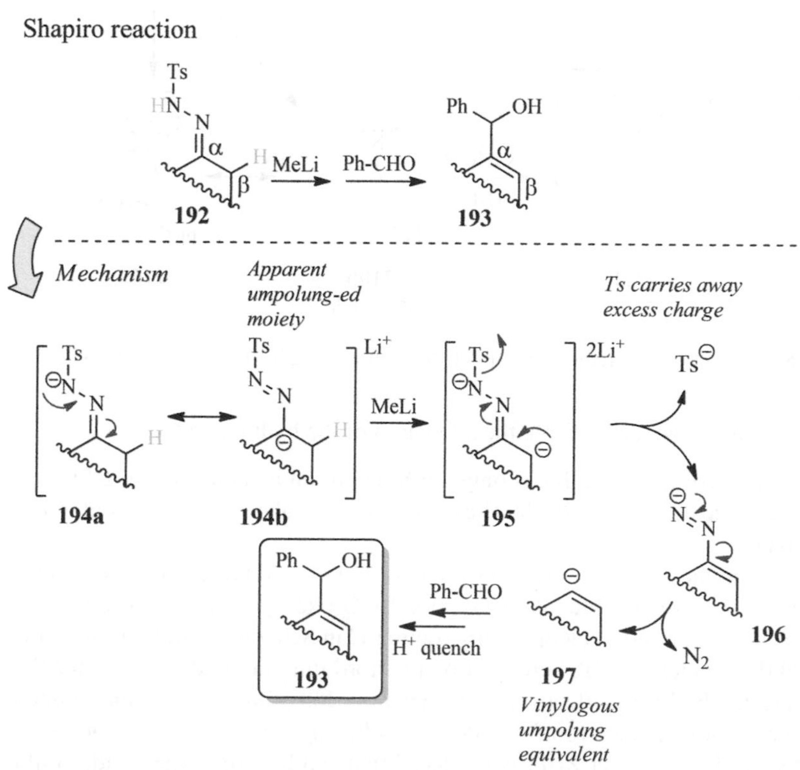

SCHEME II.35 The Shapiro reaction concept and presumed mechanism via dianion **195**.

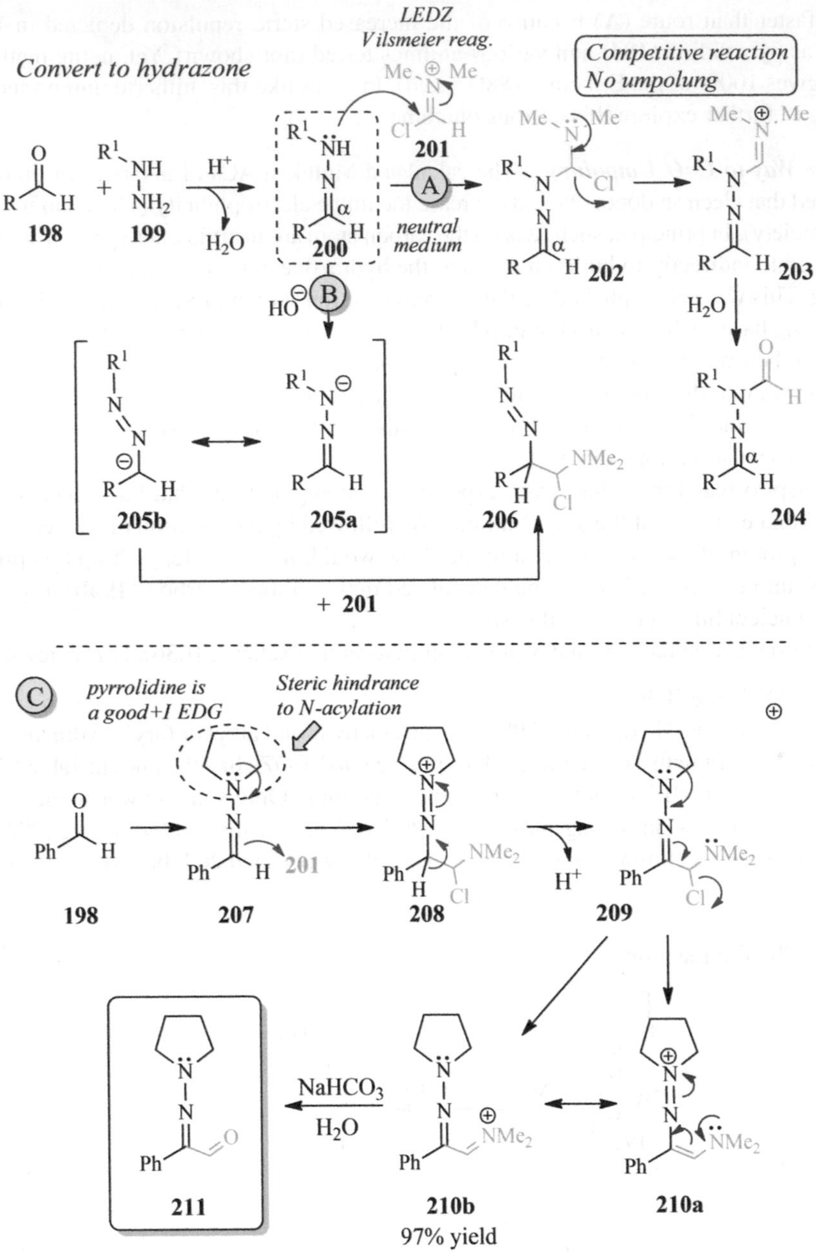

SCHEME II.36 Collection of strategies for imine direct umpolung by use of aldehyde → hydrazone conversion.

There are two ways around this obstacle to force the alkylation of the hydrazone C_α.

1. Use a moderately basic medium (route **B**): If only the N–H proton is removed and R[1] is an EDG, the resonance hybrid will be enriched in canonical structure **205b**. Thus, addition will take place at C_α preferentially to give **206** chiefly but perhaps not exclusively.
2. Increase the steric bulk at the *sp*[3] N with two substituents (e.g. *t*-butyl) to create steric congestion against the incoming electrophile; better yet if these substituents were good EDG like pyrrolidine (+*I* induction), which was successful in enhancing enamine [R_2C=C–NR_2] nucleophilicity (option **C** in Scheme II.36, bottom section).
3. Primary adduct **208** then undergoes HCl elimination to afford iminium hydrazone **210** either via regular β-elimination (not shown) or, more likely, by the electron flow shown in **208** stimulated by the ammonium ion and the sequence **209** → **210a**, a showcase of what electron flows do while obeying electromagnetic forces.
4. A nearly quantitative yield of adduct **210b** is produced that can be hydrolyzed under mild aqueous alkali to the true α-acylation target **211**.

5. The **207** → **208** step *is the key umpolung bimolecular stage*, whereas the rest are electrons and substituents standard redeployment toward isolable products.

A selected number of electrophiles other than the Vilsmeier reagent (**201**) are susceptible to C–C bond formation with **207**-type hydrazone carbonyl anion equivalents:

- Sulfonyl isocyanates (RSO_2–N=C*=O).
- Aldimines ($R_2N=CH_2$) as in the Mannich reaction.
- Acyl chlorides R–(C=O)Cl.
- Trifluoroacetic anhydride, $(CF_3–C=O)_2O$.
- Electron-poor Michael C=C and alkynes with EWGs at both ends.
- Molecular halogens.
- Sulfenyl chlorides S_2Cl_2 and aryl-SCl.
- Phosphine halides R_2PCl.

Several other electrophiles give poor yields, side products (*N*-acylation), or fail to give adducts altogether.

This abridged list, a sample of the working electrophiles in imine umpolung, can be extended to alkyl halides, alcohols, allyl acetates, aldehydes, and Michael 1,4 additions by metal complex catalysis such as those from ruthenium, copper, nickel, and iron [66], as well as reversed polarity Diels–Alder cycloadditions.

2.4.12 Issue 12: One-electron Flow

☑ *Electrons can be transferred individually during molecular transformations from radicals, triplet carbenes, and oxidizing and reducing agents.*

Up to now, we have redeployed electrons in pairs from HEDZs to LEDZs to form or dislocate σ bonds and attack or modify π bonds, retrieve or reinstall NBPs in N, O, P, or S, or cleave C–X (X = N, O, P, S, Se, halogen) bonds heterolytically. In all these cases, electrons populate available AOs and MOs in pairs of opposed spin in compliance with Pauli's exclusion principle. Frontier MO theory (FMOT) conceives HOMO to LUMO transfer of electron pairs to bring about a quantum-bonding framework to electron redeployment and energy profiles involved.

Electrons, however, can also migrate as single and independent subatomic particles and continue to comply with quantum rules of orbital occupancy. FMOT defines a specific MO to lodge these unpaired electrons: SOMO (single occupied MO). SOMO stands at the highest energy level of the occupied MOs, hence are more exposed than electrons in lower HOMOs.

As you may have anticipated, molecules with unpaired electrons, that is *atoms or molecules with an uneven number of electrons*, are *radicals*, generally very active species.

Since there are excellent treatises on this subject, we will not spend much editorial space and time in describing radical chemistry of organic compounds. Yet, since radicals crop up often enough in reaction-mechanism alternatives, a convenient summary of essential issues of radical chemistry and current updates is a convenient addition at this point.

2.4.12.1 Radicals, Reminder Takeaways Note at the start: only use single-barbed (half-headed) curled arrows for one-electron redeployment, please. This symbol is universally accepted.

Essentials:

- As you already know, radicals are atoms or molecules containing an atom with only seven valence-shell electrons instead of eight for stable structures in second-row elements, or an odd number of valence electrons.
- Single atoms of diatomic molecules are atomic entities and radicals at the same time. Classic examples: the hydrogen $H^{(\bullet)}$ and halide $X^{(\bullet)}$ atoms.
- Some simple common elementary molecules are mono or diradicals in their natural form. Examples (Lewis NBPs not shown) include:
  - O_2: O=O ↔ $^{(\bullet)}O–O^{(\bullet)}$, (highly reactive singlet oxygen).
  - NO: $^{(\bullet)}N=O$ also expressed as the dimer N_2O_2: O=N–N=O.
  - NO_2: $O=N^{(\bullet)}–O^{(-)}$; both nitrogen oxides are common air pollutants.
  - But N_2 is not a radical, hence exceedingly less reactive than O_2.

- By contrast to polar intermediates, anions, and cations, carbon and heteroatom radicals can be formed under mild conditions without the assistance of acids or bases that could interfere with other sensitive substituents in the molecule, thus offering some advantages to synthetic plans and mechanistic possibilities. Disadvantages also exist.
- Carbon radicals can be formed by:
  - Subtraction of a hydrogen atom (H•) by another radical.
  - Homolytic cleavage: C–X → $C^{(\bullet)}$ + $X^{(\bullet)}$. There are several such leaving groups including halides, Se, Sn, sulfur derivatives, peroxides, and carbon itself in strained scaffolds.
  - Transfer of one electron from an electron releasing agent, formally a reduction, furnishing an anion radical $C^{(-\bullet)}$.
  - Transfer of one valence electron from the carbon scaffold to an electron-accepting agent, formally an oxidation. A cation radical is formed $C^{(+\bullet)}$.
- Neutral $C^{(\bullet)}$s are electron deficient and generally follow the stabilization pattern of carbenium ions: $R_3C^{(\bullet)} > R_2CH^{(\bullet)} > RCH_2^{(\bullet)} > H_3C^{(\bullet)}$; however, radicals perform independently of solvent polarization since charge is zero.
- While the methyl cation $CH_3^{(+)}$ is not a comfortable proposition for an intermediate, the methyl radical is a perfectly feasible entity. During mechanism design in need of a free methyl, when in doubt, prefer the radical.
- According to atomic-orbital theory, the radical electron of alkyl $C^{(\bullet)}$ occupies the space of a sp^3 AO as it adopts a pyramidal configuration, actually a shallow pyramid with an extremely low-energy barrier for inversion of only 0.15 kcal mol^{-1} approaching the trigonal planar conformation (entry **1** of Scheme II.37-1). Consequently, the unpaired electron occupies a π orbital in MO diagrams, assuming the planar conformation with expression of the radical on both sides of the molecular plane.

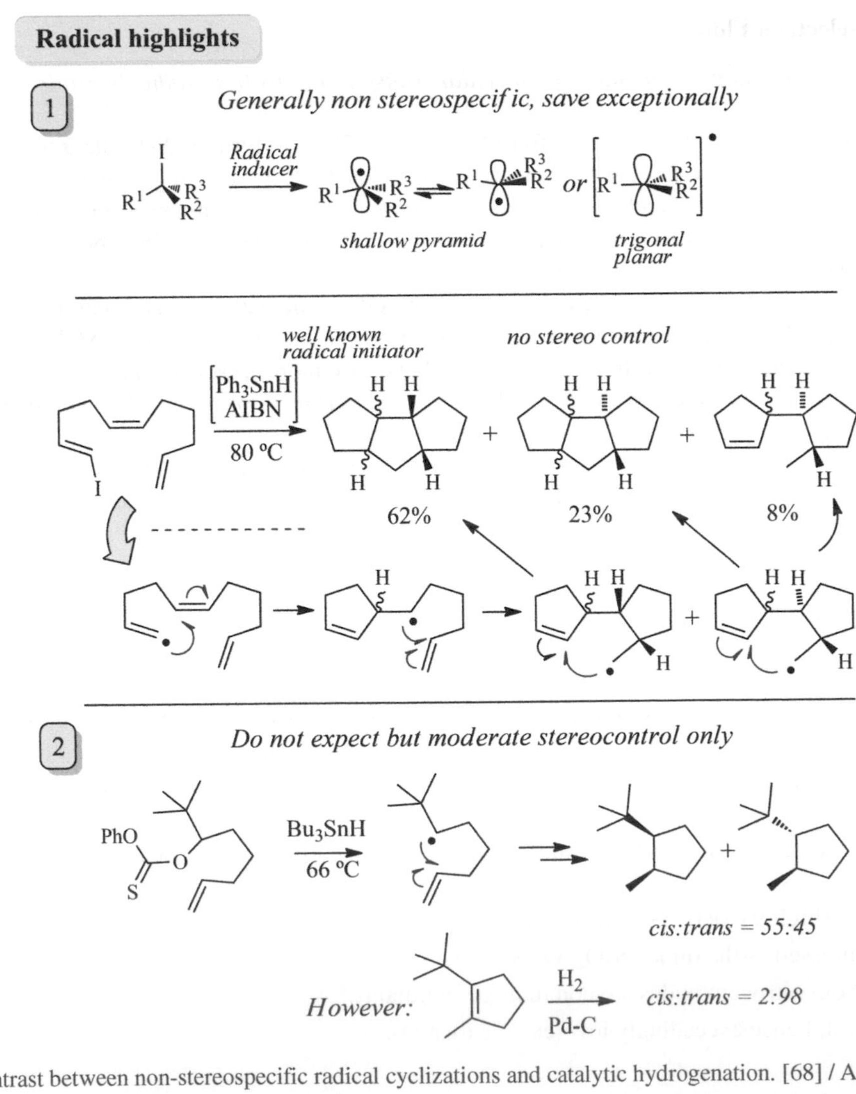

SCHEME II.37-1 Contrast between non-stereospecific radical cyclizations and catalytic hydrogenation. [68] / American Chemical Society.

- Do not expect but moderate stereospecific control in steps involving $C^{(\bullet)}$ as in the "round trip" intramolecular cyclizations shown in entry **1**, except in conformation-driven transition states or severe steric restrictions (entry **2**) [68]. Note: The mixture of Bu_3SnH – 0.1 mmol – and AIBN – 0.05 mmol – at moderate temperature is a radical initiator – see below.
- It is also worth noting the near absence of cis : trans preference in the radical cyclization (55 : 45), in contrast with the hydrogenation product of 1-*t*-butyl-2-methylcyclopentene (2 : 98) responding to thermodynamic control (entry **2**) [68].
- Trendy advances in radical chemistry are nevertheless producing an increasing number of radical-based stereospecific syntheses (Scheme II.37-2, entry **3**) [69–71].
- Once one accepts the π MO occupancy of $C^{(\bullet)}$ resonance rules for delocalization and stabilization come into effect (entry **4**). Radical delocalization is a flexible mechanism tool to explain reactivity of apparently remote atoms, be it by conjugation or H transfer (entry **5**) [72].

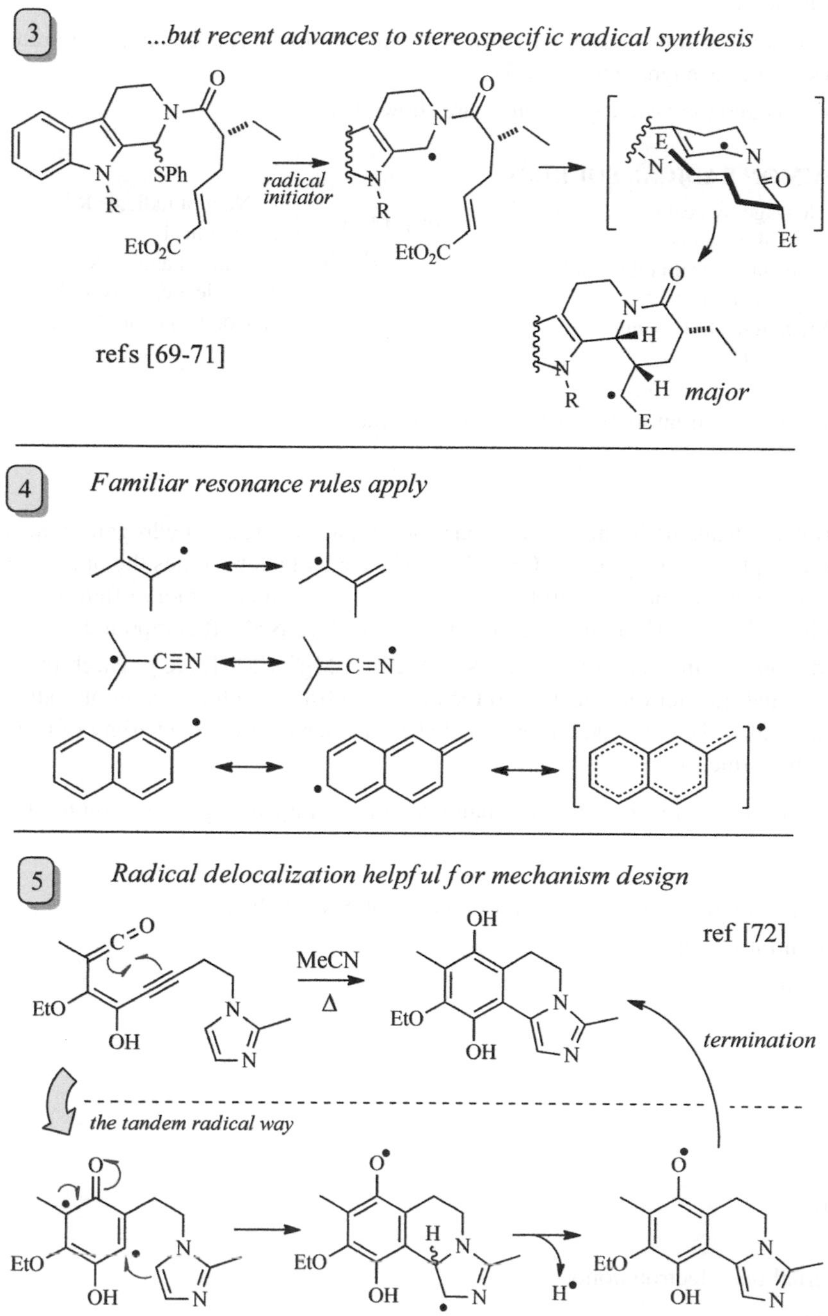

SCHEME II.37-2 The reaction mechanisms shown were Adapted from [69–71]; and [72] / American Chemical Society.

- $C^{(\bullet)}$ can develop in alkyl, aryl, vinyl, and (less often) alkynyl carbons.
- $C^{(\bullet)}$ can be stabilized by neighboring or conjugated EDG as well as EWG, a unique case among reactive intermediates in organic reactions (Scheme II.37-3, entry **6**). Known as *capto-dative stabilization*, this property is due to the establishment of radicals in specific heteroatoms at the center of EWGs and EDGs.
- Whereas radicals generally tend to rapidly complete the octet of valence electrons, *several stable, long-lived, or persistent radicals are known*, owing to the extensive delocalization of the lone electron across the molecule. One of them, a 2,2-diphenyl-1-picryldiazyl radical (DPPH) (entry**7**), is available from commercial vendors as a crystalline material in a bottle. It is extensively used, among other compounds, to test the antioxidant capacity (radical trapping) of functional foods, consumable additives, and medicinal natural products.
- The metabolism of all living things produces a vast number of radicals every day. It also copes with the intake of radical-producing environmental pollutants. To avoid their damaging reactions, known generically as oxidative stress, healthy cells produce a stream of antioxidants.
- Contrary to rare unimolecular dications or dianions, diradicals are common, produced during intramolecular homolytic C–X bond cleavage, photolysis, or rearrangement (entry **5**).
- Standard sources of radicals and the primary outcomes are shown below.

COMMON RADICAL SOURCES		
1. Cleavage by heat or radiation 2. Radical inducers 3. Propagation from other radicals 4. Triplet carbenes 5. Metal reduction agents 6. Oxidants	PRIMARY OUTCOMES ⇨	1. Neutral radicals $R^{(\bullet)}$ 2. Diradicals 3. Radical anions $R^{(-\bullet)}$ 4. Radical cations $R^{(+\bullet)}$ 5. Coupled neutral products

- Commonly used conditions and initiators of carbon radicals include:
  - High temperature.
  - UV light.
  - Compounds that furnish radicals under mild conditions and propagate through hydrogen-atom sequestration from a substrate or attack on a susceptible moiety (C=C, C=O, C=N, O_2, etc), which becomes the operating radical. There is a large collection of radical initiators to choose from [73, 74]. Among the promoters more often found in organic reactions are trialkyl or triaryl tin hydride (R_3SnH) alone or in combination with R–N=N–R compounds.
  - Dissolving metal reductions using zero valent Li^{o}, Na^{o}, K^{o}, Zn^{o}, Mg^{o}, Ca^{o}, Ti^{o} (e.g. Birch reduction) and low oxidation state Cr^{+2}, Sm^{+2}, Mn^{+3} that transfer one electron to the organic substrate to form an anion radical that can be neutralized with a suitable H source like *t*-BuOH. Radical intermediates can then be used in a variety of C–C bond-forming reactions in addition to C=C bond reduction.

Observing these conditions and/or reagents in a mechanism problem may be a good reason to propose radical-driven reaction course.

- Once radicals are formed (initiation), they evolve through a set of basic pathways:
  - Propagation (inter or intramolecular)
    - ✓ Hydrogen abstraction
    - ✓ Substitution
    - ✓ Addition
    - ✓ Rearrangement
    - ✓ Fragmentation
  - Termination
    - ✓ Diradical coupling
    - ✓ Capture of H atom
    - ✓ Reduction by external one-electron donor
    - ✓ Disproportionation

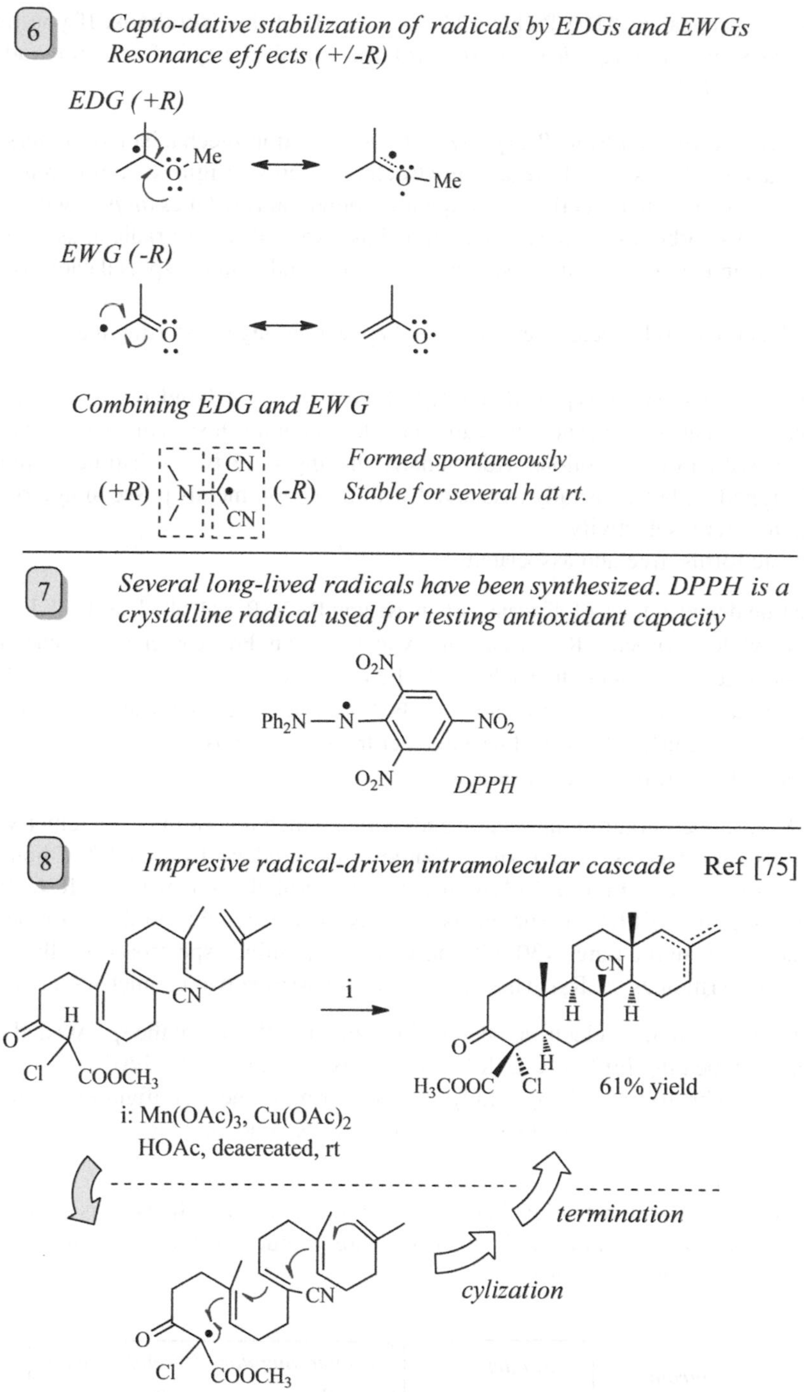

SCHEME II.37-3 Stabilization forces impinging on radicals to provide extended life and a case of radical-driven cyclization cascade with exquisite stereochemical control. [61] / American Chemical Society.

Extraordinary intramolecular *radical cascades* have been carefully designed to create regio and stereospecific molecular complexity in one-pot reactions (see for example Scheme II.37-3, entry **8**) [75], which are comparable to the more familiar cation-driven cascades (see Scheme II.26, section **A**).

Mind that the starting radical in entry **8** is *capto-dative stabilized* and should remain as such long enough for the aliphatic structure to acquire the proper conformation to cyclize rings **A** and **B** before disproportionation or H capture. The ensuing radical (not shown, it is for you to figure out) is also stabilized by the proximity of nitrile and facilitate cyclization of rings **C** and **D** (in the triterpene nomenclature). The dotted bonds at the upper end (ring **D**) of the final product depicts the observed mixture of two endocyclic and an exocyclic C=C bonds.

Examples of these reactions appear in "The Problem Chest", Part II of this workbook. If you have access to the **second edition** of *The Art of Problem Solving in Organic Chemistry*, you can gain dexterity in fascinating radical chemistry by solving Problems no. 7, 20, 23, 24, 26, and 38.

2.4.12.2 Carbenes, Overview, and Electron Redeployment As reaction-mechanism designers, we cannot do without carbenes; herein is a summary of the essentials regarding structure, energy, origin, common reactions, and some common mistakes in proposing them. Carbenes are essential and frequently found intermediates *au pair* with carbocations, radicals, and carbanions. One of the two forms carbenes adopt can be regarded as a special case of radical, as we shall review momentarily. Besides, electron movement in and out of carbenes is somewhat peculiar and requires special care to avoid falling into chemical incongruence.

Generally, carbenes, as the ephemeral species they are, do not appear among the starting materials but only their precursors.

2.4.12.2.1 Carbene Structure In the prototype carbene [R_2C:] there are two σ-bonded substituents and *six valence electrons on carbon only, with charge zero*. This is so because the carbon nucleus contains just as many protons. Most carbenes are highly reactive, short-lived species with rare exceptions. Additionally, having only two substituents reduces substantially steric requirements. Specially designed carbenes usually associated with transition metals possessing large and/or chiral ligands do show steric effects suitable for stereoselectivity.

Carbenes exist in two basic forms, free and associated.

a) *Free carbenes* are the true denuded form of divalent carbon, written R_2C**:** (R = H, alkyl, aryl, allyl, halides, heteroatoms, esters, ketones, aldehydes). Vinylidene carbenes $R_2C{=}C$: are also widely known. Free carbenes are generated *in situ* in solution from unambiguous precursors (see below) submitted to heat, UV light, or base.
b) *Associated carbenes, or metal carbenoids*, also known as Fisher carbenes, are tetravalent carbons with a $R_2C{=}MLn$ general structure, wherein M=Rh(II), Cu(II), Ru(II) and several other transition metals, and Ln is *n*-ligands on M. The carbon portion shows the gamut of reactions typical of free carbenes.

Free carbenes adopt two electronic configurations, *singlet* and *triplet*, which determine the geometry of the carbene:

Singlet carbene [$^sCR_2$] (also abridged as ^{1}:CR_2): The unshared spin-paired electrons and R–C bond electrons occupy carbon sp^2 AO in a trigonal planar configuration (Figure II.14) with a vacant p_z orbital. As a result, the R–C–R bond angle is about 102° according to electron-spin resonance (EPR) measurements but may vary depending to the steric bulkiness and mutual repellence of the R groups. Bond angles in the range 130–140° have been determined spectroscopically.

Triplet carbene [$^tCR_2$] (also written as ^{3}:CR_2): These carbenes may assume two geometries, angular and linear:

1. *Angular*: The unshared spin-unpaired electrons occupy different orbitals, one at the sp^2 AO and the other at the p_z AO. The configuration continues to be *angular* but R–C–R bond angle is now about 125–140°.
2. *Linear*: The carbene AO hybridization changes to *sp* thus acquiring a linear configuration. Unshared electrons populate the two degenerate p_y *and* p_z AOs with equal spins, and the R–C–R bond angle is 180°.

Most triplet carbenes are angular.

A companion of the more stable triplet state of carbenes is the O_2 molecule, which displays a ground-state, open-shell triplet configuration. Not a true carbene, O_2 is formally a diradical structure as much as the triplet carbene is. This feature has consequences for stabilization, resonance, and reactivity.

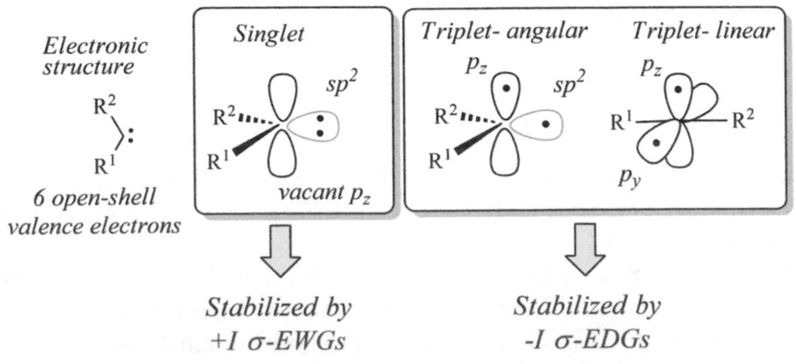

FIGURE II.14 Atomic-orbital configuration of carbenes.

Transition metal carbenoid complexes $R_2C{=}MLn$ perform as singlet, electrophilic carbenes with certain metals such as molybdenum, and triplet, nucleophilic carbenes with others such as niobium complexes according to MO calculations [75]. The $R_2C{=}M$ bond is constituted by the carbene donating to the metal its sp^2 AO electrons while the metal atom donates back its *d* AO completing four electrons in this bond.

2.4.12.2.2 Singlet and Triplet Carbene Stabilization and Two-faceted Nucleophilic-electrophilic Character [76] There is an average energy difference of 8 kcal mol^{-1} between the singlet $^{s}CR_2$ (higher) and the triplet $^{t}CR_2$ (lower) carbene with equal substituents R. Therefore, a carbene singlet *is an excited state of the triplet,* as opposed to other molecular species. Energy differences between both electronic ground states vary strongly with the electronegativity of substituents on the carbene and the order of stability.

The energy level order S > T may even be inverted in some cases. For example, the ^{s}**:**CR_2 to ^{t}**:**CR_2 energy difference $\Delta E_{s\text{-}t}$ of the methylene carbene H_2C**:** is 11 kcal mol^{-1}, thus the triplet is several orders of magnitude more abundant. By contrast, for difluoromethylene carbene F_2C**:** in which the strongest possible inductive σ-EW occurs due to F being the most electronegative of all elements, $\Delta E_{s\text{-}t} = -45$ kcal mol^{-1}, with practically total predominance of the ^{s}**:**CR_2 ground state. Therefore, the character of substituents directly bonded to the carbene atom must be considered as regards to stability/reactivity.

2.4.12.2.3 Carbenes: Electronic and Steric Effects As any other reactive intermediates of carbon, it is worth discussing briefly these two effects given their strong influence:

2.4.12.2.3.1 Electronic Effects As usual, there are three effects associated with stabilization of transient species including carbenes: through covalent bond *+I/–I* σ inductive effects, through-space field effects, and conjugation, all of them applicable to divalent carbon. Free carbenes take the ^{s}**:**CR_2 or ^{t}**:**CR_2 configuration contingent on the electronic type of substituents R (Scheme II.38, entry **1**). This branches off to four circumstances:

a) When R = EWG, inductively attractive (*–I*) (σ-EWG effect), the carbene will preferably be in the ^{s}**:**CR_2 state since these groups inductively increase the energy difference between the sp^2 and *p* AOs. The latter remains vacant.
b) When R = EDG by *+I* induction (σ-EDG effect), electron density in the sp^2 AO is reinforced and electrons escape to the lower energy level-space of separate AOs, adopting angular and linear configurations of the ^{t}**:**CR_2 state shown in entry **1**.
c) On the other hand, when the EDG has NBPs as in $-NR_2$, –OR, –SR, $-PR_2$, and halogens, resonance (*+R*) (π-EDG) is enabled with the vacant p_z AO of the carbene providing stabilization. Only the ^{s}**:**CR_2 can take an incoming NBP of its neighbor through sp^3–p_z overlap and the singlet state is preferred (Scheme II.38, entry **2**). Canonical structure **B** becomes an *ylide* whose *former carbene carbon turns into a powerful nucleophile.*
d) By contrast, when the *–R* resonance is allowed on account of π-EWG as in C=O, CN, C=NR, SOR, SO_2R, NO, or NO_2, there is a resonance conduit for carbene electrons toward the conjugated system (canonical structure **B'**) leaving a vinylogous cation. This is a strongly electrophilic unit, enhanced further by the σ-EWG property of the substituent. Again, only the ^{s}**:**CR_2 state can delocalize its unshared electrons through sp^2-π overlap by rotation of the: C–C=X σ bond and become stabilized. In rigid scaffolds where this rotation is not allowed resonance stabilization is not enabled.

By combining *+R* and *–R* effects we have the *four modes of conjugative electronic distribution of carbenes* depicted in Scheme II.38, entry **2**: π-push, π-push-pull (or capto-dative), π-push only when the second carbene substituent is π-nil (alkyl or H), and π-double push.

Notice that:

i. The first of these combinations (pull only) creates a vinyl carbocation at the carbene carbon and a dipolar unit like $R{-}C^{(:)}{-}C{=}O \leftrightarrow R{-}C^{(+)}{=}C{-}O^{(-)}$ of central importance in [3 + 2] dipolar cycloadditions of carbonyl carbenes. This reaction finds numerous applications in synthesis and crops up often in mechanism problems.
ii. The push-only mode, in turn, creates a nucleophilic ylide $X^{(+)}{=}C^{(-)}{-}R$, easily formed in heterocyclic carbenes, several cases of which appear in "The Problem Chest", Part II.
iii. The double-push mode is found in all the heterocyclic stable carbene examples selected in entry **3** of Scheme II.38. These carbenes, among which NHCs are prominent moieties, *are strongly nucleophilic*, suitable for additions to C=O. Several applications for trapping CO_2 with NHC intermediacy are being designed for fixation of this gas [77] in the hopes of scaling the procedure up to industrial use and help curbing the planet human-made atmospheric CO_2 concentrations.

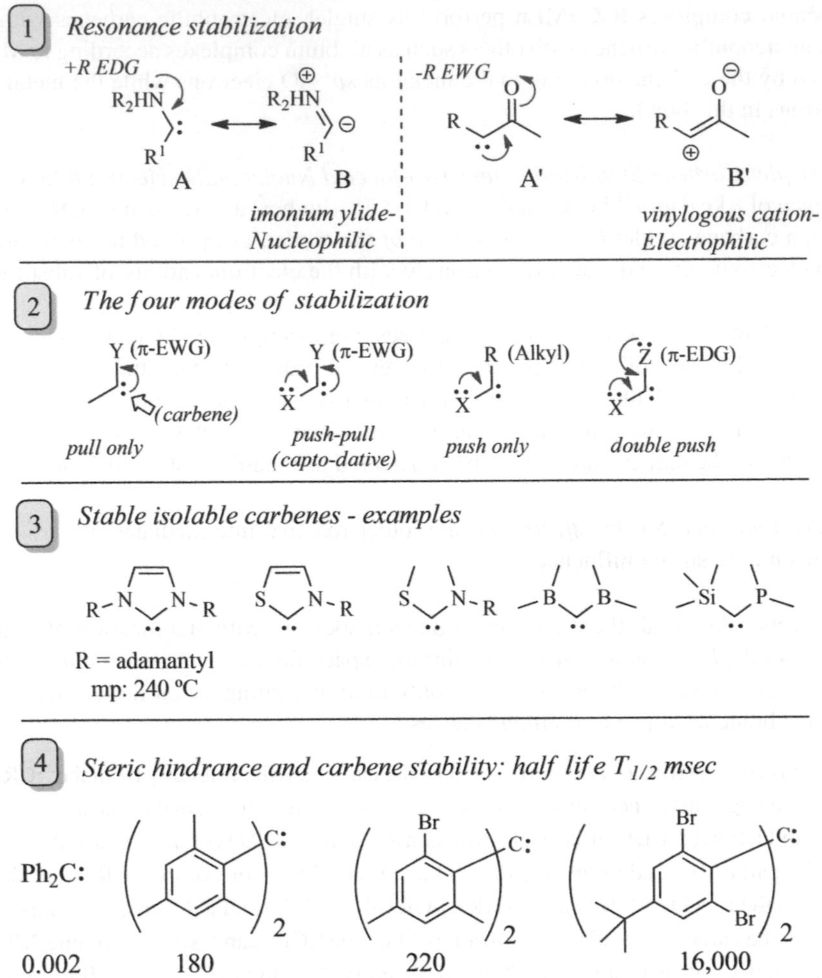

SCHEME II.38 Electronic and steric influence on carbene stabilization. See text for comments for each entry.

Based on these concepts, carbene manipulation through suitable substituents have led to the isolation of stable, or persistent, nude carbenes at room temperature, some are distillable or crystalline with high melting points, cyclic as well as acyclic (entry **3**) and can be stored in the cold for days and months as a standard reagent [78, 79]. NHCs are prominent members of this class, which were described earlier in this chapter as part of modifications of the Grignard alkylation pursuant of carbonyl umpolung synthons (see Schemes II.31 and II.32 as well as a related discussion for a timely reminder).

2.4.12.2.3.2 Steric Effects in Carbenes Steric effects influence reactivity of carbenes due to either kinetic (configuration of the TS) or thermodynamic (steric encumbrance of adducts) reasons, or both. Despite being limited to two substituents only, some stereospecificity is observed in carbene bimolecular additions. This subject is summarized in the carbene reaction section, right after the present discussion.

Large substituents such as arenes that hold sterically demanding groups impinge strongly on the half-life of carbenes as they hamper further reaction progress. Steric congestion/half-life correlations have been established (entry **4**) [80–82].

2.4.12.2.4 Carbene Sources Because carbenes, as such, do not appear among the reactants it is very convenient for the mechanism problem solver to be familiar with precursors of carbenes.

Carbenes will arise when electron configuration leaves six valence electrons on a disubstituted C, either from direct precursors or electron redeployment.

A variety of compound types produce carbenes. Key carbene sources are shown in Schemes II.39-A and II.39-B,

The Simmons-Smith reaction (Scheme II.39-A, entry **1**), discovered around the mid-twentieth century is a preferred method for the synthesis cyclopropanes from the [1+2] addition of the carbenoid to a gamut of alkenes. Later modifications have improved substantially yields and selectivity. For example, Zn° has been replaced by Zn/Cu couple prepared from Zn° and cuprous acetate, or the more reactive diethyl zinc. The introduction of asymmetric catalysts gives access to excellent enantioselectivity of cyclopropane adducts. Detailed theoretical studies of the interesting mechanism involved are available [83].

The α-elimination of methane trihalides (Scheme II.39-A, entry **2**) provides *in situ* access to an electrophilic methylene carbene under alkali, which subsequently undergoes reaction with electron-rich sources. The process responds to a universal mechanism but is constrained to very few compounds of the haloform group: iodoform, bromoform, and chloroform. After C–H insertion or cyclopropanation, the corresponding dihalides are produced.

Diazomethylene compounds (Scheme II.39-A, entry **3**), from diazomethane to complex structures, continue to be extensively used for carbene generation induced by UV irradiation, or producing metal carbenoids with a growing number of transition metals and rare earths. The sight of an $R_2C{=}N{=}N$ moiety or a tosyl hydrazone in anhydrous base (entry **4**) *are strong suggestions of a carbene taking part in the reaction mechanism.* Yet, one cannot jump into conclusions before considering the reaction medium.

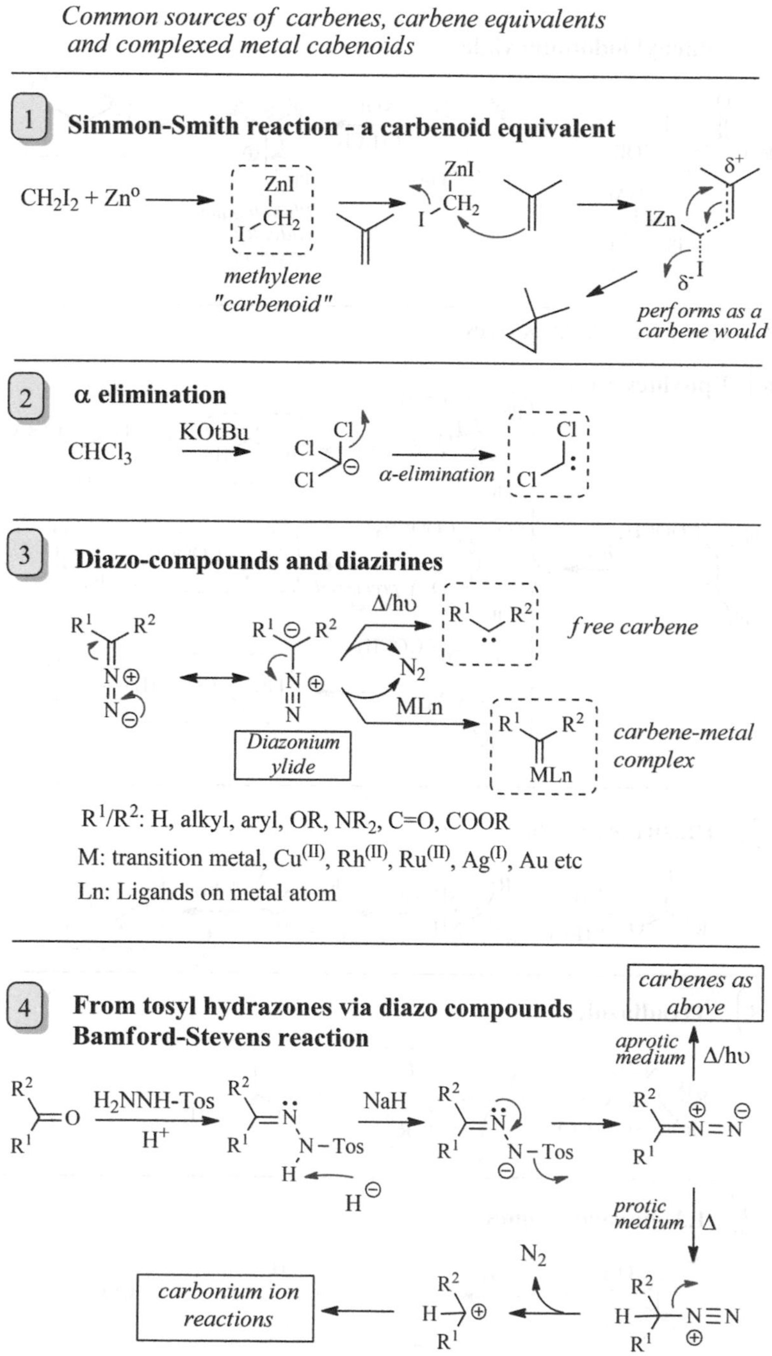

SCHEME II.39-A Preferential formation of free and metal-complex carbenes from classical sources.

If it is true that diazo compounds in aprotic neutral medium are excellent sources of carbenes/metal carbenoids as said above, they will trap a H^+ from traces of water or acid in the medium and follow the carbonium ion route (entry **4**) to S_N1 and elimination reactions.

Diazocompounds of entry **4** are one example of a few zwitterionic species that can undergo elimination of a leaving substituent on the carbene-to-be C atom. This circumstance is fulfilled in the adduct of a R–CH$^{(-)}$–R' such as diethyl malonate and phenyl iododiacetate (PIDA) (Scheme II.39-B, entry **5**). Exposure to anhydrous base (KOH, potassium hydroxide) in aprotic solvent (dichloromethane) picks up the acidic proton and hydrolyzes one acetate unit concomitantly, which forms the phenyl iodonium ylide that is isoelectronic with the diazonium ylide of entry **4** of Scheme II.39-A.

Two routes are open to this intermediate:

1. α-Elimination of iodobenzene furnishes the free dimethoxycarbonyl carbene ready for addition, insertion, and other reactions typical of carbenes to be described in the next section.

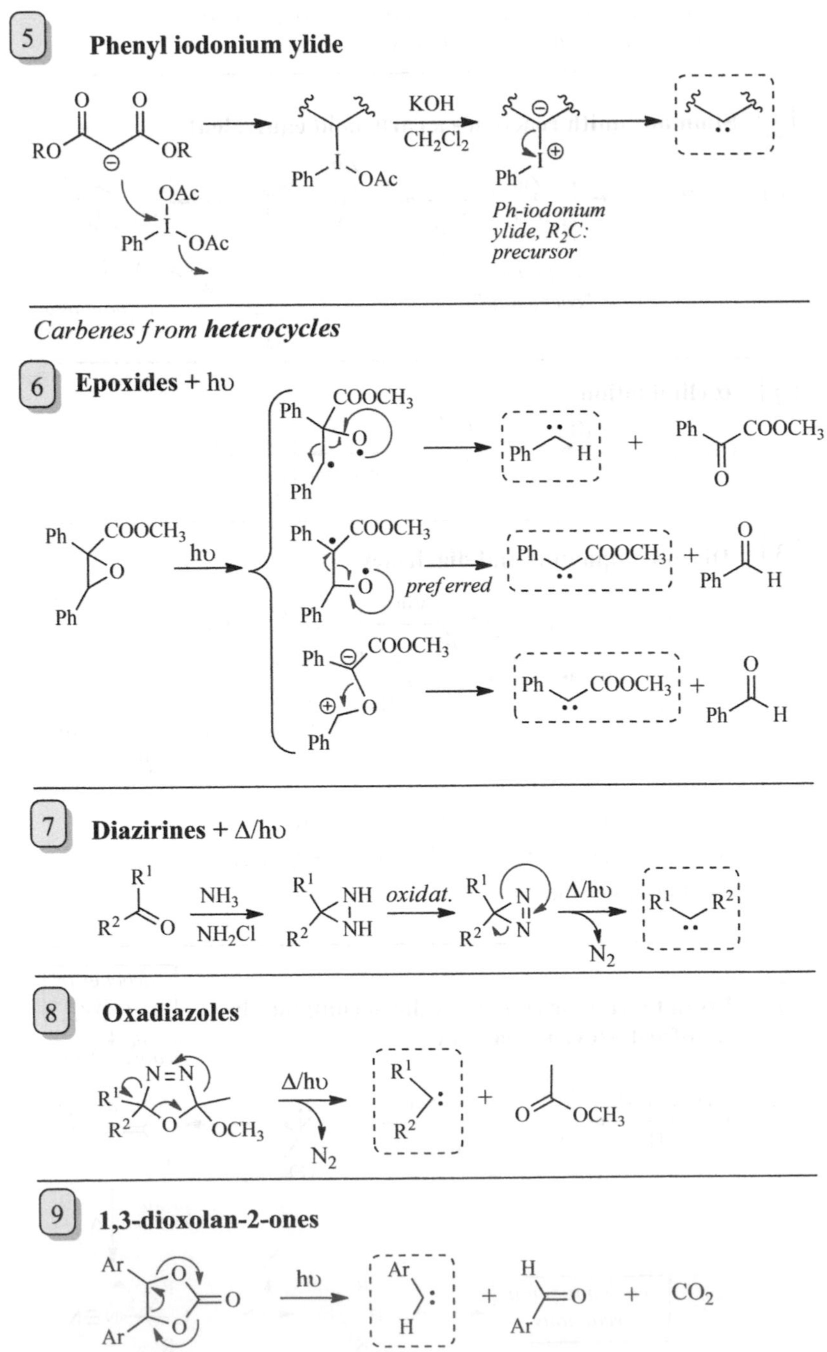

SCHEME II.39-B In situ production of free carbenes by more modern methods.

2. Conversely, addition of a transition metal catalyst, including simple Cu(I) and Rh(II) derivatives, give rise to the corresponding $R_2C{=}MLn$ carbenoids also amenable to cyclopropanation on electron-rich alkenes in 40–80% yield [84].

Other compounds, particularly O, N, and S heterocycles, produce carbenes suitable for practical applications in synthesis (entries **6–9**) via fragmentation of the parent molecule by heat or UV radiation and loss of N_2 or carbon dioxide.

In asymmetric epoxides, (entry **6**) three primary outcomes are conceivable, differing in thermodynamic stability. Authors of this discovery [85] entertained the idea of a heterolytic C–C cleavage of the ring rather than homolytic diradical from the C–O dislocation (which I prefer), to justify the more stable product. Note that a carbene can also emerge from the zwitterionic intermediate.

2.4.12.2.5 Typical Carbene Reactions As highly reactive intermediates and three diverging evolution possibilities (singlet and triplet carbenes, metal carbenoids, and diradicals depending on reaction conditions and substituents), the scope of reactions open to them is predictably generous and involve saturated and unsaturated substrates. Mixtures of by-products are therefore not rare.

Figure II.15 spans the arch of carbene reactions. The most frequent are:

✓ [1+2] cycloadditions to alkenes and alkynes furnishing cyclopropanes and highly strained cyclopropenes.
✓ C–H insertions [86].
✓ [2+3] dipolar cycloadditions [87].
✓ The Wolff rearrangement.

Yet, some control is possible to reach the other reactions of Figure II.15.

The lack of stereoselectivity of cyclopropanations used to be an annoying problem in free carbene and $R_2C{=}MLn$ species but have been solved ever since (see Chapter 3). These shortcomings are shown in Scheme II.40.

Issues to highlight in Scheme II.40 *and takeaways*:

- The reaction of interest is the cyclopropanation of ethyl diazoacetate (**216**) on a series of alkenes related to the indole structure, having S, N, and O affecting electronically the involved $C^2{=}C^3$ bond. The process is prompted by heat, Cu° (copper-bronze dust, a form of heterogeneous catalyst used commonly in many early diazocompound studies, later replaced by a breadth of soluble complexes of several transition metals already mentioned) associated to increasingly convoluted ligands.
- Diazoacetate has been a standard compound for many years in the testing field of α-carbonyl carbenes generated by heat, photolysis, or catalysis, acting upon all conceivable substrates. Few, however, explore the response of thiophenes to diazocarbonyls. Two of these are discussed in Scheme II.41 and associated text.
- Here, we observe not only poor and variable exo/endo cyclopropanation stereospecificity (why?) but also other undesirable products cropping up (e.g. **219** and **220**) in the case of benzothiophene **212**. What mechanism supports this result?

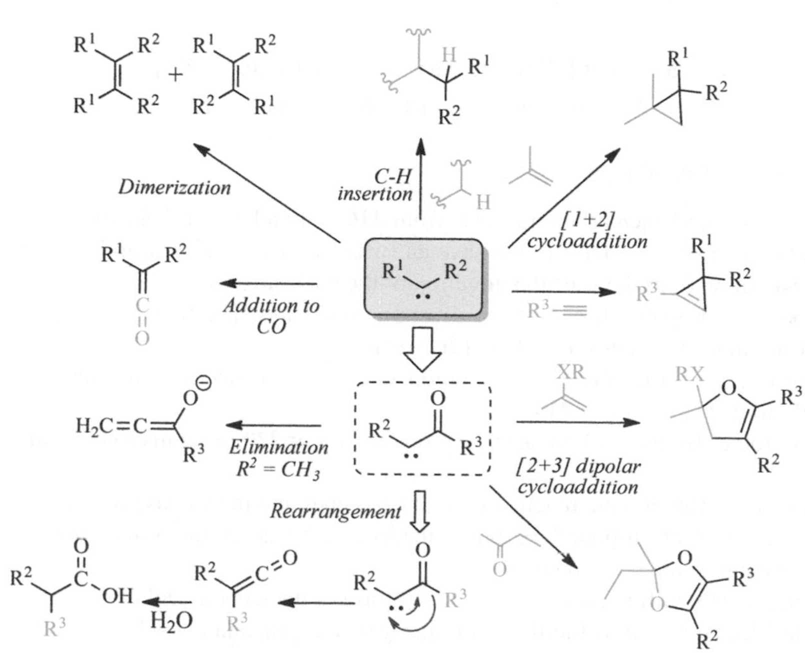

FIGURE II.15 The diversity of carbene reactions.

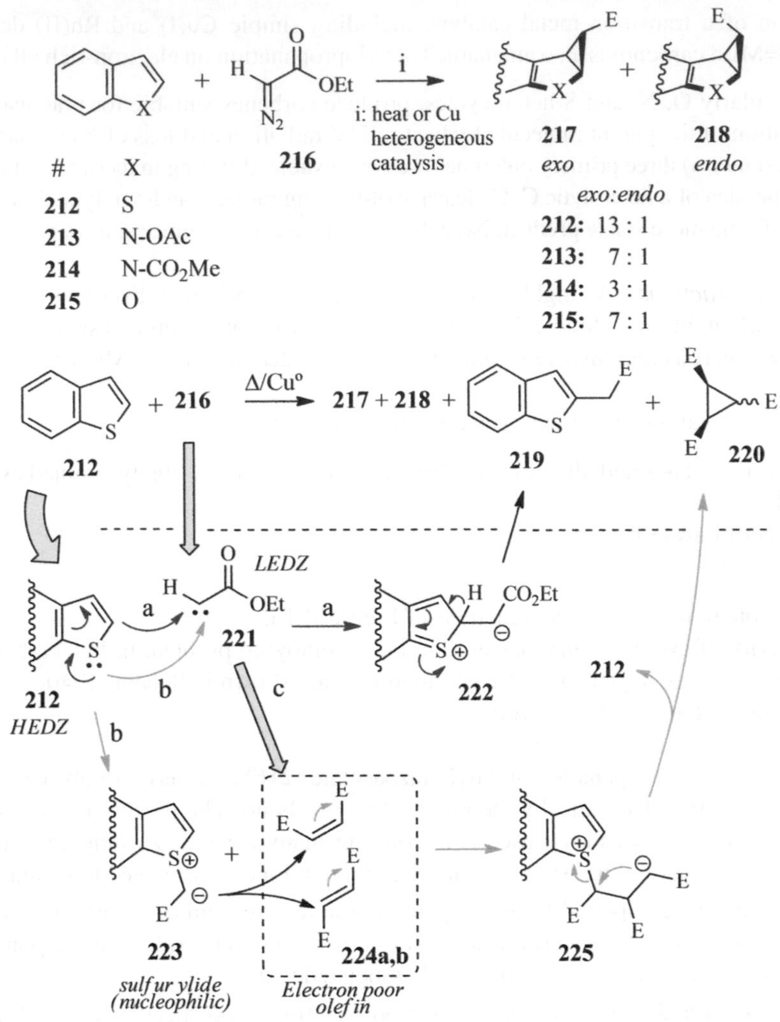

SCHEME II.40 Contrast between the cycloaddition of copper-catalyzed ethyl diazocetate and benzothiophene relative to benzofuran and indole-N-derivatives. [88] / American Chemical Society.

- Indoles **213–214** or benzofuran (**215**) do not follow the devious reaction course. Why?
- Would you take up these questions? Likely mechanism outlets are shown after the bent yellow arrow.

The prominent questions concern benzothiophene **212**:

- ✓ The first step is to create a carbene/carbenoid (**221**) from **216**, with a high probability of being an electrophilic singlet owing to the *p*-EWG effect of the ester. So, we have an LEDZ/HEDZ combination between the **221** and heterocycles **212–215**, and thus we know from the start the direction of the curly arrows.
- ✓ Thiophene **212** appears to progress through three diverging pathways. The first is cyclopropanation (standard mechanism not shown). Then there are routes **a** (red) and **b** (green).
- ✓ Route (**a**): A standard nucleophilic attack of benzothiophene C^2 on the electrophilic carbene furnishing zwitterion **222,** which evolves easily to question adduct **219**.
- ✓ Before moving on to route (**b**), have a look at cyclopropane triester **220**. It seems to be made of three moles of carbene **221**. How?
- ✓ Well, dimerization is one characteristic reaction of some carbenes and in this case it means forming maleate and fumarate esters **224a,b** with an electron-poor C=C bond, inadequate for accepting a third mole of carbene **221**. We need a *nucleophilic* carbene equivalent, enter route (**b**).
- ✓ Route (**b**): Benzothiophene's sulfur proceeds to a second form of nucleophilic addition onto carbene **221** (green arrow), yielding sulfur ylide **223**. You are now familiar with nitrogen, oxygen, and phosphorous ylides as you read this chapter,

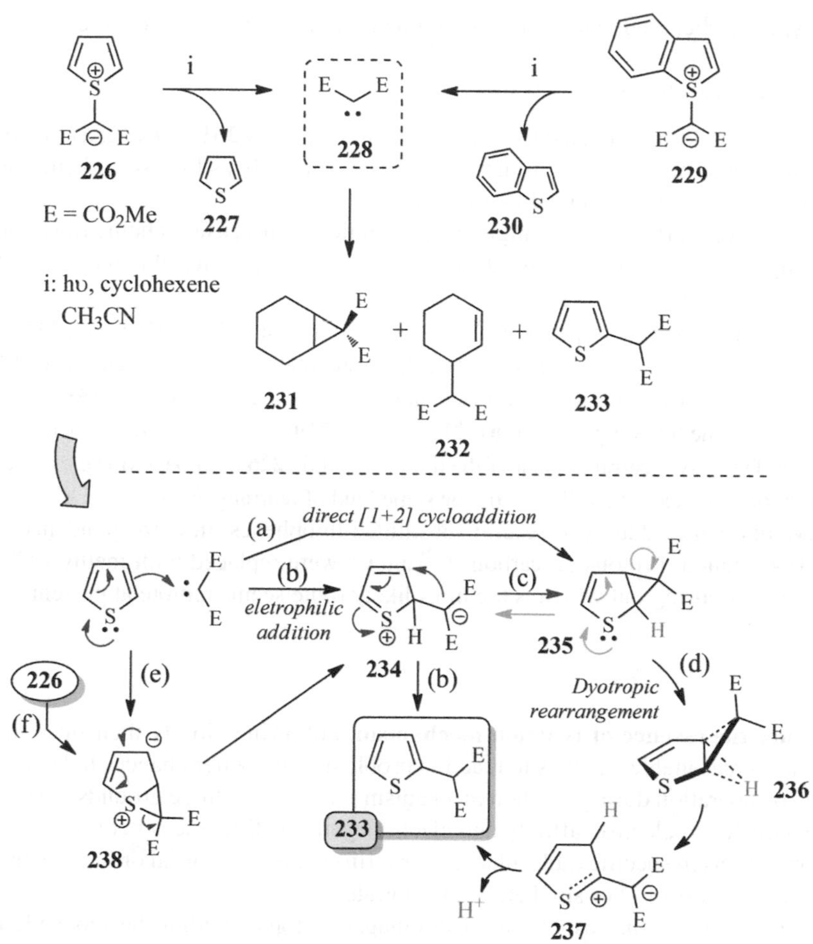

SCHEME II.41 Preparation of free electrophilic dicarboethoxycarbene **228** from the corresponding sulfur ylides **226** and **229**, and predominance of the C-H insertion (**233**) on thiophene. [89] / American Chemical Society.

and know that they are excellent cyclopropanation agents by way of a stepwise mechanism depicted in the **223** → **225** → **212** sequence.

✓ Of the other heterocycles, the indoles are *N*-substituted, sterically impeded and less nucleophilic than the sulfur homolog. As for benzofuran (**215**), the NBPs of O participate in the aromatic π cloud and cyclopropanation rather than formation of the *O*-ylide predominates.

✓ In regard to the exo : endo ratios, authors [87] did not advance any explanation probably because many more substrate models would have been necessary, and complex stereoelectronic effects participate in the exo/endo transition states contingent upon the heteroatom and its substituents. This is one instance in which the problem solver needs more information to formulate a hypothetical mechanism, should recognize it, and either ask for more or stop scrambling with the problem.

Stereo and enantioselective cyclopropanations have been developed in recent years by manipulation of the diazocarbonyl substituents and ligands of $Rh^{(II)}Ln$ forming congested $R_2C{=}Rh_2(Ln)$ complexes. Exploration of a variety of diazocarbonyls and ligands has led to a much better understanding of the stereochemical control over the [1+2] cycloadducts and the transition states involved. This subject will be dealt with in Chapter 3.

As mentioned earlier, thiophene is an oddball among the diazocarbonyl additions to heterocyclic alkenes. Scheme II.41 portrays, various thiophene ylids of dimethylmalonate ester (DMM), which are precursors of the free carbene **228** prompted by UV irradiation or heat, as suggested by the reaction of ylids **226** or **229** in cyclohexene, a standard testing racetrack for purported carbenes. As expected, cyclopropanation (**231**) and C–H insertion (**232**) were the predominant products [89].

Notes and takeaways from Scheme II.41:

Worth noting: 1 mol of the sulfur heterocycle is freed in the medium (within the solvent cage), which in principle would be available close by as substrate for the carbene. But cyclopropanation was not observed even after adding more thiophene to the reaction mixture. Similarly, no carbene adducts to benzothiophene (**230**) were detected, in contrast with the experiment of Scheme II.40.

Authors [89] put forward two reasons for this outcome:

1. While a *singlet* carbene would be expected to perform yielding adducts **231** and **232**, as that obtained from the photolysis of **226**, other thiophenes pertaining to aromatic systems would perhaps furnish carbenes of different multiplicity (*a triplet in this case*) leading to a more radical-like course.
2. Multiple possibilities arise; six in fact (**a–f**), might be in operation in this case (Scheme II.41, past the yellow arrow gate) as it occurs frequently in mechanism analysis. Deciding which is the most feasible may be quite challenging.
   a) Putting together these pathways one comes to the following:
      i) Route (**a**) is the expected cyclopropanation also accessible through path (**b**). However, **235** was not observed due to the likely decomposition back to ylide **234** (route [**c**]) or, alternatively undergo DR (path [**d**]) (see Scheme II.16 for refreshing the concept) to zwitterion **237**, which yields final thiophene derivative **233**.
      ii) Route (**b**) would be the homolog of path (**a**) (**212**→**222**→**219** in Scheme II.40), a facile entrance into target **233**.
      iii) Routes (**e**) and (**f**) involve rearrangement of the thiophene ylide **226** to thioepoxide diester **238** or direct sulfur nucleophilic addition onto the carbene followed by the same kind of rearrangement.
3. Efforts to detect cyclopropane adducts in several substituted thiophenes, in correspondence to those from benzothiophene of Scheme II.41, failed. Protons on carbons C^2 and C^5 were replaced with methyls Cl and Br to no avail. With presently available data, figuring out which is the most likely route seems remote at present.

2.5 SUMMING UP

1. **Valence electrons are the essence of reaction mechanism**. Following closely their operational rules during mechanism design guarantees reasonable answers to reaction problems with better chances to be accepted by others.
2. **Atom nuclei** need consideration during reaction mechanism primarily on three grounds: (i) to define electron shell configuration; (ii) to characterize electron affinity and electronegativity differences between bonded atoms, and thus help outline electron populations in specific molecular sections; (iii) to provide partial or full positive charge when the neighboring electron cloud decreases ED or shell electrons migrate.
3. **Electron cloud-density variations** respond to electromagnetic forces within the molecule or by close influence of nearby molecules, hydrogen bonding, Lewis acids/bases, catalytic association, metal complexation under ligand influence, and solvents.
4. **Electrons exist in probability spaces known as atomic or molecular orbitals.** Therefore, electron redeployment within or between molecules requires a certain degree of orbital overlap except in demonstrable quantum tunneling and in special material surfaces.
5. **The design of TS in mechanism problem development must contemplate reaction sites at close distances (<2.8 Å) and feasible molecular topology that grants this proximity within reasonable strain energy levels**. The concepts of early and late TS acquire prominent importance. However, molecular vibrational modes of their degrees of freedom may enhance probability of interatomic interaction at greater distances than 2.8 Å.
6. **For effective orbital overlap, electronic repulsion, also accounted for as steric repulsion, must be overcome.** TSs along the way to observed products are generally configured according to their least obstructed structure for minimum $DH^{\ddagger}$. However, there is more than one conformational energy minimum along the reaction coordinate, which can only be assessed by quantum-mechanical calculation.
7. **Electrons interact more favorably in active sectors of molecules defined by their functional group setting and their orbital occupancy in the energy sequence, from high to low:**
   $\pi^* \rightarrow \pi > \pi \rightarrow \pi > \pi \rightarrow \sigma > \sigma \rightarrow \sigma$. Molecular orbitals more favorably engaged in electron redeployment for bonding contemplate the HOMO→LUMO transition.
8. **Electrons flow favorably within the same molecule (during rearrangements for example) or between interacting molecules in mostly predictable directions, from high electron density (*HED*) to low electron density (*LED*) zones and not the reverse**. If HEDZs are not apparent in the starting materials, such centers may be created in subsequent intermediates after proton abstraction, hydride transfer, interaction with nucleophiles (substitution, addition), and one-electron transfer from a reducing agent in the reaction mixture.
9. **As important as recognizing HEDZs is the identification of LEDZs among the reactants to define the direction of electron flow.** Resonance, inductive, and field effects contribute to the electronic interplay to define HEDZs and LEDZs.
10. **If absent, an LEDZ can be generated in subsequent intermediates** by subtraction of a leaving group or cleavage of strained heterocycles (oxiranes, oxetanes, N and S equivalents, or larger hetero-alicyclics). Electrophilic attack and electron transfer to an oxidant in the reaction mixture are also contributors of LEDZ formation.

11. **The preliminary HEDZ/LEDZ assessment may be entirely reverted by transition metal complexation and umpolung operations.**
12. **In addition to carbanions and carbocations, carbon compound intermediates may contain reactive species that are not fulfilling the Lewis octet rule,** namely radicals and carbenes, both devoid of electronic charge. While diradicals may occur in the same molecule, similar dicarbenic compounds are presently unknown and should be avoided in mechanism design. However, two heterocyclic carbenic ligands on the same transition metal have been characterized [90].

2.6 ORGANIZED PROBLEM ANALYSIS WITH THE TOOLS DESCRIBED SO FAR

Having the issues exposed in Chapters 1 and 2, one practical way to organize your strategy as you approach a new mechanistic problem is shown in Scheme II.42. It is built on a binary yes/no decision-making system that allows a rapid response to the challenges of all problems. While some steps are easily and quickly solved, such as identifying the hot spots in starting materials where electron flow and molecular transformations are more likely, others take more time and considerable thinking like accounting for stereoelectronic effects in potential transition states of advance stages of your mechanism design.

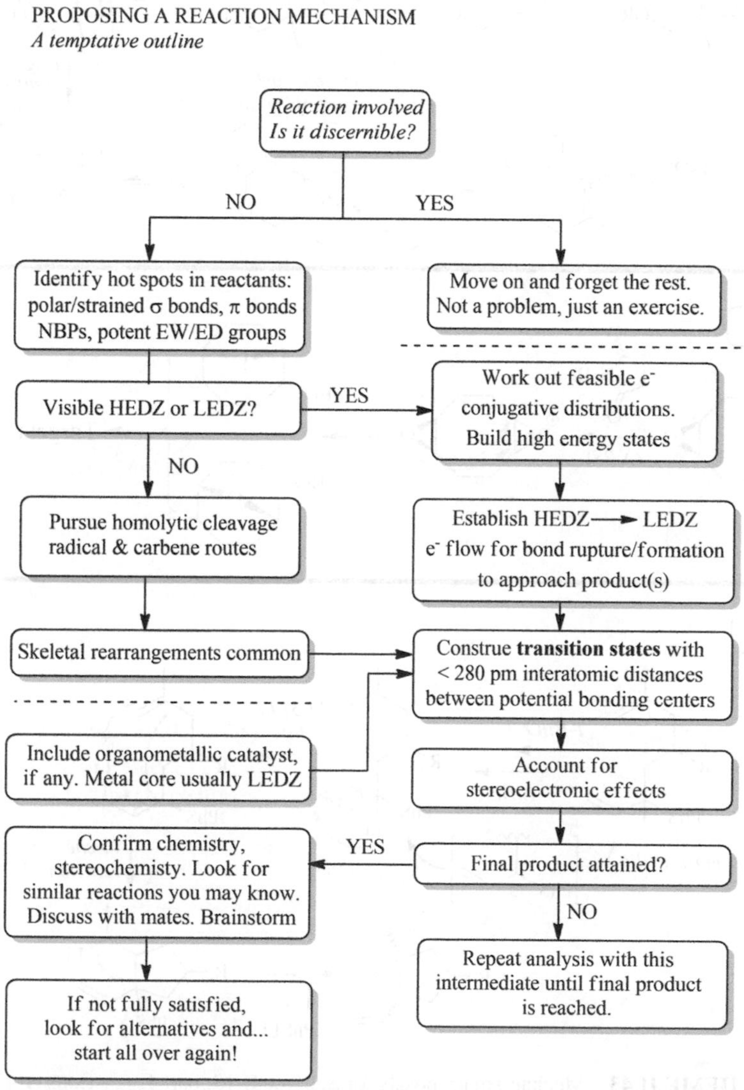

SCHEME II.42 A decision-making flowchart, useful for the analysis of reaction-mechanism problems whose solutions are not apparent after a first assessment. Follow the yes/no decision forks according to your best criteria. The chart uses the tools so far described in Chapters 1 and 2 (design: Alonso-Amelot, this book). Other analysis tools are explained in detail in the ensuing chapters, for inclusion in a deeper level of analysis.

Let me recommend that you earmark this page and have it on hand to guide your future problem-solving projects until you become familiar with the entailed logics.

2.7 SUPPLEMENTARY SCHEMES: SOLUTIONS TO PROBLEMS EMBEDDED IN THIS CHAPTER

Mechanisms of reactions of Scheme II.13 - Reactions A-D

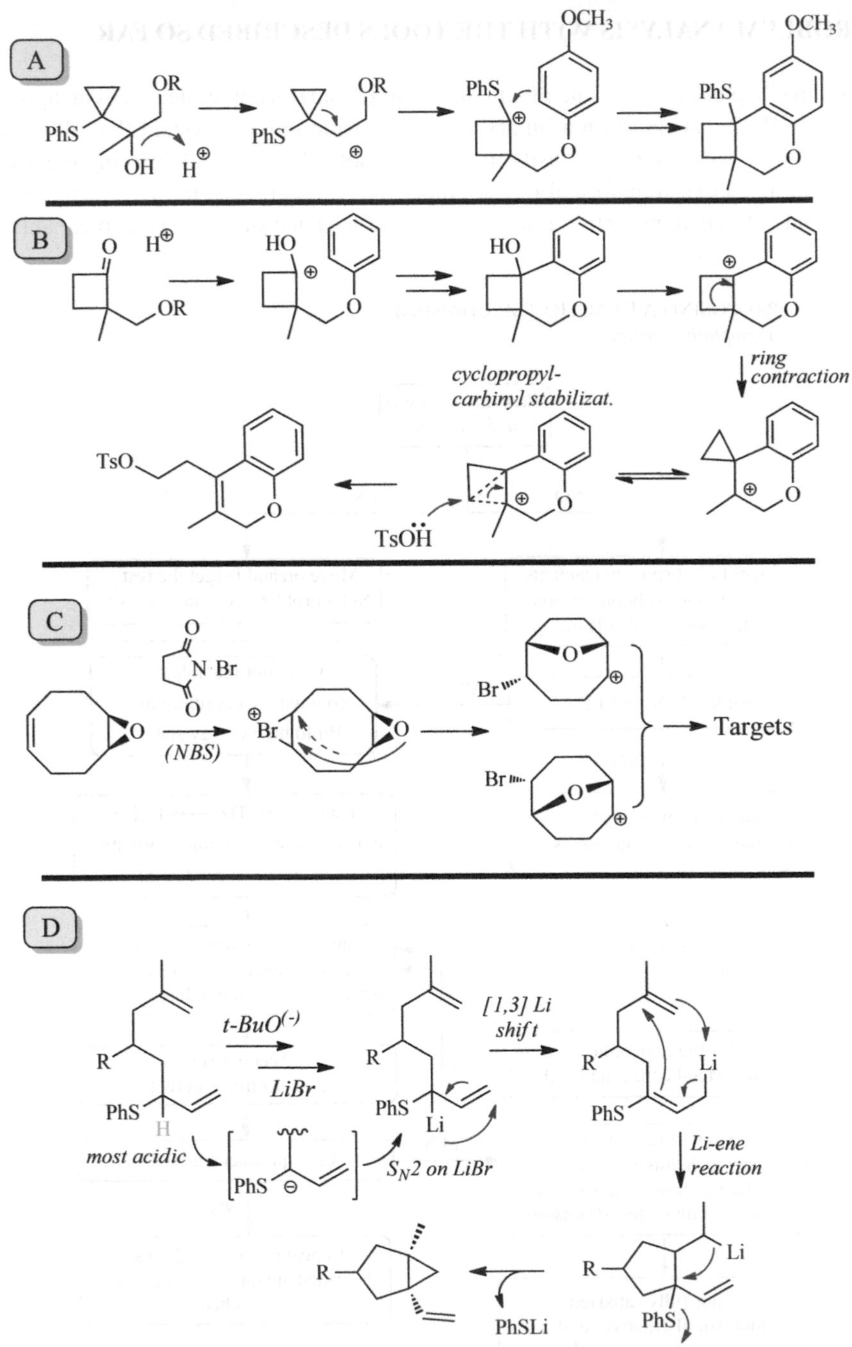

SCHEME II.43 Mechanism proposals for Scheme II.13, part 1, reactions **A–D**.

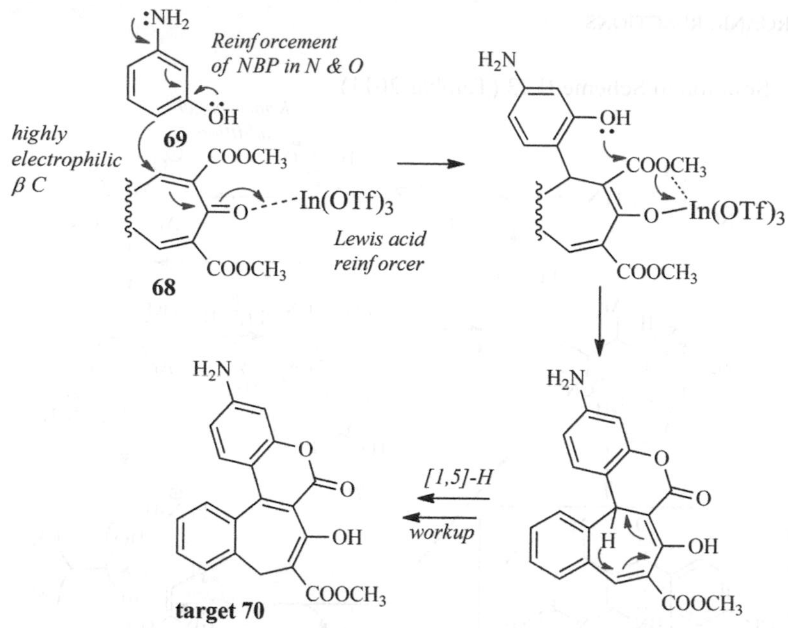

SCHEME II.44 A likely mechanism for the reaction **B** of Scheme II.18.

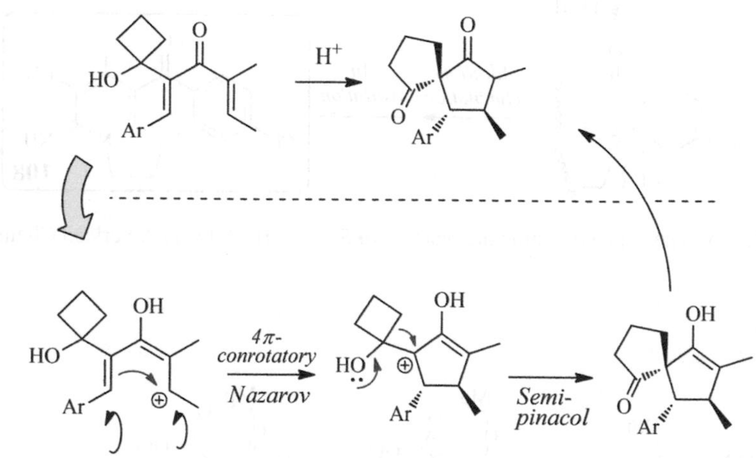

SCHEME II.45 A feasible mechanism for reaction **C** of Figure II.9. [34] / American Chemical Society.

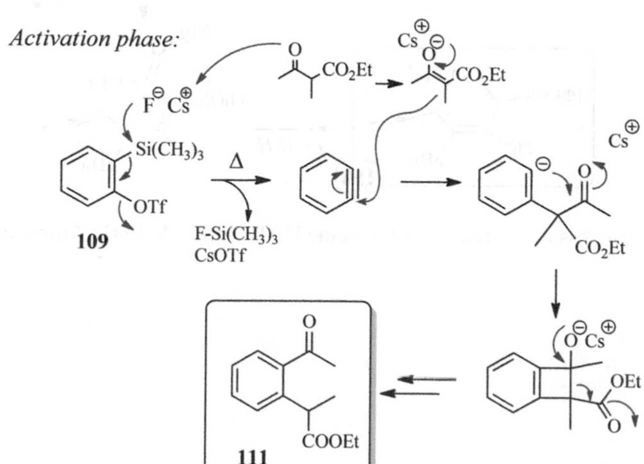

SCHEME II.46 Reaction course expected in Scheme II.24. [40] / American Chemical Society.

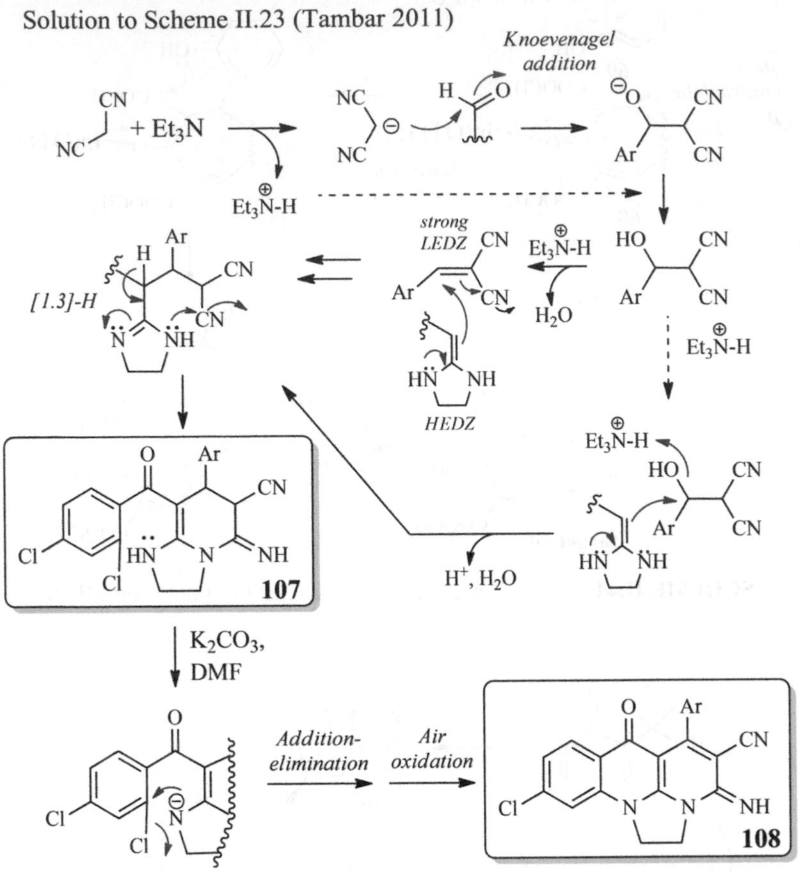

SCHEME II.47 Mechanism explaining the reaction of Scheme II.23. [39] / American Chemical Society.

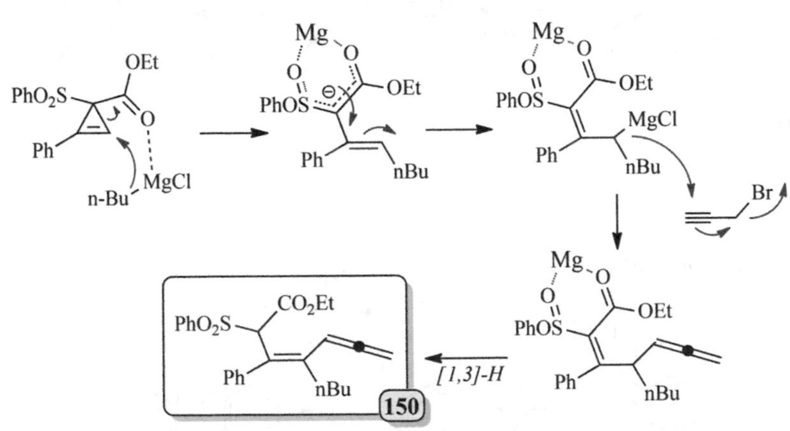

SCHEME II.48 Feasible progress of the reaction of Scheme II.31, section **A**. [54] / American Chemical Society.

NOTES

1. Few chemists realize the contribution of electrons to chemical structure and reactivity, relative to their mass proportion. Given that the mass of electrons is 1836 times smaller than that of a proton at rest (1.673×10^{-24} g) and nearly the same figure for neutrons, the mass of a glucose molecule $C_6H_{12}O_6$ is composed of 53.283% of protons, and 46.688% of neutrons (can you account for this difference?), whereas the proportion of electrons is a mere 0.029%. When accounting for the 44 electrons in the σ bonds of glucose, the mass proportion of electrons plunges to 0.013%. Therefore, the interatomic bonding that holds glucose together and all its chemical reactions arise from this very minor mass fraction. An amusing comparison of molecular mass ratios to bring the nebulous quantum physics of interatomic bonding to daily experience is to imagine a C–C bond like two African male elephants weighing 6000 kg each (the nuclei) trying to get away from each other (repulsion of positive charge) while tethered by a 15 g thread holding their tails (a pair of bonding electrons) together.
2. Concept from renowned Harvard Professor George M. Whitesides, recipient of several top awards in chemistry, in the opening lecture of the ninety-second Canadian Chemistry Conference, May 30, 2009, at Hamilton, Ontario, Canada. Mentioned in Reference [2].
3. The correct answer in Scheme II.20 is ketone **73**, due to the greater stability of the carbonium ion remaining after protonation and water E1 elimination from the *p*-chlorobenzyl carbon rather than the *p*-nitrobenzyl carbon. Then, [1,2]-H migration with attendant C=O formation finishes the sequence.
4. *about M and P nomenclature of allenes*. Read this in case you are not familiar with the M/P nomenclature of allenes (org chem. textbooks ordinarily skip this information; see Figure II.9, section **E** in this chapter for reference): Allenes carrying equal 1,3 substituents, 1,1 (different) groups plus one C^3 substituent, or all different substituents, *are chiral*. The cis/trans, S/E, R/S, syn/anti, and endo/exo nomenclatures, though, do not apply in allenes because of the two orthogonally placed C=C bonds that create confusion when assigning symmetry elements. The allene 3D structure applies well to the axial chirality concept, which can be employed in any object formed by two perpendicular planes with a center of symmetry. So, have a look at your polysubstituted allene from one end and organize the four substituents you see, using the same priorities of the R/S system, except that now we have four substituents to contend with. Select the heaviest, or more dominant group at the carbon closest to your eye and assign #1 to it. The other group on the same carbon is #2 (no matter if this is a H). Then look at the two substituents on the farthest carbon without moving the model, assign #3 to the heaviest or more dominant back there and #4 to the lower priority of the two substituents. Next, determine whether the 1,2…4 sequence is clockwise = *PLUS* = *P*. If the sequence was counter clockwise or to the left, the sequence is = *MINUS* = *M*. Because the selection of the end allene carbon was arbitrary, you should check the other end and repeat the visual process. The result should be *the same*, because the M or P is a property of the whole allenic portion of the molecule, not of a single carbon.
5. In the 1*s* atomic orbital the square radial wavefunction ψ^2, which amounts to the distribution of ED within the orbital, reaches a maximum value at 0.529 Å (52.9 pm = 1.0 Bohr radius unit [ρ]). This value decreases exponentially as a function ($e^{-\rho/2n}$), ($n = 1$ for the 1*s* orbital). The 90% probability mark is reached at 3.5 Å from the nucleus, which is way longer than the H–H bond distance (0.74 Å) in the H_2 molecule. For the 2*p* orbital, the maximum of ED is attained around 2ρ (1.06 Å) and the 90% probability edge is 10ρ (5.29 Å) away from the carbon atomic nucleus. Such a distribution of electron density should create conditions for inter-orbital interaction toward bonding at internuclear distances greater than 3–3.5 Å, as intermediate **24b** of Scheme II.8 illustrates. Nevertheless, effective π bonding is assumed to be consolidated at much shorter distances (1.34 Å). Consequently, σ and π atomic-orbital renderings in mechanism design depict lobes of about 1.0 Å long *arbitrarily*, but they should be much larger. Depicting them as such would be an impractical smear in the mechanism scheme.
6. In case you need a reminder: Formal charge = [#valence electrons] – [#electrons in NBP + ½ #of bonding electrons].

REFERENCES

1. Macias A, Alonso E, del Pozo C, Gonzalez J. *Tet. Lett.* 2004;45(24):4657–4660. DOI:10.1016/j.tetlet.2004.04.109.
2. Schmidt MW, Ivanic J, Ruedenberg K. *J. Chem. Phys.* 2014;140:204104. DOI:10.1063/1.4875735.
3. Utterbach JK, Ruzicka JL, Keller HR, Pellows LM, Dukovic G. *Annu. Rev. Phys. Chem.* 2020;71:335–359. DOI:10.1146/annurev-physchem-050317-014232. Electron transfer from electrodes to organic or biochemical receptors are favored by mediators, in some cases called "molecular wires" that attach to the electrode surface, receive electrons, and pass them on to organics or proteins in very close proximity. See for example: Creager S, Yu CJ, Barndad C, O'Connor S, MacLean T, Lam E, Chong Y, Olsen GT, Luo J, Gozin M, Kayyem JF. *J. Am. Chem. Soc.* 1999;121:1059–1064. DOI 10.1021/ja983204c. The molecular bridges through which electron transfer occurs has been observed by means of scanning tunneling microscopy in a sort of palisades and terraces of various kinds of organic molecules. For those interested see: Bumm L, Arnold JJ, Dunbar TD, Allara DL, Weiss PS. *J. Phys Chem. B* 1999. DOI: 10.1021/jp9921699. See also Section 3.4.2 of this book.
4. Wang J, Hsung RP, Ghosh SK. *Org. Lett.* 2004;6:1939–1942. DOI:10.1021/ol495624.
5. Alonso-Amelot ME, this book. Calculation method: Gaussian RHF at the 6-31G level.
6. Olah GA, Prakash GKS, Rasul G. *Dalton Trans* 2008;4:521–526. DOI:10.1039/B713188G.
7. Park KH, Kang JS. *J. Org. Chem.* 1997;62:3794–3795. DOI:10.1021/jo9703966.
8. Mamantov A. *Prog. React. Kinetics Mech.* 2013;38:1–31. DOI:10.3184/146867812X13558464799.

9. Some reactions discovered long ago continue to stir concepts not completely understood today. The benzidine rearrangement was discovered serendipitously by Nikolay Nikolayevich Zinin (1812–1880), a Russian chemist, around 1845, when scientists were debating about the structure of benzene. The underlying mechanism has elicited an extraordinary amount of research of prominent scientists for nearly two centuries.
10. Markó IE, Warriner SL, Augustyns B. *Org. Lett.* 2000;2:3123–3125. DOI:10.1021/ol006324+.
11. Makhnovskii Y, Ovchinnikov AA, Ovchinnikov YK. *Polym. Sci. USSR* 1981;23:386–396. DOI:10.1016/0032-3950(81)90179-9.
12. Zuev PE, Sheridan RS, Albu TV, Tituv V, et al. *Science* 2003;299(5608):867–870. DOI:10.1126/science.1079294.
13. McMahon RJ. *Science* 2003;299(5608):833–834. DOI:10.1126/science.1080715.
14. Greer EM, Kwon K, Greer A, Doubleday C. *Tetrahedron* 2016;72:7357–7373. DOI:10.1016/j.tet.216.09.029.
15. Castro C, Karney WL. *Angew. Chem. Int. Ed.* 2020;59:8355–8366. DOI:10.1002/anie.1201914943.
16. Nandi A, Alassad Z, Milo A, Kozich S. *ACS Cat* 2021;11:14836–14841. DOI:10.1021/acscatal.1c04475.
17. Parmar D, Matsubara H, Price K, Spain M, Procter DJ. *J. Am. Chem. Soc.* 2012;134:12751–12757. DOI:10.1021/ja30479751.
18. Patel HA, Stothers JB, Thomas SE. *Can. J. Chem.* 1994;72:56–68. DOI:10.1139/v94-010.
19. Ilyn A, Kysil V, Krasavin M, Kurashvili I, Ivachtchenko AV. *J. Org. Chem.* 2006;70:9544–9547. DOI:10.1021/jo061825f.
20. Lotesta SD, Hou Y, Williams LJ. *Org. Lett* 2007;9:869–872. DOI:10.1021/ol063087n.
21. Bernard AM, Cadoni E, Frongia A, Piras PP, Secci F. *Org. Lett.* 2002;4:2565–2567. DOI:10.1021/ol026199x.
22. Bonney KJ, Braddock DC, White AJP, Yaqoob M. *J. Org. Chem.* 2011;76:97–104. DOI:10.1021/jo101617h.
23. Cheng D, Knox KR, Cohen T. *J. Am. Chem. Soc.* 2000;122:412–413. DOI:10.1021/ja993325s.
24. Vazdar K, Margetic D, Kovacevic B, Sundmeyer J, Leito I, Jahn U. *Acc. Chem. Res.* 2021;54:3108–3123. DOI:10.1021/acs.accounts.1c00297.
25. Crespi S, Fagnoni M. *Chem. Rev.* 2020;20:9790–9833. DOI:10.1021/acs.chemrev.0c00278.
26. Nakamura M, Suzuki A, Nakatani M, Fuchikami T, Inoue M, Katoh T. *Tetrahedron Lett* 2002;43:6929–6932. DOI:10.1016/S0040-4039(02)01627-1.
27. Li W, LaCour TG, Fuchs PL. *J. Am. Chem. Soc.* 2002;124:4548–4549. DOI:10.1021/ja017323v.
28. MacKenzie JK. *J. Chem. Soc.* 1965;4646–4653. DOI:10.1039/JR9650004646.
29. Fernández I, Cossío FP, Sierra MA. *Chem. Rev.* 2009;109:6687–6711. DOI:10.1021/cr900209c. Hugelshofer CL, Magauer T. *Nat. Prod. Rep.* 2017;34:228-234. DOI: 10.1039/C7NP00005G.
30. Rablen PR. *Tetrahedron* 2018;74:3781–3786. DOI:10.1016/j.tet.218.04.031.
31. For a review of [1.j] sigmatropic rearrangements, see: Spangler CW. *Chem. Rev.* 1976;76:187–217. DOI:10.1021/cr60300a002.
32. Kielesinski L, Morawski OW, Barboza CA, Gryko DT. *J. Org. Chem.* 2021;86:6148–6159. DOI:10.1021/acs.jo0c02978.
33. Vinogradov MG, Turova OG, Zlotin SG. *Org. Biomolec. Chem.* 2017;15:8245–8269. DOI:10.1039/C7OB01981E.
34. Yang BM, Cai PJ, Tu YQ, et al. *J. Am. Chem. Soc.* 2015;137:8344–8347. DOI:10.1021/jacs.5b04049.
35. Skraba SL, Johnson RP. *J. Org. Chem.* 2012;77:11096–11100. DOI:10.1021/jo302176k.
36. Zhen L, Lin C, Du HJ, Dai L, Wen X, Xu QL Sun H. *Tetrahedron* 2015;71:2839–2843. DOI:10.1016/j.tet.2015.03.077.
37. Concellon JM, Bernad PL, Suarez JR, Garcia-Granda S, Diaz MR. *J. Org. Chem.* 2005;70:9411–9416. DOI:10.1021/jo0515096.
38. Boominathan, M, Nagaraj M, Muthusubramanian S, Krishnakumar RV. *Tetrahedron* 2011;67:6057–6064. DOI:10.1016/j.tet.2011.06.021.
39. Wen LR, Jian CY, Wang LJ. *Tetrahedron* 2011;67:293–302. DOI:10.1016/j.tet.2010.11.049.
40. Tambar UK, Stoltz BM. *J. Am. Chem. Soc.* 2005;127:5340–5341. DOI:10.1021/ja050859m.
41. Himeshima Y, Sonoda T, Kobayashi H. *Chem. Lett.* 1983;12:1211–1214. DOI:10.1246/cl1983.1211.
42. Bhojgude SS, Bhunia A, Biju AT. *Accounts Chem. Res.* 2016;46:1658–1670. DOI:10.1021/acs.accounts.6b00188.
43. CCCBDB 2020, Computational Chemistry Comparison and Benchmark DataBase, Release 21, (August 2020). NIST, National Institute of Standards and Technology. US Department of Commerce. Gaithersburg MD. https://cccbdb.nist.gov/diplistx.asp.
44. Salvador JAR, Pinto RMA, Santos RC, Le Roux C, Matos Beja A, Paixão JA. *Org. Biomol. Chem.* 2009;7:508–517. DOI:10.1039/B814448F.
45. Bian PC, Xu WJ, Gang HZ, Liu JF, Mu BZ, Yang SZ. *Int. J. Mass Spectrom.* 2017;415:85–91. DOI:10.1016/j.ijms.2017.02.003.
46. Tanuwidjaja J, Ng SS, Jamison TF. *J. Am. Chem. Soc.* 2009;131:12084–12085. DOI:10.1021/ja90522366.
47. Gaini-Rahini S, Steenek C., Meyer I, Fitjer L, Pauer F, Noltemeyer M. *Tetrahedron* 1999;55:3905–3916. DOI:10.1016/S0040-4020(99)00098-8.
48. Justus K, Beck T, Noltemeyer M, Fitjer L. *Tetrahedron* 2009;65:5192–5198. DOI:10.1016/j.tet.2009.05.005.
49. Mori K, Sueoka S, Akiyama T. *J. Am. Chem. Soc.* 2011;133:2424–2426. DOI:10.1021/ja110520p.
50. Mahoney SJ, Moon DT, Hollinger J, Fillion E. *Tetrahedron Lett* 2009;50:4706–4709. DOI:10.1016/j.tetlet.2009.06.007.
51. Seebach D. *Angew. Chem. Int. Ed.* 1979;18:239–258. DOI:10.1002/anie.197902393.
52. Wittig G, Davis P, Koenig G. *Chem. Ber.* 1951;84:627–632. DOI:10.1002/cber.19510840713.
53. Byrne PA, Gilheany DG. *Chem. Soc. Rev.* 2013;42:6670–6696. DOI:10.1039/C3CS60105F.
54. Liu Y, Ma S. *RSC Chem. Sci.* 2011;2:811–814. DOI:10.1039/C0SC00584C.
55. Matsumoto Y, Yamada KI, Tomioka K. *J. Org. Chem.* 2008;73:4578–4581. DOI:10.1021/jo800613h.
56. Knochel P, Singer RD. *Chem. Rev.* 1993;93:2117–2188. DOI:10.1021/cr00022a008.
57. Boudier A, Bromm LO, Lotz, M, Knochel P. *Angew. Chem. Int. Ed.* 2000;39:4414–4435. DOI:10.1002/1521-3773(20001215)39:24<4414::AID-ANIE4414>3.0.CO;2-C.
58. Yamada K, Yanagi T, Yorimitsu H. *Org. Lett.* 2020;22:9712–9718. DOI:10.1021//acs.orglett.0c03782.

59. Hollóczki O. *Chem* 2020;26:4885–4894. DOI:10.1002/chem.201903021.
60. Kang Q, Zhang Y. *Org. Biomol. Chem.* 2011;9:6715–6720. DOI:10.1039/c1ob05429e.
61. Wei HL, Yan ZY, Niu YN, Li GQ, Liang YM. *J. Org. Chem.* 2007;72:8600–8603. DOI:10.1021/oj7016235.
62. Vicario J, Aparicio D, Palacios F. *J. Org. Chem.* 2009;74:452–455. DOI:10.1021/jo8022022.
63. Kerr WJ, Morrison AJ, Pazicki M, Weber T. *Org. Lett.* 2012;14:2246–2249. DOI:10.1021/ol300652k.
64. Sun S, Yu JT, Jiang Y, Cheng J. *J. Org. Chem.* 2015;80:2855–2860. DOI:10.1021/jo502908v.
65. Brehme R, Enders D, Fernandez R, Lassaletta JM. *Eur. J. Org. Chem.* 2007;5629–5660. DOI:10.1002/ejoc.200700746.
66. Dai XJ, Li CC, Li CJ. *Chem. Soc. Rev.* 2021;50:10733–10742. DOI:10.1039/d1cs00418b.
67. Vilsmeier A, Haak A. *Ber. Dtsch. Chem. Ges* 1927;60:119–122.
68. Tripp JC, Schiesser CH, Curran DP. *J. Am. Chem. Soc.* 2005;127:5518–5527. DOI:10.1021/ja042595u.
69. Smith MW, Ferreira J, Hunter R, Venter GA, Su H. *Org. Lett.* 2019;21:8740–8745. DOI:10.1021/acs.orglett.9b03308.
70. Bar G, Parsons AF. *Chem. Soc. Rev.* 2003;32(2003):251–263. DOI:10.1039/B111414J.
71. Hung K, Hu X, Maimone TJ. *Nat Prod. Rev.* 2018;35:174–202. DOI:10.1039/C7NP00065K.
72. Knueppel D, Martin SF. *Tetrahedron* 2011;67:9765–9770. DOI:10.1016/j.tet.2011.08.064.
73. Sheppard CS, Kamath VR. *Polymer Eng. Sci.* 1979;19:597–606. DOI:10.1002/pen.760190902.
74. Denisov ET, Denisova TG, Pokidova TS. *Handbook of Free Radical Initiators*. John Wiley & Sons, 2003. DOI:10.1002/0471721476.
75. Zoretic PA, Fang H, Ribeiro AA. *J. Org. Chem.* 1998;63:7213–7217. DOI:10.1021/jo980518+.
76. Taylor TE, Hall MB. *J. Am. Chem. Soc.* 1984;106:1576–1584. DOI:10.1021/ja00318a007.
77. For conceptual application see:: Kayaki Y, Yamamoto M, Ikariya T. *Angew. Chem. Int. Ed.* 2009;48:4194–4197. DOI:10.1002/anie.200901399.
78. Bourissou D, Guerret O, Gabbaï FP, Bertrand G. *Chem. Rev.* 2000;39–92. DOI:10.1021/cr940472u.
79. Vignole J, Cattoën X, Bourissou D. *Chem. Rev.* 2009;109:3333–3384. DOI:10.1021/cr800549. The entire Chemical Reviews issue (2009 N° 8) is devoted to carbenes.
80. Hadel LM, Maloney VM, Platz MS, McGimpsey WG, Scaiano JC. *J. Phys. Chem.* 1986;90:2488–2491. DOI:10.1021/j100402a044.
81. Tomioka H, Okada H, Watanabe T, Banno K, Tomatsu K, Hirai K. *J. Am. Chem. Soc.* 1997;119:1582–1593. DOI:10.1021/ja9635903.
82. Tomioka H, Watanabe T, Hirai K, Furukawa K, Takui T, Itoh K. *J. Am. Chem. Soc.* 1995;117:6376–6377. DOI:10.1021/ja00128a35.
83. Nakamura M, Hirai A, Nakamura E. *J. Am. Chem. Soc.* 2003;125:2341–2350. DOI:10.1021/ja026709i.
84. Goudreau SR, Marcoux D, Charette AB. *J. Org. Chem.* 2009;74:470–473. DOI:10.1021/jo802208q.
85. Petrellis PC, Griffin GW. *Chem. Commun.* 1967;691–692. DOI:10.1039/C19670000691.
86. Doyle MP, Duffy R, Ratnikov M, Zhou L. *Chem. Rev.* 2010;110:704–724. DOI:10.1021/cr900239n.
87. Alonso ME, Jano P. *J. Heteroc. Chem.* 1980;17:721–725. DOI:10.1002/jhet.5570170419.
88. Wenkert E, Alonso ME, Gottlieb HE, Sanchez EL, Pelliciari R, Cogolli P. *J. Org. Chem.* 1977;42:3945–3949. DOI:10.1021/jo00444a034.
89. Jenks WS, Heying MJ, Stoffregen SA, Rockafellow EM. *J. Org. Chem.* 2009;74:2765–2770. DOI:10.1021/jo802823s.
90. Musgrave RA, Turbervill RSP, Irwin M, Goicoechea J. *Angew. Chem. Int. Ed.* 2012;51:10832–10835. DOI:10.1002/anie201206100.

3

STEREOCHEMISTRY AND MECHANISM OF MOLECULAR TRANSFORMATIONS

CONTENTS

3.1 OVERVIEW

For the proficient reaction mechanism designer and problem solver, consideration of the shape of molecules in three dimensions (3D) with all the symmetry, space occupation, and conformational possibilities is as crucial as electron handling proficiency. This chapter is a concentrated pill of the prominent features of molecular shape that govern reactivity, around the concept of steric effect, repulsion, hindrance, encumbrance, bulkiness, and space openings for incoming reagents.

"Stereo" comes from the Greek *stereos*, meaning "solid" among other interpretations; solid is an object occupying the three directions of the Cartesian space. Thus, stereochemistry deals with the shape of things chemical, matter composed of molecules with volume expression.

The Art of Problem Solving in Organic Chemistry, Third Edition. Miguel E. Alonso-Amelot.

Molecules display zones of greater atomic congestion than others, offering steric hindrance to incoming reagents or solvents. This hampering effect is also advantageous to the reaction design: protective groups, moieties for induction of stereoselectivity, and sophisticated control in the enantio-enrichment of products.

The relevance of stereochemistry in the modulation of the reactivity potential within and between organic molecules was presented in Chapter 2 as an essential controller of electron flow. The discussion was limited though and needs consideration of several other aspects to become a practical tool in reaction mechanism design and problem solving.

I will assume that you are acquainted with the basic stereochemistry subjects of nomenclature, asymmetry, dissymmetry, chirality, configuration, conformational analysis and engage in a higher level of discussion. At the start, however, basic questions around the peculiar features of planar organic molecules will be discussed, which become quite knotty soon enough.

A major challenge in the exploration of steric hindrance has been the separation of genuine steric effects from electromagnetic perturbation of organic substituents (electronic repulsion, dipolar effects, inductive, and field effects). Given that molecules are a tightly packed conglomerate of charged particles, for many years scientists were conscious of how difficult it was to pinpoint *pure steric effects* with surgical exclusion of electromagnetic influence of these charged components, intruding into reaction mechanism and kinetics. Bulky groups generally possess a substantial number of bonding valence electrons and in lower energy, inner shell atomic orbitals (AOs). All of them exert electromagnetic repulsion against incoming nucleophiles packed with bonding, non-bonding, and inner shell electrons of their own. Isolation of steric effects in this maze of influences is indeed a challenging task.

The evolution of this highly demanding question is rarely described in a textbook despite its utmost importance in reaction mechanism design. In this chapter, a selection of developments and setbacks is presented with appropriate abundance of examples from the literature for you to move a step forward in your way to become a better reaction mechanism problem solver.

3.2 INTRODUCTION

Molecules, no matter how small (H_2, O_2, N_2), linear (acetylene), or planar (benzene), are 3D objects, and hence, are subjected to stereochemical analysis. All of them occupy a 3D space since they are wrapped in an electron shroud of a set of van der Waals spheres of sizable diameters on the atomic scale: H = 240 pm; C = 340 pm; N = 310 pm; O = 308 pm (pm = picometers; 100 pm = 1.0 Å) [1]. The van der Waals approximation should be complemented with a molecular orbital theory background to account for the space occupied by π molecular orbitals (MOs), each one populated by as many as two electrons, thus extending the volume manifestation of molecules.

Given that covalent bonds are flexible, multicomponent molecules easily invade the third space dimension, thus becoming much more interesting objects to work with in regard to structure properties, configuration, conformation, symmetry, and reaction patterns.

3.2.1 The Question of Planar Molecules and Deviations, an Approach to Steric Effects

3.2.1.1 Planar 2D versus 3D The stereochemistry of non-chiral aliphatic compounds is limited to the cis/trans or E/Z question within the plane of the molecule. While this subject is covered in elementary courses, it also appears frequently in more sophisticated mechanism riddles.

Cyclic planar compounds pull us one notch higher on the question of the third dimension of space around molecules we need to deal with in correct reaction mechanism design. The plan for this recap section is this:

- Describe what a so-called planar molecule is and what it is not.
- How assumed-to-be planar molecules lose their symmetry elements (axes, center, planes) by grafting substituents capable of throwing molecular architecture out of symmetric arrangements and send it closer to optical activity.

The qualification of *planar* molecules only conveys the model of *their nuclei sharing a common plane* (Figure III.1). The coplanarity of atoms in compounds **1–3** stands out in the semi-profile view of stick molecular models. This design, though, disregards out-of-plane vibrations that create a small and yet significant invasion of the third dimension of space. There is also the ever-present contribution of the van der Waals radii and Connolly's volume (see Section 3.4.3) defining the electron occupation space into the third dimension.

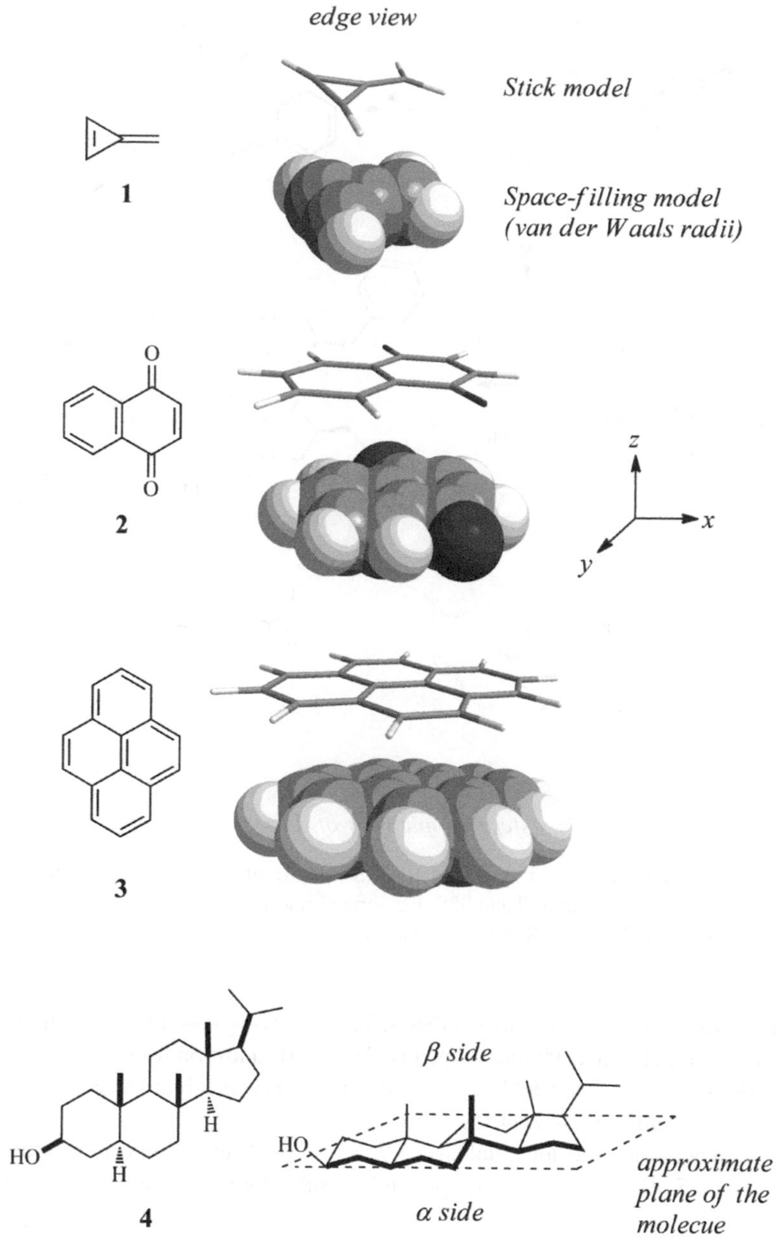

FIGURE III.1 The 3D quality of planar molecules arising from their inevitable van der Waals spheres.

In simple terms, there are three chemically relevant zones in planar molecules:

a) Above the plane (the β side by convention).
b) Below the plane (the α side, idem).
c) The edges of the plane. Edges? Yes, think of reactive atoms – aromatics for instance – or unshared electrons. Think of phenyl lithium, occupying the highly reactive nucleophilic edge; pyridine and the Lewis base quality of nitrogen's nonbinding pair (NBP) right at the edge; C=O functions in benzoquinones (**2**) representing Lewis-acid anchoring sites; electron exchange in redox operations; and so forth.

Reaction-wise, α and β sides of planar molecules are indistinguishable; for example, the well-known [4+2] cycloadditions of electrophiles on the central carbons of anthracene. The α or β side selectivity of incoming reagents can nevertheless be altered substantially by substituents grafted on the ring that may offer selective steric hindrance to one of the sides [2]. The result is enantiomeric induction.

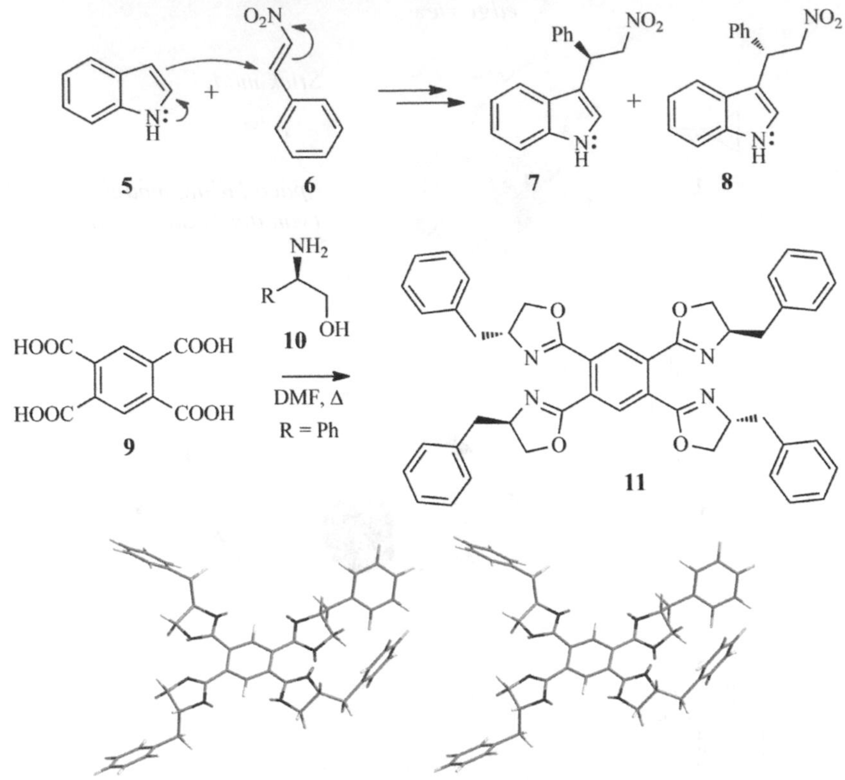

stereoscopic view of 11

SCHEME III.1 While the non catalyzed **5** + **6** coupling gave a 1:1 R:S mixture of adducts **7** and **8**, the same zinc-**11** complex catalyzed addition induced strong steric preference (98% yield of **7** and 95% EE –enantiomeric preference), due to the topological definition and steric induction of a key transition state leading to **7** and not **8**. Adapted from [4].

This approach constitutes a strong trend in organic synthesis. The following case provides an instructive illustration [3].

Indole (**5**) and 1′-nitrostyrene (**6**) perform a Michael-type Friedel–Crafts addition followed by proton elimination in a typical HEDZ/LEDZ rapport, furnishing the predicted adduct (Scheme III.1). The mechanism is simple enough to require further comment, save for the chiral carbon in products **7** and **8**.

Under these conditions (toluene, zinc chloride catalysis) a **7**:**8**, 1:1 R:S enantiomeric mixture is obtained. As expected, there is no steric preference since **5** and **6** are planar, no protruding substituents cause selective repulsion/attraction on either side of the molecular plane.

Nevertheless, adding a fourth independent component containing well-defined chiral carbons that could participate in the Zn-**5**–**6** transition state may affect the differential orientation of the reactants. Author Li [3] explored this concept by grafting four chiral *R*-benzyloxazoline moieties to a rigid benzene central core (**11**) that would serve for complexation with zinc. Earlier research explored one and two similar substituents on benzene to modify the stereochemical course of analogous A + B addition reactions.

Compound **11** was synthesized by mixing **9** and **10** in dimethyl formamide (DMF) at reflux with simultaneous removal of water to prevent decomposition back to starting materials. Although the author did not provide an X-ray structure of **11**, I allowed myself to run a computer MMFF94 molecular dynamics job to observe a possible relaxed structure at 300 K (Kelvin) shown in Scheme III.1. With some practice you can let your eyes float so that the two images merge and have a 3D look. At center, the rigid benzene ring appears surrounded by the four nearly coplanar oxazoline rings and further out to the periphery the four benzyl units bonded in the same R configuration of the starting aminocarbinol **10**, fashioning the blades of a helix in one specific direction. After attempting various conformations in the initial conditions, the same helical matrix was obtained after several energy minimization-relaxation procedures.

Turning back to the experiment, Li prepared a 2 : 1 π complex with zinc chloride (ZnCl) by heating in DMF. The ensuing addition of nitrostyrene (**6**) was assumed to yield a Zn + **11** + **6** complex of still undetermined structure, which would be the result of molecular recognition such that **6** presented only one face to the attack of indole. This latter step was carried out in toluene at −10°C, affording the R adduct **7** in a 98% yield and 95% EE. Ligand **11** had performed an excellent job in inducing a sweeping reactivity difference of the two faces of planar nitrostyrene.

Aside from mono and polycyclic fully unsaturated compounds, from cyclobutadiene to porphyrins, one also talks about the "plane of the molecule" in non-planar saturated monocyclic and polycyclic compounds such as steroids like **4** (Figure III.1, bottom), triterpenes and many other structural types containing isolated or fused 5–10-membered partially or fully saturated carbocycles. In these cases, it is customary to use the α- and β-side nomenclature to define substituent orientation and approaches of reagents.

3.2.1.2 Perturbation of Planarity by Substituents; Approaching Steric Effects Planar molecules may also steer away from planarity on account of stereoelectronic influence of substituents. Have a look at Figure III.2, which shows an unquestionable planar compound: cyclobutadiene (**12**).

A perspective view of a ball-and-stick model of (**12**) shows coplanarity of all nuclei and a presumably comfortable resonance hybrid conveying total C–C bond equivalence. All the elements of symmetry, three planes, three axes, and center of symmetry are there. However, adding the electron π cloud space of its highest occupied molecular orbital (HOMO), which is in fact the functional component of this molecule, uncovers two issues:

- The obvious third dimension stands out as the HOMO lobes, an all-important feature of every molecule, including planar, where π electrons exist.
- Because of this peculiar electron arrangement whereby the left side C=C bond is antibonding relative to the right-side C=C bond (mind the MO lobe phase) there is a distinct σ and π contribution to the carbon-to-carbon bonds. Consequently, cyclobutane exhibits two different carbon-to-carbon bond lengths: 1.32 and 1.40 Å. The perfect square symmetry is thus distorted a bit. The molecule continues to be planar, though.

Adding just one fluoride substituent introduces measurable perturbation in the square shaped cyclobutadiene. Calculation of bond distances and dihedral angles [1] reveals a distorted square structure with four carbon-to-carbon bonds of different lengths and dihedral angles away from the theoretically expected 90° for the unsubstituted **12**. The molecule is still planar considering only the atomic nuclei, but it has lost all the elements of symmetry except for the plane slicing across all atoms; just at the brink of becoming chiral.

Let us complicate matters further by looking at bis-cyclopentadienyl **14** (Figure III.3): 10 π electrons in a perfectly planar structure, and hence, are aromatic ($4n+2$ π electrons); it is intuitively rigid and maintains the planar shape at the lowest energy level due to maximum overlap of the π AOs. Not all bonds are equivalent, though; bond lengths in the rings are 1.456 Å, whereas the central C=C bond is 1.343 Å long. But this 7.7% difference does not affect the symmetry elements of **14**, it only makes the central bond more reactive in electrophilic additions.

When methyl groups are added in the cisoid configuration, as in **15** Figure III.3, steric repulsion (SRe) stemming from the short distance between the two CH_3 hydrogens (2.18 Å) (C–C distance: 3.18 Å) exerts sufficient pressure to deviate the cyclopentadienyl rings an average of 7.8° away from coplanarity (Note 1). Replacing methyls with bromine as shown in **16** increases the dihedral angle to 13° out of coplanarity (Br–Br nuclei distance: 3.38 Å). Given that the van der Waals radius of bromine is 1.85 Å, the two electron clouds are practically overlapping, which must cause considerable strain. Additionally, the dipole moment μ of the C–Br bond (1.82 D for methyl bromide) pointing in the same direction in **16** contributes not only to dipolar

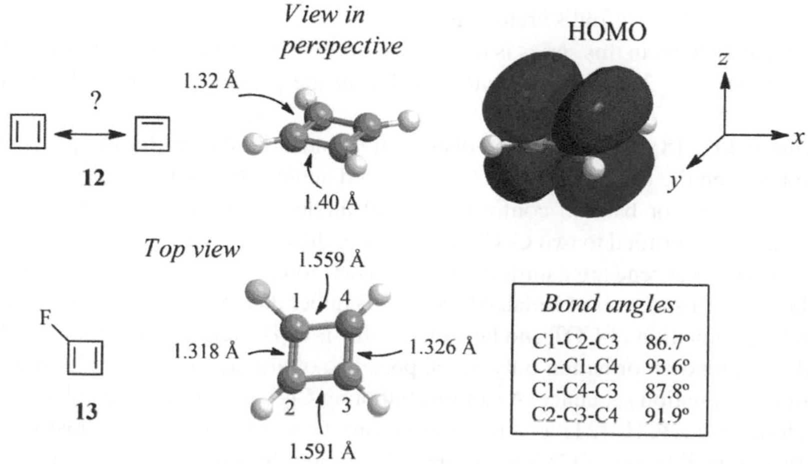

FIGURE III.2 HOMO of cyclobutadiene and asymmetries induced by a single fluorine atom. Calculation method: See Note 1 method (a).

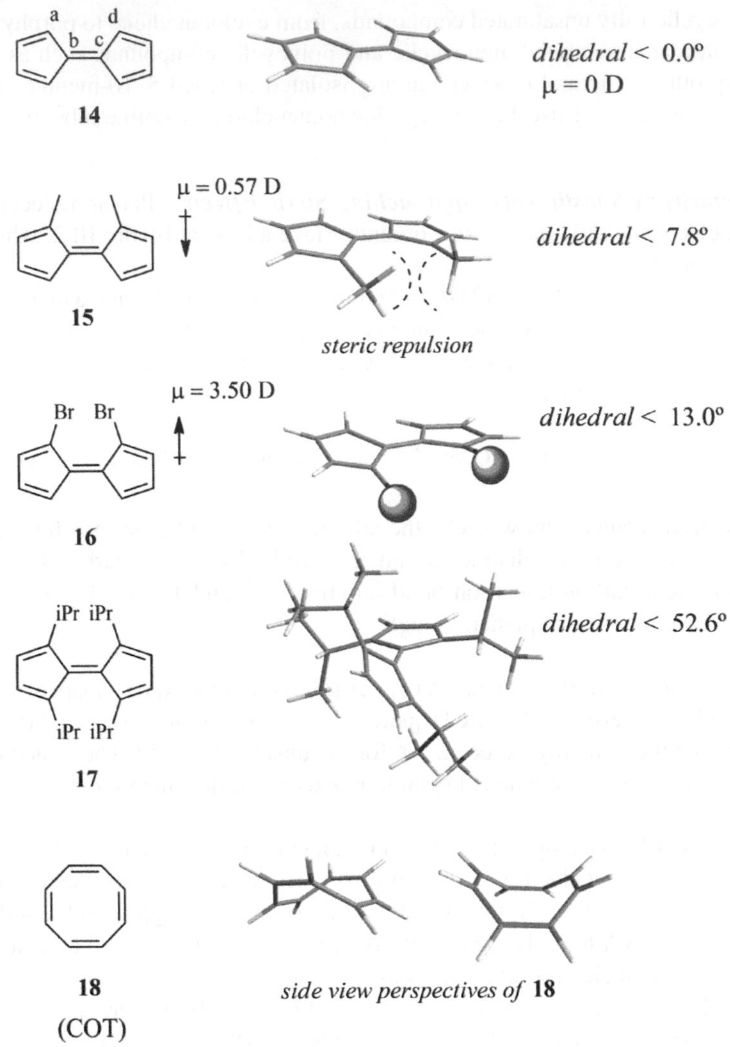

FIGURE III.3 Examples of $R_2C{=}CR_2$ out-of-plane twisting forced by steric stress (SS) of substituents. Dihedral angles and polar moments were calculated using MMFF94s energy minima at 300 K (Note 1).

field repulsion and brings a total μ of 3.50 D to this compound. These circumstances call for a combination of steric and electronic elements of distortion of the scaffold's architecture.

An extreme case of coplanarity bend in this series is the tetraisopropyl substituted model **17** of which I estimated (MMFF94 minimization-relaxation at 300 K) at 52.6°. This is clearly visible in the perspective view where the cyclopentadienyl rings appear slanted.

Finally, cyclooctatetraene (COT, **18**) exemplifies the distortion of planarity caused by the internal ring strain. This tension comes from the difference between the open chain 120° C–C–C bond angle between three sp^2 carbons and the 135° of a planar regular octagon. In adopting a boat or bathtub conformation, all angles are reduced to 127°, which decreases ring strain. Consequently, the π conjugation is modified to two C–C bond types of different lengths: 1.341 and 1.445 Å of π and σ character, respectively – C–C bond lengths in benzene are a uniform 1.39 Å. Therefore, being non-aromatic ($4n$ π electrons and a non-planar configuration) COT (**18**) performs as a set of isolated C=C bonds rather than a conjugated chain due to stereochemical strain.

The contrast of classical halogenation of COT and benzene with Cl_2 or Br_2 takes you to this peculiar feature. As you know, benzene and pyridine undergo mono-chlorination only in the presence of aluminum or iron chloride at room temperature, following the familiar addition–elimination sequence. As a plus, bubbling Cl_2 gas to a benzene solution takes halogenation all the way to the hexachlorocyclohexane (HCH) (**21**) as a mixture of various stereoisomers. Contrastingly, COT reacts with either halogen without any catalyst, at well below 0°C yielding the 1,2-dihalide (Scheme III.2) [4], very much like aliphatic alkenes. Can you provide an answer considering that:

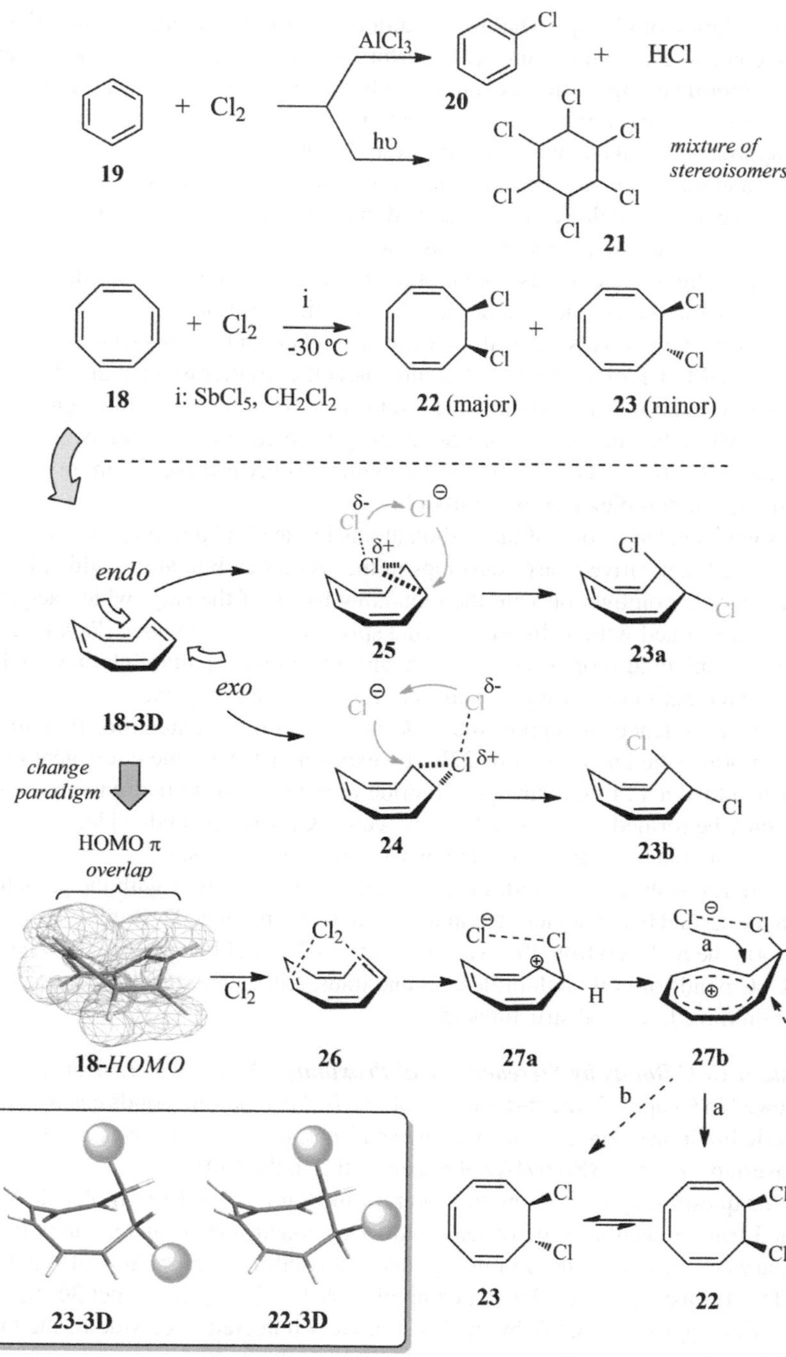

SCHEME III.2 Stark differences between chlorination of benzene and cyclooctatetraene and a feasible mechanism explaining this difference. Adapted from [4].

a) The thermodynamically favored *trans* **23** is a very minor product.
b) When bromine is used (to be able to see color changes), the solution of COT and the first mole of Br_2 remains colorless, but the addition of one droplet of Br_2 exceeding this first mole the color changes to reddish-brown [4].

Reasoning Scheme III.2:

1. Our first step after the yellow arrow is to draw a 3D rendering of COT to better observe steric effects (**18-3D**).
2. COT C=Cs resemble independent alkenes and should respond in the classical manner of halogen addition, namely: form a π–Cl_2 complex as exo (**24**) and endo (**25**) chloronium ions. Although the latter seems less favorable than the exo approach, one cannot disregard the electronic influence of the C=C bond at the opposite side of COT.

3. Following the classical evolution of C=C plus halogen addition, one would expect that the chloride anion, once detached from either endo or exo complexes, would be unable to perform the attack on the chloronium carbons from the same side as this is blocked by its chloride companion. The anion would come around and form the second Cl–C σ bond from the opposite side in an S_N2 fashion with an inversion of configuration.
4. Of note, endo and exo approaches will end up in the same trans product **23a/b** (Scheme III.2), which are indistinguishable despite our using colors, and therefore, the expected mechanism does not explain the predominance of the cis isomer **22**. In addition, this route disregards the color change observed in the solution after adding the first drop (of bromine) more than the equimolar amount of halide. This option is thus rejected:

   We need a change of paradigm since this is not an ordinary addition. The stereochemistry of COT is playing games against our rock-hard mindset in the classical model for electrophilic additions.
5. We all know that halogens and C=C bonds show affinity for each other, and the **18-3D** has two non-conjugated C=C units at the head and foot of the bathtub. Examining this structure under the frontier molecular orbital stand, one soon discovers the π HOMO forming a canopy of sorts in the endo face of the molecule (green cushion-like feature in **18**-*HOMO*), whereas the MO lobes in the exo face point away from each other; (the negative phase is light-colored to enhance the green positive MO phase). The chlorine electrophilic activity is therefore more intense in the endo side of **18**, and a chlorine–COT molecular complex (**26**) is admissible.
6. As one of the Cl atoms gets bonded to one of the carbon atoms in the C=C unit and the Cl–Cl bond is in the process of being heterolytically cleaved, a positive charge develops in the second carbon atom, which is allylic in **27a**. The empty sp^2 AO of this carbon enters in conjugation with the remaining triene of the ring, which acquires a planar seven-membered partial structure **27b** adorned with a chlorine bearing spike. This is a *homotropilium* ion, a stable structure with an *endo* chloride, an attractive and bold proposal of Huisgen and coworkers [4] although they could not stop the reaction at this stage by lowering the temperature, adding a quencher or any known complex.
7. Authors developed indirect evidence to support the *endo*-homotropilium contention: they managed to obtain the exo homotropilium mono-chloride (the enantiomer of **27b**), by exposing COT to one equivalent only of antimonium pentachloride at −40°C; then, added a 1 : 1 equivalent of chloride in the form of tetraethylammonium salt. The product was *trans* **23,** which could only be formed by the attack of the second Cl from the endo side.

   The homotropilium cation model has been adopted by COT researchers since.
8. In the progress of **26** in transit to structures **27a/b** I kept the chloro-COT complex with the partially cleaved Cl_2 to maintain the cis configuration in the second halogenation step in its way to major product **22**. If this chloride anion was fully detached it would be free to perform the S_N2 step from the exo side and *trans* **23** would be predicted to become the major product.
9. The three remaining C=C bonds of both dichlorides, cis and trans, still cause sufficient strain to maintain the tub conformation in the energy-minimized, relaxed structures [5].

3.2.1.3 Interaction of Distant C=C Bonds by Stereochemical Proximity While we are accustomed to intramolecular C=C conjugative interaction reviewed in Chapter 2, the π–π rapport of two *isolated* double bonds can also occur across space in rigid scaffolds, not only in the steric hindrance sense but also acting as if the extra C=C unit was an assisting neighboring group of sorts owing to the HOMO overlap shown in **18-*HOMO*** of Scheme III.2 in the COT case.

Scheme III.3 illustrates this question by restraining the natural flip-flop flexible COT molecule to a rigid carbon frame **30**. This compound was obtained from a peculiar type of Diels–Alder cycloaddition of maleic anhydride (**28**) on COT's distant C=Cs instead of the customary conjugated diene. In this atypical cycloaddition, the π AOs of the distant C=C bonds overlap during the transition state (TS), furnishing the cyclobutene unit observed at the top in adduct **30** [6].

Importantly, maleic anhydride approaches COT by the less sterically hindered open side of the tub-shaped COT, as shown in **29TS** of Scheme III.3. The result is a sterically well-defined carbon skeleton **30**.

In this open box-like material, three steric fields A, B, and C become well defined: the upper side "A" or lid, well exposed to exo addition, the underside "C" and the middle mouth-like section "B". Predictably, steric impedance to incoming reagents should follow the order B>C>A, given that the anhydride moiety in **30** hinders incoming species from the endo side "C". Hindrance of B explains itself.

Then **30** was submitted to halogenation using bromine, iodine chloride, and *t*-butyl hypochloride, hoping to observe preferential electrophilic attack on the "A" double bond and subsequent reactions of the halonium ion intermediates with the halogenation counterion or an added nucleophile (e.g. methanol) [6]. Scheme III.3 depicts three illustrative outcomes involving, or not, the distant "C" C=C.

Reactions occurred in two or three steps depending on halogenation conditions. Products **32, 35,** and **36** (framed structures) call for interesting mechanisms: After a first electrophilic addition of the halonium **31** occurring exclusively at the "A" C=C, the compound so activated evolves through three high yielding pathways (a), (b), or (c) contingent on the nucleophile.

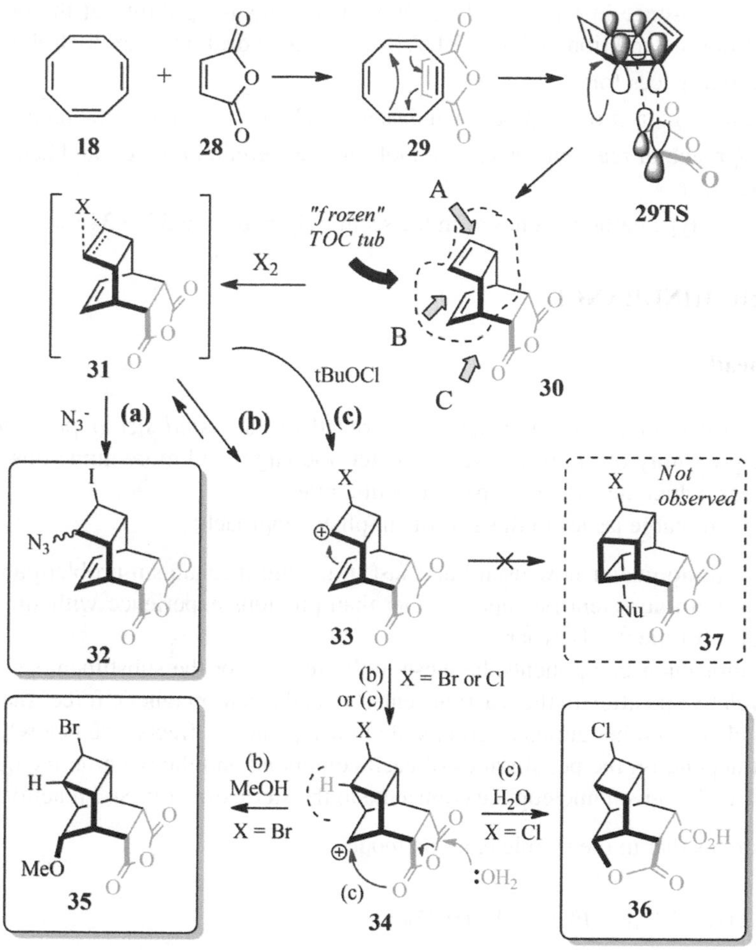

SCHEME III.3 An expected cycloaddition of cyclooctatretaene and maleic anhydride triggers a reaction cascade to awesome cage products by addition of simple reagents. [6] / American Chemical Society.

Path (a): With iodine azide, attack of the $N_3^{(-)}$ completed the standard 1,2 addition without affecting the rest of the molecule, yielding **32**. Authors [6] did not provide evidence for the cis : trans nature or ratio of this material, yet one can safely conclude that a carbonium ion or a strongly polarized halonium ion are likely intermediates. So far, no involvement of the lower C=C of the "C" area; no surprises.

Path (b): Given that halogenation was also performed in methanolic Br_2 solution at room temperature, conditions were suitable – for undetermined reasons – for the lower C=C bond to interact with the LEDZ bromonium ion **31** or carbonium ion **33**. This interaction could progress toward two secondary addition products: the direct box scaffold **37** (dashed inset) or the crossed structure **34**. The nuclear magnetic resonance (NMR) spectra ruled out **37** in all cases but reasons for this preference remain obscure. The difference in ring strain substantiates this outcome.

In route (**b**), the secondary cation in the strained cage structure **34** is susceptible to exo and endo attack by methanol. Both approaches find steric hindrance by either the succinic anhydride moiety or the angular proton highlighted in green. That the isolated product was **35** only (38% yield) suggested kinetic control due to steric hindrance differences of both approaches.

Path (c): followed the same course of (**b**) as far as cation **34** in acetonitrile/water, 1 : 1. In this instance, water did not operate as a nucleophile on this cation; the succinic anhydride (in green) oxygen atom concurred first, aided by a hydrolytic cleavage **34** → **36**. Reasons for this selectivity are unknown but may respond to TSs including the halonium-lower C=C bond-succinic anhydride moieties acting concurrently.

Scheme III.3 *takeaways*:

- The distant C=C bonds in COT operate in diverse manners depending on the side of the conformational tub: compare **26** → **22** of Scheme III.2 and **18** + **28** → **30** in Scheme III.3.

- The 3D architecture is of the essence in qualitatively defining reaction susceptibility of the two represented moieties, the HEDZ (C=C) toward electrophilic addition and the LEDZ (succinic anhydride) toward hydrolysis.
- Rigid structures like **30** define spaces for:
  - intramolecular interactions, e.g. π–π overlap between distant C=C bonds forced by the native configuration.
  - allowing the selective approach of reactants or solvent molecules at certain moieties, and hampering (e.g. zone "C" of **30**) others by SRe (e.g. zone "B").
  - furnishing unexpected cage-type carbon frames as in the seemingly nonsense **33**→**34** transition.

3.3 MEASURING STERIC HINDRANCE

3.3.1 The Roadblocks Ahead

The elementary treatment of steric hindrance is generally restricted to the *visual size appearance* of substituent groups R directly bonded to, or in the proximity of, a given reaction center: the larger and more numerous substituents offer increased impedance to reaction progress, which negatively impacts reaction rates.

However, there are two questionable issues in this sort of simplistic approach:

1. The eye-based appraisal, no matter how useful and fast in getting mechanism problem answers, is only a superficial qualitative assessment with no scientific support other than previous experience with similar bulky (a perfectly non-quantitative estimate) substituents elsewhere.
2. The steric effects of molecular components, be these in the scaffold or the substituent set, are a combination of sheer substituent volume, relative position to the reaction center(s), and electromagnetic forces that are modulated by through-bond and field channels from other atoms or groups of atoms, quantum effects (see below), and solvation. Their nature and distribution will depend on the peculiarities of each compound and the reaction medium considering that solvent molecules associated with a given nucleophile contribute to the steric effect in S_N2 reactions.

Recent research [7] demonstrates that in the simple reaction model:

$$F^{(-)}(H_2O)n + CH_3I \rightarrow FCH_3 + I^{(-)}(H_2O)m$$

where $n=0$–3 and $m=$undetermined.

The native non-solvated reaction is strongly quenched upon increasing microhydration of fluoride, adding a complexity factor to a far from simple problem of steric hindrance.

3.3.2 Steric Requisites for Building σ Bonds

For two AOs or MOs involved in electron flow for σ bonding four conditions need to be met:

1. Adequate distance as discussed in Chapter 2.
2. None, minimal, or moderate steric impediment of vicinal groups in the way of bonding.
3. Sufficient molecular drive or kinetic energy to overcome or take advantage of stereoelectronic influence from nearby α or β bonds and NBPs, including control of partially inhibited resonance or conjugation due to steric deformation of the required coplanarity of conjugated AOs.
4. Compliance with symmetry rules and requisites of interacting HOMOs and lowest occupied MOs (LUMOs).

Molecular steric characteristics, namely occupation of space and spatial distribution of its components, have four major consequences for reaction mechanism:

1. Cause hindrance of reaction rate (*).
2. Elicit acceleration of reaction rate (*). See reminder Scheme III.4 for the dominion of SN reactions.
3. Define the stability of molecular conformations and/or configurations, equilibrium constants, and reaction outcomes.
4. Serve as a conduit for stereo and enantioselectivity.

(*) *depending on the location of slow step along the reaction coordinate.*

Each one of the points above is discussed next with the aid of a selection of published study cases.

3.3.3 Reaction Rate Retardation Due to Steric Hindrance

Bimolecular reactions in the landscape of bimolecular nucleophilic substitutions (S_N2), nucleophilic or electrophilic alkenyl, aryl, or C=X (X=O, N) additions, and bimolecular eliminations (E2) respond most conspicuously to steric hindrance negatively impacting reaction rate. The old gold rule of reactivity in such cases: primary > secondary ≫ tertiary (if at all) of receptor carbons, and the so-called α series: methyl > ethyl ~ propyl > isopropyl > isobutyl ≫ *tert*-butyl = neopentyl in S_N2 reactions, applicable also to nucleophiles in some cases (but not others), has been a most respected *qualitative* rule with consistent results not devoid of certain limitations of group size and substitution pattern.

Despite their apparent straightforwardness, steric encumbrance is just **one of five effects** through which substituents modify reaction courses and measurable rates (Box). Recent research adds quantum effects to this list.

Some of these effects operate in opposite directions. For example, in the alkyl bromide series CH_3Br, EtBr, *i*-ProBr, and *t*-BuBr there is an increasing inductive *+I* effect by substituents that accelerates as far as electronic influence the S_N2 rate, especially if there is a significant increase in δ^+ charge at the reaction center by asymmetric distancing of the bromide nucleofuge relative to the incoming nucleophile. At the same time, the growing steric encumbrance across the series hampers it, added to the lesser polarization R→C(δ^+)–Br(δ^-) owing to the same *+I* electron density pumping of the R group onto the reacting **C**. Taken together, the R series of substituents offer a give-and-take game impinging strongly on the reaction rate of the S_N2 reaction. One wonders if these opposed effects can be quantified experimentally.

Substituent effects have been grouped in two classes: Through space and through bonds. A third class is currently emerging: *quantum effects*.

Through space:
- Steric (relative size, location, subjected to variation during bonding)
- Field
- Solvation

Through bonds:
- Polar, inductive, hyperconjugation
- Conjugation and resonance

The latter two are collectively treated as *bonding effects*.

Quantum effects:
- Quantum mechanics studies reveal that alkyl branching–as in isopropyl and *t*-butyl–may cause deformation of the branching carbon, changes in electron density distribution in the molecular frame, volume, and energy [8, 9]. Thus, steric encumbrance of a given substituents also modifies the electron distribution of the scaffold section it belongs in, modifying its reactivity.

The macroscopic result is a reduction in the reaction rate, i.e the steric influence dominates over the *bonding effects* of induction and hyperconjugation. However, this is not always true. A case-by-case methodology employing systematically built series of compounds sharing a given reaction is needed to understand and quantify the performance of specific nucleophile–electrophile systems and the interference of steric hindrance. This is what the free energy relationships research did, precisely, as we shall see presently in abridged form.

3.3.4 The Saga of Purely Steric Effects

Attempts to isolate steric effects cleanly for quantitative assessment from this turbulence of electronic authority have been a challenging task for several outstanding scientists, ever since they progressed from visual judgment to numerical evaluation in the 1930s and the following 50 years, while employing increasingly refined reaction rate measurements. Subsequently, quantum mechanics took the lead up to today. The influential work of heavy weight chemists–Hammett, Taft, Streitwieser, Ingold, Mayr, MacPhee, Okamoto, DeTar, Shorter, and several others–produced mathematical expressions to estimate and discriminate some of these effects. (See Note 2) [10–16].

Among others, Professor Robert W. Taft [11] ran extensive kinetic studies of substituted aryl or alkyl ester hydrolysis and comparisons with the *pKa* of the corresponding carboxylic acids and nucleophilic substitution. He produced Equation 3.1, one of several similar linear free energy relationships of which the first version was proposed by Hammett in 1935 [10]:

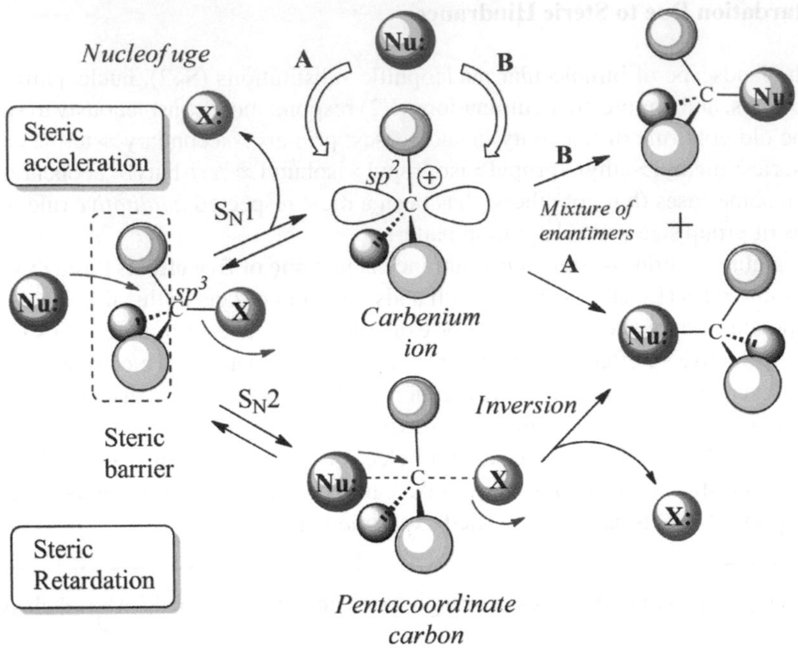

SCHEME III.4 Schematic *reminder* of the two main sequences of nucleophilic substitution reactions and contribution of steric effects.

$$\log(k/k_o) = \rho^*\sigma^* + \delta E_s \tag{3.1}$$

whereby (*as a reminder*):

- ✓ Taft's ρ^* is the reaction constant, or sensitivity of a given reaction to the effect of substituents in the vicinity of the reaction center or transmitted through an aromatic ring, as a result of partial or effective electronic charge transmission in the TS or intermediate related to the slow step of the reaction; ρ^* is calculated from the slope of a linear correlation plot of $\log(k/k_o)$ (for example *pKa* of a series of para-x-benzoic acids in the abscissa relative to benzoic acid) with $\log(k/k_o)$ in the plot's ordinate (rate of esterification or hydrolysis of a similar series of para-x-benzoate derivatives). The same relationships can be established for many reaction types of a variety of substrate molecules undergoing a common transformation.
- ✓ σ^* is the substituent constant, established as $\sigma^* = \log(k/k_o)$ for each substituent in a reaction series; k_o is the rate constant of an unsubstituted member of the series.
- ✓ E_s is the *steric factor of a given substituent.* E_s can be determined assuming that the electrostatic (polar, field, resonance) factors included in $\rho^*\sigma^*$ are either negligible, constant, or whose contribution to the measured rate can be calculated from a different set of reactions in compound series containing the same collection of substituents.
- ✓ δ is the slope of the solitary set of E_s in a homologous reaction series *when the electronic effects have been canceled.* The term ρ_s is also used instead of E_s.

Over time, Equation 3.1 has been modified and expanded to include other effect terms in addition to E_s, the elusive steric contribution. Criticism has also been very active, to the chagrin of problem solvers forcibly buried under piles of data and contradicting conclusions [15]. In this regard, some authors have replaced E_s with other sets of values that include additional effects. There is no space here to put together this endless flow of dissatisfaction. For those interested in a deeper insight, illuminating reviews [14, 15] are available. The flow of discussions has not stopped since; lately in the hands of computer-equipped quantum mechanics theoreticians [8, 9, 16, 17].

In this section, we will navigate through these troubled steric waters by examining closely five selected study cases and constitute the present saga (See Note 3).

Intuitively, SRe influences primarily the rate constants of second-order reactions involving bimolecular TSs responsible for the slow step along a reaction coordinate. The selected reaction types herein are substitutions and addition–eliminations of a variety of substrates and nucleophiles. Comparison of results from different studies will demonstrate that, curiously, the steric effect does not operate equally in electrophilic and nucleophilic moieties and varies with the type of nucleophile, making vis a vis comparisons of different reaction types difficult or invalid.

3.3.4.1 Study Case 1: Substitution (S_N2) of Alkyl Bromides by Sodium Methoxide Professor Okamoto and coworkers in Japan [18] studied with admirable oriental patience the second-order reaction constants of a long series of primary alkyl bromides of increasing classically accepted steric bulkiness in the bimolecular substitution:

$$R-Br+NaOCH_3 \rightarrow R-OCH_3+NaBr$$

Quantification was performed by measuring the rate of 34 S_N2 reactions using an assortment of nucleophiles including lithium and sodium halide salts, amines, alkyl, aryl, and naphthyl alkoxides, and others [18]. The *reactivity constant r(R)* for each alkyl group R was defined by the expression:

$$r(R)=(\log k_{RX}-\log k_{EtX})=\log\left(k_{RX}/k_{EtX}\right) \tag{3.2}$$

where:

k_{RX} = the average second-order rate constant of compound R–Br with sodium methoxide in methanol.

k_{EtX} = average second-order rate of Et–Br, selected as the comparison standard. Et–Br was the best candidate, probably because it represents an intermediate size electrophile without substituent branching and the smallest alkyl halide in which elimination could compete (it did not under the employed conditions).

Governed by Equation 3.2, *r(R)* is the *relative rate* of each member of the series against the rate of Et–Br. Table III.1 portrays the recorded data: the second-order rates (Table III.1-A) and reaction constants defined by Equation 3.2 (Table III.1-B).

Note that rate comparisons can be done in longitudinal and transversal terms; that is, contrasting first, the performance of the receptors (electrophiles) of the S_N2 reaction (the alkyl halides) against one specific nucleophile ($NaOCH_3$, second column) and second, the behavior of each electrophile with the four transversal nucleophiles, from CH_3ONa to pyridine (all columns).

In the first two columns of Table III.1-A one observes a regular decrease in the reaction rate of the S_N2 process covering two orders of magnitude from CH_3Br to *i*-butyl-Br. This is reproduced in the first two columns of Table III.1-B and Figure III.4 in reference to the relative rate (RR) according to Equation 3.1. The declining trend runs in parallel with the *apparent* steric hindrance of the R group assessed qualitatively.

A scrutiny of Table III.1-A across the whole matrix of electrophiles/nucleophiles is also consistent with a marked decrease of nearly four orders of magnitude from the $CH_3Br/NaOCH_3$ cell at the top left, to the *i*-butyl bromide cell at the bottom right.

TABLE III.1 Second-order reaction rates ($mol^{-1}s^{-1}$) (Table A) and reactivity constants *r* (Table B) of a series of alkyl bromides with sodium methoxide in methanol at room temperature, calculated from Equation 3.1. Data from [18]

TABLE III.1-A: Second-order rate constants

Electrophiles	Selected nucleophiles			
R–Br	**NaO CH_3**	**NaOPh**	**Dimethyl aniline**	**Pyridine**
CH_3–Br	8.14×10^{-02}	2.84×10^{-02}	2.75×10^{-05}	2.71×10^{-05}
EtBr	9.06×10^{-03}	6.21×10^{-03}	1.78×10^{-04}	1.52×10^{-04}
***n*-Propyl Br**	3.35×10^{-03}	2.89×10^{-03}	8.67×10^{-05}	8.42×10^{-05}
***n*-Butyl Br**	3.34×10^{-03}	3.08×10^{-03}	9.03×10^{-05}	ND
***i*-Butyl Br**	6.75×10^{-04}	5.12×10^{-04}	3.10×10^{-06}	2.54×10^{-06}

TABLE III.1-B: Reaction constants *r(R)* relative to ethyl bromide

R–Br	**NaO CH_3**	**NaOPh**	**Dimethyl aniline**	**Pyridine**
CH_3–Br	0.953	0.660	−0.811	−0.749
EtBr	0.000	0.000	0.000	0.000
***n*-Propyl Br**	−0.432	−0.332	−0.312	−0.257
***n*-Butyl Br**	−0.433	−0.305	−0.295	
***i*-Butyl Br**	−1.128	−1.084	−1.759	−1.777

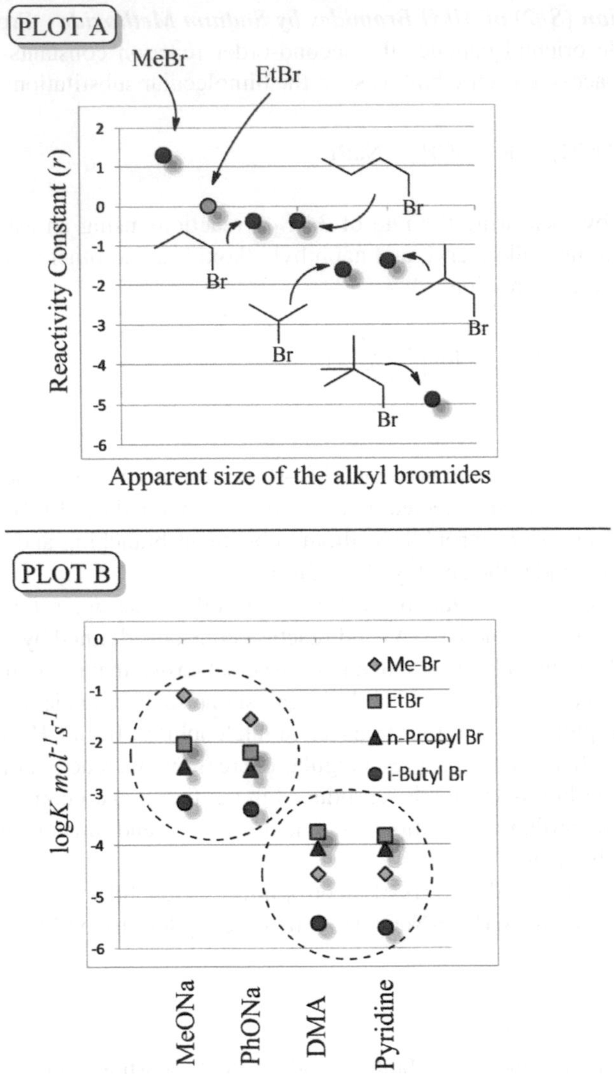

FIGURE III.4 Plot **A**: Reactivity constants *r(R)* (see Equation 3.1) of a series of alkyl halides of increasing steric hindrance in the bimolecular substitution with sodium methoxide in methanol. Steric hindrance is organized along the *x*-axis only as a categorical variable, according to *apparent* encumbrance. Plot **B** organizes measured second-order rates of a series of alkyl bromides reacted with equimolar amounts of nucleophiles of increasing steric bulkiness. DMA = 2,6-dimethylaniline. The *r* and *k* data were extracted from Drawing by Alonso-Amelot, Data from [18].

At first glance, all rate constant numbers are in keeping with the general idea that as substituents get larger, increasingly strong SRe hampers the S_N2 reaction progress.

All things considered, the data plots of Figure III.4 unveil a number of interpretation difficulties, some of which were mentioned by the Japanese group [18]:

1. Both plots use categorical (non-numeric) variables in the abscissa axis (small, medium, and large substituents). This sort of treatment can only provide steric hindrance in *qualitative* terms.
2. While it is comforting to find that CH_3Br is much more reactive than EtBr as plot **A** shows, the nearly similar rates of *n*-propyl and *n*-butyl bromides in the S_N2 reaction with the methoxide anion suggests that *k* is nearly independent of the carbon chain size after the second carbon if ramifications are ignored. Only *sec*-bromide (isopropyl) and β-substituted (isobutyl) bromides offer more resistance to the attack of methoxide.
3. The neopentyl bromide appears at the bottom of the scale (plot **A**) five orders of magnitude slower than methyl bromide, despite that the *t*-butyl group is distanced from the primary carbon supporting the leaving group.

4. Drawing an imperfect straight line with a negative slope across the points of plot **A** would be unproductive and hardly useful as predictor for other analogous reactions. No quantification of this slope is possible.
5. Two groups of data emerge in plot **B** (dashed circles). In the first group, alkyl halides follow the same trend when exposed to methoxy and phenoxide alkoxides, and a reduced k in the more sterically demanding phenoxide. However, in the second group, the reactivity of methyl drops under ethyl and n-propylbromides for unexplained reasons.
6. Replacing the oxygen anion (upper circle) by neutral amine nucleophiles (lower circle) brings about electronic effects not accounted for in the provided data. Conclusions on the contribution of the steric effects, independently of other factors, cannot be assessed unless other experimental designs are considered.

Conclusively, the absence of a linear free energy relationship prevents a dependable quantitation of E_s and casts doubt on Taft's reported values when applied to Okamoto's nucleophilic substitution model.

3.3.4.2 Study Case 2: Hydrolysis of Esters and Esterification of Carboxylic Acids Professor DeTar in Florida used the concept described for Equation 3.1 above by the separate estimation of electronic and solvent effects embedded in the ρ*σ* term [19, 20]. The selected model reaction was the hydrolysis of carboxylic esters under acid and alkaline catalysis, and the esterification of similar alkyl carboxylic acids.

These reactions occur in two steps: nucleophilic addition on the C=O unit to give a tetrahedral, and hence a more crowded intermediate, followed by elimination to reconstitute the carboxylate unit. Predictably these are sensitive to steric effects from both R and R' of the RCO_2R'electrophile. In the case of the acid hydrolysis of esters the reaction course is:

$$\mathrm{R-COOR'(trigonal)+H_2O+H^+} \rightleftarrows \mathrm{R-C(OH)_2-OR'(tetrahedral)} \rightleftarrows R-COOH\,(\mathrm{trigonal})+HOR'$$

Reaction rates of many alkyl carboxylates and their esters were measured in four water-soluble solvents (acetone, methanol, ethanol, dioxane) at 25 °C and the enthalpy change ΔH obtained.

To determine the contribution of the steric effect *per se*, one needs to isolate it from other substituent electromagnetic influences, which equally impinge on reaction rate. There are smart ways to neutralize selectively these influences (reader: think of what you would do to achieve this goal before checking the following set of strategies).

These are:

a) **Solvent contribution**, by comparison of the kinetics of two similar reactions. The solvation influence is expected to be small in bimolecular interaction with little charge development. Running kinetic measurements in two or more solvents of varying polarity should reveal solvent influence on reaction rate.
b) **The bonding effect** wrapped by the ρ*σ* factor of Equation 3.2 can be evaluated once *Es* values are known from a comparable set of reactions.
c) **Resonance effects** can be reduced to zero using saturated substituents to block conjugation, and having the same atoms bonded to the reaction center.

After gathering the critically analyzed published data obtained from similar hydrolysis/esterifications of earlier researchers in addition to his own observations, Professor DeTar calculated E_s from the rate data using Equation 3.3 *for each alkyl substituent*:

$$E_s = a + b \log k \tag{3.3}$$

This is the math expression of a straight line of intercept a and slope b and is statistically validated by deviations from linearity according to the correlation coefficient (r^2) and the sum of the differences from the experimental points and the best-fit correlation line. In addition, the standard deviation (SD) of the calculated E_s in a number of replicates (four to five in this case, for each reaction) was taken into account for reproducibility.

DeTar reported *excellent values for all these parameters strengthening* E_s thus obtained. After averaging the validated results, the quantitative scale of new steric effects shown in Figure III.5 was developed [20]. It is relevant to note that the abscissa axis of this plot is, once more, categorical: the *expected* steric hindrance (small, regular, so-so, big, bigger, biggest of the series). The actual value of E_s of an alkyl series and others previously determined for similar hydrolysis by researches Taft in 1956 [11] and MacPhcc in 1980 [12] arc availablc.

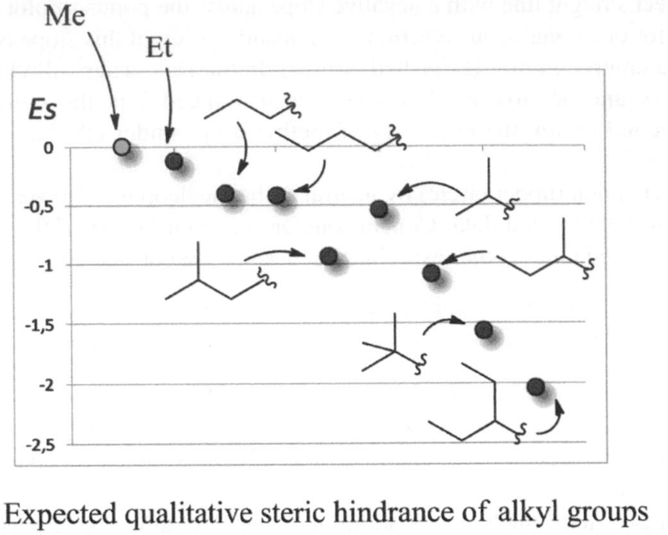

FIGURE III.5 Exploration plot by author Alonso-Amelot of steric effects E_s of the first series of nine common alkyl substituents obtained from the rate constants of acid and base catalyzed hydrolysis/esterifications of a series of R–COOR' (R' = alkyl, H, respectively) Adapted from [19]. The E_s of R = methyl (blue dot) was used as reference. Note that the abscissa is not numerical but categorical; hence, any approximation to a linear correlation is meaningless and shows a qualitative trend only.

Lessons from Study Case 2 and some shortcomings:

1. DeTar's work included many rate constants and *Es* from a variety of published and validated sources of RCOOR' hydrolysis or esterifications of the corresponding carboxylic acids. Some of the data were outlier points (away from the plots correlations or computations) and had to be discarded. Explanations for these abnormalities were not substantiated.
2. Deviations of DeTar E_s values from Taft's [11] and MacPhee's [21] were observed in some commonly found alkyl groups, including *n*-propyl, *n*-butyl, and isopropyl. However, these differences did not invalidate the general sequence.
3. The increasing steric impact on the four monitored reactions expressed as the steric factor E_s (Figure III.5) is reasonably compatible with the *visual* assessment mentioned at the beginning of this section. This fact contributes to place the E_s quantification in our comfort zone as problem solvers of organic reaction mechanisms and are *intuitively* acceptable.
4. At any rate, if electromagnetic forces (polar, resonance, field), which in the case of alkyl substituents can be of minor importance relative to steric impediment, they constitute a major factor in nucleophiles or electrophiles bearing electron attractor/withdrawer groups near the nucleophilic/electrophilic center, a common occurrence.
5. In spite of DeTar's advances and those of others, the scientific community of the early 1980s was not fully satisfied, and further studies continued to appear, seeking to revise, expand, and improve our understanding of the origins of the E_s. Besides, DeTar numbers cover only a small fraction of alkyl substituents, were obtained from a very narrow set of addition–elimination reactions in very simple molecular substrates and have not been accepted universally.
6. The E_s estimation role in increasingly complex molecules as those found in contemporary organic chemistry, materials science, chemicals-by-design, and so forth, *constitute an unquantifiable factor*. This is so because the bulky substituents influence the molecular scaffolds in various ways, in addition to the electromagnetic sway. Among others, steric forces lead to twisting molecular planes – see Figure III.3 – that decrease (or enhance) conjugation affecting reaction centers in other sections of the molecule, modify the equilibrium between two or more conformations, induce regio-, stereo-, and enantioselectivities, and can also exert their control from a different ancillary molecule, a ligand of organometallic complex, or purely organic catalyst as many examples demonstrate – see Scheme III.1.
7. The plots of Figures III.4 and III.5 are semiquantitative correlations only. Physical organic chemistry, however, requires *robust quantitative correlations* between two numeric variables from separate sources to establish whether they are linearly associated to statistical satisfaction, they are supported by one or more common factors, and have mechanistic meaning.

These considerations boil down to finding a counterpart of E_s, either experimental or computed, related to substituent size, which may be contrasted with the steric factor furnishing linear associations with high correlation coefficients.

The following study case focuses precisely on this objective.

3.3.4.3 Study Case 3: Connolly's Molecular Volume In addition to the physical or computer-generated molecular models a few descriptors of numerical expression have emerged as an aid to appreciate key features. The most common descriptors are:

- ✓ Solvent accessible molecular area (SAMA)
- ✓ Solvent exclusion molecular area (SEMA)
- ✓ Solvent excluded molecular volume (SEMV) or solvent exclusion volume (SEV)
- ✓ Molecular ovality (MOv)

3.3.4.3.1 Measuring Surface Area and Volume of Molecules or Molecular Sections

-A Story of SAMA, SEMA, and SEV

Over 30 years ago, Professor Michael Connolly in California presented a computation algorithm to measure both area and volume of low and high molecular weight (e.g. proteins) compounds based on a simple and ingenious concept [22]. Think of a static collection of atoms forming a molecule as a set of hard spherical bearings, each with a size corresponding to van der Waals radii (vdWR) of every atom. A solvent molecule (water), assumed to be a sphere 1.5 Å vdWR, is rolled over the entire surface of the tested molecule so it never passes twice over the same spot.

Connolly's algorithm follows the solvent ball course and converts it into a calculated surface in two forms:

1. The SAMA measured in square angstroms ($Å^2$).
2. The SEMA described by the virtual contact of the water sphere with the set of rigid bearings representing the molecule.

From the solvent trajectory in 3D, the SEV is calculated ($Å^3$). Here, exclusion means that the molecular body not accessible to the water sphere surface: crevices and interstices of the molecule. In turn, this defines a molecular surface with protrusions and depressions whose area and contained volume constitute SEMA and SEV.

The method is accurate within a 0.001% error and is sensitive to conformational differences. The top section of Figure III.8 portrays the concept.

SEV is a major intuitive (thus attractive) and quantifiable component of steric hindrance, regardless of stereoelectronic influences. An additional advantage is that SEV can also be split into molecular sections suspected of wielding a major steric effect.

Professor Connolly published computer programs in the C language to calculate SAMA, SEMA, and SEV that have been adopted in a variety of molecular handling softwares accessible to non-computer specialist organic chemists [23]. This software can be retrieved or accessed online to perform these calculations reliably. It is like measuring the space occupied by a knot of high voltage electric wires, independently of whether there is electric current in the wires (electrons in bonds, resonance) or not.

The lower section of Figure III.6 collects SEVs of primary and secondary carbinols used as testing esterification agents of carboxylic acid kinetic studies described in the preceding case and with several other substrates. Predictably, SEVs increase with molecular weight and branching. *tert*-Butyl alcohol falls off the trend in my application of Connolly's method, but the SEVs of methyl-isopropyl and methyl-*tert*-butyl ethers (two last entries in Figure III.6) account correctly for the extra methyl group of isopropyl versus *tert*-butyl (16.2–17.9 $Å^3$) recorded when comparing ethanol and isopropanol.

3.3.4.3.2 SAMA, SEMA, and SEV: Definitions in Practical Terms **SAMA:** Connolly's algorithm calculates the surface described by *the center* of the water rolling ball. This surface is 1.5 Å above the actual surface described by the vdWr of nuclei. What you get is the solvent accessible area SAMA in $Å^2$. In regards to the molecular volume, it includes a 1.5 Å thick wrapping mantle.

SEMA is obtained with the same procedure except that it is the area measured by the point where the *solvent ball touches the molecular van der Waals model*. Theoretically the solvent (if it was water) cannot get any closer to the molecule; hence, SEMA is the smallest area accessible to it. SEMA is also measured in $Å^2$.

SEV is obtained likewise, representing the volume ($Å^3$) of the enclosed SEMA. As said, the 1.5 Å pebble cannot get into crevices and interstices of the space-filling model, which are inaccessible as well to external reagents unless the kinetic energy of the incoming reagent overcomes the electrostatic repulsion offered by the van der Walls balls protecting the depression.

Here, we add another parameter related to molecular 3D shape: MOv.

MOv (molecular ovality) comes from the oval shape of the majority of organic molecules as the name suggests. MOv is the ratio of the surface of this oval and of a sphere of the same volume.

Table III.2 puts together relevant descriptors of a few cyclohexyl derivatives. Keep in mind that steric descriptors are sensitive to different configurations and conformations. Predictably, they are also roughly proportional to molecular weight (MW). In addition, compounds with equal MW may have different SAMA, SEMA, SEV, or MOv, as Table III.2 illustrates. Also note the SEMA and SEV differences of *i*-butanol and *t*-butanol, reflecting a more compact configuration of the latter while both have the same MW.

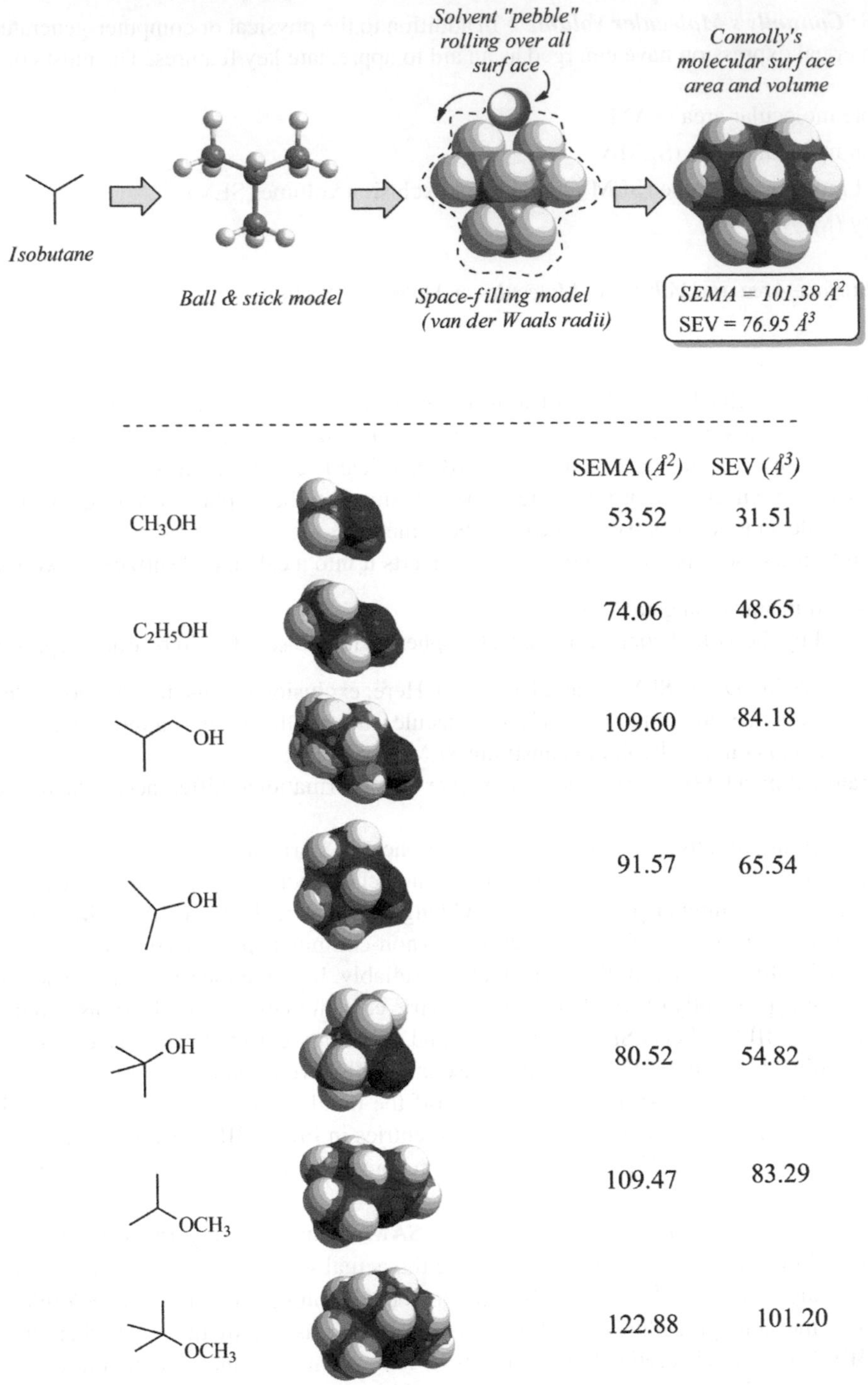

FIGURE III.6 (Connolly method: Isobutane) Schematic illustration of Connolly's algorithm to determine SEMA and SEV of molecules in Å^2 and Å^3 units, respectively. The blue envelope in the space-filling model at right is a visual representation of the calculated surface where the probe solvent ball would not be able to enter. Connolly's SEMA and SEV values are shown for a few carbinols and encumbered ethers (calculated through application of Connolly's computation algorithm, Alonso-Amelot, this book).

SAMA, SEMA, and SEV, and MOv are suitable numeric parameters defining general features of molecular shape. In addition to providing clues related to stereochemistry's role in reaction mechanism conditioning.

Connolly's SEVs can also be tested in several other series. For example, by fixing the alkyl group while changing the halide from fluorine to iodine, alongside other fundamental parameters related to size (Table III.3). The SEV of methyl iodide is twice that of methane, underpinning the steric effect of iodine.

TABLE III.2 Example of selected steric descriptors. For acronym definitions see text. Green ovals emphasize the general oval shape of each molecule to illustrate the molecular ovality (MOv). Calculations: Alonso-Amelot, this book

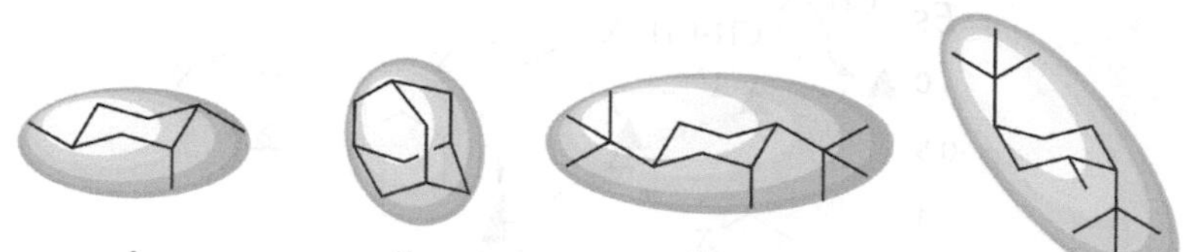

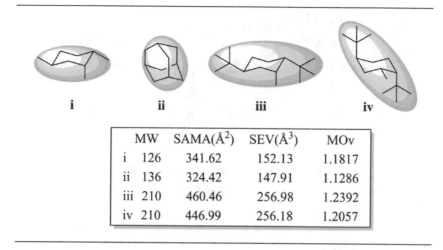

	MW	SAMA(Å^2)	SEV(Å^3)	MOv
i	126	341.62	152.13	1.1817
ii	136	324.42	147.91	1.1286
iii	210	460.46	256.98	1.2392
iv	210	446.99	256.18	1.2057

TABLE III.3 Selected dimensions, including C–H and C–halide bond distance, Connolly's SEV (Å^3) of the methyl halide series calculated after MMFF94 energy minimization (Alonso-Amelot, this book)

Compound	C–H (Å)	C–halide (Å)	SEV (Å^3)
CH_4	1.113	–	24.339
CH_3F	1.103	1.362	27.439
CH_3Cl	1.108	1.780	38.631
CH_3Br	1.111	1.943	44.394
CH_3I	1.113	2.149	52.047

3.3.4.3.3 Testing SEV in a Kinetic Reality Having access to SAMA, SEMA, and SEV steric descriptors, one can address the question of whether there is any relationship between SEVs and reaction rates in bimolecular events. I have been unable to find any such work in the literature, but we can explore this putative correlation by selecting published kinetic data of bimolecular reactions from a systematically designed series of nucleophiles according to their steric hindrance and explore correlations with Connolly's parameters calculated herein. The selected reaction is the hydrolysis of esters.

Our first goal is to calculate SEVs of the alkyl substituents of carboxylic esters described in the previous section [19, 20] (Figure III.5) and plot SEVs against Taft's E_s [11]. Results are portrayed in Figure III.7.

Interpretation of Figure III.7:

The analysis of plot **A** in Figure III.7 allows one to contend that:

1. Generally speaking, E_s and SEV appear correlated. The trend of steric hindrance exerted by alkyl groups announced by E_s is roughly accompanied by computed SEVs based only on the molecular volume. Therefore, the DeTar study succeeded in accounting for the electromagnetic contribution of the examined alkyl groups and isolating it from the assessment of the steric hindrance parameter E_s.
2. In the ester hydrolysis/esterification sequence (plot **A**), all but one data point are associated reasonably well by an exponential expression (after putting aside the starkly deviated *n*-pentanoic methyl ester point, dashed circle). Considering that E_s has negative values as these are obtained from log(*k*) of increasingly hindered reactions – hence, negative log numbers – the exponential correlation suggests that SEV impacts with increasing severity the course of the hydrolysis/esterification reaction rates.
3. The outlier point of *n*-pentanoic methyl ester can be interpreted as an overrating of SEV relative to the steric hindrance demonstrated by the measured reduction of the reaction rate (reflected in E_s). The increase in molecular volume of the *n*-carbon chain *once passed three carbons in a straight chain does not further affect the rate*, because the growing molecular appendage of the substituent is located further away from the reaction center to have a proportional steric influence

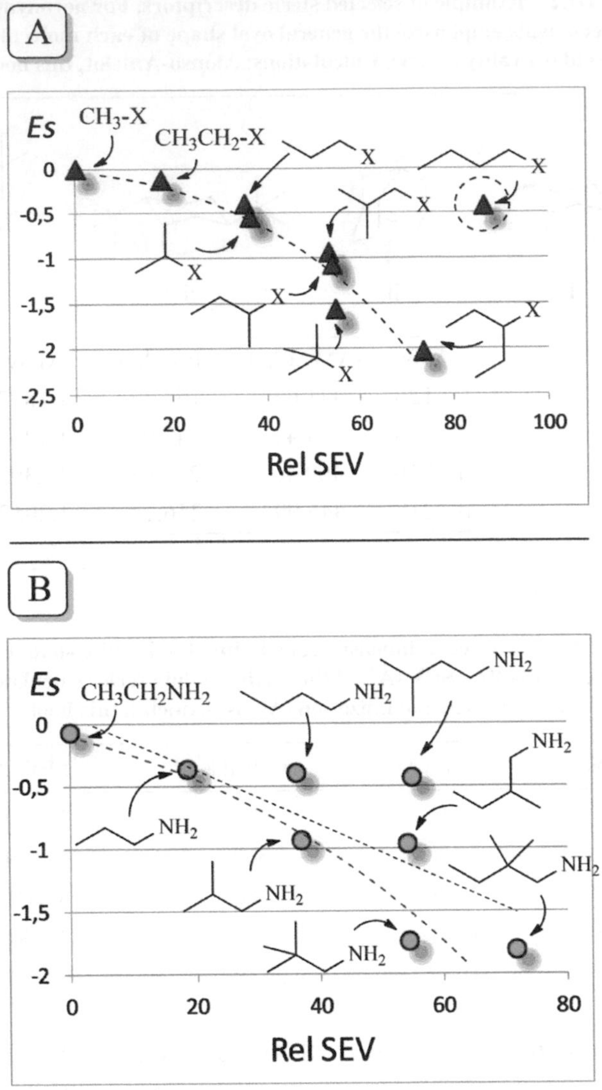

FIGURE III.7 Attempts to correlate the steric factor E_s and the SEV in two different reactions have in common a bimolecular mechanism. Plot **A**: Data from the hydrolysis/esterification of RCO_2R' (R' = alkyl or H, respectively) of the same series of R-substituted esters and carboxylic acids, [19]. Plot **B**: Data from the S_N2 amination of allyl bromide by a series of primary amines [20]. Connolly's SEV values in both plots are relative to methyl acetate (plot **A**) and ethylamine (plot **B**, the smallest of each series). SEVs were computed (Alonso-Amelot, this book) for the all the molecules of interest. Dotted lines represent adjusted curve fitting to validated data. Adapted from [19]; and Adapted from [20].

on the incoming nucleophile. This is also compatible with the aliphatic chain conformational flexibility responding to electronic self-repulsion of the chain methylenes. Were it a little flexible, cyclopentane this conformational escape would not be possible.

4. The data points on or close to the correlation line beyond the SEV value of the *n*-butyl ester/carboxylate are branched at the α carbon and the inhibition effect is very substantial. A similar influence is noticed in substituents on β carbon atoms, like isopentanoic ester/acid. However, having only one data point, no trend can be envisioned, of course.

Plot B of Figure III.7 **is far less consistent**. Note that it portrays a different reaction, an S_N2 amination of allyl bromide to yield alkyl-allyl amines E_s and SEV follows a very coarse general trend in the same direction.

1. *However*, tracing a linear correlation such as the black dashed line shown in Figure III.7-**B** is more a daring exercise of imagination than anything statistically meaningful (see Note 3). The correlation factor of the whole set of data, $r^2 = 0.609$, is just too poor.

2. Despite the data points of the linear *n*-propyl, and β branched amines isobutyl, and isopentyl derivatives, and 2,2-dimethylbutylamine (bottom right) might appear associated with the exponential red line quite nicely, such appraisal would leave three amines (*n*-butyl-, isopentyl-, and 2-methylbutylamine), nearly 40% of the sample, as orphan data points; too many for any validation pursuant of quantitative assessment.
3. Author DeTar concedes that E_s values of *sec*-butyl and isobutylamine, one of the grossly deviated points in *my* plot, were interpolated from the correlation of Taft's steric constants [20]. One cannot surmise whether this practice affects the experimental correlation.
4. Conceivably, the allyl moiety of the receptor electrophile interacts with the incoming alkyl amine attendant on variations in the structure of its alkyl chain related to branching and preferred conformation in the TS, not only on SEV.

Amines are electronically active nucleophiles that may accommodate bimolecular substitution behavior in models other than S_N2. Study Case 4 describes such a study with an ample series of secondary amines.

3.3.4.4 Study Case 4: The Anilines Arylsulfonyl Chloride Model We move on nearly four decades ahead of DeTar's work. In a recently published study (2017), researchers in Ivanovo, Russian Federation [24] exposed a series of *N*-alkylanilines to an arylsulfonyl chloride and studied in detail the kinetics in tetrahydrofuran (THF) and other polar organic solvents, pursuant to assessing solvent participation in the TS of the slow step Scheme III.5.

The *m*-nitro substituent enhanced the reactivity of the sulfonyl group **38** as an electrophilic receptor. In principle, this is a bimolecular substitution passing through transition state TS_{38+39} subjected to a governing influence of the steric effect of the N–R substituents of the studied anilines **39a–h**. Given that all variable Rs are aliphatic, their electronic influence should be limited to a moderate *+I* inductive effects. I disregarded the aryl group in the calculation of the solvent exclusion molecular volume SEV as a common factor to all amines. Selected results for anilines **39a–h** are collected in Figure III.8.

Figure III.8 *highlights*:

1. Predictably, a negative linear correlation exists between *k* and SEV for most of the alkyl nucleophiles tested (red points). Only two amines (blue points) fell off the trend line.
2. Elongation of the *n*-alkyl chain reduces *k* significantly (**39a,b,d**). But it does not follow a consistent pattern, when broken abruptly by **39c**. Notably, elongation by one β-methyl (**39c**) has a greater impact on reducing *k* than lengthening the *n*-aliphatic chain to four carbons (**39d**).
3. Branching at the β carbon pulls the trend off the line to lower *k* values (blue points) that could not be predicted by the corresponding SEVs (**39c, 39e**). The rate constant of anilines **39e, 39g,** and **39h** respond also to branching of the β carbon.
4. Comparison of **39d** and **39e** have nearly the same SEV (~89 Å^3) and yet $k_{39d}/k_{39e}=3.96$. This implies that *SEV alone cannot predict kinetic behavior given major impact of substituents on the α and β carbons.*

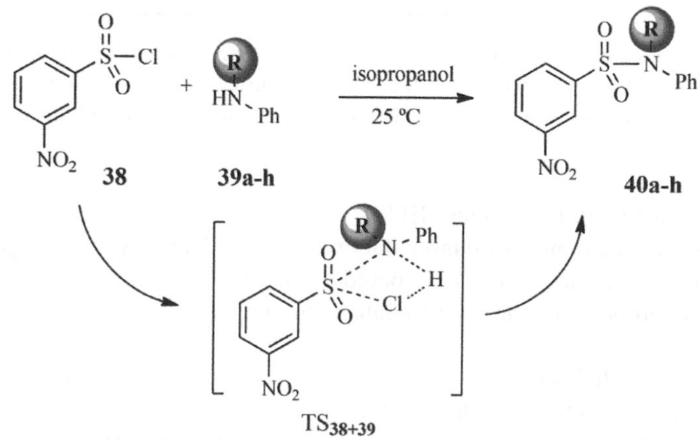

SCHEME III.5 Reaction of *m*-nitro-arenesulfonyl chloride and secondary amines showing the expected transition state (TS_{25-26}) where nucleophile, leaving group and substrate form a congested activated complex. Specific amines are shown in **Figure III.8** (Kochetova), top section. [24] / Springer Nature.

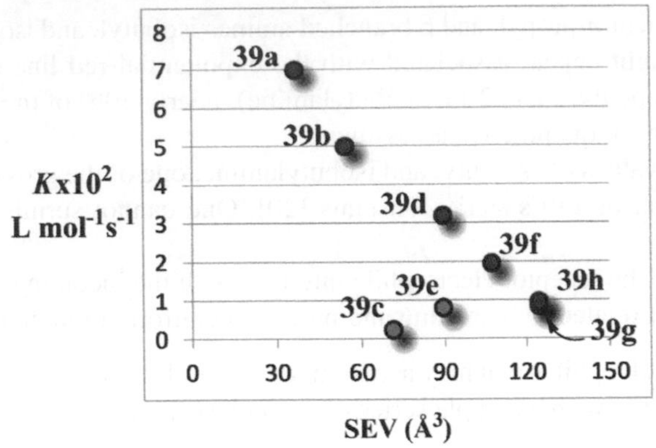

DATA:

Nº	Amine	$K \times 10^2$ L mol^{-1}s^{-1}	SEV Å^3
39a	MeNHPh	7.03	36.31
39b	EtNHPh	5.02	53.72
39c	 NHPh	0.25	71.31
39d	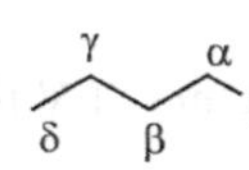 NHPh	3.21	88.87
39e	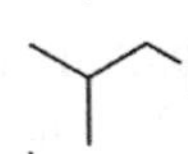 NHPh	0.81	89.51
39f	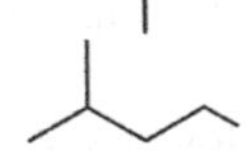 NHPh	1.98	106.86
39g	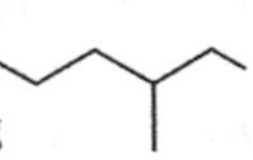 NHPh	0.86	124.08
39h	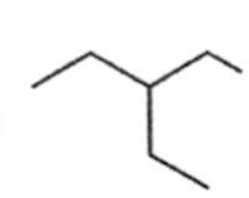 NHPh	0.98	124.07

FIGURE III.8 Second-order rate constants (×10^2) of the amination of *m*-nitroarylsylfonyl chloride (**38**) (Scheme III.5) with a series of *N*-alkyl anilines **39a–h** in isopropanol at 25 °C, contrasted with the solvent exclusion molecular volume SEV (Å^3). SEVs were calculated by Alonso-Amelot, this book, using Connolly's algorithm [22], and assuming a measuring solvent sphere of 1.4 Å. Data from [24].

Are there electronic effects of the alkyl groups in Figure III.8*?*

Alkyl groups are well-known electron donors through inductive effects (*+I*) and hyperconjugation in addition to the recently introduced concept of quantum effects. These effects are based on the electron density changes of carbon atoms induced by alkyl appendages [25, 26]. These properties contribute to enhancing the amine nucleophilicity and constitute a confounding factor for the *k*/SEV relationship

The strategy to elucidate this primordial question was the classical grafting of the alkyl groups on the aromatic ring of aniline in the meta and para positions, whereby the steric influence on the transition state TS_{38+39} was expected to be nil [24]. The reaction rate was measured using a primary (*p*/*m*–R–Ar–NH_2) and a bulkier secondary (R–Ar–NH–*i*–Bu) amine. Their results are condensed in Table III.4.

Takeaways from Table III.4:

1. All secondary anilines were less reactive than primary anilines by a factor of 22 to 36 (compare columns three and four), regardless of the enhanced electronic *N*-nucleophilicity induced by the isobutyl group. Thus, steric hindrance was the major controlling factor represented in this case by the *N*-isobutyl group, in line with the general nucleophilic order primary > secondary ≫ tertiary pattern of substituents (with exceptions).

TABLE III.4 Second-order rate constants of the reaction of Scheme III.5 using a series of 3- and 4-substituted aniline derivatives. Electron donors (entries 1–4) and withdrawers (entries 5–7) were monitored. Aniline and isobutylaniline served as standards of comparison. Data from [24]

Entry	R–Ar–NHR'	$k \times 10^2$(*) R' : isoBu	$k \times 10^2$ (*) R' : H	k relative to H–Ar
1	4-CH_3–Ar	3.48	93.7	3.21
2	4-Et–Ar	3.38	76.2	2.61
3	3-CH_3–Ar	1.69	45.3	1.55
4	3-Et–Ar	1.54	47.6	1.63
5	3-CH_3O_2CAr	0.43	3.83	0.13
6	4-CH_3O_2CAr	0.22	0.51	0.02
7	4-EtO_2CAr	0.16	0.4	0.01
8	H–Ar	0.81	29.2	1.00

(*) k units = $\mathrm{L\,mol^{-1}\,s^{-1}}$

2. Table III.4 clearly separates two sets of substituents influencing rate constants k: (a) electron donors (entries 1–4) and (b) electron withdrawers (entries 5–7). Group (a) boosts k well above the unsubstituted phenyl (entry 8) owing to inductive electron donation from alkyl groups across the phenyl ring to the nucleophilic center N. Predictably, strong electron withdrawers in the (b) group impact k in the opposite direction.
3. One can also anticipate that methyl and ethyl groups in the meta carbon do not offer steric hindrance, as the k of entries 3 and 4 of the primary and secondary amines (columns three and four) are very similar. The S_N2 rate enhancement in meta is nonetheless significantly *less* pronounced than in the para position, despite that the electron density donation (+*I*) is usually more strongly felt from meta than para substituents.

This first contradiction suggests the involvement of other factors affecting the proposed TS depicted in Scheme III.5, a feature that should not be ignored by the proficient problem solver. Indeed, authors [24] report two additional observations. In short:

a) *The reaction constant* k *is influenced strongly by the solvent polarity*. When water is added (from 20 to 30% by weight) to the organic solvent (dioxane) the kinetic parameters change substantially: k increases from 5.1 ± 0.3 to $17.4 \pm 0.9\,\mathrm{L\,mol^{-1}\,s^{-1}}$; the enthalpy of activation $\Delta H^\ddagger$ decreases from 10.28 to 7.89 kcal mol^{-1}; and the entropy of activation $\Delta S^\ddagger$ is more negative, from −19.59 to −35.85 kcal mol^{-1} K^{-1} *implying a reduction in the degrees of freedom caused by partial bonding to water-solvent in the TS*.
b) The reaction constant k also varied with the solvent employed: isopropanol, THF, dioxane, or dimethyl sulfoxide (DMSO), in the presence or absence of water. After detailed theoretical calculations, authors contend that the *alkyl anilines in such solutions create water-aniline-solvent complexes that participate in facilitating the encounter with arenesulfonyl chloride as these build the TS en route to product* **27** [24]. Both observations are consistent with the organization of a water-solvent complex along the reaction coordinate.

Once again, however, *the specific contribution of steric effects by themselves to the reaction rate, independently from electronic perturbation of the reaction center, continued to be elusive except for the manifest influence of primary versus secondary nucleophile and the* α *and* β *carbon ramification.*

3.3.4.5 Study Case 5: Other Sources of Evidence After nearly a century of systematic exploration of the steric effect, the flow of experiments to clear the subject continues to this day, providing new evidence of use to the knowledgeable mechanism problem solver.

New insights arise from Banert and Seyfert at Chemnitz, Germany in 2019 [27] who addressed the reaction of allenyl isothiocyanate (**41**) and increasingly sterically encumbered dialkylamines **42a–k** (Figure III.9) furnishing aminothiazoles of type **43** (Scheme III.6).

The experiments required a great deal of sophistication relative to earlier research. Allenyl isocyanates are extremely reactive relative to alkyl and aryl isothiocyanates. These compounds form intractable polymers in the absence of the amine or other nucleophiles and require specialized manipulations in their synthesis: vacuum flash pyrolysis or vapor spray at high temperature followed by rapid condensation of products at low temperature. The isothiocyanates thus produced are exposed to a very

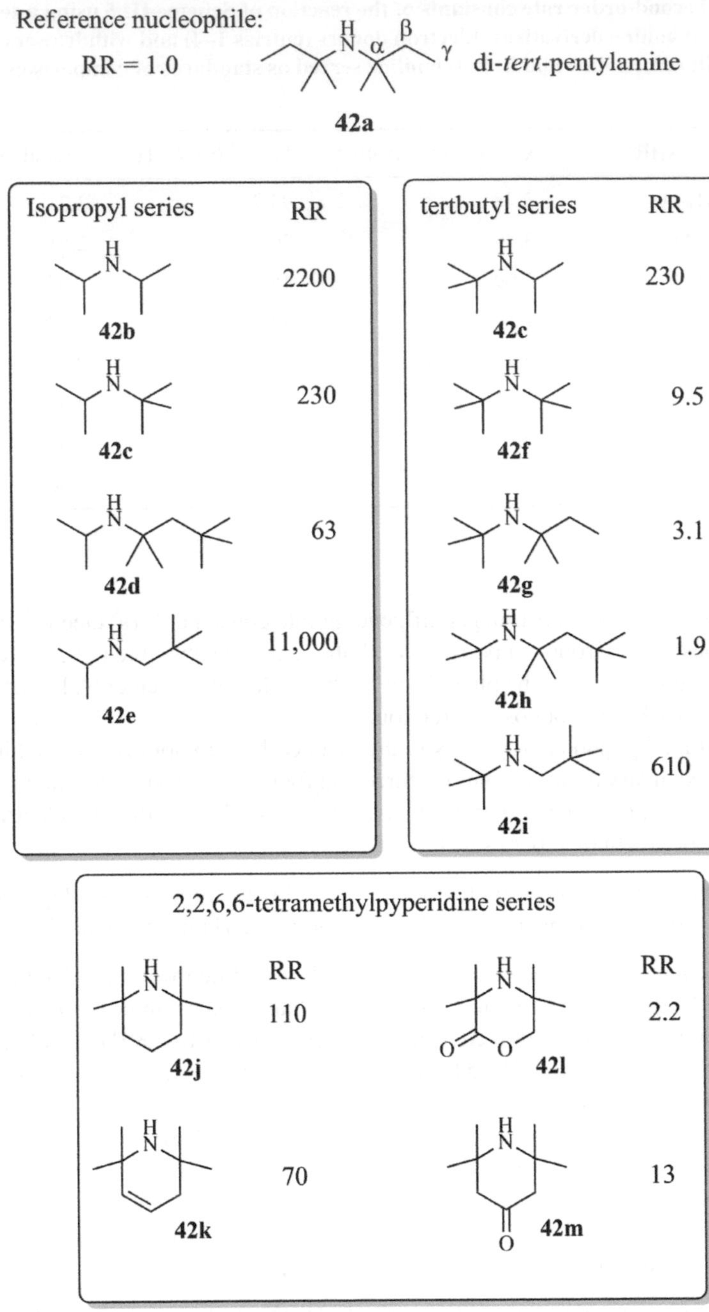

FIGURE III.9 Relative reaction rate constants (RR) of several secondary amines recorded in the addition-cyclization of allenic isothiocyanate **41** (see Scheme III.6) in tetrahydrofuran. Di-*t*-pentylamine was used as standard of comparison. Data is organized according to common substituents, isopropyl, *t*-butyl, and alicyclic by permission with [27] / Royal Society of Chemistry.

hot environment for a very short period only. Then, the cold material is diluted in solvent and converted smoothly in a few minutes to the heterocycles even with poorly nucleophilic amines.

The temperature range for the kinetic measurements was 0 °C to room temperature. When amines were very bulky though, the reaction took several days at moderate temperature (30 °C), suggesting from the start a strong dependence of the reaction kinetics on steric encumbrance of the nucleophile.

To correlate the steric bulk of a given reactant and reaction rate, the most likely mechanism must be elucidated first from other sources of evidence or resort to an educated guess. The lower panel of Scheme III.6 describes my line of reasoning that

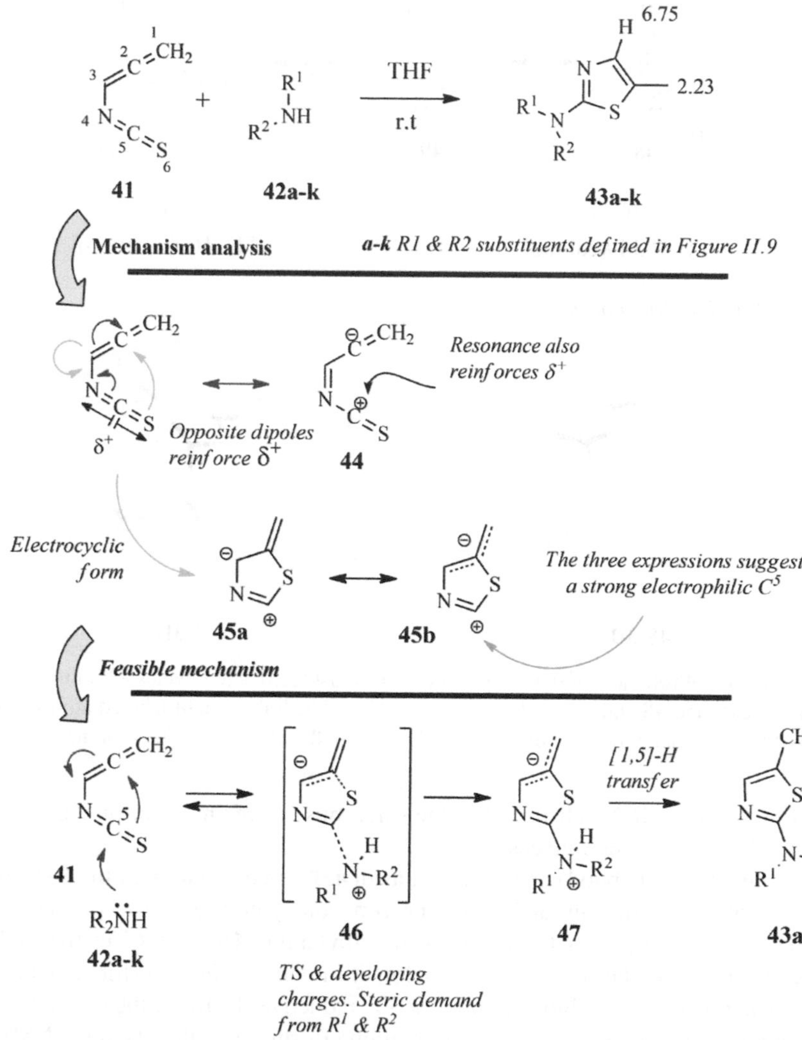

SCHEME III.6 Top section: Nucleophilic addition-cyclization of allenyl isothiocyanates with secondary amines of increasing steric bulkiness. Numbers in adduct **43** are key ^1H NMR signals for the dimethylamine derivative (R^1=R^2=CH_3) reported by authors for kinetic measurements [27]. Lower section: mechanistic reasoning and proposals. Atom numbers in **41** are arbitrary to facilitate discussion. [27] / Royal Society of Chemistry.

leads to the primary electrophilic character of the isothiocyanate carbon C^5 flanked by N and S. Polar effects and resonance (red arrows) converge to flag C^5 as the most favorable for nucleophilic attack. Reversible cyclization of **41** to zwitterion **45a–b**, stabilized by delocalization and charge separation, buttresses the electrophilic character of C^5.

The approach of a sec-amine (third panel) evolves to a likely TS with ample charge separation while R^1 and R^2–N substituents resist the consolidation of TS and subsequent collapse to adducts of type **43**. Therefore, this mechanism anticipates a dependence of the rate of reaction primarily in accordance with the steric hindrance of the amines. Assuming a nucleophilic addition of amines to C^5, the expected TS (**46**) anticipates a bimolecular process governing the reaction and hence, a second-order reaction rate susceptible to the steric effect of the N substituents.

Banert and Seifert [27] carried out a long series of **41** + **42a–k** additions monitoring the formation of adducts of type **43** by the C^4 methyl singlet at δ 2.23 ppm in the ^1H–NMR spectrum. This signal was shifted gradually downfield to 2.40 ppm as the amine steric encumbrance increased. Similarly, the chemical shift of the C^3–H underwent a substantial shift from δ 6.75 to 7.19 ppm with increasing steric effects of amines.

Can you explain this? Please give it a try and then carry on reading.

Figure III.10 provides feasible answers. In the less hindered nucleophiles, the AO of the non-bonding pair of the amine N atom in adduct **48** is nearly parallel to the π MO of the aromatic aminothiazole nucleus (see model **48-3D**). Unimpeded

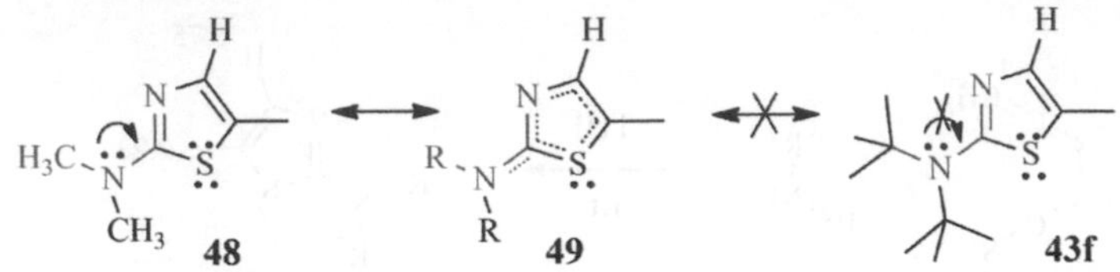

48 **49** **43f**

DA: 1.6º
TE: 0.34 *k*cal mol^{-1}

DA: 60.6º
TE: 10.4 kcal mol^{-1}

3-D molecular models

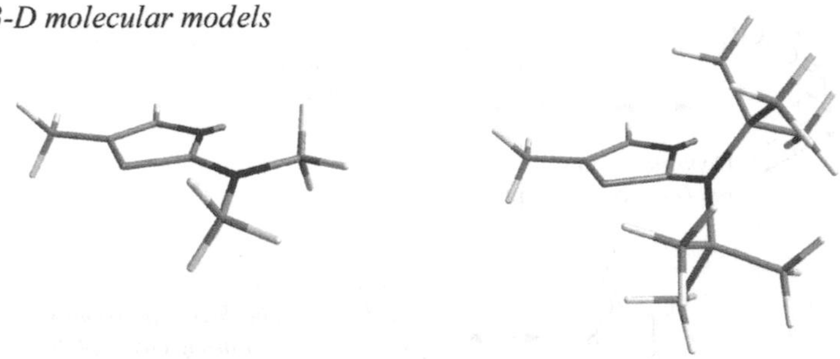

48-3D **43f-3D**

FIGURE III.10 Energy minimization of rotamers of sterically demanding substituted heterocycles **43f** of the Banert and Seyfert series [27] / Royal Society of Chemistry (see text) and **48** and selected parameters: DA = dihedral angle of labeled atoms (red); TE = torsional energy. Results suggest that resonance of the external N non-bonding pair is allowed in **48**, whereas it is predominantly inhibited in **43f**.

resonance allows electron flow from the R_2N substituent to the ring, thus enhancing electron density of the ring π current. A shielding effect over atoms on the plane of the ring ensues.

By contrast, placing large substituents (*t*-butyl) on the amine, as in **43f,** forces rotation of the R_2N-group, throwing the non-bonding pair *p* AO out of the plane of the ring and inhibiting *p*-π conjugation as steric pressure increases. The dihedral C–N–C=N angle (out of the ring plane) passes from 1.6º in dimethylamine-**48** to 60.6º in the di-*N*-*t*-butyl derivative **43f** according to MMFF94s minimization calculations (Alonso-Amelot, this book). At the same time, torsional energy also increases substantially (Figure III.10). The resonance inhibition allows the –*I* (inductive) factor of the amine to take effect, withdrawing electron density from the aromatic ring and therefore, field deshielding of ring substituents in the NMR spectrum.

Despite these considerations, the ^{1}H-NMR or ^{13}C-NMR shifts were not linearly correlated with a quantifiable measure of the substituent steric demand of adducts. Instead, the integration of the methyl singlet relative to a standard in the mixture was employed as analytical criterion for the variation in concentration of adducts **30a–k** with time in competition kinetic runs using equimolar amounts of two amines in the same reaction flask.

Yields were good in most cases (80–90%), reactions were finished within 10 minutes at 0°C but had to be extended up to 9–10 hours when very bulky amines were employed. Rate retardation by steric influence is obvious. Figure III.9 provides amine structures and second-order rates relative to di-*tert*-pentylamine **42a**, the apparently most congested nucleophile of the lot. All employed amines were secondary and could be grouped in three classes (framed groups in Figure III.9): symmetric, asymmetric, and aliphatic alkyl substituents versus alicyclic amines. Reasons for this design are now explained.

Interpreting the results:

At the start:

1. All amines furnished heterocycles of type **43** without bothersome side products, paving the way to brilliant conclusions.
2. Two aspects, (a) and (b), suggest a strong intervention of steric congestion.
3. A third aspect, (c), provided insights into the pervading electronic effects.

a) *The reaction rates were very sensitive to the enlargement of the alkyl substituents*. The data of Figure III.9 (R_2NH steric effect) [27] are organized in three blocks that show structures and the reaction rates *k* relative to k_o of sterically encumbered amine **42a** shown at the top. The upper left group of the isopropyl amine series shows a sharply decreasing reactivity in the first three compounds. This ought to be the result of the additional bulkiness of the di-*tert*-pentylamine **42a**. The increasing volume in the series **42b** < **42c** < **42d** is compatible to the recorded sharp rate decrease.

The rate declining sequence in the second block of Figure III.9 (R_2NH steric effect) follows the same trend with the additional feature of a generally lower k range relative to the first block. This is in line with the larger steric impediment of the *t*-butyl substituent, common to all amines of this group.

Surprisingly, however, the introduction of a neopentyl substituent, namely a *tert*-butyl group on the α carbon of **42e** and **42i**, which is equivalent to saying that three methyl groups occupy the β carbon of the chain, was not hampering the reaction but actually pushing it well over three orders of magnitude *greater* than the standard in the isopropyl group (RR = 11 000) and 610 times faster in the *tert*-butyl set than the already bulky standard and about 5 and 2.7 times, respectively; a case of steric acceleration perhaps?

This observation is not only contrary to computed Connolly's SEVs (160.85 Å^3 of **42e;** 124.85 Å^3 of **42b**) and to visual perception of substituent molecular volume, but also in sharp contradiction with the case of electrophiles (e.g. alkyl halides) in S_N2 reactions, wherein the neopentyl substitution hinders the reaction much more effectively than isopropyl substituents. Beta-elimination becomes a highly competitive reaction in these cases.

b) In highly sterically hindered adducts like **43h,** *attempts to purify them by silica gel chromatography failed; instead, products* **50** *and* **51** *were isolated from* ***43h*** (Scheme III.7) [27]. This is an indication of excess steric compression, which leads to a thermodynamically controlled release of the bulky group.

Banerts and Seyferth [27] do not venture a *mechanism* but I will momentarily: The densely structured adduct **43h** contains a methyl hydrogen of the N chain and a thioimidazole N with an NBP on a *p* AO orthogonal to the ring's aromatic π MO, hence localized and highly basic, at six atoms distance. This is suitable for proton elimination through a thermally allowed [1,5]-H shift shown in **52** with liberation of the sterically demanding alkyl chain as terminal alkene **50**.

c) *The question of electronic effects in the alkyl groups and the alicyclic series*: The third block in Figure III.9 shows the nucleophilic rate (RR) relative to the standard **42a** of six-membered N heterocycles on the reaction of Scheme III.6. In principle, the stereochemical impact of the fully saturated 2,2,6,6-tetramethyl piperidine **42J** should be comparable with di-*tert*-butylamine **42f**, from the *+I* influence of α carbons, but it is not. There is a 11.6-fold increase in $k_{(\mathbf{42J})}$ relative to $k_{(\mathbf{42f})}$. This increment is likely due to:

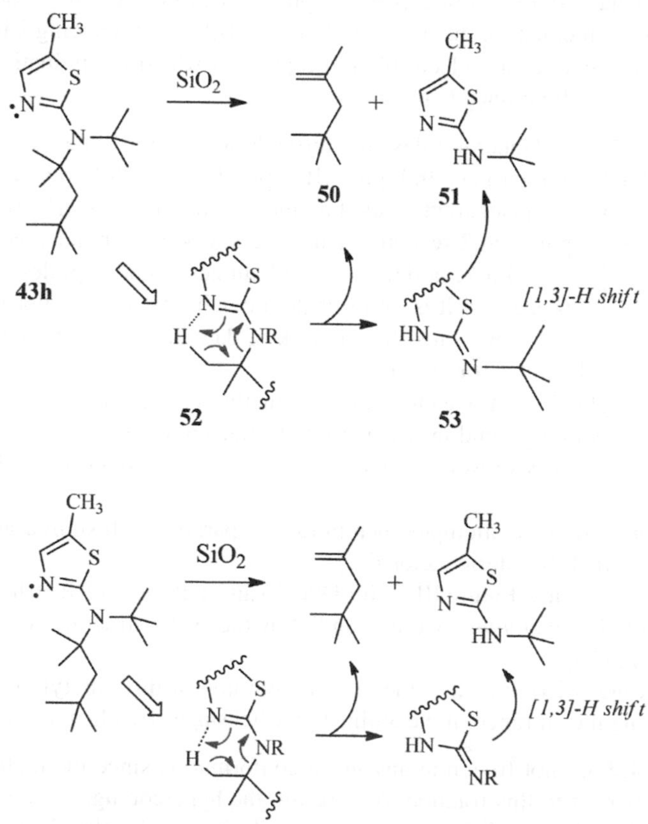

SCHEME III.7 Fragmentation of sterically encumbered adduct **43h** upon contact with silica gel at room temperature, which prevented silica gel chromatographic purification [27] / Royal Society of Chemistry. A feasible mechanism based on the Hoffman elimination of the longer *N*-alkyl chain may proceed through a six-electron [1,5]-H transfer. Mechanism proposed in this book.

✓ The added +*I* effect of the methylene cyclic scaffold of **42J**.

✓ The electron density withdrawn from the ring through C–C bonds generating CH_2–CH_2 dipoles with a vector component reinforcing the N NBP density via through space F (field) effects. Mind that energy-minimized conformations of these cyclohexylamines display the N NBP–directly involved in the nucleophilic attack–in the equatorial conformation; hence, this orbital is coplanar with the ring where C–C +*I* vector effects are aligned, and hence, more effective.

On the other hand, lactone and ketone carbonyl units in **42l** and **42m** operate in the opposite direction, as the potent electron withdrawers they are, decreasing sharply N electron density and nucleophilic power.

Is there a relationship between k *or* k/k_o *and molecular volume?*

I computed SEVs for the entire set of aliphatic dialkylamines and plotted them against RR but no correlation was apparent after various clean-up considerations such as grouping by block (Figure III.9) or by α or β substituent load. It is unfortunate that SEV does not take us out of the woods, even though at times it seems to work just fine.

***Lessons from these experiments*:**

Have Figure III.9 on hand to grasp the meaning of the following conclusions. RRs are shown in brackets:

1. The isopropyl substituent induces a more pronounced reduction of *k* than the neopentyl group:
   → Compare **42b** [2200] and **42e** [11 000].
2. Increasing the α substitution in isopropyl-N-R by one methyl furnishes a stark reduction of reactivity:
   → Compare **42c** [230] with **42f** [9.5],
   or **42c** [230] with **42d** [63],
   or **42e** [11 000] with **42h** [1.9],
   or **42e** [11 000] with **42i** [610].
3. Therefore, the steric effect of α branching in *k* is more potent than double β branching in the aliphatic series of amines.
4. The α/β branching in amines follows *the opposite trend* of alkyl halides. In these compounds, the isopropyl derivatives experience S_N2 reactions much more rapidly than the neopentyl halides. Benert and Seyfert [27] contend that the secondary alkyl amines of their study operate as nucleophiles, with the benefit of the through-bond electron density donation (+*I* effect) and reinforcement of the basicity of the N NBP, thus enhancing *k* in the opposite direction of steric hindrance; whereas the alkyl halides are electrophile receptors in which the inductive +*I* effect of alkyl substituents work in parallel with their steric hindrance to hamper *k*.

In short, steric effects of alkyl nucleophiles compose a different scenario than alkyl electrophiles in binuclear reactions.

General rules are difficult to enforce when contradicting results are used for comparison. In Figure III.7, plot **B**, you will find a poor correlation of the reactivity of monosubstituted alkyl amines, in their *nucleophilic* role over allyl bromide as electrophilic S_N2 receptor. This reaction is conceptually not far from Benert and Seyfert's model in regard to the role of amines as nucleophiles and bimolecular character of both processes. It is unfortunate, though, that in this plot isopropyl and *t*-butyl amines are not represented for the lack of data in the original paper, but *n*-propyl-, isobutyl- and neopentyl amines serve as homolog substitutes for our discussion.

In Figure III.7, in addition to ethylamine on one end and 2,2-dimethyl butylamine (bottom right) on the other, these compounds appear well organized in a declining trend in reaction rate (reflected in the steric factor E_s). This series of five amines drives one to conclude that *the greater the β-substitution the slower the rate* k, as opposed to the isothiocyanate case we examined presently.

The failure of the RR-SEV correlation we attempted herein runs in parallel with similar association problems recorded by Benert and Seyfert in connection with Taft's steric factor E_s.

Alicyclic amines of the cyclohexyl series (Figure III.9, third block) also pull-down RR strikingly, relative to aliphatic amines in the first block. But there ought to be an *enhancing* effect on RR by the cyclic structure when you:

→ Compare **42f** [9.5] with **42J** [110].

The 2,2,6,6-tetramethylpiperidine (**42J**), which is the closest homolog of di-tertbutyl amine (**42f**) in expected steric hindrance, is less effective at hindering the entrance of the amine to the TS stage. Two factors may be invoked.

1. The *tert*-butyl groups of **42f** are not frozen in any given conformation since the methyls rotate around the C–N axis, despite the small energy barrier to this rotation. As a result, the hyperconjugative and vector field effects of these six methyls–both induce modest electron donation and increase the basicity/nucleophilicity of N–are the vector sum of all C–H and C–C components. The same argument holds for all other aliphatic amines.
2. By contrast, in the cyclohexyl chair conformation of **42J** the C^2–C^3 and C^5–C^6 dipole vectors are fixed and aimed in the same direction as the N NBP, which lay on the plane of the ring, thus contributing to reinforce basicity/nucleophilicity by a **42J/42f** factor of 110/9.6 = 12.6.

The same effect can justify the much-enhanced reaction rates of tetramethylpyrrolidine and dicyclohexyl amine, as Figure III.11 shows quite dramatically. The polar factors involved in the decrease in RR recorded for alicyclic amines of the third block of Figure III.9 are depicted in the top part of Figure III.11, which also portrays the large differences of these amines' influence over RR.

3.3.5 Steric Acceleration of Reaction Rates

After our previous dissertation about steric hindrance of bimolecular processes, to expect enhanced reaction rates caused by steric acceleration from bulky substituents seems total nonsense. This would be true *unless* steric strain is released in the TS of the slow step–lower $\Delta H^{\ddagger}$ and higher rate constants–and in end products–thermodynamically favored relative to the starting material. Think of unimolecular nucleophilic substitutions (S_N1), eliminations (E1), even addition–elimination reactions in which a bulky group, say a tosylate, a tributylsilyl, or a *t*-butyl carbinol from an ester were the leaving groups. As the nucleofuges are cleaved, the energy associated with SRe decreases to zero and the molecular scaffold can breathe in a more relaxed environment.

Studies of *rate acceleration* by steric strain (SS) have nevertheless been demonstrated in very few cases and special reaction conditions. In many instances, researchers studying the SS of alkyl substituents on a β carbon, where they impact reactivity more effectively, give more conceptual weight to the *+I* effect firmly supported by Taft's σ^* correlations than to an SS influence.

This subject has been examined for the past 70 years, but a full comprehension is not with us yet (2022). All the same, to honor the challenging concept of steric acceleration a few foundational works are now reviewed, including key experimental features for problem-solving readers to have a first-hand appreciation of their importance to mechanism design in the dominion of steric effects.

3.3.5.1 Steric Acceleration in SN1/E1 Competition during Solvolysis In the present case one asks whether SS of substituents on the β position relative to the reaction center of an aliphatic scaffold influence competition between substitution and elimination reactions as well as the type of olefin produced therefrom (Zaitsev versus Hofmann). Substitution and elimination reactions, be they uni- or bimolecular, are prototypical cases of competing reaction mechanisms, a warning sign for mechanism problem solvers.

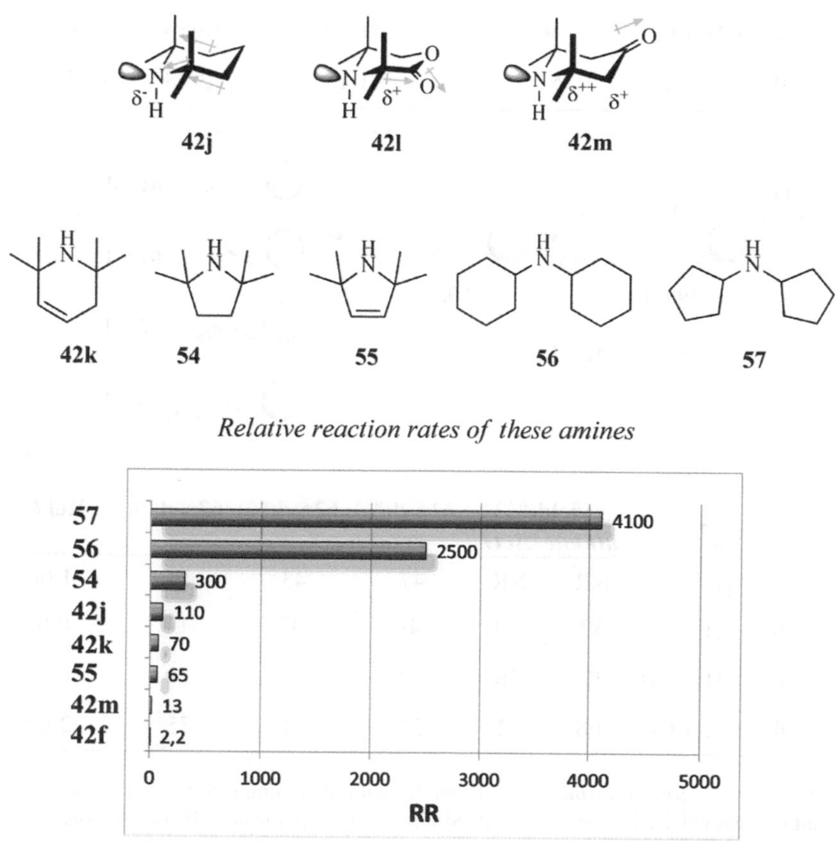

FIGURE III.11 RRs of alicyclic amines and isothiocyanate **41** described in Scheme III.6. Dipoles are shown as blue arrows. Di-*tert*-pentylamine (**42a**) (see Figure III.9, top) was used as reference, like in the rest of this research. Adapted from [27].

The $S_N1/E1$ competition was addressed by Professor Herbert C. Brown and coworkers in the United States in a triad of papers, two of which are extracted herein [28, 29]. Vintage chemistry? Yes, but perfectly valid today and an instructive analysis of a thorny mechanistic question–steric acceleration.

Here is the problem: Under the hydrolytic conditions of Scheme III.8, a set of tertiary bromides **54a–d** with increasing SS is assumed to form tertiary carbenium ions **55a–d** and then yield three possible outcomes **56a–d** to **58a–d** [28]. Carbinols **56a-d** were not observed so the only sources of evidence to work with were the olefin yield, rate of elimination (first order), and the proportion of 1- versus 2-alkene (**58 : 57**) in each case. While the latter alkene is thermodynamically stabilized by C–H hyperconjugation, the cis group enhances SS in this isomer in proportion to its size, which favors a $\mathbf{TS}_{(55\text{–}58)}$ en route to the terminal olefin.

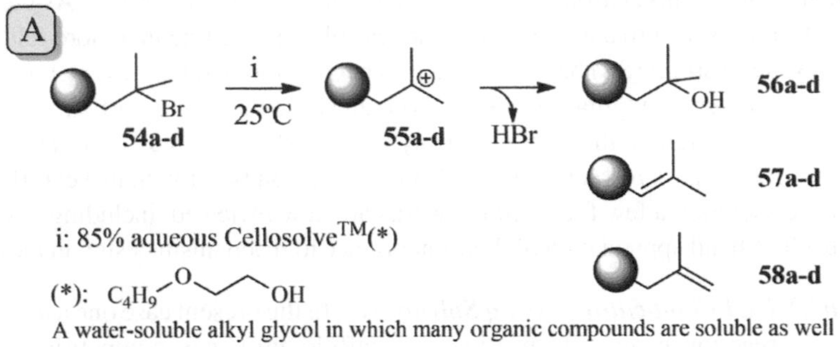

	57+58 yield	58:57 ratio	*k* (s⁻¹)x10⁴	Rel *k*
a CH_3	27%	21:79	1.075	1.00
b C_2H_5	33%	29:71	0.825	0.77
c $(CH_3)_2CH$-	46%	41:59	1.936	1.80
d $(CH_3)_3C$-	57%	81:19	13.08	12.17

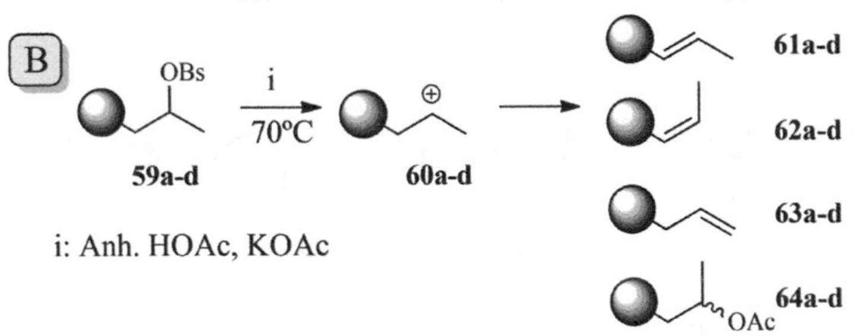

	yield(%) *alkene*	*AcO*	61a-d(%)	62a-d(%)	63a-d(%)	Rel *k*
a CH_3	NR	NR	47	43	10	1.00
b C_2H_5	32	50	49	35	16	0.95
c $(CH_3)_2CH$-	47	36	53	27	20	0.89
d $(CH_3)_3C$-	68	12	24	1	75	2.64

SCHEME III.8 Part **A**: Reaction of a series of *tertiary* alkyl bromides with increasing steric strain from β alkyl substituents, under hydrolysis conditions. The dominant process is E1 elimination. Part **B**: Similar analysis in the acetolysis of *secondary* brosylates (OBs). NR = data not reported. Adapted from [28, 29].

*Scheme III.8 part **A** interpretation*:

1. Steric strain (SS) emerges as a clear pattern (table in the scheme).
2. Under the same reaction conditions, after 60 hours the yield of isolated olefins increases significantly in proportion to SS. One might attribute this trend to a growing hyperconjugation and *+I* stabilization of the incipient $C^{(+)}$ in the preceding TS as the substituent provides a greater number of C–H bonds. But, in the authors' own words, it is believed that such hyperconjugative assistance will not be enhanced by the changes in the alkyl β substituents. This tenet is arguable though, but the kinetic data (third column) brings support to this contention.
3. Rate constants show a modest increase in the first three bromides **54a–c** up to isobutyl (right column, Scheme III.8). Then, there is a sudden sevenfold jump in the neopentyl case **54d**, relative to isobutyl **54c**. Why is that? This is attributable to the SS of the *t*-butyl group and the release of this strain as the bromide is cleaved from **54d** to a much greater extent than in any of the other models studied. In kinetic terms, *rate acceleration occurs on account of steric strain*.
4. Along the same lines, the growing ratio of 1- to 2-olefins **58 : 57** from less to more sterically hindered bromides, calls for a mounting SS in the TS of the H elimination step from $C^{(+)}$ **55**, as the methyl and the β substituent become aligned to form the trisubstituted C=C in **57**;.the larger the β substituent, the higher the energy of this TS. By contrast, the said alignment does not apply in the TS leading to the terminal alkene. Thus, *the growing SS swerves the regiochemistry of the elimination in the direction of the Hofmann product* **58**.

These results and conclusions drive us to a second question:
→ Does the preceding analysis hold for alkyl moieties with nucleofuges in *secondary* carbons?

Scheme III.8 part **B** portrays the studied reaction [29]: same carbon scaffolds as in part **A** save for one less methyl to create a secondary position, *p*-bromosulfonate (brosylate) as the nucleofuge, acetic acid/acetate as the reaction medium, and a higher temperature to have the reaction going. As in part **A**, olefins were distilled from the reaction vessel as they were formed and collected at time intervals to avoid isomerization. Three olefin types **61**, **62,** and **63** were collected in the distillate; their composite yield and proportion were determined. The corresponding acetates **64a–d** were isolated and characterized at the end of the reaction and yields determined. Thus, data to feed the kinetics of acetolysis were not available, only the final yield.

*Scheme III.8 part **B** interpretation*:

1. Steric strain (SS) seems to find expression here but in a softer dimension.
2. Given the solvent polarity, heat, and the ease of brosylate departure, the intermediacy of a secondary $C^{(+)}$ is assumed, an important element of judgment to postulate the SS of the β substituents. Reaction rates are first order in total olefin produced, which is the measured variable.
3. As opposed to experiment **A** on the tertiary bromides, the secondary brosylates furnished a substantial fraction of acetates of type **64** from nucleophilic substitution (S_N1). Hence, S_N1 versus E1 competition emerges.
4. The total olefin yield in cases **b–d,** which can be interpreted as the efficiency of the E1 branch of the reaction, correlates well with the bulk of the β substituent. This trend matches the decrease in the yields of acetates **64b–d**.
5. However, *the relative* k *values* derived from olefin time-dependent recovery by concomitant distillation *appear unaffected by substituents in* **59a–c** *and only moderately so in* **59d**. This latter value is only a fraction of the relative *k* of **54d** in the tertiary bromide series (part **A**). Therefore, the SS of the β group on the TS or the $C^{(+)}$ (**55a–d**) is much less significant in the secondary series (part **B**).
6. Authors explain this by stating that SS exerts a marginal rate acceleration only in the largest member of the series (*t*-butyl **59d**) and the increase in olefin yield must result from a decrease in the rate of substitution *ks*, as is in fact the case (table of part **B**). This would arise from the steric hindrance offered by the large β substituent to the approach of HOAc or $AcO^{(-)}$ to the $C^{(+)}$ and the subsequent formation of a more sterically strained tetrahedral acetate product.
7. This picture is put together in Scheme III.9, showing specifically the most sterically demanding case of the neopentyl methyl brosylate **59d**. Given the ramification of the tentative reaction diagram and the number of partial rates at the gates of the several steps with their corresponding TSs, it is evident that the kinetics of the olefin production only measure the overall reaction course exploiting the slowest step but cannot validate the intermediate stages, activation, and Gibbs energies. Only a theoretical quantum density functional theory (DFT) calculation or *ab initio* equivalent might provide a deeper insight. To the best of my knowledge this study has not yet been performed.

In any event, based on the evidence of Professor Brown's experiments [28, 29], Scheme III.9 suggests:

✓ The departure of the leaving brosylate relaxes the SS of the starting material to a flat trigonal $C^{(+)}$ **60d**.

✓ The evolution of this species splits in two pathways that will eventually define de S_N1/E1 competition reflected in the ratio k_2/k_4. Based on acetate/olefin yields alone, $k_2/k_4 > 1$ for less sterically demanding β groups, represented by **59a** and **59b** (evidence provided for the latter compound only).

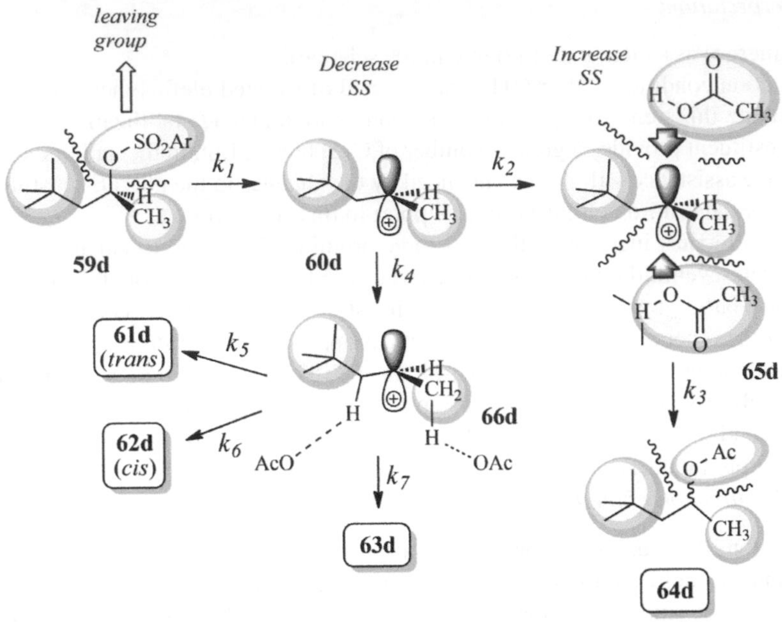

SCHEME III.9 Combined reaction model to explain conclusions of the experiments of Scheme III.8.

- ✓ However, the ratio becomes inverted to $k_2/k_4 < 1$ as the steric requirements of the β groups expand by one and two methyls (**59c,d**) with a preponderance of olefinic products, whereas the acetate yield wanes to hardly significant values.
- ✓ This increased steric demand for the S_N1 sequence is clearly expressed in the molecular encounters of TS **65d** on either side of the $C^{(+)}$ by the contributions of the methyl and neopentyl groups; the k_3 contribution becomes less and less important to the overall set reactions.
- ✓ Acetate performance changes from nucleophile to base under the circumstances. The contrastingly less sterically demanding cationic TS **66d** conceivably exhibits a more distant acetate acid–base rapport, rendering TSs down the line (not shown) en route to cis and trans Zaitsev 2-alkenes **61d**, **62d**, and the Hofmann terminal 1-olefin **63d**.
- ✓ Following the classical picture, the cis and trans Zaitsev alkenes are equally competitive and prevail over the terminal Hofmann alkene in cases **a** and **b** (Scheme III.8 table of part **B**), but the *trans* olefin becomes the dominant product in the 2-alkene **59d** → **61,62** elimination, reducing the cis isomer to a mere trace (1%). At the same time, the Hofmann 1-alkene **63d** becomes the major product of the acetolysis as a result of the much lower sensitivity to the steric demands of the more distant neopentyl section to the methyl proton abstraction in **66d**.

In a nutshell, the progressive change of S_N1 to E1 products yield, rate of solvolysis (especially in the tertiary cases of part **A**), and cis/trans/terminal olefin composition are consistent with steric strain of the β groups, in addition to the fact that a *substantial rate acceleration in tertiary models (and a modest increase in secondary models), is supported by these experiments.*

Additional convincing evidence for steric acceleration of reaction rates in liquid phase has appeared more recently in other reaction models like the $Br_3^{(-)}$ oxidation of ortho-substituted benzyl alcohols to benzaldehydes [30].

3.3.5.2 Steric Acceleration in the Gas Phase Kinetic studies in solution cannot escape from the solvent intruding in the reactive species under scrutiny [31]. However, solvent cages, hydrogen bonding, and substrate-solvent dipolar and field interactions are often ignored when proposing TS successions from experimental data or quantum mechanical calculations designed in the gas phase.

Modern sophisticated techniques such as electrospray mass spectrometry (ESMS) have been developed to study transient species of a growing number of reactions in solution inside the electrospray-ionization chamber of a mass spectrometer [32, 33].

Another approach being developed is to do away with solvents altogether by carrying out kinetic studies in solvent-free reactions in liquid phase, with or without catalysis [34]. Serious limitations arise though as reactants are consumed and new products accumulate; hence, reaction conditions change sharply, cumulative concentration is always very high, and the reacting system is far removed from the ideal solution conditions required for getting dependable kinetic data.

On the other hand, reactions performed in the *gas phase* at low pressure are free from solvent interference and are measured under nearly ideal gas conditions. Although the kinetics of many unimolecular reactions compose the major portion of experimental gas phase studies, some bimolecular or intermolecular reactions of volatile mixtures have been examined [35].

One of the works that addressed the question of steric strain of β substituents on the first-order rate of elimination in the gas phase was authored by late Dr. Gabriel Chuchani in Venezuela [36], who devoted most of his scientific life to studying a variety of reactions in such a system. The selected model was the HCl elimination of a series of β-alkyl substituted primary alkyl chlorides depicted in Scheme III.10.

In the absence of solvent or anchimeric assistance of neighboring substituents (which can also be detected in the gas phase [37]), one would expect a concerted proton and chloride detachment with little or no charge development. Two sources of evidence, however, suggested a discrete δ^+ buildup in this series: a linear correlation of Taft's σ^* and log k_R/k_o on the one hand and $\rho^* = -1.81$ on the other.

Given that σ^* is a measure of the *+I* inductive electron donation capacity on the rate enhancement, the data would be symptomatic of a small measure of electron demand at C^1, giving credit to a TS like **67a–e**. If the electron demand is low, though, the *+I* effect on demand will also be discrete and the SS of the β substituent would have room to contribute to the observed *rate acceleration* trend.

While the data does not allow a valid separation of *+I* and SS contributions, I anyway calculated the Connolly's SEV of the β substituents R in alkyl halides **66a–e** and plotted them against k_R. An excellent linear correlation emerged (Scheme III.10) ($r^2 = 0.9871$). It not only showed a strong rate acceleration, comparable to the one registered for the *t*-butyl group in the SN1/E1 competition (Section 3.5.1, and Scheme III.8), but also suggested the involvement of SS in the expulsion of chloride.

3.3.6 Summary of Steric Hindrance and Reactivity Takeaways

1. Steric effects (SE) are born in the natural electromagnetic repulsion of electron clouds between non-bonded atoms at a short distance (<3–4 Å). SEs can be applied in three landscapes: *repulsion, stress*, and *hindrance*. These terms are often used as synonyms of the same hampering effect of bulky substituents, but more precisely, each refers to a distinct situation.
2. *Steric repulsion* (SRe) results from the mutual rejection between two zones of the same molecule. In rigid scaffolds such as cage compounds, many [x.y.z…n] bicyclic or polycyclic structures and *cis*-aliphatic alkenes this repulsion determines *steric stress or strain* (SS). Compare, for instance, *cis* and *trans* 2,7 bisubstituted bicycloheptanes **69a,b** and **70a,b** (Figure III.12). The calculated ΔG^o_r energy differences between cis and trans isomers **69a** and **70a** were estimated, relative to the Gibbs energy of the unsubstituted bicycloheptane and then compared cis and trans [38]. Not only the dimension of SS arising from the 7-methyl ⟷ isobutyl interaction is significant ($\Delta E = 4.60\,\text{kcal mol}^{-1}$) but the contribution of the additional methyl in the *t*-butyl substituent contributes a $\Delta\Delta E = 4.60 - 2.47 = 2.13\,\text{kcal mol}^{-1}$ to the SS caused by these two groups.
3. SS within the same order of magnitude occurs in cis/trans models **71/72** and **73/74**, with the added σ-bond rotation of the phenyl ring in the cis isomer, as shown in model **74-3D**. While $\Delta E_{74\text{-}73} = 5.92\,\text{kcal mol}^{-1}$ the dihedral angles ϕ indicated in Figure III.12 reduce the coplanarity of the styrene section, whereas the CH_3–CH=C–C(Ar) fragment remains essentially planar ($\phi = 179°$) [38].

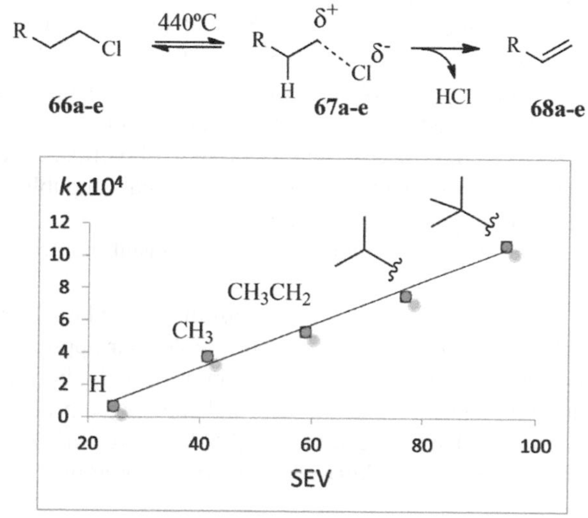

SCHEME III.10 Unimolecular elimination of HCl in the gas phase. Reaction rates used for the correlation plot were Adapted from [37]. SEV calculations of R substituents by M. E. Alonso-Amelot, this book.

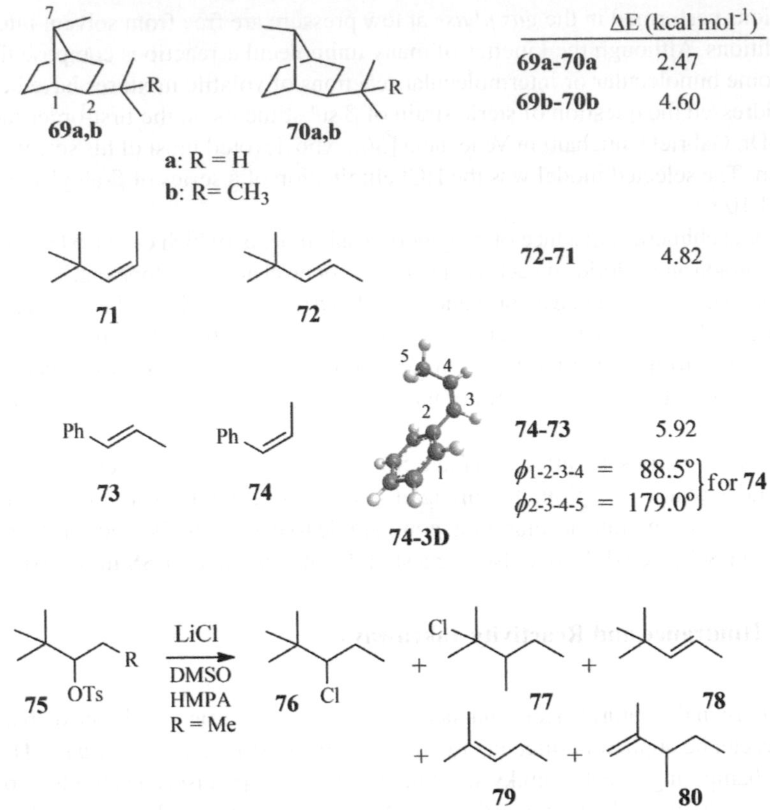

FIGURE III.12 Selected models showing steric strain energy differences depending on configuration Alonso-Amelot (Author).

4. In di-, tri-, and tetrasubstituted alkenes, the E ⇄ Z isomerization equilibrium is displaced to the thermodynamically favored isomer, and may include the terminal olefin as well to achieve the minimum SS. Structural rearrangements may also be driven by the decrease of SS.
5. The release of SS is also observed during a reaction course as a result of the cleavage of a large leaving group, formation of less encumbered cations, elimination, and addition–elimination reactions. In such cases, steric rate acceleration may be observed, but developing fully convincing evidence continues to be a challenging task.
6. The term *steric hindrance* (SHI) is better applied to the bimolecular TS built from an attacking reagent upon a sterically compromised reaction center. SHI can also be called upon in the mechanism design of large molecules in which an intramolecular rapport between an HEDZ and an LEDZ is subjected to the influence of sterically hindering moieties. Interference of SHI in substitution reaction courses in which the intermediacy of $C^{(+)}$ is undesirable to prevent possible Wagner–Meerwein rearrangements constitute a problem. Among others, the neopentyl group is particularly prone to rearrangement side products (Scheme III.12, bottom). In an attempt to expand a synthesis method of pinacolyl chloride **75** R = H [39] to higher homologs (R=CH_3) like **76**, a portentous mixture of olefins, a rearranged product, and two chlorides were characterized in the distillate in various yields, all totally impractical (unpublished, of course); a good warning sign for mechanism problem solvers facing large SHI near reaction centers.
7. As opposed to this, SHI is also widely used in organic synthesis for installing protecting groups and forcing regio- and stereoselectivity.
8. During problem solving, visual appraisal of SHI can still be a useful tool to build reasonable TSs leading to product targets when substituent barrier differences between two or more molecular approaches are very obvious. The sequence primary > secondary≫ tertiary carbons bonded to leaving groups or neighboring atoms in the S_N2/E2 competition dilemma, and addition–elimination in regard to reaction rates and anticipated products, continues to be valid.
9. To the good fortunes of reaction mechanism addicts, recent (2022) *theoretical calculations on the S_N2/E2 competition converge on the classical concepts of SS and SHI linked to visually apparent bulk of substituents* [40].

10. Steric or stereoelectronic influence of substituents on reactivity in bimolecular S_N2 and E2 mechanisms vary according to the nature of reacting compounds, electrophilic or nucleophilic, as well as the acting nucleophile atom, C, N, O, or S and the electrophile moiety, C=C, alkyne, or arene.
11. The SEV, as defined by Connolly [22, 23], of some alkyl substituents in the simple β-substituted moieties as well as enzyme–substrate associations are suitable to explain hindrance and even acceleration of reaction rates in specific reactions. A reduced number of such studies have been published.
12. SEV and other steric descriptors are helpful, no doubt, but their impact on reaction rate varies considerably with charge development at the reaction center or TSs as a result of the electromagnetic effects impinging on their stabilization or destabilization, in addition to quantum influence of substituents on atomic charge density.
13. Isolating steric from electronic effects of substituents and molecular sections near the reaction center with regards to inductive, field, and hyperconjugative factors, or in conjugation, continues to be a challenging problem.
14. The question of early and late TSs in bimolecular reactions is proposed frequently at the discretion of the scientist interpreting the rate data, even though DFT at a suitable level can provide more precise models to better define the geometry of TSs and the associated energy.

3.4 APPLICATIONS TO STEREOCHEMICALLY COMPETENT REACTION MECHANISMS

Readers should be convinced by now that once the energy orbital compatibility requirements are fulfilled electron flow, as discussed earlier, and the tridimensional shape of the molecules they belong in constitute the two pillars of reaction mechanism. In this section, these two conceptions will be treated together in a selection of reactions that would not proceed without a careful scrutiny of the stereochemical features of intervening molecules.

3.4.1 A Case of Regio and Stereoselective Reaction in a Sterically Simple Compound

Let us get back to the workbook spirit starting with a simple case and building complexity from the bottom line up: *trans* decalin (**81**), fully saturated, uncomplicated stereochemistry, undergoes a well-known Friedel–Crafts acylation by simply mixing it with aluminum chloride and acetyl chloride in dichloromethane (DCM) at room temperature (Scheme III.11). In spite of no distinguishable HEDZ/LEDZ in **81,** an assortment of compounds was isolated and characterized [41a] (only one depicted: alkene **82**).

Scheme III.11 portrays a set of reactions whose mechanism I would like you to devise. There are three questions:

1. Explain the genesis of **82** from **81.**
2. Consider **82** → **83** *with the indicated stereochemistry.*
3. Why does **84** behave differently, affording **85** instead?

After you work out a proposal come back to this line, and I will later drop a confounding element.

Feasible mechanisms:

A quick reminder at the start: For many years the structure of the AcCl·$AlCl_3$ complex has been a matter of discussion: was it a donor–acceptor complex (**86**) or an ion pair (**87**) (Scheme III.12, part **A**) equivalent to an acyl cation? Contradictory results arose depending on the survey technique used. ^{1}H and ^{27}Al NMR experiments of their mixture in DCM finally settled this dilemma in favor of the donor–acceptor complex concept in solution [42]. In this model, the C=O unit in the complex retains a potent LED character that attracts weak nucleophiles.

Having this information on hand, the subtraction of a hydride from decalin by the complex **86,** as in part **B** of Scheme III.12, is perfectly feasible. The angular proton is selected since the developing carbonium ion grows on the only tertiary carbon. This calls for a late, product-like TS. The primary products of cation **89** are the E1 elimination alkenes **82** and **84**.

We continue with part **C** of Scheme III.12, which explains itself, furnishing equatorial carbinol **83**. The key to the stereo-controlled addition of water is the neighboring group participation of the pivotal acetyl unit installed in **90**.

Now, the confounding factor: what if I told you that *trans* carbinol **83** *was **not** the reported product* but the *cis*-acylcarbinol instead [41b]? Let me add that **85** was identified as the cis derivative as well. In addition, it was an outcome of the same alkene **82.** Take this rather light manner of treating hard scientific data for the benefit of opening mechanistic possibilities in your mind. Think for a while before checking out Scheme III.22 at the end of this chapter.

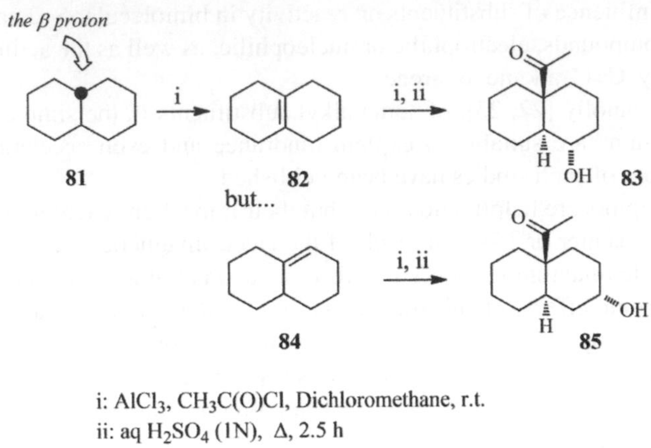

SCHEME III.11 Building molecular complexity from a hydrocarbon without identifiable high or low electron density zones. [41b] / Royal Society of Chemistry.

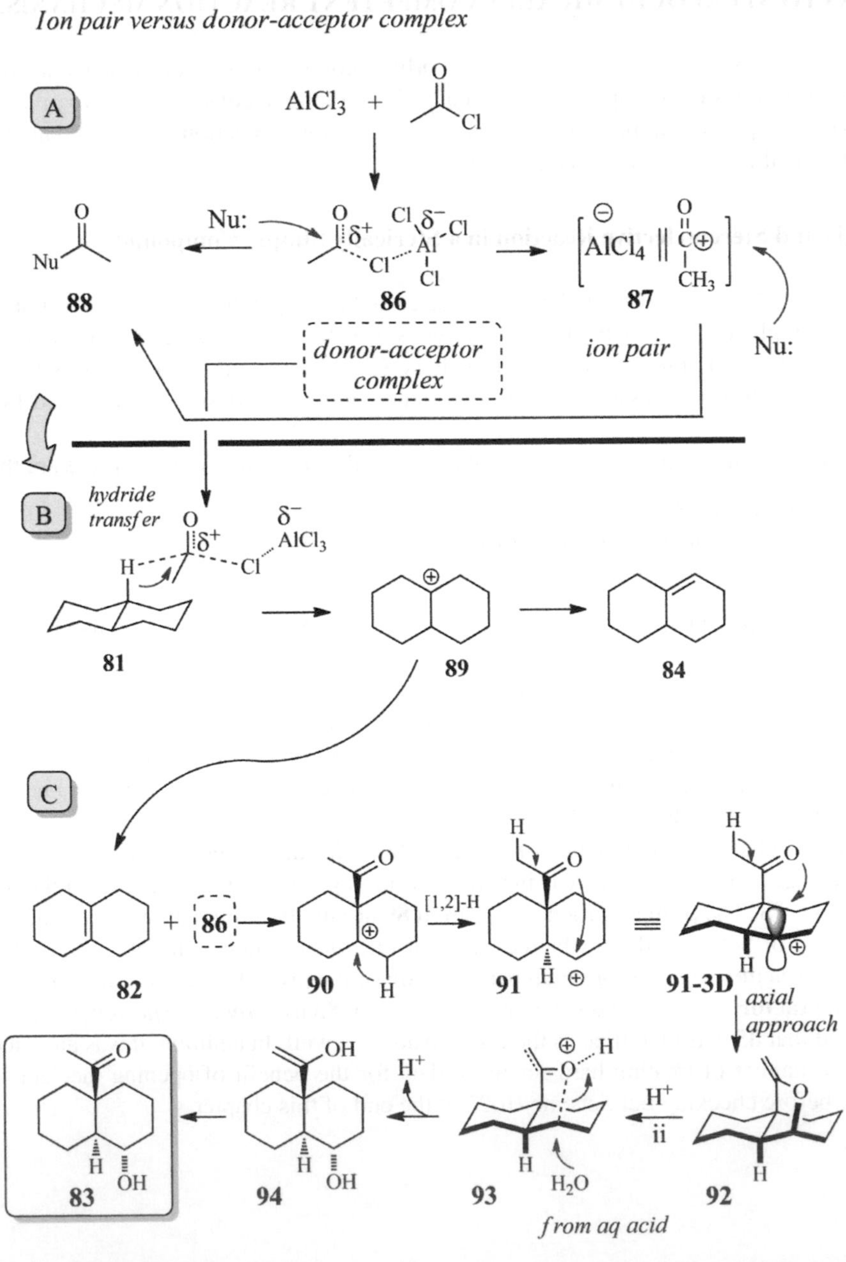

SCHEME III.12 Stepwise activation of a remote C-H bond and regio and stereospecific a-hydroxylation through control of the angular acyl group. [41b] / Royal Society of Chemistry.

3.4.2 The Role of Steric Umbrellas

In addition to the discussed effects of SHI on reaction rate and as group protectors, a third consequence is their use as steric control to create chiral centers during nucleophilic substitutions. The stereochemical component of the analysis of reaction mechanism should consider SHI when enantio-control becomes a product requisite. The following example, based in the well-known Meinwald rearrangement (MR) is an instructive application.

The MR is a typical reaction of epoxides by virtue of their tendency to create a carbonium ion or a close relative upon partial or complete ring opening initiated by Lewis- or protic-acid catalysis (Scheme III.13, part **A**). In this circumstance, α substituents may undergo [1,2]-migration to the developing $C^{(+)}$ while the carbonyl consolidates the final ketone **96** or aldehyde **97**. This is contingent upon two conditions: i) the most favorable $C^{(+)}$ position and ii) the ability of the α-migrating group. Note that the combined circumstances might create up to four primary products. However, given that **97** includes a quaternary carbon, the MR finds use in the synthesis of enantiopure products.

A case in point is the work of researchers in Kharagpur, India [43], pursuant of (*R*)-α-cuparenone (**101**) (Scheme III.13). Cuparenone has two vicinal quaternary carbons, one of which is stereogenic upon which the bioactivity depends. This feature "*poses a dreadful synthetic challenge*" in the words of the authors and a mechanistic trial for us. The heart of this synthesis strategy appears framed in blue, an MR. From there on, the absolute configuration of C^3 must be preserved during the remaining 10 steps to **101**.

The success of the retrosynthesis (section **B**) was contingent on two factors: i) the direction of the epoxide ring opening upon attack of the Lewis acid, and ii) which of the three substituents in the epoxide would migrate. Scheme III.14 shows five isolable products (**106–110**) (dotted area).

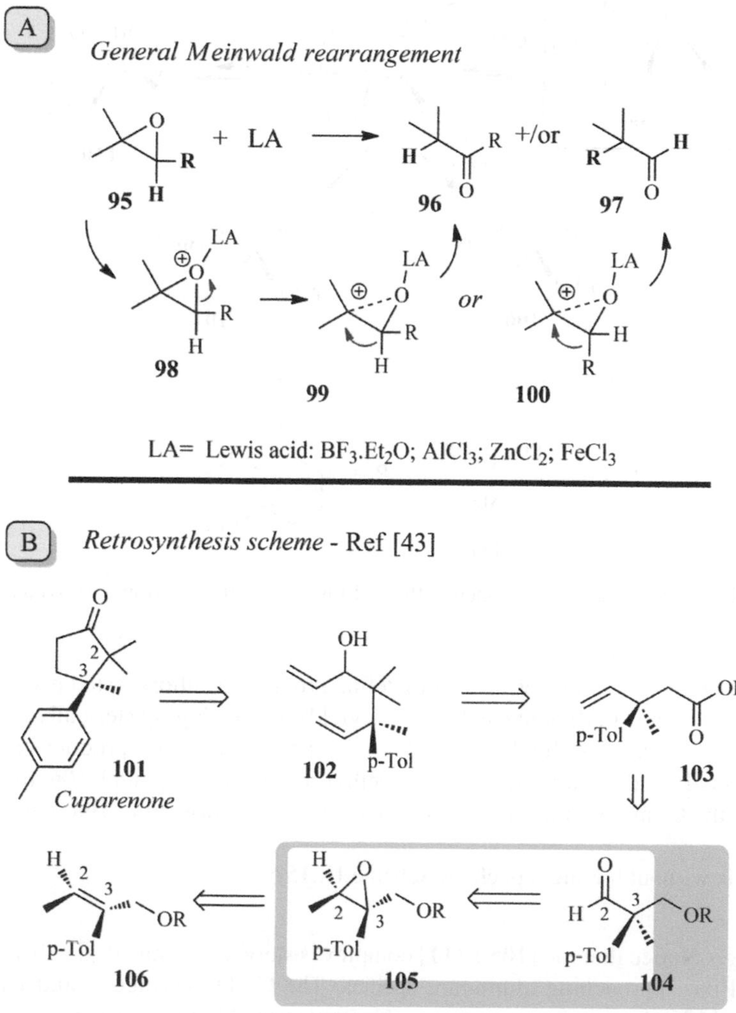

SCHEME III.13 Part **A**: basic mechanism of the MR. Part **B**: abridged retrosynthetic analysis of cuparenone (**101**). Open arrows point in the direction of preceding compounds. [43] / Elsevier.

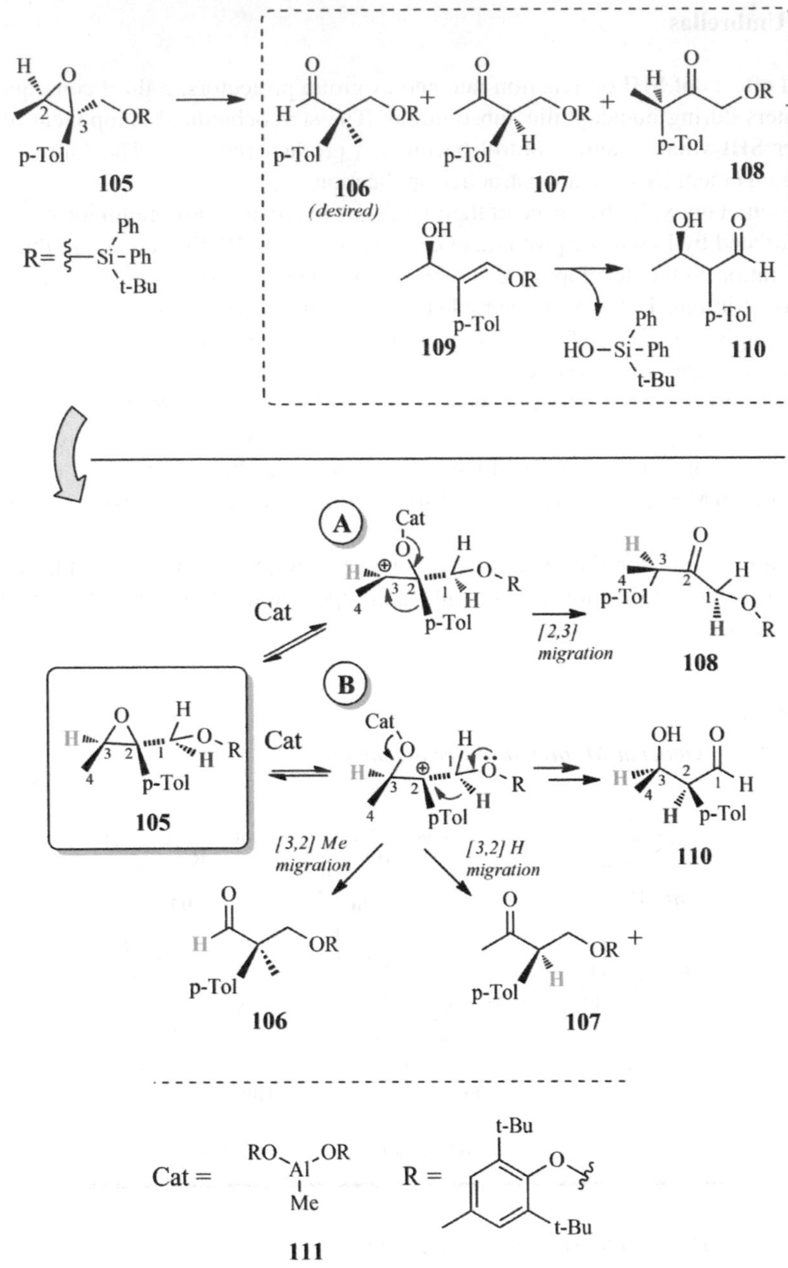

SCHEME III.14 Possible mechanisms open to the MR involved in the reaction portrayed in Scheme III.13.

The ring opening direction of epoxide **105** was critical to define products. Authors tested a series of Lewis-acid catalysts at very low temperature (−78 °C). The most efficient catalyst (95% yield) was the highly sterically hindered methyl aluminum bis phenoxide **111**, whereas all other much less hindered catalysts furnished rearranged product in the 30–55% range. In every case, however, the only detected product was the desired methyl-migrated compound **106** from the selective C^3–O bond cleavage. A second issue was the C^2 methyl migration to C^3 to the back side furnishing **106**, which would involve a forbidden *antarafacial* shift.

Can you explain why this is without having a peek on Scheme III.15?

Reasoning:

1. *Selective C–O cleavage*: Notice that the [**105** + **111**] complex is sterically strained (SS) heavily on account of the silicon tolyl substituents and the approaching aluminum species. The C^2–O cleavage would lead to increased SS as these groups come closer in **112**, whereas the opposite would occur in the C^3–O breakage.

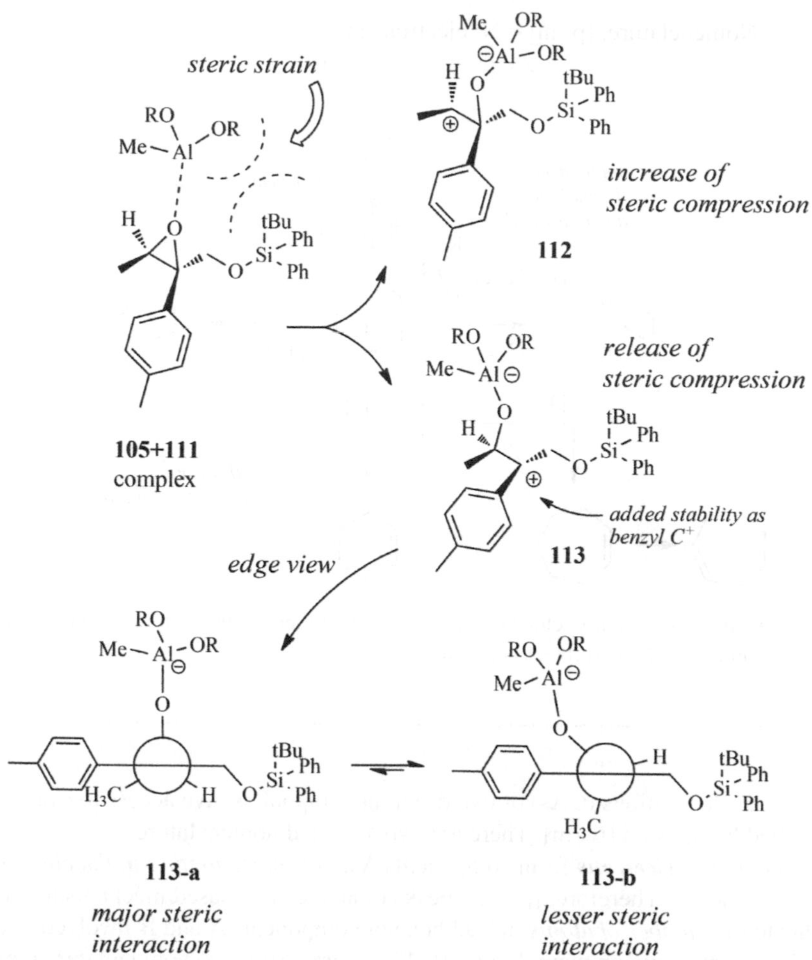

SCHEME III.15 Implication of steric strain in the regio and stereoselective course of the MR.

2. The *antarafacial* [1,2] methyl migration requires the rotation of the C^2–C^3 bond such that the C^2-methyl bond is parallel to the *p* orbital of the carbonium ion at C^3. This is depicted in rotamers **113-a** and **113-b**, the latter showing less SS from the increased distance between the large groups, thus favoring the now suprafacial methyl migration.
3. The double steric umbrellas of the trisubstituted aluminum and silicon units offer sufficient SS to allow the only suitable conformation for the methyl migration and the construction of the stereogenic quaternary carbon C^3.

As you can surmise, the stereochemistry assessment of this reaction is essential for the understanding of the involved mechanism and the accurate structuring of the chiral quaternary C^3.

3.5 STEREOCHEMISTRY IN BIMOLECULAR REACTIONS: CYCLOADDITIONS

Among many others, the all-important intermolecular cycloaddition reactions are a showcase for sterically demanding transition states: two molecules approach at short distance in their way to a TS in which two σ bonds are formed on account of π–π interactions. The TS tight configuration is thus expected to be strongly influenced by the steric and electronic effects of substituents in the interacting components.

Running the risk of repetitious insistence, a note of the designation of cycloadditions is timely to avoid the common confusion of what [p+m] cycloadditions mean (see box alongside Figure III.13).

Nomenclature: [p+m] = N° electrons in TS
[p+m] = N° backbone atoms in TS forming the ring

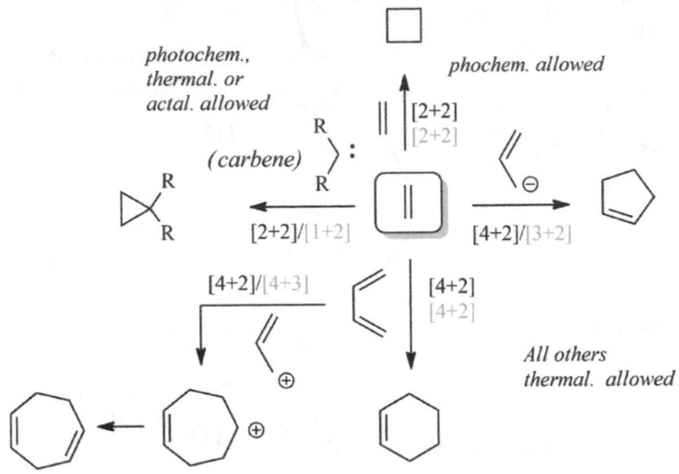

FIGURE III.13 Common types of bimolecular cycloaddition reactions and their nomenclature based on the number of electrons being redeployed in the TS from each component [p+m] (black) or atoms (green).

> ***A brief reminder***
>
> 1. Common intermolecular cycloadditions (CAs) of two different compounds give access to rings of various sizes, from three to seven, established by the sum [p+m]. There are two accepted nomenclatures:
> 2. [p+m] represents the *number of* ***electrons*** from components A and B *participating in the electron redeployment* of the CA, generally involved in the TS. Therefore, [p+m] speaks of an electron-based mechanism in this case.
> 3. [p+m] does also indicate the *number of* ***atoms*** in backbone of components A and B involved in the TS leading to the future ring, not counting substituents on them. Figure III.13 illustrates the two nomenclature types for the most common CAs.
> 4. CAs may also include heteroatoms, particularly N and O and less commonly S and Si.
>
> Use whichever nomenclature you like as long as you let people know whether you speak about electrons or atoms in the S.

3.5.1 The Diels–Alder Cycloaddition: the CA Prototype

The [4+2] CA widely known as the Diels–Alder cycloaddition (DACA) has been an intensely studied model for many years at the laboratory and industry scales used in countless syntheses of challenging natural products and many other compounds [44]. All undergraduate chemistry students have been exposed to textbook level DACA. But there is much more to it than dienes and dienophiles.

DACA may be fashioned between separate diene and dienophile carbon species (alkenes, alkynes, aromatics) or $R_2C{=}X$ (X=N, O, S) known as heteroDACA, which brings about conceptual density. Intramolecular combinations are often employed as well. The advanced problem solver should profit from another round of DACA insights at a deeper level.

The simplest prototype combination–butadiene and ethylene–requires high temperature and pressure to give cyclohexene [44], but electron donor substituents in the diene, electron withdrawers in the dienophile, and Lewis-acid catalysis facilitate greatly the cyclization and is generally performed under standard laboratory conditions as well as even low temperature.

DACA is also regio- and stereoselective and subjected to effects of substituents away from the reaction center. Because DACA is the foremost 4π–2π, namely [4+2] cycloaddition and probably number one in experimental, synthetic, and theoretical exploration of its class, anything concerning the details of this reaction affects the rest of the concerted cycloadditions in the catalog and is thus a learning showcase.

This chapter is focused on stereochemistry, and by now you should know that it never operates separately from electronic effects. This section is presented as a reminder in these two fundamental aspects: the stereochemical constraints and the stereoelectronic effects brought by substituents on the π–π components.

3.5.1.1 *DACA Steric Domain: Essential Takeaways* Please, check out these issues while keeping an eye on Scheme III.16:

1. Because a total of four sp^2 carbons in the diene and dienophile (if alkene) become tetrahedral, the occurrence of substituents are potential sources of cis : trans conflict in the adducts. The use of acetylene dicarboxylate ester (**114**) in section **A1** of Scheme III.16 reduces the conflict to the appended substituents in the diene. Adducts **116** and **118** are expected to be the major products of **115** and **117**, respectively. Steric control is easily achieved.
2. When we translate the stereochemical definition potential to the dienophile, as in **119** of section **A2**, adduct **121** mirrors the stereochemistry of both components **119** and **120**. We are talking about the expected *major* products, but these may be accompanied by minor quantities of stereoisomers like **122** (dashed frame).
3. This is also the case of reaction **B**. Here, we fix the all-trans configuration of the diene and explore the cycloaddition of maleate and fumarate esters **123** and **124**. Both are archetypical dienophiles in DACA explorations. The self-explained section **B** illustrates the expected outcome: cis gives cis, trans affords trans. Period.
4. As you can surmise, there is nothing to squeeze your memory about to handle with proficiency these simple stereochemical issues. Be reminded though that all these bimolecular encounters in **A1/2** and **B** reactions arise from a concerted – albeit not fully simultaneous – formation of the two new C–C bonds, wherein there is little opportunity, if at all, for C–C bond rotation of the transforming dienophile in the TS.

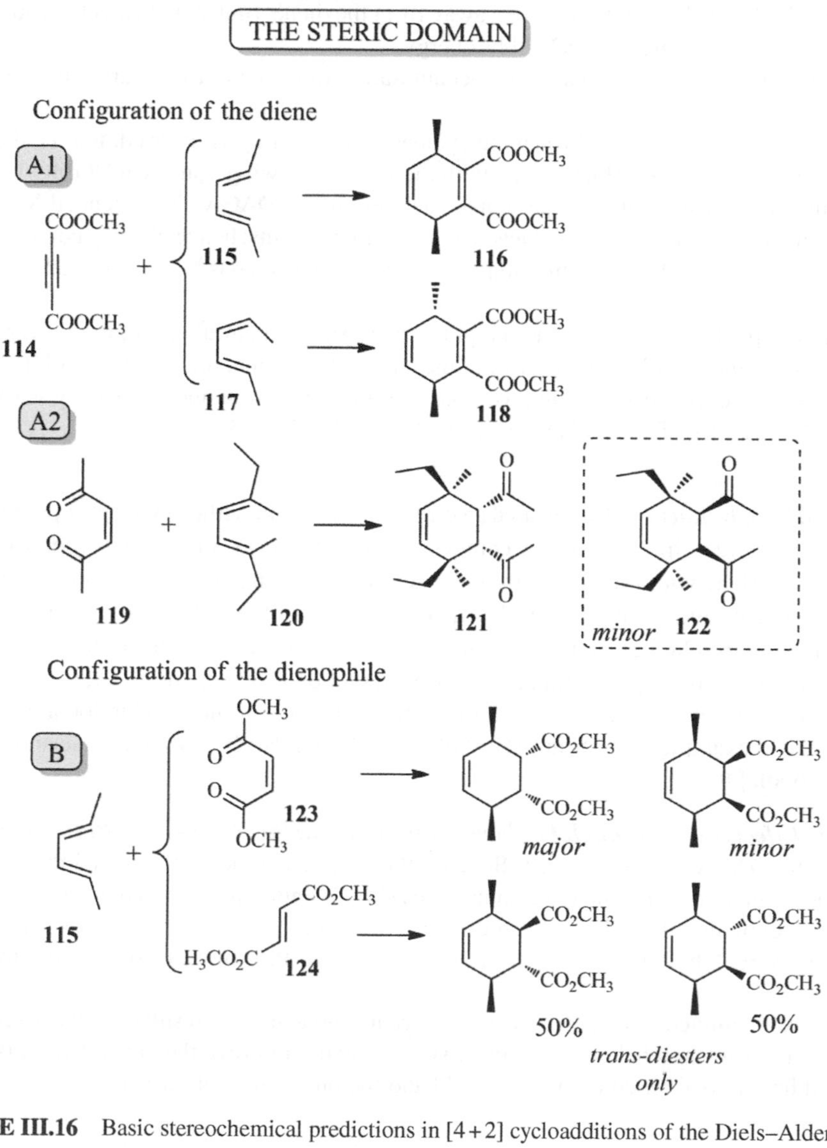

SCHEME III.16 Basic stereochemical predictions in [4+2] cycloadditions of the Diels–Alder model.

3.5.1.2 DACA Stereoelectronic Domain: Essential Takeaways Please, check out these issues while keeping an eye on Scheme III.17:

CAs are harmonious reactions with little polarization marked by two σ bonds being formed from sp^2 elements and no leaving groups; even though the diene-dienophile rapport is concerted, partial charges emerge because of inhomogeneities in their electron density configuration and delays in bond rate formation due to SRe in the tight TS.

3.5.1.2.1 The Frontier Molecular Orbitals Model

1. All cycloadditions involve HOMO/LUMO interactions that may lead to increased di-π transitory interactions in the TS overcoming moderate SS.
2. Usually, the LUMO is assigned to the electron-deficient dienophile. Its LUMO, which is electron-**U**noccupied, opens orbital space for electron redeployment toward it from the electron-**O**ccupied HOMO of the (generally) electron-rich diene.
3. This π orbital interaction is depicted in the simple terms of **TS1** in reaction A of Scheme III.17. The diene's HOMO is π bonding in $C^1{=}C^2$ and $C^3{=}C^4$ but π* (antibonding) in the C^1–C^4 conjugation. Therefore, the AO phase lobes appear inverted in $C^3{=}C^4$ relative to $C^1{=}C^2$. (In this case, Huckel's MO models appear confusing and were skipped.)
4. By contrast, the LUMO of the dienophile expresses a fully antibonding π* across the conjugated ketone moiety. Convincingly, the diene and dienophile coupling fits perfectly the **TS1** model where the C=O bond also overlaps the diene's *p* AOs, lowering the **TS1** Gibbs energy. This configuration affords **125** exclusively, whereas **126** arises from an alternative **TS2** in which the C=O unit is oriented away from the diene, (not shown, it is for you to draw) thus decreasing the π–π interaction and increasing the **TS2** Gibbs energy.
5. This picture confirms the principle of maximum accumulation of centers of unsaturation, namely, major π–π interactions in the TS.
6. This HOMO/LUMO rapport is reversed when the cycloaddition polarity is inverted, that is, the dienophile is electron-rich, and the diene is electron-poor. (Please play with these ideas drawing your own HOMO/LUMO models.)
7. Additional π-conjugated substituents also participate actively in the DACA-TS. Extended MO overlaps may furnish a higher yield of the most sterically stressed adducts within certain SS limits; hence, SS may be overcome by electromagnetic forces lowering the activation $\Delta H^{\ddagger}$. This question is far from simple and has been examined from various postulates and theories in recent times (see below).
8. Not much attention is paid to the fact that to adopt the necessary structural configuration for the [4+2] TS, aliphatic *trans* dienes must adopt the cisoid rotamer form. In the case of butadiene, this σ 180° rotation costs 6–8 kcal mol^{-1}. The cis-concerted TS → product process is more favorable than the stepwise sequence starting with the diene trans + dienophile → TS1 → 108° rotation → TS2 → adduct by around 10 kcal mol^{-1} [45].

3.5.1.2.2 The Polarization by Substituents Conjugated electron donating and withdrawing groups (EDGs and EWGs) in diene and dienophile, respectively, create dipoles that may compensate SRe in the TS and adducts therefrom (section **B** of Scheme III.17). There is no need to overload your memory with this material. All you need to do is locate the expected δ^+ and δ^- of each component by applying your wits and what you have learned so far about EWG, EDG, and conjugation, put it together in the suitable configuration [brackets in scheme], and carry on to products. If the polarity is inverted, the reaction will go anyway but might require heat and catalysis. It is for you to fill in the spaces in section **C** of Scheme III.17.

Cycloadditions occur not only between bimolecular arrays but often in the intramolecular domain, a fruitful tool for building molecular complexity. Related examples appear in "The Problems Chest", Part II as well as in several challenging problems of the second edition of this book [46].

3.5.1.3 Assessing Steric Effects (SEs) through Products Configuration: the Endo Alders Rule Preference for endo adducts was observed soon after the DACA was discovered. Because the majority of adducts showed the endo configuration, it was thought to be a general phenomenon and came to be named the **Alder's endo rule**. Not having a frontier orbital theory at hand in those days (1937) [44] to explain things, the endo products seemed to conceal some sort of mystery since they were counter-intuitive: the exo products were "obviously" less sterically hindered and thus more favorable, but they were minor products only. Why?

This dilemma has been re-examined systematically over the years and a few key results are discussed next for you to develop a solid criterion about this relevant stereochemical issue. Two approaches to prove the general validity (or the opposite) of the Alder's rule are presented herein, one based on product yield and the other on reaction rates.

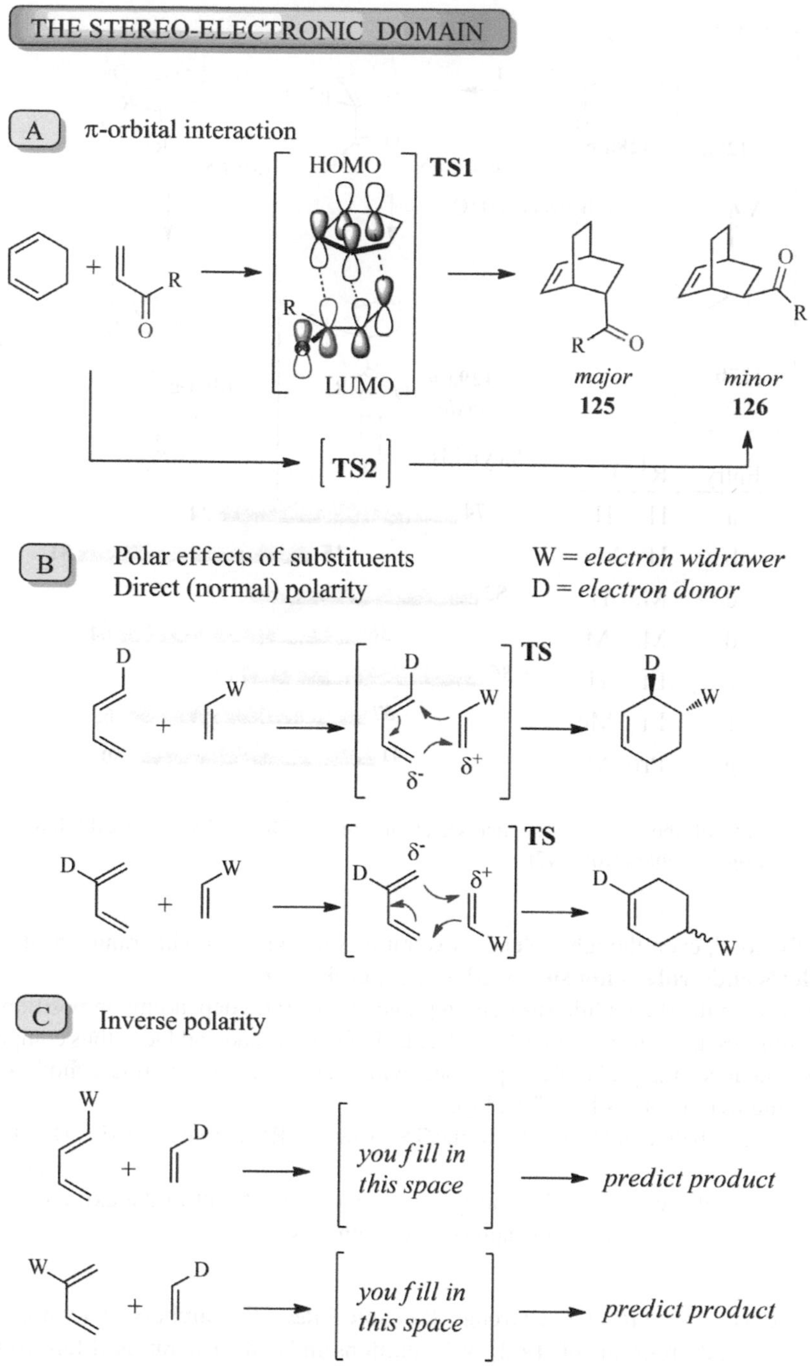

SCHEME III.17 Dominant electronic effects in [4+2] cycloadditions and regiospecific outcomes. Although the rules shown here apply in most cases, secondary interactions between substituents may participate in defining the regio- and stereochemistry of adducts.

3.5.1.3.1 Proof Based on Product Yield Researchers in England tested this question in a series of methodically selected dienophiles (α,β-unsaturated ketones) on cyclopentadiene [47]. Condensed results are put together in Scheme III.18. This chart is divided into two parts. The top section depicts the reaction under study, the endo and exo TSs and the end products. Endo and exo adducts **129** and **130** yields are shown in the bar plot.

Relevant issues and takeaways of Scheme III.18:

1. Cyclopentadiene (**127**) (CPD) is one of the classical testing dienes of DACA. It is a potent *dienophile*, so much so that what you have in the CPD flask is its dimer (endo by the way). CPD is transformed back to the monomer by heating it just before or during exposure to the dienophile.

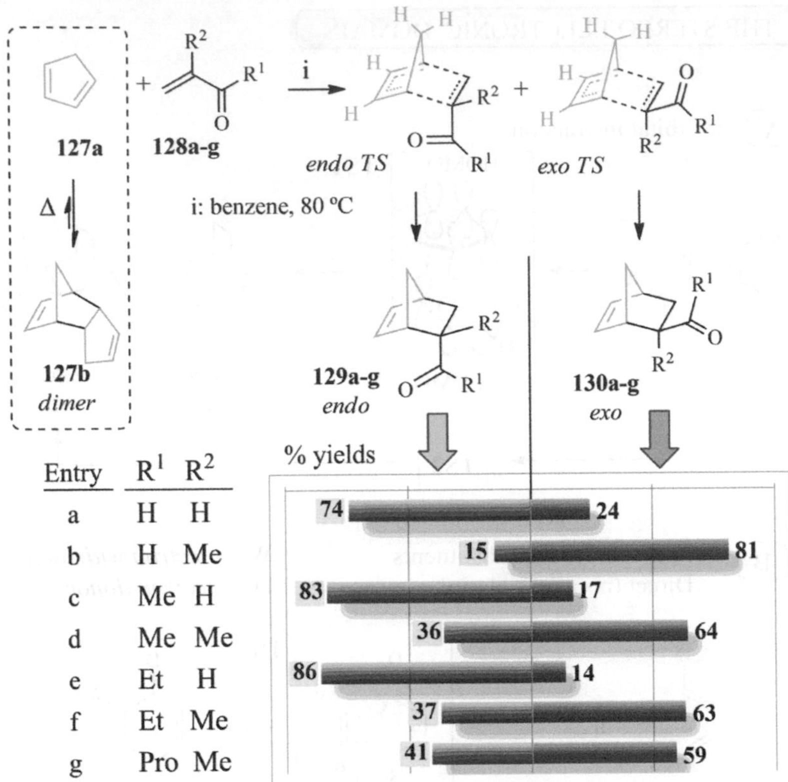

SCHEME III.18 Results of the *exo/endo* adduct selectivity in the Diels-Alder cycloaddition depending on the dienophile substitution pattern (**128a-g**). Adapted from [47].

2. Yields are generally good, even though the endo : exo ratios vary within a wide range, as the bar plot shows. The data confirms that **Alder's endo rule is not sustained as a general tenet**.
3. Substituents R^1 and R^2 in the dienophile strongly impinge on this ratio depending on position and size:
   a) R^1(H, CH_3, Et) (entries a, c, e in Scheme III.18) decidedly favor the endo adducts, thus complying with the Alder's rule.
   b) By contrast, R^2 enhances the yield of exo products with increasing size (compare modest exo yields H of entries a, c, and e against methyl in entries b, d, f, and g).
   c) Competition for space between R^1 and R^2 in the TS grows as R^1 increases in size (R^1 : Et, *n*-pro; R^2 : CH_3, bottom entries).
4. This feature reveals *SRe* of the bridge CH_2 and R^2 in the endo TS, absent in the exo configuration. Distal protons of cyclopentadiene do not resist the dienophile approach in both TSs.

3.5.1.3.2 Assessing Steric Effects (SEs) in DACA through Reaction Rates Researchers in Kyoto, Japan [48] developed this approach by measuring the second-order rate of [4 + 2] cycloadditions and endo : exo ratios of 1,4-disubstituted cyclopentadiene and vinyl carbonyl derivatives **135a–h** at 35 °C without catalysis (Scheme III.19). Strongly reactive maleic anhydride (**132**) was used as reference.

Relevant issues and takeaways from Scheme III.19:

1. There was a steady decrease in the endo preference:

Dienophile	*Endo %*
maleic anhydride (**132**)	99
ethyl-vinyl ketone (**135a**)	83
acrylaldehyde (**135b**)	75
methyl acrylate (**135c**)	73
α-*n*-butyl methyl acrylate (**135h**)	31

2. Surprisingly, when the second-order rates relative to acrylate methyl ester (**135c**) were compared with the recorded endo : exo ratios a linear correlation emerged after log-transforming the data (correlation coefficient of data points 2–9: $r^2 = 0.9725$) with the exception of entry **1**: maleic anhydride (**132**) (plot in Scheme III.19). The relative cycloaddition rate of **131** (R^3=H) + **132** was 3.1×10^5 times faster than that of butyl ester **135h**.
3. Entries 2, 3, and 4 demonstrate considerably higher endo : exo RRs with unsubstituted cyclobutadiene, independently of the aldehyde, ketone, or methyl ester moiety of the dienophile. However, *RRs were impacted by SS*, due to increasing steric encumbrance at the TS stage.
4. The more reactive the dienophile, the higher the endo stereoselectivity. Authors [48] supported these results by the pivotal concept of *secondary orbital interaction* between diene and dienophile in the TS, as mentioned earlier (Scheme III.17). Continuing research has confirmed this tenet.

3.5.1.4 The Exo : Endo Ratio and Lewis-acid Catalysis Lewis-acid catalysis, typically aluminum chloride, is known to increase substantially the reaction rates of carbonyl-endowed dienophiles in DACA and other cycloadditions. The acid–base interaction decreases the electron density of the C=C bond, brings the LUMO energy level closer to the HOMO of the diene, and lowers the TS $\Delta G^{\ddagger}$.

Better adherence to the Alder's endo rule would be expected in DACA under Lewis catalysis, given the relationship between reactivity and endo preference are related, This question was put to experimental test with and without catalysis in benzene at 30 °C. The abridged results appear in Figure III.14 [49]. Enhanced endo selectivity is clearly evidenced in all tested reactions.

Relevant issues and takeaways from Figure III.14:

Authors [49] discussed their results on three grounds:

1. *Steric stress in the TS*: A larger steric compression would be expected in the dienophile-$AlCl_3$ complex than in the non-catalyzed reaction. However, given that the endo : exo ratio was already high in most of the non-catalyzed reactions the enhanced steric effect on account of the catalyst is perhaps of secondary importance at least in this series of dienophiles. More recent developments using an expanded succession of Lewis acids have given rise to not only substantial

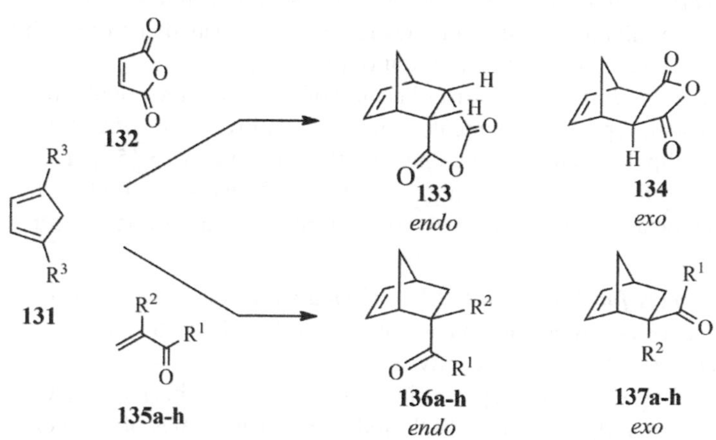

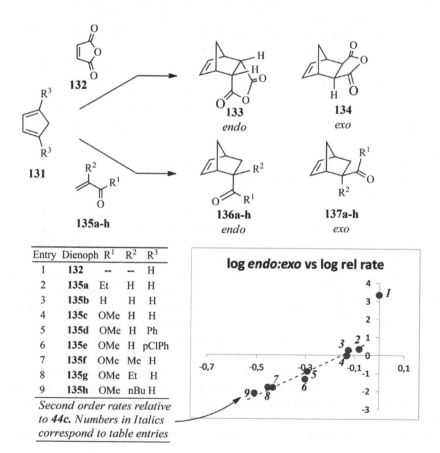

Entry	Dienoph	R^1	R^2	R^3
1	**132**	--	--	H
2	**135a**	Et	H	H
3	**135b**	H	H	H
4	**135c**	OMe	H	H
5	**135d**	OMe	H	Ph
6	**135e**	OMe	H	pClPh
7	**135f**	OMe	Me	H
8	**135g**	OMe	Et	H
9	**135h**	OMe	nBu	H

*Second order rates relative to **44c**. Numbers in Italics correspond to table entries*

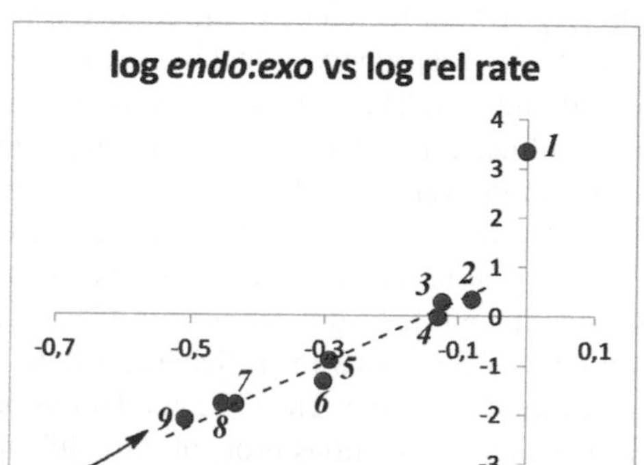

SCHEME III.19 Study of the *exo/endo* stereoselectivity and reaction rates in the Diels-Alder cycloaddition of some diene and dienophile combinations. Adapted from [48].

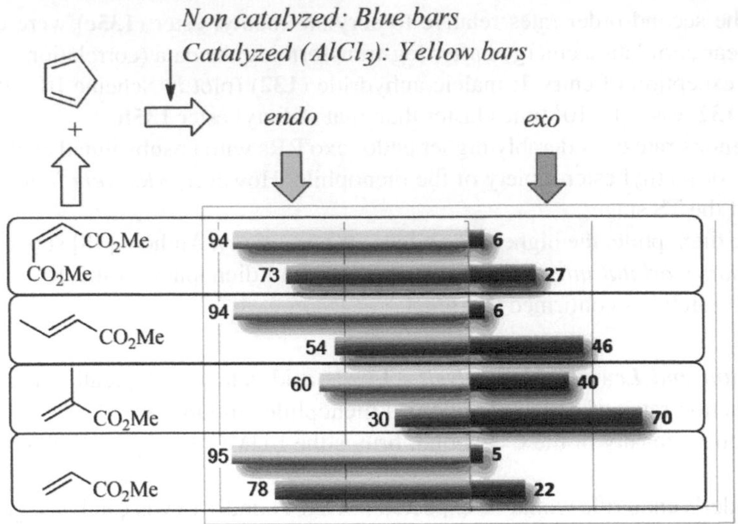

FIGURE III.14 [4+2] cycloadditions of dimethyl maleate and three acrylate esters with cyclopentadiene, with and without aluminum chloride catalysis. Adapted from [49].

improvements in stereoselectivity associated with catalyst steric congestion [50] but have reached the stage of chiral catalysts of aluminum, boron, titanium, nickel, copper, lanthanides, and other metals capable of inducing not only very high endo : exo ratios or the opposite but also enantio-control in DACAs on the order of >96% *ee* (enantiomeric excess). The subject has been reviewed [51], comprising new mechanistic insights and applications.

2. *Greater π–π rapport in the endo TS than in the exo TS through secondary orbital interactions.* Authors [49] contend that this interaction is enhanced in the aluminum-ester complex relative to the bare diene-dienophile TS according to models and the influence of the increased reaction rate on account of the catalyst.
3. *A polar effect of the dienophile-catalyst complex* arises because of C=C electron redeployment by the C=O–$AlCl_3$ complexation. This is consistent with reports of augmented endo : exo ratios *as the solvent polarity is increased.* Authors contend that there is a larger dipole moment in the endo TS than in the exo TS [52]. This tenet has led to the recent introduction of water as solvent for Lewis acid-catalyzed DACAs [53] and several other environmentally friendly solvents, e.g. glycerol and gluconic acid, polyethylene glycol, and supercritical carbon dioxide [54].

3.5.1.5 Current and Future Prospects for DACA and Other Cycloadditions DACA research is far from outmoded. A stream of current research, theoretical or experimental, update concepts of C–C bond-forming reactions and even turn upside-down some paradigms thought to be well settled until recently.

The role of secondary orbital interactions (SOIs) depicted in Scheme III.17 has been criticized and is a subject of current debate. The concept has migrated from a fully convincing SOI model [55], that is, AOs of C=O or other electron withdrawers of the dienophile and the distal *p* AOs of the diene in the endo TS, to a complete disregard for SOI to explain endo selectivity. An alternative model states that SOI is a primary orbital interaction (POI) to which *electrostatic attraction* of these two sections is added [56] favoring the endo adduct. This can be pictured (Scheme III.20, top part) as a re-interpretation of the endo TS wherein the light hyphened red lines represent this non-bonding electrostatic attraction while the dark dashed lines mark the AOs in the process of bond-forming overlap, in the way to the endo adduct proposed by the frontier molecular orbital model.

The long-held model of little or no charge separation in the DACA TSs, as proposed by the frontier molecular orbital theory (FMO), has been debunked, even in highly polar dienophiles (Scheme III.20) by DFT scrutiny. Nucleophilic–electrophilic interactions between diene and dienophile lead to the notion of a polar Diels–Alder (P-DA) TS based on global charge transfer (GCT) or electron redeployment in our terms [57] (Scheme III.20, bottom part). The dimension of this charge transfer is a function of the EWG power of the dienophile. The GCT model has been introduced to replace the harmonious concerted C–C bond formation [57]. Such reagent combinations brought a modification of cycloaddition nomenclature: $[4+2^+]$ to indicate a cationic dienophile or $[4^+ +2]$ for the diene.

Taken together, a new DACAs classification is proposed: non-polar (N-DA), polar (P-DA), and ionic (I-DA) (Scheme III.20, bottom part) [57, 58].

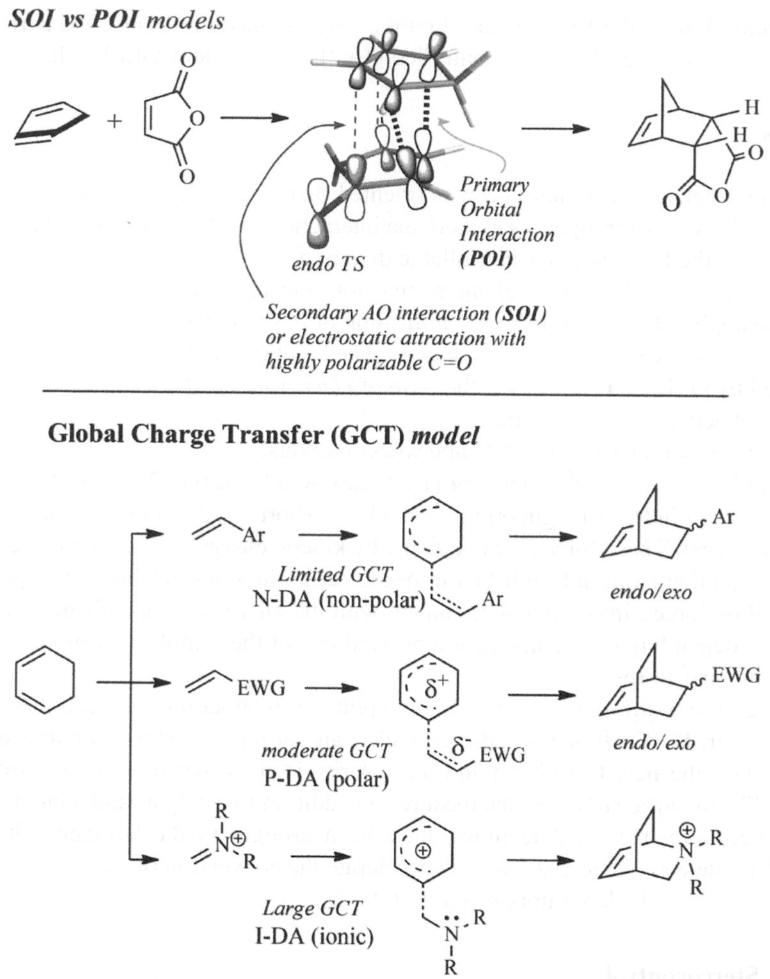

SCHEME III.20 Novel TS models for Diels–Alder cycloadditions. Illustrations are inspired in References [55–58].

Additionally, the various models of cycloadditions involving heteroatoms that yield new C–X bonds (X=N, O, S, Si) or diene/dienophiles bearing boron, phosphorous, and other appended substituents bring about new mechanism challenges and products never seen before [59]. These advances are nutritious additions to interesting reaction mechanism problems.

3.6 THE ULTIMATE STEREO- AND ENANTIO-CONTROL: ORIENTED EXTERNAL ELECTRIC FIELDS (OEEFS)

Electrons and thus chemical reactions respond to local electromagnetic fields created by neighboring molecules. Given that electric fields possess vector properties, and thus change directions at every turn of molecules, before these come close enough for these local fields to have any effect on others and build up a local organized molecular array, the four rotational degrees of freedom and displacement of each individual molecule in all possible directions of space at great speed turn the dipole collection in gas or liquid state into a characteristically chaotic mess. Consequently, in such an unruly environment, only a small fraction of molecules per unit time (say picoseconds) in bimolecular reactions will achieve viable TSs in the direction of products.

Enquires on Alder's endo rule in DACA and the tailing controversies have discussed the prolific subject of specific orientations according to SOI, POI, N/P/I-DAs, or other intermolecular forces and TS structure proposals, working on the locally and momentarily organized TS amidst a surrounding electrostatic field (EF) chaos in the rest of the reaction body.

Now, it is time for a change in paradigm, and the familiar DACA and stereochemical overtones are adequate to guide our steps through it.

Accepting that a certain amount of dipole moment (μ) assists either diene, dienophile, or both, and therefore, their susceptibility to local EFs, what if one applies an oriented *external* EF (OEEF) to the reaction vessel to put some order in the natural generalized EF pandemonium? Would the EF so focused on the reaction medium induce the correct positioning of a dipolar dienophile close to the diene to get a specific stereo-controlled [4+2] adduct? Does OEEF achieve this experimentally?

3.6.1 How OEEF Works

EF possesses vector properties, hence directional. It can be oriented sharply in a desired direction of the Cartesian XYZ space. Because the dienophile μ displays vector properties as well, the interaction of EF + μ should result in the orientation of all polar or polarizable molecules within the EF reach in a controllable direction.

Moreover, charge transfer species $X^{(\delta-)} \rightarrow Y^{(\delta+)}$ along the reaction coordinate, as suggested in Scheme III.20 [57, 58], would be influenced as well by the applied OEEF furnishing stabilization or destabilization of the corresponding coupling or decoupling transient assemblages and decrease the activation energy. The kinetics and thermodynamics of reactions of a potentially broad scope may be affected by OEEFs. This leads to the birth of *electrostatic catalysis* [60], a concept nature developed eons ago alongside the evolution of active sites of enzymes.

This concept is graphically shown in Scheme III.21 and works like this:

Imagine a set of polar molecules $A^{(\delta+)}$–$B^{(\delta-)}$ moving chaotically in a Cartesian XYZ space (part **A**, left rendering). Each dipole generates a local EF, which induces neighboring molecules to short-lived field interactions (H-bonding, dipole–dipole, van der Waals, and London forces). These forces are overcome by kinetic energy driving molecules away.

Then, an electric field is applied externally (OEEF) in one specific direction of this XYZ space (part **A**, right drawing). Instantly, the molecules will be forced into ordered alignment with the OEEF vector. This new organization may or may not have chemical consequences depending on the interaction possibilities of these molecular dipoles, such as head–tail bonding, S_N2, addition, polymerization, and so forth.

Let us examine a mixture of two different compounds susceptible to interactions between them, water and methyl acetate, for example (Scheme III.21, part **B**). Each one is endowed with μ and/or may constitute a polarized TS. This TS gets aligned with the OEEF co-axially with the new C–O bond and the polarity is such that it favors electron traffic. Bimolecular rate acceleration should result: *Electrostatic catalysis*, the most recent addition to catalysis and a budding field at this time.

This concept, and the necessary technical facilities, have been brought to the laboratory in a handful of organic and biochemical systems [60, 61]. Indeed, some enzymes can accelerate the conversion of substrates docked in their active site up to 20 orders of magnitude relative to the laboratory reaction [60, 62].

3.6.2 OEEF and DACA Stereocontrol

The orientation of one of the (polar) components relative to the other reagent may define the stereochemistry of the final adduct. This is precisely the case of DACAs where stereoisomers are formed. If the required orientation of diene and dienophile is induced by the OEEF, there would be control over the topology of the TS.

Has it been realized in the laboratory? Yes, just a few times up to now in organic and organometallic combinations at the molecular scale employing unorthodox equipment, a great deal of quantum theoretical treatment and ingenuity [63–65]. Two of these are of instructive and mechanistic relevance as well as state-of-the-art of cycloadditions. While this experimental array is momentarily designed at the molecular scale only, it may attain micromolar scales not far in the future.

3.6.3 The Experimental Array

Using a sophisticated facility known as scanning tunneling microscopy (STM) to observe *single molecule performance*, (see Figure III.15) a team of researchers in Spain and Australia managed to probe a DACA under a unidirectional EF [63]. As opposed to molecules in solution moving in every possible direction and colliding in such a way to produce several stereoisomers, it is nowadays possible to use surface chemistry techniques to affix compounds rigidly to a surface and force their reaction in a pre-established direction (a vast development in the aging model of solid–liquid phase hydrogenation under palladium–charcoal catalysis).

The setting can be briefly described as two tiny gold pieces separated by a gap a few angstroms wide (Figure III.15, top part); the upper one (plate *i*) is a very sharp wire mounted on a frame that can be moved with extreme precision over the lower surface (plate *ii*); the device was designed to map plate *ii* surfaces at the atomic scale. There are protruding, inverted, pyramid-shaped, extremely fine apexes at the tip of plate *i* comprising a few gold atoms. Plate *i* is chemically treated to load molecules of reactant **A** through a thiol-gold (S–Au) bond, hoping that one individual molecule will be tethered at the very end of each inverted pyramid.

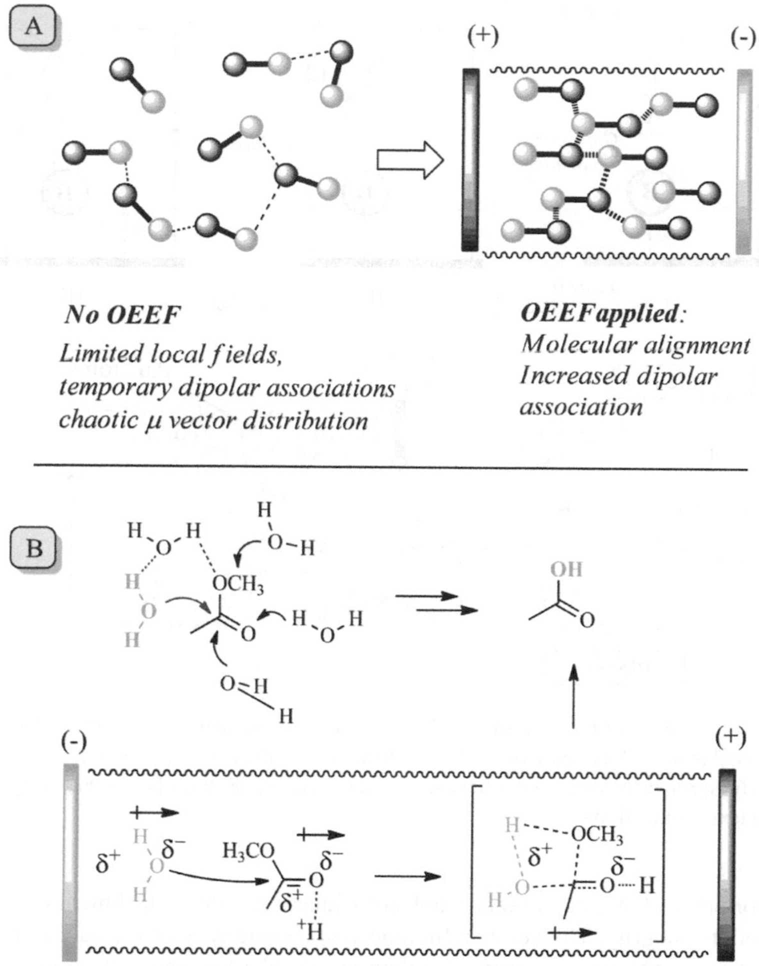

SCHEME III.21 Hypothetical model depictions of (part **A**) dipolar molecules moving randomly, only controlled by local electrostatic fields (EF) of molecular dipoles, and the effect of an oriented external electrostatic field (OEEF). Dipolar molecules become oriented instantly in this field and the expression of dipolar interactions is much enhanced. Part **B** portrays a simple model of ester hydrolysis occurring in the absence of OEEF. Water molecules approach from various fruitless directions except for one (green water molecule). In principle, as OEEF is applied, water and ester dipoles should become aligned to the OEEF, ineffective alignments of water disappear and the effective backside attack on C=O is much enhanced. The actual experiment has not been carried out at the time of printing.

In this case, **A** is 2-methyl-3-thiomethyl furan, our diene. Because it is singly bonded to the Au tip, the molecule can swirl around freely. In so doing it exposes two different faces to the dienophile. This feature will have stereochemical consequences as we shall see presently.

The dienophile, a norbornene derivative **B** is chemically affixed to the bottom gold surface (plate *ii*) through *two trans* S–Au bonds. **B** rises above the gold surface in the direction of **A,** a bit tilted from the perpendicular line of the plate *ii* surface (the *z*-axis). Unable to rotate, **B** is rigidly held exposing its distal C=C bond to the diene **A**.

The gold plates are sections of a direct current circuit. They are moved with extreme precision until **A** and **B** are at near TS distance (Figure III.15 top section, left image **I**). Once there, a small voltage is applied to create an enveloping oriented electric field, the desired well-focused OEEF, which results from the pointed inverted pyramid of the top electrode plate (center image **II**). The electric field is pointed along the reaction *z*-axis.

The ± polarity of the current can also be inverted to modify the electron flow between molecules **A** and **B** in the TS. According to quantum calculations, authors report that a polarized TS was induced by the OEEF, which evolved to the Diels–Alder adduct [63]. After passing through the TS, the bonded product becomes an electric wire, a *molecular junction*, that connects the two electrode plates. This was registered as a brief current spike or *blink*, which was interrupted as the furan-tethering sulfur bridge collapsed soon after to open this molecular section of the circuit and end the blink (image **III**). A blink sparked

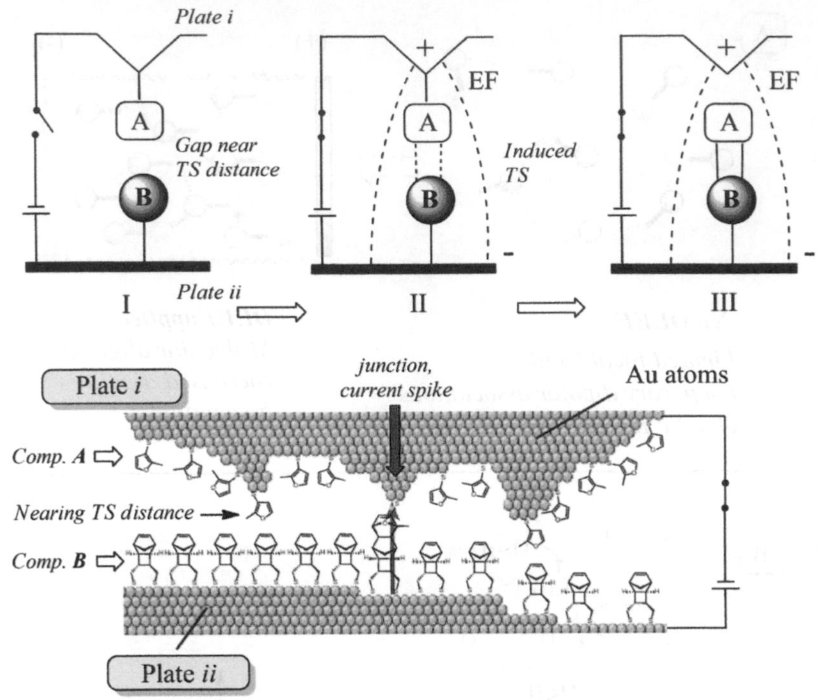

FIGURE III.15 Simplified schematic representation of a molecular scale arrangement adapted to a STM facility to investigate a Diels–Alder cycloaddition at the molecular scale Adapted from [63]. The technique is explained in the text.Color rendering reproduced from the original Reference [63]. © 2016, Springer Nature, under license no. 5284101239568, with labels added by this book's author (M Alonso-Amelot) for compatibility with text explanations.

every time a junction was formed and stopped in other molecular junctions after detachment of the upper sulfur bridge while the rest of the adduct retained its integrity. The number, lifetime, and conductance of the current blinks were monitored at various voltages at direct or reversed polarity as several bonding events like the one just described occurred in other similar face-to-face sites of the electrodes in the same experiment.

Therefore, the experimental data of the reaction is not yield, purity, NMR/infrared/mass spectra of adducts, and bulk reaction rates as we, wet chemists, are used to but electric blinks from sophisticated equipment (commercially available by the way) and a great deal of quantum theoretical modeling.

Outcomes, interpretation, and takeaways from this OEEF experiment:

1. *Compounds at play*: In the actual experiment, **A** and **B** were furan **131 (F)** and norbornene **132 (NB)** (Figure III.16). Combining the asymmetry of **131** and the exo/endo coupling with **132** a set of six major stereoisomers (**133a,b–135a,b**) may be formed. This many products complicates any interpretation. One can cross out the endo-cis adducts **135a,b,** since **131** and **132** are not aligned from their respective sides of the plates. Likewise, the TSs leading to *endo-trans* **134a,b** require a closer imbrication along the furan–norbornene approach axis. This leaves the two *exo* **133a,b** as the most likely products to work with.
2. *Polar versus non-polar TSs*: Once a selection of more likely products is made, a TS can be built. There are three assumptions regarding furan–norbornene electron density transfer induced by the OEEF focused in the direction of the reaction axis:
   a) Given that **132** is a non-polar, electron-poor dienophile, little if any polarization is expected to arise in the TS, and a concerted cycloaddition follows. This assumption disregards the stepwise GCT model of Scheme III.20, which still awaits universal acceptance. Partial charge transfers can be postulated but would be minor contributors to the concerted TS **II** in the absence of OEEF.
   b) However, if **131** is a clear-cut HEDZ molecule, **131** → **132** charge transfer in the TS is normally directed but may be reversed owing to the lack of EWGs in **132**.
   c) Given the –*I* effect of sulfur and oxygen in **131,** the opposite electron density drift direction can be proposed.
4. The three possibilities are resonance structures of the same TS but their sensitivity to the polarity effect of OEEF; hence, their contribution to the resonance hybrid, would display different energy levels depending on the strength and polarity of the applied OEEF.

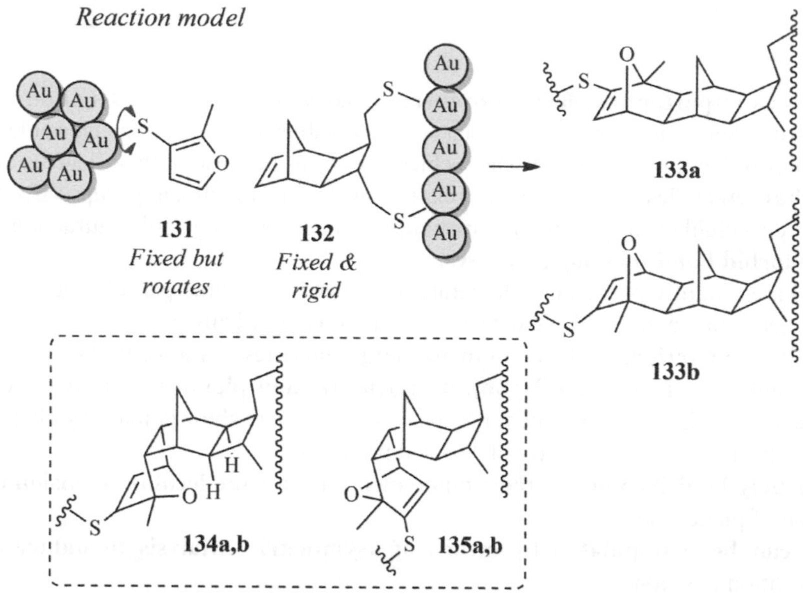

FIGURE III.16 Reaction model adopted for the OEEF study [63] and possible stereoisomers. [63] / Springer Nature.

5. *Charge transfer* stemming from a one electron transfer would be furthered by the OEEF according to the polarity of the field.
6. *Activation energy*: a plot of calculated $\Delta E^{\ddagger}$ ($kJ\,mol^{-1}$) as a response of the strength of the OEEF in the range of −1.0 to +1.0 $V\,nm^{-1}$ showed a smooth positive correlation in **133a**. The lower $\Delta E^{\ddagger}$ values corresponded to a field direction stabilizing structure and increased its contribution to the hybrid. Electrons flow from the negative bottom plate through **132**, and up to **131**, which is the opposite direction one expects from an uncharged array (no field). Quantum calculations reveal that the variation of $\Delta E^{\ddagger}$ between −0.8 and +0.8 $V\,nm^{-1}$ was 91.6–93.2 $kJ\,mol^{-1}$ in stereoisomer **133a**. The difference is minor and yet significant enough to be capable of altering the reaction rate.
7. For reasons difficult to explain, similar $\Delta E^{\ddagger}$ calculations of **133b** showed it to be practically insensitive to the variations in EEF strength, remaining at an average $\Delta E^{\ddagger} = 92.5\,kJmol^{-1}$. From this data, at EEF < 0 the more favored product is **133b**. The opposite occurs at EEF > 0.4, whereas the **133a/b** ratio is ~1 at EEF +0.1 to +0.3 $V\,nm^{-1}$.
8. *Blinks lifetime meaning*: The average lifetime of blinks was 0.4 seconds. This is the time the adduct remains connected to both plates and responds to a stable molecular junction (red arrow, Figure III.15). This is only possible if a *robust molecular wire exists between the plates*. Authors could show that blinks cease not because of C–C bond cleavage in the molecular backbone but after Au–S rupture. Thus, DACA results when **131** and **132** are brought together under an OEEF.
9. *Reaction rate*: This is possibly the major discovery of this research. Considering the blink frequency per unit time and other quantum tunneling microscope techniques, authors observed a substantial increment of molecular junctions (sending an electric blink), namely increased DACA rates, as the surface field potential changed: 252 adduct molecules out of 6000 attempts at −0.05 V to 1118 adduct molecules out of the same number of attempts at −0.75 V, a 4.4-fold rate increase [63]. This is not only consistent with the $\Delta E^{\ddagger}$ calculations as a function of field strength, but the first-time demonstration that a C–C bond-forming reaction rate is accelerated by an OEEF.

Translating this ultra powerful microscope arrangement to the preparative scale is still beyond the horizon of today's lab scale preparative technology but may be forthcoming in the near future. Advances in theoretical calculations of the effects of EFs on the dipolar orientation of molecules approaching cycloaddition TS (and other reactions) promise effective control of complete stereo and enantioselectivity in DACA and other cycloaddition models by selecting properly the direction and polarization of the applied EF [64].

To our satisfaction as reaction mechanism designers, the OEEF experiments would not be possible without previously devising detailed pencil/paper reaction models.

3.7 SUMMING UP

1. **Stereochemistry is as important as electron properties** and MOs for the analysis of organic reaction mechanisms. The redeployment of valence electrons in molecular scaffolds subjected to reaction is feasible only in specific shapes of interacting molecules in the ground state or other achievable forms because of their degrees of freedom.
2. **Stereochemical characteristics** (scaffold shape, relative position of functional groups, SEs, hindrance, encumbrance, or assistance through neighboring group participation, hydrogen bonding, polar attraction of appropriately situated groups) **enable or forbid bond-forming processes.**
3. Organic molecules are a tightly packed conglomerate of charged subatomic particles occupying space and generating highly interactive electromagnetic fields. What we know as **steric hindrance** or congested molecular zones **constitute negatively charged field barriers against of approaching molecules or molecular sections of the same compound**.
4. **Individual atoms and all molecules are 3D objects, no matter how planar they may seem on paper.** Several atoms in each molecule can nevertheless share a common plane and yet, may show potentially chemical properties associated to space with consequences in the symmetry of TSs and products.
5. **Planar molecules may be driven out of their plane owing to intramolecular repulsion** opening opportunities for differential reaction of plane sides.
6. **Steric repulsion can be manipulated by means of asymmetric catalysis to induce chirality** in bimolecular substitution and addition reactions.
7. **The stereochemistry of unsaturated carbocycles of seven atoms or more forces seemingly conjugated C=C bonds out of conjugation by a severe distortion of the molecular plane,** as in cyclooctatetraene (COT). The pattern of well-known and predictable reactions of such compounds undergoes substantial changes, leading to novel reactivity paradigms (Scheme III.3).
8. **Molecular steric characteristics have four major consequences for reaction mechanism**. They:
   a) Cause hindrance of reaction rates.
   b) Cause acceleration of reaction rates.
   c) Define the stability of molecular conformations and/or configurations, equilibrium constants, and reaction outcomes.
   d) Serve as conduits for stereo and enantioselectivity.
9. **Whereas steric retardation has been experimentally observed very frequently, cases of formal demonstration of rate acceleration by steric congestion are rare.** Although one cannot postulate generalizations on such a small set of evidence, one can call upon liberation of SS passing from a starting compound to a rate determining TS, aided by an overall exothermic reaction. The cleavage of a large, sterically congested group would be a typical case of SS release influencing reaction rate.

   Qualitative appraisal of SS in each compound is perfectly admissible in problem analysis of organic reactions and continues to be useful for determining relative stability of rotamers, conformers, enantiomers, and diasteromers. This assessment is also adequate for predicting favorable spatial approaches of two or more molecules en route to feasible TSs and explaining the target products. **This approach is nevertheless much less suitable for predicting energy levels of the TS, activation enthalpies, and comparison of overall ΔG^o change from reactants to products**.
10. **The general reactivity rule primary > secondary ≫ tertiary carbon substitution patterns continues to be valid in bimolecular substitutions and additions** in regards to the carbon center undergoing attack. Unimolecular reactions follow the opposite order. **Increasing steric congestion at the α and β carbons has a greater impact on the reaction center than groups located beyond the β carbons**. Quantitative studies have discovered exceptions to this appraisal.
11. While the qualitative assessment of steric hindrance or interaction is a valuable tool in problem analysis, applying **quantitative steric factors of substituents like Taft's E_s gives additional support to visual appreciations**.
12. **Evaluation of purely SEs isolated from the concomitant electronic influences (inductive, polar, field effects) entails considerable difficulty to prove experimentally.**
13. **Efforts to bring SS to quantitative terms independently of electronic effects has been achieved with mixed success** by means of detailed kinetic studies of systematic series of compounds submitted to well-known reactions. However, the quantitative steric factors of substituents are not of universal application to all types of bimolecular processes. **Case-by-case scrutiny alongside quantum chemical calculations have proved a more powerful tool to account for SEs.**
14. **Other descriptors to evaluate potential SS have been developed** and can be accessed through computers. Connolly's solvent exclusion surface area (SEMA) and volume (SEV) based on the van der Waals radii of all atoms in the molecular conglomerate provide well-defined figures in $Å^2$ and $Å^3$ dimensions, representing the area and space occupied by molecules, respectively, from simple organic compounds and subgroups to proteins and active sites of enzymes.

15. SEMA and SEV offer clues of stereochemistry's role on reaction mechanism conditioning, solvent accessibility-solubility, intermolecular non-bonding interactions, and conformation in medium-size carbocycles.
16. **Attempts to correlate kinetically determined steric factors (E_s) with Connolly's SEV meet mixed success only** but continue to be a reassuring tool when correlations approach linearity. Alkyl groups located in α and β carbons of alkyl halide electrophiles during S_N2 reactions do correlate well with experimental E_s although *n*-alkyl chains of primary alkyl halides do not. Translating SEV to the nucleophile in S_N2 reactions furnish poorer correlations E_s/SEV, casting doubt on their universal applicability in the assessment of quantitative steric influence on reactivity. Cautious interpretation is required.
17. **SEs are often used to the advantage of reaction synthesis design:** Protective moieties of reactive groups not only replace reactive protons but also operate as steric umbrellas to fend off reactants aimed at other molecular sections or as adjuvants to induce stereo and enantioselectivity.
18. **Cycloaddition reactions have been a showcase for the intervention of SEs in reactants impinging on the stereochemical outcome.** From a diene/dienophile single pair as many as eight stereochemically differentiated products are obtained. Rules exist to rationalize the preference of some adducts over others (endo versus exo plus cis/trans selection, for example), but **these canons are easily violated** depending not only on relative SS but EW/ED effects of substituents, their specific location in the interacting π systems, and Lewis-acid catalysis.
19. **Application of externally applied electric fields oriented in specific direction of space over polar or polarizable reacting species can exert considerable control over their positioning in bimolecular cycloadditions as they approach to form the TS.** Enantiospecific products can be obtained with high specificity.

3.8 SUPPLEMENTARY SCHEMES

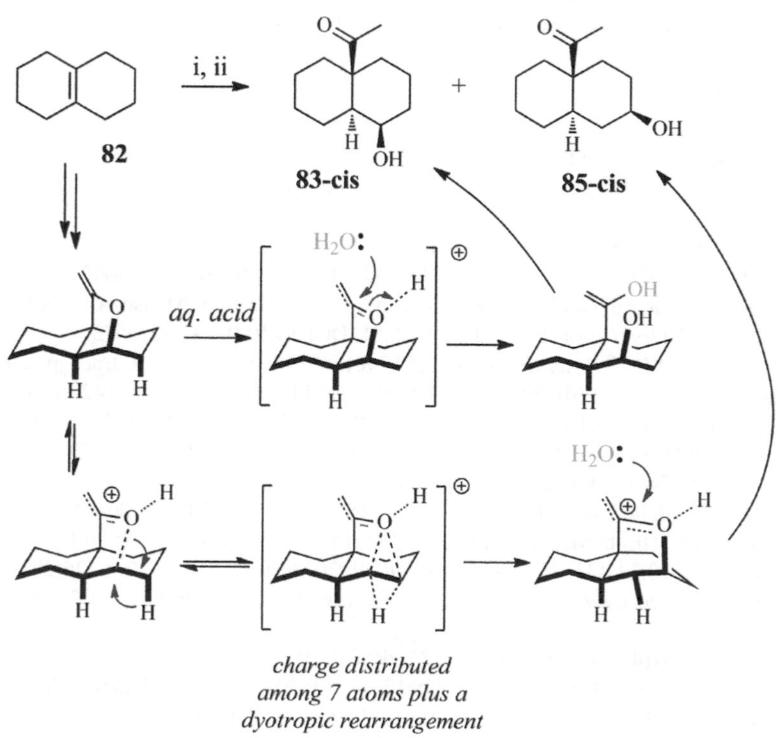

SCHEME III.22 Sequence explaining the correct stereochemistry of the reaction proposed in Scheme III.11.

NOTES

1. Calculation methods employed: (a) B3LYP/6-31G level; (b) MMFF96s; (c) HF/6-31G level. Alonso-Amelot ME, this book.SEs of substituents have been explored by several methods. The most well-known are Taft's steric substituent constants *Es* and their modified values of other constants developed ever since. Taft's prominent criteria of classification are the RRs of ester formation, hydrolysis of aliphatic carboxylic derivatives, and analogous o-substituted aromatics. Taft found that steric hindrance is more relevant in β than in α substituents, which has been confirmed many times. For guidance, see References [10–15].
2. *On the big problem of isolating the steric effect on reactivity.* In this regard, there has been a string of papers over the years attempting to separate steric and electronic effects of substituent groups, given their variety, elemental composition, configuration conformation, dipolar interaction, and electromagnetic fields they create in their surrounding environment. One example of such efforts among many others is that of Professor Kunio Okamoto and colleagues in Japan. They attempted to correlate the reaction constants r of alkyl halides with the steric factor E_s developed earlier by Professor Robert Taft who created a quantitative dimension to SS or hindrance of several substituents in various reaction types (see Note 1). Okamoto's data was based on many kinetic measurements (34) of a variety of S_N2 reactions that they patiently carried out in the laboratory with excellent correlation coefficients. His r results would not give the expected r versus E_s linear correlation that would have supported the congruence of a unique criterion for the quantification of the steric effect regardless of other factors. Professor Okamoto spent several years of his career pursuing this complex subject. See: Okamoto K, Nitta I, Imoto T, Shingu H. *Bull. Chem. Soc Jp.* 1967;40:1905–1908 and ensuing papers.
3. Kinetic studies of selected organic reactions are carried out with clean, simple, and pure compound/solvent combinations under well controlled experimental induction conditions, concentration, temperature, time, and fully tested/accepted analytical methods. Thermolysis in the gas phase and low pressure further debug the reaction system of solvent effects, often decisive. Under these conditions, very few fortuitous phenomena, like unexpected side reactions, intervene, thus small standard deviations of replicates and data associations and trends showing high correlation coefficients $r^2>0.8$ are not uncommon. By contrast, quantitative studies of biological systems in cell, animal and human biology and biochemistry, ecology, disease, population stats and others involving animals and their organs are filled with the so-called confounding (unknown, imponderable) factors. These untraceable interferences tend to spread data points widely in correlation plots with $r^2<0.5$; yet many scientists draw imaginative conclusions and claim experimental demonstration to working hypothesis, which are based on preliminary observations only. For this reason, outlier data points in chemical laboratory experiments should be put aside from correlation lines and studied individually in search of reasons for off-the-line performance.

REFERENCES

1. Bondi A. *J. Phys. Chem.* 1964;68:441–451. DOI:10.1021/j100785a001.
2. Rodríguez-Esrich C, Davis RL, Jiang H, Stiller J, Johansen TK, Jørgensen KA. *Chem. Eu. J.* 2013;2932–2936. DOI:10.1002/chem.201300142.
3. Li WJ. *Catal. Commun.* 2014;52:53–56. DOI:10.1016/J.catcom.2014.04.013.
4. Huisgen R, Boche G, Huber H. *J. Am. Chem. Soc.* 1967;89:3345–3346. DOI:10.1021/ja00989a042.
5. Calculation method: MMFF94 molecular dynamics at 300 K and energy minimization. Alonso-Amelot ME, this book. 2022.
6. Sasaki T, Kanematsu K, Kondo A. *J. Org. Chem.* 1974;39:2246–2251. DOI:10.1021/jo00929a025.
7. Liu X, Xie J, Zhang J, Yang L, Hase WL. *J. Phys. Chem. Lett.* 2017;8:1885–1892. DOI:10.1021/acs.jpclett.7b00577.
8. Turovtset V, Orlov YD. *Russ. J. Phys Chem.* 2010;84:965–970. DOI:10.1134/S0036024410060142. Accessed: 03/17/2021.
9. Turovtset V, Orlov YD. *Russ. J. Phys Chem.* 2010;84:1174–1181. DOI:10.1134/S0036024410070174. Accessed: 03/17/2021.
10. Hammett LP. *J. Am. Chem. Soc.* 1937;59:96–103. DOI:10.1021/ja01280a022. Accessed: 03/21/2021.
11. Taft Jr. RW. *J. Am. Chem. Soc.* 1952;54:3120–3128. DOI:10.1021/ja01132a049.
12. Panaye A, MacPhee JA, Dubois J-E. *Tetrahedron* 1980;36:759–768. DOI:10.1016/S0040-4020(01)93691-9.
13. Komatsuzali T, Akai I, Sakakibara K, Hirota M. *Tetrahedron* 1992;48:1539–1556. DOI:10.1016/S0040-4020(01)88712-3.
14. Shorter J. *Quart. Rev. Chem. Soc.* 1970;24:433–453. DOI:10.1039/QR9702400433. Accessed: 03/26/2021.
15. Shorter J. In: Chapman N, Shorter J. (ed.). *Advances in linear Free Energy Relationships*, Chapter 2, pp 71–118. Plenum Press, London, New York, 1972.
16. Uggerud E. *Pure Appl. Chem.* 2009;81(4):709–717. DOI:10.1351/PAC-CON-08-10-03.
17. Santiago CB, Milo A, Sigman MS. *J. Am. Chem. Soc.* 2016;138:13424–13430. DOI:10.1021/jacs6b08799.
18. Okamoto K, Nitta I, Imoto T, Shingu H. *Bull. Chem. Soc Jp.* 1967;40:1905–1908. DOI:10.1246/bcsj.40.1905.
19. DeTar DF. *J. Org. Chem.* 1980a;45:5166–5174. DOI:10.1021/jo01313a029.
20. DeTar DF. *J. Org. Chem.* 1980b;45:5174–5176. DOI:10.1021/jo01313a030.
21. MacPhee JA, Panaye A, Dubois JE. *J. Org. Chem.* 1980;45:1164–1166. DOI:10.1021/jo01294a052.
22. Connolly ML. *J. Am. Chem. Soc* 1985;107:1118–1124. DOI:10.1021/ja00291a006. Since this work was published, a string of improvements and applications, especially in protein and polymer properties, have appeared and continue at present.
23. Connolly ML. *J. Molec. Graph.* 1993;11:139–141. DOI:10.1016/0263-7855(93)87010-3.
24. Kochetova LB, Kustova TP, Kuritsyn LV, Katushkin AA. *Russ. Chem. Bull. Int. Ed.* 2017;66:999–1006. DOI:10.1007/s11172-017-1846-0.
25. Liu S, Hu H, Pedersen LG. *J. Phys Chem.* 2010;114:5913–5918. DOI:10.1021/jp101329f.

26. Turotsev VV, Orlof YD. *Russ. J. Phys Chem.* 2010;84:1174–1181. DOI:10.1134/S0036024410070174.
27. Banert K, Seifert J. *Org. Chem. Front.* 2019;6:3517–3522. DOI:10.1039/c9qo00312f.
28. Brown HC, Nakagawa M. *J. Am. Chem. Soc.* 1955;77:3610–3613. DOI:10.1021/ja01618a055.
29. Brown HC, Nakagawa M. *J. Am. Chem. Soc.* 1955;77:3614–3619. DOI:10.1021/ja01618a056.
30. Kumar A, Sharma PK, Banerji KK. *J. Phys. Org. Chem.* 2002;15:721–272. DOI:10.1002/poc541.
31. Slakman BL, West RH. *J. Phys. Org. Chem.* 2019;32:e3904. DOI:10.1002/poc.3904.
32. Cech NB, Enke CG. *Mass Spect. Rev.* 2001;20:362–387. DOI:10.1002/mas.10008.
33. Wang HY, Zhang JT, Zhang SS, Guo YL. *Org. Chem. Front.* 2015;2:990–994. DOI:10.1039/C5Q00154D.
34. Balland L, Mouhab N, Cosmao JM, Estel L. *Chem. Eng. Procs.: Procs. Intens.* 2002;41:395–402. DOI:10.1016/S0255-2701(01)00164-7.
35. Kondratiev VN, Nikitin EE. *Gas phase reactions: kinetics and mechanisms.* Springer-Verlag, Berlin, 1981. DOI:10.1007/978-3-642-67608-06.
36. Chuchani G, Hernández JA, Avila I. *J. Phys. Chem.* 1978;82:2767–2769. DOI:10.1021/j100515a003.
37. Chuchani G, Martin I, Bigley DB. *Int. J. Chem. Kinetics* 1978;10:649–652. DOI:10.1002/kin.550100610.
38. Alonso-Amelot ME. Calculation method: MMFF94s interface energy minimization + molecular dynamics, relaxation at 300 K. This book.
39. Alonso ME. *J. Org. Chem.* 1976;41:1410–1412. DOI:10.1021/jo00870a026.
40. Gallegos M, Costales A, Martín Pendás A. *J. Phys. Chem. A* 2022;126:1871–1880. DOI:10.1021/acs.jpca.2c00415.
41. (a) Baddeley G, Wrench E. *J. Chem. Soc.* 1959;1324–1327. DOI: 10.1039/JR9590001324; (b) Baddeley G, Heaton BG, Rasburn JW. *J. Chem. Soc.* 1960;4713–4719. DOI: 10.1039/JR959600004713.
42. Wilinsky J, Kurland RJ. *J. Am. Chem. Soc.* 1978;100:2233–2234. DOI:10.1021/ja00475a045.
43. Kumar R, Halder J, Nanda S. *Tetrahedron* 2017;73(6):809–818. DOI:10.1016/j.tet.216.12.072.
44. Otto Diels and Kurt Alder discovered this reaction in 1928 at the University of Kiel, northern Germany, and published their findings in 28 papers between then and 1937, an epoch of deep social, economic, and political unrest in the country. They shared the Nobel Prize in Chemistry in 1950 for their landmark discovery.
45. Cui C-X, Liu Y-J. *J. Phys. Org. Chem.* 2014;76:652–660. DOI:10.1002/poc3133. In addition to dealing with the mechanism details of the cycloaddition of butadiene and ethylene, this article reviews updated (up to 2014) issues around the Diels–Alder addition worth collecting. It also shows how far a mechanism can be brought to a realm of great complexity, contradiction, and uncertainty.
46. For interesting and challenging cases of [2+4] carbon cycloadditions see second edition of this book, Problem no. 6, 11, 21, 27, 28, 31, 37–39; and hetero-[2+4] cycloadditions, problems 37, 38, and 44.
47. Mellor et al: (a) Mellor JM, Webb CF. *J.C.S. Perkin Trans.* 1974;2:17–22; DOI:10.1039/P2970000017; (b) Cantello BCC, Mellor JM, Webb CF. *J.C.S. Perkin Trans.* 1974;2:22–25. DOI:10.1039/P29740000022.
48. Seguchi K, Sera A, Otsuki Y, Maruyama K. *Bull. Chem. Soc. Jp.* 1975;48:3641–3644. DOI:10.1246/bcsj.48.3641.
49. Inukai T, Kojima T. *J. Org. Chem.* 1966;31(6):2032–2033. DOI:10.1021/jo01344a543.
50. Yepes D, Pérez P, Jaque P, Fernández I. *Org. Chem. Front.* 2017;4:1390–1399. DOI:10.1039/C7QO00154A.
51. Dias LC. *J. Braz. Chem. Soc.* 1997;8:389–332. DOI:10.1590/S0103-50531997000400002.
52. Berson JA, Hamlet Z, Mueller WA. *J. Am. Chem. Soc.* 1962;84:297–304. DOI:10.1021/ja00861a033.
53. Fringuelli F, Piermatti O, Pizzo F, Vaccaro L. *Eur. J. Org. Chem.* 2001;2001(3):439–455. DOI:10.1002/1099-0690(200102)2001:3<439AID-EJOC439>3.0.CO;2-B.
54. Soares MIL, Cardoso AL, Pinho e Melo TMVD. *Molecules* 2022;7:1304(+46p). DOI:10.3390/molecules27041304.
55. Arrieta A, Cossio FP, Lecea B. *J. Org.Chem.* 2001;66:6178–6180. DOI:10.1021/jo0158478. Authors conclude: "*...we have found that SOIs do exist and are responsible for at least an important part of the observed stereocontrol. Although the absolute magnitude of this kind of interactions is small, its impact on the stereochemical outcome is very important, because of the exponential relationship between the differences in activation energies and the kinetic product ratio.*" Indeed, the magnitude of the interaction energies of endo and exo adducts at a ratio 80:20 in DACA at 25 °C is only 0.8 kcal mol^{-1}.
56. García JI, Mayoral JA, Salvatella L. *Acc. Chem. Res.* 2000;33:658–664. DOI:10.1021/ar0000152. These scientists conclude that "*... the hypothesis of SOI is not necessary to explain the stereoselectivity results found in pericyclic reactions. We believe that a combination of well-known mechanisms (such as solvent effects, steric interactions, hydrogen bonds, electrostatic forces, and others) can be put forward in their place.*"
57. Domingo LR, Sáez JA. *Org. Biomol. Chem.* 2009;7:3576–3583. DOI:10.1039/b909611f.
58. Domingo LR, Aurell MJ, Pérez P. *RSC Adv* 2014;4:16567–16577. DOI:10.1039/C3RA47805J.
59. Welker ME. *Molecules* 2020;25:3740(+23p). DOI:10.3390/molecules25163740.
60. Warshel A, Sharma PK, Kato M, Xiang Y, Liu H, et al. *Chem. Rev.* 2006;106:3210–3235. DOI:10.1021/cr0503106.
61. Lai W, Chen H, Cho KB, Shaik S. *J. Phys. Chem. Lett.* 2010;2082–2087. DOI:10.1021/jz100695n.
62. Fried SD, Bagchi S, Boxer SG. *Science* 2014;346(6216):1510–1514. DOI:10.1126/science.1259802.
63. Aragonés AC, Haworth NL, Darwish N, Campi S, Bloomfield NJ, et al. *Nature* 2016;531(7592):88–91. DOI:10.1038/nature16989.
64. Wang Z, Danovich D, Ramanan R, Shaik S. *J. Am. Chem. Soc.* 2018;140:13350–13359. DOI:10.1021/jacs.8b08233.
65. Joy J, Stuyver T, Shaik S. *J. Am. Chem. Soc.* 2020;142:3836–3850. DOI:10.1021/jacs9b11507.

4

ADDITIONAL TECHNIQUES TO POSTULATE ORGANIC REACTION MECHANISMS

CONTENTS

4.1 OVERVIEW

In the previous chapters the basis for the identification of real problems in organic reaction mechanisms, and the all-important electron flow and stereochemistry, were discussed. These concepts, however, still need to benefit from additional practical techniques and strategies developed by this author over the years to fully rationalize reaction mechanisms. The supplementary strategies described in this chapter, some of them novel to problem solvers not acquainted with previous editions of this book, can be applied more successfully during the preliminary analysis of reactants and products in preparation for the assault of the mechanism itself, following the guidelines of Scheme II.42 relative to electron redeployment and the steric descriptors of Chapter 3.

The Art of Problem Solving in Organic Chemistry, Third Edition. Miguel E. Alonso-Amelot.

These methods and attitudes are:

- ✓ Do not rush through reaction schemes seeking a quick answer, there may be better ones.
- ✓ Avoid drawing messy structures, clear molecular renderings are part of the answer.
- ✓ Bookkeeping: maintain atom and molecular budgets in starting materials and products.
- ✓ Look at molecules from more than one perspective.
- ✓ Apply fragmentation analysis (FA): the system behind dissecting products into parts resembling starting materials.
- ✓ Apply the functionality number (FN) concept: Track changes in the oxidation level of reactants and products.
- ✓ Combine FA and FN.
- ✓ Refine your mechanism options with quantum-mechanical calculation methods if you have access to them.

4.2 TAKE YOUR TIME

Some people contend that the longer you sit learning something, the longer this information will stay with you. Others argue just the opposite. Scientists of the cognoscitive mysteries do not know the answer, they are convinced that humans are diverse animals subject to an individual perception and interpretation of the environment.

Problem solving requires a lot of concentration to get the essentials, discard the chaff, build reasonable and successful answers, and sieve these through the fine mesh of logics, chemical or otherwise; for some, better results are found when sharing with brain-storming fans; out-of-the-box inspiration crops up occasionally. All this takes time. On the other hand, rare geniuses prefer approaches of their own design far removed from systematics. Answers emerge after an inspiring blink of an eye, why not.

4.3 USE CLEAR AND INFORMATIVE MOLECULAR RENDERINGS

Molecular drawings weigh heavily in the language of organic chemistry (Scheme I.1). Molecular structures, reaction sequences, branching mechanism trees, electron redeployment at the tip of curly arrows, stereochemistry, this is all graphical language. So, try your best in learning how to draw neat molecular renderings. In this era of electronic screens, there is still no substitute for pencil/paper or chalk/board to draw organic reaction-mechanism schemes and convey a great deal of information. Drawing molecules by hand is a bit of an art. Computer calculations come afterward for refinement, to inform about interatomic distances, bond angles, conformational preferences, rotating 3D structures of tremendous visual impact, getting all-important energy parameters, molecular properties, and so forth. But all this can only be done once you know what to look for from your hand renderings (see Scheme IV.1).

This may seem naïve and old-fashioned, but students and meeting attendants expect advanced organic-chemistry students and professionals to devise and explain reaction-mechanism sequences on a piece of paper or blackboard in their own words, not by clicking automatically generated, pre-drawn structures with a mouse.

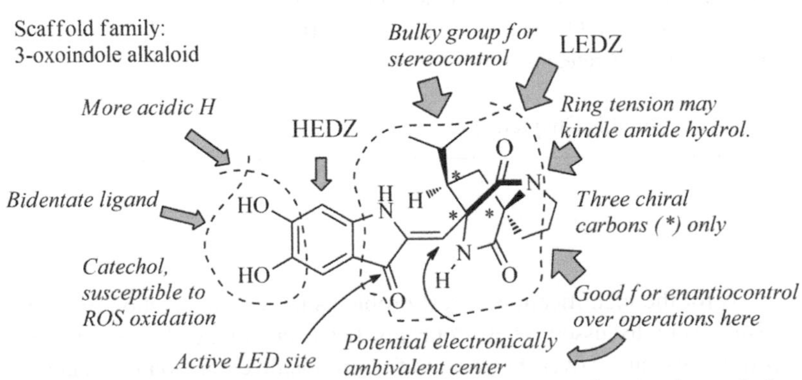

SCHEME IV.1 Ultra-fast information retrieval from a good molecular rendering.

4.4 ELEMENT AND BOND BUDGETS

Organic reactions involve changes in the number and/or type of bonds and/or number and distribution of elements. Valuable information can be extracted by simply comparing starting materials and target product(s) in terms of elementary composition and type/number of interatomic bonds.

Take, for example, the reaction of Scheme IV.2 [1]. We shall examine it through the proposed element and bond budgets only and see how far it takes us, after a preliminary bird's eye view.
Reasoning points:

✓ Is this a problem or a simple exercise? Tidy clues can be discerned in the final target, but the structure of intermediate **3** remains undetermined, thus obscuring the mechanism. It is thus a problem.
✓ The reaction involves considerable molecular enlargement; hence, multiple new bonds are expected. In the absence of skeletal heteroatoms, these bonds should be C–C and C–H.
✓ Given that OBn and *p*-methoxyphenyl (OPMP) are moieties of **1** and **2,** respectively, and both appear in **4**, the target must result from their coupling. This facilitates the element budget by counting them as groups instead of individual atoms/bonds and save precious time.
✓ Self-explanatory *element* and *bond budgets* bring valuable information shown in Table IV.1.

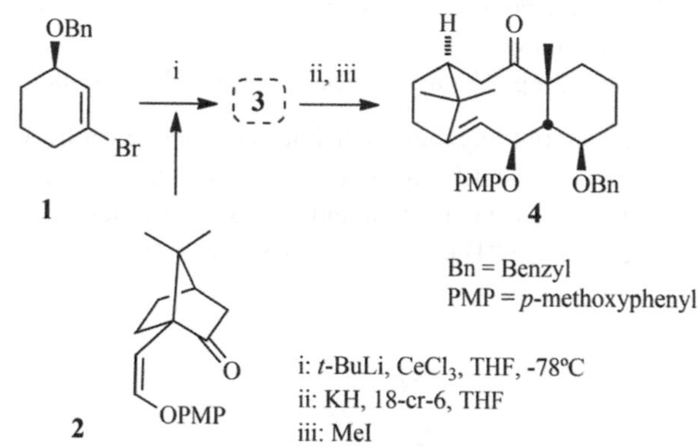

SCHEME IV.2 Test problem to be solved by element and bond budgets of starting materials and product alone. [1] / American Chemical Society.

TABLE IV.1 Element (Part A) and bond (Part B) budgets of components in the reaction of Scheme IV.2. Empirical formulas are abridged by excluding common protective groups, which appear in parenthesis. In Part B, as reactants 1 and 2 are coupled, they will be counted together

Part A

Compound	Empirical formula
1	C_6H_8Br (OBn)
2	$C_{11}H_{15}O$ (OPMP)
1+2	$H_{17}H_{23}OBr$ (OBn)(OPMP)
4	$C_{18}H_{26}O$ (OBn)(OPMP)
4−(1+2)	$+CH_3$, −Br

Part B

Compound →	**1+2**	**4**	**4−(1+2)**
C–C	17	19	+2
C=C	2	1	−1
C–O	2	2	0
C=O	1	1	0
C–Br	1	0	−1

Conclusions from Table IV.1

I) *Part A: Element budget*

1. Target **4** results from the coupling of **1** and **2** without loss of carbon's hydrogens.
2. Bromide is lost, most likely displaced by *n*-BuLi in step *i*. Thus, the newly created C–Li turns **1** into a powerful high electron density zone (HEDZ), a suitable complement to the low electron density zone (LEDZ) in compound **2**.
3. The balance **4**−(**1**+**2**) also detects that a methyl group is incorporated. Where from? You can visualize it easily among the reactants.
4. A second look at **4** reveals four chiral centers, which demand careful stereochemical considerations of the coupling and methylation reactions.

II) *Part B: Bond budget*

1. The accounting balance of the right column indicates that two σ C–C bonds are created while there is a net loss of one C=C. The two events are probably connected. C=C bonds may have migrated to be used in building new C–C bonds.
2. The second C–C bond must be related to methylation.
3. C–O and C=O bonds are preserved all along. The assistance of one C=O to the methylation can be anticipated.

On these grounds and a bit of logic, one can put together a feasible sequence. A pathway is shown in Scheme IV.3. Element and bond balance proved to be a quick way of identifying other important features: specific bond breaking, intramolecular hydrogen and group shifts, stereochemical features, and rearrangements. Some of these processes have occurred in the sequence of Scheme IV.3. This strategy is used repeatedly in "The Problem Chest", Part II.

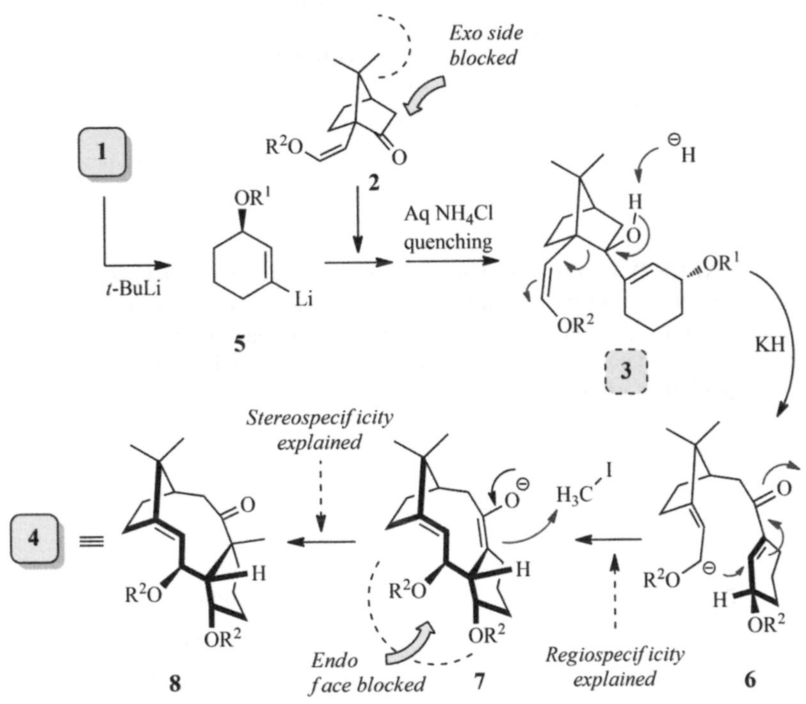

SCHEME IV.3 The previously described budgets discussed in the text unveil the structure of **3**; from there a reasonable reaction mechanism can be reasoned. [1] / American Chemical Society.

4.5 LOOKING AT MOLECULES FROM DIFFERENT PERSPECTIVES

Among others, there are two molecular-drawing techniques worth commenting on herein: the convenience of planar versus 3D renderings to smooth the mechanism design of organic transformations and reshaping starting materials as close as possible to the target compounds form.

1. To the advantage of the problem solver, organic compounds should be observed from all the sides that one requires to comprehend their architecture, considering their 3D character, then compare it with other homologs and grasp their

transformation possibilities. Drawing 3D structures may not be easy in complex manifolds, but there are ways to make this easier. Take compound **9** (Figure IV.1) as a moderately complex molecule. All the components of this dibasic alkaloid, *scholarisin VI* [2], are shown in depiction **A**. The 2D rendering **A** shows five stereogenic centers, while a 3D drawing allows us to assign R or S configurations.

Building a 3D version that solves this question and several others can be performed step by step:

First, remove all functional groups from the 2D drawing and concentrate on the molecule's manifold comprising C–and N in this case–and twist it to a half profile view as in **B**. Then follow the steps shown in Figure IV.1 (top part).

It takes a bit of practice to get the knack of this, but it pays off. What if you were asked whether structures **10, 11,** and **12** (Figure IV.1, bottom) represent different compounds or not? Check out first the carbon-scaffold consistencies, then the position of functionalities, and finally, the stereochemistry in this order.

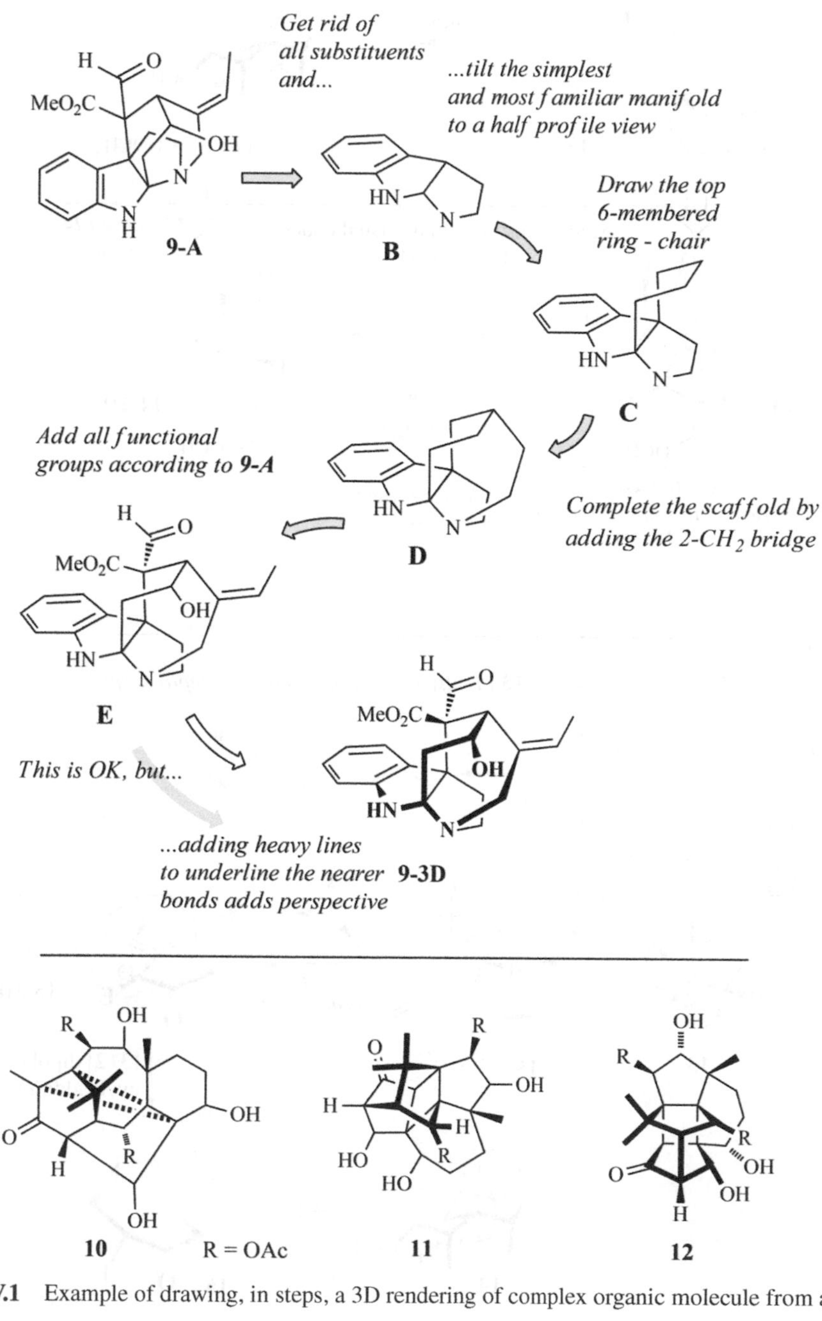

FIGURE IV.1 Example of drawing, in steps, a 3D rendering of complex organic molecule from a 2D figure.

The three renderings show the same *canataxapropellane* [3], a convoluted box product of the taxane group of complex natural compounds with anti-cancer properties, isolated from *Taxus canadiensis*. These substances are so successful that their extraction from the bark of the *Taxus* trees threatens this species due to over-exploitation of the slow growing plants native to northwestern United States and Canada. Synthetic substitutes containing only the active part have eased the exploitation pressure.

Three-dimensional views allow assessing distances between functionalities, steric roadblocks, through-space field interactions, the apparently remote H that gets removed at some point of the reaction coordinate, and possibilities of interaction with external agents. Turning around hand-held molecular models and computer software is not only very useful but also exceedingly fun.

In some instances, however, 3D structures may bring about confusion rather than analysis improvement. The **13** → **14** conversion of Scheme IV.4 is a case in point [4]. Being a thermolysis, one suspects isomerization at the start, but structure **14-3D** includes too much information for a preliminary analysis. So let us run the analysis by working with **14** in 2D, compare

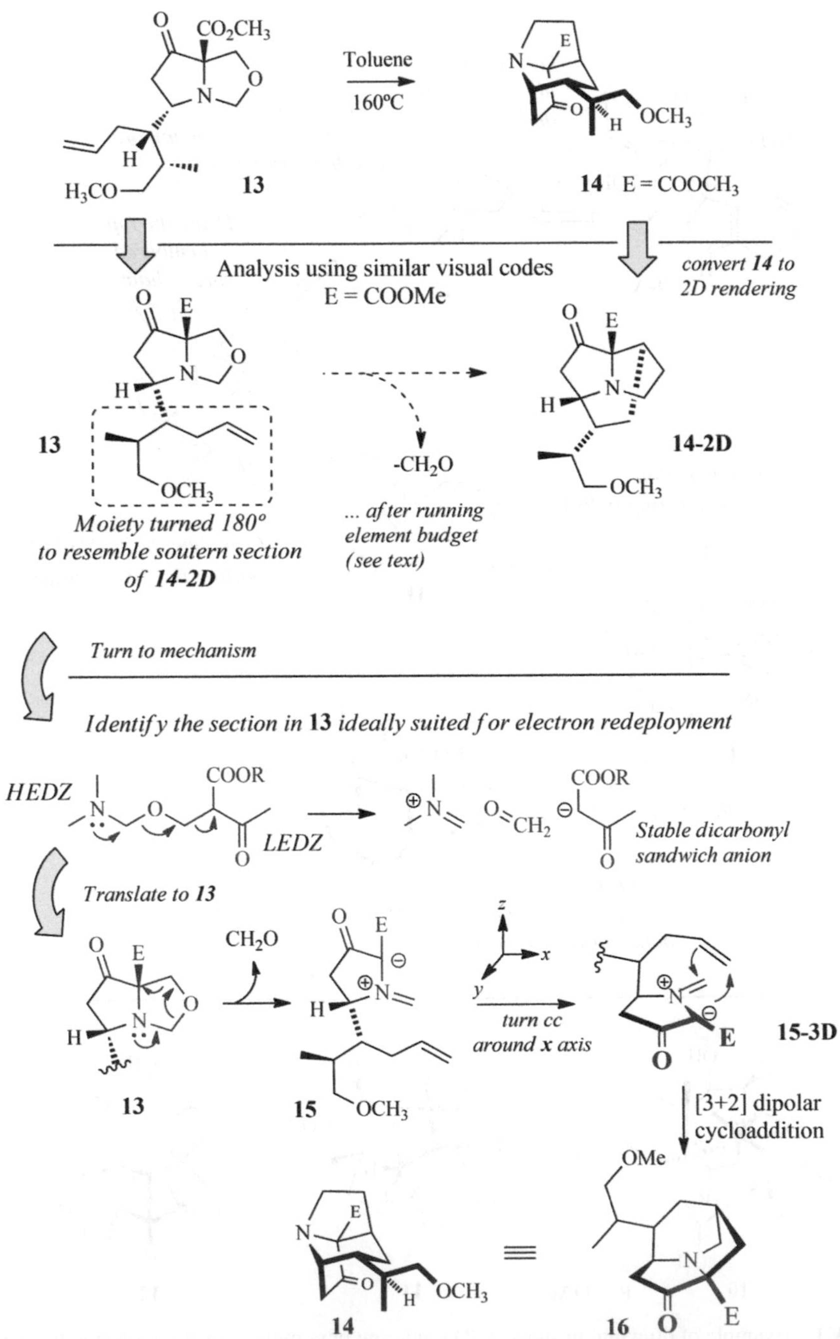

SCHEME IV.4 Illustration of the advantage of converting starting compound(s) and target(s) in a similar visual pattern, bidimensional views (2-D) in this case. The mechanism analysis and proposal become more accessible. [4] / Elsevier.

13 and **14** drawn under similar 2D codes and find similarities and differences (Scheme IV.4). These become clearly visible after the **3D → 2D** transformation.

The element budget, keeping common functional groups locked up in parenthesis is this:

Compound	Empirical formula (abridged)
13	$C_{13}H_{19}NO_2$ $(CH_3O)(CO_2\ CH_3)$
14	$C_{12}H_{17}NO$ $(CH_3O)(CO_2\ CH_3)$
14–13	C (−1), H (−2), O (−1)

The element content balance shows the loss of a CH_2O. The only feasible compound with this elementary composition is *formaldehyde*, in the absence of other reagents or catalysts. Thus, this is not an isomerization, nor a true skeletal rearrangement. Activation must come entirely from **13** without involving ester and methoxy units or cleavage of the tethered southern section. In fact, new σ bonding seems to engage the C=C bond after some commotion occurs in the heterocyclic moiety associated with the loss of CH_2O. That said, the mechanism unfolds as shown in Scheme IV.4, following the yellow curved arrow, once you realize the HEDZ/LEDZ combination contained in the blue sector of **13**.

What happens with the iminium ylide **15** thus formed is difficult to realize unless you go back to its 3D rendering. I chose turning the molecule counter clockwise around the *x*-axis about a 110° to be able to put the terminal C=C poised above the highly active ylide. Both sections can be positioned to execute a [3+2] dipolar cycloaddition from which the five/six-membered fused-ring structure **16** results. It is for you to be convinced that **16** is identical to the target product **14.** During these maneuvers, chiral carbons and stereochemistry in general need special precautions to avoid undue inversions of configuration that will wreck the entire mechanism.

4.6 REDRAW REACTANTS SUCH THAT THEY RESEMBLE PRODUCTS

Solving reaction-mechanism problems, especially those in the category of true riddles, necessarily requires the meticulous comparison of product(s) and starting material(s). Up to this point, we have shown several reaction analyses doing just that (element and bond budget just revised is but one of them) and will continue to do so until the last problem in this book.

There are other strategies of special value to extract fruitful information. In this section, *redrawing starting materials and/or products so both structures look as similar as possible* is described.

This is relatively easy to do in structures with two or more σ bonds since they can be rotated at will on paper/pencil sketches. This idea is based on the principle of minimum movement of molecular components between a pre-transition state (TS) array, the TS itself, and the target compound. There may be more than one TS as it happens in cascade (also known as domino) reactions. Take, for example, Scheme IV.5 where I have sketched indole-oxadiazole **17** without thinking of any reaction possibilities and I have also deleted any hints. In this spirit, I outlined the reported product (**18**) of heating **17** at 230 °C in TIPB (triisopropylbenzene) [5].

There are plenty of σ C–C bonds and one σ C–N linkage in **17** to play with once we identify the amide-oxadiazole-carboxylate ester axis in **18**. The dinitrogen section is lost as N_2 on the way to the product and contributes somehow to the new C–C bonds appearing in **18**–even though at this point we do not know how or where in the reaction coordinate. Rotating a few bonds as shown in the **A → B → C** sequence of Scheme IV.5 takes you to a sterically admissible conformation of **17** that is much closer to target **18** than the original depiction. This simple operation greatly facilitates the design of a feasible reaction mechanism, which appears in Scheme IV.6.

This is an enticing one-pot thermal tandem intramolecular cycloaddition cascade that combines an inverse-electron-demand [4+2] Diels–Alder-type (**17 → 19**) followed by a polar [3+2] cycloaddition (CA) **21 → 18** [5]. Once the oxadiazole heterocycle is redrawn and functional groups combining LEDZ/HEDZ at short distance are recognized, the [4+2] CA is easily grasped. The all σ-bond chain that tethers heterocycle and enol ester units allows a comfortable arrangement of the two moieties to furnish a forcedly regiospecific cycloadduct.

Furthermore, the approach of the oxadiazole ring from above the molecular plane reproduces the trans geometry of the alkene in **19** while molecular nitrogen is forced out by the ensuing retro-Diels–Alder electron redeployment.

Zwitterion **20** can be expressed as carbonyl ylide **21** which confronts LEDZ and HEDZ sections at short distance. This gives rise to the second [3+2] cycloaddition and the observed product in 44% yield. Authors [5] found that adding a methoxy group on C^7 on the aryl section of indole and changing the configuration of the enol ester to *cis* Et-OBn enhances the yield to more than 95%, furnishing the cis adduct homolog of **20** and **18** as would be expected by the intervening Diels–Alder CA.

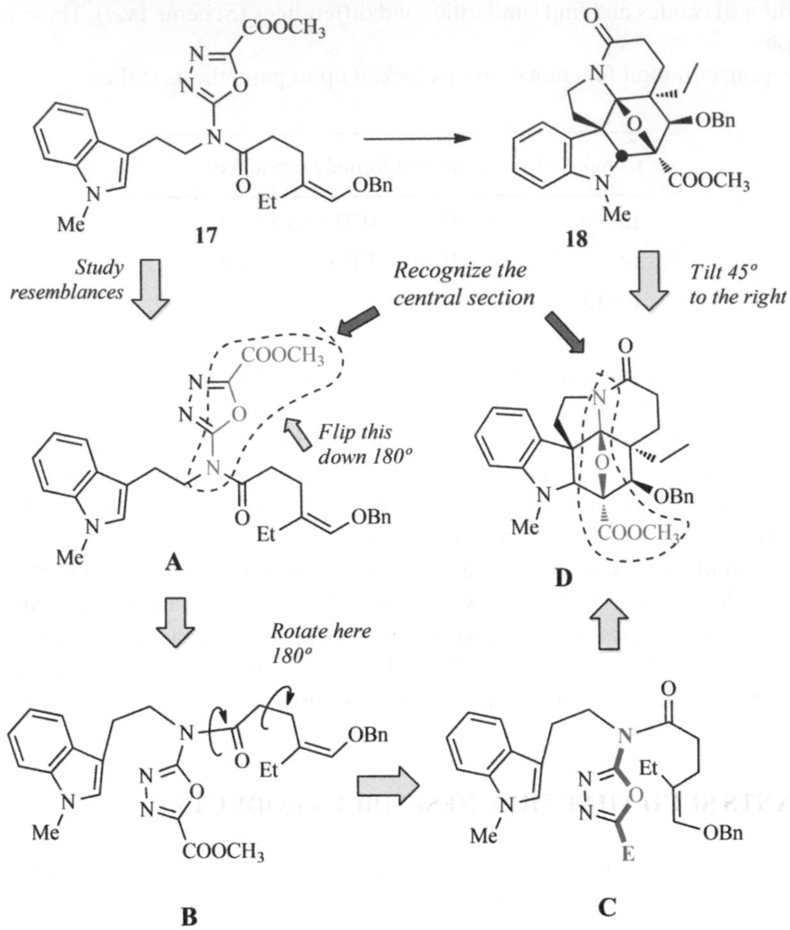

SCHEME IV.5 An example of moving molecular renderings to detect similarities between initial and final compounds.

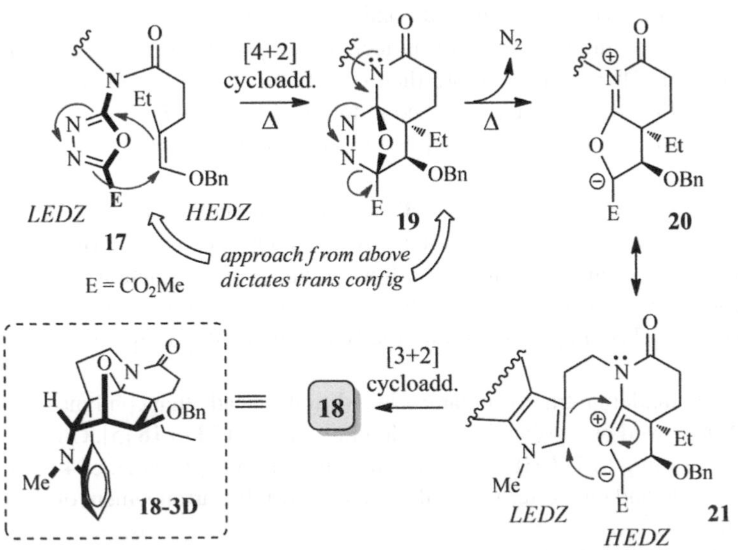

SCHEME IV.6 Illustration of the advantage to move or rotate the molecular sections to closer distances pursuing to build the required bonds leading to target **18**. [5] / American Chemical Society.

It is worth noting that the indole–ylide closing in **21** occurred with excellent control of the endo stereochemistry that is apparent in the 3D rendering **18-3D** (dotted frame, Scheme IV.6).

In the end, we were able to tackle a relatively complex reaction sequence by placing the incumbent sections at the right positions through σ-bond rotation and careful consideration of the stereochemical features [5].

4.7 FRAGMENTATION ANALYSIS (FA): DISSECTING PRODUCTS IN TERMS OF REACTANTS

4.7.1 The Fundamental Proposition

By now it should be clear that problem analysis profits from all sources of dependable information on both sides of the equation. This strategy is applicable to many other situations of professional endeavor: what you presently have (starting materials) and the precise profile of your ends (the target product).

In this section are the details of inverting the reasoning order, that is from end to start: observe the target compound and find all possible structural relationships with starting materials by identifying common sections and pulling them apart as one would in a puzzle. This is an *FA of reaction mechanisms.*

FA is a practical technique to identify sections of the product that also appear in starting material(s) – not excluding some reagents – one by one until most, if not all, sections are accounted for. Molecular segments one cannot associate with visible fragments will be termed here as *orphan* groups.

There may be functional group transformations and skeletal rearrangements in which the groups' identity may seem to have been erased. However, footprints of their occurrence remain in the product – unless they get removed by cleavage – such as more reduced or oxidized functional groups because of bond-forming or bond-breaking operations, respectively, incorporation or exclusion of active sections, protective groups, intra or inter nucleophiles/nucleofuges, and others.

One can enrich the information from this dissection by postulating the electronic charges that result from deconstructing bonds. For example:

$$\text{ROC-C-CR}_2 \text{—} \;\rightarrow\; (\textit{deconstruct}) \;\rightarrow\; \text{ROC}^{(+)}\,\text{C}^{(-)}\text{-CR}_2 \text{—} \;\rightarrow\; \text{etc}$$

And then establish – if this were the case – HEDZs and LEDZ fragments as deconstruction progresses on the way back to the starting materials that could thus operate in the way forward.

A standard technique, each zone of the puzzle in the target that is also found or traceable in the starting materials is marked by a dotted-line field. Bond connections/disconnections are found readily *at the boundary of the dotted fields in the drawing*. One or more hypothetical mechanisms can be advanced from this basis.

This analysis strategy, first proposed in the first edition of this book in 1987, is not to be confused with the brilliant retrosynthetic analysis of Professor Stuart Warren [6], although both approaches could well be cousins.

Given that FA is a most valuable resource for reaction-mechanism design and reasoning, we will resort to it in combination of other strategies described in this workbook as required. For a better understanding of FA, three examples of growing difficulty are now presented.

4.7.2 Study Case 1

Have a look at Scheme IV.7, top section [7]. Although you may feel a bit stymied by the imposing structure of target **24**, recognizing the traits of starting materials with only a few modifications smoothens the first impact. To make things easier, it is not necessary to worry about reaction conditions at the stage of FA. Consideration of reagents becomes important later when devising the forward mechanism to reassemble the pieces.

By examining the composing pieces of target **24** that resemble components **22** and **23,** one finds two large sections we call northern and southern that can easily be traced to the starting materials (Scheme IV.7, lower part).

As said above, bond connections (for the forward mechanism) are determined readily at the boundary of the dotted fields in the FA sketch. As we do this in target **24** the 2-acyl-indole segment at the southern side is easily recognized (Scheme IV.7, bottom section). The northern section is nothing but a six-membered piperidine with a methyl propionate ester appendage that can easily be traced to *N*-methylpyridine derivative **23**.

You may have noticed a third section at the top that does not appear in the equation. (an orphan). It does exist though, hidden among the reactants added in step ii.

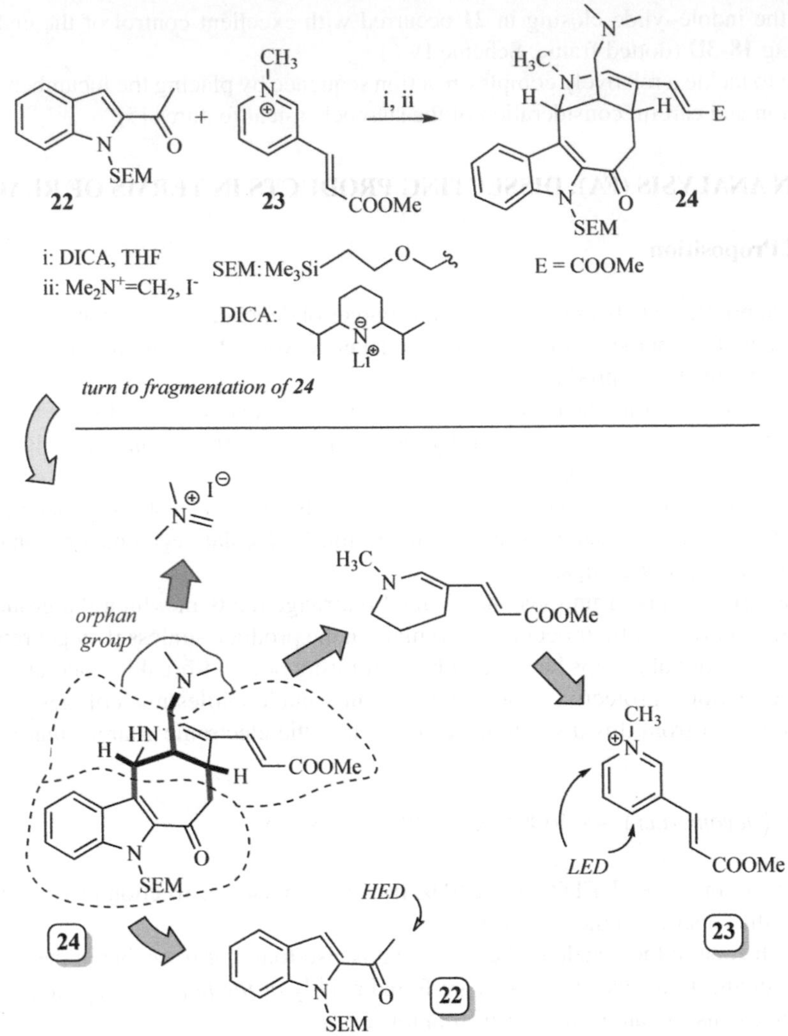

SCHEME IV.7 A feasible deconstruction of target **24** into fragments as close as possible to starting materials **22** and **23**. [7] / American Chemical Society.

In summary, FA discovered three C–C bonds (in purple) that will be formed in the forward direction.

As a final touch, we now add the expected electronic density property to each of the fragments, whether LED or HED, and check if the C–C bonding determined by the dashed fields results from HED/LED combinations or there are incongruous linkages like LED-LED or HED-HED. In such cases, one resorts to non-polar or radical couplings.

With this information at hand, the conversion **22 + 23 → 24** becomes feasible. Considering the specific reaction conditions:

Step *i*: strong base (DICA) in aprotic solvent (THF), followed by dimethyl iminium ylide iodide, origin of the orphan section, in step *ii*. A reasonable explanation is portrayed in Scheme IV.8.

4.7.3 Study Case 2

Despite its successful applications, FA is not bullet proof against error. To start with, FA design is personal; each problem solver is free to assign molecular fragments and draw dissection lines according to their best criterion. The target can be dislodged into as many fragments as deemed necessary to resemble sections in the starting compound(s). This molecular surgery is basically ruled by perception and logics.

The present case (Scheme IV.9, top) conceals some steps of more difficult comprehension than the previous example, even though at first glance nothing seems especially thorny.

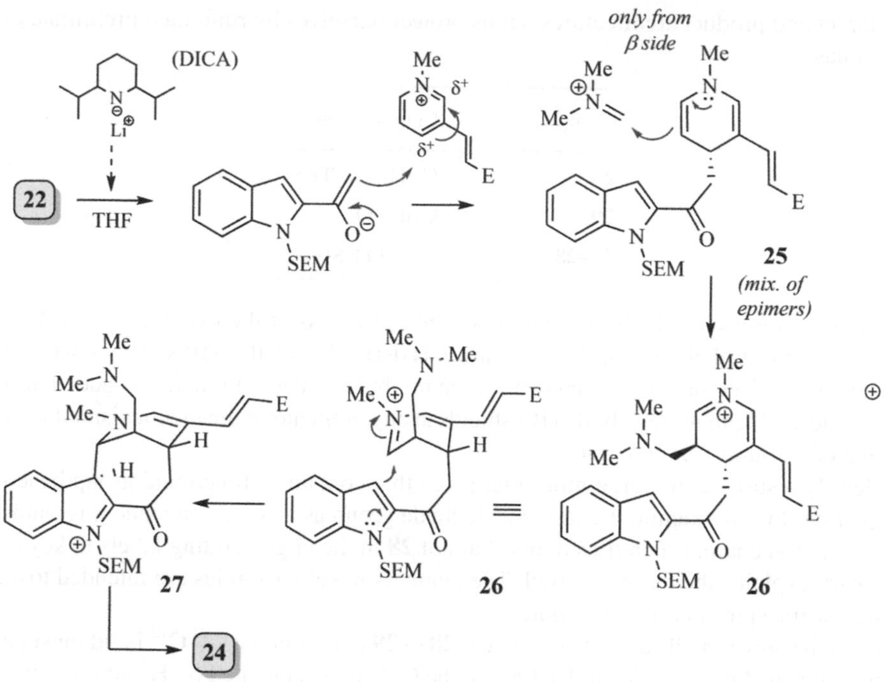

SCHEME IV.8 A reasonable mechanistic sequence based on the fragmentation analysis shown in Scheme IV.7. [7] / American Chemical Society.

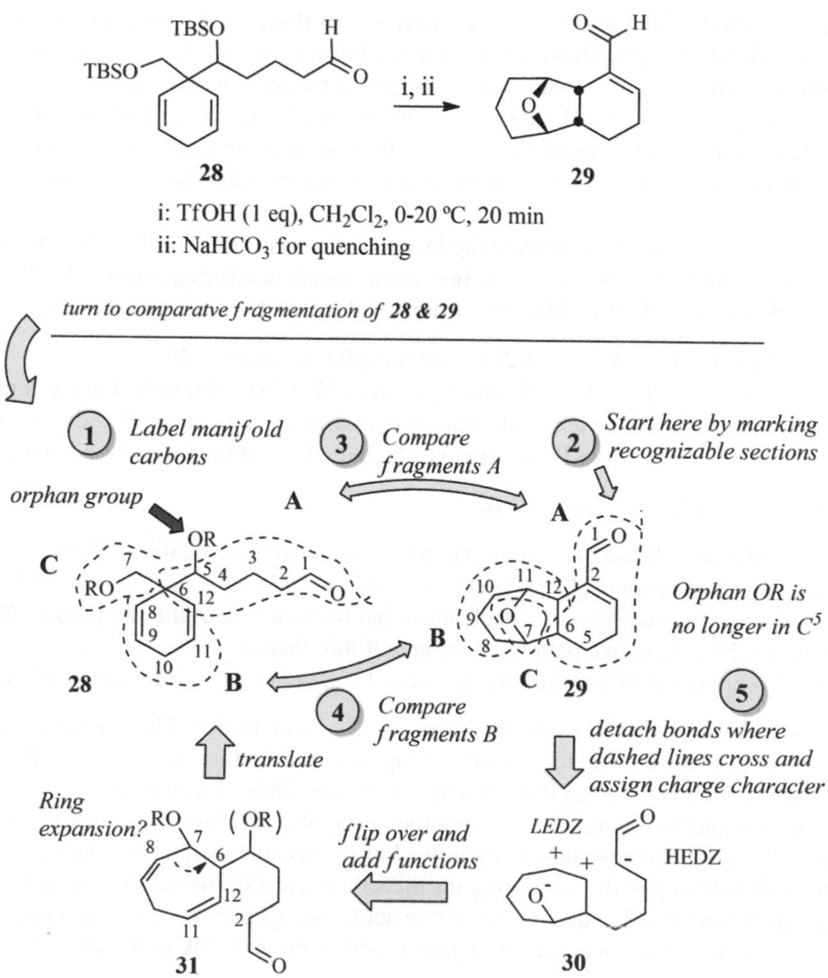

SCHEME IV.9 A first fragmentation analysis applied to target 29 seeking to propose a reaction mechanism based on it. [8] / Elsevier.

To better articulate target and product architectures, let us protect ourselves by running a preliminary element balance from abridged empirical formulas:

Compound	Composition
28	$C_{12}H_{16}O_3$ (TBS)$_2$
29	$C_{12}H_{16}O_2$
29–28	–O, –(TBS)$_2$

Such a simple accounting exercise (and a bird's eye view) shows the loss of the two TBS (*t*-butyldimethylsilyl) protecting groups and one oxygen atom, most likely one of the two carbinols derived from the TBSs. The other O should be connected in **29** at the bicyclic oxygen bridge. The only other oxygen source might be water, although it is added at the workout stage. The strong acidic medium (triflic acid) can remove both TBS shields at any point along the mechanism, thus the *t*-butylsilyl (OTBS) sections can also be handled as bulky OH functions.

A good idea is to identify a sufficiently large molecular piece that contains a functional group beacon and appears at both ends of the reaction equation. In this example I chose the aldehyde group as this beacon since it remains unchanged in **28** and **29** at the end of a $(CH_2)_4$ carbon chain; I then traced this chain of **28** in the target, adding labels to key carbons to keep track.

Scheme IV.9, lower part, explains the reasoning trail. The numbers in yellow circles are intended to dissipate the reasoning-order haze, so please follow the signs one step at a time.

A key dashed-field sketch shown in **29** suggests that in the **28** → **29** conversion a C^2–C^{12} bond must be formed at some early stage of the forward mechanism. An aldehyde enol acting on the C=C bond activated by H^+ addition in **28** as suggested by the HED/LED combination in **30** would accomplish this goal. Additionally, fragmentation structure **31** also recommends ring expansion in the way forward involving a C^7–C^8 bond. Because **28** → **29** occurs in strong acid (TfOH) one may assume the intervention of $C^{(+)}$, a powerful driving force for a variety of reactions that provides mechanistic flexibility. Following these leads, pathway **A** of Scheme IV.10 emerges. However, it soon reaches a dead end in **33**, since playing further bonding (feel free to do so) does not furnish anything similar to the desired target but unrecorded side products.

Get back to **32** and try the *green* pathway **B** by inverting the order of electron redeployment; reassemble the aldehyde appendage to C^{12} by a [1,2] alkyl shift to **34**, run ring expansion **35** → **36**, and carry on "uneventfully" to target **29**. *Green* **29's** architecture is identical to the desired product depicted in Scheme IV.9 and should fulfill our expectations for you to try a mechanism now. Or does it?

No, it does not. You may realize this after comparing the labels of the *green* **29** with the labels we assigned during our first FA (see the *yellow* **29**, in dashed inset of Scheme IV.10). Instead of a predicted formation of a C^2–C^{12} to guide the rest of the reaction we got a C^2–C^6 bond in *green* **29**. This difference is not at all trivial, because it carries the notion of:

1. Ring expansion through the C^8–C^6 → C^8–C^7 [1,2] C shift redeployment (**35** → **36**).
2. Preceded by the C^5–C^6 → C^5–C^{12} [1,2] C shift rearrangement (**32** → **34**). Although density functional theory (DFT) or other calculations would be necessary to assess the energy course of green pathway **B** and its feasibility, one can advance that breaking two C–C bonds along route **B** is energy costly and other pathways might be more likely.

There are two ways out of our self-created quicksand.

1. Using the same fragmentation scheme, design an alternative route from key cation **32,** going through the obliged ring expansion but skipping the troublesome [1,2] C^5 → C^{12} shift that would retain the allocated carbon labels. Be my guest to try this as your present assignment. One such mechanism that furnishes the right end product **29** with the correct labels is provided in Scheme IV.20 (alternative path **C**) at the end of this chapter.
2. A much more interesting approach in line with the power of FA that one intends to show in this problem.

Since there are no oxygen labels, we do not know the origin of the oxygen bridge. This circumstance enables an *alternative* **28** → **29** FA involving two pieces instead of three without leaving an orphan group on the wings. However, some of you may consider this an incomplete disconnection, given that one oxygen atom of **28** is absent in **29**.

Taking advantage of the new partition, one can reassign the atom labels–arbitrarily–such that the five-carbon aldehyde chain materializes in the oxabicyclooctane portion of **29** instead of the six-membered ring. The C^1 aldehyde is no longer the best beacon; quite an abrupt change of paradigm spurring the didactic interest of the reaction of Scheme IV.9.

The high functional group density of **28,** enhanced by strong acid, opens a variety of reaction opportunities that give us the chance to think out of the box, literarily. Scheme IV.11 puts together the new FA and carbon label sequence. An entirely

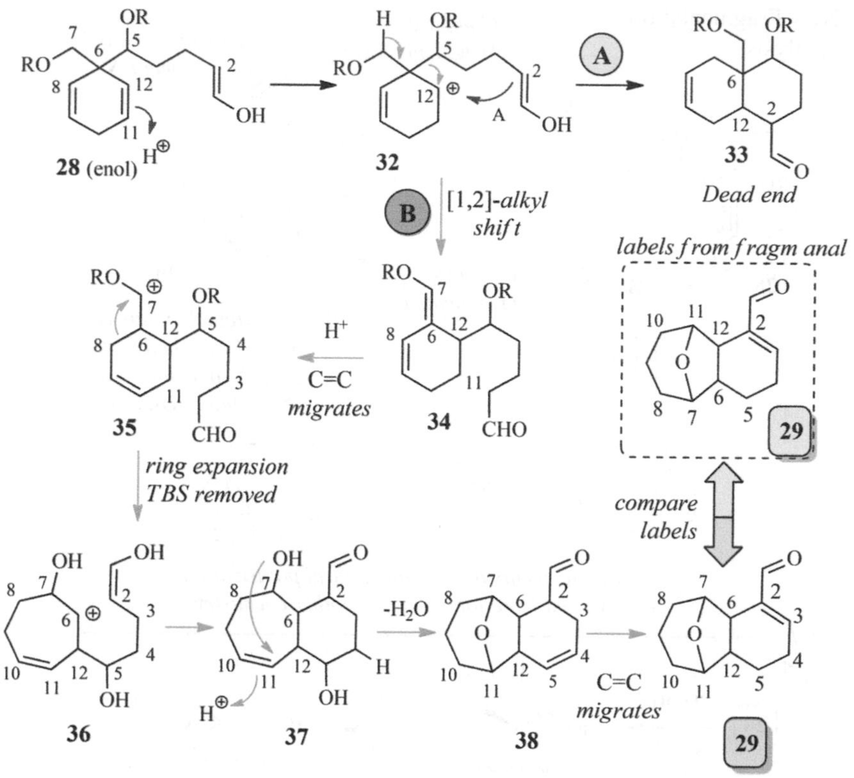

SCHEME IV.10 A first reaction mechanism of **28** to **29** based on the fragmentation analysis (FA) shown in Scheme IV.9, lower part.

different and more reasonable mechanism emerges from it, which includes stereochemical considerations. There are substantial differences with the mechanisms proposed earlier:

1. The aldehyde in **29** is not C^1 but C^7, quite unexpectedly.
2. The new σ bonds are C^5–C^{12}, C^1–C^{11}, and C^5–O.
3. There is no need for any C=C bond migration.

The analysis of the reaction in Scheme IV.9 is flexible enough to allow for a third mechanism alternative (see Scheme IV.19, at the end of this chapter).

4.7.4 Learning Lessons and Takeaways from FA

1. Paying attention to details in reactants and products to find clues of the mechanism linking them is rather obvious. However, attention to detail **is contingent on the type of feature you have used to observe molecules**. Functional group changes, loss or gain of scaffold elements, ring expansion/contraction, changes in oxidation level, and stereochemical differences are most commonly scanned by problem solvers. Important as these are, insufficient information is often extracted from them in problems beyond the average difficulty.
2. One can supplement the above reactants/products survey with a fast and efficient **FA of the target in terms of reactants, a *deconstruction strategy* to understand targets in their composing sections as a function of the starting compound(s).**
3. **FA consists of identifying similar or somewhat modified molecular sections, emphasizing the scaffold elements common to starting materials and product.** These are marked as fields.
4. If properly selected, **the boundary lines of molecular fields indicate sites of bond formation or loss that must be accounted for in the forward mechanism.**
5. In some cases **"orphan" functionalities**, which are not equal to or identifiable in any field of either reactants or products, **become apparent during comparisons,** and should be interpreted as potential leaving groups in the starting compound, or highly modified functions in the forward reaction. For example, primary carbinol in reactants versus a carboxylate in

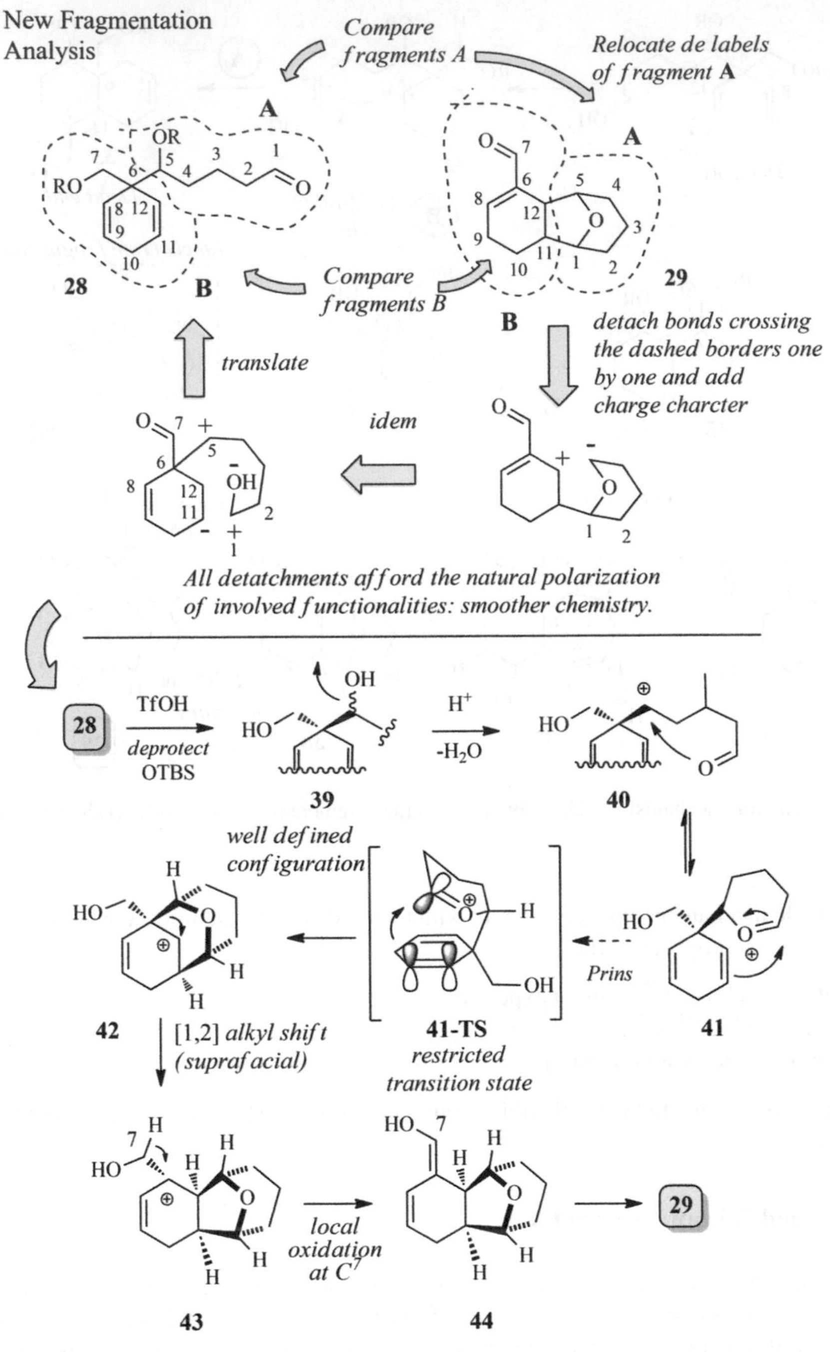

SCHEME IV.11 Top part: A second FA of the reaction of Scheme IV.9. Lower section: a mechanism based on an entirely different approach.

the product (oxidation), an aldehyde versus amine (addition–elimination followed by N-alkylation or reduction), and very many others. **Location in a given field provides clues about their relationship.**

6. **Once the molecular sections** or fields **have been defined**, one proceeds to **separate the boundary scaffold bonds in steps and assign HEDZ or LEDZ character, if possible.**
7. **When HEDZ/LEDZ assignments are not reasonable, chances are that homolytic bond breakage, diradical coupling of non-polar cycloadditions, or retrocycloadditions are responsible for the examined deconstruction step.**

8. **Deconstruction continues until all the starting materials or sections are accounted for.** Once there, the construction mechanism (or fragmentation, as in bond-breaking thermolysis, pyrolysis, or other decompositions), can be developed following the various leads provided by the previous deconstruction sequence.
9. **Adding labels** to key scaffold atoms helps keep track of deconstruction and the reaction mechanism in the forward direction.
10. *Keep in mind that FA is a planning strategy that always* ***works backward from target to starting materials;*** *the actual physical reaction coordinate points in the opposite direction.*
11. **There are no stringent rules in FA;** hence, opinion outweighs other considerations so long as it is chemically sound.
12. As it happens with other strategies of problem analysis, **FA is not infallible and may lead to preposterous hypotheses.**
13. Because the mechanism design is misguided by the wrong FA and scaffold labeling, it becomes mandatory to review the deconstruction sequence to detect inconsistencies or propose a different fragmentation and envision a different mechanism from the start.
14. FA is of limited applicability in reactions heavily marked by deep-rooted skeletal rearrangements or, on the other end, hydrocarbon scaffolds denuded of significant labels (chiral centers, functional groups, deuterium labels, and so forth).

4.8 OXIDATION LEVELS AND MECHANISM

Oxidation and reduction are among the mainstay reactions of organic compounds (and everything else chemical). For chemists, it is natural to associate the oxidation or reduction of organic compounds with the presence of oxidants or reducing agents among the reactants. There are other forms, independent of the presence/absence of redox agents, though. Heterolytic C–C cleavage gives rise to reduced [$C^{(-)}$] and oxidized [$C^{(+)}$] carbons with consequences for the oxidation level of each molecular section thereafter, depending on the individual course of these sections if they become separate entities

It is of interest to mechanism design learning whether a given reaction involves oxidation or reduction to draw plans for electron transfer. This planning should rest on identifying the oxidation level of reagents and target products.

There are a few methods to estimate oxidation levels like *oxidation state or number*, and *degree of unsaturation (DU)*. The latter is much easier to calculate and is taught in introductory courses of general chemistry. Yet, it fails to fulfill expectations with regards to being a measure of the oxidation level changes in many instances. Take the reaction of Scheme IV.12 as one of many examples. 3-Methyl indole (**45**) was subjected to iron(III) chloride oxidation and the isolated product was characterized as **46** [9]. Years later a similar transition metal oxidation [Cu(II) salts, pyridine, and oxygen] of **45** was confirmed by nuclear magnetic resonance (NMR) studies [10]. More recently, the reaction was revised once again using a horseradish-peroxidase enzyme and the earlier proposed structure [9, 10] was found to be in error and corrected [11]. For the present assessment of oxidation level changes, the incorrect structure will fill our purpose first and then the revised compound will be used as a check-up and a mechanism challenge.

From visual inspection and hydrogen peroxide/peroxidase in the reaction medium, compound **45** underwent oxidation. I used DU to estimate oxidation level changes, calculated as shown in Scheme IV.12 and found that $\Delta DU = DU_{46} - 2 \times DU_{45}$ (since 2 mol of indole 45 were consumed) equals zero. Thus, ΔDU does not reflect the expected change in oxidation level and does not provide a notion of electron transfer.

For the benefit of mechanistic problem analysis we need another system to estimate oxidation levels and how they are modified during organic chemical conversions.

Enter FN: the functionality number.

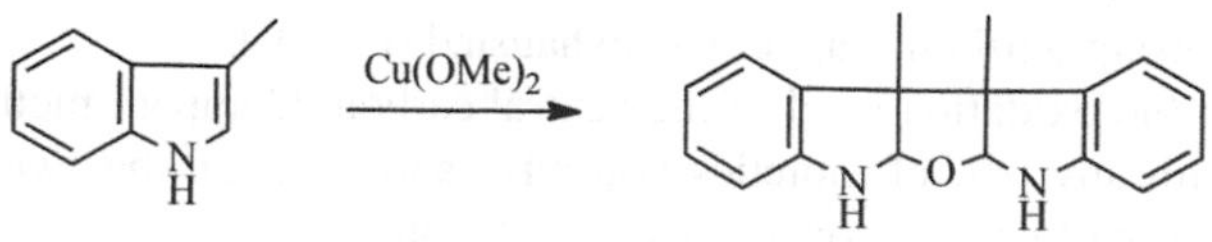

45

C_9H_9N

DU = 9

46

$C_{18}H_{18}N_2O$

DU = 18

ΔDU = 18 - 2 x 9 = 0

SCHEME IV.12 Early interpretation of 3-methyl-indole oxidation, as a test for validation using the degree of unsaturation (DU) as oxidation criterion. DU failed to record it. [11] / Elsevier.

4.9 THE FUNCTIONALITY NUMBER (FN)

4.9.1 What Exactly is FN?

FN is an integer number designed to illustrate the comparative oxidative status of individual carbons within functional groups as well as in whole molecules [12]. FN is derived from the algebraic sum of local FNs corresponding to individual carbon atoms. Modifications in FN [$\Delta FN_{(p-r)}$] (p = products, r = reactants) during chemical conversions of organic compounds mirror changes in oxidation status, thus providing key information about electron whereabouts during conversions.

The FN concept is straightforward: take a single carbon atom within a functional group marked C*, an alcohol R_3C^*–OH, for example. Then, ask yourself how many C*–C or C*–H σ bonds can be formed from R_3C^*–OH to attain the minimum oxidation level of the carbinol carbon (R_4C, where R = C or H). *This figure is the FN.* In this example, $FN_{(carbinol)} = 1$. For an aldehyde, $FN_{(aldehyde)} = 2$, and so forth.

Repeat this reasoning at a higher sp^3 oxygen substitution: trimethyl *ortho*-formate H<u>C*</u>$(OCH_3)_3$. Potentially, a maximum of three σ C–H or C–C bonds can be built on the central C*, one for each CH_3O; hence, $FN_{(orthoformate)} = 3$. Now, replace trimethyl *ortho*-formate with formic acid and ask the same question: $FN_{(formate)}$ is also three, two new σ bonds on account of the C=O and a third from the C–O. Therefore, ortho-*formate and formic acid, despite their belonging to disparate functional group sets, share the same FN.*

4.9.2 Organizing Carbon Functionalities in FN Groups

This very simple interpretation can be extended to the dearth of carbon functional groups you know. By organizing them according to this concept, a table categorized in FN classes (integers from zero to four) takes shape (Table IV.2). CH_4 and any other fully saturated alkane carbon will occupy the "FN = 0" column whereas *ortho*-carbonate and CO_2 at the high oxidation level end belong in the "FN = 4" column. The growing unsaturation state from top to bottom is added as organizing criterion. There is no need to memorize this table as it is easy to place any carbon-containing functionality in its proper FN cell. It is anyway helpful to become familiar with this classification system and learn how to estimate FNs of individual groups.

As might be expected, there are exceptions, chiefly R_3C–H moieties susceptible to remote C–H activation, which are thus capable of initiating electron flow in the "FN = 0" group to furnish one C–C, C–Si bond, or of higher oxidation levels as C–X (X = N, O, S, P, halogens).

4.9.3 Main FN Groups Properties

There are a few rules, and several FN properties that can be extracted from Table IV.2 by simple logics.

A word on notation:

- FN_i: individual carbons.
- $FN_{(CONHR)}$: of a specified group in the subindex, an amide carbon in this case.
- FN_t: carbons of all functional groups in the same molecule = ΣFN_i.
- $\Delta FN_{(p-r)}$: change of FN between products (p) and reactants (r).

The most important FN features are:

1. FN is an integer as it responds to a simple question: the number of new bonds – an integer – that can be formed from the central carbon of a functional group until this capacity is exhausted (FN = 0).
2. The greater the FN, the higher the oxidation level of the central carbon. Compare methanol (FN = 1) and CO_2 (FN = 4). Run the same comparisons with any pair of moieties you wish and estimate the oxidation level distance, that is, the number of valence electrons you will need to convert one into the other.
3. Functional groups in the same FN group possess a similar oxidation level: aldehydes, ketones, acetals, acylals, dithianes, α-dihalo alkanes, imines, and diazocompounds; they all belong in the FN = 2 class. By the same token, functions within the same FN group can be interconverted without necessarily resorting to oxidations or reductions. Check this against Table IV.1 as you read.
4. Conversion of functional groups from lower to higher and the opposite occurs in FN = ±1 increments in redox reactions. This is the result of the participation of two electrons, at most, for each step. When designing redox reaction mechanisms, it is always advisable to execute this in one or two electron stages. One may call this *the minimum oxidation level change*

TABLE IV.2 Common functional groups organized according to their FN category of the principal or C* labeled carbon atom. Adapted from [12]

FN	−2	−1	0	+1	+2	+3	+4
				$R_3C–OR'$	$R_2C–(OR')_2$	$RC–(OR')_3$	$C(OR)_4$
				$R_3C–NR_2'$	$R_2C–(NR_{2'})_2$	$RC–(NR_{2'})_3$	$C(NR_2)_4$
			R_4C, $R=C, H$	$R_3C–SR'$	$R_2C–(SR')_2$	$RC–(SR')_3$	$C(SR)_4$
		$R_3C–SiR_3$		$R_3C–X$	$R_2C–X_2$	$RC–X_3$	CX_4
				$R_3C–[N{\equiv}N]^+$, $R_3C–[N_3]^+$			
	$R–C^{(-2)}SO_2R'$	$R_3C^{(-)}$		$R_3C(+)$			
				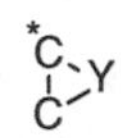 Y (O, NR, S)	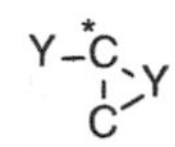 Y (O, NR, S)		
					$R_2C=O$	$RC(=O)YR'$ (N,S)	$RO–C(=O)–OR$
					$R_2C=NR'$	$RC(=NR)YR'$(N,S)	$R_2N–C(=O)–NR_2$
	$R_2C^*=PR_3'$				$R_2C=N=N$		
					$R_2C=S$		
	$RO–C=C^{(-)*}R_2$	$RO–C=C^*R_2$		$R_2C^*=CR_2$	$RO–C^*=CR_2$		
	$R_2N–C=C^{(-)*}R_2$	$R_2N–C=C^*R_2$			$R_2N–C^*=CR_2$		
	$RS–C=C^{(-)*}R_2$	$RS–C=C^*R_2$			$RS–C^*=CR_2$		
					$X–C^*=CR_2$		$X–C(=O)–X$
					$R_2C=C^*=CR_2'$	$R_2C=C^*=O$ (N,S)	$Y=C=Y$ (N,S)
	$RO–C{\equiv}C^*R$				$RC{\equiv}CR$	$RO–C^*{\equiv}CR$ (N,S)	
						$N{\equiv}C–R$	$N{\equiv}C–OR$ (N,S)
							$N{\equiv}C–X$
							$^{(-)}C{\equiv}N^{(+)}–X$

rule. Practically speaking, if you wish to form a carboxylate ($FN_i=3$) in the final product, look for a ketone or aldehyde as a priority source ($FN_i=2$) in the starting compound, rather than functions with lower FN like alkenes, halides, and ethers, or other functions removed two or more FN levels away from your target carboxylate. Less energy will be required.

5. FN and HEDZ/LEDZ: When $FN>0$, the central carbon occupies the positive end of a dipole and is consistent with a LEDZ. Thus, C is the *receptor* of bonding electrons from the incoming $C^{(-)}$ or $H^{(-)}$, other anionic moiety, or electron source – HEDZ by definition. All electrophilic carbons fall in this $FN>0$ category. The quintessence is the carbocation-bearing functions ($FN_i=+1$) as well as more advanced groups like N and O ylides or acylium ions $O=C^{(+)}R$ ($FN_i=+3$).
6. By contrast, $FN<0$ responds to C carrying anionic character (as in carbanions, enolates) or a pseudo-anionic profile (neutral nucleophilic forms of C, enol acetates, enamines, $R_3Si–C$, for instance). At this point, only groups with an FN up to −2 are known, including double anions, oxyvinyl anions, and Wittig-type ylides. All are strong HEDZs.
7. Importantly, FN is not limited to individual carbons but collections of them in carbon scaffolds. The algebraic sum of individual carbon FNi (ΣFN_i) corresponds to the relative oxidation level of neutral compounds with two or more functional groups. This number is FN_t (t = total).

8. Oxidation level changes during reactions can be assessed by the comparison of ΣFN_i of products and reactants. That is: $\Delta FN_{(p\text{-}r)} = \Sigma FN_{i(p)} - \Sigma FN_{i(r)}$. The result, whether positive or negative, is equivalent to the oxidation level change:

$[\Delta FN_{(p-r)}] > 0$	oxidation
$[\Delta FN_{(p-r)}] < 0$	reduction
$[\Delta FN_{(p-r)}] = 0$	no net oxidation or reduction

A few study cases illustrate FN applicability and the extracted information.

4.9.4 Study Case 1

This first example examines determining whether overall oxidation or reduction has occurred in a given reaction, and from there, devise a feasible mechanism.

Let us first use ΔDU to predict whether any oxidation occurs in **45** → **46** as one might expect from Cu(II) oxidation by one electron transfer (Scheme IV.12). However, ΔDU between target and reactants is *zero*; it does not record any oxidation level change. Application of FN change, however, detects correctly the oxidative process by ΔFN = +2 (Scheme IV.13, top section), which allows us to propose a radical oxidative sequence shown below the yellow arrow.

4.9.5 Study Case 2

Mechanism analysis through FN is not limited to a superficial comparison of reactants and products seeking to establish the occurrence of oxidations or reductions. FN is also a tool to provide information of individual reaction steps. Have a look at the **49** → **50** conversion (Scheme IV.14). You are welcome to propose a mechanism while figuring out local FN changes.

One can rationalize this reaction using the FN perspective:

a) $\Delta FN_t = 0$; hence, no net change in oxidation level. This fits well with the absence of redox agents or net C–C bond cleavages.
b) However, C=O is oxidized to COOH while the C–Cl bond (FN = 1) in **49** is no longer present in **50**. This is equivalent to saying that the ketone is oxidized at the expense of the C–Cl reduction to C–C or C–H, even though the two occupy non-conjugated and distant positions. How can this take place?
c) According to the minimum oxidation level change rule (# 4 above), we should look at the ketone carbon as the source of the carboxylate: a *local* oxidation caused by a neighboring C–C cleavage.
d) Direct bonding interaction between CH_2Cl and C=O is not an option [both are LEDZs (FN > 0)]. We need to find a potential HEDZ within **49** from existing functional groups and the contribution of reagents in the medium. The α carbonyl carbon (via enolate) is the only HEDZ candidate.

Scheme IV.15 portrays the reaction mechanism buttressed by FN analysis. Each FN_i is shown in circled numbers. Local oxidations and reductions emerge smoothly through ΔFN balance at each step. The FN interplay allows one to spot the effective intramolecular oxidation level transfer in **51** → **52** and the ensuing Haller–Bauer ketone cleavage. This sequence can be envisioned as a homo-Favorskii rearrangement, an extension of the classical Favorskii rearrangement comprising a transient cyclopropenone intermediate instead of cyclobutenone **62**.

From Scheme IV.15 one concludes that **heterolytic C–C bond *formation* will be reductive at the FN > 0 carbon** (the LED end). Conversely, **heterolytic C–C bond *cleavage* will be oxidative at the FN>0 carbon.** The opposite holds true for FN < 0 carbons.

4.9.6 Study Case 3: Heterolytic C–C Cleavage and the Electron Sink

The impact of the conclusion in the box on reaction-mechanism design is paramount to the direction of electron traffic during the heterolytic bond cleavage. Even though electron redeployment can be determined with relative ease once HEDZs and LEDZs have been identified, this flow can follow unexpected courses that the ΔFN_t analysis may anticipate. Every time you

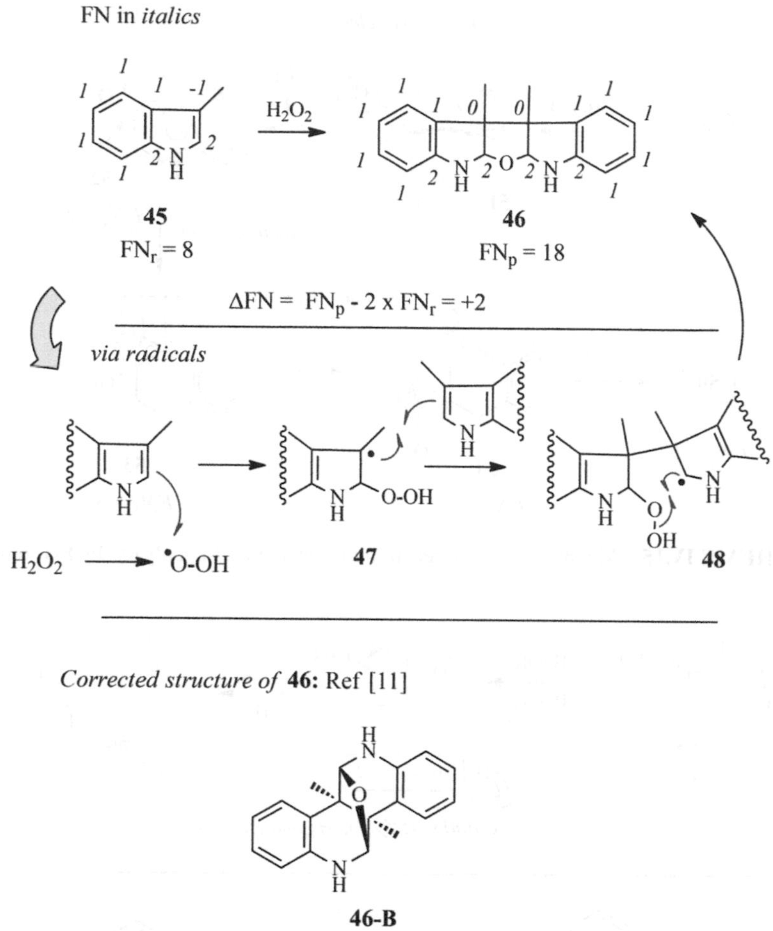

SCHEME IV.13 Functionality number (FN) analysis of an accessible reaction believed to yield an oxygen-bridged 3-methyl indole dimer and a feasible mechanism. The structure of **46** was later corrected to **46-B**. Please provide a reasonable mechanism to explain **46-B**. A solution is given in Scheme IV.20 at the end of this chapter. [11] / Elsevier.

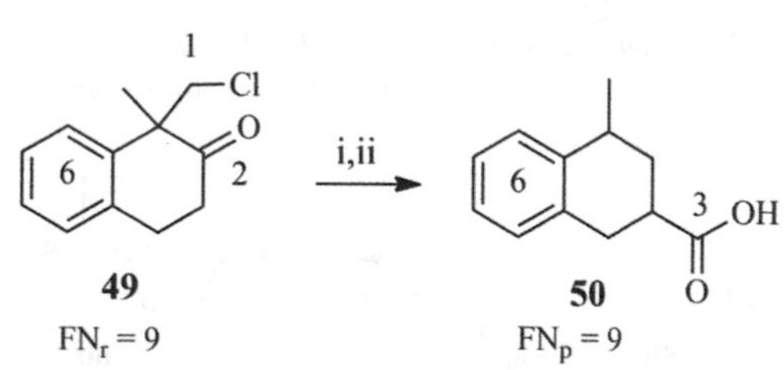

i: t-BuOK. t-BuOH
ii: H_3O^+

SCHEME IV.14 Modification of oxidation level of individual carbons in the dehalogenation of **49** assessed by partial functionality numbers FNi and their addition. Adapted from [13–14].

suspect an oxidation, look for the electron sink in your starting material(s) or redeploying electrons away from the target. During reductions, look for the electron source. FN helps in their identification.

Compare reactions **55**→**56** and **57**→**58** in Scheme IV.16 [15]. The first one with a net $\Delta FN_t = +1$, meaning one notch (two e^- oxidation) is prompted by t-BuO$^-$K$^+$/t-BuOH, a non-oxidative mixture; hence, a net oxidation occurs on account of C–C cleavage only, which means no electron ejection from **56** and therefore no electron sink.

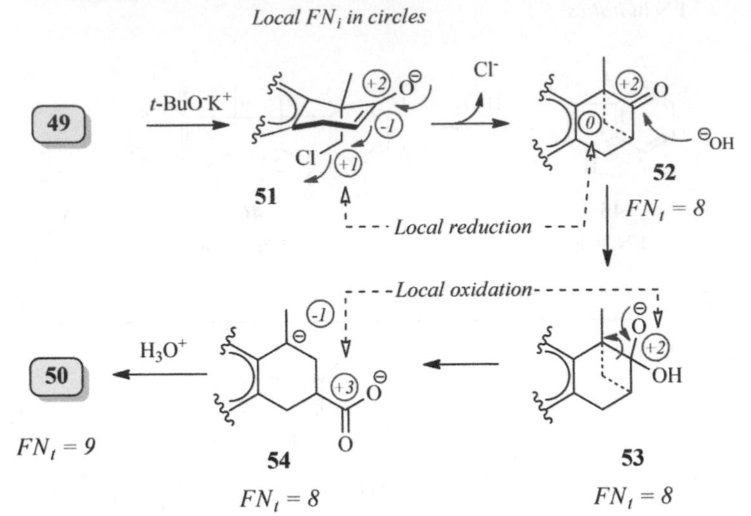

SCHEME IV.15 Mechanism of the reaction of Scheme IV.14 analyzed by FN logics.

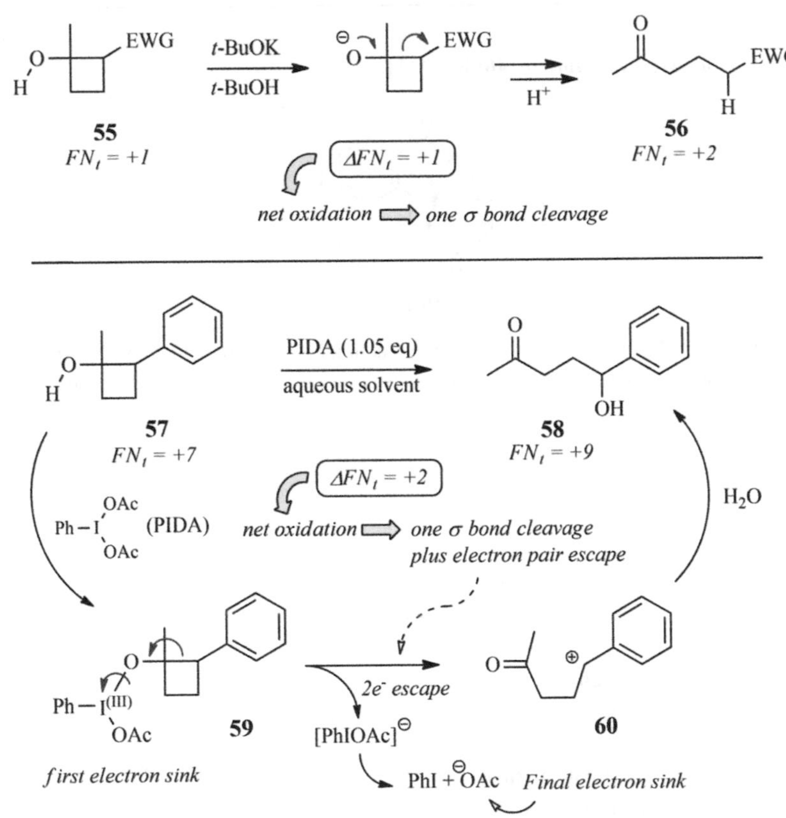

SCHEME IV.16 Discovering opposing electron flows in similar cyclobutane carbinols through FN analysis. [15] / Elsevier.

By contrast, in the second reaction of the cyclobutyl carbinol product **58** is two FN notches higher in the oxidation ladder (ΔFNt=+2). C–C *cleavage alone cannot account for this change; thus, one electron pair needs to be estranged from* **57**. An external oxidant is required. Phenyl iodine diacetate (PIDA) fulfills this role by carrying away one electron pair on iodine and eventually acetate as the end electron sink. Notably, *there is a complete inversion of the electron flow* in the oxycyclobutane ring fracture (**59** → **60**) relative to the **55** → **56** conversion above, giving rise to an entirely divergent reaction mechanism. Let me emphasize that the mechanism course could be predicted based on ΔFN_t data analysis alone.

4.10 COMBINING FRAGMENTATION ANALYSIS AND FUNCTIONALITY NUMBERS

Supplementing reverse electron redeployment in retro-mechanistic analysis with the FN scrutiny becomes a potent tool for reaction-mechanism design. It is especially productive for the reconstruction of functional groups demanded by our fragmentation of reaction products in terms of starting materials.

Scheme IV.17 (top part) sketches in a self-explanatory manner a first straightforward case [16] for you to grasp the general idea, in preparation for the real life mechanistic problem **61** + **62** → **63** of the lower section of Scheme IV.17 [17]. As you may have surmised already, functional groups with opposite FN signs in these and all other schemes define unambiguously the identity of HEDZ and LEDZ and consequently, the direction of electron flow for bond formation. Likewise, equal FN signs do not warrant bond assembly between these carbons unless there is a previous umpolung operation or radical intermediates.

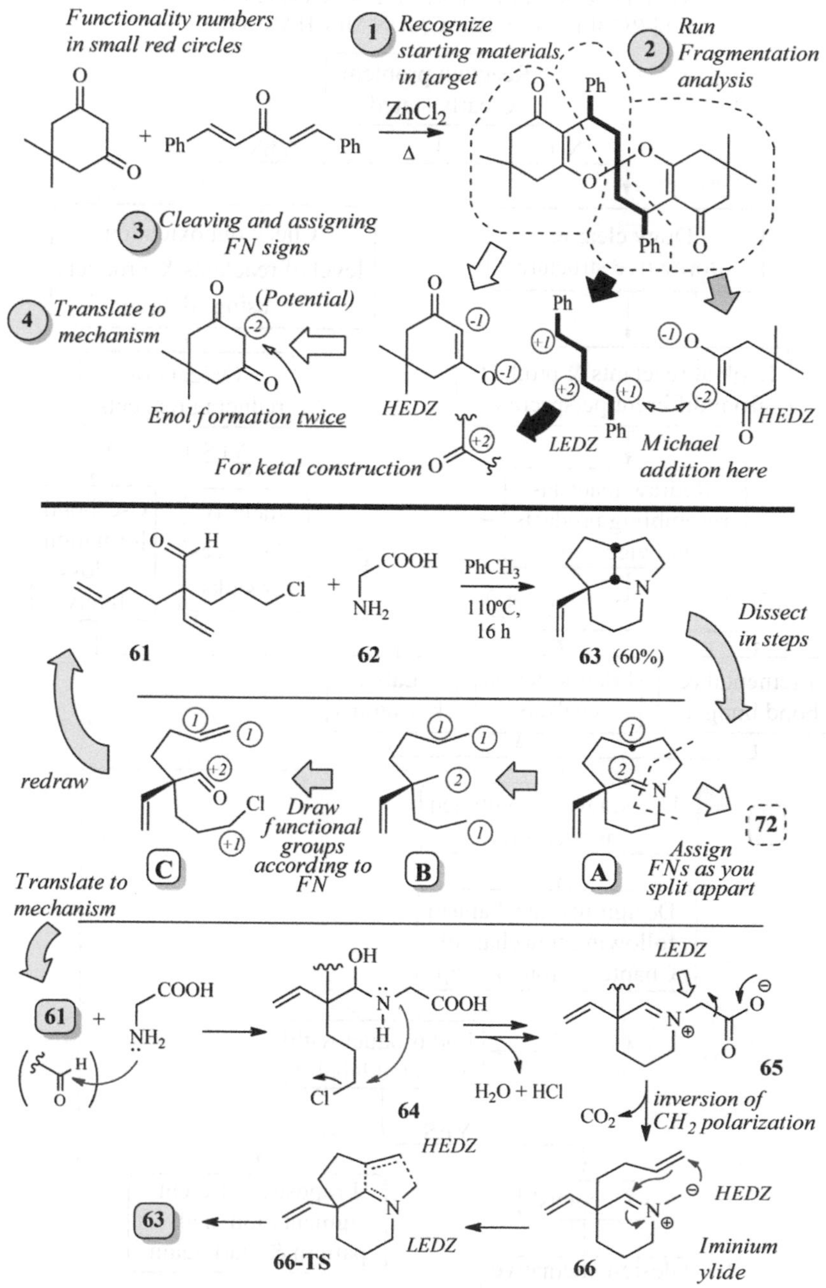

SCHEME IV.17 Top part, [16] / Elsevier; Lower part [17] / Beilstein Institute for the Advancement of Chemical Sciences.

4.11 A FLOWCHART TO ORDERLY EXPLOIT THE STRATEGIES OF THIS CHAPTER

As with other chapters in this book, ideas, techniques, and strategies for problem analysis and drawing solutions may get a little fuzzy because of their abundance. Profusion, however, also means a wealth of resources you may have seen for the first time while reading these chapters. These are now in your possession to proficiently tackle reaction mechanisms with confidence. The analysis chart of Scheme IV.18 is just one approach to take advantage of the strategies described in this chapter and by no means replaces similar decision-making flowcharts of previous chapters. It is up to you to decide which analysis flow better suits each problem you may come across. You may also overlap this chart's suggestions with similar diagrams shown at the end of previous chapters herein.

FLOWCHART:
APPLYING SUPPLEMENTARY STRATEGIES
TO PROBLEMS OF REACTION MECHANISM

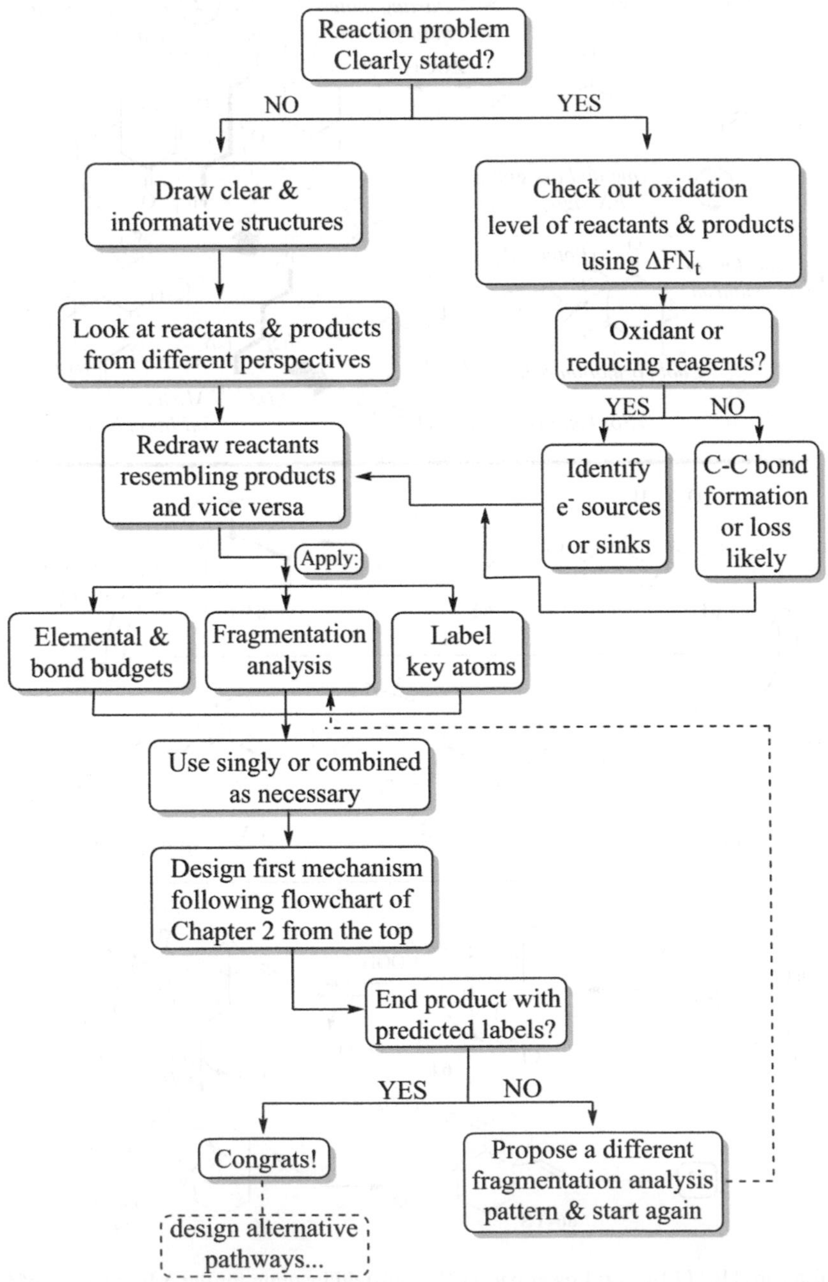

SCHEME IV.18 Recommended problem analysis flow diagram for using the strategies described in this chapter.

4.12 SUMMING UP

1. Devising a clear picture of electron redeployment by first defining HEDZs and LEDZs in starting materials or alternatively, sites for advantageous formation of radicals and one electron acceptor–donor sites, is a productive tactic to start problem analysis.
2. Personal attitude toward deep-minded concentration is highly productive not only for finding better and more thorough problem solutions and understanding chemical conduct, but also for extracting robust and long-lasting learning from the chemistry involved in each problem.
3. For most people, quickly garnered knowledge tends to be forgotten much faster than gradual and deliberate uptake. Take your time.
4. Proficiency in molecular rendering with pencil/paper or chalk/board, using computer software as complementary material, as well as the ability to draw and study molecular structure viewed from various angles are exceedingly valuable tools for the mechanism design of complex organic transformations. Drawing partial structures of compound zones subject to change saves time and provides neatness to reach mechanistic solutions.
5. Cutting target products into molecular pieces as a function of starting materials in a retro-mechanistic or deconstructive manner, here named *FA*, greatly facilitates the conception of mechanistic routes open to the building of these targets. This is so valuable that it is the most frequently used strategy in "The Problem Chest", Part II, of this workbook.
6. Frequently, compounds undergo redox transformations without the concourse of reduction or oxidation reagents, because of net carbon-to-carbon or heteroatom bond formation or cleavage and loss of molecular fragments alone. These hidden redox operations, of critical consequence to reaction mechanism, can be traced with a few tools including unsaturation assessment. The FN provides a more thorough and fast analysis as it semi-quantitatively estimates oxidation level changes for individual atoms as well as entire compounds and their mixtures through a comparison of reactant and target product oxidation status in a very simple manner.
7. Combining FA and FN analyses provides a potent instrument to devise reaction mechanisms and predict functional group status for participating components at various stages of the reaction coordinate.

4.13 SOLUTION TO PROBLEMS EMBEDDED IN THIS CHAPTER

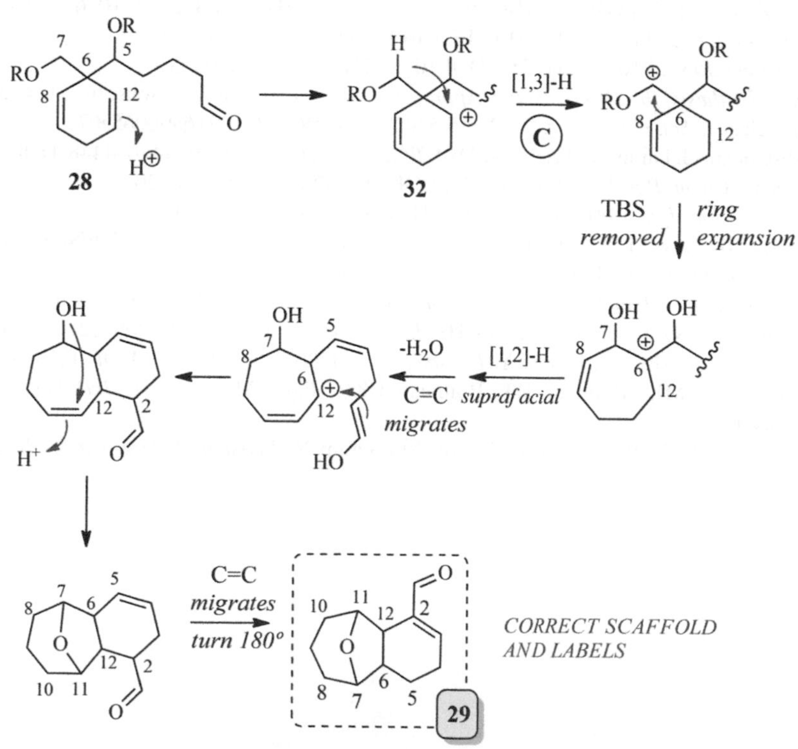

SCHEME IV.19 Alternative pathway to proposed routes **A** and **B** to the reaction **28**→**29** of Scheme IV 9. See also Schemes IV.10 and IV.11 for comparisons. Numbers in structures are scaffold carbon labels.

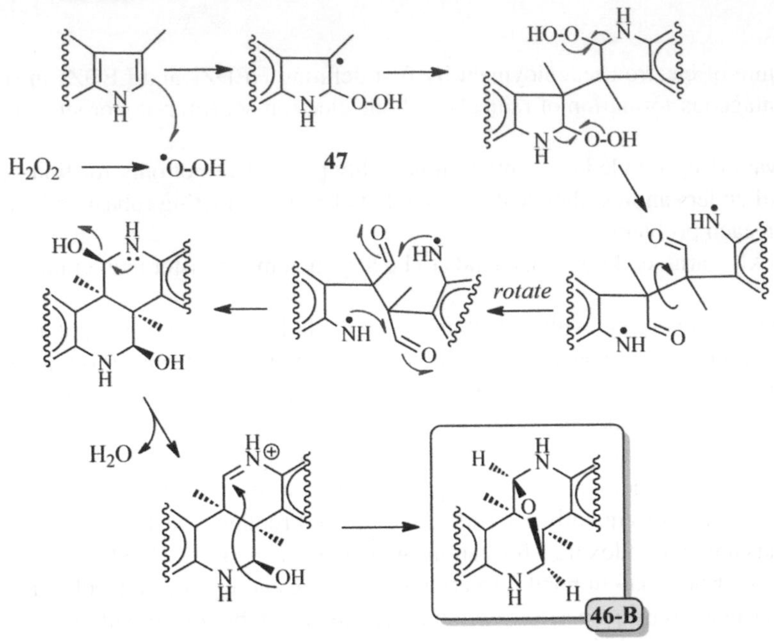

SCHEME IV.20 Solution to the reaction of Scheme IV.12, corrected target structure **46-B** [11] / Elsevier.

REFERENCES

1. Paquette, LA, Huber SK, Thompson RC. *J. Org. Chem.* 1997;58:6874–6882. DOI:10.1021/jo00076a058.
2. Rossi-Ashton JA, Clarke AK, Taylor RJK, Unsworth WP. *Org. Lett.* 2020;22:1175–1181. DOI:10.1021/acs.orglett.0c00053.
3. Huo CH, Su XH, Yang YF, Zhang XP, Shi QW. *Tetrahedron Lett.* 2007;48:2721–2724. DOI:10.1016/j.tetlet.2007.02.063.
4. Dietz J, Martin SF. *Tetrahedron Lett* 2011;52:2048–2050. DOI:10.1016/j.tetlet.2010.10.038.
5. Sears JE, Boger DL. *Acc. Chem. Res.* 2016;49:241–251. DOI:10.1021/acs.accounts.5b00510.
6. Warren S, Wyatt P. *Organic Synthesis, the Disconnection Approach*, 2nd ed. John Wiley & Sons, Inc., UK, 2008.
7. Bennasar ML, Vidal B, Bosch J. *J. Am. Chem. Soc.* 1993;115:5340–5341. DOI:10.1021/ja00065a074.
8. Butters M, Elliott MC, Hill-Cousins J, Paine JS, Westwood AWJ. *Tetrahedron Lett.* 2008;49:4446–4448. DOI:10.1016/j.tetlet.2008.05.022.
9. Von Dobeneck H, Lehnerer W. *Chem. Ber.* 1957;90:161–171. DOI:10.1002/cber.1957.0900203.
10. Tsuji J, Kezuka H. *Bull. Chem. Soc. Jpn.* 1981;54:2369–2373. DOI:10.1246/bcsj.54.2369.
11. Ling KQ, Ren T, Protasiewicz JD, Sayre LM. *Tetrahedron Lett.* 2002;43:6903–6905. DOI:10.1016/S0040-4039(02)01623-4.
12. Alonso-Amelot ME. *J. Chem. Educ.* 1977;54:568–570. DOI:10.1021/ed054p568.
13. Wenkert E, Bakuzis P, Baumgarten RJ, Doddrell D, et al. *J. Am. Chem. Soc.* 1970;92:1617–1624. DOI:10.1021/ja00709a033.
14. Wenkert E, Bakuzis P, Baumgarten RJ, Leicht CL, Schenk HP. *J. Am. Chem. Soc.* 1971;93:3208–3216. DOI:10.1021/ja00742a020.
15. Fukuoka H, Komatsu H, Miyoshi A, Murakai K, Kita Y. *Tetrahedron Lett.* 2011;52:973–975. DOI:10.1016/j.tetlet.2010.12.032.
16. Ahmed MG, Ahmed SA, Uddin MK, Rahman MT, Romman UKR, Fujio M, Tsuda Y. *Tetrahedron Lett.* 2005;46:8217–8220. DOI:10.1016/j.tetlet.2005.09.103.
17. Coldham I, Burrell AJM, Guerrand HDS, Watson L, Martin NG, Oram N. *Beilstein J. Org. Chem.* 2012;8:107–111. DOI:10.3762/bjoc.8.11.

PART II
THE PROBLEM CHEST

In this section, there is a collection of 50 selected reactions extracted from published research in mainstay journals with mechanisms of increasing difficulty that are, yet, accessible to most advanced students and professionals in organic chemistry. Each case is first presented as a problem for you to study and solved by the readership to practice and hopefully enjoy how solutions crop up after application of the strategies described in Part I. Then, it is analyzed in depth using one or more of the toolbox strategies, which greatly facilitate the design of feasible ways from starting material(s) and product(s), the idea being to serve as a comparison with your own musings and solutions. Feel free to discuss each problem with your classmates, in seminars or group brainstorming sessions, or present them as part of advanced courses if you happen to be a lecturer.

As said repeatedly herein, more than one solution is provided in the discussion in addition to the one conceived by authors of the original research. These solutions arise as I took the liberty of providing my own opinions, developed over five decades of practice in this fascinating game of exploring the inner workings of reaction mechanisms. I hope you enjoy these challenges and grow further in your career and curiosity for scientific discovery and organized reasoning.

PROBLEM 1

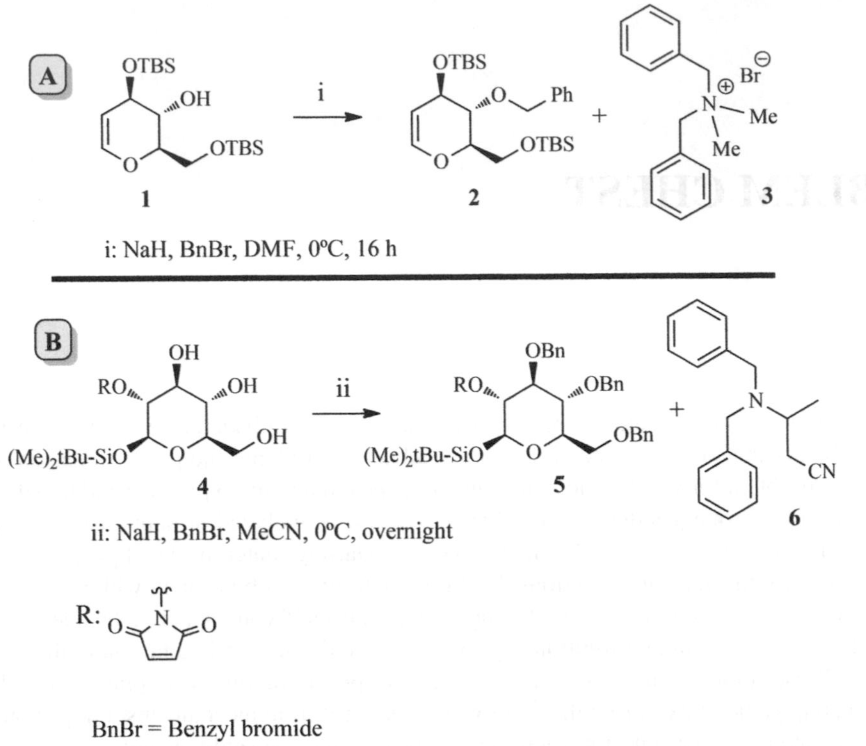

SCHEME 1.1 [1] / American Chemical Society.

PROBLEM 1: DISCUSSION

1.1 Overview

Scheme 1.1 illustrates two closely related cases of unexpected complications in an otherwise easily predictable performance of NaH (sodium hydride): two problems in one, so to speak. The authors' plan was to use NaH as a base in aprotic dimethyl formamide (DMF) to pick up OH (hydroxide) protons in **1** and **4**, promote the ensuing bimolecular nucleophilic substitution (S_N2) O-alkylation of benzyl bromide and furnish the corresponding O-protected products **2** and **5**, respectively. There is no mystery here, so you do surmise that "the problem" is not explaining these products.

Simple reactions such as these, however, are often marred by low yields and the formation of one or more side products that consume reagents and complicate isolation and purification of desired products.

Explaining these side reactions leading to **3** and **6** is the actual problem, if you think they constitute "a problem" in the terms indicated in Chapter 1.

1.2 Analysis of Reaction A

Product **3** has no relationship with **1** or **2**. We need clues to the starting materials of this unexpected product. In such cases, molecular dissection is the selected strategy despite the quite simple structure before us. Dissecting **3** (Scheme 1.2) shows two fragments strongly reminiscent of benzyl bromide (BnBr) and a dimethylamine, obviously coming from dimethylformamide (DMF) used as solvent.

The question turns to determining the way BnBr and DMF interact under relatively mild conditions. The difficulty is that the benzyl carbon and N in DMF are bonded to electron-withdrawing groups (EWGs) and are therefore, lousy candidates for N–C bonding. We need a previous step to turn one of the two into a high electron density zone (HEDZ) for prompting electron flow. The lack of O in **3** suggests a previous *reductive removal from* DMF, a job only attainable by NaH among the reaction components. How? Nucleophilic addition of hydride on C=O. Thereafter, two pathways (**a** and **b**, Scheme 1.2, bottom) are open to the first NaH + DMF intermediate **7** → **8**. Both converge on the tertiary amine **10** after the removal of formaldehyde. The passage to the ammonium salt **3** is probably enhanced by the Na^+-assisted departure of bromine.

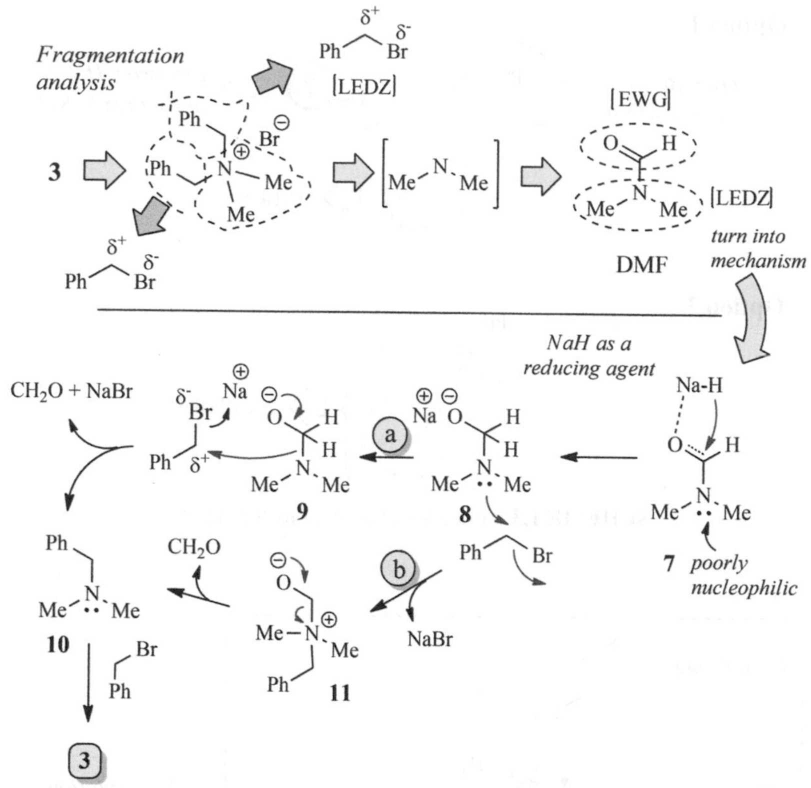

SCHEME 1.2 Dissection of target compound **3** and electronic properties of fragments.

An earlier investigation claimed that when a base-free mixture of DMF and methyl bromide was heated at 80 °C, well below its boiling point (bp; 153 °C) for six days in a sealed tube, carbon monoxide and tetramethylammonium bromide were obtained [2]. However, this disproportionation reaction, namely oxidation and reduction of the same compound, DMF in this case, could not be ascertained by ^{1}H nuclear magnetic resonance (NMR) monitoring, but the addition of NaH gave the annotated ammonium salt [1].

1.3 Analysis of Reaction B

The use of acetonitrile also proved troublesome in regard to low yields and the formation of by-products. The goal of reaction **B** was very much like that of **A**: protect three OH units of glucose derivative **4** with benzyl bromide and NaH to give **5**, a trivial operation. Again, explaining **6** is the problem.

Fragments of **6** can be established as before following first a very obvious lead: On the east side of the molecule the traits of acetonitrile appear clearly delineated in **Option 1** (Scheme 1.3), even though it leaves a CH–CH_3 unit up in the air. As this book also deals with making mistakes and learning from them, this option entails a miscalculation perhaps, since we did not assign any pertinence to the N atom in the dibenzyl western block. You may probably have been misled (I did this on purpose) by the reasoning of reaction **A** in which DMF was the solvent, while it is absent in **B**. This blunder calls for a second fragmentation pattern: **Option 2**.

By engulfing N alongside the mysterious CH–CH_3 unit, one realizes right away that this is a highly reduced form of acetonitrile; hence, 2 mol of CH_3–CN participate in the C_4 frame. This reduction from nitrile to amine (ΔFNt = 2) is the result of two C–C bonds being formed. In addition, the N atom performs as a double nucleophile against 2 mol of benzyl bromide. Therefore, the centerpiece of the mechanism plan is the self condensation of 2 mol of acetonitrile driven by base. After you try your own sequence down to product **6**, check Scheme 1.4 for a plausible string of reactions.

Authors [1] used CD_3CN (trideuteroacetonitrile) and various reaction conditions to support a more solid argument in favor of the mechanism of acetonitrile self-condensation route. They also managed to isolate various quantities of adducts **15** and **16** you can easily gather where they come from. And yet there is one point of doubt. In the actual experiment of reaction **B**, NaH faces two distinct protons with regard to their *pKa*: ROH (10) versus CH_3CN (acetonitrile, 25). One may surmise that being 15 orders of magnitude less acidic than the alcohols of **4**, acetonitrile stands no chance to compete for the hydride anion base. One can nevertheless argue that there is an enormous molar excess of acetonitrile since it is the solvent, after all; hence, NaH–CH_3CN collisions are far more frequent and have a chance to react and yield self condensation immediately after.

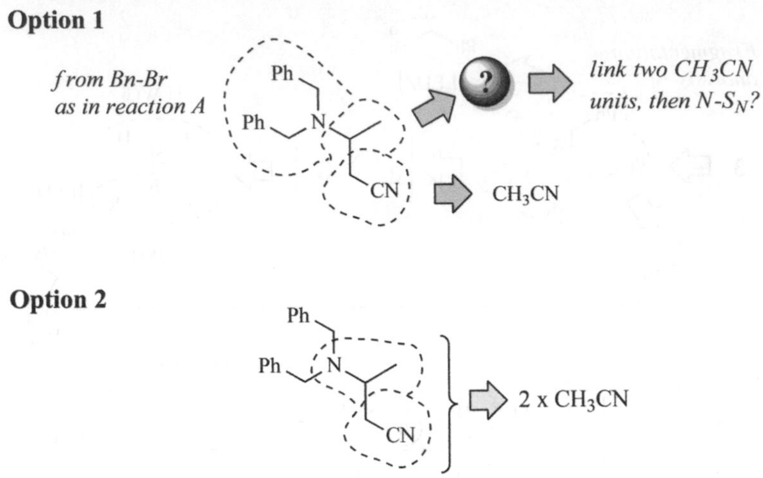

SCHEME 1.3 Fragmentation options for target **6**.

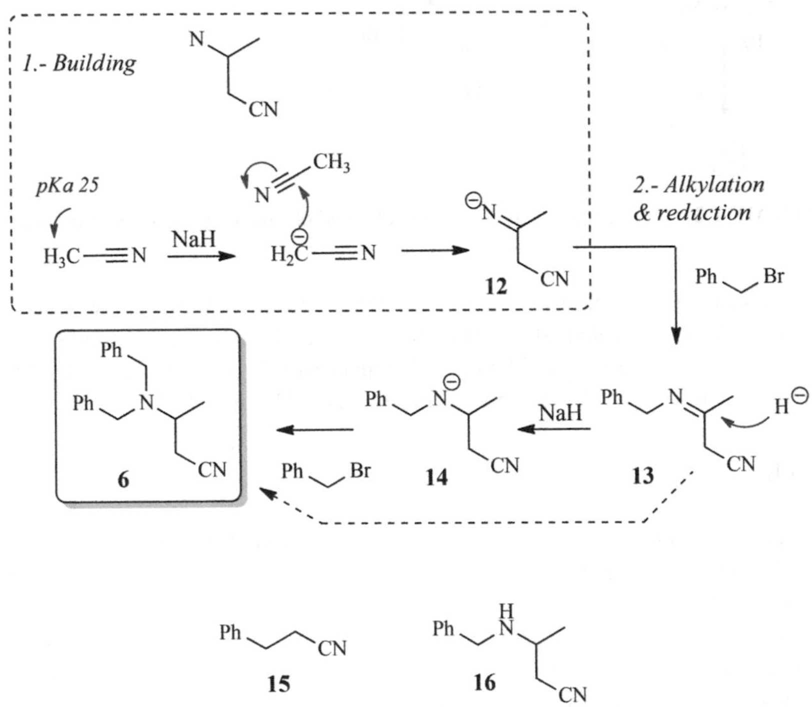

SCHEME 1.4 Double role of NaH as base and reducing agent to explain the reaction mechanism to target **6**.

It should be clear by now that *NaH and LiH can operate as both strong bases and reducing agents* in the same reaction sequence. So, pay attention to their dual personality in future problems.

REFERENCES

1. Hesek D, Lee M, Noll BC, Fisher JF, Mobashery S. *J. Org. Chem.* 2009;74:2567–2570. DOI:10.1021/jo802706d.
2. Neumeyer J, Lahtinen M, Busi S, Nissinen M, Lochmainem E, Rissanen K. *New J. Org Chem.* 1961;26:4681–4682. DOI:10.1021/jo01069a507.

PROBLEM 2

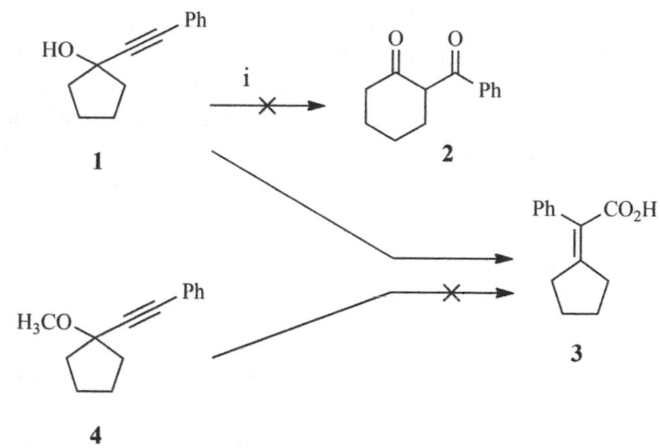

i: m-CPBA*, (1.5 eq)
Cl_3CCO_2H (1.5 eq), MeCN, r.t. overnight
Quenched with aq. $Na_2S_2O_3$

*: m-chloroperbenzoic acid

SCHEME 2.1 [1] / Thieme Medical Publishing Group.

PROBLEM 2: DISCUSSION

2.1 Overview

Acetylenes are exceedingly useful building blocks for organic compounds, some of them of great structural complexity. One expects substituted acetylenes to behave decently under a large variety of reagents, though not all of them; for example, the well-known peroxyacid oxidation furnish mixtures of products and poor yields of the anticipated oxidation of the end carbon to a carboxylic acid or ester [2]. Worse comes to worst, when disubstituted acetylenes are submitted to *m*-chloroperbenzoic acid (*m*CPBA) oxidation the number of products expands uncontrollably [3] (Scheme 2.2).

Scheme 2.1 portrays one of these devious reactions. The anticipated product from **1** and *m*CPBA was cyclohexanone **2. Not a trace of this compound was obtained but** only **3** in 20% yield. This value varied considerably with substituents in the aryl group [1]. On the other hand, the methoxy derivative did not react at all under the same conditions.

Therefore, this problem encompasses three questions:

1. Why did authors expect the unseen cyclohexanone **2**?
2. What mechanism explains the **1** → **3** reaction?
3. Why does the OCH_3 hamper the oxidation of the alkyne.

A fourth question will be added once the first three have been satisfied.

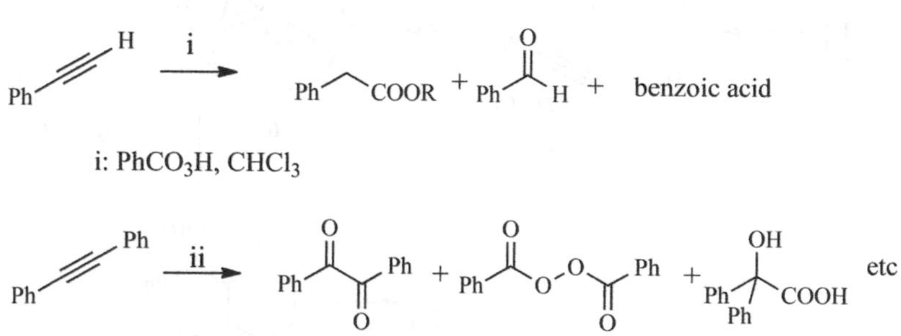

i: $PhCO_3H$, $CHCl_3$

ii: m-CBPA, CCl_4

SCHEME 2.2 Examples of uncontrolled oxidation of monosubstituted versus disubstituted acetylenes with peroxyacids.

2.2 Problem Analysis

2.2.1 Question 1 Mechanistic analysis is quite straightforward given that ring expansion stands out strongly suggesting a *pinacol-type rearrangement*. The tertiary carbinol is ideally suited for promoting this, so long as there is an electron-deficient carbon to collect the electron pair of the [1,2] migrating carbon of the expanding ring. The alkyne is not the group in question since it is a typical HEDZ. We need the concourse of a markedly low electron density zone (LEDZ) represented here by *m*CPBA owing to the weak and polarized O–O bond. This synthon ought to be associated with the alkyne for the oxidative process. Putting both HEDZ and LEDZ together leads to the first complex **5**, and ensuing oxirene **6**, setting the stage for the **semi-pinacol rearrangement**. The electron flow evolves like a breeze to target **2**. Unfortunately, **2** was not to be seen.

2.2.2 Question 2 This is more complicated than Question 1 and demands some analysis before rolling down a mechanism. Given that no ring expansion occurs, and the phenyl group has migrated, labeling scaffold carbons may bring leads and ideas (Scheme 2.4, top part). All the action occurs in the C^1–C^6–C^7 axis, including the loss of OH, probably as water. Because there are no α protons other than the ring methylenes, the extrusion of OH must take place owing to electron redeployment from whatever happens in the C^6–C^7 section. Also, the [1,2]-phenyl migration should accompany the alkyne oxidation at some stage. Thus, let us assume that transition state (TS) **6a** gives rise to **8** either directly or by an actual oxirene **6b** conduit. Ring strain and a highly basic oxygen characterizes this species. In the presence of *m*-clorobenzoic acid generated from *m*CPBA within the solvent cage, **6b** would be immediately protonated as soon as it is formed and cleaved to **8**, a proposal advanced in early mechanistic studies of the peracid oxidation of acetylenes [3]. Once cation **8** would be produced, we can work it out from there by changing the direction of the electron flow, as shown in Scheme 2.4 middle part.

The ensuing key electron redeployment **8** → **9** leaves the α carbon (C^6) with six valence electrons only. This is a carbene, a suitable intermediate for the desired phenyl migration known as **Wolff rearrangement** [4]. Similar shifts of hydrogen, alkyl, and backbone carbons next to C=O also undergo this sort of rearrangement.

You may wonder if the Wolff rearrangement **9** → **10** phenyl migration and carbene (presumably singlet) electron redeployment for fashioning the π bond occurs in steps or in a concerted manner (see note in Reference [5]). The ketene **10** thus formed is susceptible to hydrolysis after the previous exit of the C^1–OH, as we had anticipated in the mechanism design not resorting to the usual water elimination but by the intervention of another functionality in the scaffold (as in **11**).

2.2.3 Question 3 Methoxy propargylic derivative **4** (Scheme 2.1) hampers the oxidation of the triple bond. This is a bit peculiar since the oxidation of the triple bond is achieved by other oxidants. Gold and zirconium catalysis, among others, enhance the process to ketenes and unsaturated ketones [6] (Scheme 2.5, top part). These precedents lead to the notion that contrary to the C^1–OH, the C^1OCH_3 interacts differently with *m*CPBA in the construction of the TS. If so, this takes us back to **TS 6** with a few modifications depicted in Scheme 2.5, bottom section. Authors of the original paper [1] suggest that the lack of the H-bonding capacity in **TS6c** from **4** is a probable cause of the lack of reactivity against *m*CPBA. How would you demonstrate this hypothesis experimentally?

A note in closing: Oxirenes (**6b**-type) have never been isolated in a reaction vessel, distilled, crystallized, or chromatographed, but only detected as transient species in laser flash photolysis of α-diazoketones during their Wolff rearrangement course, photolysis of ketenes, and so forth, and postulated in the peroxidation of acetylenes as this problem describes. High power quantum calculations using large basis sets have concluded that oxirane exists not in a peak of the energy profile along

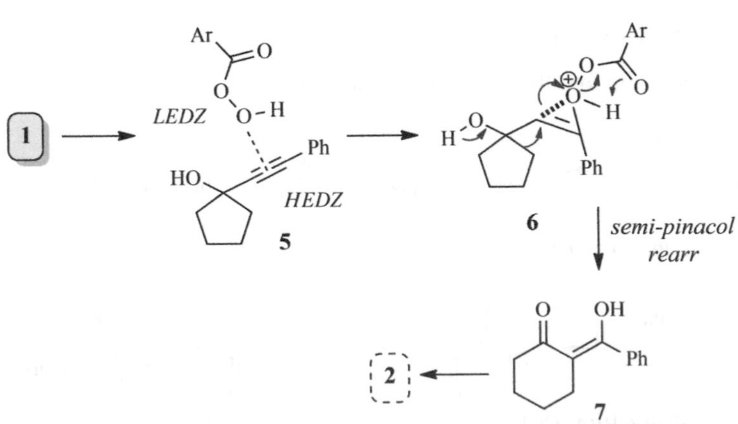

SCHEME 2.3 Answering Question 1.

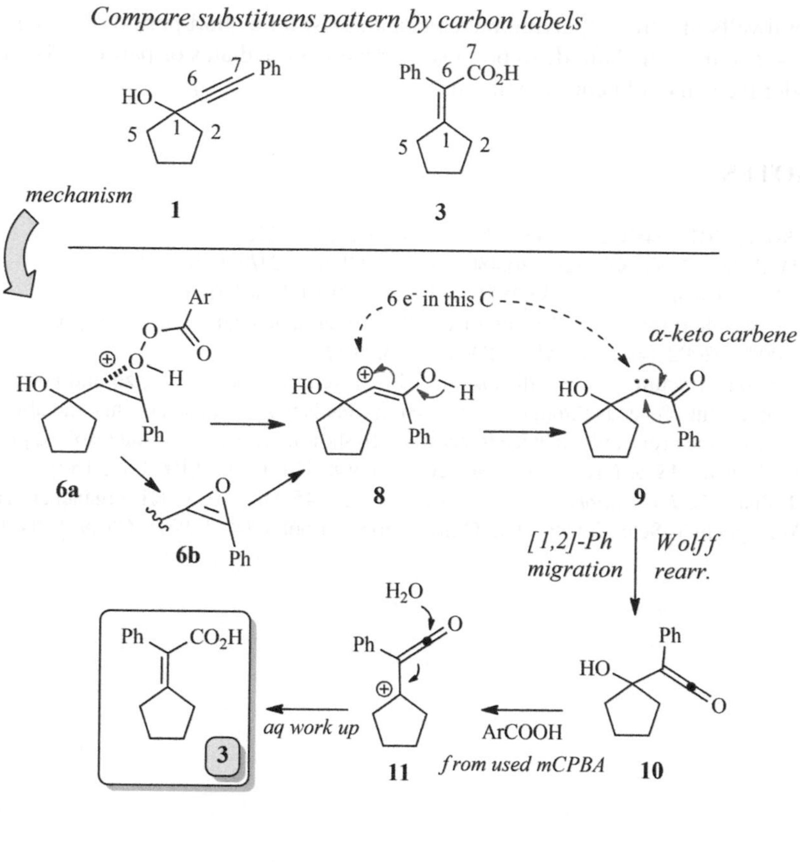

SCHEME 2.4 Answering Question 2.

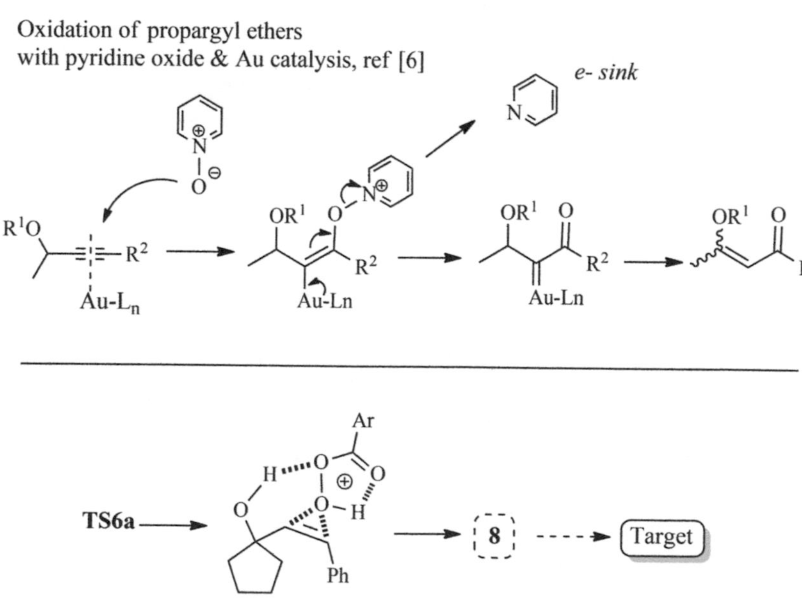

SCHEME 2.5 [6] / Elsevier.

the reaction coordinate but dwells in a trough, giving it a chance to exist for a short period of time [7]. In terms of mechanism proposal, these calculations give us some latitude to propose oxirane intermediates or part of TSs en route to products but not as isolable compounds under the standard bench conditions.

REFERENCES AND NOTES

1. Rodríguez A, Moran WJ. *Synlett* 2013;24:102–104. DOI:10.1055/s-0032-1317711.
2. McDonald RN, Schwab PA. *J. Am. Chem. Soc.* 1964;86:4866–4871. DOI:10.1021/ja01076a028.
3. Stille JK, Whitehurst DD. *J. Am. Chem. Soc.* 1964;86:4871–4876. DOI:10.1021/ja01076a029.
4. For an authoritative review, see: Kirmse W. "100 years of the Wolff rearrangement." *Eur. J. Org. Chem.* 2002;2002(14):2193–2256. DOI:10.1002/1099-0690(200207)2002:14<2193::AID-EJOC2193>3.0.CO;2-D.
5. The answer is contingent on the structure of the α-diazoketone. In alicyclic ketones, the diazo and keto groups are locked in the syn configuration and the rearrangement proceeds through a concerted mechanism. As opposed to this, the aliphatic diazoketones and esters adopt a flexible conformation and the preferred sequence is stepwise. To show this ultra-fast, time resolved photochemical techniques were employed. See: Burdzinski G, Platz MS. *J. Phys. Org. Chem.* 2010;23:308–314. DOI:10.1002/poc.1601.
6. Ji K, D'Souza B, Nelson J, Zhang L. *J. Organomet. Chem.* 2014;770:142–145. DOI:10.1016/j-jorganchem.2014.08.005.
7. Vacek G, Galbraith, JM, Yamaguchi Y, Schaefer HF, et al. Oxirene: to be or not to be. *J. Phys. Chem.* 1994;98:8660–8665. DOI:10.1021/j100086a013.

PROBLEM 3

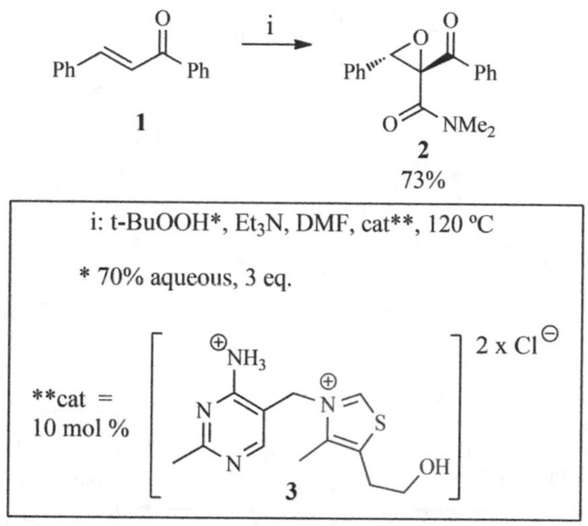

SCHEME 3.1 [1] / American Chemical Society.

PROBLEM 3: DISCUSSION

The preceding problem was a warmup of what this reaction entails.

3.1 Problem Analysis

It seems quite simple, even though the number of components and the catalyst structure may be a bit overwhelming:

1. Product **2** shows the same carbon backbone of **1**.
2. Modification of the latter is reduced to epoxidation with conservation of the E steric configuration of the chalcone moiety and amidation.
3. Chalcone (**1**) and DMF, the obvious amidation reagent, are both LEDZ, so their chances of undergoing C–C bond formation in the way to **2** through polarity-driven electron redeployment are low.

3.2 Attempting a First Mechanism

The aforesaid circumstances suggest a reaction pathway governed by *radical chemistry*. In actual fact, authors found that, on one hand, adding a radical scavenger stops the reaction. On the other hand, *t*-butyl peroxide is a good candidate for spreading radicals around, especially if there is no light protection [2], not to speak of the 120 °C employed. Would you mind drawing such a mechanism? My all-radical version is portrayed in self-explainatory Scheme 3.2.

Formally, Scheme 3.2 is a sound exploit of radical performance for oxidative C–O, a C–C bonding. I managed to oxidize the α,β-unsaturated ketone, perform the acylation of the ketone's α carbon **7** radical-wise and oxidize the resulting aldol **9** following allowed H shifts and obtain the desired target **2**. And yet, Scheme 3.2 does not respond to the experimental realities of Scheme 3.1. Notice that no specific role is assigned to the catalyst or the base (Et_3N, triethylamine). Therefore, my sequence is nothing more than an exercise of imagination.

3.3 A Second Alternative

More data are needed. Indeed, the authors provided a good batch as they improved conditions and tested the scope of this reaction [1]. Check out Scheme 3.3 for a few hints before you start working this one out. Not all entries explain feasible mechanisms.

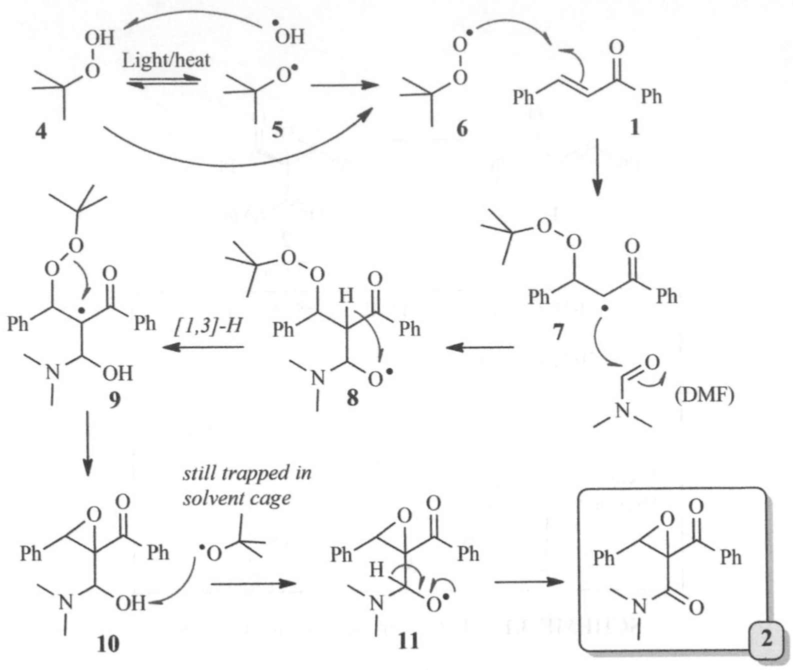

SCHEME 3.2 The radical-only $1 \rightarrow 2$ way; not the making of authors of [1] / American Chemical Society but of Alonso-Amelot for this book.

After studying Scheme 3.3 for suggestions, and given that the reaction will not proceed in the absence of catalyst, (maximum product yield efficiency at 10 mol%), key issues crop up:

✓ What is the role of the catalyst? The mechanism scheme must account for its *regeneration* to have it available for another catalytic cycle.
✓ What is the purpose of the base (Et_3N) considering that the reaction does not proceed in neutral medium?
✓ Is this epoxidation–amidation a concerted double addition or a stepwise (tandem) sequence? Either option should have consequences for the mechanism.
✓ If your choice is the tandem model, which comes first, epoxidation or amidation?

3.3.1 A Word or Two About the Catalyst in the Way to a Mechanism Catalyst **3** is a double salt that your own body stores (25–30 mg in a healthy adult): thiamine, better known as vitamin B1. If deprived of it, you would die painfully, stalled in a few hours. The catalytic sector of thiamine is the blue block shown in Scheme 3.1. This sub-structure should ring a bell if you paid attention to Section 2.4.11.2, regarding NHCs (nitrogen heterocycle carbenes). A prominent feature of **3** and congeners is the acidic character of the single proton in the heterocycle. Upon removal by base (Et_3N in this case) and electron redistribution of the zwitterion, a *nucleophilic carbene* (**12b**) is formed. As an LEDZ, chalcone **1** is susceptible to this carbene (Scheme 3.4, top section). Their coupling prepares the ketone for the amidation step. After this, the actual tandem radical-driven oxidation (epoxidation and reconstruction of the amide) occurs followed by the liberation of thiamine for another catalytic cycle (Scheme 3.4, blue arrow from **16**). Worth noticing, the independently synthesized epoxide (entry **7**, Scheme 3.3) remained unchanged in hot DMF, putting amidation ahead of epoxidation.

3.4 Open Questions and Partial Answers

Mechanism proposals, no matter how well adjusted to experimental data, often leave unsolved questions. A few:

✓ *The radical inhibitor issue* (entry **5**, Scheme 3.3): When authors [1] added a radical scavenger, the reaction was inhibited. This strongly suggested that a free radical must have been formed early in the sequence, but the first radical intervenes at the $\mathbf{14} \rightarrow \mathbf{15}$ stage (Scheme 3.4) when the DMF coupling was been accomplished. The side pathway shown in the dashed box would be an escape route to stable chloride salts **18** and **19** or ketone **20** identified as deoxy-Breslow adduct after the

Hints

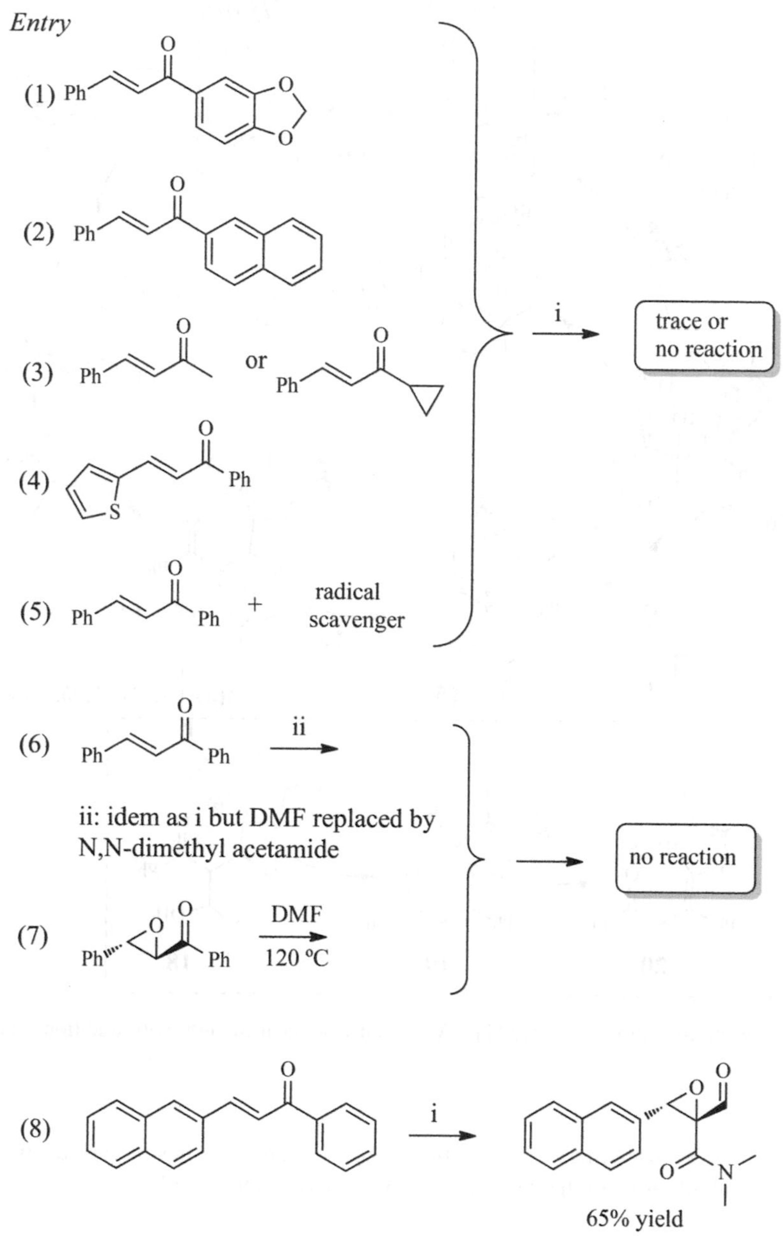

SCHEME 3.3 Clues worked out by the authors during detailed optimization of reaction conditions and scope [1] / American Chemical Society.

proposal of this author in the early days of NHC chemistry [3]. Given that these products were not observed, the only explanation for the no-reaction scene is the reversibility of **1**+**2** (+cat **3**) ⇄ **13** or even **14** through a retro-aldol cleavage.

- ✓ *Steric effects*: Enolate **13** is a congested structure, no doubt, and DMF condensation may be hampered further by the steric hindrance of groups larger than benzene on the C=O end (entries **1** and **2**, Scheme 3.3). Along the same lines, replacing DMF by *N,N*-dimethylacetamide (entry **6**) completely inhibits this condensation. By contrast, placing much larger substituents on the β carbon of **1** such as a naphthyl group (entry **8**), which offers nullifying resistance on the other end, does not interfere with the NHC+**1** addition, even though thiamine **3 is a sterically bulky carbene**; nor does an *o*-bromobenzene substituent relative to the *p*-bromo derivative but only marginally (not shown).
- ✓ *Alkyl ketone inability to yield products* (entry **3**), it probably results from a decreased reactivity of **12a,b** toward the β carbon of **1** by the *+I* effect of the alkyl group, rather than a steric effect.

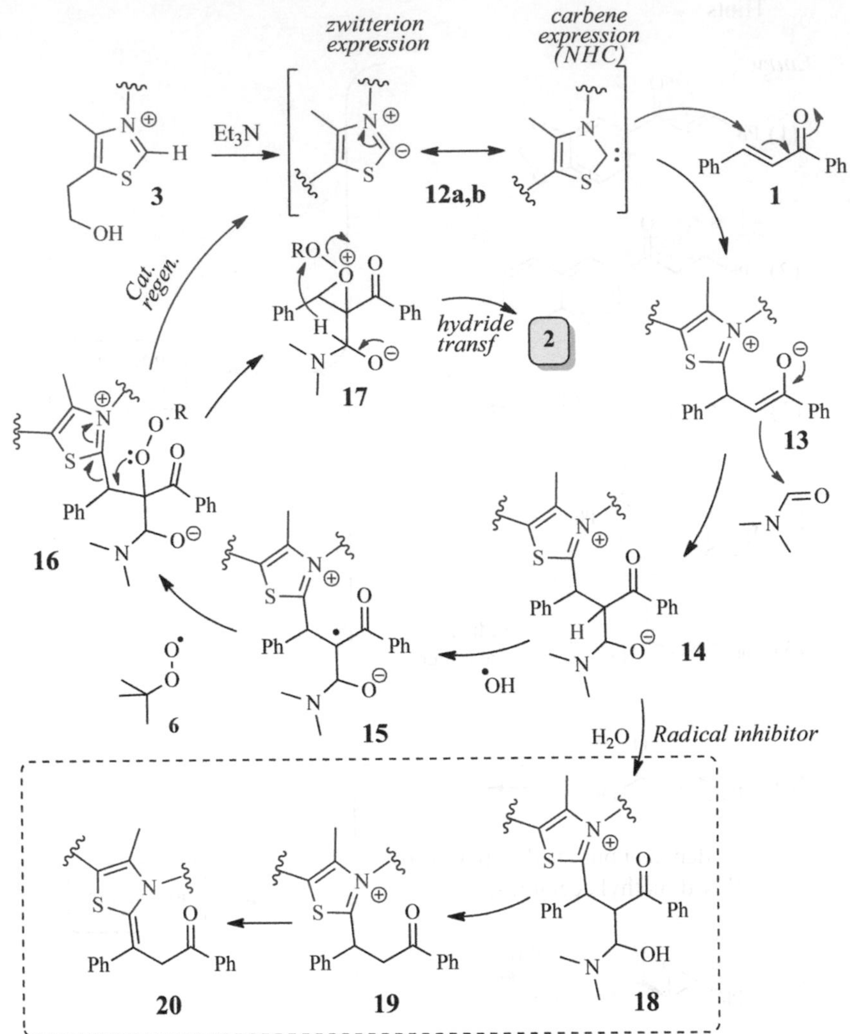

SCHEME 3.4 Tandem amidation–oxidation sequence [1] / American Chemical Society with additions. The catalyst (blue) structure is abridged for clarity.

A more informative method to explore these uncertainties might be kinetic studies of various model compounds used in this study or quantum theoretical calculations in the density functional theory (DFT) style.

REFERENCES

1. Sankari Devi E, Pavitra T, Tamilselvi A, Nagarajan S, Sridharan V, Maheswari CU. *Org. Lett.* 2020;22:3576–3580. DOI:10.1021/acw.orglett.0C01017.
2. Solar and UV light induce the homolytic cleavage organic hydroperoxides to peroxyl-, alkoxyl-, and hydroxyl radicals. This phenomenon has been observed in natural or synthetic organic hydroperoxides formed by terpene ozonolysis, which occurs in the atmosphere. Dispersed in aerosol particles, they furnish radicals five times faster than hydrogen peroxide when exposed to UV light under laboratory conditions and may contribute to cloud water and aerosol chemistry in the atmosphere. See: Badali KM, Zhou S, Alkawhari J, Antiñolo M, Chen WJ, Et al. *Atmos. Chem. Phys.* 2015;15:7831–7840. DOI:10.5194/acp-15-7831-2015.
3. Breslow R. *J. Am. Chem. Soc.* 1958;80:3719–3726. DOI:10.1021/ja01547a064.

PROBLEM 4

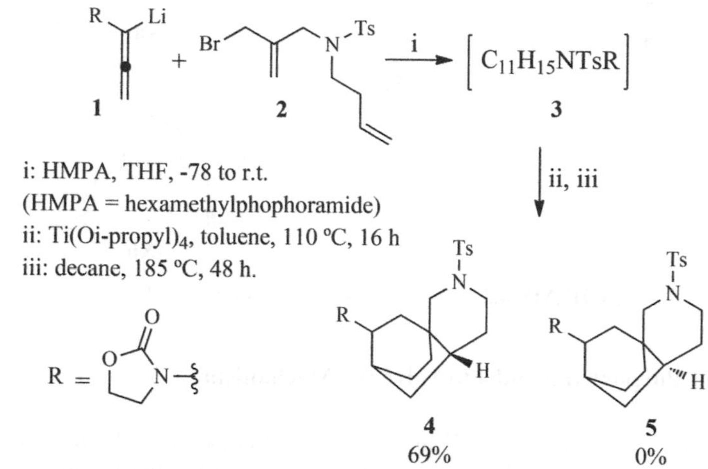

SCHEME 4.1 [1] / American Chemical Society.

PROBLEM 4: DISCUSSION

4.1 Overview

While compound **3** was not isolated, authors knew its structure thanks to their previous work [2] and similar reactions in their present paper [1]. Let us imagine, however, that you were a new student to this laboratory and nobody told you that or allowed you to read the previous paper, and so the structure of **3** is an open question. Once you solved this first challenge, proceed with what appears to be a true problem: explaining feasible **3** → **4** mechanism(s) and reasons for not observing **5**.

4.2 Unveiling the Structure of *3*

Because there is little to hold to, we have only two elements of judgment to work with.

a) Abridged element balance that keeps R and Ts (tosylate) as such for simplicity, given that both are maintained in starting materials and products, as explained in Chapter 4:

Compound	Abridged molecular formula
1	C_3H_2Li**R**
2	$C_8H_{13}BrN$**TsR**
3	$C_{11}H_{15}$**NTsR**
3− (1+2)	Li, Br

From this elementary balance we can say this much: **1** and **2** unite their carbon scaffolds to create **3** without altering their unsaturation.

b) The lithium allene is clearly an HEDZ, whereas **2** is a typical LEDZ as far as the C=C–C–Br section. One can easily predict a **1**+**2** direct S_N2 reaction on C–Br accompanied by the vinylogous S_N2' process, which furnish **Xa** and **Xb**, respectively (Scheme 4.2). Now you have two compounds to contend with for the ensuing step, don't you? Or do you? The answer is no, because **Xa** and **Xb** are identical although sketched differently. The molecular formula of **X** is $C_{11}H_{15}$**NTsR**, which is the same assigned to intermediate **3**. Thus, there is a chance that **X** equals **3**, but you were given insufficient information to confirm this. The only way is to relate **3** to the next **3** → **4** thermolysis, then work this out backward from the latter by establishing feasible mechanisms.

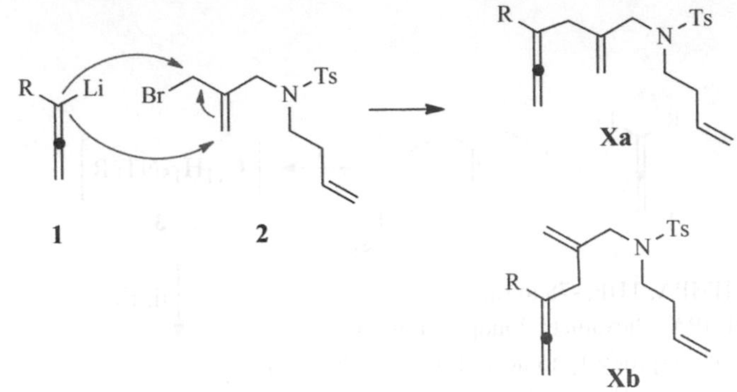

1 2

SCHEME 4.2 Proposing a structure of **3** from **1** + **2**.

4.3 Comparative Analysis of Purported 3 and 4 to Solve the Mechanism

Apparently, structures **3** and **4** seem quite disparate. Which analysis strategy would be more effective to study this reaction, considering that it is a thermal isomerization under Lewis-acid catalysis [Ti(O–*i*–Pr)$_4$]? I would suggest assigning atom labels supported by the recognition of the equal relative positions of R and *N*-tosylates (*N*Ts) groups in both compounds. From there, bond accounting may provide clues about bond forming/cleaving and skeletal rearrangements, if at all. This strategy is worked out in the top section of Scheme 4.3. Prompted by the substantial bond construction in **Xa** → **4**; additionally, there are no functional groups for intramolecular interaction other than the C=C bonds; hence, it is a good idea to redraw **Xa** by moving closer these functionalities. In so doing, one realizes the intimate relationship between **Xa** and target **4**. Carbon tags reveal three C–C σ bonds (marked in green): C^1–C^6, C^2–C^{11}, and C^5–C^{10}.

4.4 Analysis of Route A

(See Scheme 4.3, middle section.) Based on the assumption that **Xa** is the unknown intermediate, it soon faces two conceptual errors:

1. Having selected the formation of the C^1–C^6 link first, this approach forces the allene and alkene into bonding, taking for granted that the allene C^1 is an LED carbon. A tertiary $C^{(+)}$ develops at C^5, thus opening the $C^5 \rightarrow C^{10}$ future bonding. On the allene side, an sp^2 $C^{(-)}$ is formed concomitantly and undergoes stabilization by the +*I* inductive effect of the oxamido group R. Although in principle this would favor the completion of the **6** → **7** dipolar cycloaddition, the unshared electron pair atomic orbital (AO) at the C^2 anion required for bonding to C^{11} is in the pseudo plane of the six-membered ring facing away from C^{11}, and hence, is perpendicular to the *2pz* AO of the latter carbon. Drawing your own 3D model of the purported TS will convince you of his. Therefore, the second cycloaddition is hardly realistic.
2. The steric strain (SS) from the bridge head C^2=C^3 would contribute to increase the ring strain and the activation energy to reach this intermediate. Approach **A** is thus discarded.

4.5 Analysis of Route B

(See Scheme 4.3, bottom section.) The C^1–C^6 bonding may be secured first by transforming the allene to a conjugated diene involving C^4. All it takes is a C^4–H → C^2 [1,3] proton shift and π electrons redistribution thereafter. A triene **9** results. For allylic functions, symmetry rules anticipate an *antarafacial* H transfer, a physically unfeasible proposition under thermal conditions. The AO distribution of allenes, nevertheless, allows the 1,3 shift given that the terminal C=C π orbital is perpendicular to the inner π C=C allenic bond. Therefore, one of the C^4 hydrogens may face the C^2 *2pz* AO at a suitable distance to execute the *suprafacial* [1,3]-H shift with symmetry adequacy as shown in sketch **8**.

Authors had reported earlier a series of trienes from similar thermolysis of allylic allenes grafted with substituted oxamides on C^3 [2], thus supporting **the identity of X as compound 9 rather than allene Xa**. (See Section 4.7 and Scheme 4.4 for more details.)

Continuing with Scheme 4.3, the C^1–C^6 linkage develops from the familiar thermal 6 π pericyclic reaction, inasmuch as the [1,3]-H shift furnishes the E C=C to situate the three double bonds in the right position for the electrocyclic process. Product **10** thus obtained is ideally suited for a second intramolecular cyclization, a Diels–Alder type, passing through **TS10**.

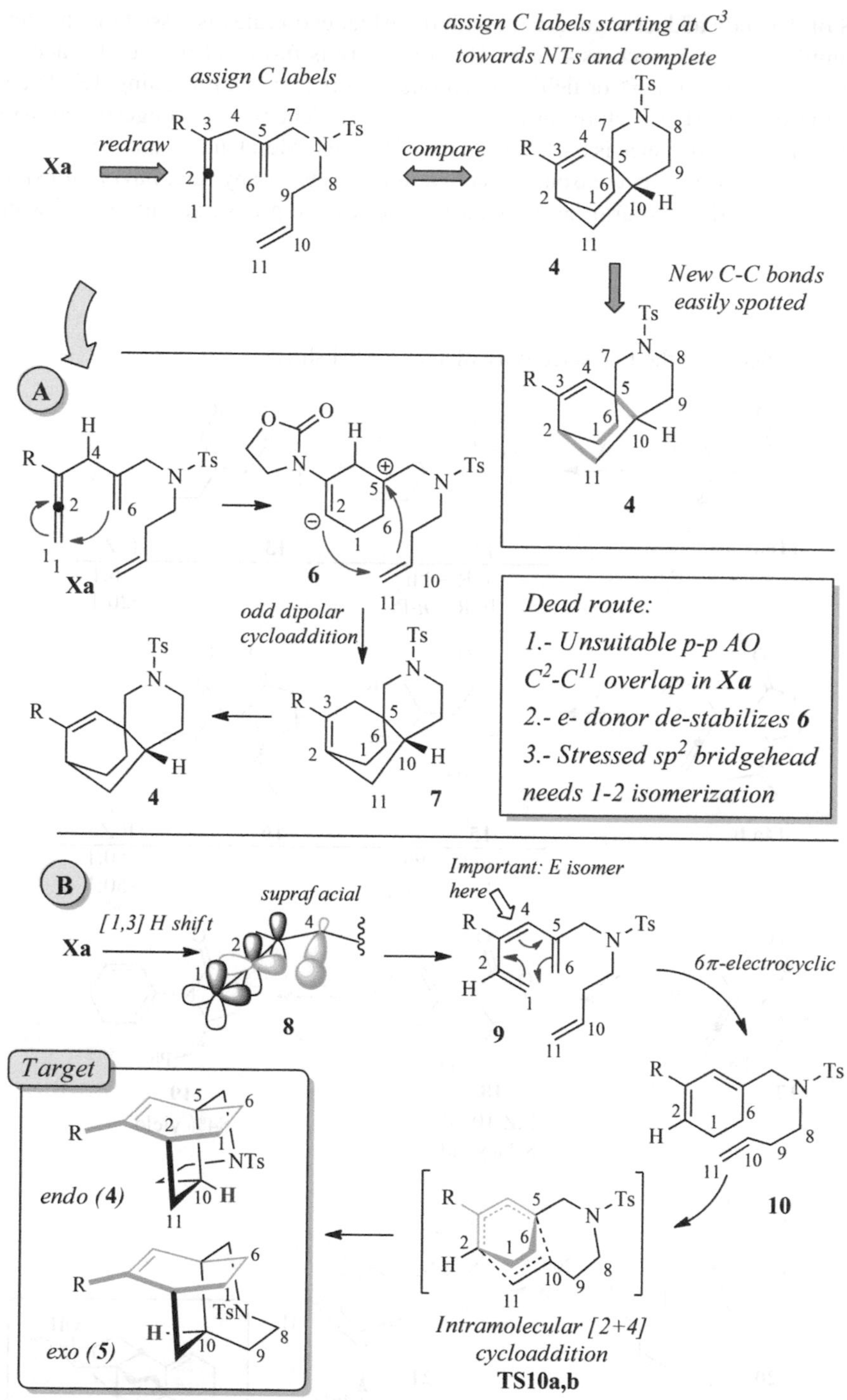

SCHEME 4.3 **Top part:** Comparing purported intermediate **Xa** and target **4**, assigning carbon labels and revealing new bonds to form the product. **Bottom section:** Possible mechanism routes **A** and **B** with entirely different feasibilities.

4.6 Understanding the Stereochemistry of Target *4*

The absolute configuration of C^{10} defines the stereochemical arrangement of the diene-dienophile components of the intramolecular Diels–Alder cycloaddition (DACA). This coupling is nearly impossible to understand using a 2D rendering of **4**. With a bit of skill, one can translate **TS10** and the resulting adducts into 3D endo and exo depictions (framed structures, Scheme 4.3). In endo (**4**), the C^1–C^6 bridge points in the direction of the C=C moiety, whereas these two sections point in opposite directions in the exo adduct **5**.

Preference for the TS of the endo adduct is congruent with the Alder endo rule discussed in Chapter 3, but it is more difficult to visualize in a complex structure such as **4** or **5**. This difficulty is furthered by the absence of polar substituents in $C^{10}{=}C^{11}$ that might help in proclaiming π–π* or field interactions favoring *endo* **4** preceding **TS**. The reason might rather be steric. A second look at the framed 3D structures at the bottom left of Scheme 4.3, suggests repulsion between C^1 and C^9 methylenes in *exo* **5** where the 1–3 H distance is only 2.33 Å in a MMFF94 (Merck molecular force field) minimized structure. This interaction is absent in *endo* **4**, thus decreasing the activation energy. At any rate, having no stereogenic atoms in precursor **9**, the intramolecular cycloaddition furnishes two diasteromeric endo products, only one of which appears depicted in Scheme 4.3.

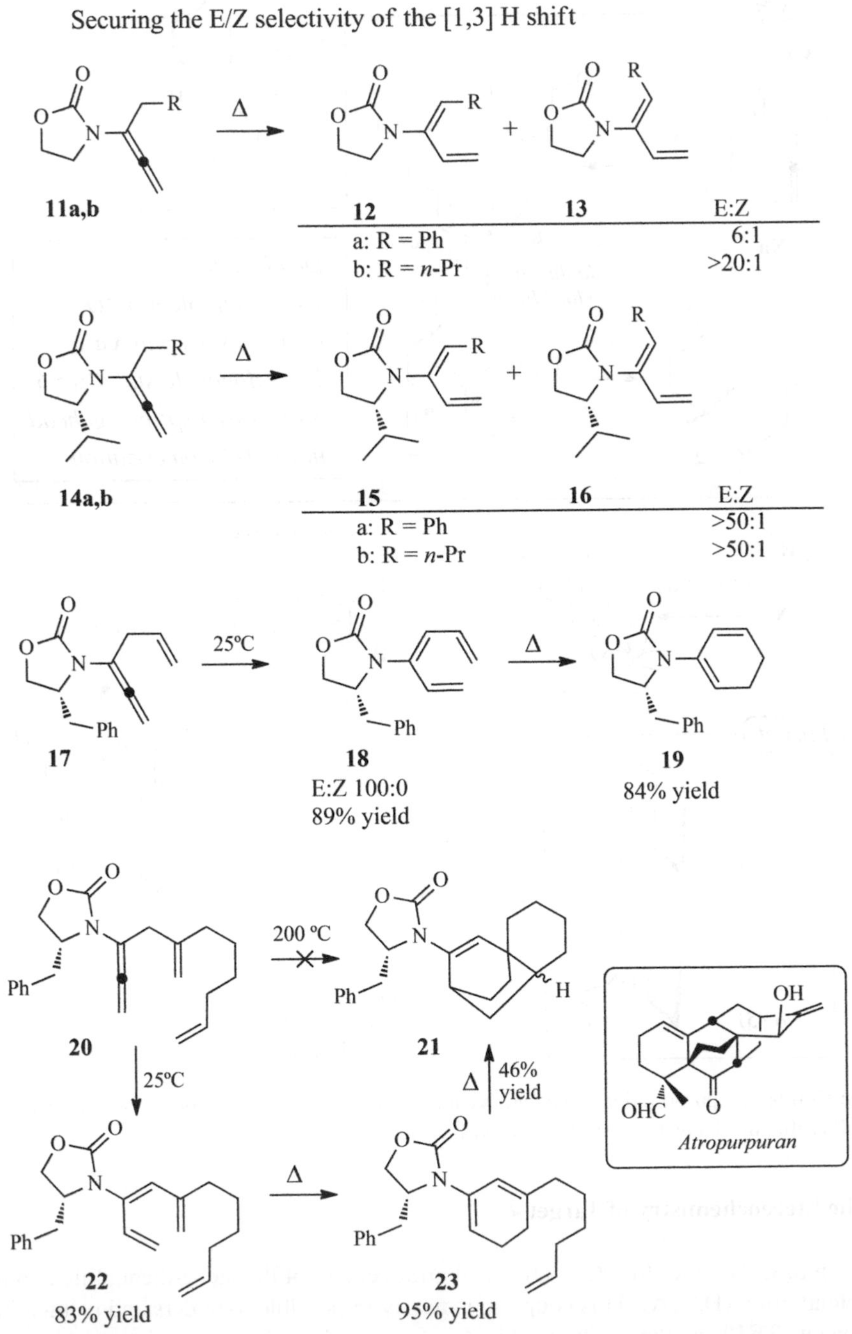

SCHEME 4.4 Earlier approximations to the suitable structure for the tandem cyclizations. [2] / American Chemical Society.

4.7 Background Reactions and Takeaways

Discoveries in chemistry often come in steps from a seedling, foundational paper. Scheme 4.1 is the result of earlier developments by the same research group [2–4], based on conceptually supporting References [5, 6] in search of higher structural complexity inspired in the always admirable and frequently convoluted chemistry of living organisms [7] (see inset, Scheme 4.4). This scheme collects just a few of the reactions paving the way toward the said complexity, those that failed or succeeded after careful thinking of the reaction mechanism. Among others, three issues stand out:

1. *The question of the* [1,3]-*H shift* discussed above leads to two possible outcomes: E and Z olefins **12** and **13**. Only the former is of any use if an electrocyclic reaction is desired to build the six-membered ring, as in **18** → **19**. The steric effect of the oxamido moiety must have controlled the TS in the way to *E isomer* as indicated by the enhancement of the E : Z ratio in **12a/12b** and notably **15a/15b**.
2. The triene from the [1,3]-H shift could be intercepted before electrocyclization by controlling the temperature (25 °C).
3. The shortcut from the amidoallene **20** to tricycle **21** by heating and Lewis-acid catalysis would not work but following the aforementioned steps, allowing better control and higher yield, after exploring models ever closer to the target atropurpuran scaffold.

REFERENCES

1. Hayashi R, Ma ZX, Hsung RP. *Org. Lett.* 2012;14:252–255. DOI:10.1021/ol203030a.
2. Hayashi R, Hsung RP, Feltenberger JB, Lohse AG. *Org. Lett.* 2009;11:2125–2128. DOI:10.1021/ol900647s.
3. Hayashi R, Hsung RP, Feltenberger JB. *Org. Lett.* 2010;12:1152–1155. DOI:10.1021/ol902821w.
4. Hayashi R, Feltenberger JB, Lohse AG, Walton MC, Hsung RP. *Beil. J. Org. Chem.* 2011;7:410–420. DOI:10.3762/bjoc.7.53.
5. Hsung RP, Wei LL, Xiong H. *Acc. Chem. Res.* 2003;36:773–782. DOI:10.1021/ar030029i.
6. Krohn K. *Angew. Chem. Int. Ed.* 1993;32:1582–1584. DOI:10.1002/anie.199315821.
7. The beacon of this synthesis sequence is the pentacyclic motif of the alkaloid *atropurpuran* from *Aconitum hemsleyanum* (inset, Scheme 4.4).

PROBLEM 5

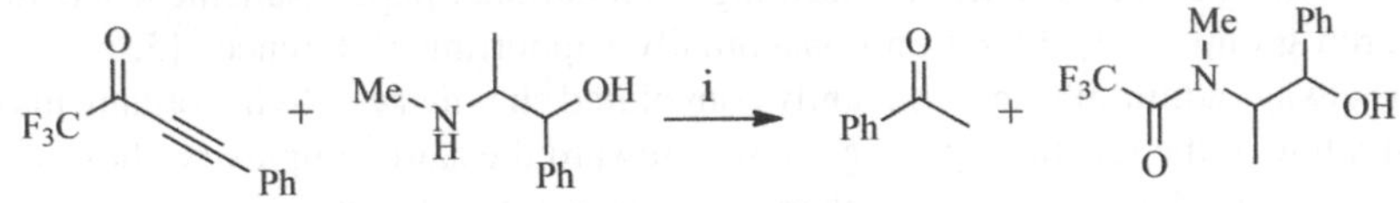

1 2 3 4

i: EtOH, 80 °C, 12 h, >50% yield

SCHEME 5.1 [1] / Elsevier.

PROBLEM 5: DISCUSSION

5.1 Overview

The reaction of Scheme 5.1 is a development of earlier investigations aimed at exploring the addition of nucleophilic ethanolamines and 1,3-diamines to acetylene (propargyl) ketone acceptors with EW substituents [2]. The functionally dense adducts were prone to C–C fragmentation in similar terms of Scheme 5.1. Reaction conditions are very modest, no fancy reagents or catalysts, just moderate heat in ethanol and water (96% ethanol), convenient proton sources and a hydrolytic reserve. In a nutshell, an organic chemistry's bare bone performance. This problem is one example of the diversity of reaction mechanisms open to apparently accessible transformations, a teaching tactic extensively used in this workbook.

5.2 Problem Analysis

At first glance, one notices bond dislocations after comparing products and reactants; hence, some sort of backbone rearrangement is expected. In such cases, fragmentation analysis (FA) is an efficient strategy to identify molecular pieces in the target reflecting the starting compounds and from there the likely origin of these sections. Our first FA is sketched in Scheme 5.2. It shows that:

1. Fragments **A** and **B** are assigned without question.
2. Dashed lines crossing bonds in the product (**4**) imply C–C bond formation in the forward mechanism sequence. In this case, this involves the secondary amine pseudoephedrine attacking the trifluoromethyl ketone moiety, but it does not reveal at which point of the reaction mechanism it occurs. Also, this attack would leave a deactivated alkyne to further react by hydration furnishing acetophenone. Hydration of acetylenes requires Lewis- or protic-acid catalysis but not this time over. In other similar experiments, authors employed CuCl [copper(I) chloride; wait until Scheme 5.4, please].
3. The origin of acetophenone (fragment **C**) is not clear. If the phenyl acetylene moiety is your choice (Scheme 5.2, fragmentation **B**) two questions arise: i) Where does the oxygen atom of the C=O genesis come from (green arrow [**a**])? ii) If you prefer choice (**b**), wouldn't this entail a methylation-fragmentation of the ethanolamine derivative **2**?
4. In regard to the propargyl ketone **1**, FA indicates that the C–C bond between Ph–C=O and the vicinal acetylene unit gets cleaved since the dotted line of fragments **B** and **C** in option (**a**) crosses this bond.

> *Reminder: FA dotted line crossing X–Y bonds of products indicates the site of formation of this bond. The FA dotted line crossing the Y–Z bonds of starting compound indicates the breakage of this bond during the reaction.*

Addition–elimination may therefore account for this breakage, which calls for an inventory of high electron density (HED) and LED centers available. These are depicted in the lower section of Scheme 5.2, opening various possibilities of **1**+**2** interactions to work with. Please try your own sequence before checking out my mechanistic musings in Scheme 5.3.

Not one but four pathways are conceivable to explain Scheme 5.1.

5.2.1 ***Route A, Scheme 5.3*** The natural electron density gradient of **1** from the terminal phenyl to the CF_3C–CO moiety is put to work for the **1**+**2**→**5** addition. Having two nucleophilic centers in **2**, I selected the carbinol for the first Michael-type 1,4-addition on the alkyne to keep the amine end available for the necessary 1,2-addition on the C=O, and thus be able to form

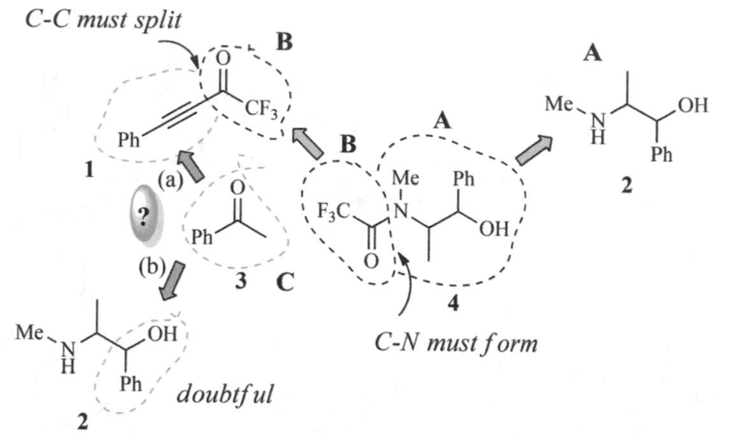

HED and LED centers:

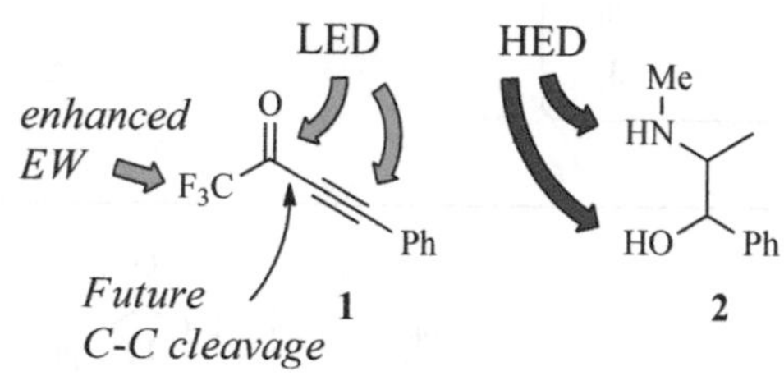

SCHEME 5.2 A first scrutiny by fragmentation analysis (FA), HED/LED identification.

product **4**. A *suprafacial* [1,3]-H transfer ensues in **5**, allowed in allenes as described in Problem 4. If properly positioned by the Z configuration of **6**, the N end of the appendage provides access to the semiaminal **7**. After enol ether protonation, conditions are ripe for the desired C–C bond cleavage announced in our FA by way of a pseudo-retro aldol condensation in **8**. Terminal enol ethers like **9** are unstable in hot aqueous alcohol, which gives rise to C–O cleavage that is necessary for freeing the target acetophenone **2**. Route **A** offers a reasonable entry into the observed targets leaving little, if any, space for non-sense chemistry. (Authors chose a different route, read on.)

5.2.2 Route B, Scheme 5.3, with Interference by Crashing Side Road C Given that intermediate **6** is a suitable receptor for a second Michael addition, and having the amine end adequately positioned for intramolecular attack, it is tempting to form oxazolidine **10** in preparation for the anticipated Cα–Cβ scission (relative to C=O). In so doing, however, intermediate **10** can take the "green turn" (route **C**) affording unrecorded products **11** and **13** locked up in the dashed frame. This disgraceful event can nevertheless be avoided by a second mole of pseudoephedrine (**2**) acting on the C=O of **10**, getting us closer to the construction of the N–C=O section revealed by FA. Steps marked by the sequence **2**+**10**→**14**→**targets 3** and **4** are self explained in Scheme 5.3, lower part. By the way, the second mole of pseudoephedrine **2** is recovered after the iminium ion hydrolysis of **16** near the end; hence, in principle only, a small molar excess of **2** is required.

5.2.3 Route E, the Authors' Version The authors of this work were in possession of privileged information from their own previously published studies of 1,4 diamine, ethanolamine, and propargyl ketone additions [1,2]. By exposing a much less electrophilic propargyl ketone **17** to pseudoephedrine **2** (Scheme 5.4, upper section), they isolated the first Michael addition product **18** before it went further to the cleaved products [2]. Only the Z alkene **18** was formed, possibly due to the H bond, according to authors. The reaction was slow (several hours), especially with sterically hindered ethanolamines. After adding CuCl as catalyst and pyridine to the hot solvent mixture, the cleaved compounds **19** and **20** were isolated. Similar conditions induced the tandem transformation in a one-pot reaction (**17**→**19**+**20**).

By careful handling of reaction conditions, authors managed to stop the sequence at the vinylogous amide stage [1]. Upon heating in ethanol, conversion to cleaved products was achieved in reasonable yield. Given that **18** was a likely intermediate, route D emerged from the authors' mechanism mind [1] (Scheme 5.4, lower section). Obviously, a second mole of ethanolamine was required for the aforementioned reasons in our pathway **B**→**D** structure **14** in Scheme 5.3. As in that sequence, the

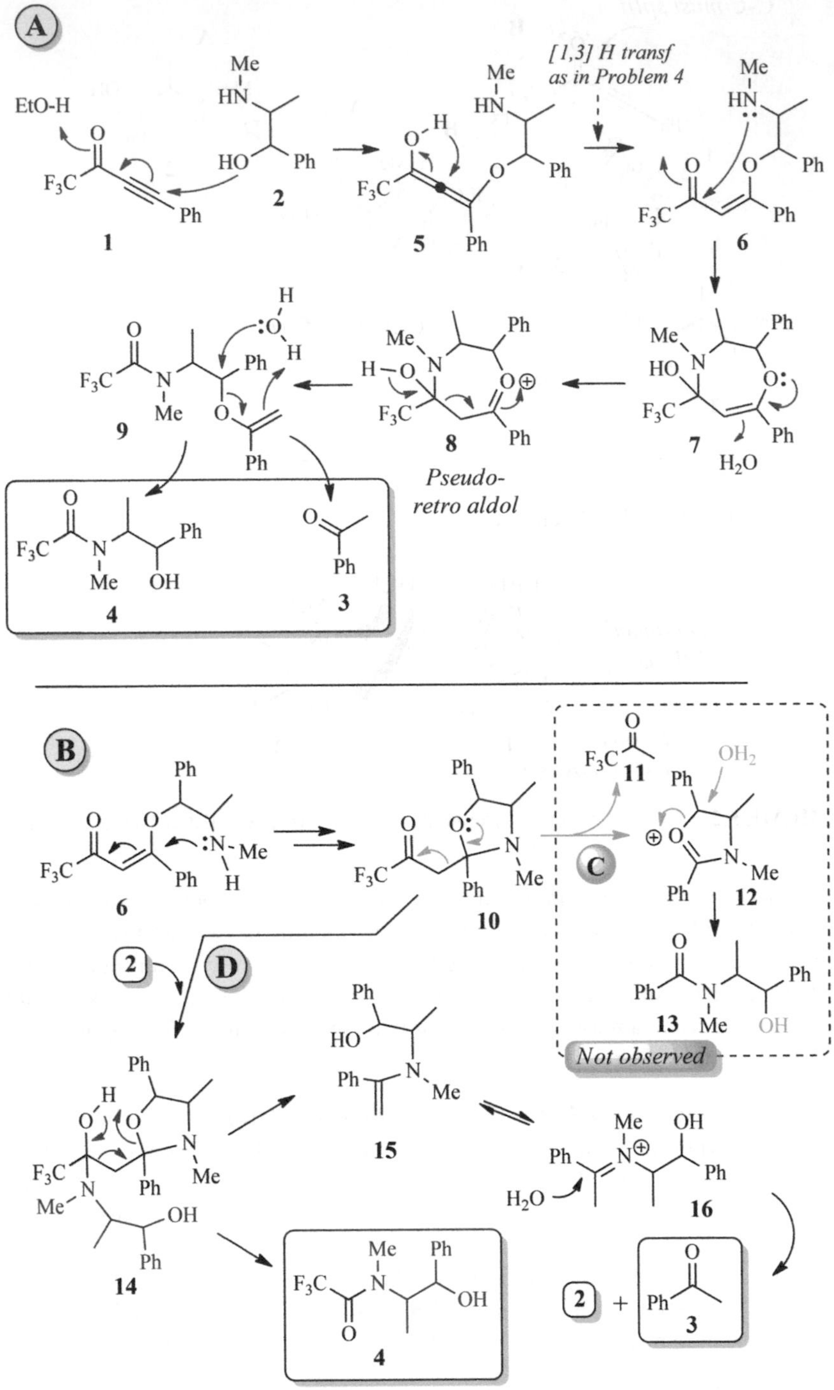

SCHEME 5.3 Alternative reaction sequences to feasibly explain targets **3** and **4**.

ethanolamine derivative upon which this sequence depends is regenerated in both pathways **D** and **E**, as the purple renderings show in the latter.

Finally, the hydroxyl unit in the ethanolamines employed appears to play an important role, since authors report the failure of these reactions when alkylamines are employed. The key step is the H transfer in **23**, which facilitates the C–C cleavage to **24** and prevents the build-up of negative charge at the enamine sp^2 end. If there is no proton source, as for example in anhydrous dioxane, the reversibility of the **21** → **23** step leads the reaction to a stationary point with no product being produced.

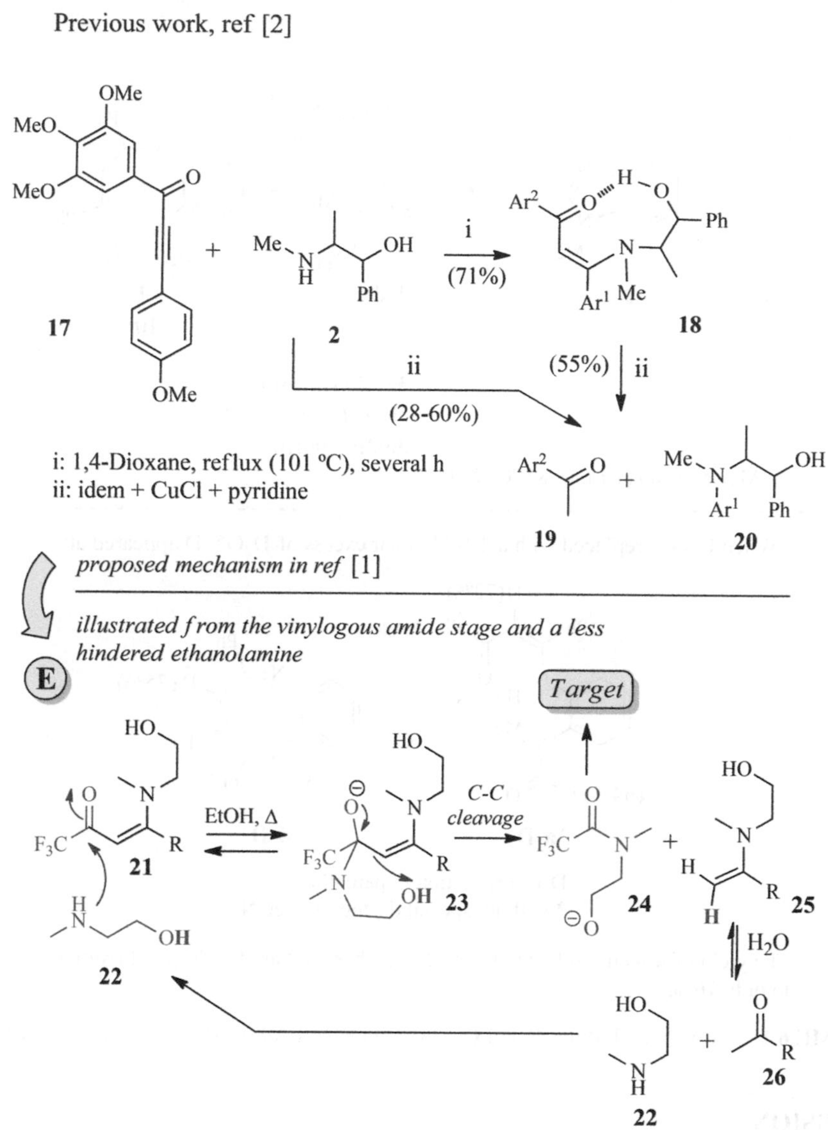

SCHEME 5.4 [2] / Elsevier.

REFERENCES

1. Davydova MP, Vasilevsky SF, Nenajdenko VG. *J. Fluorine Chem.* 2016;190:61–67. DOI:10.1016/j.jfluchem.2016.08.008.
2. Vasilevsky SF, Davydova MP, Mamatuyk VI, Pleshkova NV, Fadeev DS, Alabugin IV. *Mendeleev Commun.* 2015;25:377–379. DOI:10.1016/j.mencom.2015.09.021, and references cited therein.

PROBLEM 6

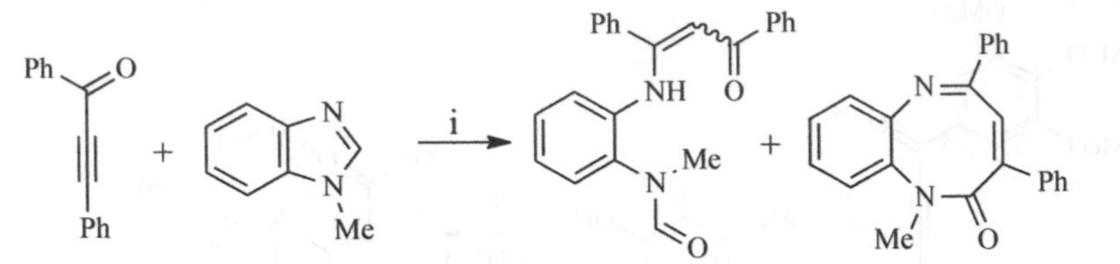

1 2 3a,b 61% 4 10%

3a: *Z-(syn) isomer*
3b: *E-(anti) isomer*
3a:3b = 90:10

i: MeCN, H_2O (1 mol), 82 °C, 70 h

When H_2O is replaced with a 4-fold molar excess of D_2O*, D appeared at:

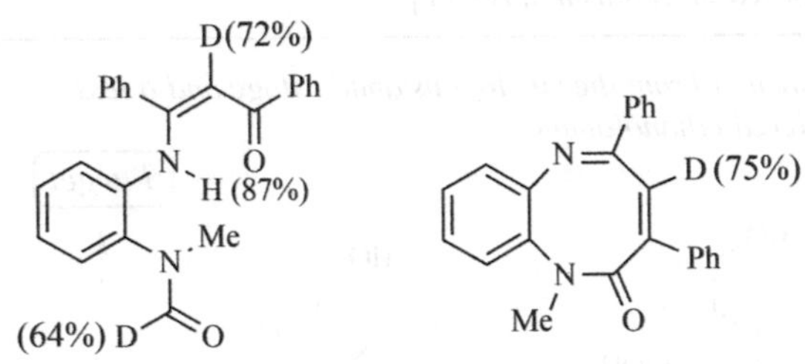

3a-D **4-D**

D-incorporation in parenthesis
* without special drying of MeCN

The yield of **4** went up from 10% to 28%, whereas that of **3** dropped from 61% to only 16%.

SCHEME 6.1 Problem and routes **B** and **C** shown in the body of the discussion were [1] / Elsevier.

PROBLEM 6: DISCUSSION

6.1 Overview

Scheme 6.1 depicts one of the many novel reactions of heterocycles and electron deficient alkynes developed for nearly half a century by distinguished Professor Boris Trofimov, at the University of Irkutsk, Russian Federation. Many of these reactions are performed in simple media: organic solvent, water, and unsophisticated base, acid, or none of these. Such a set of conditions leaves organic compounds sort of by themselves to perform their mutual chemistry unaided but by heat and solvent effects to stabilize transient polar species, if any. I find this didactically attractive. Solving this mechanism also takes advantage of what you have learned in the past few problems, making it readily accessible. It nonetheless involves an unanticipated variety of solutions.

6.2 Preliminary Analysis

The problem is split in two: The top part of Scheme 6.1 portrays the products obtained after heating in acetonitrile and a 1 : 1 mole ratio of water : propargyl ketone. Under the dashed line, there is additional data in regard to product composition when water is replaced with D_2O. Let us take up these two parts one at a time.

6.2.1 The Reaction in CH_3CN and Water One rapidly gets the point: There is a crystal-clear combination of an HEDZ heterocycle and an LEDZ alkyne whose thermally induced combination is a Michael-type *N*-alkylation adduct as the primary product. The *problem* starts after this first obvious step because it can branch out into several diverging conduits and products. In addition to one more coupling type, these side reactions are relevant in that they may have consumed part of the starting materials (yields are not impressive), even though the corresponding outcomes were not observed among the column chromatography fractions. In my own experience, when reactions are carried out at a small-scale minor products may escape detection. Three routes are described here, one of which splits into the authors' and my own proposal as a minor deviation from the designed mechanism. Scheme 6.2 and our annotations at the end provide key details. As will be demonstrated, electron flow in benzimidazoles is diverse and brings about flexibility to reaction mechanisms.

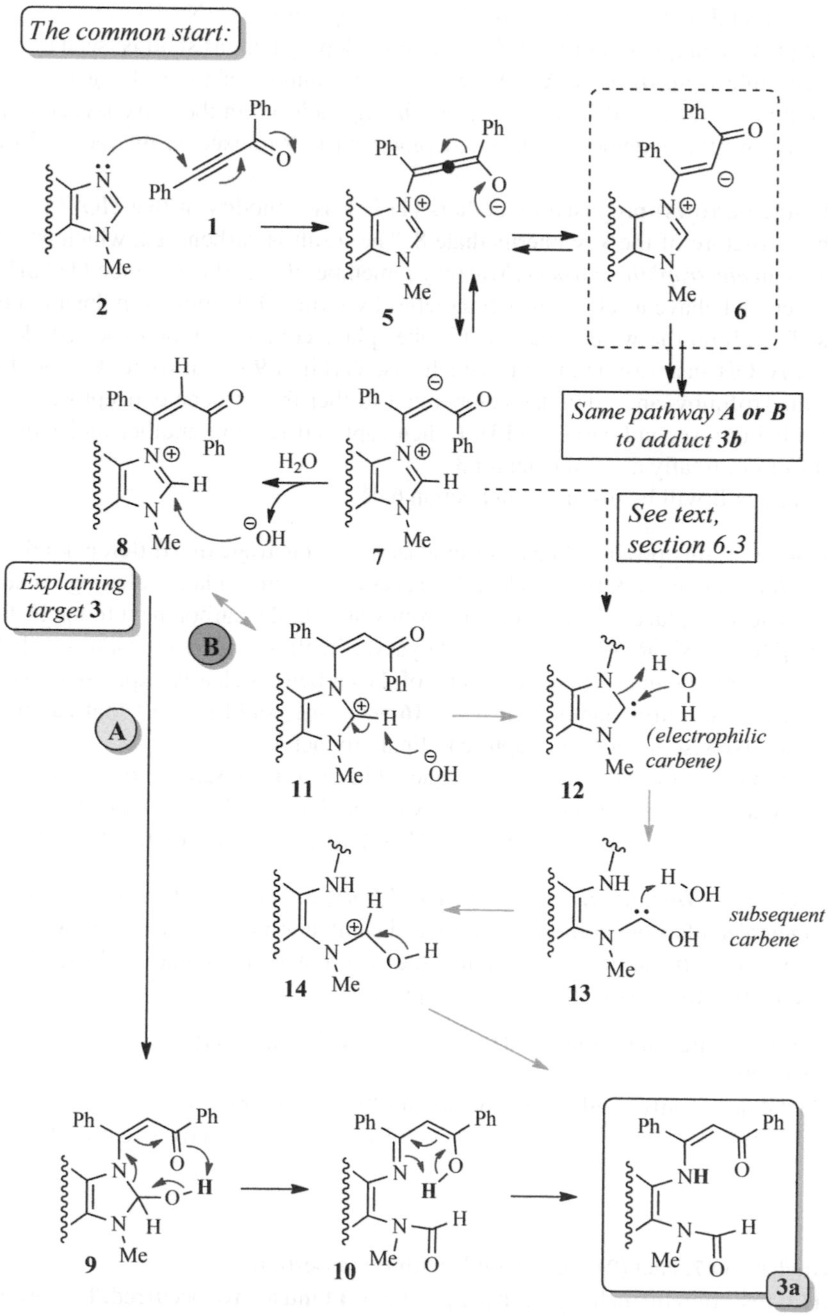

SCHEME 6.2 Feasible routes to target **3**.

6.3 Comments and Takeaways

Having 1 mol of water in the medium, one assumes carbanions are very short-lived species only or a previous manifestation of HED carbons enabling certain steps. N and C cations are admissible in the reaction medium. Nucleophilic forms of water are expressed here as hydroxyl anions for emphasizing proton transfer but never ignoring that carbocations and $HO^{(-)}$ cannot coexist in the same medium. Besides, anions and cations are taken as passing species trapped in the peculiar ambient of a polar solvent cage, which isolates in a bubble of sorts these intermediates momentarily from the rest of the proton sources (water).

The no non-sense entry contemplates the straightforward **1**+**2** coupling perceived as the first step. In the absence of other influences, E and Z α,β-unsaturated ketones **6** and **8** are the primary adducts. Preference for **8** may be the result of electrostatic attraction in allenic enolate intermediate **5**.

6.3.1 Explaining Target 3 The choice of **8** as a strategic structure for the rest of the problem is based on: i) the scaffold of **8** is strongly reminiscent of target **3**. All it takes is the oxidative cleavage of the $C{=}N^{(+)}R_2$ bond; ii) benzimidazoles are highly functional frameworks that provide ample opportunity for electron redeployment as shall be seen momentarily.

Two pathways are conceivable from **8:** route **A** (my own) takes advantage of the $HO^{(-)}$ generated during proton transfer (water is the only protic source) with the added advantage of it being enclosed in the solvent cage of polar **8**. Attack of $HO^{(-)}$ on the iminium ion carbon allows the extended electron redeployment in **9** to execute the required ring cleavage **9**→**10** and [1,5]-H transfer to target **3**.

Route **B** (Scheme 6.2, green arrows) proposed by authors [1] is less orthodox in that $HO^{(-)}$ picks up the angular proton of **11**, which is a resonance structure of the key intermediate **8**. The result is carbene **12**, which is attacked then by a second $HO^{(-)}$ and hence, operates as an *electrophilic carbene*. You may remember that carbenes embedded in heterocycles of this type, known as NHCs (see Chapter 2), behave as electron-rich moieties by virtue of the double *p* donation of adjacent non-bonding pairs (NBPs) of N atoms. Therefore, the water attack in **12** takes place contrary to the expected electron flow. This carbene might nevertheless act in a H–OH insertion manner in which case carbinol **9** (from route **A**) would emerge and carry on to target **3**. Anyway, Belyaeva, Trofimov, and coworkers elaborated further the carbene concept by proposing the **12**→**13** ring cleavage, the latter species being a second carbene (**13**). It then captured H from another molecule of water to form highly stable cation **14** and C=O from it, finally accessing target **3**.

The validity of route **A** versus **B** will be discussed in Section 6.4.

6.3.2 Explaining Target 4 Forming product **4** involves imidazole ring cleavage of a different kind as the heterocycle evolves to an eight-membered ring. As with target **3**, that product **4** is accessible from oxidative cleavage is suggested by the carbonyl group C=O. This implies the need to place an oxygen atom (from water) at the carbon next to the N–CH_3 section and promote a retro-Mannich reaction: [HO–C–$N^{(+)}R_3$→O=C+NR_3]. Belyaeva, Trofimov, and coworkers [1] departed from NHC **12** (Scheme 6.3, pathway **C**), whereby the nucleophilic character of the carbene is clearly expressed. The ensuing cation is water-hydroxylated in preparation for the retro-Mannich cleavage **16**→**4**. One could suggest that cation **15** might be trapped as epoxide **17** also, which would likewise induce ring rupture to final product **4**.

A hypothetical pathway bypassing the carbene stage is depicted in route **D** of Scheme 6.3, which illustrates the mechanistic possibilities of non-bonding pairs. It also branches off from key cation **8** and evolves by way of the fused bicyclic structure **18**, now expressed as a dication. Electron flow along the **19**→→**21**→**4** sequence lands on the desired product.

6.3.3 Which Pathway is Correct? Insights from Experimental Data, Scheme 6.1 Lower Section As with many reactions without systematic experimental studies or submitted to advanced computer models, there is no way to know. One fundamental question, though, was addressed by the authors [1] regarding the occurrence of carbene **12**. When D_2O (deuterium oxide) was used instead of water (Scheme 6.1, lower part) three things happened.

1. D was incorporated at three sites in product **3a-D** (see Scheme 6.1, lower part: incorporation: 15 to 20%).
2. D appeared in product **4** at the only available vinyl carbon (20% incorporation).
3. The yield of **4** went up from 10% to 28%, whereas that of **3** went down from 61% to only 16%.

Meaning:
Fact (1) indicates:

a) The existence of vinyl anion **7**. Fact (2) is compatible with this assertion.
b) The D_2O-carbene (**13**) insertion furnishing aldehyde precursor **14** must have occurred. This gives credence to the carbene **12** route by extension but does not discard the non-carbene route **A** as a competing pathway.
c) The occurrence of $DO^{(-)}$ attack on **8**, which opens my route **A**. This would be the only entry of D to the vinylogous amide N.

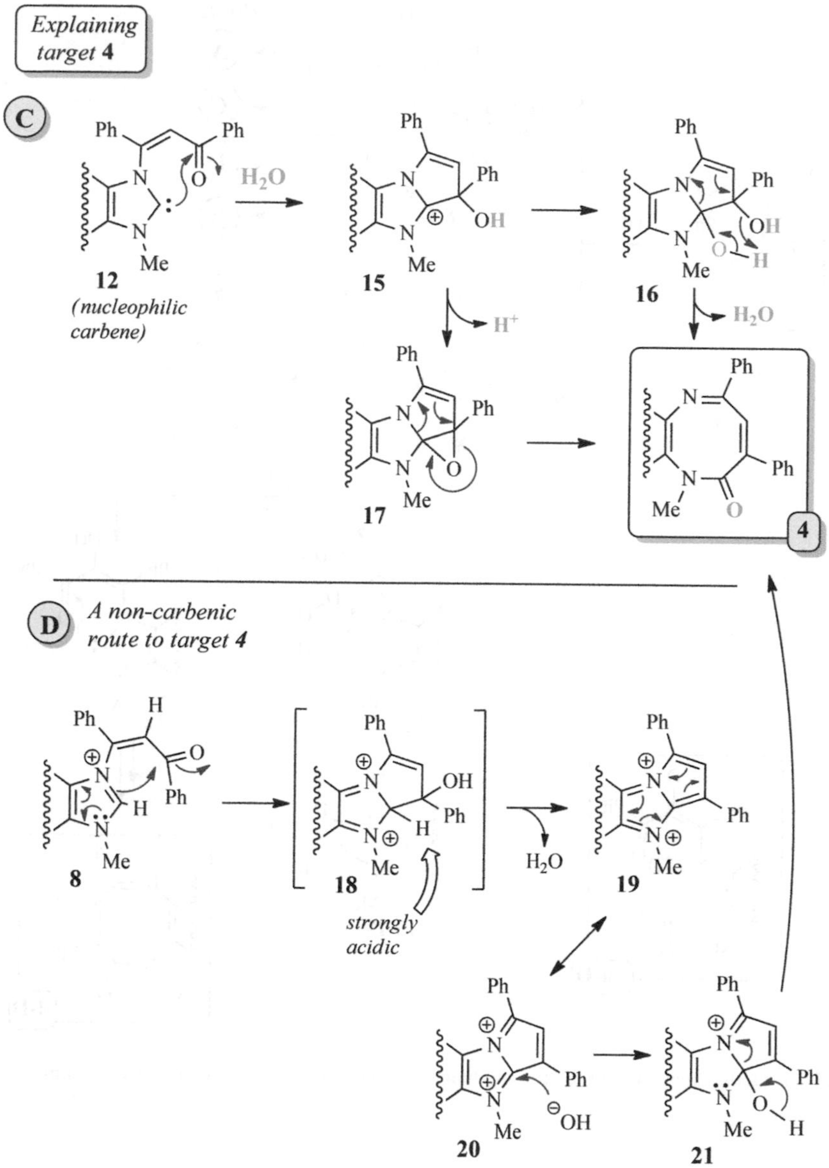

SCHEME 6.3 Carbenic and non-carbenic routes to target **4**.

Fact (3) has more belabored implications:

The increased mass of D relative to H decreases the rate of proton transfer. This *primary isotope effect* may have two consequences:

a) The overall reaction rate decreases if the D transfer participates in the slow step of the reaction (this is not the case as kinetics studies were not performed).
b) The yield of products may be impacted. If a certain intermediate may evolve along two reaction pathways, each leading to different products, their yields will be affected by the slowing effect of the D transfer.

This takes us first to the **7** → **8** protonation. In D_2O, this is the only hydrogen atom source, but the limited deuteration of anion **7** (only 72%) suggests a second proton supply. Which? Authors assign this supply to the imidazolium moiety in competition against D_2O (Scheme 6.4). The intramolecular H transfer favors route **B** toward the construction of carbene **12**. However, **12** can still evolve to **13** by H–O (or D–O) insertion and eventually to ring open target **3a** continuing on green route **B**, or take route **C** (Scheme 6.3) to macrocycle target **4**. The D effect to increase the yield of the latter compound ought to occur *after 12 is formed*. The crucial step is thus the slower D–O insertion of the carbene in **12** → **13-D** (relative to the H–O insertion). This circumstance favors the alternative route **C** (Scheme 6.3), which affords an increased yield of target **4** at the cost of **3**.

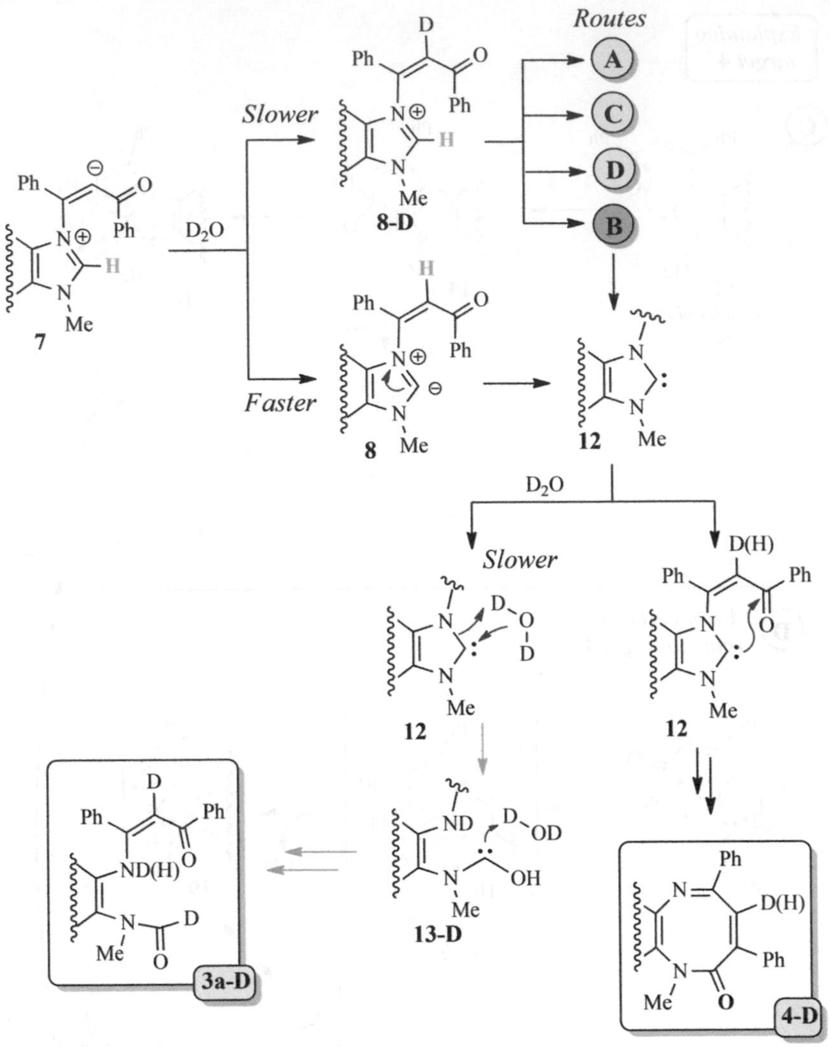

SCHEME 6.4 Reaction mechanism testing by deuterium incorporation performed by the authors, [1] / Elsevier.

Conclusively:

The deuterium oxide experiment supports the carbene routes to the target compounds. However, non-carbene pathways **A** and **D** cannot be ruled out altogether since deuterium incorporation at the vinyl position occurs prior to the outlets to the various routes. While route **A** explains deuterium incorporation at vinyl and vinylogous amide N sites, it does not account for deuterium at the terminal amide.

REFERENCE

1. Belyaeva KV, Andriyankova LV, Nikitina LP, Bagryanskaya IY, Afonin AV, Ushakov IA, Mal'kina AG, Trofimov BA. *Tetrahedron* 2015;71:2891–2899. DOI:10.1016/j.tet.2015.03.056.

PROBLEM 7

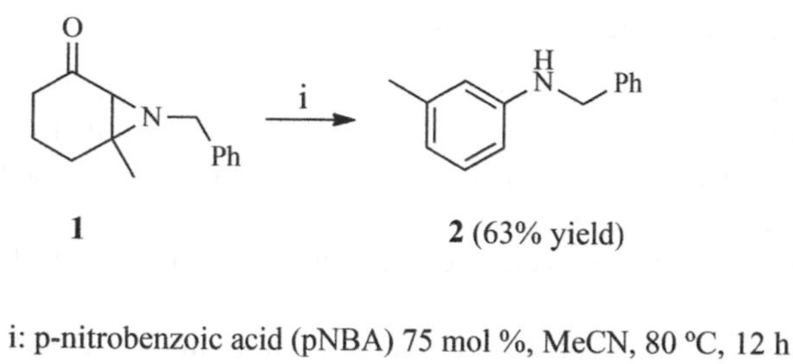

SCHEME 7.1 [1] / Elsevier.

PROBLEM 7: DISCUSSION

7.1 Overview

In science, annoying results often turn into interesting discoveries when they get the attention they deserve. The birth certificate of the Scheme 7.1 reaction is the irritating appearance of a side product when scientist pursued a vastly more interesting goal: the synthesis of bromoxone (**8**) from an aziridine precursor [2, 3]. The synthesis strategy of researchers from the Universidade Nova in Lisbon, Portugal, was based on the addition–elimination reaction of *p*-methoxybenzyl amine (**4**) on α,β-unsaturated 2-iodocyclohexanone **3** (Scheme 7.2), followed by conversion of aziridine **5** to epoxide **7** and the desired potent antitumor and antibiotic bromoxone (R = acetyl) **8** by standard reactions. The aziridine moiety was used as protective group of the enone C=C while the enantioselective installation of epoxide and 4-carbinol functions were undertaken. During the aziridine grafting on **3** with other benzylamines, compounds of type **6** cropped up persistently. Portuguese researchers decided to explore this further by expanding the scope of benzylamines and ketones and found this to be a general, synthetically useful reaction of which Scheme 7.1 was selected here as your next assignment.

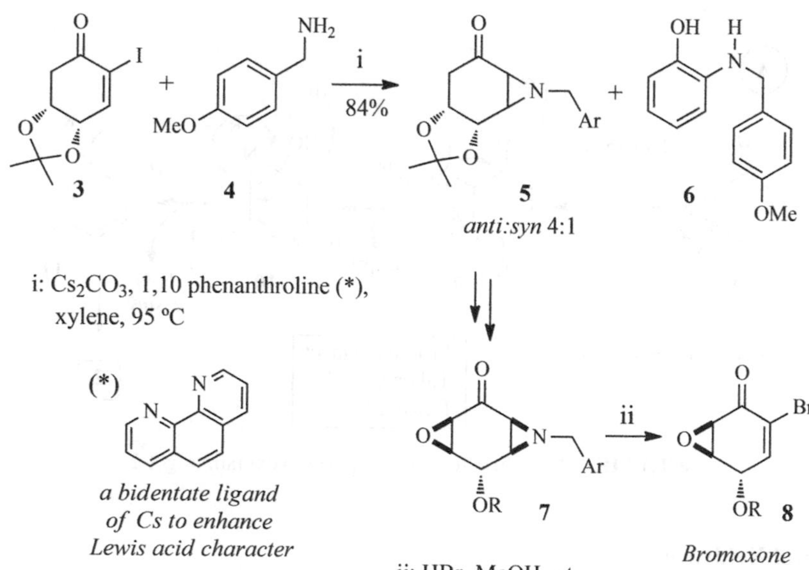

SCHEME 7.2 [2] / American Chemical Society.

7.2 Problem Analysis

Comparison vis a vis of **1** and **2** provides a mixture of questions as well as leads:

1. Following the principle of minimum change, the tolyl group of **3** originates from the cyclohexanone moiety of **1**.
2. However, **1** does not have the potential for aromatization. The functionality number (FN) tells us that: FN_t(**1**)=4 (disregarding the benzyl group) while FN_t(**2**)=7 (idem). This is a local oxidation but no oxidant in sight. Aromatization should involve an H-shift in the carbocycle at some point.
3. The relative positions of CH_3 and NH–Bz groups change from germinal or 1,2 in aziridine, to 1,3 in the tolyl block. Does this imply a 1,2–CH_3 shift?
4. How is the oxygen atom removed? A primary amine would, but we have none among the reagents. It ought to be created at some point, form an imine on the C=O, and be removed thereafter helping in the aromatization step.
5. The aziridine removal like in **7** (Scheme 7.2) by HBr/CH_3OH, to liberate the primary benzylamine could fulfill the requisite of issue (**4**) if we only found a nucleophile in our reaction mixture equivalent to HBr.
6. Let us have a second check up of reagents in Scheme 7.1. Do you grab it? There is *p*-nitrobenzoic acid (*p*-NBA), a strong proton donor (pKa=3.44), thus a weak nucleophile, weaker than $Br^{(-)}$, in any case. Having no other nucleophile (Nu), we might use it in the sense of point (**5**).
7. However, *p*-NBA does not appear among the products and only 75 mol% is added, which suggests a catalytic role. If so, one must account for a catalytic cycle to recover *p*-NBA in the mechanism proposal.

With these presuppositions at hand there are two **1**→**2** routes. These are presented next and then validated, or not, according to further experimental evidence.

7.3 First Proposal

This route (Scheme 7.3, part **A**) is based on the absence of any free amine since it is not among the available reagents. It therefore rejects the formation of an imine at the start as proposed in item four of the previous analysis and takes *p*-NBA as a catalytic proton source only. Imine **11** is built after a sort of walking aziridine rearrangement guided by proton stimulus across the **1**→**9**→**10** sequence: proton borrowing from *p*-NBA and subsequent proton release from the carbon scaffold. In the end, proton balance is even, the aromatic ring was put together without oxidation (although there is an electron sink: water in step **10**→**11**); hence, *p*-NBA operates solely as a Brønsted catalyst. A simple, uncomplicated (even elegant) sequence I was able to create.

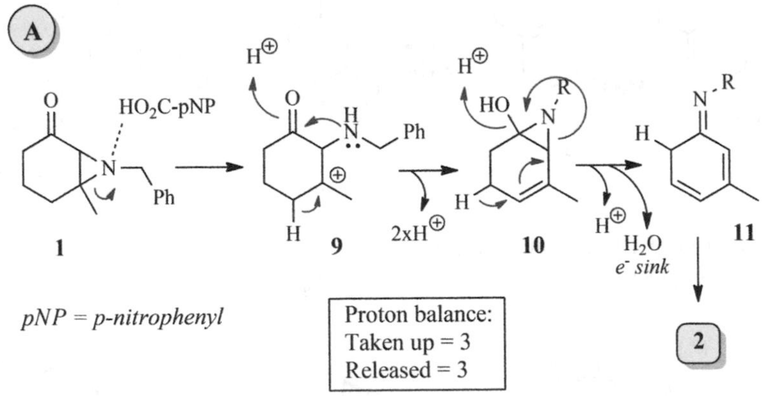

SCHEME 7.3 A first feasible proposal to explain target **2**.

7.4 Adding Experimental Evidence

In their efforts to get a deeper insight into the mechanism, Barroso and coworkers [1] designed two key reactions (Scheme 7.4, top section) and observed that:

When exposing aziridine **12** to 1 equivalent (eq) of benzylamine and excess (1.5 eq) of *p*-NBA, not only the anticipated *N*-phenylbenzylamine (**13**) was obtained but also, and most importantly, enol ester **14** (framed structure). This meant that *p*-NBA was acting as nucleophile on the aziridine effectively at the α-carbonyl carbon, ejecting the amino moiety by subsequent β–elimination. They submitted the isolated enol ester to benzylamine again employing *p*-toluenesulfonic acid as a strong proton source but a non-nucleophilic base after deprotonation, ketoester **14** reacted within a few hours furnishing *N*-phenylbenzylamine (**13**) again in addition to products **15** and **16**. This result was solid evidence that enol ester **14** was indeed a valid intermediate in Scheme 7.1, and therefore, mechanism **A** (Scheme 7.3) was not supported by this evidence. The **1**→**9** step, no matter how logical, does not contribute significantly to the reaction progress.

7.5 A Second Proposal Attained to Experimental Evidence

Once the previous evidence has been reviewed, one must recognize a temporary incorporation of *p*-NBA to aziridine **1**. *p*-NBA must be reclaimed after completing its deed for another reaction cycle. The authors' proposal, which accommodates these facts, is depicted in Scheme 7.4, mechanism **B** [1].

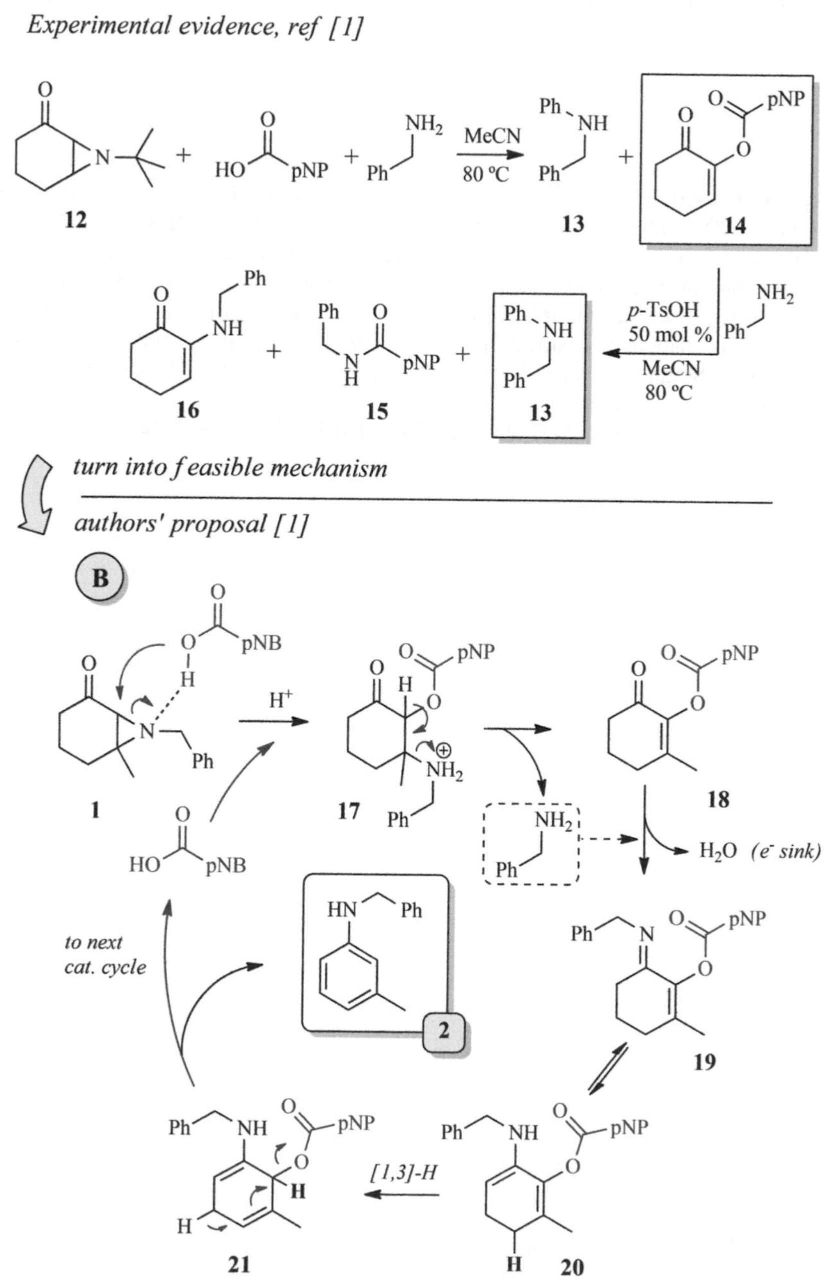

SCHEME 7.4 Experimental evidence developed by Barros et al. [1] and their mechanism proposal (section **B**) to explain the reaction of Scheme 7.1. [1] / Elsevier.

7.6 Lessons and Takeaways

a) Our qualms about reacting aziridines with a weakly nucleophilic carboxylate is unjustified. Ring opening at either C–N bond is a feasible and often a high yielding reaction with several nucleophiles. These include a variety of alcohols, water, cyanide, halides, and several carboxylic acids [4].
b) Although it is of no consequence for this problem, aziridines respond as S_N2 receptors, leading to *trans* 1-amino-2-Nu products. However, in unsymmetrical aziridines, the primary versus secondary regioselectivity of the Nu varies with conditions, Nu, and steric constraints, which may result from the contribution of a unimolecular nucleophilic substitution (S_N1) [5].
c) The tandem set of reactions involved in **1**→**2** is a curious case of self feeding of one of the components in **1** (benzylamine), which is liberated early in the sequence and recaptured later immediately after to create the imine-enamine moiety.
d) In both mechanisms **A** and **B**, water acts as electron sink in the oxidative process from cyclohexanone to benzylamino benzene.

REFERENCES

1. Barros MT, Dey SS, Maycock CD, Rodrigues P. *Tetrahedron* 2012;68:6263–6268. DOI:10.1016/j.tet.2012.05.054.
2. Barros MT, Matias PM, Maycock CD, Ventura MR. *Org. Lett.* 2003;5:4321–4323. DOI:10.1021/ol035576i.
3. The source of bromoxone is quite remote and worth an adventure story. In short, this metabolite, apparently a secretion substance, was isolated along with several other unusual compounds from *Ptychodera* acorn worms, a primitive lifeform, collected from a deep-sea cave in the Hawaiian island of Maui. This family of sea worms is found in several coasts of the Atlantic, Indian, and Pacific oceans. Higa T, Okuda RK, Stevens RM, Scheur PJ, He CH, Changfu X, Clardy J. *Tetrahedron* 1987;43:1063–1070. DOI:10.1016/S0040-4020(01)90042-0.
4. Kumar M, Gandhi S, Singh Kalra S, Singh VK. *Synth. Commun.* 2008;38:1527–1532. DOI:10.1080/00397910801928723. Prasad BAB, Sekar G, Singh VK. *Tet. Lett.* 2000;41:4677–4679.
5. Concellón JM, Riego E, Suárez JR. *J. Org. Chem.* 2003;68:9242–9246. DOI:10.1021/jo0350514.

PROBLEM 8

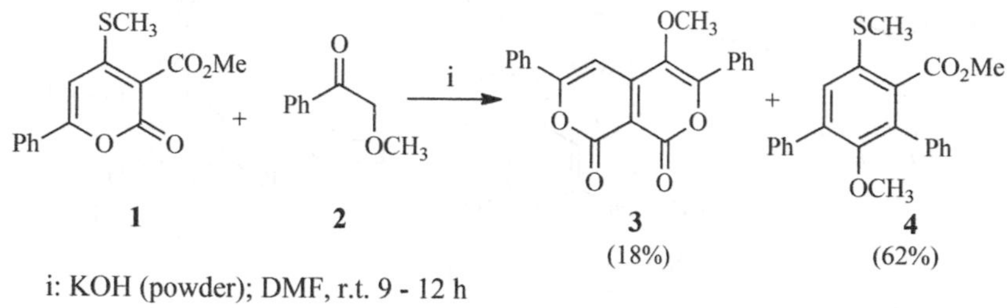

SCHEME 8.1 [1] / Elsevier.

PROBLEM 8: DISCUSSION

8.1 Overview

Fungal metabolites are often built on peculiar carbon architecture. Compounds of type **4** are among these (isolated from the fungus *Telephoraceae* family), showing a characteristic series of non-fused aromatic rings in succession that contain EWGs and/or electron-donating groups (EDGs). These compounds are potent activators or inhibitors of several animal cell enzymes associated with disease and antioxidants [2]. As they occur only scarcely in nature, synthetic methods are most desirable. The work of Professor Goel and collaborators in Lucknow, India, presented herein [1, 3], pursued a facile synthesis of model compound **4** and some derivatives in the best possible yield using very simple and inexpensive experimental conditions seeking to replace the use of expensive palladium catalysis employed in previous syntheses. For this workbook, it was of interest to present the **1**→**4** reaction as a setting for mechanism design and the unanticipated appearance of compound **3** added a bit of more accessible pizzazz to it.

8.2 Explaining Compound 3

As the most accessible of the products, let us apply one of my favorite problem scrutiny strategies that will serve as constructive illustrations. FA seems suitable and shows the division of **3** into two blocks closely resembling **1** and **2** (Scheme 8.2, top part) that should form C–C and C–O bonds at the dotted line crossings. The most relevant issue here is that both starting materials contain EW functions that make them preferential LEDZs but the desirable **1**+**2** coupling in basic highly polar medium (e.g. DMF) should be HEDZ + LEDZ. This is easily circumvented by converting one of the blocks into a HEDZ. All one needs is an anion-prone active methylene group. Our best and only candidate is **2**. The expulsion of SCH_3 (dimethylsulfide) calls for an addition–elimination reaction, portrayed in Scheme 8.2 (lower part). Additionally, one CH_3O must be expelled as well, probably the ester alkoxy unit. There seem to be no contradictions, difficulties with stereochemical issues, or anion/cation unbalance to account for side product salts.

8.3 Explaining Compound 4

The reactivity constraints that limit the **1**+**2** interaction with regards to HEDZ/LEDZ rapport are pretty much the same in the route of **4**. But there are a few complications that enhance a bit the mechanistic and instructive interest. Once again, *FA* seems a most informative strategy (Scheme 8.3, top). Sections **A** and **B** of target **4** are easily found in starting materials **1** and **2**, but fragments in purple are excluded. The fragmentation scheme also indentifies two C–C bonds to be formed.

A feasible mechanism is shown after the yellow arrow, split into two possibilities, a stepwise 1,6 addition, preferred by authors [1] and/or a DACA with reversed polarity passing through TS **10TS**. Both converge in **8**, illustrated here as the endo isomer in case the DA route was followed. The retro-DA cycloaddition excludes CO_2 and water thereafter.

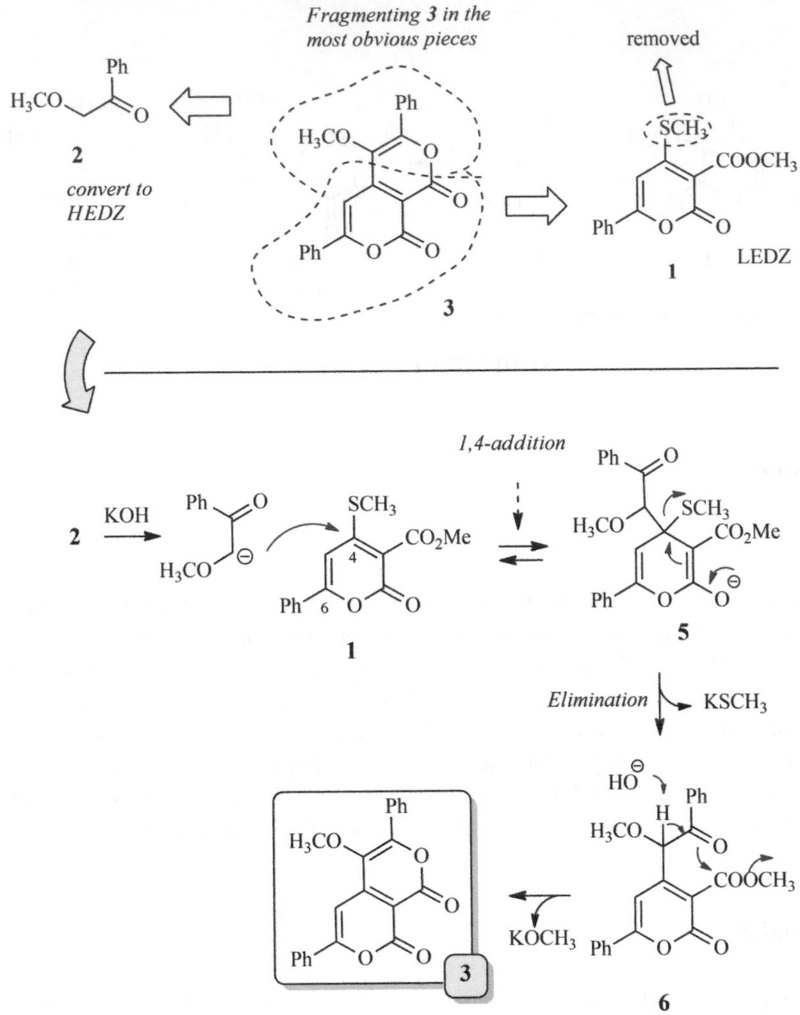

SCHEME 8.2 Fragmentation of target **3** as a function of the reactants structure and a suitable mechanism therefrom.

8.4 Lessons and Takeaways

1. The reactions of this problem are simple enough to require a special discussion to determine the most probable pathway. Both targets **3** and **4** can be explained by a carbanion performing 1,4 and 1,6 additions on a diene polarized by an EWG at the end (the lactone). These additions then progress to products by elimination.
2. Given that both addition types occur in the same reaction medium, competition between the two pathways is likely. The yield difference of **3** and **4** may be due to the ratio of equilibrium constants of the 1,4 and 1,6 additions. The long reaction time and moderate temperature provide conditions for thermodynamic control.
3. The difference between routes **I** (authors [1]) and **II** (my alternative proposal) may have been assessed if compounds **8** or **9** had been intercepted and characterized, or observed by running the reaction in a suitable spectrometer−^{13}C NMR for example−or drawing samples at regular times, clean up, and extraction in deuterium-organic solvent at low temperature. The diastereomers of **9** may have provided insights on the endo/exo characteristics of **8** and support a DA-type TS. In the end product, **4**, all the stereochemical features disappear except the regioselectivity of the 1,6 addition.
4. Route **II** responds to a DA cycloaddition of inverted polarity, therefore slow and likely to take place in steps, formally similar to route **I**.
5. Given that solid KOH is insoluble in DMF, the proton abstraction of the methylene in **2** occurs at the solid–liquid interface, a complex environment.

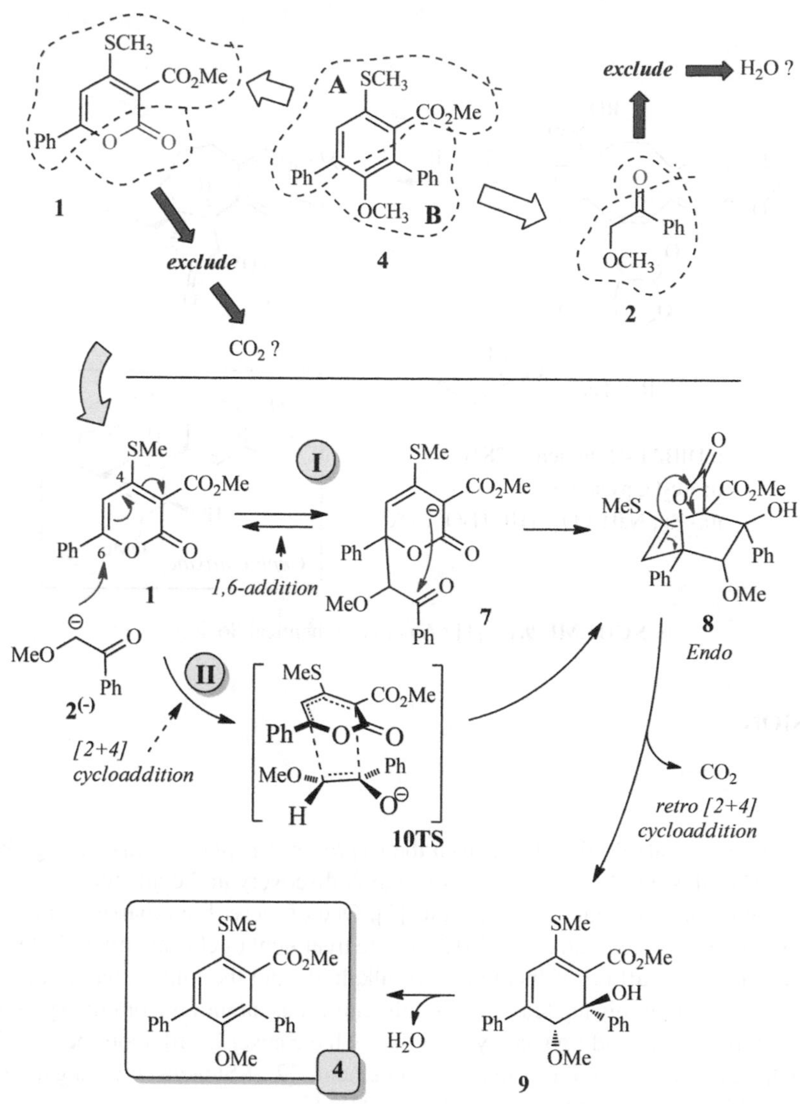

SCHEME 8.3 Fragmentation analysis of target **4** and two feasible mechanism routes to explain it.

REFERENCES AND NOTES

1. Goel A, Verma D, Singh FV. *Tet. Lett.* 2005;46:8487–8491. DOI:10.1016/j.tetlett.2005.10.018.
2. Yang, WM, Liu JK, Hun L, Dong ZJ, et al. *Zeitschrift Naturforsch* 2004;59C:359–362. DOI:10.1515//znc-2004-5-612.
3. The title of Reference [1] was: A *vicarious* synthesis of unsymmetrical *meta-* and *para*-terphenyls from 2*H*-pyran-2-ones. Formally speaking, a *vicarious reaction* in aromatics is defined as the result of a nucleophilic substitution on the ring with loss of an aromatic proton instead of other substituents (leaving group) in the ring. In the discussion of this problem, Scheme 8.3 can be said to include such a vicarious synthesis because SCH_3, a potential leaving group, is retained in the final product. An H^+ is lost in step **9**→**4** as well as CO_2 in the previous reaction.

PROBLEM 9

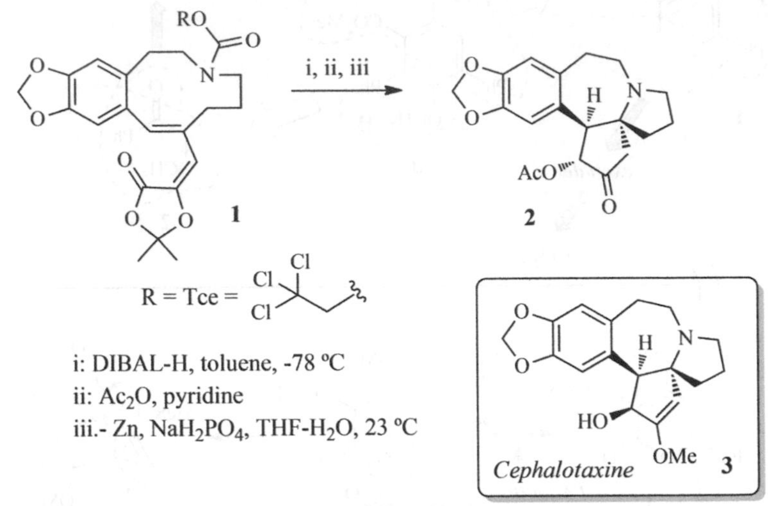

SCHEME 9.1 [1] / American Chemical Society.

PROBLEM 9: DISCUSSION

9.1 Overview

Cephalotaxine (**3**) (inset) is the major alkaloid isolated from the *Cephalotaxus* plant genus. A large body of research has been devoted to the several alkaloids of this family, particularly, since their discovery in the nineteenth century, owing to their cancer cell antiproliferative activity at the nano-molar concentrations [2]. One of these, homoharringtonine, is a U.S. Food and Drug Administration (FDA)-approved pharmaceutical drug (2012) for the treatment of chronic myeloid leukemia. For organic chemists and biochemists, the pentacyclic scaffold of this family of alkaloids and the numerous variants is particularly intriguing with three stereogenic carbons. Building this nucleus has brought enormous attention from the synthesis stand [3] and a gamut of reaction types have been employed, including some you probably have never heard about before: 3-aza-Cope/Mannich cascade, transannular Mannich reaction, Ireland–Claisen rearrangement, [2,3]-Stevens rearrangement, Parham-aldol domino cyclization, and others, applied just in the past five years (2016–2021) [3].

The present problem deals precisely with one peculiar and accessible approach to the pentacyclic structure under quite simple reaction conditions, the name of which will not be revealed until after the mechanism is solved herein.

9.2 Problem Analysis

It is hard to avoid the feeling that this problem is too easy: A cursory comparison of **1** and **2** provides clear-cut clues (Scheme 9.2, top section). One is thus driven to throwing a thoughtless mechanism into the frying pan without much doubt (pathway **A**, lower section).

However, there are a several potholes in route **A**:

1. From start to finish, all of it appears accountable to DIBAH (diisobutyl aluminum hydride) only, a reducing agent, save for the final acetylation.
2. *O*-trichloroethyl (*O*Tce) is an N and S protecting group, unreactive to DIBAL–H; hence, it should remain in place as far as DIBAH, thus blocking N activity.
3. The *O*Tce-protected nitrogen's NBP is part of an amide with no capacity toward addition to a C=C, even though it is five atoms away potentially forming a favorable five-membered cyclic TS. *O*Tce should be removed first.
4. The electron flow shown in intermediate **4** would require Lewis or protic acid activation, but there is none in the reaction mixture.
5. No role was specified for zinc sodium phosphate; authors used it for a reason.
6. No consideration was given to the stereochemistry, a crucial concept for the organic synthesis design of chiral organics like cephalotaxine.

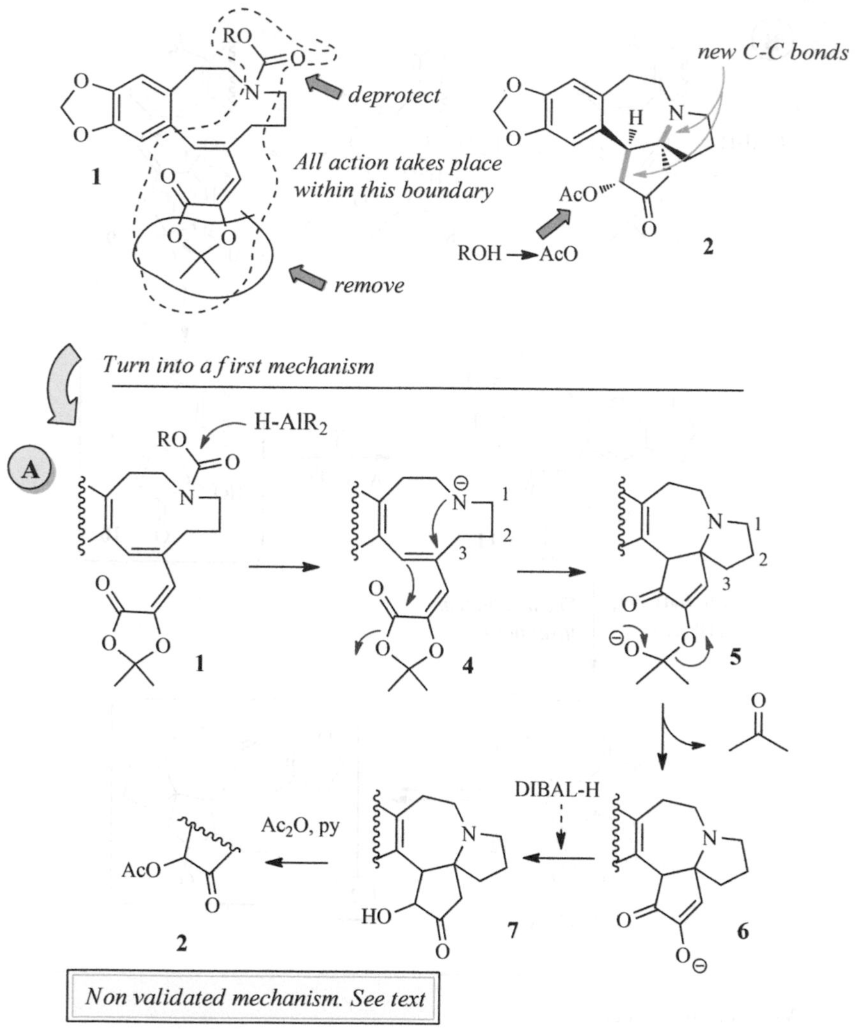

SCHEME 9.2 Comparative analysis and a first proposal.

As you suspected, I created route **A** to show that:

- Electron flow alone does not always account for all factors involved in forming/cleaving operations in tandem or domino schemes.
- Swiftly conceived answers often drag us into mistakes despite the apparent conduit to the target product.
- The ugly head of wrong answers that look like the correct ones appears once again.

So let us think again, and take one step at a time.

9.3 A Second Better Thought Proposal

To determine how far the DIBAH reduction can drive compound **1** toward the target, four issues need to be considered in reverse order: from target **2** to starting compound **1**.

- ✓ *Step iii*: The authors used *zinc-sodium phosphate* for a reason: To remove the N-protective group, which is the third step of the sequence. The mechanism of this deprotection is shown in Scheme 9.3, bottom section. If so, cyclization to the heterocycle must be postponed after the southern section undergoes the DIBAH activity.
- ✓ *Step ii*: Given that *acetic anhydride and pyridine* are the trademarks of acetylation of primary and secondary carbinols, an alcohol with all the attributes of the southern cyclopentanone moiety must have been created previously.

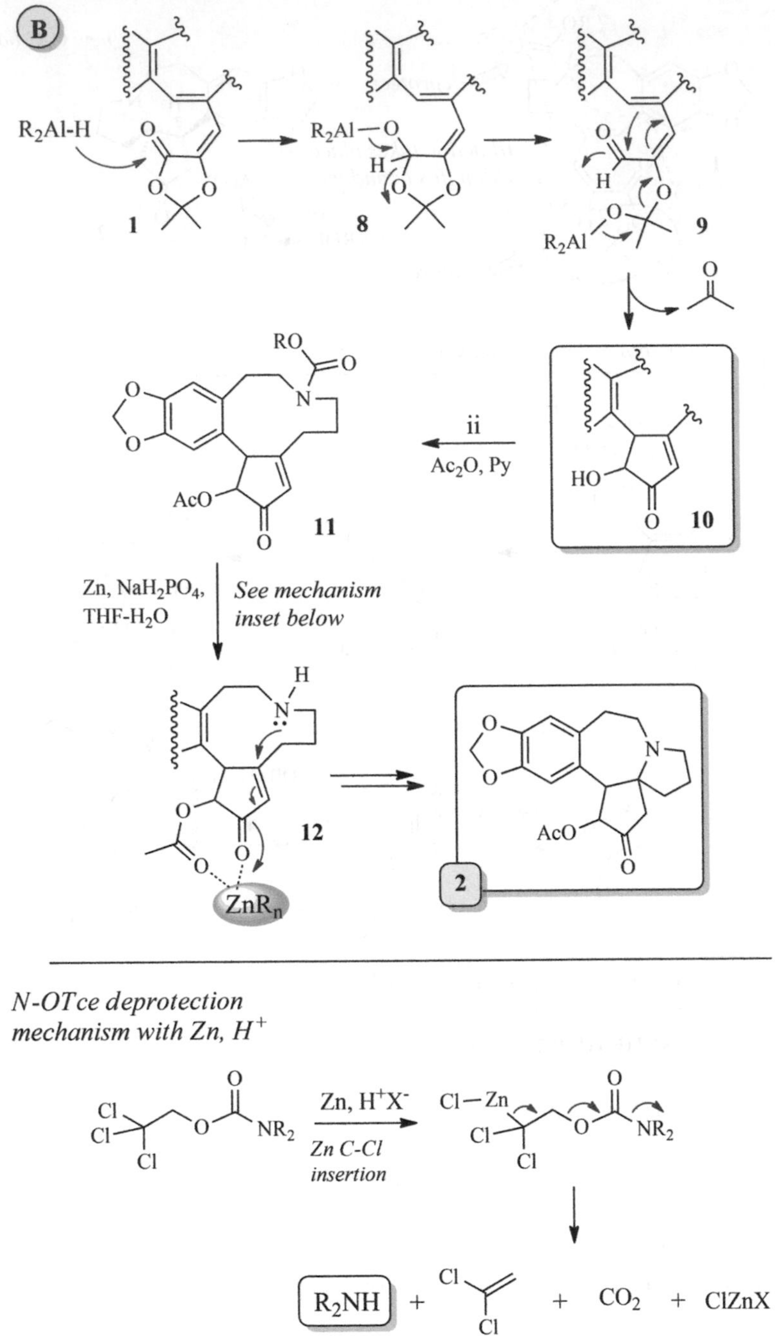

SCHEME 9.3 A second proposal.

- ✓ *Step i: DIBAH is responsible for the construction of the cyclopentanone-2-ol* of the southern section while the northern side remains unchanged.
- ✓ Once the electron flow and C–C bonds (green linkages in Scheme 9.2) are solved, the stereochemical issues should be treated.

Route **B** seems to have corrected most of the shortcomings of route **A**. It relays entirely in the reductive **1**→**8** step, as a homolog of the DIBAH reduction of lactones to lactols after work up [4]. Dioxolane carbinols like **8** are perhaps not as stable as aluminate and their cleavage is expected. Aluminum species **8** furthers the electron flow shown in **9** for the fragmentation and ensuing electrocyclization that yields **10**. After acetylation, deprotection of the amine by Zn° insertion in the Cl–C bond liberates the N NBP for the final construction of the tricyclic sector of the target alkaloid.

The validity of pathway **B** cannot be assessed with the information so far examined, because there are other ways to explain this mechanism. Its roots go back to 1984 when the DIBAH cyclization was discovered as part of a continuing investigation of dioxolanones by scientists in Manchester and Liverpool, United Kingdom led by Professor Robert Ramage [5]. Among the several reactions studied, those that interest us here constitute examples of fundamental chemistry worth checking (Scheme 9.4, top section). Take note: although authors [5] do not venture a mechanism interpretation and they used an inseparable mixture of **13a** and **13b**, they report that the DIBAH reduction of *trans* **13b** afforded ketoaldehyde **14** (enol form), whereas our familiar cyclopentanone **15** *came from the reduction of the* cis *13a isomer*. This result calls for a similar mechanism attending the **1** → **2** reaction of our present problem. By the way, it should be easy for the readership to propose a feasible route for the alkaline hydrolysis of the **14** + **15** mixture to **16**.

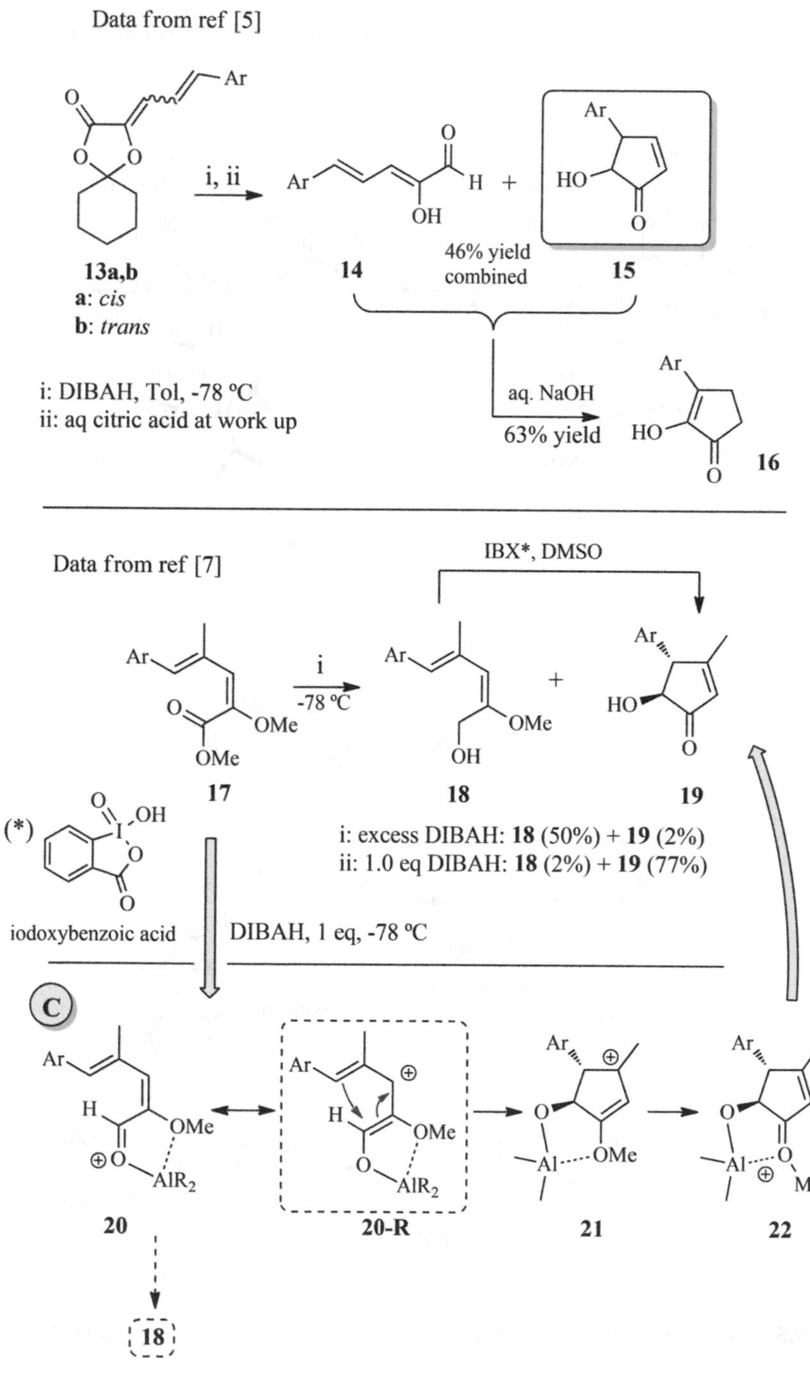

SCHEME 9.4 Top section: Adapted from [5][6]. Proposal **C** Adapted from [7].

The key cyclization to cyclopentanone is then re-examined by the same group, Professor Wei-Dong Li's in Chongqin, China (2017) [6] from whom I selected this problem [1]. They used simple models tailored to solve the mechanism question (Scheme 9.4, lower section). Please mind the experimental details shown there. The iodoxybenzoic acid (IBX)–DMSO oxidation of carbinol **18** (this mixture affords aldehydes without further oxidation to carboxylates) and the concomitant conversion to cyclopentanone **19** suggested that aldehyde **20** (shown as the aluminum complex) was indeed an active participant in the genesis of the cyclized product. Excess DIBAH would reduce it to carbinol **18** and become the dominant product with almost complete inhibition of ketone **19**. But limiting DIBAH gives **18** a chance to follow route **C** (Scheme 9.4, bottom). Expressed as resonance structure **20-R**, this intermediate fills the requisites of an iso-*Nazarov 4π electrocyclization* [7]. Having the extra oxygen atom, it becomes a case of **oxi-Nazarov** cyclization in the course **20-R** → **21** → **22** → **19** after aqueous acid work up. As a result, the yield of the latter jumps to 77% while carbinol **18** drops to nearly nothing. This is evidence of competition between pathways **20** → **18** versus **20** → **21**.

You can adapt this scheme to the case of the dioxolanone **1** adopting the concepts exhibited in routes **B** and **C**.

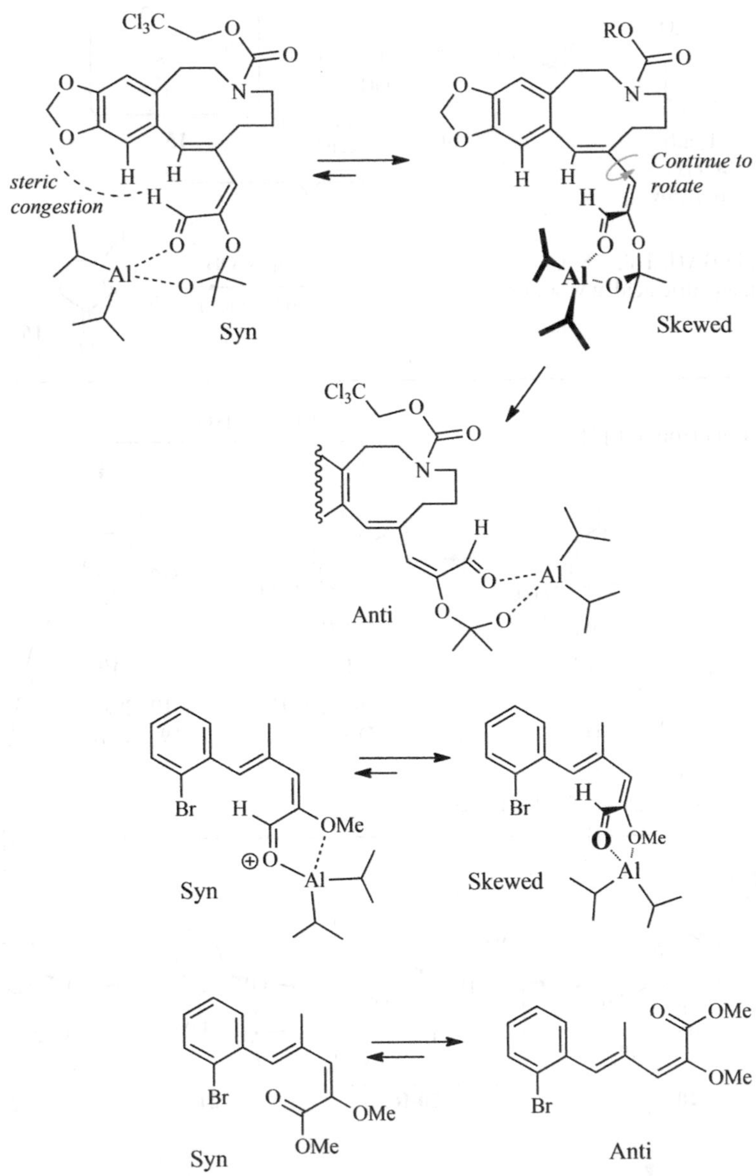

SCHEME 9.5 Stereochemical considerations of DIBAH controlling the configuration of intermediates.

9.4 Which Route, B or C, Better Explains the Mechanism?

Both routes are sound with regard to electro flow pulled by the LEDZ of the southern portion. Doubts in both pathways emerge from conformational preferences. There are three asymmetric carbons in **2**, formed enantiospecifically during the tandem cyclizations. In the first of these, mind that the southern dioxolane moiety is tethered to the main scaffold by a σ C–C bond. In either pathway model (**B** or **C**) the reacting diene elements need to be coplanar for suitable atomic orbital overlap. But the SS caused by the bulky diisobutyl aluminum and C=O against the aromatic ring of **1** forces this section to rotate away from the endocyclic C=C (Scheme 9.5) turning the cyclization unfeasible. There is no dipolar force, hydrogen bond, solvent interaction (toluene), or aluminum complex to hold the two functions together at the correct *syn* position for the 4π interaction that leads to cyclization. The other models show much preferred anti-rotamers as well. Therefore, the energy difference between the two rotamers is expected to be large enough to retard or hamper the reaction. Because theoretical calculations of the energy profile of reaction models such as **17** or **20** are not available presently, the **B** or **C** hypothesis cannot be supported unmistakably. We can only hold on to the experimental evidence.

In regard to the *second* cyclization, once the N-protecting group is removed, models (not shown) put the 10-membered ring in a boat form such that the N atom sits above C(β) of the α,β-unsaturated cyclopentanone in intermediate **11** (Scheme 9.3) furnishing a unique enantiomer and a trans-ring junction as in the targeted natural product.

REFERENCES

1. Li W-DZ, Duo WG, Zhuang CH. *Org. Lett.* 2011;13:3538–3541. DOI:10.1021/ol201390r.
2. Chang Y, Meng FC, Wang R, Wang CM, Lu XY, Zhang QW. In: Atta-ur-Rahman (ed.). *Studies in Natural Product Chemistry*, Chapter 10: chemistry, bioactivity, and structure-activity relationship of Cephalotaxine-type alkaloids from Cephalotaxus sp, vol. 53, pp. 339–373. Elsevier B.V, 2017. DOI:10.1016/B978-0-444-63930-1.00010-7.
3. For a review of the most modern synthetic approaches, see Jeon H. *Assian J. Org. Chem.* 2021;10:3052–3067. DOI:10.1002/ajoc.202100543.
4. Winterfeldt E. *Synthesis* 1975;613–630. DOI:10.1055/s-1975-34049.
5. Ramage R, Griffiths GJ, Shutt FE, Sweeney JNA. *J. Chem. Soc. Perkin Trans.* 1984;1:1531–1537. DOI:10.1039/P19840001531.
6. M. E. Alonso-Amelot is author of articles published in the Royal Society of Chemistry (RSC) and given permission as such to reproduce material from RSC journals.
7. For the classical Nazarov cyclization, the substrate is a cross-conjugated enone (the C=O sits between two C=C bonds). Strong Lewis- or Brønsted-acid catalysis creates a (A)-O-$C^{(+)}$ cation in conjugation with both C=C bonds. In the present case, the C=O occupies the end of a dienyl moiety. In both cases, however, the five atomic orbitals involved are quite similar and furnish the five-membered ring likewise.
8. Chen Y, Li W-DZ. *Tet. Lett.* 2017;58:248–251. DOI: 10.1016/j.tetlet.2016.12.020. Accessed: 12/12/2018.

PROBLEM 10

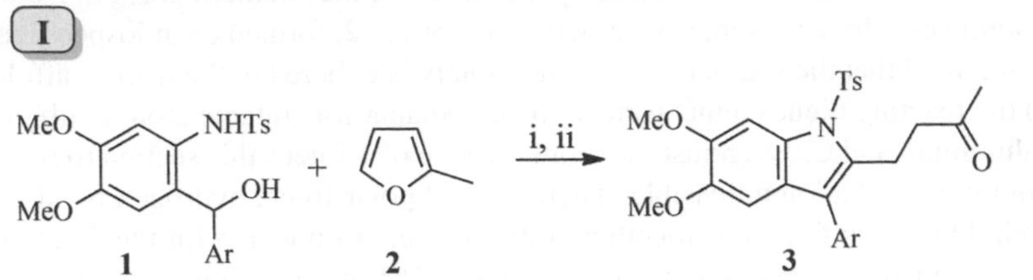

i: pTSOH, CH_2Cl_2, reflux 3-5 min, then $NaHCO_3$
ii; HCl, EtOH, reflux, 15 min

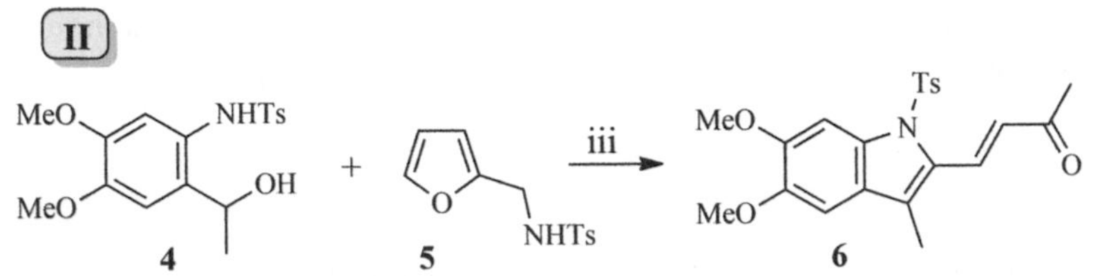

iii: HOAc, H^3PO_4, reflux 11 h, then $NaHCO_3$

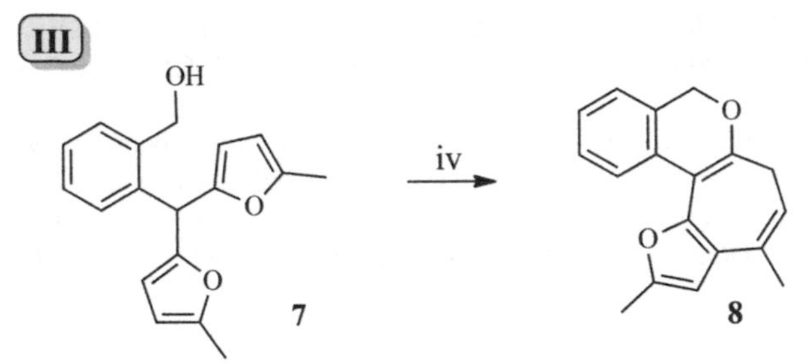

iv: HCl, EtOH, reflux, 20 min

SCHEME 10.1 This and the other Schemes of this problem have been adapted from Reference [1]: © 2001, Elsevier Science Ltd., used with permission, license no. 5311791156469; Reference [2]: © 2005, Elsevier Science Ltd., used with permission, license no. 5311800679159; and Reference [3]: © 2005, Elsevier Science Ltd., used with permission, license no. 5311810727766.

PROBLEM 10: DISCUSSION

10.1 Overview

The chemistry of indole synthesis and furans chemical exploits have been highly attractive subjects for well over a century. When both motifs get together, interesting products are created by relatively simple and unsophisticated routes that all the same furnish instructive mechanisms. As part of a 20-year effort pursuing furans as equivalents of 1,4 dicarbonyl compounds and other molecular blocks useful for the synthesis of a number of heterocyclic scaffolds (see, for example, References [1–4]), Professor Alexander V. Butin and his collaborators in Krasnodar, Russian Federation developed the reactions of Scheme 10.1. These were added to "The Problem Chest", Part II, as three closely related processes, each with peculiar features and open questions. For example, why do products **3** and **6** appear with different oxidation levels even though they are outcomes of apparently very similar reactions? We will discuss reactions **I** to **III** in this order, although by the time you solved reaction **II**, the third should be a breeze.

10.2 Discussing Reaction I

For some readers, this reaction is probably not a problem but an exercise; other less well-prepared students may find them a bit more challenging. A short comparative analysis of target and starting compounds reveals that all carbon atoms of **1** and **2** appear in the scaffold of **3**, while the ketone appendage is likely to be formed from the furan nucleus. There is one missing component: the hydroxyl group in **1**. Given that the medium is strongly acidic and polar, and the carbinol is bis-benzilic, a $C^{(+)}$ in this position would be not only feasible but mandatory to trigger the attack of 2-methylfuran, an HEDZ moiety. This hypothesis is developed in Scheme 10.2.

There is one fork in the route at intermediate **13**, though. Authors propose pathway **A** based on the assumption of proton elimination driving the cleavage of the dihydrofuran ring, although I would expect the *pKa* of this proton to be quite high for the conjugate base of the acid used (chloride anion). Instead, green route **B** would take advantage of i) the strong acidic medium and ii) the known instability of the RN–C–OR' hemiaminal function in acid. Proton elimination would be fueled by the EW effect of iminium ion **16**. The exocyclic to endocyclic C=C bond migration would follow the preferred thermodynamic profile thereafter.

Although authors explored the scope of this reaction with eight different compounds having substituents of different electron donor/acceptor capacities in the aromatic section, yields fell within a similar range, so conclusions regarding electronic effects impinging on product preference or yield cannot be drawn.

10.3 Discussing Reaction II

The acidic and polar medium promote a carbonium ion-marked pathway very much like reaction **I**, yet product **6** is, as suggested at the start, in a higher oxidation level (the extra C=C bond in the aliphatic ketone appendage). This difference may have something to do with the additional tosylamino function in 2-methylaminofuran **5** relative to 2-methylfuran **2**. Try your mechanism design now, compare it with Scheme 10.3 and come to conclusions.

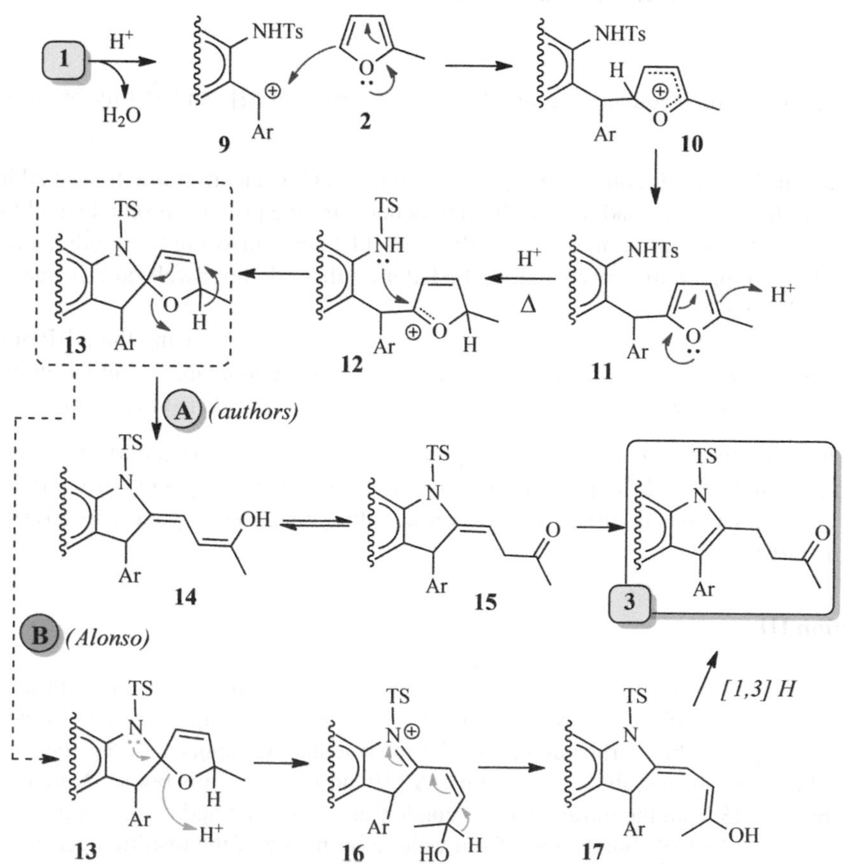

SCHEME 10.2 Likely mechanisms of reaction **I**.

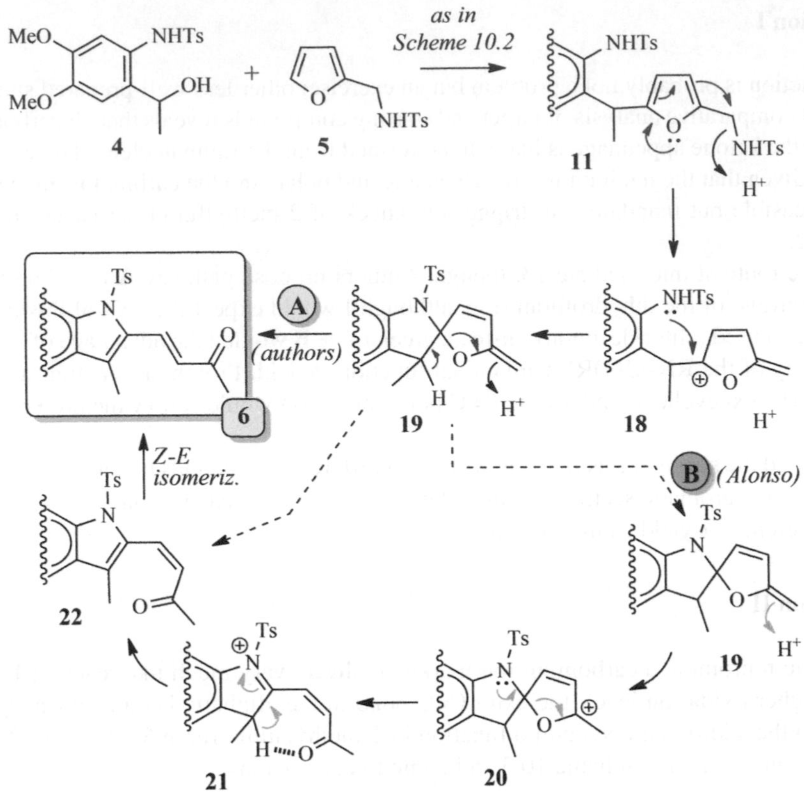

SCHEME 10.3 Discussing options for reaction **II**.

Scheme 10.3 also shows two feasible sequences, route **A** (authors' proposal [2]) and **B** (my own) with some fundamental disparities:

1. Once they arrived at spirobicyclic **19**, authors propose concomitant C=C electrophilic proton addition and β-elimination reaching directly target **6**. After a second look at the stereochemistry the primary product should be Z isomer **22** (dotted arrow). Under acidic conditions and long exposure (heat for 11 hours–favoring thermodynamic equilibria) the Z→E isomerization should set in, predominantly yielding the latter (authors do not dwell on this question, but this is a workbook, and we must underline it).
2. Alternatively, in green route **B** the dihydrofuran ring cleavage goes in steps by using the NBP of N for the same reasons annotated for target **3**. The Z- intermediate **21** is then sterically suitable to enable elimination of the acidic α-iminium-benzyl proton, thus arriving at Z-**22** as the primary product followed by Z → E isomerization.

There is little doubt about the trans C=C in the isolated product, given that authors report two unequivocal signals at 6.01 and 8.09 ppm (d, $J = 16.5$ Hz), in the ^{1}H NMR spectrum of a similar derivative with a phenyl group in lieu of the indole methyl [2]. As you know, C=C cis protons display distinctively lower coupling constants (typically 6–10 Hz). Conclusively, Z → E isomerization precedes the target product.

10.4 Discussing Reaction III

In comparing target **8** and difuran **7** (Scheme 10.1, reaction III), the four-carbon unit of the eastern furan is discernible forming part of the seven-membered ring of isochromene **8**. This implies a furan ring opening homolog of reactions **I** and **II**, followed by addition of an active form, preferably carbocationic, of the intermediate C_4 appendage by the subsisting southern furan. Again, there are two feasible mechanisms, the authors' route **A** [3] and my own (this book). Note that in route **A** the seven-membered carbocycle is built in **25** from the intramolecular condensation of furan and ketone under Brønsted-acid catalysis, a well-known reaction in hot acid. The C=C bond comes from water elimination of the resulting carbinol.

Green route **B** branches off from the same **23** (dotted frame) by the spontaneous cleavage of the ketal substructure and ensuing $C^{(+)}$ (**27**), which maintains the cis configuration of the appendage's C=C bond. This arrangement places furan and $C^{(+)}$

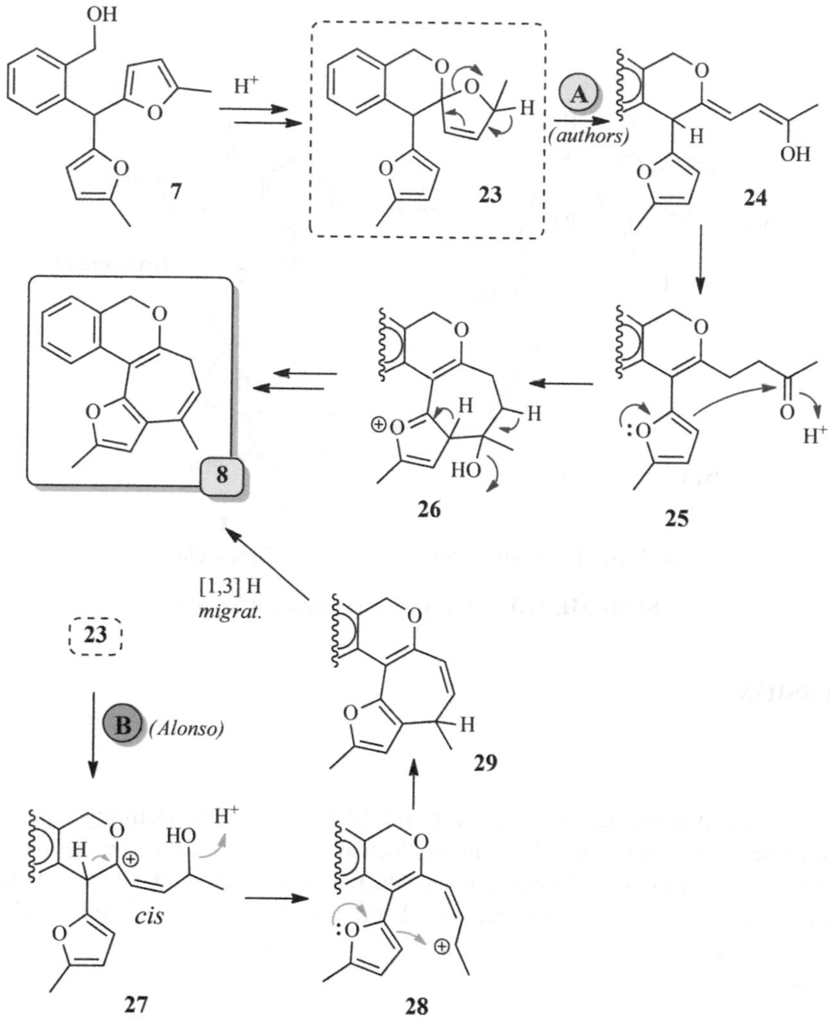

SCHEME 10.4 Two independently conceived sequences to explain target **8:** Sequence **A** by authors of the original work [3] and **B** by this book author.

at close range in the cisoid rotamer **28** (much closer in fact than in **25**) in preparation for a facile cyclization to **29**. The thermodynamic conditions (heat in polar acid), allow the C=C migration via [1,3]-H shift to the more stable cycloheptatriene target **8**.

10.5 Which Route, A or B, of Schemes 10.3 and 10.4 is More Valid? A Takeaway

Experimental evidence with the models tested cannot establish the preponderance of either route. When designing hypothetical mechanisms, it is advisable to create one or more alternatives and pathway branching at key points according to previous experience, homologation of earlier information, or a downright out-of-the-box thinking. This is always interesting, instructive, materials for debate, and entertaining, no doubt.

REFERENCES

1. Butin AV, Stroganova T, Lodina IV, Krapivin G. *Tet. Lett.* 2001;42:2031–2033. DOI:10.1016/S0040-4039(01)00066-1.
2. Butin AV, Smirnov SE. *Tet. Lett.* 2005;46:8443–8445. DOI:10.1016/j.tetlet.2005.09.057.
3. Butin AV, Abaev VT, Mel'chin VV, Dimitriev AS. *Tet. Lett.* 2005;46:8439–8441. DOI:10.1016/*j.tetlet*.2005.09.056.
4. Uchuskin MG, Molodtsova NV, Lysenk SA, Strel'nikov VN, Trushkov IG, Butin AV. *Eur. J. Org. Chem.* 2014;2014(12):2508–2515. DOI:10.1002/ejoc.201301762

PROBLEM 11

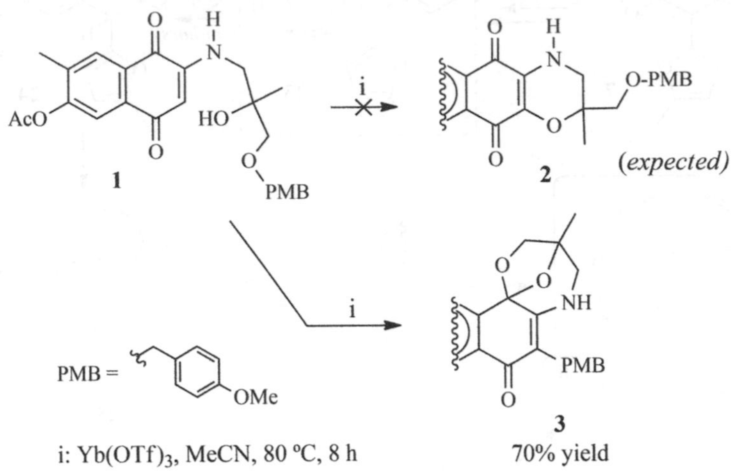

SCHEME 11.1 [1] / American Chemical Society.

PROBLEM 11: DISCUSSION

11.1 Overview

Here, we have a heavily substituted aminonaphthoquinone **1** with EDGs and EWGs defining separate zones within the same molecule. This combination seems perfectly suited for intramolecular additions such as the one required to engineering morpholine **2.** In principle, exposure to a Lewis acid would enhance the reactivity of the C=C quinone bond in **1** to direct the nucleophilic attack of the carbinol and give a one step access to **2**. Predictions, however, failed. Could you please advance an explanation for this frustrated outcome?

I will give you two reasons:

1. The intended nucleophile in **1** is not only a tertiary carbinol; hence, a sluggish C–O-bond forming oxygen atom, but the neighboring *O*-benzyl substructure is bound to offer added steric encumbrance to an already congested compound.
2. In the reaction zone there are five Lewis bases that could link the Lewis acid. Two of these sites would operate as bidentate ligands, one at the northern amino-ketone section, the other at the southern appendage. The first of these would decrease the electron density at the β C=C carbon, fueling the desired reaction, but the nucleophilic power of *O*-benzylglycol moiety, also a bidentate ligand, would be obstructed by the Lewis-acid nucleus.

Things are easily justified once they occur but are otherwise difficult and risky at the time of predicting unmapped molecular behavior. Scientist Laguishetti and coworkers at Chongqing and Huandong in China [1] were especially surprised to find in structure **3** that the *p*-methoxybenzyl fragment was re-crafted on an unexpected site. This is, in fact, the centerpiece of this mechanism problem.

Some pieces of experimental information [1] may also be helpful.

- Several acid catalysts at 70 mol% in acetonitrile were tested with compound **1**.
- The following failed to give products:
  - Proton catalysis ($HClO_4$; perchloric acid)
  - $Ni(OTf)_2$ (nickle triflate) at room temperature
  - AgOTf at 80 °C
  - $Zn(OCl_4)_2$ (zinc perchlorate) at 80 °C
- The **1** → **3** reaction succeeded (55–82% yield) with triflates of Cu, Zn, Yb, and Sb at 80 °C. The best among these Lewis-acid catalysts proved to be $Zn(OTf)_2$ (82% yield).
- Derivatives of **1** having O-protecting groups other than electron donors such as *p*-methoxy-benzaldehyde (PMB) did not react.
- Compound **2** was never seen.

11.2 Problem Analysis

Given that the western sector does not undergo any change, render, or withdraw electronic charge (apparently) to still unknown intermediates, only partial structures of **1** and **3** are compared in Scheme 11.2 (top section). Analysis strategies:

a) Fragmentation considering only the acting blocks, now shown in color to avoid the confusion of dotted lines in congested structures. Because doubts will not all be cleared by fragmenting the puzzle, we will also use:
b) Atom labels
c) Element balance

Results:

Fragmentation: Three blocks are clearly discernible: the fixed scaffold (black), the PMB migrating moiety (purple), and the green section undergoing the most of the action, which is bond cleavage to allow the transfer of the purple section and the construction of the ketal sector.

Atom labels: I selected C^4 as the anchor point since the methyl group on it retains its position relative to carbons C^3 and C^5, thus it does not migrate, and bears witness to the key ketal formation taking place just nearby. After placing labels in the most active molecular section, one easily spots formed and cleaved bonds. The question remains, though, with regard to the identity of the second ketal oxygen atom as there are two candidates, O^8 or O^9.

Element balance: Two hydrogens and one oxygen atom are lost in **1** → **3** associated with the ketal assembly: water elimination likely, although zinc may also get involved.

11.3 A First Approach to the Mechanism

Based on these ideas, a first mechanism plan can be devised from the hypothetical maneuvers of the organic components of the reaction, leaving aside the details of the catalyst, which will be added later. A feasible electron flow moves this approach forward and allows organizing the order of the two main events: ketal genesis and ejection–recapture of the PMB block, or the reverse (Scheme 11.2, mechanism **A**). In this model, O^6 and O^9 constitute the ketal function while O^8 is taken up by Zn to create a tertiary $C^{4(+)}$. Two $H^{(+)}$ are released as well in **5** → **7** and **8** → **3**, hence completing the anticipated element balance. Fragmentation of the PMB moiety (**6**) in **4** → **5** and reincorporation at C^1 at **7** are separate and independent events. Migration of PMB and survival as a stabilized benzyl cation is made possible by the solvent cage within which all these processes take place.

An only organic model with the inverse order of events (detachment–migration of PMB followed by ketal construction) is not feasible without the intervention of the catalyst. Besides, as commented at the start, the reaction is strictly dependent on the kind of catalyst used, and so we must take it into account necessarily. This is not an easy task. Having five core heteroatoms with non-NBP serving as Lewis bases for the catalyst, and at suitable distances, substrate **1** may serve as a bi-and tri-dentate ligand of $Zn(OTf)_2$. Authors of this work [1] offer two feasible sequences depicted in Scheme 11.3, (**B** and **C**–red and green curled arrows, respectively). It is worth noticing that neither **B** nor **C** follow the organic electron deployment of Scheme 11.2 owing to the central role of the catalyst in controlling electron flow. In both **B** and **C** intramolecular alkylation takes precedence to ketal formation. It is also noteworthy that this model builds the ketal moiety with O^6 and O^8, as opposed to model **A**.

When carefully considering the catalyst recovery cycle and the stoichiometry of TfO(-)/TfOH (trifluoromethylsulfonate), the **12** → **13** → **3** steps are difficult to reconcile since O^9 departs as a zinc triflate oxide, which is not the original catalyst that requires recycling. Sequence **D** (Scheme 11.3) offers a feasible catalyst recovery by using the two moles of HOTf produced in steps **1** → **9** and **11** → **12**, which impinge on the latter complex to free the glycol unit. This section completes the ketal synthesis by regular Lewis-acid catalysis and classical steps passing through oxo-cation **14**.

11.4 Conclusions and Takeaways

1. Given that no reaction could be observed at room temperature and only after heating at 80 °C (bp of acetonitrile used as solvent), the C–C bond cleavage of the PMB group is likely the rate limiting step of the domino sequence.
2. When authors tested various other compounds similar to **1** having substituents other than PMB, no conversion was observed after heating for several hours and zinc triflate catalysis [1]. This is compatible with the intermediacy of a duly stabilized $C^{(+)}$ such as **6**, or a late **9** → **11** TS in which the Zn–O – $C^{\delta+}H_2$–Ar was well developed and influenced by electron donation from the aryl group. Substituents in this position unable to afford this stabilization would not proceed to the C^1 alkylation by either route **A–C**.

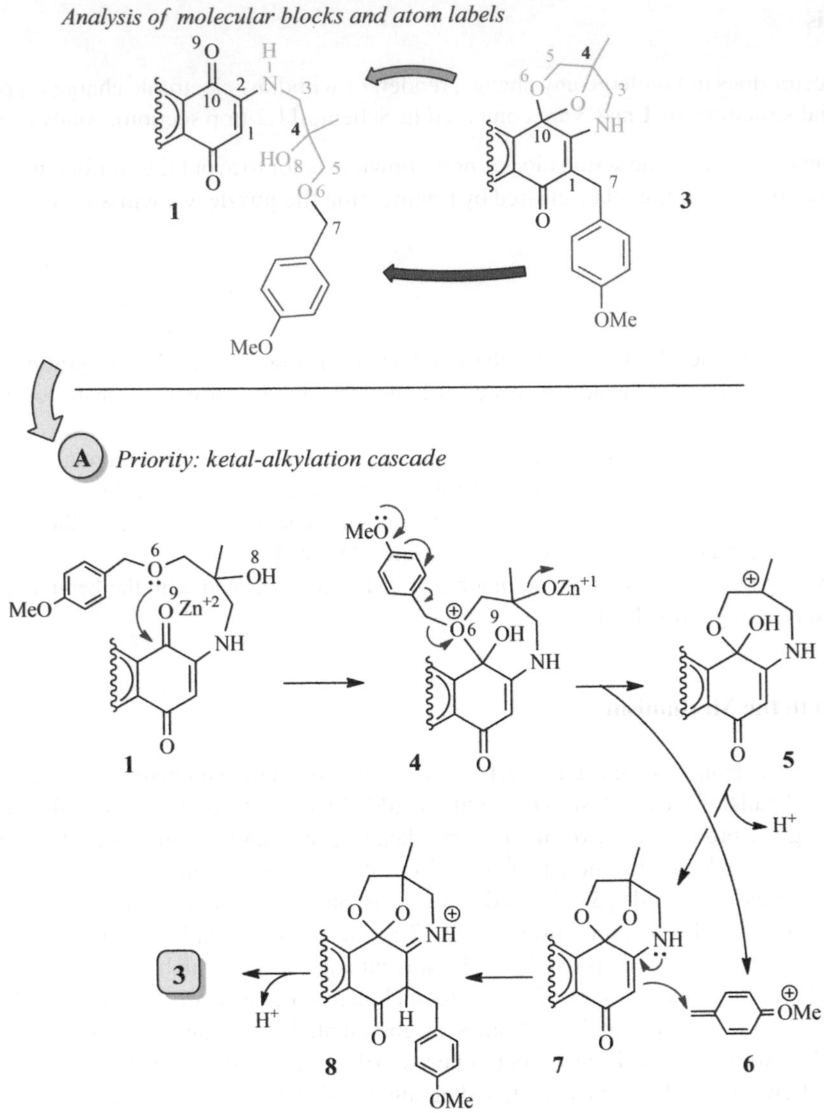

SCHEME 11.2 A first reaction model.

3. A variety of EW and ED substituents on the western aromatic section did not significantly alter the yield of **3**-type products. This suggests little influence on the ketal construction or a marked difference in the rate constants of alkylation/ketalization such that $K_{alk} \ll K_{ketal}$.
4. When no O-protective group was installed in the primary carbinol of **1**, the compound could not be forced to build the bicyclic ketal moiety and remained unchanged [1]. Clearly, PMB cleavage must be the first step of the cascade. It therefore provides support to the order of sequences **B** and/or **C**. Model **A** is thus rejected.
5. In regard to sub-model **D**, it explains satisfactorily the recycling of the catalyst.

Now, if you were to shed light on these mechanisms by attaching a deuterium in **1**, where would you place it and what would you expect to observe?

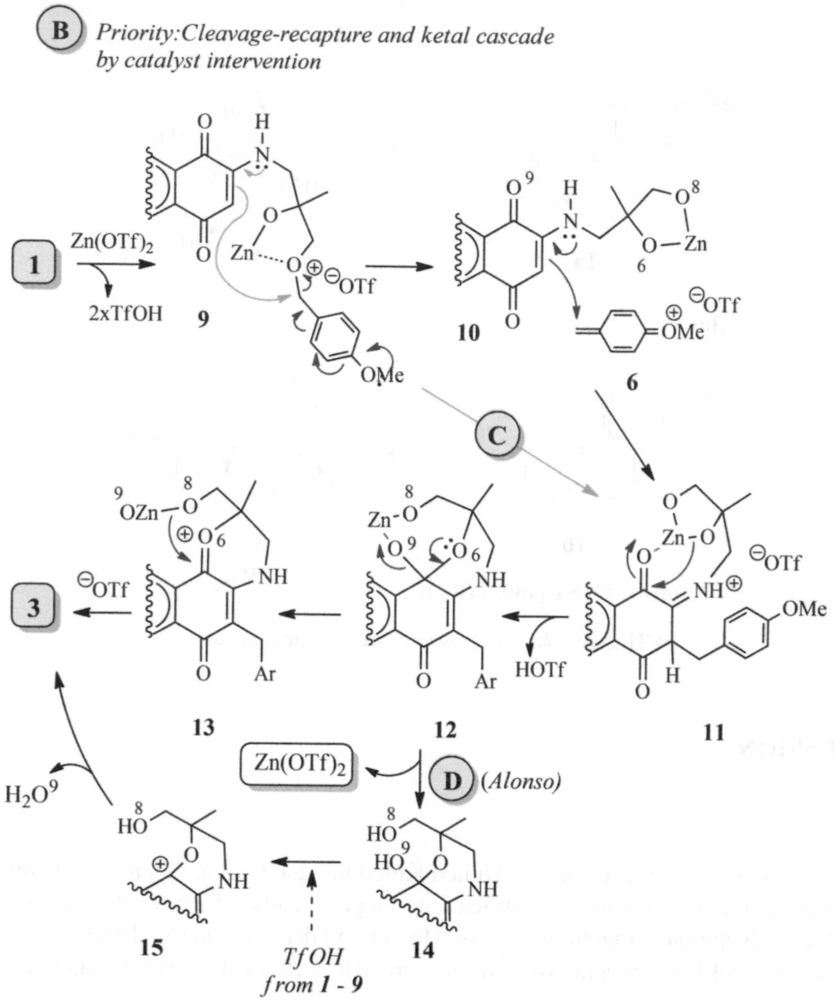

SCHEME 11.3 Mechanism sequences (**B** and **C**) provided by authors of [1] / American Chemical Society.

REFERENCE

1. Lagishetti C, Banne S, You H, Tang M, Guo J, Qi N, He Y. *Org. Lett.* 2019;21:5301–5304. DOI: 10.1021/acs.orglett.9b01912. Accessed: 03/01/2021.

PROBLEM 12

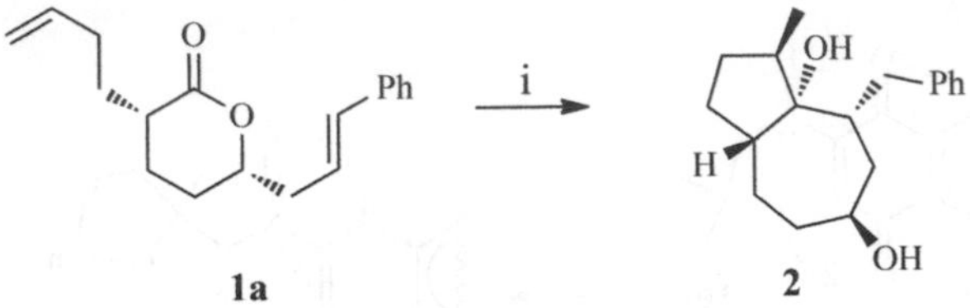

1a **2**

71%

But:

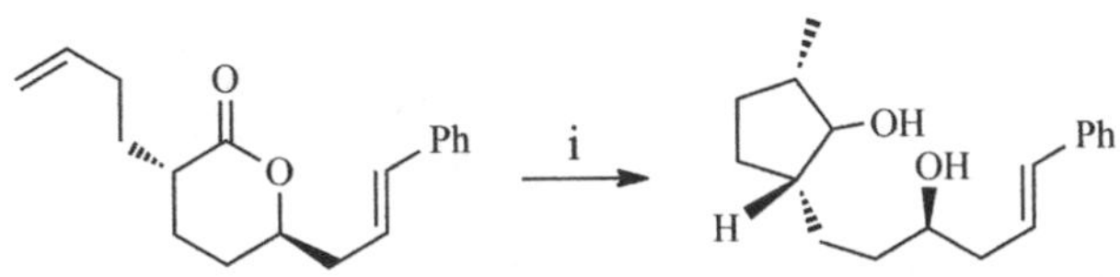

1b **3**

75%

i: SmI_2-H_2O (8 equiv), THF, rt

SCHEME 12.1 [1] / American Chemical Society.

PROBLEM 12: DISCUSSION

12.1 Overview

This reaction developed by researchers in Manchester (United Kingdom) and Osaka, Japan [1] was selected here as an example of the great impact in product outcome of a minor modification of a given carbon framework, even though both enantiomers **1a** and **1b** are endowed with exactly the same functional groups, (for a reminder, see issue 5: “Electron traffic and stereochemistry”, Scheme II.9, Chapter 2 in this book). The mechanism was not provided there to allow you to figure it out.

12.2 Problem Analysis

Two levels of analysis will be applied here: first, the balance of bond formation and cleavage in **1** → **2** disregarding temporarily the stereochemistry for the sake of simplicity. The influence of stereochemistry is treated next when considering the closely related **1b** → **3** process.

12.2.1 The 1 → 2 Conversion By momentarily ignoring the chiral centers, visual comparison of target and starting material rapidly exposes the bonding changes (Scheme 12.2, top part) as one identifies three-carbon subunits (different colors) in **1**. At the centerpiece of the molecular commotion is the lactone group in the black section. Mind that the capacity for forming C–C bonds of C=O in lactones is three as a functional electrophile, in agreement with the *functionality number* [FN_i = +3] (Chapter 4). In the present case, it goes as far as two new C–C bonds with a remnant tertiary carbinol (FN_i = 1). This FN change ($\Delta FN = -2$) indicates a reduction. These green bonds obviously come from the double addition of the C=C linkages in sections **A** and **C**. On first impressions, activation of the C=O is derived from some form of acid–base Lewis association with samarium diiodide (SmI_2; however, see next section). Here crops up a fundamental feature: These additions are *both anti-Markovnikov*, which do not occur customarily under the command of $C^{(+)}$ intermediates. This is a loud call for free radical chemistry.

The balance of molecular formulas between **1** and **2** (**1:** $C_{12}H_{17}O_2$-ph → **2:** $C_{12}H_{21}O_2$–ph) points in the same direction, given that four hydrogens are added in **2** relative to **1** without loss of other elements. Such a process requires the transfer of four electrons to the organic substrate. Where do electrons come from? The SmI_2–H_2O complex is the only and efficient provider of electrons. Indeed, SmI_2 is used lavishly in the reactions of Scheme 12.1 (8 eq), at room temperature [1].

12.2.2 *The Uniqueness of Samarium Diiodide–Water Complex for Electron Transfer, in a Nutshell*

✓ Known as the Kagan reagent [2], it is one of the electron transfer (one or two) reagents of choice for the selective reduction of ketones, esters, carboxylic acids, **lactones**, and cyclic diesters to alcohols [3].

✓ The reduction potential of SmI_2 can be modulated by solvents including hexamethylphosphoric triamide (HMPA), alcohols, and especially water. The acidity of water is increased when complexed with SmI_2 and becomes an effective proton donor in quenching the reaction once the electron transfer and subsequent processes have taken place.

✓ Water is so essential in the activation of SmI_2 that some reductions of ketones furnishing a greater than 98% yield of carbinols fail completely in its absence.

✓ The stereo- and regiocontrol have been extensively exploited in the synthesis of several natural products of complex structure and adorned with several functional groups [3].

12.3 Explaining the Mechanism of 1a → 2

(Scheme 12.2, lower two sections.) Since we have two active C=C bonds in the appendages of the lactone nucleus, one can in principle conduct the cyclizations using first the styryl C=C bond and then the 1-butenyl C=C as in **4** → **5** or in reverse order as in **4** → **5**. We do not have enough information yet to distinguish either one. Wait until Scheme 12.5 to learn the decisive data. The key point is radical **4**.

12.4 Focusing on the Stereochemistry Domain

Once in possession of Scheme 12.2, you should be able to work out the **1b** → **3** reaction by yourself and turn your attention to explaining the impact of changing the absolute configuration of just one carbon in **1a** versus **1b** on the regio- and stereoselectivity of this reaction. The problem boils down to this: What stereochemical feature hampers the cyclization of the methylene-styrene appendage onto the lactone (or the ketone in a second step) in *trans* **1b**?

The answer ought to be in the TSs involved in the cis and transforms due to: i) The feasibility of putting together the intervening functions C=C and ex-lactone radical of type **4** at accessible distances, and ii) the activation energy required to reach each TS.

a) **Accessible distances**:
Because this is your first encounter with a stereochemical challenge in "The Problem Chest", Part II, you may find Scheme 12.3 a bit overwhelming. Let us move in steps through this maze that is so packed with information.

1. The centerpiece: a radical, preceding the cyclization of the styrene appendage. A chair conformation is assumed by authors reasonably enough [1]. The C–[O–Sm–$(H_2O)_n$] bulk can be equatorial (**13a**) or axial (**13b**). Both are in equilibrium owing to the radical inversion, but **13a** is the major component due to the greater diaxial steric stress found in **13b**. Additional stabilization of **13a** is brought by the anomeric effect of axial substituted radicals next to an oxygen atom in a pyranose [4–6].
2. Pathway choices **A** or **B** open from each configuration furnish different outcomes.
3. The AO of the radical is projected in the pseudoaxial direction, in parallel with the methylene-styrene chain in **1a**, whereas the 1-butenyl arm is equatorial. For the terminal C=C to reach this radical, inversion of O-C• is mandatory. This inversion places the [O–Sm–$(H_2O)_n$] complex to the axial position, thus exerting 1,3-diaxial SS shown in **13b**.
   a. ***Route A***: In **13a** the equatorial Sm complex enables the styryl C=C to reach the one-electron axial AO to perform the anti-Markovnikov addition to **14**, while water transfers an H atom to trap the residual benzyl radical furnishing **5**. To achieve this, the styryl appendage adopts an endo conformation (**13a**) without significant steric compression by other substituents on the pyranose ring.
      i) Cleavage of **5** in the manner explained in Scheme 12.2 reconstitutes the C=O, which is reduced by a second equivalent of SmI_2. Again, axial (a) and equatorial (e) [O–Sm–$(H_2O)_n$] complexes coexist in equilibrium in **15**. Models show that the *a* O–Sm configuration should be preferred since steric stress caused by the e,e configuration with the *e*-benzyl group is greater than in the e,a distribution, as shown in **15**.
      ii) [O–Sm–$(H_2O)_n$] thus located enables the e-approach of the 1-butenyl chain via an anti-Markovnikov TS affording target **2** in the correct *trans* CH_3–C–C–OH configuration.

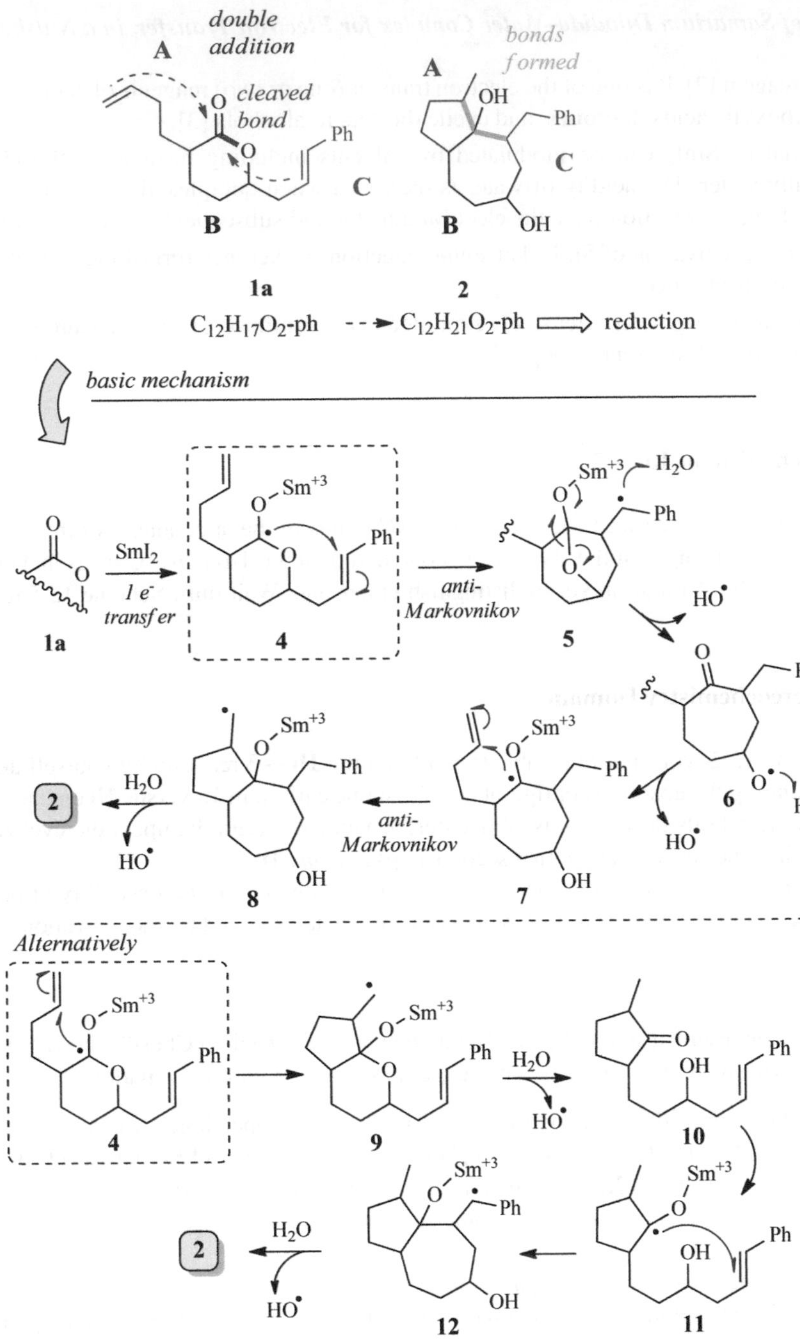

SCHEME 12.2 Preliminary problem analysis of the **1a** → **2** conversion (see text for details) and a basic electron flow of the tandem cyclizations, stereochemistry not withstanding momentarily.

b. ***Route B***: The pathway enters the dotted frame zone. For the C=C to radical addition interacting groups face SS as in TS **17**. If at all, sequence **B** leads to *unobserved* hemiacetal *cis* CH_3–C–C–OH **19a**.
c. ***Route C***: The 1-butenyl side chain can adopt a reverse TS as in **20**, with less sterical conflict to afford the *trans* CH_3–C–C–OH configuration of **19b**. However, this hemiketal was not observed by authors in this and similar treatments with sibling compounds.

4. Therefore, routes **B** and **C** can be discarded.

Scheme 12.3 thus explained enables a better understanding of the pending **1b** → **3** reaction exhibited in Scheme 12.1. In this instance, the order of cyclizations exposed in our basic mechanism scheme is inverted; the 1-butenyl chain takes the lead as a result of the styryl chain positioned equatorially, hence away from the [C(•)O–Sm–$(H_2O)_n$] complex (Scheme 12.4). Once pentanone **24** is produced, excess Sm–$(H_2O)_n$ yields the corresponding carbinol passing through complex **25**, which might offer

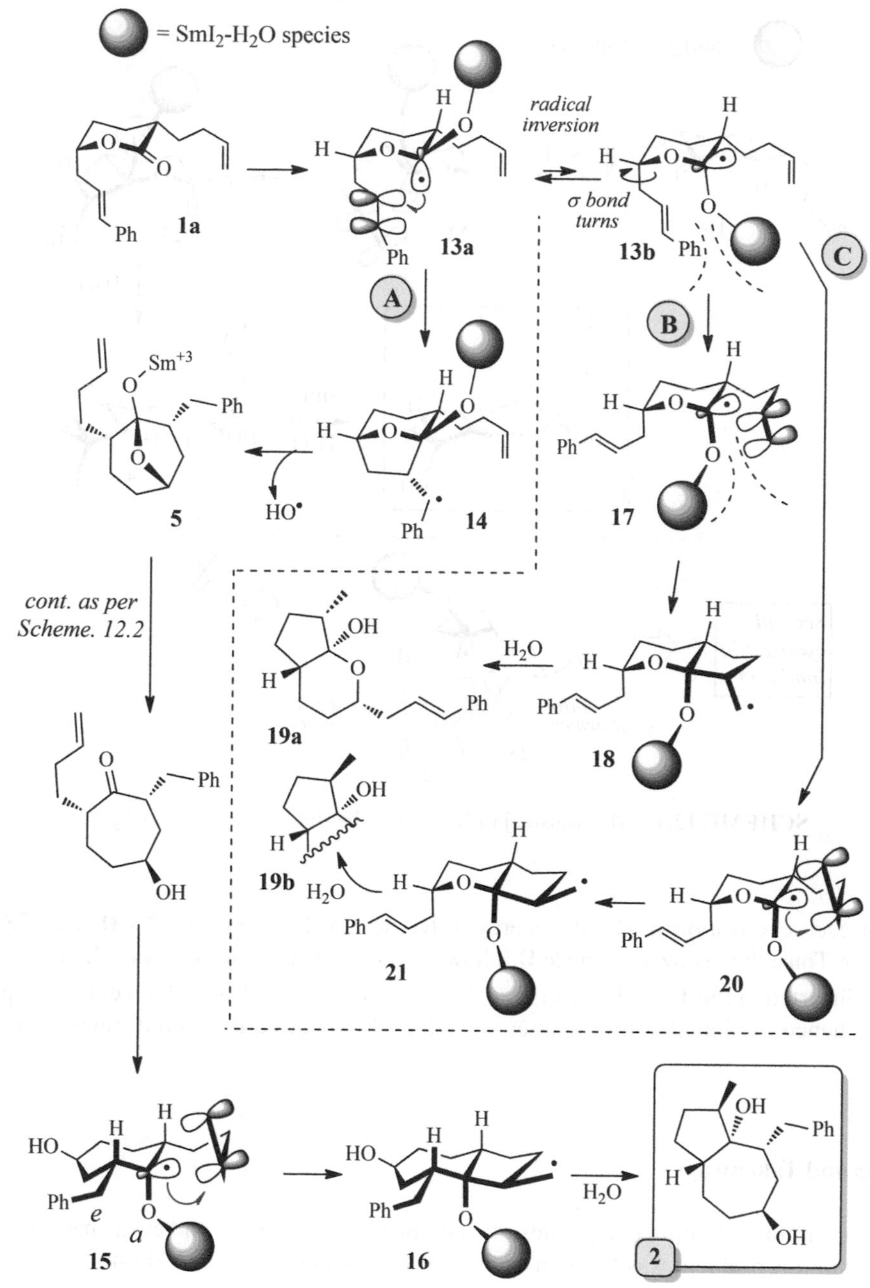

SCHEME 12.3 Detailed description of the stereochemical control to reach target **2**.

the opportunity for the second cyclization. However, this is hampered by steric encumbrance found in the required endo approach of the styryl C=C, and hydrogen transfer from water first finishes the reaction at the carbinol stage observed by researchers [1].

b) **Difference in energy levels affects the order of the two tandem cyclizations.**

Authors [1] studied this issue using a theoretical model compound **I** that contains the same traits of *cis* **1a** and two side chains with terminal C=C bonds [7]. The difference between the two appendages was chain length and relative position in the ring. For accessibility to the calculation method employed, the $SmI_2{\cdot}H_2O_n$ component was replaced by an alkoxide anion. Scheme 12.5 shows the results. In short:

✓ Preference was given to radical anion II due to the stabilization of the anomeric effect [4].

✓ A radical inversion equilibrium **II** ⇄ **III** is established but displaced in favor of **II**. The **II/III** ratio is the first determinant of cyclization preferences.

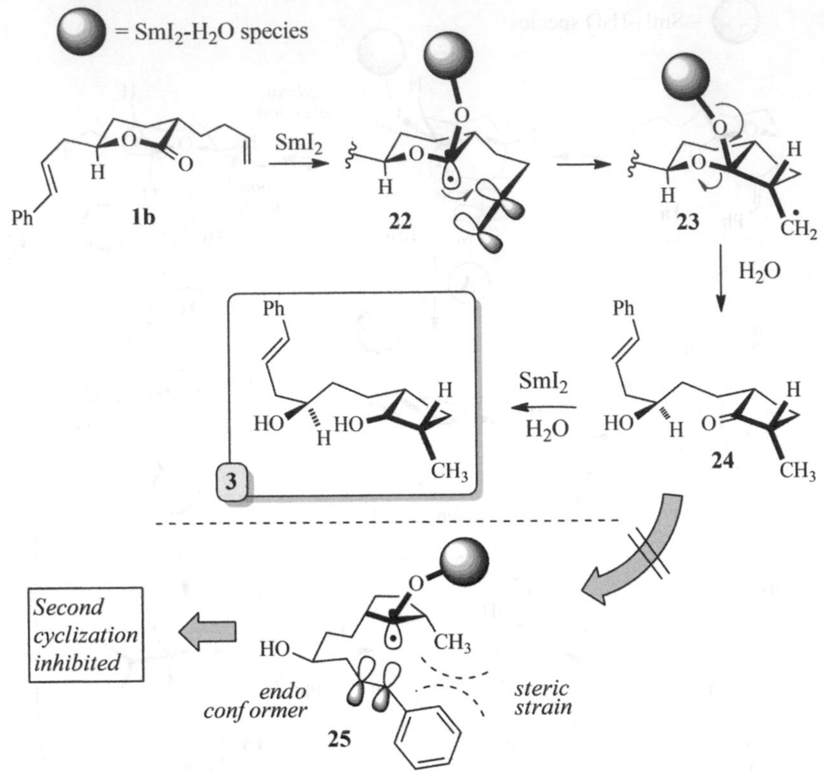

SCHEME 12.4 Mechanistic justification of the stereochemistry of target **3**.

✓ Down the line of each cyclization mode, the energy difference of the respective TS(**II**) and TS(**III**) is 5.2 kcal mol^{-1} favoring the former. Thus, the cyclization mode **B** following routes **B** and **C** of Scheme 12.3 is a very minor contributor.

✓ If this situation is inverted in the **1b** → **3** conversion, it is because the styryl side chain extends equatorially in the plane of the pyran ring; hence, it is unable to fold over and adopt the required endo conformation to reach the axially oriented radical AO.

12.1.5 Some Lessons and Takeaways

1. Stereochemical constitution of chiral compounds often defines regio-, stereo-, and even enantioselectivity.
2. While the analysis of this problem was far from easy because of the use of several 3D structures, it must have been good training in this respect for future problems.
3. In cases where stereochemistry dealings may be potentially complex, it is advisable to start the problem analysis by defining electron flow involved in bond formation/cleavage and redox balance, momentarily ignoring the stereochemical features. On this basis, move on to consider these features using 3D renderings when necessary. Developing your drawing ability of 3D structures gives you a professional edge over those who do not.
4. It takes considerable experience to know beforehand the chemical properties of transition metal and rare earth complexes, samarium iodide in this case. Hints about their performance can anyway be inferred from the manner the organic framework is responding (the anti-Markovnikov addition mode, for example).
5. Computer calculations of reaction sequences, including energy levels of intermediates and TSs, are published with increasing frequency. A visit to the references provided in "The Problem Chest" reactions in difficult problems or when a choice between various apparently feasible reaction mechanisms are on the table is always a good idea.

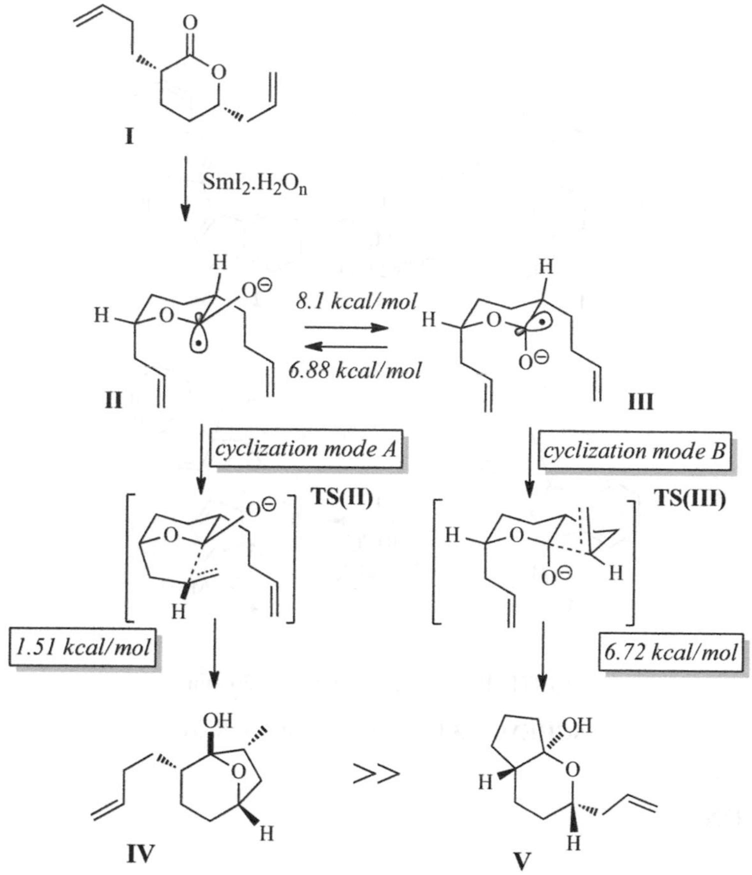

SCHEME 12.5 Energy profile of key transition states (TS) of the two cyclization modes of either side chain in a theoretical model **I.** Theoretical calculations by authors of this research [1, 7]. [1] / American Chemical Society.

REFERENCES AND NOTES

1. Parmar D, Matsubara H, Price K, Spain M, Procter DJ. *J. Am. Chem. Soc.* 2012;134:12751–12757. DOI:10.1021/ja3047975.
2. Girard P, Namy JL, Kagan HB. *J. Am. Chem. Soc.* 1980;102:2693–2698. DOI:10.1021/ja00528a029.
3. For a review, see: Szostak M, Spain M, Parmar D, Procter DJ. *Chem. Commun.* 2012;48:330–346. DOI:10.1039/c1cc14252f.
4. The anomeric effect refers to the axial energy preference of an electronegative substituent, or the AO of a radical intermediate, adjacent to the ring oxygen of a pyranose ring. The NBP AO of the adjacent O interacts with the radical or electron deficient axial AO that provides electron density. It has been examined in the light of *ab initio* calculations [5] and a more modern view poses complex reasons for it [6].
5. Delbecq F, Lefour JM. *Tet. Lett.* 1983;24:3613–3616. DOI:10.1016/S0040-4939(00)88182-4.
6. Wiberg KB, Bailey WF, Lambert KM., Stempel ZD. *J. Org. Chem.* 2018;83:5242–5255. DOI:10.1021/acs.joc.8b00707.
7. Authors [1] calculated the energies on MP2/aug-cc-pVDZ/ MP2/ -cc-pVDZ with solvation in tetrahydrofuran (THF), for those of you who understand the meaning of this.

PROBLEM 13

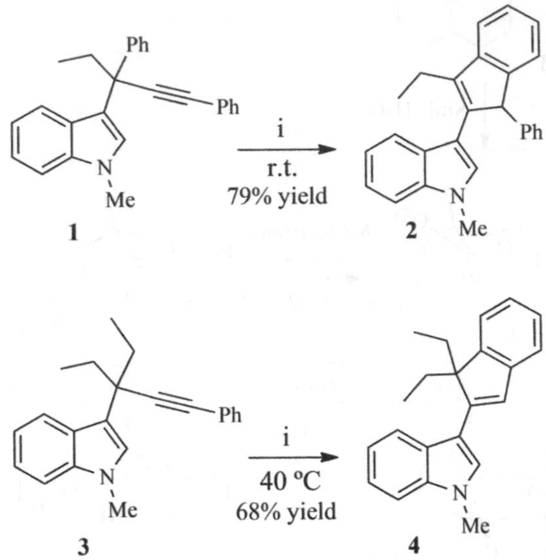

i: $AuNTf_2(Ph_3P)$ 5 mol %, CH_2Cl_2, 30 min

SCHEME 13.1 [1] / John Wiley & Sons.

PROBLEM 13: DISCUSSION

13.1 Overview

This is our first encounter with gold catalysis, which has gained considerable momentum in organic synthesis and the discovery of novel reactions based on the activation of alkenes, alkynes, and other synthons [2]. A large number of propargyl indoles were submitted to the Au(II) complex of Scheme 13.1 by this team of researchers from three universities in northern Spain [1]. All substrates reacted similarly under mild conditions underscoring the general character of the **1** → **2** reaction. However, dialkyl indole **3** was the only one tested requiring higher, albeit still mild, temperature. Product **4** had a different substitution pattern. Your job is to provide a reasonable mechanism for both reactions, to justify the different outcomes and justify the need of a higher temperature for the **3** → **4** reaction to proceed.

13.1.2 Understanding the 1 → 2 Reaction

By dividing structures **1** and **2** in blocks comparatively, one can easily recognize three common sections in the target and starting compounds (Scheme 13.2). If the phenyl group on the acetylene maintains the C–C bond link to the rest of the scaffold, one observes the following:

- The aryl substituent on the quaternary carbon forms a C–C bond with one of the *sp* carbons.
- The indole block migrates from the quaternary carbon to the vicinal alkyne center carbon via C^3. These new bonds are marked in green in Scheme 13.2.
- However, considering the top phenyl alkylation as an aromatic electrophilic addition, the electron-rich alkyne and the well-known tendency of indole's C^3 to operate as a nucleophile, we have a set of three HED sections whose intramolecular interaction for enabling electron flow is unlikely, unless an electrocyclic reaction sets in. But this would need a high temperature and/or activation by LED/HED substituents. Having none of these, we have two options: i) a radical-commanded process or ii) another factor, external to compound **1** that changes the electron density framework.
- This is where the gold catalyst breaks this stationary situation. A growing body of literature supports the model of Au(II) forming π complexes with alkenes and alkynes, as other transition metal elements do (e.g. Ag, V, Sc, Ti, Fe, Pd, Pt, and others) affording a fallout of reactions [2].

These observations drive a tandem sequence triggered by an Au-alkyne complex turning this section into the required LEDZ and the ensuing indole migration (Scheme 13.2, middle section). The mechanism model is based on a transient cyclopropane initiated by an indole nucleophilic attack on the electron deficient gold-alkyne complex **5**. Electron redeployment leads to an electrophilic gold-carbenoid **7**. Worthy of note: In the latter, indole and the Au complex appear in a cisoid conformation to decrease the phenyl-indole steric stress of the initially formed transoid conformation. Given that this change occurs by rotation of the adjacent *s* bond, here we talk about a torquoselective ring opening of the cyclopropylidene group (**6**→**7** step).

Considering the partial positive charge in the five carbon dienyl substructure, the setting is an iso form of a Nazarov-type cyclization. The *metalla-iso-Nazarov* denomination for the **7**→**8** step grows from the participation of gold. The sequence ends after the C protonation of the C–Au$^{(I)}$Ln bond and cleavage that regenerates the catalyst. Having an aprotic medium, this proton must emerge from the preceding elimination step, bringing proton stoichiometric balance.

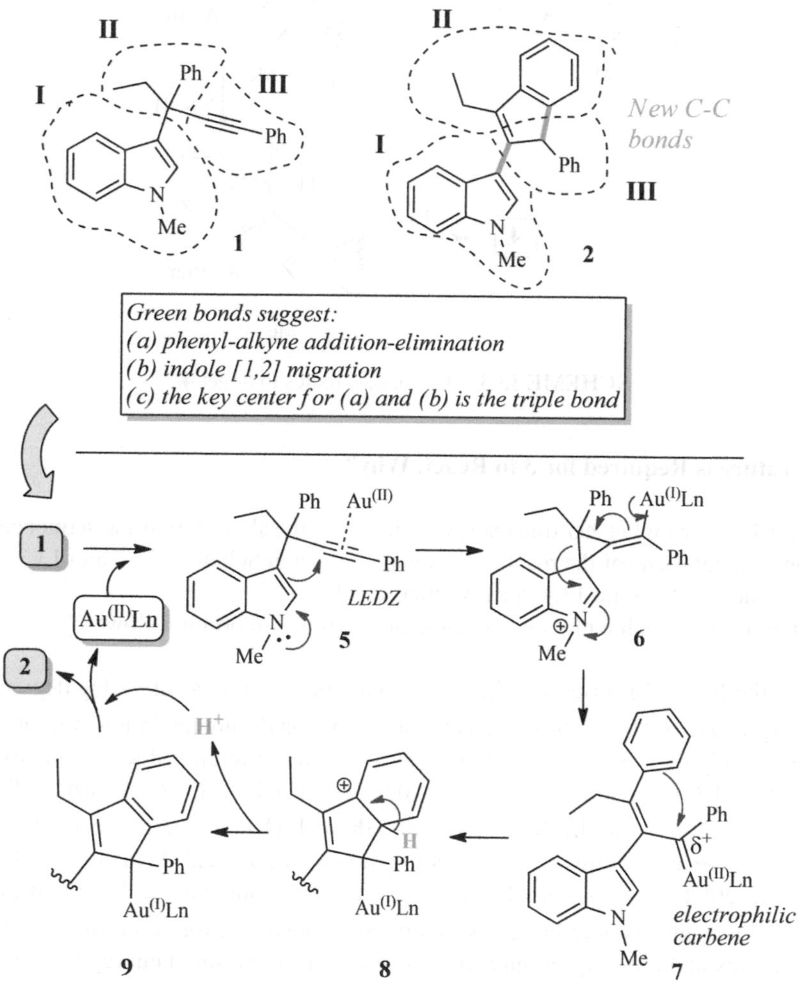

SCHEME 13.2 Problem analysis by segmenting target and starting compound in the **1**→**2** reaction. [1] / John Wiley & Sons.

13.1.3 Understanding the 3→4 Reaction

Just one aspect marks the difference between target structures **2** and **4:** The end phenyl substituent gets incorporated in the indene moiety and is no longer a σ-bonded substituent. Other facets are quite alike as both **1** and **3** seem to undergo the 1,2-migration of the indole section, the unique hitherto unknown process unveiled by this research. We may thus assume that **4** also stems from a similar cyclopropane **10**. This is where Scheme 13.3 begins. As in Scheme 13.2 (not shown there), two reaction modes are likely; both are marked by green and red curly arrows and self explained in the renderings. Both converge in cation **12** and a brief evolution pathway to target **4**.

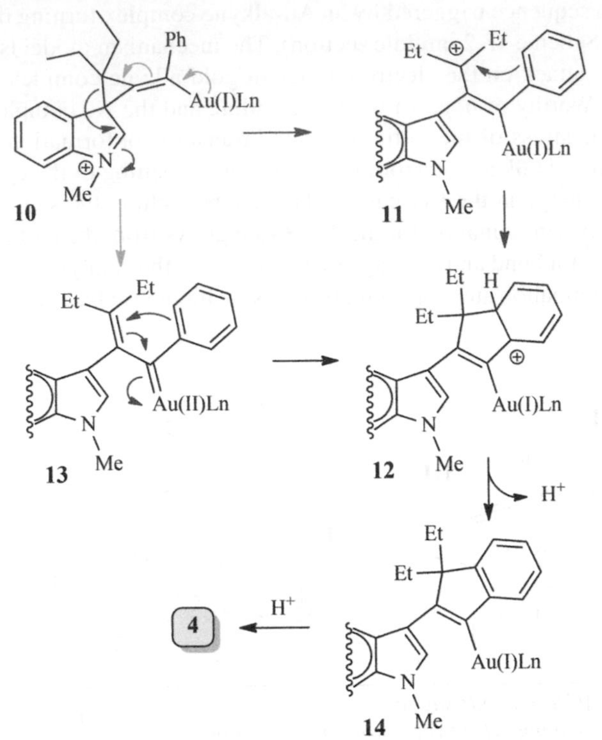

SCHEME 13.3 Two access routes to target **4**.

13.1.4 A Higher Temperature is Required for 3 to React. Why?

Higher temperatures generally imply higher ΔG free energy in the TS of the slow step of the sequence. The Spanish authors [1] included DFT computational calculations of the reaction pathways shown in Schemes 13.2 and 13.3 for two related models (**Ia** and **Ib**) replacing ethyl with methyl. Scheme 13.4 portrays their results.

Among the several comments that such a plot elicits, some of the most important follow:

- ✓ The energy profiles of the **Ia** and **Ib** series are different as anticipated but to a surprising degree.
- ✓ The energy demand expressed as $\Delta G^{\ddagger}$ of the crucial cyclopropanation during the indole migration is less favorable for the gem-dialkyl series **b** but only by less than 2 kcal mol^{-1}. Such a small difference does not justify the need for higher temperature to force compound **3** into action. Another more demanding TS must occur down the line.
- ✓ This rate limiting step is found in the **TS2** between **IIb** and **IIIb**, namely the cyclopropylidene ring cleavage ($\Delta G^{\ddagger}$ = 9.48 kcal mol^{-1}). In the **a** series this step progresses smoothly ($\Delta G^{\ddagger}$ = 2.58 kcal mol^{-1}). This argument seems to be at the core of the lack of reactivity of gem-diethyl compound **3** at room temperature, thus requiring moderate heating.
- ✓ The plot of Scheme 13.4 may also be interpreted as a thermodynamically favored return from cyclopropylidene **IIb** to the first alkyne-gold π complex and recovery of unreacted material, given the small energy barrier of the reverse reaction via **TS1**.
- ✓ The displacement of the indole unit toward the phenyl, being nearly quite similar in **IIIa** and **IIIb** does not explain the 10 kcal mol^{-1} $\Delta G^{\ddagger}$ difference. Computed models showed the rotation of the allyl C–C bond coupled with the ring cleavage here as the substructure adopts helical conformation, thus reducing indole-phenyl SS.
- ✓ The resulting gold-carbenoid structures **IIIa,b** undergo considerable relaxation thereafter.
- ✓ Further C–C bond turning is required in the **IIIb** → **IVb (TS3)** to place the terminal phenyl ring in the proper position for the ensuing Nazarov-type electrocyclization. This rotation creates pressure between the rotating phenyl and the gem-dimethyl section, which is reflected in a $\Delta G^{\ddagger}$ = 9.5 kcal mol^{-1} in the **b** series.
- ✓ In the **a** series the reaction follows a different metallo-iso-Nazarov electrocyclization with its own stereoelectronic requisites involving the phenyl → C = AuLn carbenoid coupling, and equally requiring a $\Delta G^{\ddagger}$ (8.30 kcal mol^{-1}) of the same order as in the **b** series.

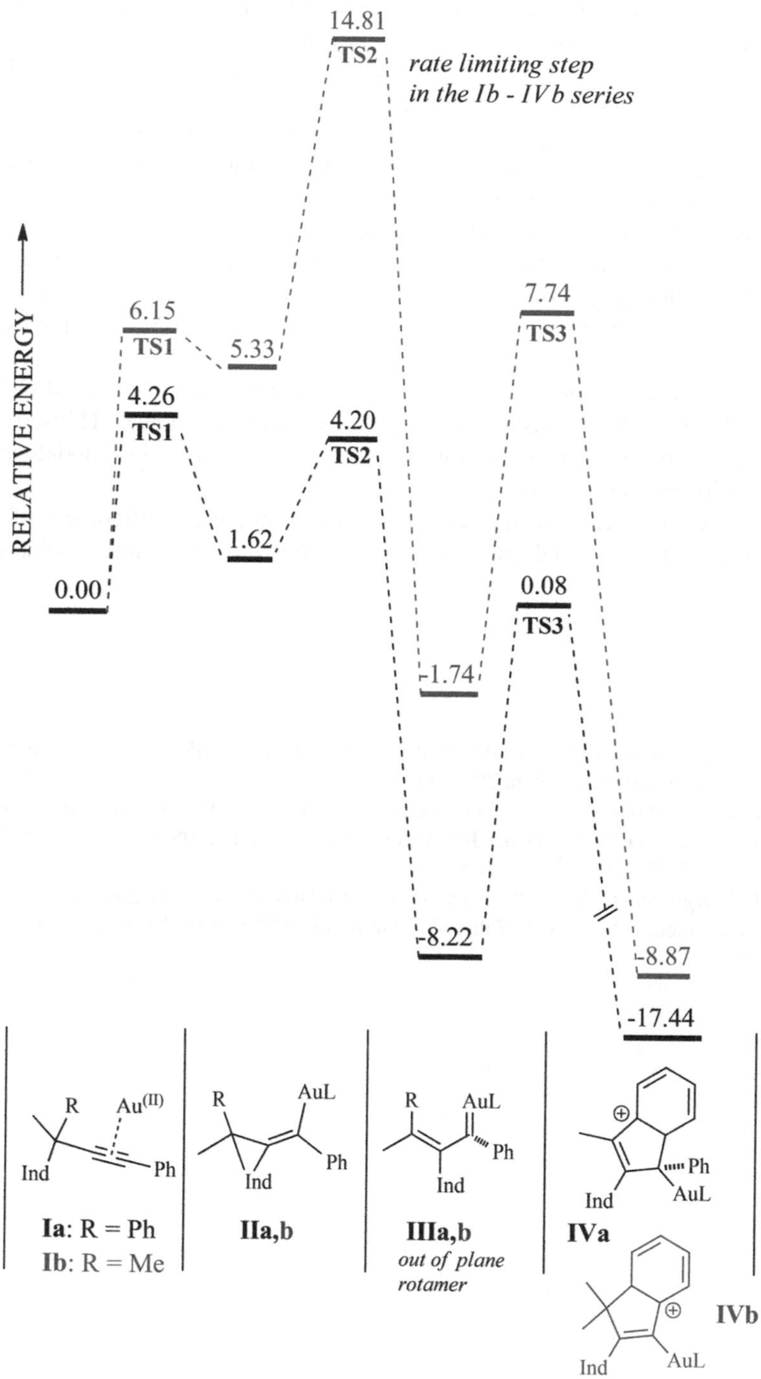

SCHEME 13.4 Results of DFT computations of two model compounds **Ia** (R=phenyl) and **Ib** (R=CH_3, in purple) representing starting indoles **1** and **3** of this problem (Scheme 13.1). Values shown are free energies ΔG (kcal mol^{-1}) of intermediates shown at the bottom and the activation energies $\Delta G^{\ddagger}$ of the TSs between the shown structures; all these are relative to the starting model compounds. The position of energy levels in this picture are not to scale. Drawing by Alonso-Amelot based on table form data in [1] / John Wiley & Sons.

13.2 LEARNING POINTS AND TAKEAWAYS

1. Gold(II) is a relative newcomer element among precious metal catalysts such as Ag, Rh, Pd, and Pt. This popularity grows from remarkable versatility, different reduction potential, reactivity toward alkenes and alkynes, and the number of reactions with other functional groups from the alkene and alkyne-gold complexes.
2. The essence of the majority of the gold-alkyne complexes reactivity is the strong electrophilic character, thus enabling C–C bonding with intra and intermolecular carbon nucleophiles.
3. The intramolecular addition of a Nu to the gold-alkyne complex can cyclize by a C–C attack on the closer alkyne carbon (as in this problem) to form a three-membered ring. More distant intramolecular nucleophiles can also cyclize on the terminal alkyne carbon, as other reactions in this chapter illustrate.
4. The 1,2-migration of indole acting as a nucleophile on the complex via a cyclopropylidene is the third case of such shifts by the time this discovery was published. Previous migrations were observed for acyl (RCO_2) and thio (SR) groups.
5. The Au=C carbene is electrophilic (e.g. structure **7** of Scheme 13.2), short-lived and highly reactive like other metal carbenes (MLn=C) are, and susceptible to intra and intermolecular C, N, and O nucleophiles, which offers a gamut of synthetic possibilities.
6. Although gold-carbenes imitate other metal carbenes in cyclopropanation of alkenes and C–H insertion reactions, they are less efficient in furnishing the characteristic cyclopropyl derivatives that show C–H insertion as the major product [3, 4]. Exploitation of gold-carbenes is more interesting in other less trivial reaction models as the one selected for this problem and others you will find in this chapter.
7. The DFT results of Scheme 13.4 clearly demonstrate that substituent pattern differences of apparent irrelevance bring about often unanticipated and substantial changes in the energy profile of intermediates and TSs along reaction cascades.

REFERENCES

1. Sanz R, Miguel D, Gohain M, García-García P, Fernández-Rodríguez MA, González-Pérez A, Nieto-Faza O, de Lera AR, Rodríguez F. *Chem. Eur. J.* 2010;16:9818–9828. DOI:10.1002/chem.201001162.
2. A number of reviews are available. Among the most cited, see: Jiménez-Núñez E, Echevarren AM. *Chem. Rev.* 2008;108:3326–3350. DOI:10.1021/cr0684319. Michelet V, Toullec PY, Genêt J-P. *Angew. Chem. Int. Ed.* 2008;47:4268–4315. DOI:10.1002/anie.200701589. Shapiro ND, Toste FD. *Synlett.* 2010;675–691. DOI:10.1055/s-0029-1219369.
3. Diaz-Requejo MM, Pérez PJ. *J. Organomet. Chem.* 2005;690:5441–5450. DOI:10.1016/j-jorganchem.2005.07.092.
4. Fructos MR, Belderrain TR, de Frémont P, Scott NM, Nolan SP, Diaz-Requejo MM, Pérez PJ. *Angew. Chem. Int. Ed.* 2005;44:5284–5288. DOI:10.1002/anie.200501056.

PROBLEM 14

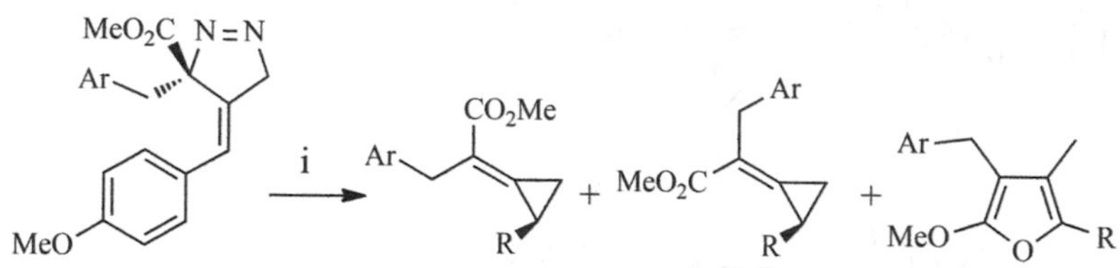

1 **2** **3** **4**

83% 9% 8%

Ar = *p*-NO_2-Ph
R = p-MeO-Ph
i: Benzene, 80 °C, 4h

SCHEME 14.1 [1] / American Chemical Society.

PROBLEM 14: DISCUSSION

14.1 Overview

The mechanism of this thermolysis is not particularly difficult if one limits oneself to moving bond electrons from here to there pursuing to break and form bonds with curly arrows. After all, the chemistry of alkylidene pyrazolines, of which **1** is one more example, has been studied for quite some time [2]. However, the extensive work developed by Japanese researchers from Osaka as they report this novel transformation [1] is far more interesting once the variety of reaction possibilities and stereochemical issues are added to the mix.

Important data to consider at the start:

- ✓ All four compounds are functionally dense opening electron flow possibilities. Indeed, upon thermolysis or photolysis, alkylidene pyrazolines are known to produce many cyclopropanes and olefins.
- ✓ Extending the reaction time from 4 to 12 hours furnishes the following proportions of products: **2** = 84%; **3** = 0%; and **4** = 16%.
- ✓ When a free energy relationship study of seven siblings of **1** (in which the *p*-CH_3O substituent was replaced with H, CH_3, Ph, *p*–ClC_6H_4, *p*–$NO_2C_6H_4$ and two phenyls at the exocyclic alkene), at 60 °C, $\rho = -0.62$ emerged from a linear ($R^2 = 0.97$) Hammett–Taft reaction rate association. The range of the Gibbs energy of activation $\Delta G^{\ddagger}$ was 23.56–27.19 kcal mol^{-1}, implying a moderate energy level of the TS.
- ✓ The major product (**2**) is the less stable due to steric hindrance offered by the syn di-arene substituents.

As you can see, the profile of this problem is far more attractive and enjoyable.

14.1.1 Discussion The basic electron redeployment of type-**1** heterocycles includes the obvious extrusion of N_2 through two modes: concerted (Scheme 14.2, mode **I**) and stepwise (modes **II** and **III**). The latter form is modulated by replacing substituents **A** or **B** or both with EWG favoring the heterolytic cleavage to a carbanion followed by C–C bond rearrangement. Both modes may theoretically produce type **2** and **3** products. As Scheme 14.2 shows, the boundary between these thermolysis modes may fall within similar activation energy ranges of TSs leading to mechanistic competition. The number of cyclopropylidene isomers (CPs) thus produced is considerable. While such an outcome may elicit theoretical interest, it is mechanistically confusing and synthetically nightmarish. Besides, all types **I–III** offer feasible routes to the cyclopropyl alkylidene structure of targets **2** and **3**. Thus, electron deployment per se as shown in Scheme 14.2 brings about more questions than answers. Yet, Scheme 14.1 describes a relatively clean result, suggesting a dominant mechanism. Your job is to fish it out from the pond of several hypothetical routes.

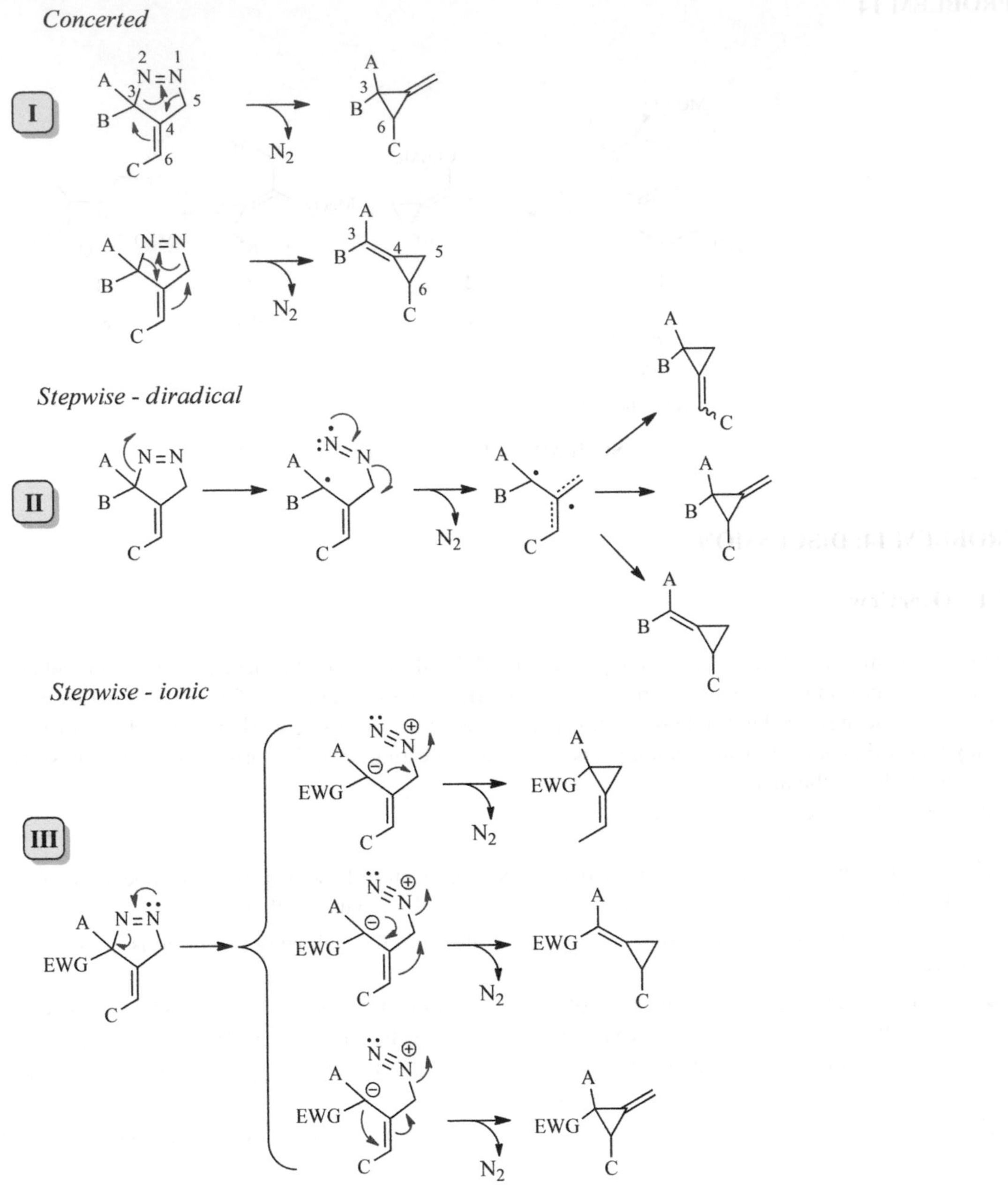

SCHEME 14.2 Modes of electron redeployment in five-membered N_2 heterocycles with exclusion of a nitrogen molecule.

14.2 Concerted versus Stepwise

Substituents in pyrazoline **1** provide a rigid configuration, which is altered in isomeric products **2** and **3**. In principle, this might suggest a stepwise mechanism passing through intermediates possessing a free-rotating σ C–C bond to accommodate the syn- and anti-isomers **2** and **3**. However, mind that in the major isomer **2** the two larger vicinal substituents Ar–CH_2– and *p*–NO_2Ph– are syn to each other offering more steric hindrance than the anti-isomer. This observation is not compatible with the freely rotating stepwise model.

Furthermore, the stepwise route should furnish several CPs, as opposed to the results of Scheme 14.1.

If the mechanism followed a stepwise route, it would call for a second question: radical or anionic?

C^3 of **1** includes one EWG (methyl ester) reducing electron density at the tertiary carbon C^3. And the N=N bond influencing electron density at C^5 as well. We know from the overview (Section 14.1) that researchers determined that $\rho = -0.62$ at 60 °C for a homolog series of compound **1**. This moderate negative value is suitable with the build-up of positive charge or a decrease in electron density at the reaction center. This result impinges on **1** with regard to the role of the nucleophilic character of the C=C bond. Indeed, EW substituents on the *p*-X-Ph group of the homologs decrease the rate of reaction and require higher temperatures to yield CP products. In this condition, however, the process would not evolve in a stepwise fashion.

14.3 A Feasible Mechanism for 1 → 2 + 3 Not Devoid of Difficulties

Given the priority of the concerted pathway and the production of the CPs with the tetrasubstituted C=C bond, it appears as if the second option of the concerted process (**I**), Scheme 14.2, was more feasible. Of note, the combined effects of an EDG in C^6 and two EWGs in C^3 provides a framework of electron density shift passing across the N=N bridge, as shown in Scheme 14.3,

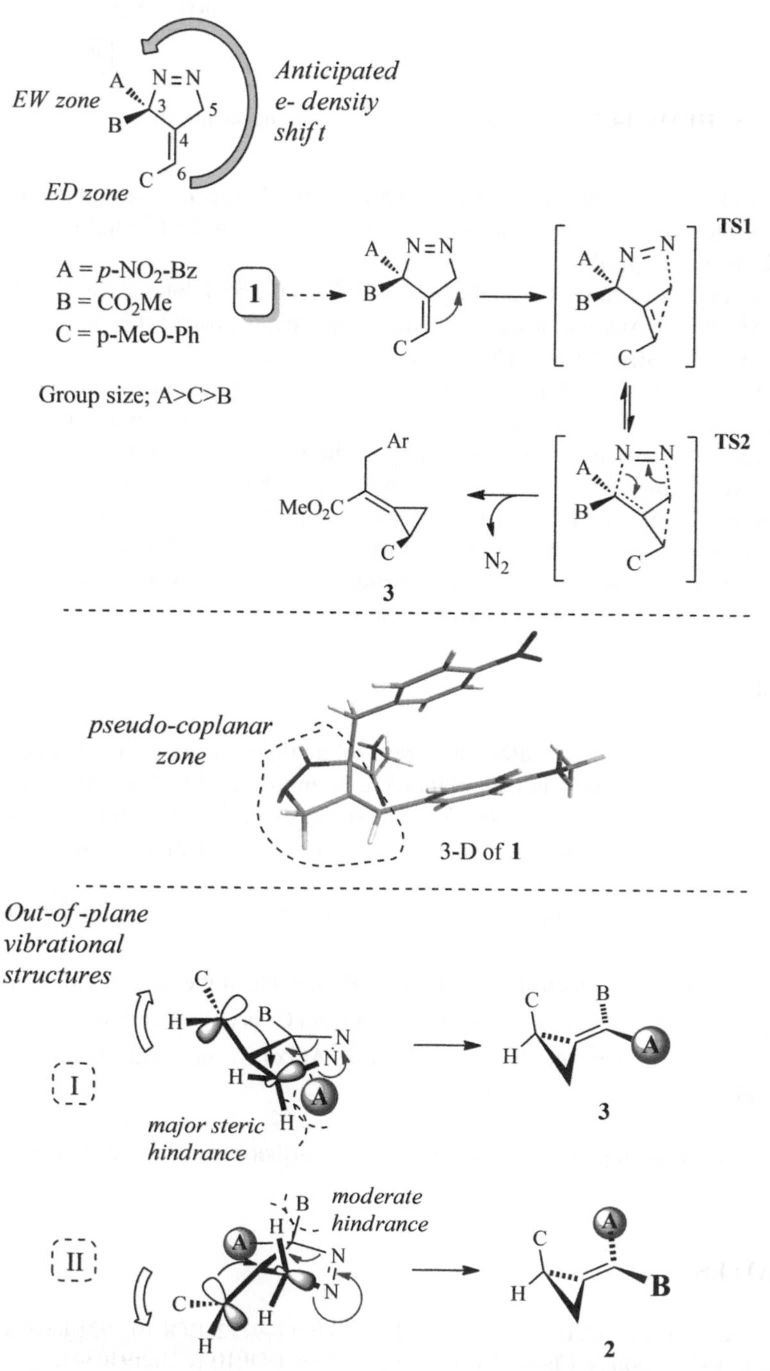

SCHEME 14.3 A schematic rendering of anticipated electron movement explaining target **3** and a more elaborate treatment of stereochemical issues to account for the prevalence of **2** over **3.** Acronyms for substituents **A, B,** and **C** are used for simplicity. The 3D model of **1** was computer energy minimized Adapted from [4].

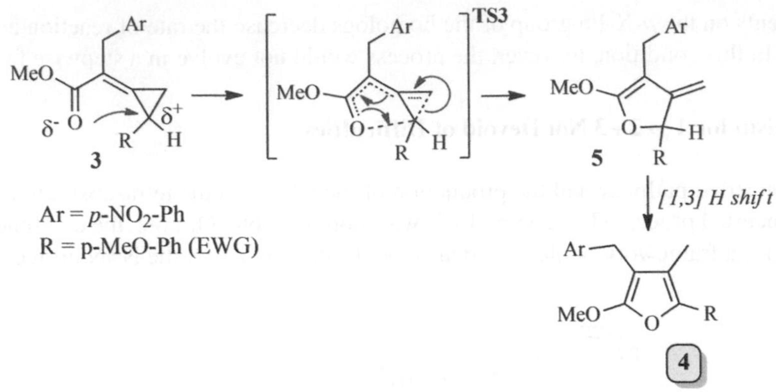

SCHEME 14.4 A sigmatropic concerted rearrangement explains target **4**.

upper section. The proposal of two transition states, **TS1** and **TS2**, is entirely speculative and depicts the idea of a concerted but not simultaneous cleavage of the two N–C bonds, and consolidation of the new C=C bond in **3**. The concerted mechanism so depicted does not explain **2**, the major product.

The construction of **TS1** requires overlap between the *p* AO of C^6 and the back lobe of the σ C–N bond in this intramolecular nucleophilic displacement. The acting AOs are nevertheless orthogonal in the ground state because of the vinylidene pyrazoline atoms lying on the same plane [3]; hence, **TS1** or **TS2** would be unfeasible.

In this book, it has often been emphasized that molecules are not rock solid objects, less so after considering their vibrational degrees of freedom; in **1** out-of-plane vibration of the vinyl tail may escape the said coplanarity to forms **I** and **II** (Scheme 14.3 bottom) and provide conditions for the required orbital overlap leading to **TS1**. Substituents on the ring adopt pseudoaxial and pseudoequatorial positions accordingly. In **I** the large substituent **A** (*p*–NO_2–Bz) undergoes 1,3 diaxial interaction with H, whereas this SS now involving CO_2CH_3 and H is much less significant in **II**. Cyclopropanation by the vinyl carbon nucleophilic attack to displace N will progress through a **TS1** of higher activation energy in **I** than **II**; therefore, **3** is the minor product despite that the steric repulsion of substituents **A** and **C** is greater than **B** and **C,** as in target **2**. These results indicate that the **1**→**2**+**3** conversion runs under kinetic rather than thermodynamic control.

14.4 Explaining Furan 4

Furan-like products are infrequent outcomes of alkylidene pyrazoline thermolysis, which suggests a secondary origin. In the overview section the disappearance of **3** and a concomitant increase in the yield of **4** as the heating period was extended from 4 to 12 hours was pointed out. This is consistent with a **3**→**4** rearrangement. Contrastingly, the yield of **2** continued to be high under these conditions. Four characteristics are associated with **3**, favorable for ring closure to furan:

- ✓ The *p*-CH_3O-Ph group on C^6 withdraws electron density from this CP carbon, favoring intramolecular nucleophilic attack by the carbonyl oxygen atom.
- ✓ The α,β-unsaturated ester facilitates electron flow during CP cleavage at the distal C–C bond.
- ✓ In the ester *syn* configuration of **3**, the distance between C=O and C^6 is only 3.16Å in the energy-minimized structure [4].
- ✓ The distal CP bond points in the opposite direction of the sp^2 AO of oxygen non-bonding electron pair, giving access to the forming O–C^6 bond.

Isomer **2** is ill suited for such rearrangement and does not react further once formed. Scheme 14.4 puts these ideas into a short logical sequence.

REFERENCES AND NOTES

1. Hamaguchi M, Nakaishi M, Nagai T, Tamura H. *J. Org. Chem.* 2003;68:9711–9722. DOI:10.1021/jo035321j.
2. For a review example, see: Brandi A, Goti A. *Chem. Rev.* 1998;98:589–636. DOI:10.1021/cr940341t.
3. Hamaguchi, et al. also demonstrated the planarity of this structure by single crystal X-ray diffraction [1].
4. Alonso-Amelot ME, this book, conformational energy minimum calculated using MMFF94G, root mean square (RMS) gradient at 0.100 and 5000 iterations.

PROBLEM 15

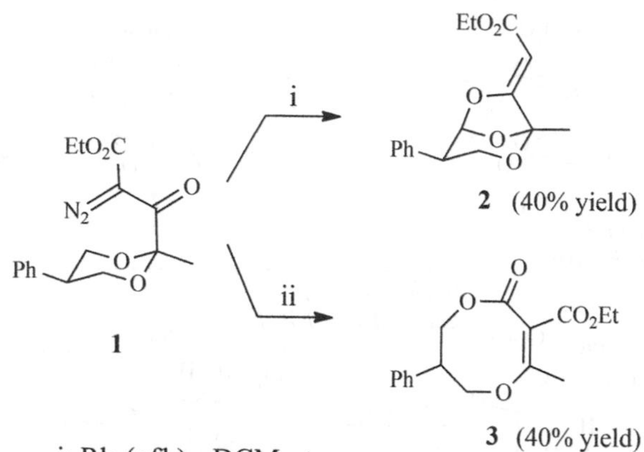

i: $Rh_2(pfb)_4$, DCM, rt.
pfb = perfluorobutyrate
ii: $Rh_2(AcO)_4$, benzene, 80 °C

SCHEME 15.1 [1] / Elsevier.

PROBLEM 15: DISCUSSION

15.1 Overview

This is one of many examples of functionally dense compounds: functional groups packed around a small molecular frame. Distances between active atoms are short, transition metal catalysis, Rh(II) in this case, increase reactivity, and centers of high and low electron density exists from the start or develop further upon catalyst influence. As it has been reviewed in Chapter 2 of this workbook, diazo compounds $C{=}N_2$ are practical sources of carbenes when exposed to Cu, Rh, Pd, Au, and other metal complexes. These highly reactive species evolve rapidly to other active species or final products through four diverging channels. Scheme 15.1, developed by scientists in Durham, UK and Benin, Nigeria [1], illustrates two of these pathways from the same starting material, contingent upon the ligands of the catalyst and temperature. It is worth noting that in either case only one product is obtained, so this is not a question of reaction cascades producing side products, or equilibrium between outcome species, but reactions going separate ways to isolated products: two problems in one set.

15.2 Solving 1 → 2

A quick comparison of dioxolanes **1** and **2** easily reveals the β-ketoester moiety in both (Scheme 15.2, top section). However, there are two less comfortable features:

1. The new C–O bond (green) seems to be the product of activation of a relatively inactive H leading to an effective O-alkylation of the enolate. This activation is momentarily obscure.
2. Considering the previous chemistry of $C{=}N_2$ under Rh(II) catalysis one would expect a major participation of the metal carbene carbon (gray ball), but all it does is to take refuge in **2** as an uneventful section of a 1,3-dicarbonyl moiety. Nonetheless, notice there is a vinyl H there, which in the absence of any other proton source (mind the aprotic medium) should have been removed from the CH_2–O in the dioxolane ring and reinstated elsewhere.

These two issues point in the direction of an initially formed metal carbene, triggering first the H removal from the pertinent CH_2–O; hence, turning this carbon into a strong electrophilic atom ready to be trapped by the enol form of the ketoester. A mechanism thus unfolds in three steps (Scheme 15.2, second part). The most relevant issues here are:

- ✓ An Rh carbene is first formed. Ligands (perfuorobutyric acid) are strong EW inducing electron density deficit at the divalent carbon.
- ✓ In the depicted syn conformation, the axial H and the carbene are six atoms away, a very suitable arrangement for interaction. Hydride transfer creates a carbonyl ylide **5**, which serves for the required O-alkylation of the enolate. The Rh catalyst is thus recovered.

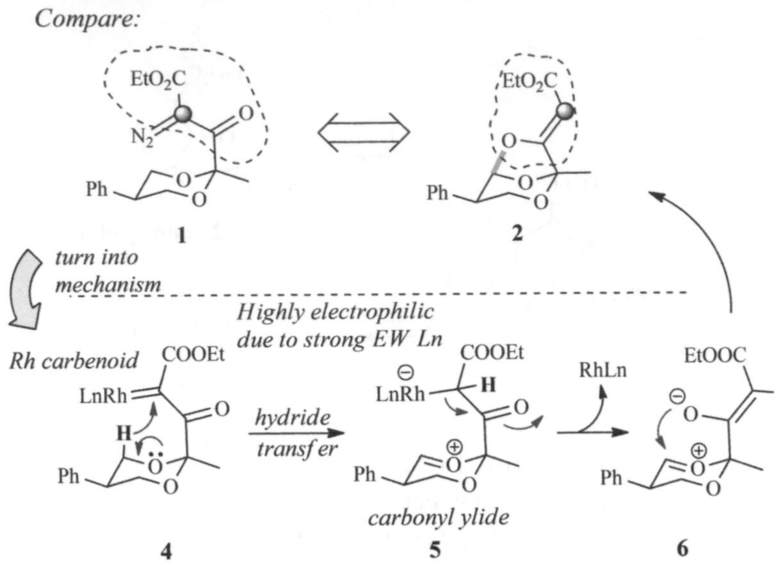

SCHEME 15.2 Comparing starting compound **1** and target **2** provides easy access to the design of this mechanism.

15.3 Solving 1→ 3

By the same comparison strategy used for **2**, two apparent incongruences now crop up.

The dotted areas identify the β-ketoester section including the methyl substituent we employ here as a flag (by assuming that it remains in this position all along). But while there are five carbons in the selected section of **2** there are only four in **3**. By tagging these carbons, one realizes that C^3 has abandoned the ketoester section leaving a C^2=C^4–O bond. This can only be explained by skeletal rearrangement. Carbenes are prone to one particular such bond redistribution, the Wolff rearrangement [2]. Electron redeployment forces the 1,2-migration of one of the substituents, the ester unit in this case, while a ketene results (as in **8** → **9**, Scheme 15.3).

The six-membered dioxane ring expands to an eight-membered cycle involving a C^3–O bond (green in Scheme 15.3, top section). The O atom performs as a nucleophile, operating on electrophilic C^3. Thus, bond disruption at two places is involved, calling for reactions in sequence (Scheme 15.3, lower section).

15.4 Learning Issues and Takeaways

✓ Reaction conditions i and ii differ in two basic features: catalyst ligands and temperature.

✓ The pfb ligands are strong electron withdrawers that affect the electron density of the metal carbenoid through the intimate bonding of carbon and metal-ligand entities. As a result, the naturally electrophilic character of the metal carbene is enhanced by these ligands and the hydride transfer is thus feasible. This is aided by the six atom H to carbene distance attainable in the syn conformation of dioxolane and Rh units. Of note, there is a higher strain energy in the syn- versus anti-rotamers due to the steric hindrance of the hefty Rh_2Ln block on the two axial C–H bonds. Yet, the sequence cannot progress toward target **2** without hydride transfer.

✓ By contrast, OAc ligands are much less EW and hydride transfer does not occur. Authors indeed report the recovery of starting compound **1** when using $Rh_2(OAc)_4$ at room temperature.

✓ When the temperature was raised to 80 °C (refluxing benzene) using the latter catalyst, diazocompound **1** was observed to decay and product **3** could be isolated. High temperature was instrumental for the Wolff rearrangement.

✓ In a parallel experiment, authors performed the reaction at 70 °C only for 15 minutes in a microwave reactor by mixing **1**, $Rh_2(pfb)_4$, and toluene instead of adding **1** dropwise. After cooling to room temperature, silica gel was added to the flask, solvent evaporated under vaccuo, and the powder material was purified by flash column chromatography to yield lactone **3**. Experimental details may be boring to some, but this is valuable information furnishing hints to other feasible mechanisms if one reads between lines. For example, silica gel may carry water, and provide a protic acid medium during solvent

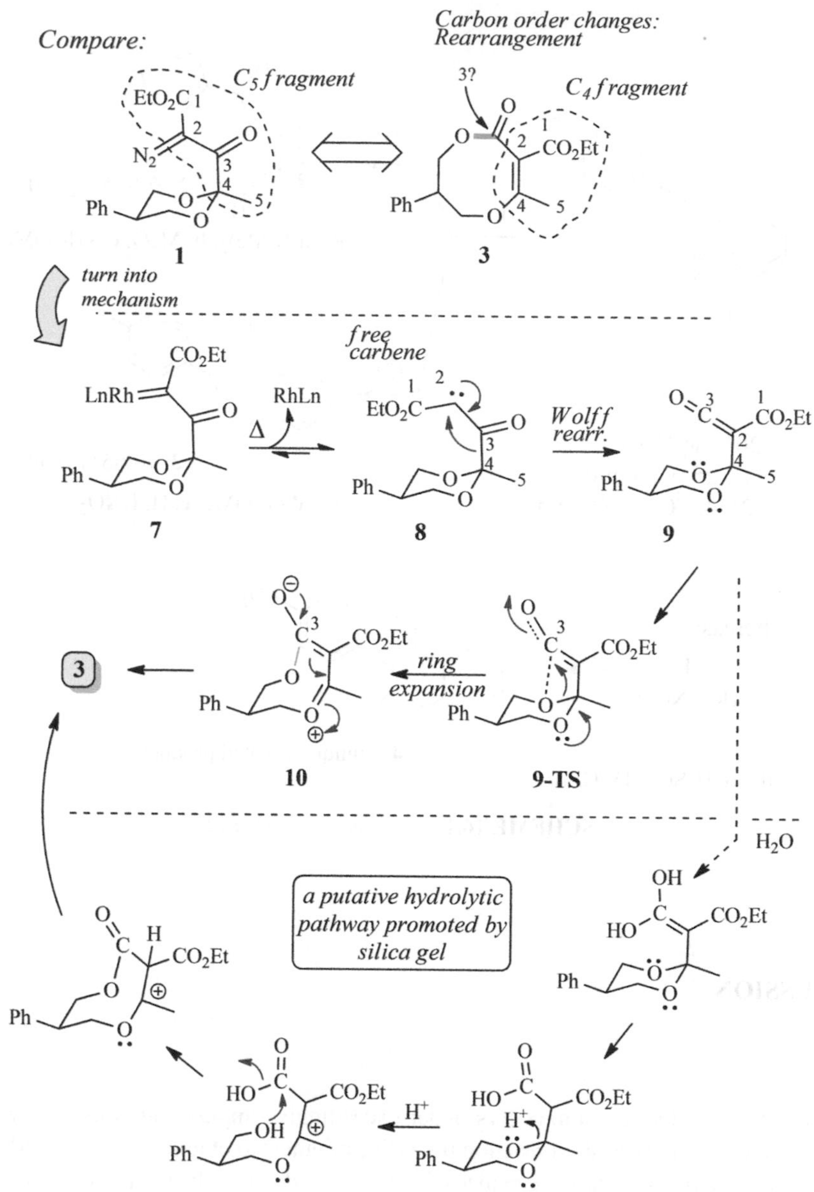

SCHEME 15.3 Two feasible mechanisms to explain the construction of target **3** from **1**.

evaporation. This brings hydrolytic conditions for an alternative pathway starting from ketene **9**, which might thus be the primary product of this reaction and the lactone an artifact (see Scheme 15.3, bottom section).

✓ Authors tested this possibility by the obvious control experiment: fishing out the product from the reaction mixture prior to adding silica gel. They found product **4**, not ketene **9**; hence, sequence **7** → **10** → target **4** gets solid support.

REFERENCES

1. Erhunmwunse MO, Steel PG. *Tetrahedron Lett.* 2009;50:3568–3570. DOI:10.1016/j.tetlet.2009.03.025.
2. For a review of the first 100 years of the Wolff rearrangement, see: Kirmse W. *Eur. J. Org. Chem.* 2002;2002:2193–2256. DOI:10.1002/1099-0690(200207)2002:14<2193::AID-EJOC2193>3.0.CO;2-D.

PROBLEM 16

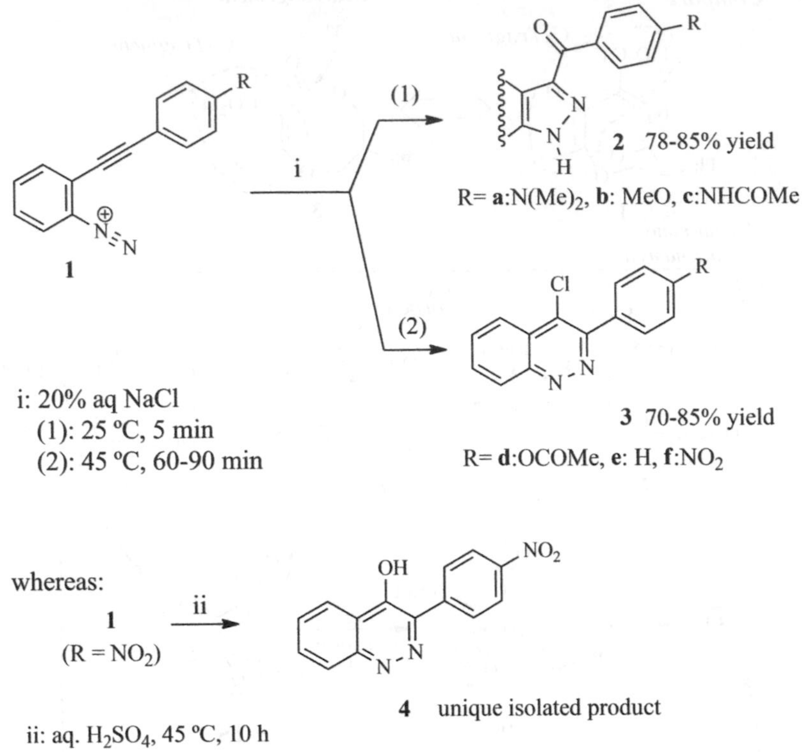

SCHEME 16.1 Adapted from [1] & [2].

PROBLEM 16: DISCUSSION

16.1 Overview

This conversion, known as the Richter reaction, occurs under exceedingly simple conditions: salt water, room temperature, no sophisticated catalysts or reagents, and very short reaction time (five minutes) to obtain **2** in high yield. This calls for an uncomplicated mechanism–to be discussed below–until one realizes that rising moderately the temperature and extending the reaction time furnishes a structurally different product line (**3**). In addition, product composition was strongly contingent on the R substituent. Electron donors gave benzophenone-type products **2a–c**, while moderate electron acceptors or H yielded chloro-derivatives **3d–f**.

Scientists at Novosibirsk, Russia who improved this well-known process [3] thought there was much more than meets the eye in these transformations and decided to undertake a deeper mechanistic study using strategies at their reach including Mulliken partial charges (MOPAC) quantum mechanical calculations [1,2]. So, your assignment is not just deploying electrons according to the identification of HEDZ and LEDZ sections, to build and break bonds but render a reasonable explanation of such changes in chemical performance.

16.2 Going After the Bottom Line: 1 → 2

Comparison of **1** and **2** (Scheme 16.2, top section) enables the quick identification of the alkynyl carbons in the target and the expected direction of electron redeployment. All one needs is a nucleophile to spark this and either chloride anion or water can fulfill this role. This Nu would attack the C^2 alkynyl carbon by virtue of the π-conjugation with the diazonium salt. The geometrically difficult interaction of two linear functions in **4** → **5** may be bypassed by vinylic anion **6**. The rest of the sequence is standard chemistry (Scheme 16.2, middle section).

In a separate experiment, authors [1] discovered a somewhat diverse course of the one-pot diazotization-cyclization in a differently decorated compound **7**. *Gem*-dichloride **8** was obtained as the only product in high yield after just two minutes (Scheme

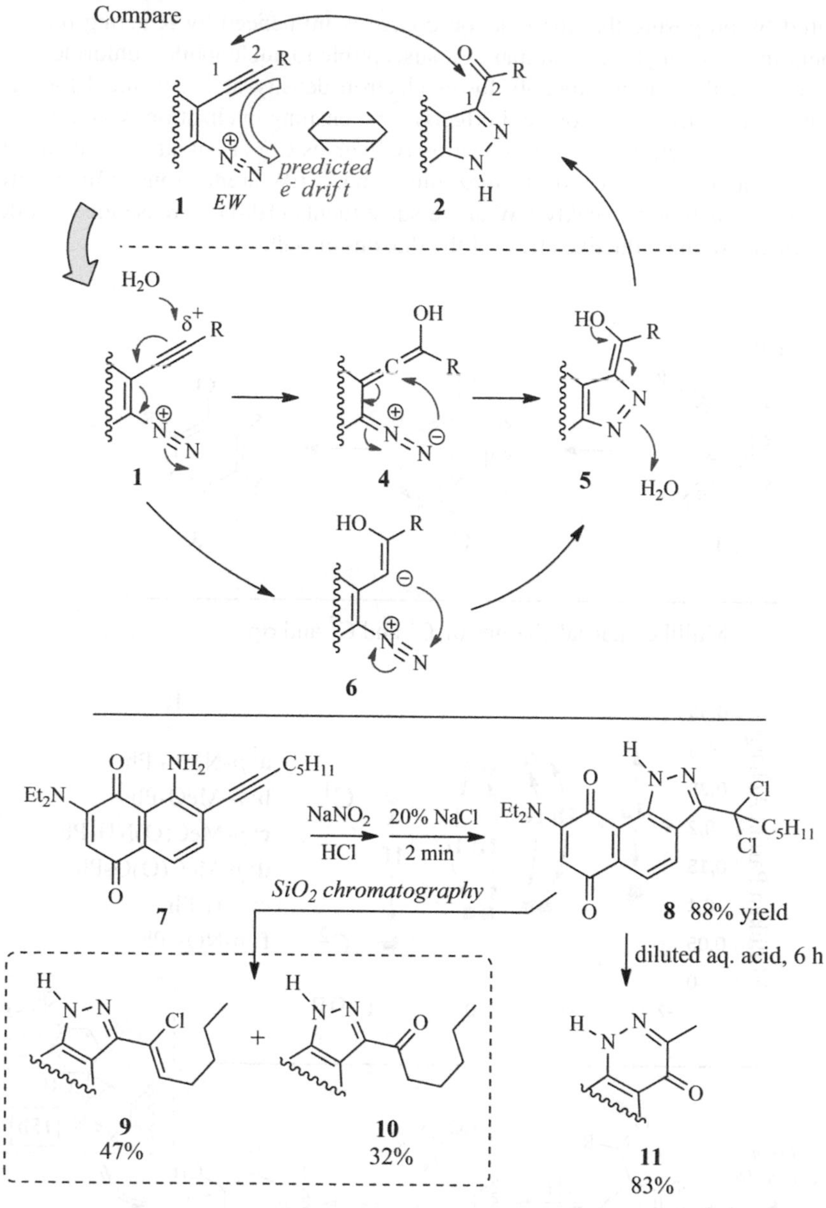

SCHEME 16.2 Disparate behavior of similar diazo-arene alkynes discovered by the authors [1].

16.2, lower section). The question is why such a compound came about from substrate **7** and not from **1**. Try a feasible mechanism and then check Scheme 16.4 at the end of this discussion.

On the other hand, **7** proved unstable as it was converted to **9** and **10** during silica gel chromatographic purification. β-elimination of HCl (hydrochloric acid) and hydrolysis via Cl–$C^{(+)}$ aided by the large excess of water justifies this. By contrast, applying the rapid dilution with 20% sodium chloride to **7** as above and stirring it for six hours at room temperature afforded the six-membered keto pyridazine **11** instead. This event takes us closer to the mechanism furnishing **3d–f**.

16.3 Explaining 1 → 3 and 4

Having a reasonable pathway for **2** there are two crucial questions as regards to targets **3** and **4:**

1. While compound **2** is obtained only when substituents R are +M electron donors (**1a–c**), **3** proceeds from milder electron donors or attractors (**1d–f**).
2. From the comparison of **1** and **3**, one observes that $Cl^{(-)}$ attacks on C^1 of the alkyne function rather than C^2 as in the case of target **2**. Why?

The first doubt is cleared by proposing that the reaction course is influenced by R acting on C^2, which suggests partial positive charge development in the alkynyl area that makes it susceptible to nucleophiles, chloride, or water.

About the second uncertainty, there is presumably lower electron density at C^1, in the **1d–f** cases, stemming from the proximity to the diazonium cation (stronger –*I* or field effects). The ensuing cyclization would be much like to **1**→**6**→**5** in Scheme 16.2 (see Scheme 16.3, top section). The site selection (C^1 versus C^2) of Nu attack boils down to two EW competing influences: the *p*-x-arene on C^2 and the –*I* effect of the diazonium on C^1. EW arene groups (**1a–c**) force the regioselective Nu attack on C^2, whereas electron donating or weakly EW arene substituents (**1d–f**) increase electron density in C^2 and the Nu addition on C^1 is preferred owing to the inductive effect of the diazonium salt.

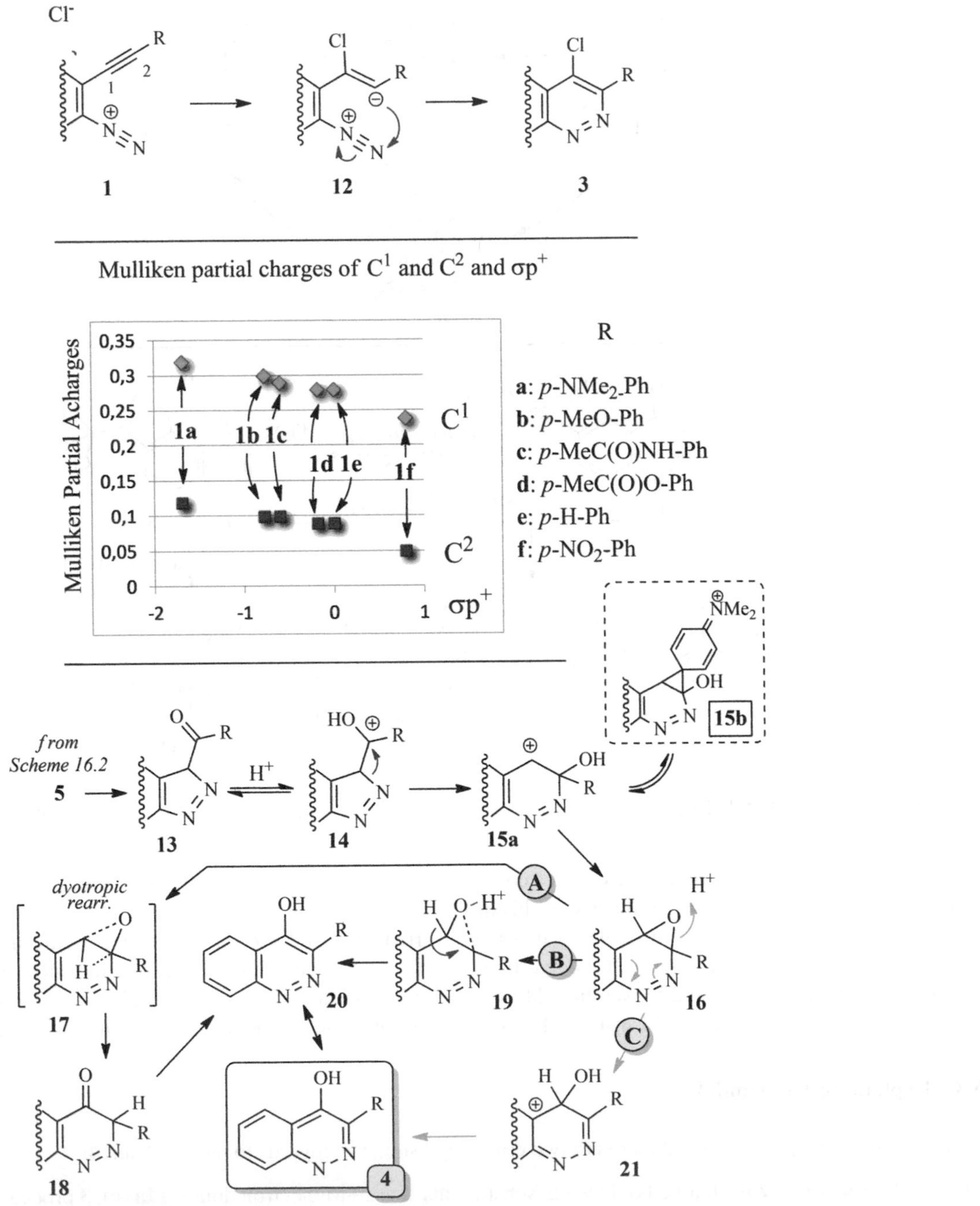

SCHEME 16.3 Mechanisms explaining the **1**→**4** conversion. The graph depicting Mulliken partial charges of atoms C^1 and C^2 of the alkynyl moiety and σp^+ of the arene substituents was drawn [4] based on data of [2] / Elsevier.

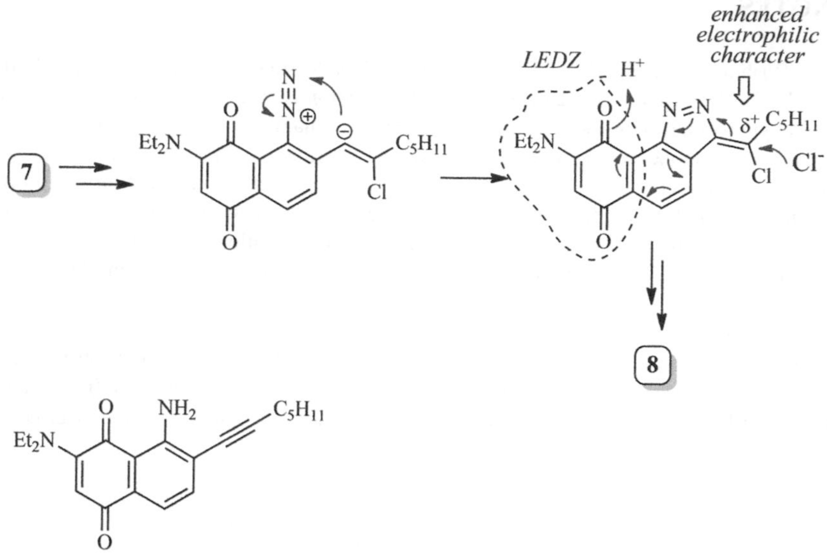

SCHEME 16.4 A feasible sequence for **7** furnishing *gem*-dichloride **8.** The quinone moiety is not only an HEDZ due to the two C=O dipoles but also their π-conjugation with the distant vinyl chloride carbon, which is expected to decrease electron density at this point.

This line of thought seems to have solved the problem. However, Lidiya Fedenok and her coworkers were not fully satisfied and opted to explore this further by solving the C1 versus C2 selectivity by quantum MOPAC method in the alkyne carbons, for one, and studying the performance of **1a–f** in aqueous sulfuric acid at two temperatures (25 and 45 °C) to compare reaction rates. We now examine the first and more significant of these two.

Partial charges: Based on the authors' data [2], I plotted Mulliken partial charges of C^1 and C^2 against the Taft σp^+ electrophilic constants of the R substituents in **1a–f** (Scheme 16.3, middle section). In a nutshell, one concludes that:

✓ Predictably, the less ED is R, the less negative will be the partial charges in both C^1 and C^2, as the negative correlations show.
✓ Both C^1 and C^2 show net negative charges in the ground state.
✓ But C^2 is less negative than C^1, thus being more electrophilic. Chloride anion and water should prefer C^2 over C^1.

This debunks the C^1 preference **1** → **12** → **3** of Scheme 16.3 top section. Therefore, the six-membered heterocycle in **3** results from the ring expansion of an initially and rapidly formed five-membered cycle of type **2**. Extending the reaction time after this first step and warming allows the **2** → **4** conversion. Scheme 16.3 (bottom section) depicts the multi-stage sequence.

16.4 Worth Noting and Takeaways

1. Cation **14** is reinforced by the EWG (*p*–NO_2–Ph), promoting ring expansion.
2. The ensuing cation **15a** is postulated to be stabilized as cyclopropyl structure **15b** (Scheme 16.3, inset) in the case of strong ED groups as in **1a** [2].
3. Cation **15** is assumed to evolve to epoxide **16**, a novel contribution to the mechanism of the Richter reaction.
4. In aqueous protic acid this key epoxide (**16**) may follow three separate routes: **A**, **B**, and **C**; the prevalence of either one cannot be established or ruled out based on existing data. It is fair to say that authors selected route **B**, but my proposals **A** and **C** might as well justify reasonably well target **4** [5].
5. A second cation on C^2 is avoided in the subsequent epoxide ring opening by way of all three routes.
6. Qualitative assessments such as the competition between inductive and field effects of the diazonium salt and arene groups in R on the polarization of the alkyne bond may fail in predicting reaction mechanism pathways (Scheme 16.3, top section compared with middle section and text discussion). Whenever possible, it is advisable to perform accessible quantum mechanical calculations in simplified model systems when computer power is limited (a laptop with suitable software).
7. Apparently simple problems like this one may provide unexpected complications and a great deal of instructive chemistry when examined critically in detail.

REFERENCES AND NOTES

1. Fedenok LG, Barabanov II, Bashurova VS, Bogdanchikov GA. *Tetrahedron* 2004;60:2137–2145. DOI:10.1016/j.tet.2003.12.058.
2. Fedenok LG, Shvartsberg MS, Bashurova VS, Bogdanchikov GA. *Tetrahedron Lett.* 2010;51:67–69. DOI: 10.1016/j.tetlet.2009.10.078.
3. The diazotization of aminoalkynyl arenes with sodium nitrite was known for many years. The established conditions, however, required a high mineral acid concentration in acetone–water. When the mixture was warmed to induce cyclization, these conditions were too harsh, and decomposition occurred. Fedenok et. al [1] changed the method by allowing the diazotization to go briefly (one minute) at room temperature in water–acetone–HCl, and then rapidly diluted the mixture with a large amount (10 to 30-fold) of water–sodium chloride to reduce acid concentration and then applied moderate heat if necessary. The diazotization and cyclization stages were thus separated in the same flask and could be monitored by color changes of the solution and thin-layer chromatography (TLC). Cyclization occurred in just two to five minutes.
4. Plot by Alonso-Amelot ME, this book, based on data of Reference [2].
5. Fedenok et. al also calculated enthalpies of formation for the **b** and **f** series [2], and found that, on the one hand, the overall Richter reactions examined are exothermic, and ΔHs of the **f** sequence are 40–70 $kcal\,mol^{-1}$ higher than the **b** equivalent series along the reaction coordinate of route **B** of Scheme 16.3. For the sake of space saving, readers are welcome to visit the original work [2] for details.

PROBLEM 17

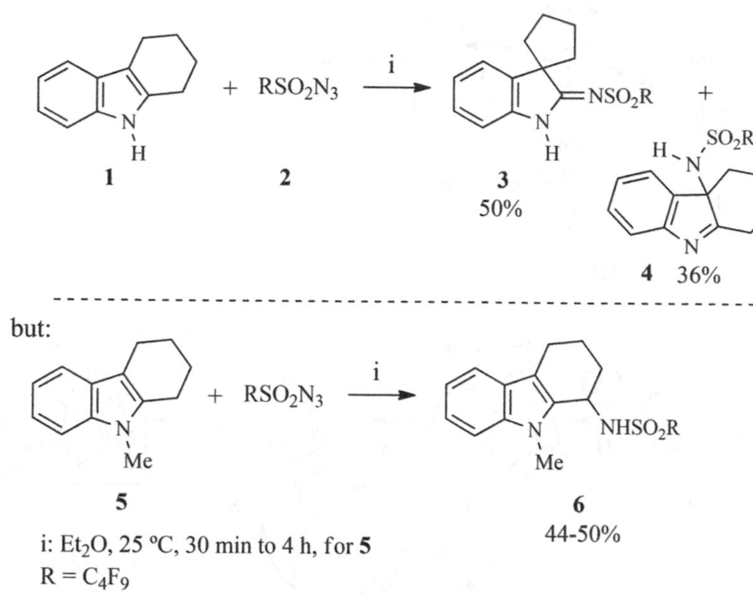

SCHEME 17.1 [1] / Elsevier.

PROBLEM 17: DISCUSSION

17.1 Overview

The carbazole scaffold, shown in **1** in its tetrahydro form, is the nucleus of a family of plant alkaloids with numerous medical applications, including cancer. Synthetic chemists work hard to craft the widest possible gamut of functional groups to enhance this bioactivity, as researchers He and Zhu from Shanghai and Guangzhou in China [1] did by taking advantage of the enamine properties of carbazole. The outcome of their efforts was unexpected, though, adding much interest to carbazole chemistry from the mechanistic as well as synthetic stands.

For this discussion it is worth noting that:

✓ Authors tested six azide models and observed that the longer the R chain, the longer the time needed for reaction completion. In the **1** model, the $R=C_4F_9$ case required the shortest time to reach completion (30 minutes), while $R=Cl–C_8F_{16}–O–C_2F_4$ required four hours.

✓ For *N*-methyl carbazole under the same conditions, the reaction was complete in less than 10 minutes.

✓ Tosyl azide resisted carbazole reaction after three days, even though tosyl azide forms adducts with indole in DMSO at 50 °C and 12 hours under air exposure [2].

17.2 A First Approach to Mechanism

Given the variety of outcomes, let us first develop the apparently more accessible $\mathbf{1}+\mathbf{2}\rightarrow\mathbf{4}$ conversion.

Comparison of **1** and **4** clearly indicates two elements of judgment: loss of N_2 and operation of carbazole as a traditional C^3 nucleophile on the remaining N atom in the sulfonamide section. Sulfonium azides resonance structures unveil versatile reactivity possibilities (Scheme 17.2, top section). Nitrene **IV** will be put aside and select the N-electrophilic electron distribution **II** as a first approximation to **4**. Scheme 17.2, option **A**). This elementary sequence is compatible with: i) The growing steric effect of substituents R in retarding the bimolecular $\mathbf{1}+\mathbf{II}\rightarrow\mathbf{7}$ step, which may well be rate limiting; and ii) the strong EW effect of fluoride substituents in R increased reactivity of canonical form **II**. However, it assumes a hardly stabilized dipole **7** when negative and positive ends could well be linked as in **9** and create a bridge to eliminate molecular nitrogen in the way to target **4**.

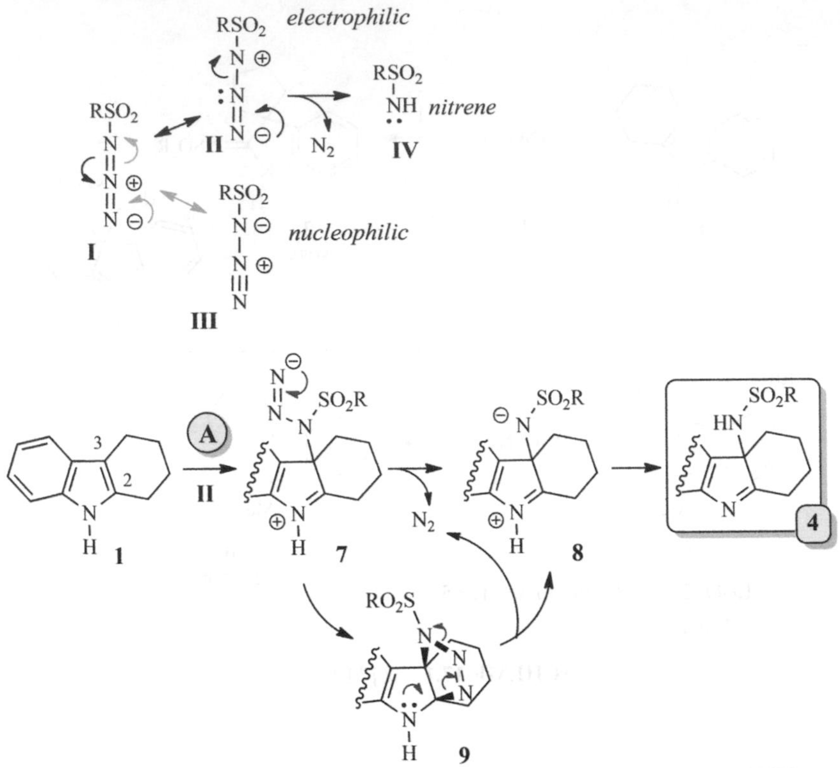

SCHEME 17.2 Top section: multiple electronic expression of the diazo function. Lower section: a first mechanism to explain target **4**.

17.3 A Second Approach in View of This Mishap

In spite of the correct usage of electronic properties of **1** and **2** in the canonical form **II** in Scheme 17.2, previous experience with indole contradicts with this design [2] (Scheme 17.3, top section). Product **12** cannot be understood without binding the diazonium end of tosyl azide to indole C^3. The scope of this reaction comprised 16 indoles with as many *N*-substituents furnishing similar adducts and yields in the 25–70% yield range [2]. Thus, the addition is so consistent that our present problem should also follow a similar course, albeit with different final products.

To this end, we now apply a more interesting model for azide **2** (canonical form **III**) for the first **1** + **2** step. The azide operates as a 1,3-dipole while carbazole **1** behaves as a dipolarophile (Scheme 17.3, lower section). However, as one follows the natural polarization of both components, the N–SO_2R group ends up on carbazole C^2 instead of C^3 in primary adduct **13**. This moiety needs to be moved to C^3 to account for target **4**. The 1,2-shift involved can be performed by a "walking aziridine rearrangement." The first step is to disengage the heterocycle into diazonium and N–SO_2R sections (**13** → **14** shown in 3D for clarity). Since diazonium salts are excellent nucleofuges, molecular nitrogen is released, which leaves a resonance-stabilized tertiary cation **15**. Interestingly, two pathway options open up from **15:** the announced aziridine route (red) to **4** via **16** and the green route involving ring contraction to target **3**, a somewhat unplanned but welcome outcome.

17.4 *Explaining Product* 6

In connection with target **6**, the relevant feature is the aminosulfonate group bonded to one carbon removed from the C^2–C^3 axis on the cyclohexyl scaffold; previous activation of the allyl methylene carbon of carbazole is a requirement. This is accessible through two routes (Scheme 17.4).

a) *The dyotropic rearrangement route*: Having *N*-methyl carbazole **5** as our starting substrate, zwitterion **17** can be produced by application of Scheme 17.3. Obeying the stimulus of the strong LEDZ created in the indole cation western fraction a [1,2]-H shift can be accompanied by a concomitant 1,2 migration of the aminosulfone group via TS **18-TS**, a case of type 1 dyotropic valence isomerization we have come across several times in this workbook. Once in the angular carbon, this proton is eliminated to restitute the carbazole C^2=C^3 bond of target **6**.

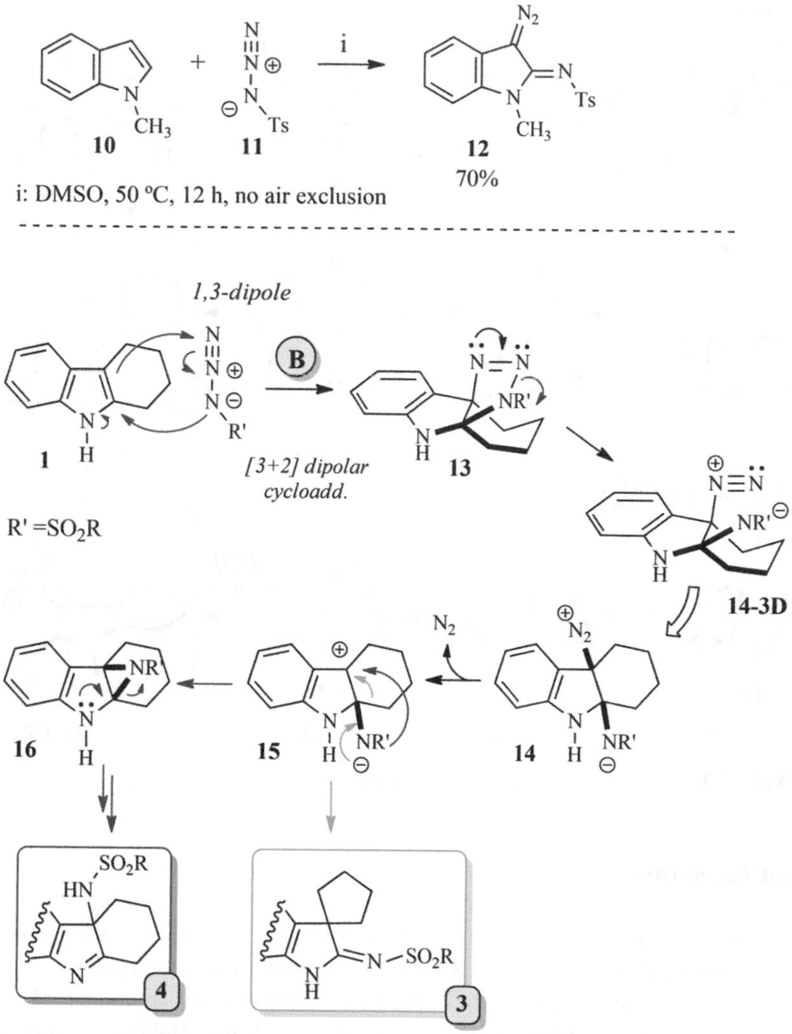

SCHEME 17.3 Upper section [2] / American Chemical Society. Lower section: application of the [3 + 2]-dipolar cycloaddition concept to carbazole, proposed earlier for indole-tosyl azide combinations [2].

b) *[1,3]-hydride shift route*: the hydride would migrate to $C^{3(+)}$ as a nucleofuge of the S_N aminosulfonate attack from the opposite side, hence forming novel and conceivably unstable amino- aziridine **21**. Ring opening and β-elimination would yield **6**. Unless the crucial hydride migration tales place in two steps (hydride release and then bonding to $C^{3(+)}$ in a second stage), [1,3]-H shifts are electronically allowed but physically prohibited. According to the Woodward–Hoffmann rules the [1,3] transition is antarafacial.

And yet, in my view both routes suffer from a stumbling drawback. The initial [3 + 2] cycloaddition that creates eventually zwitterion **17** is sterically less favorable than in the case of **1** + **2** by virtue of the steric repulsion of the bulky R group in the sulfonyl azide and the *N*-methyl of carbazole **5**. Mind that this appears as the rate-limiting step, which substantially extends the reaction time and reduces the adduct yields, *in addition to the fact that only target* **6** *was obtained from N-methyl carbazole*. Therefore, an alternative must be designed, and the authors of the original paper offer one, without comment [1]: the [1,3]-shift of the entire aminosulfone section from C^3. This move deserves a few lines:

Let us go back to intermediate **15** (Scheme 17.3) in the *N*-methyl carbazole version (**24**, Scheme 17.4). β-elimination would prepare the scene for the desired [1,3] shift in **25**. Although the concerted shift is prohibited by the impossible antarafacial course, a closer view of the involved AOs clears the way by considering that the *N*-sulfonyl anion can establish a σ–π interaction with the distal *p* orbital, following a modified version of the accepted model for the sibling [1,3]-alkyl migrations with inversion of configuration, as shown in **26-TS**. Target **6** is created without further transitions.

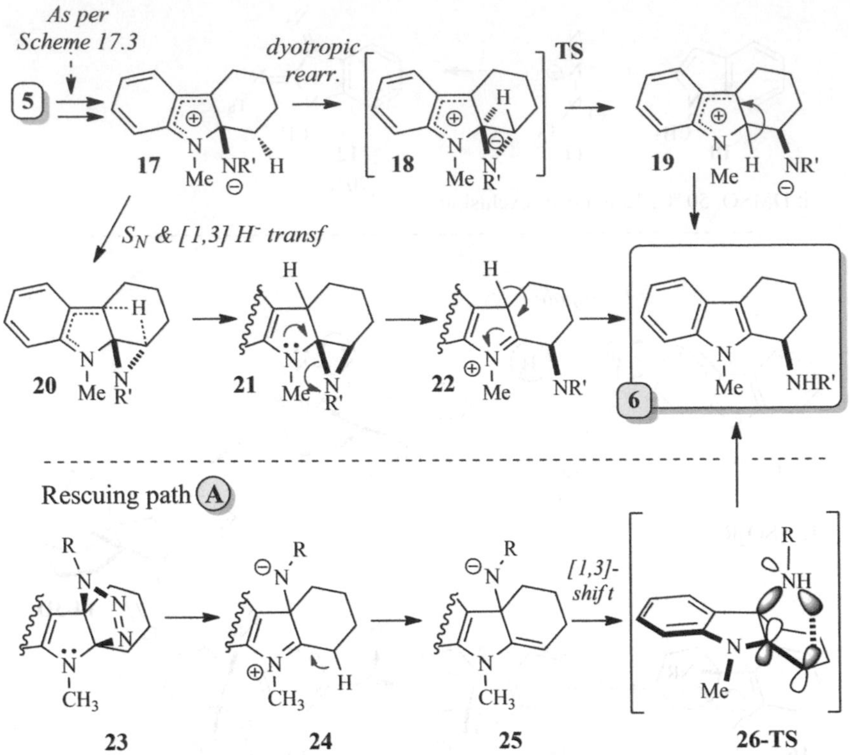

SCHEME 17.4 Target **6** can be accessed from **5** through more than one mechanism sequence.

17.5 Learning Issues and Takeaways

1. The abundance of conjugated π and NBP electrons of alkyl and sulfonyl azides offer a gamut of reactivity opportunities in the thermal and photochemical dominions. A rich literature supports this contention.
2. A conceptual issue divides approaches **A** and **B** of this problem, a stepwise versus concerted [3+2] dipolar cycloaddition. As it happens, the two routes have been proposed depending on the nucleophile, an important issue determining the regiospecific performance of the resulting triazoles (e.g. **13** in Scheme 17.3).
3. For instance, alkyl and aryl sulfonyl azides and acetylenes are both low polar compounds with no clear definition as nucleophiles or electrophiles. Their interaction is therefore limited. For this reason, metal catalysis, particularly copper(I), has customarily been employed to observe triazoline products. As metal-acetylene complexes are formed, their electrophilicity increases and interaction is possible at moderate temperature (e.g. Scheme 17.5, Section 1) [3].
4. Likewise, when alkynes are replaced with enamines that have natural nucleophilicity polarization, cycloaddition to triazoles occur albeit at high temperature (100 °C, several hours) (e.g. Scheme 17.5, section **2**) [4]. The wide scope of this investigation attests to the general behavior of this reaction.
5. Importantly, these reactions yield 1,4-substituted triazoles regularly, in contrast with the reactions of Scheme 17.1 at the start of this problem; that is, the only adduct exhibits substituents at opposite ends of the heterocycle less exposed to steric hindrance in the TS (cases **1** and **2** of Scheme 17.5).
6. On the other hand, when the alkynyl group acquires a much enhanced nucleophilicity as an anion following proton removal by *n*-BuLi, 1,5-substituted triazoles (both substituents on vicinal atoms, Scheme 17.5, section 3) are formed consistently, as shown through a number of derivatives by professor M.P. Croatt and coworkers from North Carolina, United States, who exhibit a long research record in this area [5]. Their mechanism proposal is not a concerted [3+2] aside-alkyne cycloaddition but a stepwise process in which the alkynyl carbanion attacks the RSO_2–N atom exploiting canonical form **II** (Scheme 17.2) as I first proposed therein. Others have come to similar conclusions based on theoretical calculations [6]. There are stern differences between tetrahydrocarbazol (**1**) and alkynyl carbanion moieties to be sure, but Croatt's proposal and experimental support opens mechanistic possibilities that require further studies in the quantum chemical calculations field, currently under development in Brazil and other places [6, 7].
7. In the end, what looked like an accessible mechanism turned into a set of pathways subjected to stereoelectronic effects. It is thus more difficult to select a unique route furnishing all the answers.

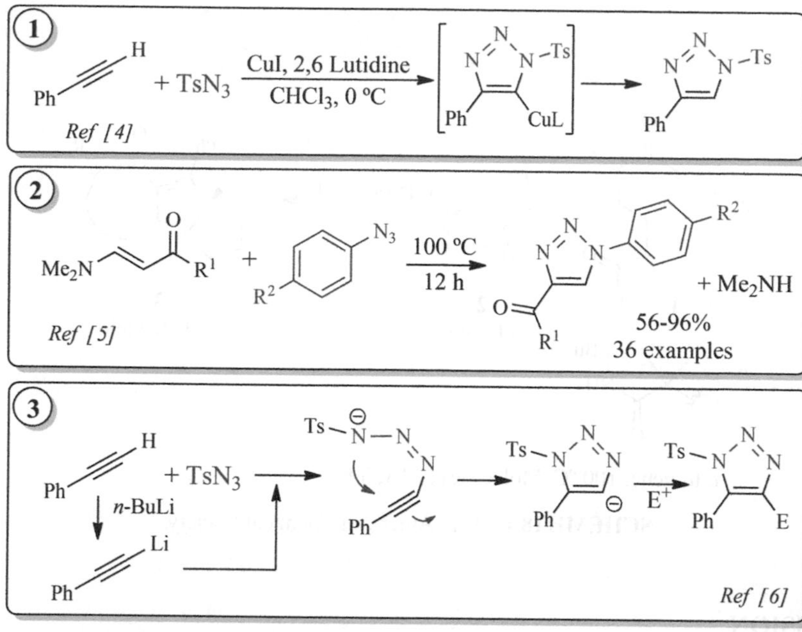

SCHEME 17.5 Selected research on cycloadditions of aryl and sulfone azides on alkene/alkyne substrates that impinge on the present mechanism discussion. Adapted from [4], [5] & [6].

REFERENCES

1. He P, Zhu S. *Tetrahedron* 2005;61:12398–12404. DOI:10.1016/j.tet.2005.08.100.
2. Sheng G, Huang K, Chi Z, Ding H, Xing Y, Lu P, Wang Y. *Org. Lett.* 2014;16:5096–5099. DOI:10.1021/ol502423k.
3. Yo EJ, Ahlquist M, Kim SH, Bae I, Fokin VV, Sharpless KB, Chang S. *Angew. Chem. Int. Ed.* 2007;46:1730–1733. DOI:10.1002/anie.200604241.
4. Thomas J, Goyvaerts V, Liekens S, Dehaen W. *Chem. Eur. J.* 2016;22:9966–9970. DOI:10.1002/chem.201601928.
5. Meza-Aviña ME, Patel KM., Lee CB, Dietz TJ, Croatt MP. *Org. Lett.* 2011;13:2984–2987. DOI:10.1021/ol200696q. License Am Chem Socc by permission.
6. López SA, Munk ME, Houk KN. *J. Org. Chem.* 2013;78:1576–1582. DOI:10.1021/jo302695n.
7. Gaspar FV, Azevedo MFMF, Carneiro LSA, Ribeiro SB, Esteves PM, Buarque CD. *Tetrahedron* 2022;in press. DOI:10.1016/j.tet.2022.132856.

PROBLEM 18

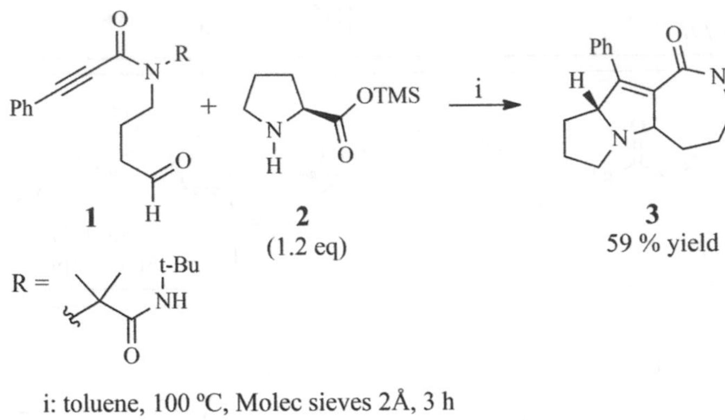

SCHEME 18.1 [2] / American Chemical Society.

PROBLEM 18: DISCUSSION

18.1 Overview

This contribution by researchers from Uttar Pradesh in India, is a development of a reaction type first reported in 1984 pursuant of the synthesis of bicyclic azaheterocycles possessing potent bioactivity. The strategy of putting together most of the structural elements, especially the ring scaffold with grafted substituents suitable for further derivatization to final products is an excellent and frequent entry into medicinal compounds. Aside from the primordial mechanism, it is interesting to note that derivatives of **1** with R = alkyl or benzyl groups do not yield the desired product, do not react under stringent conditions, or furnish complex mixtures instead. Replacing R with the tertiary amide shown in Scheme 18.1 offers conditions for conducting the process in the desired direction; an additional issue to be addressed in this problem.

18.2 A Preliminary Analysis

Target **3** visibly shows the traits of **1** and **2** as one applies the FA technique (Scheme 18.2, top section). One can start with the pyrrolidine ring at the west end, which is reminiscent of proline (**2**). The other fragments stand out easily and reveal three new bonds (in green): two C–C and one C–N linkages stemming from **1** + **2**. In addition, an orphan section remains, which calls for hydrolytic trimethylsilyl (TMS) removal and decarboxylation (but where does this water come from? Molecular sieves were added to swap up any water from the reaction medium). The bonds marked in green indicate the action scene.

18.3 A Feasible Mechanism

The disintegration of target **3** into its composing parts (Scheme 18.2, top section) suggests that the sequence begins by bonding **1** and **2**. The latter displays chiefly electrophilic carbons as a result of carbonyl-directed polarization. One therefore expects O-protected proline (**2**) to be the electron-rich zone under the command of the amide N atom. The aldehyde at the tethered butyl chain is the anticipated attack site, but in the absence of protic or Lewis acids to enhance the electron deficit of the C=O carbon, the addition of **2** would be rather sluggish. The success of this first **1** + **2** rapport is contingent on the substituent pattern of the propargyl amide section, particularly the nature of substituent N–R, since authors observed that alkyl and benzyl derivatives were unreactive or gave complex mixtures [1]. After recording the fine reactivity of **1** having an additional amide function in R, they suggested a "neighboring group participation" effect of this amide but did not offer further reasoning. We will come back to this issue in the next section.

In the absence of protic acid, the **1** + **2** addition might have ceased at the stage of amino-carbinol **4**. In the mechanism suggested by authors [1], they assume *the loss of water* to drive the formation of iminium ion **5**. In my view, the proton stoichiometry does not comply with this as one more proton is needed. They later use this mole of water for the hydrolytic cleavage of the trimethyl silyl function and allow the desired decarboxylation to account for the orphan group. Yet, molecular sieves in the

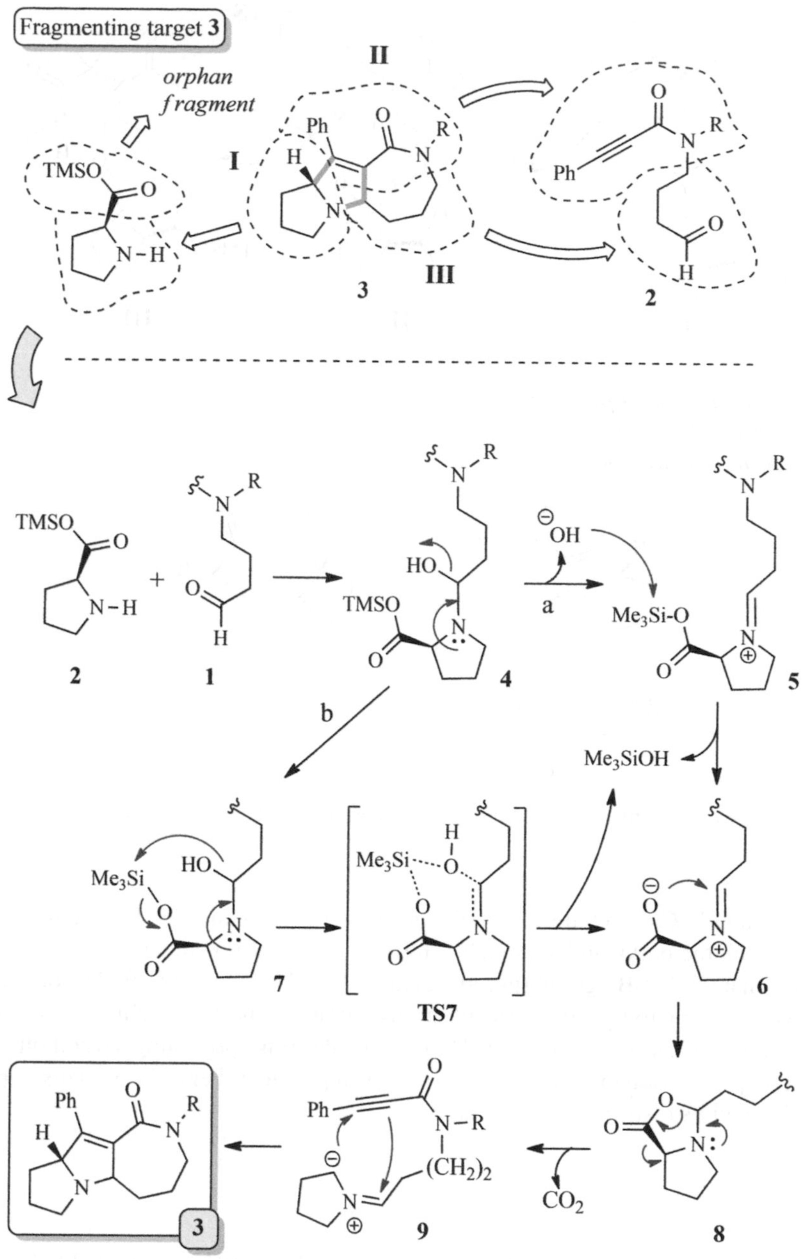

SCHEME 18.2 Fragmentation analysis of target **3** as a function of starting compounds **1** and **2,** and two feasible mechanisms leading to the desired product according to this analysis.

reaction mixture would trap this water. There is a bypass for the water usage, in fact two: i) Simply kick off the hydroxide anion aided by N's NBP (**4**→**5**) and ignore the fact that $HO^{(-)}$ is a lousy leaving group. ii) Use the departing hydroxide for the simultaneous cleavage of the trimethylsilane group located nearby (**4**→**7**→**TS7**) to **6** and steer clear of the need of water in the stoichiometric balance. The ensuing decarboxylation follows an interesting course ending at the key iminium ylide **9**. This highly reactive dipole is ideally suited and sterically located for the [3+2] dipolar cycloaddition to the acetylene group, which is the core of this synthetic strategy.

18.4 Dealing with the Neighboring Group Participation (NGP) of the Bulky Amide R

After developing the entire mechanistic sequence, one concludes that **1**+**2**→**5** (Scheme 18.2, lower section) is the most likely slow step; hence, it is subjected to hampering or enhancing effects including NGP. Upon studying various conformations that

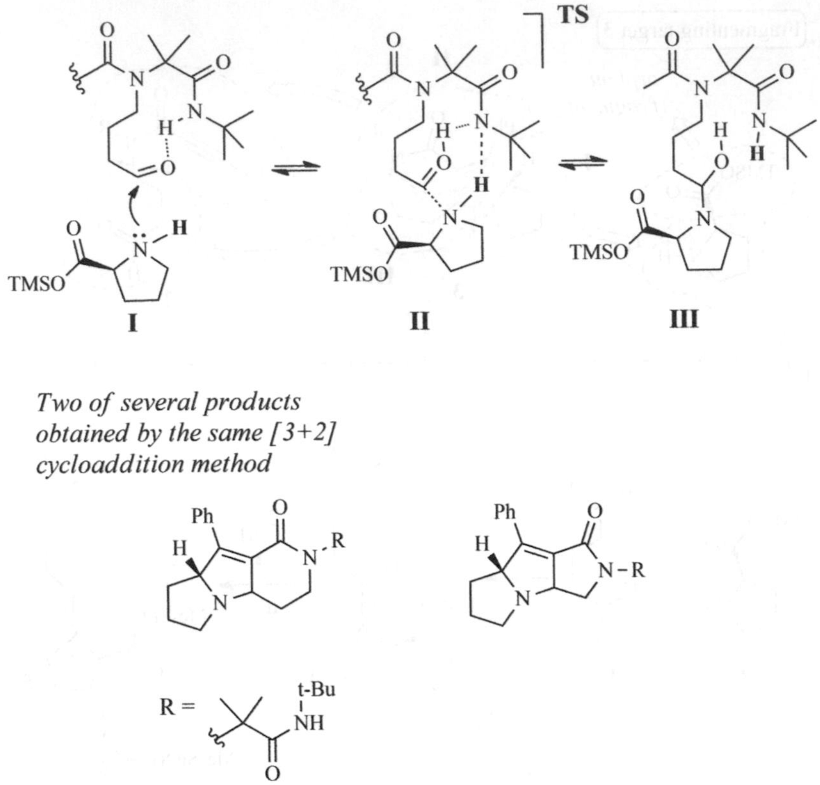

SCHEME 18.3 Explaining the neighboring group participation (NGP) of the tethered *t*-Bu amide group in the reaction mechanism.

would facilitate the expected amide-C=O addition to yield an iminium ion such as **5**, I came up with a cyclic TS (sequence **I**→**II** of Scheme 18.3), in which the bulky amide branch R forms a pseudo 10-membered ring boat. Molecular models of this temporary architecture show that the *N*-*t*-Bu group (pseudoequatorial) and the *gem*-dimethyls would be oriented away from the approaching (also bulky) *O*-trimethylsilyl proline, permitting its entrance, and thus, achieving successfully the first addition step. Similar TSs can be fashioned (please try your own TSs) for smaller rings preceding several other compounds synthesized by Professor Srivastava's group in reasonably good yields (two examples in Scheme 18.3). This hypothesis may be tested by using an *N*-deuterium proline and its distribution in the target.

REFERENCE

1. Ghoshal A, Yadav A, Srivastava AK. *J. Org. Chem.* 2020;85:14890–14904. DOI:10.1021/acs.joc.0c01539.

PROBLEM 19

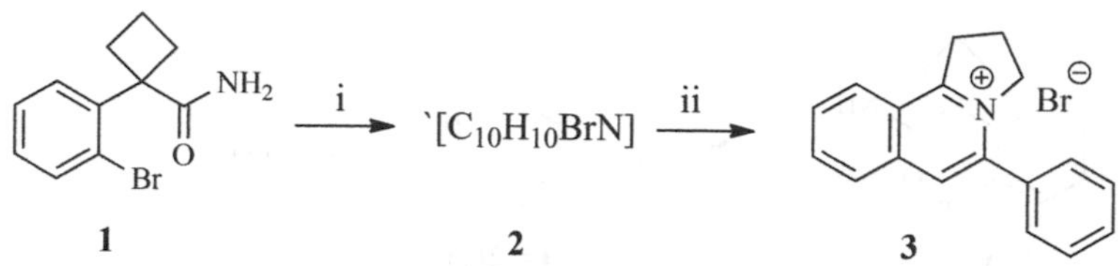

i: bis(trifluoroacetoxy)iodobenzene (BTI) 2.5 equiv. CH_3CN:H_2O 3:1, r.t., 12 h

ii: phenyl acetylene, $Ni(Ph_3P)_2Br_2$ 0.05 equiv. Zn^0 0.5 equiv, THF, 80 °C, 10 h. air and water excluded.

SCHEME 19.1 [1] / John Wiley & Sons.

PROBLEM 19: DISCUSSION

19.1 Overview

A current trend in organic synthesis is the design of cascade, domino, and multiple component reactions in one single step. When this is not possible, scientists find ways to build molecular complexity in the least number of steps pursuing higher yields, reduced time-consuming isolation procedures, and environmental pollution caused by side products. Planning requires a great deal of chemical know how, including a good handling of reaction mechanism to avoid side reactions and unexpected failures. In this problem, two reactions in a row are presented with sufficient information for you to come up with an adequate answer, in preparation for other more complex problems in more complex reaction cascades further down in this collection.

19.2 Problem Analysis: Determining the Structure of 2

A reaction mechanism is entirely speculative if the structure of the target remains hidden. So, our first task is finding ways to unveil this structure and then pass on to mechanism considerations. Only the molecular formula of **2** is provided; therefore, let us squeeze the most from it. The molecular formula differences in the **1** → **2** conversion are:

Compound	Molecular formula
1	$C_{11}H_{12}BrNO$
2	$C_{10}H_{10}BrN$
2 – 1	$-1\times C; -2\times H; -1\times O$

The best guess to account for this loss is the departure of the C=O of amide **1.** As well as:

- ✓ N remains as an amine.
- ✓ Br is bonded to the aromatic ring. There is no reason to believe the contrary.
- ✓ If no other process occurs, the structure of **2** emerges as the cyclobutyl amine **4**. This sort of reaction is known as the Hofmann rearrangement of alkyl and aryl amides to the corresponding amines (Scheme 19.2).
- ✓ However, as depicted, **4** is a $C_{10}\mathbf{H_{12}}BrN$: two extra H atoms. Hence, this structure is inaccurate. To reach the correct atomic composition we need an oxidizing agent, bis(trifluoroacetoxy)iodobenzene (BTI) in the present case, but we cannot put it to work until the structure of **2** is determined.

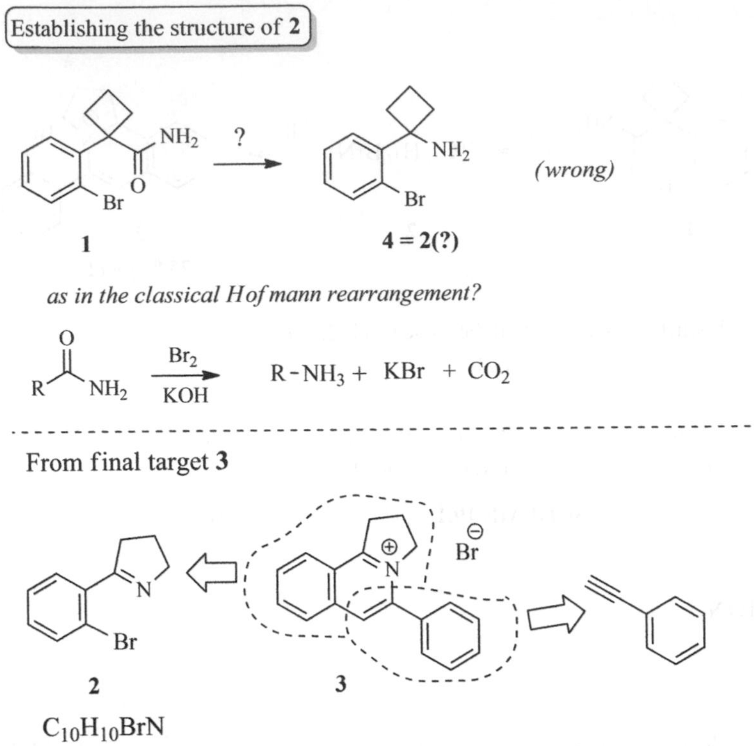

SCHEME 19.2 Does the Hofmann rearrangement of **1** explain intermediate **2**? The deconstruction of **3** provides a more feasible answer.

To this end, one can try a deconstruction of the final target **3** of Scheme 19.1 and reason it backward, as the second section of Scheme 19.2 shows. The key is identifying phenyl acetylene in **3**. Once we detach it, having no other carbon source, the rest of the core should emerge from the purported intermediate, whose structure reveals itself, with the correct $C_{10}H_{10}BrN$ atomic composition.

19.3 Solving 1 → 2

Ring expansion of the cyclobutane section and decarbonylation are the two features defining this part. As the amide N is engulfed during this expansion, decarbonylation should occur first, followed by, or simultaneously with, the apparent NH_2 migration to the benzylic carbon. Scheme 19.3 portrays the reasoning hierarchy.

All this commotion cannot take place by the release of molecular forces in compound **1** no matter how much you heat it (the experiment was performed at room temperature). The mechanism design follows these lines:

1. For the [1,2]–CH_2 shift on N to the ring expansion, one thinks in terms of a hetero- Wagner–Meerwein (W–M) transition. This in turn requires a low electron density in N, equivalent to a good electron withdrawer and leaving group substituent on N.
2. Obviously, PTI plays this role here. This trivalent iodine hyperoxide is often used in Hofmann-type rearrangements of amides in a variety of syntheses including natural product approaches [2]. For a quick and instructive look at the way this reactant works, check Scheme 19.3, middle section [3].
3. Electron density pull by both trifluoroacetate groups reduce electron density in the polarizable electron cloud of iodine, thus becoming attractive to nucleophiles, amide N in this case.
4. The S_N2 of amide **2** on iodine ensues with trifluoroacetic acid (TFA) as nucleofuge.
5. Unstable **5**, thus produced, undergoes the Hofmann rearrangement proper, which involves the redeployment of 10 electrons. Despite the $4n+2$ number of electrons involved, one cannot presume an aromatic TS, since part of the electronic commotion occurs in an exocyclic section. This step is more likely under thermodynamic control owing to steric stress release in **5** → **6**.
6. The instability of ketenimines like **6** yields carboxylates in aqueous medium (the solvent here is acetonitrile : water). If spontaneous decarboxylation occurred in carbamate **7**, the sequence would die in **8**, which is none other than our first (wrong) guess for intermediate **2**.
7. But a second mole of PTI, which was appropriately added in excess by authors, opens the gate to the desired reduction in electron density of N exposed above. Ring expansion ensues.

19.4 Solving 2→3

With previous experience in a past problem of this collection, it will be easy for you to surmise that the overall addition of phenyl acetylene to imine **2** follows traditional lines, formally a [4+2] cycloaddition. Alkynyl groups undergo substantial electrophilic enhancement by coordination with transition metal complexes, nickel(II) in this case (Scheme 19.3). This is probably a stepwise process since the imine-acetylene addition step is unrelated to the removal of bromide. This is procured once the NBP sp^3 AO of N is parallel to the aryl ring π MO as in **12** and extrusion of $Br^{(-)}$ can be achieved. The first addition is also regioselective since the two bulky phenyl groups are away in the presumed TS (Scheme 19.4).

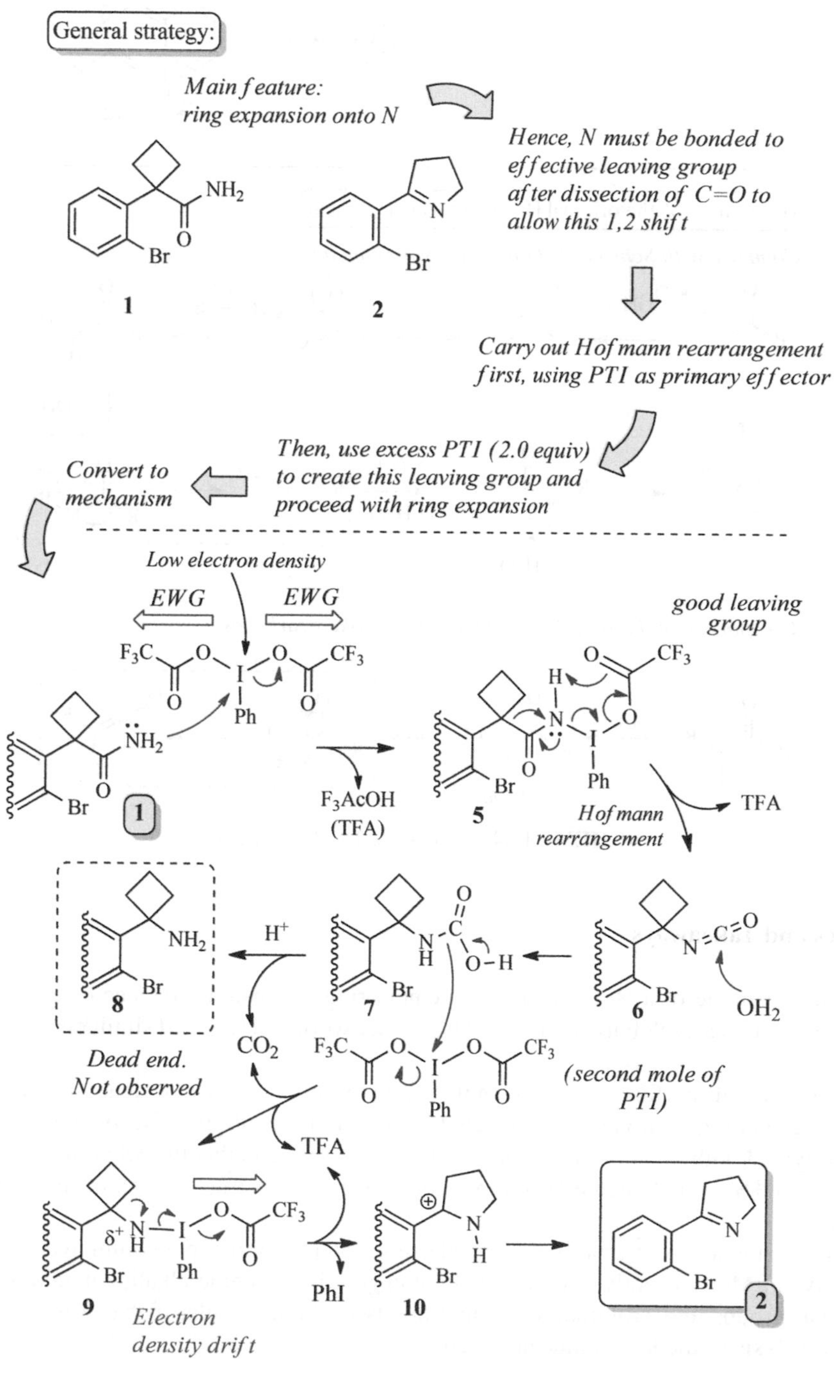

SCHEME 19.3 A feasible **1**→**2** mechanism.

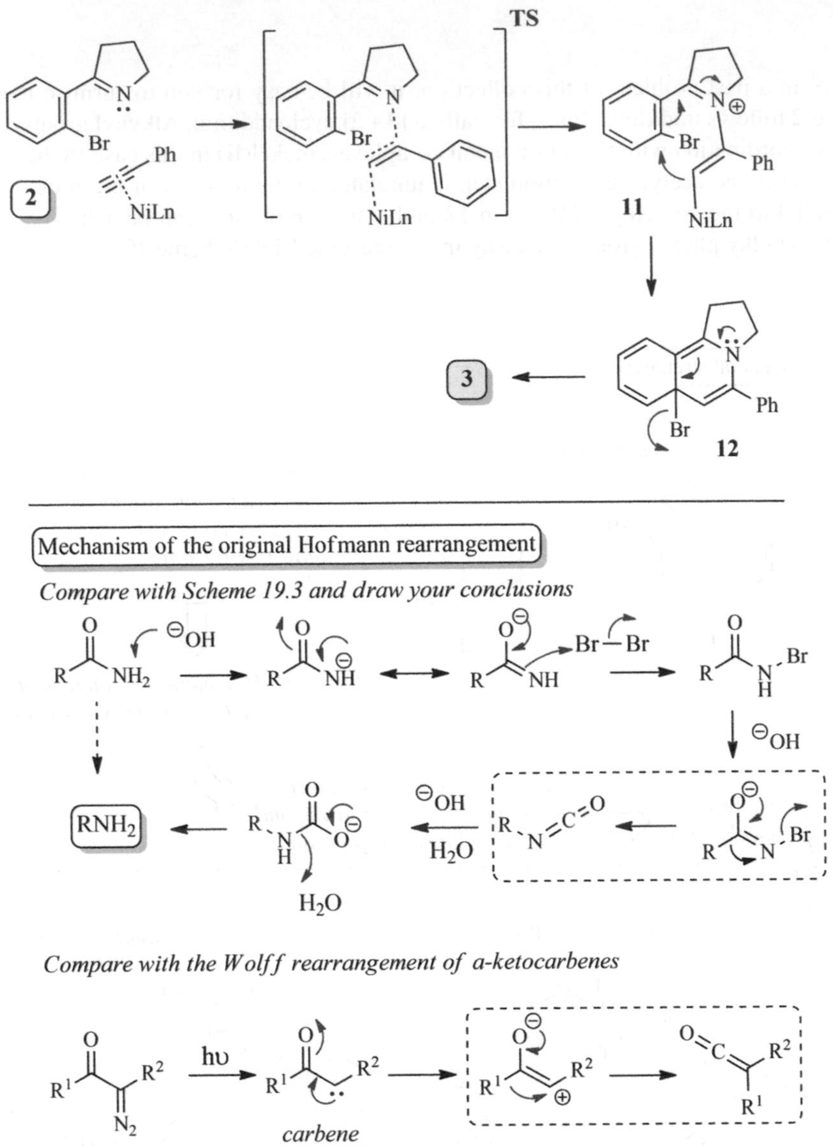

SCHEME 19.4 A feasible **2** → **3** mechanism.

19.1.5 Learning Issues and Takeaways

1. This work is the first example of a Hofmann rearrangement-ring expansion sequence or cascade (in the same reactor) observed for the first time in a cyclobutyl derivative [1]. Earlier work had reported similar ring expansions in cyclopropyl amides.
2. This is not about an odd-ball performance of a single example; the scope was widened by authors to other substituted amides and acetylenes getting high yields of adducts [1]. Among these, diphenylacetylene also gave the corresponding diphenyl adduct of type **3** with an 85% yield from **2**. This result suggests that the SS argument used to explain the regioselective addition **2** → **TS** → **11** (Scheme 19.4, top section) does not obstruct this TS or the ensuing nucleophilic attack on the aryl bromide.
3. The Hofmann rearrangement was originally developed using water, bromine, potassium hydroxide, and a suitable water-soluble organic solvent. More recently, PTI replaced the aggressive bromine-alkaline medium (that would hydrolyze a variety of functional groups and brominates multiple bonds in the molecule, if that was the case). Yet, water is still required for the hydrolysis of the ketenimine intermediate.

4. Conceptually, both mechanisms differ chiefly in the first step whereby the amide releases a proton to KOH to enable the S_N2 attack on molecular bromine, while the neutral amide can perform a similar substitution on the halide in PTI.
5. The captured halogen, be it bromide or trifluoroacetoxy-phenyl iodide, serve as efficient leaving groups for the crucial Hofmann rearrangement that furnishes the *N*-substituted ketenimine. You can appreciate the similarities by following the accepted mechanism of the alkaline Hofmann rearrangement (Scheme 19.4, lower section).
6. The Hofmann and Wolff rearrangements of amides and α-ketocarbenes, respectively, are conceptually similar (Scheme 19.4, lower section).

REFERENCES

1. Huang H, Yiang Q, Zhang Q, Wu J, Liu Y, Song C, Chang J. *Adv. Synth. Catal.* 2016;358:1130–1135. DOI:10.1002/adsc.201501071.
2. For a recent review, see: Debnath P. *Curr. Org. Chem.* 2019;23:2402–2435. DOI:10.2174/1385272823666191021115508.
3. For the oxidative rearrangements of hypervalent iodine reagents, see: Singh FV, Wirth T. *Synthesis* 2013;45:2499–2511. DOI:10.1055/s-0033-1339679.

PROBLEM 20

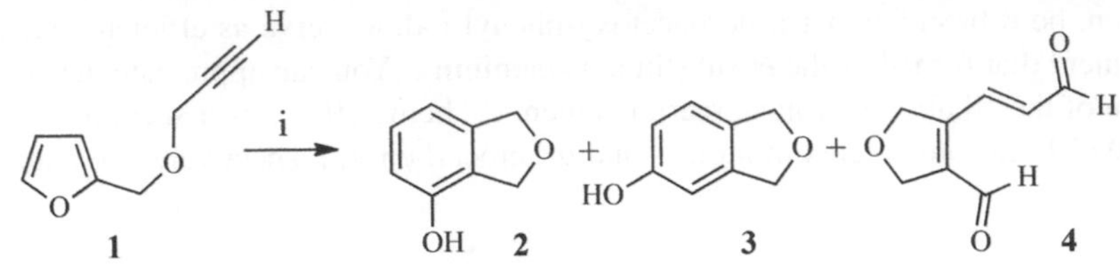

2 & **3** major products, **4**: trace

But:

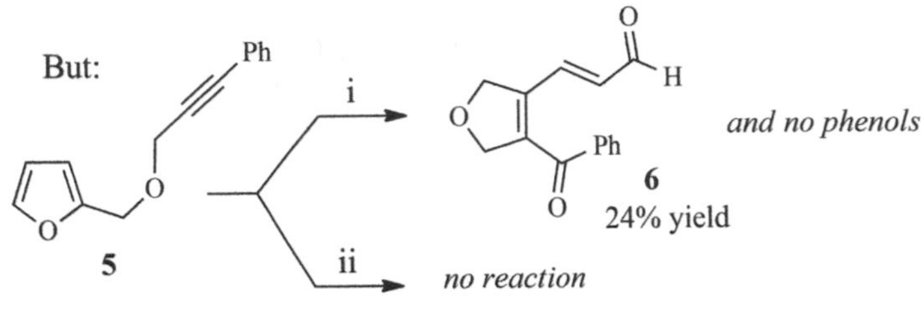

i: $PtCl_2$, acetone/water, reflux 15 h
ii: $PtCl_2$, acetone, reflux 15-17 h

SCHEME 20.1 [1] / John Wiley & Sons; and [2] / American Chemical Society.

PROBLEM 20: DISCUSSION

20.1 Overview

The complexation of alkynes with a number of mono and divalent metal catalysts is an efficient conduit for turning multiple bonds into electrophilic carbon centers, which makes them susceptible to a variety of nucleophiles. From this first encounter the primary species can evolve in several ways depending on the metallic element, the ligands of the complex, the presence of water, and the particular structure of the alkynyl organic compound [1,2]. In addition to classical homogeneous (soluble in organic solvents) metal catalysts of Pt, Ag, and Rh, gold (Au) has entered the race of product diversification and regio- and stereocontrol; intra- and intermolecular addition yielding phenols are known as well [3]. The reaction shown in Scheme 20.1 developed by a group of scientists in Madrid at the time of publication is just one of many that can be observed in the particular combination of substituted furans with an appended propargilic ether moiety [4]. A gamut of mechanism routes emerge from this apparently simple reaction, with lots of chemistry to learn.

20.2 Feasible Mechanisms: A First Appraisal and Solution in Classical Terms

A quick view of Scheme 20.1 shows eight carbons in the starting furan and all products **2–4**; therefore, no fragmentation or condensation occurs. Substituted furans are also manageable sources of unsaturated mono and dicarbonyl compounds by classical methods, aqueous acid for example, suggesting platnium–water promoted furan unraveling as an eventual source of target **4** Yet, the 3,4-substituted furan moiety of these targets arises from the tethered terminal propargyl unit by way of C–C bonding from the alkyne to the original furan.

There are two ways to conceive such bonding, depending on the orientation of the side chain of the starting furan: orientations **A** and **B**.

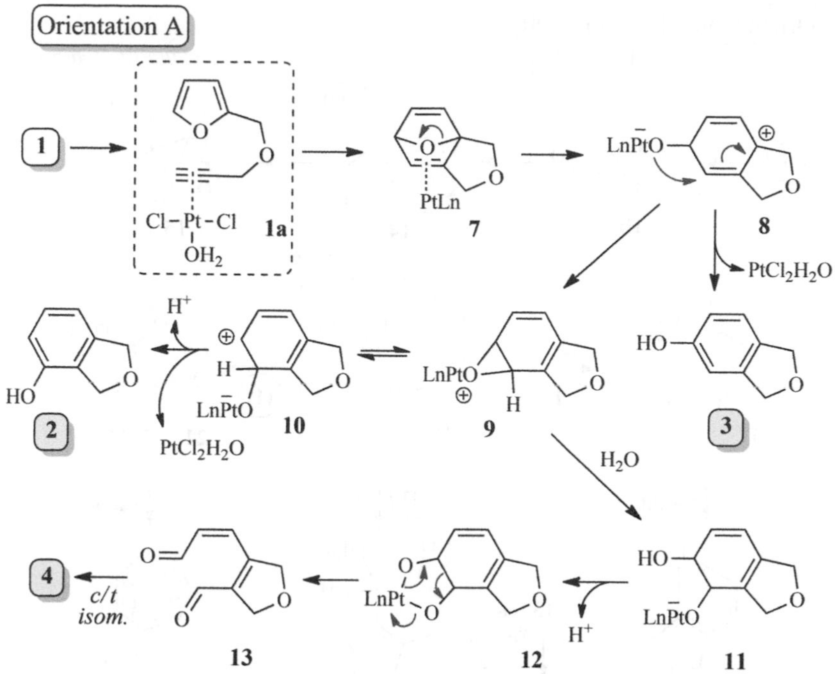

SCHEME 20.2 A first sequence that rationalizes all targets of Scheme 20.1 based on an initial Diels–Alder cycloaddition (DACA).

Orientation A: **1a** confronts electron-rich diene and electron deficient dienophile in an irresistible DACA manner, giving access to phenols **2** and **3** as well as the dialdehyde **4**, Scheme 20.2 explains the sequences. Note that we have resorted to an oxide walking mechanism **8** → **9** → **10**, a hypothesis proposed by researchers in a related context many years ago [5]. Importantly, authors indicate that when water is excluded from the reacting system (condition ii), only phenols of type **2** and **3** were observed in the several substrate-catalyst combinations they tested [1,2], thus showing that water ought to be involved in the production of **4** (mind the **9** → **11** branch of the sequence and the cleavage of the Pt bidentate complex **12**). One exception was there: phenyl acetylene derivative **5**, to be explained later. In the end, the Diels–Alder approach seems to explain the formation of all three targets.

Orientation B: In this conformer the alkyne-Pt complex is exposed unsymmetrically to the furan π MO, which operates as a conventional nucleophile on the electron deficient terminal acetylene-Pt as in **1b** (Scheme 20.3). Electron redeployment brings about a first spyrocycle cation **14** and from there a putative cyclopropyl Pt carbene **15**. The highly unstable scaffold is cleaved to aldehyde-carbene **16**, a central intermediate from which three pathways **I–III** are reasonably conceivable.

Pathway I: Proposed by authors, the [2+2] intramolecular cycloaddition was designed as a conduit to epoxide **18** in preparation for diol **19** or its Pt complex equivalent upon attack by water and a trace of acid. Phenols **2** and **3** would develop easily from this intermediate.

Pathway II: Divalent carbon metal carbenes are electron deficient and sensitive to nucleophilic attack (water in this case). This enables **20** → **19** ring closure and the phenols outcome. Alternatively, **20** can follow the route to target **4** by cleavage of the C–PtLn bond via **13** shown in Scheme 20.2.

Pathway III: Inspired by an early proposal of Professor Albert Padwa in a somewhat different framework [6] and a later hypothesis linking oxacycloheptatrienes (oxepines) as intermediates in the transition metal catalyzed synthesis of phenols [7], authors of the work presented here [1,3] offered this interesting alternative channel, adding that intermediates **22** and **23** probably coexist until water and a trace of acid converts the latter epoxide to diol **19** and desired targets thereof.

You may wish to learn that dialdehyde **13** (Scheme 20.2) and phenols **2** and **3** may also be connected by the reductive coupling–known as cross-pinacol coupling–to the glycol (**19**) in anticipation to the target aromatic derivatives (see Scheme 20.3, bottom). For this reaction to proceed, one needs an electron source to chose from, metallic elements Mg, Al, Zn, Mn, SmI_2, and others, with or without catalyst [8]. In the present case, $PtCl_2H_2O$ or any of the intermediary metallic species would have to perform as an electron donor for the corresponding transfer. However, this option was not considered by any of the papers focusing on similar reactions to that of Scheme 20.1, and logically so, as the platinum catalyst would be consumed as Pt^0.

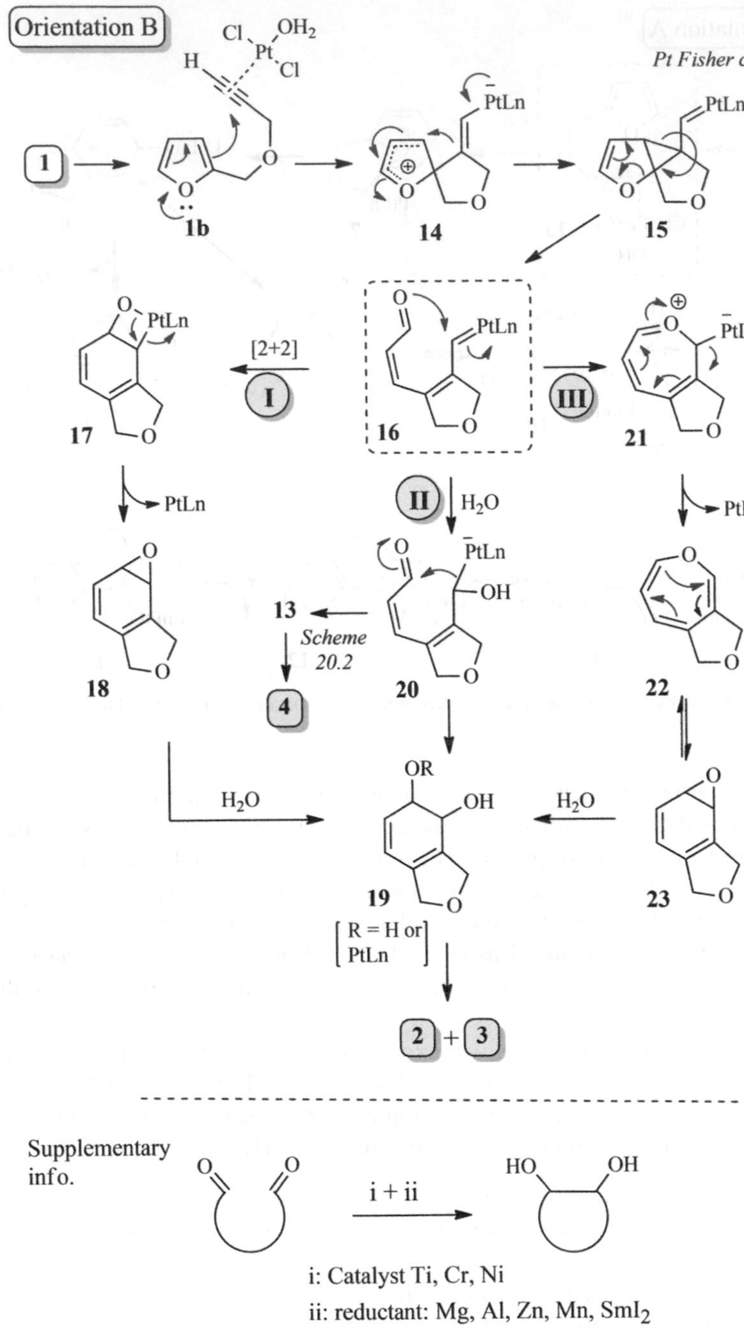

SCHEME 20.3 An alternative mechanistic tree growing from a different orientation of the tethered propargyl moiety relative to the furan ring. Adapted from [1] & [2].

20.3 Selecting the Most Feasible Mechanism

This is often the final dilemma in most reaction mechanism problems. By fortune, authors [1] carried out computer calculations (DFT) to sweep away some pathways and accept others based on activation energy differences. The condensed results were:

1. The Diels–Alder approach, Scheme 20.3, orientation **A**: In a closely related model, the activation energy of the TS between **1a** and cycloadduct **7** was calculated (DFT) at 30.2 kcal mol^{-1}, a bit too excessive. Given that this is the very first step of the entire mechanistic tree of Scheme 20.2, the whole of it must be tossed quickly into the waste basket, no matter how elegant and classic, if other less energy-costly alternatives exist. Other sources of key intermediate **9** would be preferable, as parts of Scheme 20.3 offer.

2. Scheme 20.3, orientation **B**: The furan section acts as classical furans do, affording the strained spirocyclopropyl scaffold **15**. Attack of the ring occurs from the opposite side of the Pt complex. The $\Delta G^{\ddagger}$ relative to complex **1b** of the advanced TS is only 9.4 kcal mol^{-1} (DFT). Along this first section of the pathway, other relative ΔG values were: (**15**) = −3.4 (exothermic), (**15** → **16**) (TS) = +7.6, and (**16**) = −9.8 (exothermic). One may conclude that the **1b** → **16** route is kinetically and thermodynamically favorable as far as free energy profile and surmountable stereochemical barriers. The evolution of key intermediate **16** is our next decision point since options branch out from it.
3. The [2+2] cycloaddition approach Scheme 20.3, orientation **B**, option **I**: According to a simplified model system of cyclic structure **17**, DFT afforded an activation energy of 31.4 kcal mol^{-1}, high enough to put it aside to consider other options. Besides, this imaginative alternative has no precedent in the literature (which means nothing, there is always a first case).
4. Scheme 20.3, orientation **B**, option **II**: Nothing against it, except that phenols are the only products of **1** in the absence of water. Thus, option **II** may be operative when using $PtCl_2H_2O$ complexes but not in dry $PtCl_2$. By fortune, there is a feasible way for option **II** to yield target dialdehyde **4** through the **20** → **13** → **4** side road (Scheme 20.3, center).

   Having discarded the first part of Scheme 20.2, and options **I** and **II** in anhydrous media, we are left with option **III** to product the target phenols.
5. Scheme 20.3, orientation **B**, option **III**: Several authors of closely related reactions propose the oxepin ⇆ epoxide route to explain the construction of phenols [3, 6, 7], and computer calculations disclose a low $\Delta G^{\ddagger}$ for the **16** → **21** TS (2.9 kcal mol^{-1}) [1].

We may conclude, in a very tight nutshell, that:

✓ All targets are reasonably well explained through orientation **B**, options **II** and **III**.

20.3.1 Learning Issues and Takeaways Sufficient discussion and reference materials have been brought up in the previous discussion to review these again. I assume you can do this yourself and discuss it with coworkers/classmates.

REFERENCES

1. Martín-Matute, B.; Cárdenas, D. J.; Echavarren, A. M. *Angew. Chem., Int. Ed.* 2001;40:4754–4757. DOI:10.1002/1521-3773(20011217)40:24<4754::AID-ANIE4754>3.0.CO;2-9.
2. Martín-Matute B, Nevado C, Cárdenas DJ, Echavarren, AM. *J. Am. Chem. Soc.* 2003;125:5757–5766. DOI:10.1021/ja029125p.
3. A closely related intermolecular gold catalysis example driven by a similar mechanism: Huguet, N.; Leboeuf, D.; Echavarren, A. M. *Chem.—Eur. J.* 2013;19:6581–6585. DOI:10.1002/chem.201300646.
4. For a review see: Dorel R, Echavarren AM. *J. Org. Chem.* 2015;80:7321–7332. DOI:10.1021/acs.joc.5b01106.
5. Kasperek JG, Bruice PY, Bruice TC, Yagi H, Jerina DM. *J. Am. Chem. Soc.* 1973;95:6041–6046. DOI:10.1021/ja00799a035.
6. Padwa A, Kassir JM, Xu SL. *J. Org. Chem.* 1997;62:1642–1652. DOI:10.1021/jo962271r.
7. Ohe K, Yokoi T, Miki K, Nishino F, Uemura S. *J. Am. Chem. Soc.* 2002;124:526–527. DOI:10.1021/ja017037j.
8. for a recent review of this and other methods with similar dicarbonyl couplings see: Takeda M, Mitsui A, Nagao K, Ohmiya H. *J. Am. Chem. Soc.* 2019;141:3664–3669. DOI:10.1021/jacs.8b13309.

PROBLEM 21

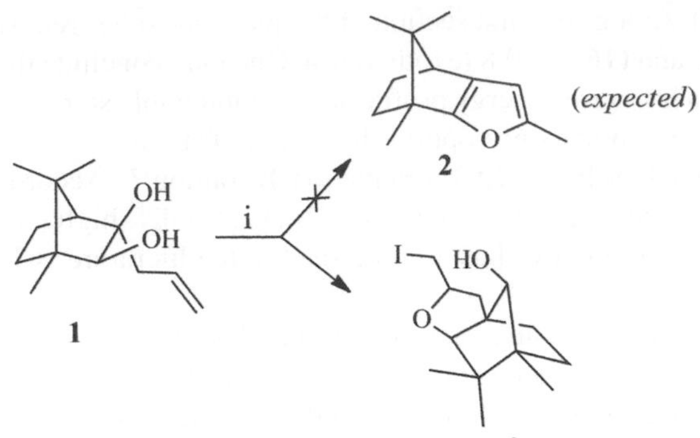

2

1

3

i: I_2 (50% molar excess),
1,4-dioxane, r.t. 24 h

88% yield

SCHEME 21.1 [1] / Royal Society of Chemistry.

PROBLEM 21: DISCUSSION

21.1 Overview

Plant polycyclic monoterpenes are a compact arrangement of 10 carbons and bonded hydrogens in often sterically stressed scaffolds. In response, a variety of skeletal rearrangements occur, driven by enzymes in the natural matrix. These give rise to a great diversity of monoterpene structures found in the essential oils. Many such rearrangements have been reproduced in the laboratory under nonenzymatic conditions, and camphor is perhaps the quintessential compound for studying such rearrangements along W–M transposition criteria. By contrast, the rearrangement of camphor quinones such as **15** (Scheme 21.3, top part) are rarely exploited as valuable tools for organic synthesis and associated mechanisms. The reaction of Scheme 21.1, developed by a research group in Puducherry, India[1] is one such case, although the outcome was rather unexpected. It is portrayed here in a peculiar manner for reasons you will understand soon, considering the workbook characteristic of this edition.

21.2 Reasons for Expecting Furan 2

Scheme 21.1 is an off shoot arising from a wider scope research [2, 3] pursuant to the synthesis of polyaromatic furans (PAFs) displaying interesting spectroscopic properties. PAF synthesis was based on the coupling of enolizable ketones and aryl 1,2 diketones under ferric chloride catalysis [2]. Among several examples of this type, Scheme 21.2 (top section) shows the smooth access to PAFs **6**, **8**, and **9** from phenanthrene quinone (**4**) and two standard ketones in satisfactory yields. Be my guest to try a feasible mechanism. Two solutions are portrayed in Scheme 21.4 at the end of this discussion.

Non-PAH 1,2 diketones also serve as suitable substrates for accessing trisubstituted furans with high yield from a different approach (Scheme 21.2, lower section). Allyl and propargyl diols were prepared from a variety of alkyl and aryl α-hydroxyketones and the corresponding propargyl or allyl halide and zinc dust or metallic indium. This procedure, known as the Barbier reaction, is reminiscent of the Grignard reaction but may be performed in the presence of water, thus the nucleophilic organometallic species is much less reactive. The addition products (e.g. **11** and **13**) are then exposed to iodine or bromine in a suitable solvent at room temperature or with heating as required, furnishing the desired trisubstituted furans (e.g. **12** and **14**).

In view of such a consistent background, while authors attempted to expand the scope of this reaction to terpenic ketones like camphor, they had every reason to expect furan **2** from allyl glycol **1** submitted to the same procedure. To their delight, however, a novel twist of the compact carbon scaffold cropped up to afford **3** in high yield; a 2-alkyl substituted furan, no doubt, but in a tetrahydro version and at the unexpected site.

---

[1]Dr Ahana Saha, the coauthor who first observed the reaction selected for this problem, passed away recently, a victim of COVID-19.

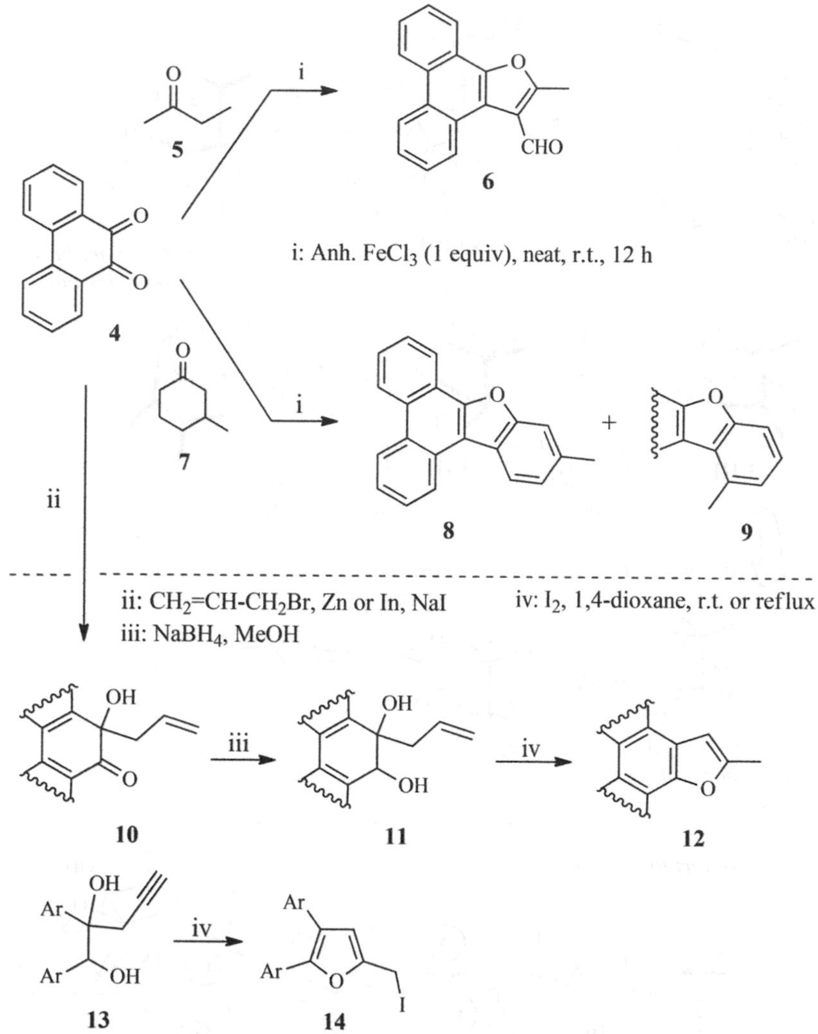

SCHEME 21.2 Selected examples of furan synthesis from phenanthrene quinone developed in previous research by Rao and coworkers [1, 2]. The propargylic glycols case (**13**) was adapted from [3] / Elsevier.

21.3 Feasible Mechanisms for 1→ 3

21.3.1 A Word or Two about Synthesis of 1 and Stereochemistry In contrast with other substrates submitted to the Barbier reaction, camphor sibling **1** is adorned with six substituents on a rigid [2.2.1] bicycloheptyl scaffold. Alkyl groups in 3-ketocamphor **15** (Scheme 21.3) define accessible and inaccessible molecular zones by opposing steric hindrance. Owing to this condition, keto camphor could be handled regio-, chemo-, and stereoselectively along the synthesis sequence to yield one single, well-defined product **1** that possesses four stereogenic centers [1]. Two of these carbon atoms are in the reactive section of the molecule. Entrance of the allyl nucleophile in the Barbier addition and C^2-ketone reduction take place from the endo side only (Scheme 21.3, top section). As a result, after rearrangement elicited by iodine, target **3** turned out in a sterically pure form with five chiral centers, a true accomplishment for a low molecular weight compound.

21.3.2 An Apparent Confusion After a cursory comparison of **1** and **3** (Scheme 21.3), one may conclude that a deep-seated skeletal rearrangement involving a true cascade has taken place. Before jumping into despair, let us remember that it is always wise to observe 3D objects from different angles for in-depth perspective [4, 5]. Scheme 21.3 depicts two rotations pursuing a view of the carbon backbone of **3** as close as possible as that of **1**, so we put into practice the least movement of involved atoms.

As in previous problem analysis, we select a framework section that is less likely to undergo cleavage in **1** (highlighted in black) and identify it in **3**. Structure **3-I** arises. Yet, it does not allow us to follow a pathway linking starting diol **1** and target **3**.

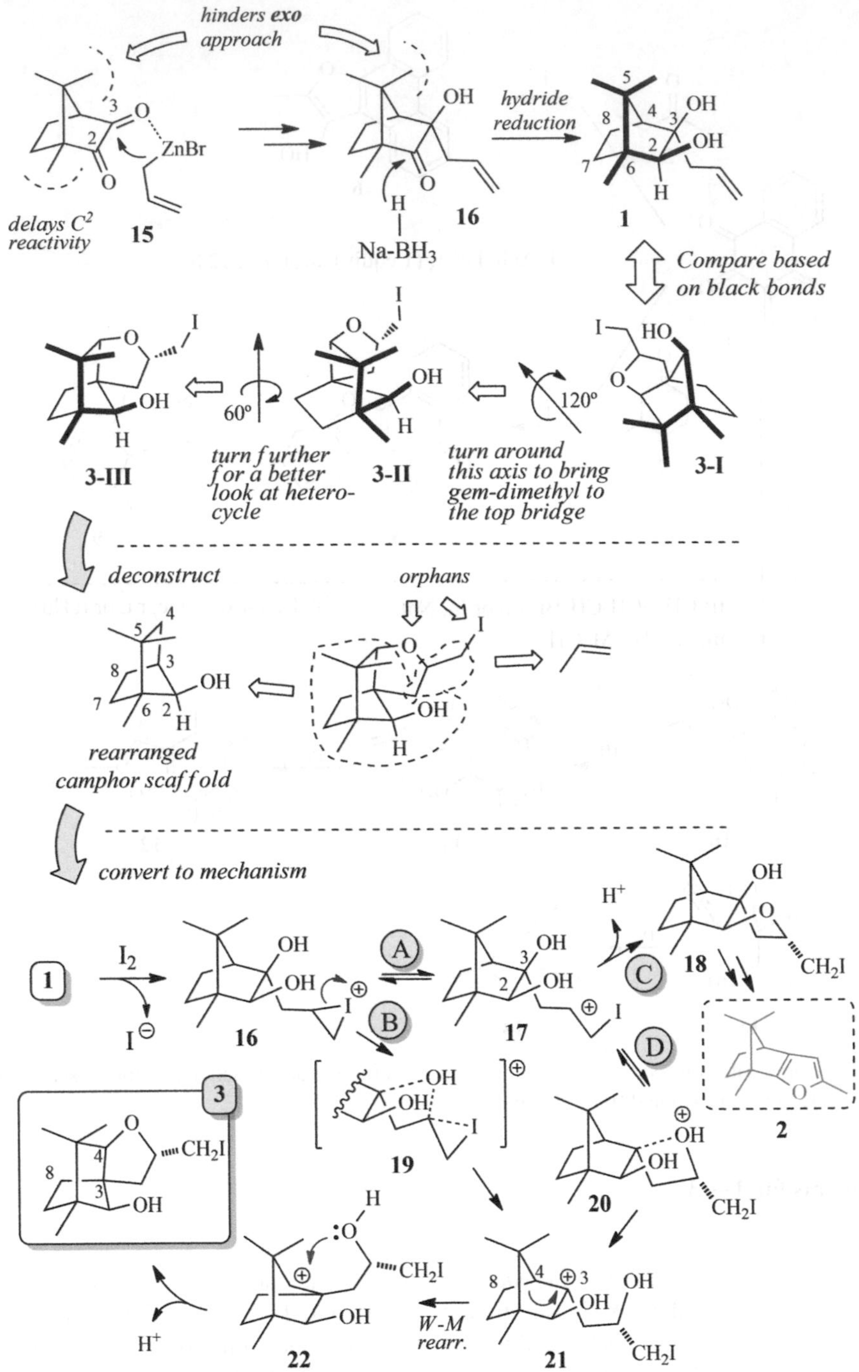

SCHEME 21.3 Fragmentation analysis of **3** as a function of **1** after re-orienting the target to observe similar structural features. Then, the reaction mechanism becomes accessible.

Next, we do some twisting (as shown in Scheme 21.3) until both highlighted sections look alike (**3-I → 3-II → 3-III**). Once there, one can proceed with fruitful comparisons and our customary *deconstruction* of the target as a function of **1**.

After recognizing two molecular sections in **3**, one realizes that the monoterpene part has undergone a modest skeletal rearrangement. Once atom labels are added following the numbering of compound **1** in Scheme 21.3, it becomes clear that:

✓ The two orphan elements have separate origins. Iodine arises obviously from the Markovnikov addition of I_2 to the terminal alkene whereas the heterocyclic oxygen is probably the result of migration from C^3–O, possibly associated with the backbone rearrangement.

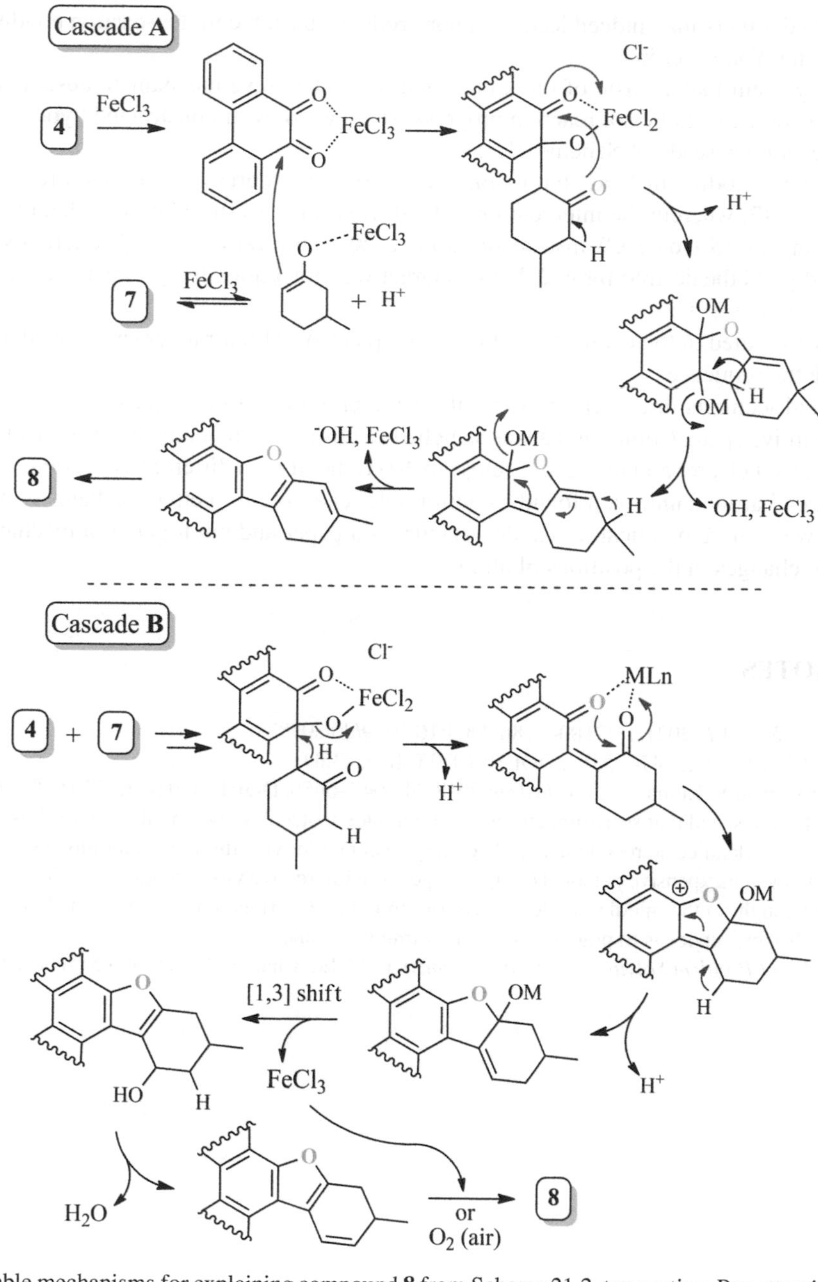

SCHEME 21.4 Reasonable mechanisms for explaining compound **8** from Scheme 21.2, top section. Pay attention to the difference between the two reaction cascades in regard to the origin of the oxygen atom in the final furan.

- ✓ The camphor rearrangement involves cleavage of C^4–C^8 and formation of C^8–C^3.
- ✓ This model calls for a W–M rearrangement, which requires a carbocation in the terpenic scaffold or one that can be transferred to it.

Scheme 21.3 (lower part) shows a feasible mechanism based on these assumptions.

21.4 Learning Issues and Takeaways

1. As you take your time wading through Scheme 21.3, it will be apparent once again that conversion pathways are rarely a straight line but rather prone to forks and unforeseen outcomes. Some of these results are mere speculative thoughts that

rarely appear published, others may indeed lead to minor products, unstable in the reaction conditions or in contact with chromatographic media. You never know.

2. Even though this may seem like a waste of time, proficiency in alternative mechanism design is exceedingly useful in preparing quantum theoretical studies of reaction progress where good, well-constructed hypotheses are necessary.
3. Noteworthy in the reaction cascade of Scheme 21.3:
   a. After complexation of iodine in C=C, the iodonium ion **16** can progress in various directions. Route **A** furnishes commanding cation **17**, wherein the intervention of C^2–OH seems less stringent than that of C^3–OH since it forms a five-membered ring in **18** (route **C**) instead of strained **20** (route **D**) or through alternative **B**. In fact, preferred pathway **C** would yield the desired furan **2**, but this target was not among the products. It is not clear (at least to me) why route **D** prevailed over **C**.
   b. Once cation **21** was seated in the camphor backbone, a typical W–M rearrangement ensued, opening C^4 for bonding in **22** and closing the sequence.
4. Seasoned students and lecturers draw certain molecular structures from a given perspective out of habit, especially in polycyclics like [i.ii.iii.iv…] etc. Compound **3** was sketched in this problem not as the original authors portrayed in their paper [1] (which was a lot clearer and instructive, close to **3-III**), but from a different viewpoint. I did this for you to work out ways to turn target **3** around until you found a comfortable rendering for better mechanism insights [4, 5].
5. In any event, it is always more practical to handle structures on paper and put together a mechanism sequence with the minimum number of changes in the positions of atoms.

REFERENCES AND NOTES

1. Rao HSP, Saha A, Vijjapu S. *RSC Adv* 2021;11:7180–7186. DOI:10.1039/d0ra09839f.
2. Rao HSP, Vijjapu S. *RSC Adv* 2014;4:25747–25758. DOI:10.1039/C4R02435D.
3. Rao HSP, Vijjapu S, Kanniyappan S, Kumari P. *Tetrahedron* 2018;74:6047–6056. DOI:10.1016/j.tet.2018.08.032.
4. In previous editions [5], I have insisted that structural attributes of complex molecules – potential reactive sites, spatial relations to predict intramolecular interactions, and detailed stereochemistry – become apparent after viewing molecular models from various angles. Several computer apps accessible to even inexpensive personal computers perform impressive visual renderings of complex compounds in motion. Yes. Yet, you may often be caught in the middle of a test, class, or job interview armed only with a pencil or a piece of chalk while your answer is eagerly awaited. So, my advice is to practice molecular turning on paper.
5. Alonso-Amelot ME. *The Art of Problem Solving in Organic Chemistry*, 2nd Ed. Chapter 3, section 3.5, Wiley, New York, 2014.

PROBLEM 22

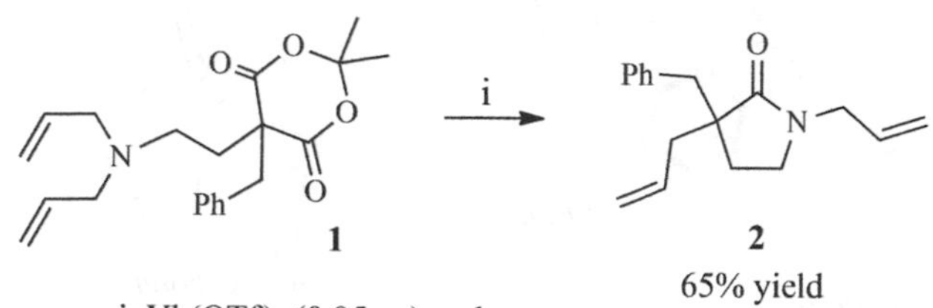

i: $Yb(OTf)_3$ (0.05 eq), xylene
150 °C, microwave

SCHEME 22.1 [1] / Elsevier.

PROBLEM 22: DISCUSSION

22.1 Overview

Gamma lactams, of which compound **2** is one of many examples, have attracted the attention of synthetic chemists for many years, motivated by the deluge of such bioactive compounds found in natural sources and the legion of derivatives from these [2]. One crucial episode in the chemical evolution of these compounds occurred in 1986, a time when pathogenic bacteria were showing resistance to penicillin, a four-membered β-lactam ring. Because the activity was in the amide rather than the strained ring, scientists of the time thought that expanding the β to γ lactam core would open a new line of antibiotic strategies against resistant bacteria [3]. They were right. Since then, the synthetic literature has exploded with imaginative methods to procure many derivatives of this scaffold. The drive persists as scientists Xu and coworkers from Changzhou, China demonstrated in their recent contribution of a domino-type reaction shown in Scheme 22.1. While the reaction course may be short, it involves a molecular commotion of sorts since sections of compound **1** disappear and others migrate. Your job is to turn this upheaval into fine chemistry.

A couple of hints:

No reaction is observed when just heat is applied (159 °C) to **1** in xylene overnight.
Adding a catalytic amount of scandium triflate [$Sc(OTf)_3$] improves the result to 50% yield after four hours in the same solvent and temperature but fails to induce reaction in polar solvents (CH_3CN, DMF, DMSO).
Some yield improvement (65%) is achieved by using ytterbium triflate [$Yb(Otf)_3$] as the catalyst in xylene at 150 °C for four hours.

22.2 Preliminary Analysis

Given that **1** → **2** is not a trivial conversion for many readers, we will resort two strategies of analysis described in "The Toolbox", Part I: an ultra-fast element balance, the identification of molecular sections of **2** that are observable in **1**, add labels to critical C and N atoms, and determine whether there are moieties of opposed polarization (HEDZ versus LEDZ).

a) Element balance: Ultra-fast means counting C, N, and O plus phenyl as it remains undisturbed all along, skipping H altogether to save time. Thus:
   1 [$C_{15}H_xNO_4$–(Ph)] → **2** [$C_{11}H_xNO$–(Ph)]:
   Difference = loss of C_4O_3, which might as well correspond to a cleavage yielding acetone (C_3O) and CO_2.
b) In this occasion, marking dotted areas (Scheme 22.2, top section) for every molecular section creates visual complications, but adding color to selected parts highlights the pieces much better. After some meditation, one can assign the green zone from the allyl amine to one C=O of the ring in **1**, which can be observed in **2** without significant change. However, one of the *N*-allyl appendages (purple) is cleaved to be relocated on the α-carbonyl carbon of **2**, likely involving rearrangement. The origin of acetone and CO_2 are easily spotted in the black section of **1** while the thick black bonds underline cleavage in coincidence with the dotted line fields.

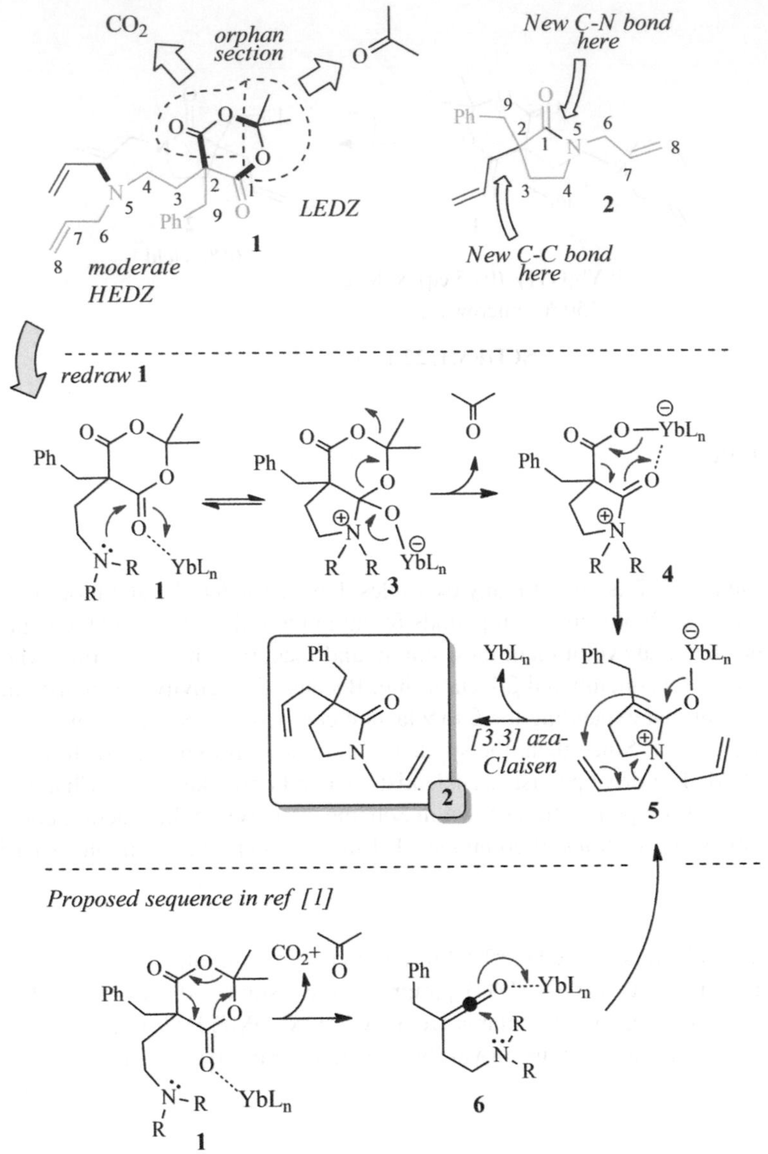

SCHEME 22.2 The proposed sequence in the lower section is that proposed in [1] / American Chemical Society as stated in Scheme 22.1.

c) Selected atom labels provide a still clearer picture: The departure of the black section in **1** must give rise to new bonds (C^2–C–allyl and C^1–N^5) to form **2**.
d) We can further observe an LEDZ in the dicarbonyl area to the east of **1**, which contains all the components of Meldrum's acid, a well recognized strong electron withdrawer, and a moderate HEDZ in the green tether expressed in the tertiary amine. These two opposed zones are poised to cyclize to afford the γ-lactone.

22.3 Feasible Mechanisms

Two reaction mechanisms can be proposed on this basis by adding ytterbium triflate as the necessary Lewis-acid catalyst for C=O activation (Scheme 22.2). Both designs must account for the restitution of the catalyst (only 0.05 eq. was used) [1]. The first of these (my own, middle section) is a prudent, stepwise lengthy sequence designed to offer a detailed account of electron redeployment. Mind the relocation of the allyl tether in **5** by way of a [3,3] aza-Claisen sigmatropic rearrangement (SR).

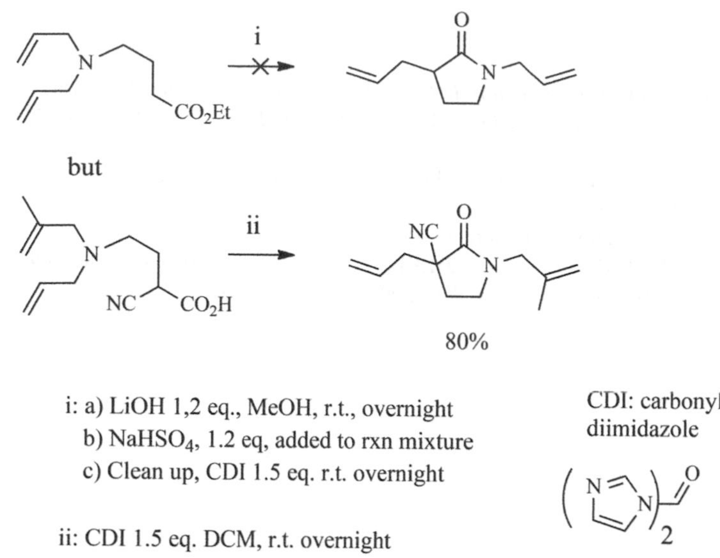

SCHEME 22.3 [4] / American Chemical Society.

Authors of this research [1] opted for a bold thermolysis of the entire Meldrum's acid section to acetone and CO_2, which yields ketene **6** (Scheme 22.2, see lower section) [3]. The Lewis acid prompts cyclization of lactone precursor **5**, which then follows the pathway exposed in the previous sequence.

22.4 Learning Issues and Takeaways

1. Authors had proposed a similar progression in *N*-allyl disubstituted amino acid esters (AAEs) in a previous work [4]. However, the simple AAEs required further EWGs to yield products. The question is exposed herein in the form of *an additional mechanism problem in* Scheme 22.3 *that you are welcome to solve* based on what you have learned in this discussion. Consideration to other issues including the curious DIC (diimidazolcarbonyl), while there was no heat applied.
2. Ketenes and ketene enolates often crop up as highly reactive LEDZ species that stem from a variety of sources (to create ketenes *in situ*) with numerous applications in synthesis [5]. The good mechanism designer should keep this synthon in mind.
3. Microwave heating (see experimental conditions, Scheme 22.1): This form of energy input constitutes a current trend in static and flow reactors, compound extractions from natural matrixes, distillations, hydrodistillations of plant essential oils, and other chemical processes. One of the main advantages over the traditional heating mantle is the much faster energy transfer from the heating source (the magnetron) to the receptor (the reaction or distillation medium) through energy-transporting microwaves than the contact heat transfer from mantle surface to the glass vessel to the reaction medium. This is particularly critical in kinetic studies. In the old days of heating mantle, sand, or silicon oil baths, one had to preheat the whole setup to the desired temperature and let it stabilize, then dip in a small quantity of solution under study while hoping for a fast heat transfer from the inner glass surface. Microwaves, however, go directly for the solvent wavelength that captures the photonic energy with nearly instant response. One may also speculate about the heterogeneous catalysis of overheated glass surfaces in heat-transfer devices or metal elements trapped as nanoparticles in seasoned or involuntarily doped reactors (which I have often observed), a problem that one can appease by pre-treating the vessel with trimethylsilyl chloride and a solvent wash. As far as mechanism musings, the vessel glass intervention cannot be overruled but is rarely mentioned in the literature.
4. On the other hand, if one is concerned about global warming, the electrical power required by the heating mantle is less than the microwave oven to reach the same temperature, as shown with plain water not long ago [6].

REFERENCES

1. Lin XW, Han M, Shen MH, Zhu CF, Xu HD. *Tetrahedron. Lett.* 2022;99:article153816. DOI:10.1016/j.tetlet.2022.153816.
2. For a recent rich review see: Caruano J, Muccioli GG, Robiette R. *Org. Biomolec. Chem.* 2016;14:10134–10156. DOI:10.1039/C6OB01349J.
3. Meldrum's acid derivatives are known to undergo decarboxylation–decarbonylation to ketenes. For a review, see: McNab H. *Chem. Soc. Rev.* 1978;7:345–358. DOI:10.1039/CS9780700345.
4. Shen MH, Han M, Xu HD. *Org. Lett.* 2016;889–891. DOI:10.1021/acs.orglett.5b02843.
5. For a hefty review, see: Allen AD, Tidwell TT. *Chem. Rev.* 2013;113:7287–7342. DOI:10.1021/cr3005263.
6. Devine WG, Leadbeater NE. *Arkivoc* 2011;5:127–143. DOI:10.3998/ark.5550190.0012.552.

PROBLEM 23

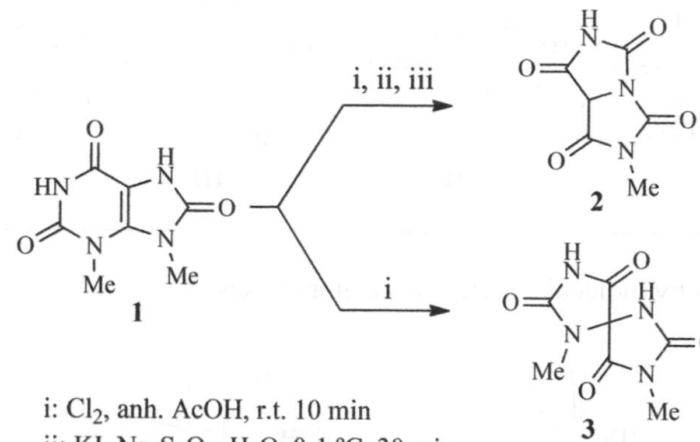

i: Cl_2, anh. AcOH, r.t. 10 min
ii: KI, $Na_2S_2O_3$, H_2O, 0-1 °C, 30 min
iii: 4N HCl, r.t.

SCHEME 23.1 These reactions and Scheme 23.3 have been adapted from [1] / American Chemical Society.

PROBLEM 23: DISCUSSION

23.1 Overview

The reaction of Scheme 23.1 is not only a mechanism problem but a challenge for researchers in purine chemistry. Purines, siblings of the oxopurine **1**, exist in large quantity in all living things, forming part of a number of biological monomers (ADP, ATP, AMP, GMP, and so forth), biopolymers (RNA, DNA), and many natural products including universally familiar caffeine [2] (see Problem 24). These compounds garner high heteroatom to carbon ratios, and are subject to a number of reactive agents such as sugars (e.g. ribose), methylation, and redox substances. These reactions modify substantially the biological function of purines; hence, knowledge of their chemistry has been crucial for fundamental aspects of cell life, disease, and survival. Aside from simple alkylations and glycosilations, the mechanism associated with such processes may be more complex than anticipated or difficult to study due to the instability of primary products (cascades often occur) that hamper their isolation or direct spectroscopic characterization, and the few carbons that witness the transformation exhibit a fine-tuned spectral response (sharp signals in ^{13}C– and ^{1}H–NMR for example).

Many of these reactions are studied in the laboratory in nonbiological conditions for a deeper insight, and yet the molecular structures products thus inferred are a question of debate. Professors Poje from Zagreb, Croatia [1] embarked on a detailed investigation of one such reaction, chlorination of oxo-purine **1** by molecular chlorine and acetic acid under anhydrous conditions. In the early literature, researchers had offered potential structures **I**, **II**, and **III** (Scheme 23.2) organized along the chlorination reaction coordinate ending in IV after reductive dechlorination [3]. Hydantoin **3** was also reported as an easily produced side product. However, Poje and Poje found these structures questionable and critically submitted **1** to scrutiny by carrying out the reaction under spectral surveillance. This is a case of the fruitful combination of the experimental capture of intermediates and linking them through a reasonable mechanistic hypothesis.

To this end, the reaction of Scheme 23.1 was performed in a 5 mm NMR tube and time-dependent monitoring of key spectral signals with an emphasis on C^5 (see numbering in Scheme 23.2) using a ^{13}C-labeled derivative of **1**. Authors concluded that **2** was the most likely product according to spectral data, as they ruled out the previous structures one by one (details in the original article). The centerpiece was the impossible reconcilement of purported structure **IV** and that of rearranged target **2**, which was verified from spectral data. Potassium iodide and sodium thiosulfate in water was used to reductively remove chloride, followed by strong mineral acid hydrolysis of an intermediate species that you are supposed to propose.

23.2 Problem Analysis, 1 → 2

Comparison of **1** and **2** using atom labels, abridged molecular formulas, and identifiable fragments provide the following keys (Scheme 23.2, lower section).

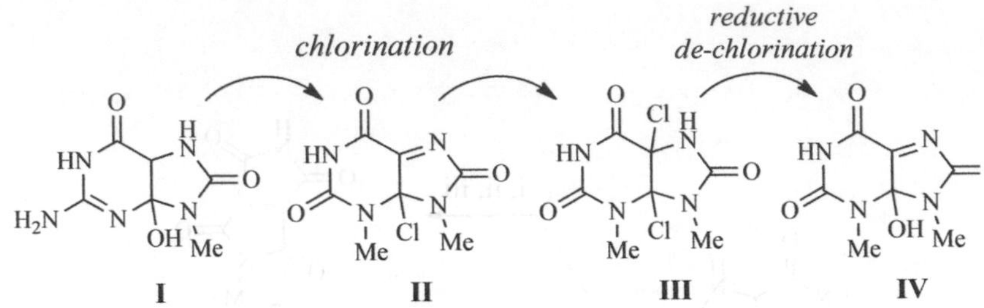

Analysis by molecular sections and atom labels

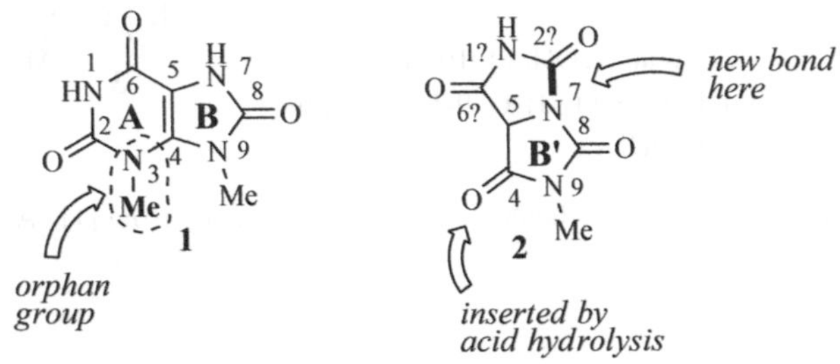

SCHEME 23.2 Top section: [3] / John Wiley & Sons. Mechanism analysis of **1**→**2**.

23.2.1 The Visibly Obvious Skeletal rearrangement occurs alongside ring contraction of ring **A** at the west of the molecule, an unusual outcome in the purine reactivity patterns, while on the other hand, backbone rearrangement examples in purines are well known. Meanwhile, the substitution pattern of ring **B** undergoes modification (to **B'**), possibly involving hydrolysis near or at the end of the sequence (mineral acid hydrolysis).

23.2.2 Abridged Atomic Balance

Compound	Abridged molecular formula
1	$C_5H_2N_4O_3$–$(CH_3)_2$
2	$C_5H_2N_3O_4$–$(CH_3)_1$
2–1	C_0H_0–N_1 +O_1 –CH_3

The **2–1** account indicates that the C and H balance is maintained, one *N*–CH_3 unit is lost (as methylamine?) and one O atom is incorporated. The latter two may be connected to the last hydrolysis, suggesting a methyl iminium ion as immediate precursor. It should therefore be conserved along the reaction coordinate until this ending step occurs.

Furthermore, the preservation of the C and H balance means that: i) In the absence of external carbon sources (CO_2, CO, or methylating agents), all carbons in **2** are part of **1**. Hence, ring contraction does not involve the loss of C, and ii) given the role of chlorine to spur the reaction while there is no chloride in **2**, this element should be incorporated in **1** at the earliest stage to be extruded reductively later (by KI and $Na_2S_2O_3$ and water).

Atom labels: Using the International Union of Pure and Applied Chemistry's (IUPAC) labeling system for purines in **1**, the elements in ring **B'** of **2** are easily assigned, but the rearranged ring at the north offers some uncertainties even though we try to keep the order assuming minimal cleavage and re-bonding. Moving clockwise from C^5 in the new ring, and assuming that the C^5–C^6 is preserved, it is unavoidable to recognize a new bond (N^7–C^2=O) in target **2**. Additionally, N^3–CH_3 in **1** is an orphan group since it is absent in **2**. This ought to be the methyl iminium ion that is extruded by hydrolysis during the final step and would give rise to methylamine.

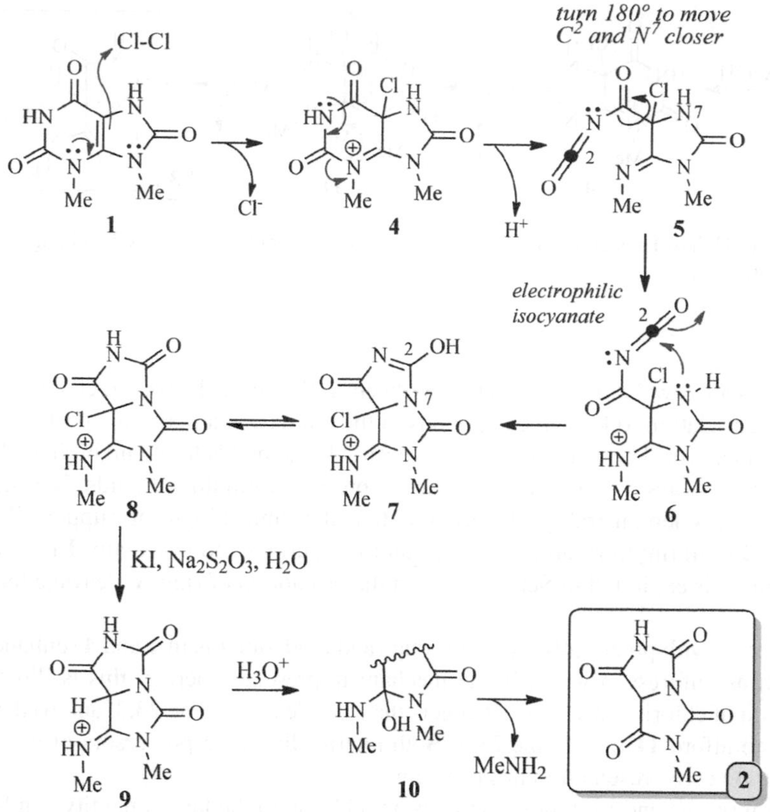

SCHEME 23.3 A feasible sequence to explain target **2** that temporarily incorporates a chloride atom.

23.3 Reaction Mechanism 1 → 2

Although the previous analysis is competent to provide a plan, there are still two open questions: i) Where does chlorine enter the sequence, and ii) what sort of intermediate is needed for the new amide N^7–C^2=O construction since N^7 and C^2 are so far apart in **1**, a rigid molecule? The previous disengagement of ring **A** to liberate C^2 as part of a freely rotating appendage is required for this to happen. This cleavage must involve chloride incorporation to the purine scaffold. These issues are put to work in Scheme 23.3. Mind the **4** → **5** fragmentation that leads to the isocyanate moiety, which shows a highly electrophilic C^2 that is stereochemically enabled after ring **A** cleavage to approach N^5.

23.4 A Reaction Mechanism 1 → 3

As authors [1] demonstrated unequivocally the structure of **2**, the sequence shown at the top of Scheme 23.2 became unfeasible. There was still the question of the relatively facile formation of spirodihydantoin **3** reported earlier [3]. This framework visibly bears witness to a different rearrangement characterized by ring **A** contraction. Justification is accessibly conceived through halogenation-unsymmetrical addition of the C=C–oxidative rearrangement cascade of **1** (Scheme 23.4). Because this sequence departs from a common intermediate **4** (Scheme 23.3), both branches compete against each other. Therefore, cleavage to isocyanate **5** seems to outweigh the addition of acetate and rearrangement to **3**.

23.5 Learning Issues and Takeaways

1. Nitrogen-rich heterocycles can be regarded as a packed collection of polar bonds and NBPs of electrons, often in π-conjugation, which is susceptible to multiple hydrogen bonding, protic and Lewis-acid catalysis, nucleophilic and electrophilic, oxidative and reductive agents, hydrolysis, acting as metal ligands, and so forth. These structures are a true playground for electron redeployment in various directions within and outside the molecular core and a source of many derivatives.

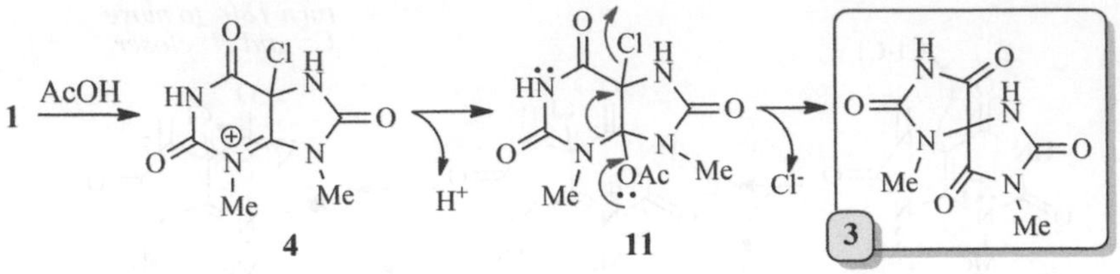

4 11 3

SCHEME 23.4 The addition of HCl to **1** instead of molecular chlorine as in Scheme 23.3 markedly changes the sequence direction toward the second target **3** via ring contraction.

2. Skeletal rearrangements are few, but most of them have been well studied. However, characterization of molecular structures, carried out by degradation to known compounds in the old days, and current spectroscopic methods often cannot provide an unequivocal answer. The ultimate and most desirable method is to obtain high quality monocrystals for X-ray diffraction, but this is not always possible in many purines prone to form thick liquids. Devising a molecular mechanism in parallel to spectral data is demonstrably a helpful tool to add chemical logics to support likely structures.
3. The (C=O)–N(R)–C=O [*N*-formylformamide] or urea motif in purines and especially **1** is a particularly effective source of reactivity. One of these is exploited in Scheme 23.2 in the promotion of ring **A** cleavage that furnishes the unexpected isocyanate **5**.
4. Adorning the basic purine with potentially leaving groups and oxidation (as in **1** and **4**) enhances reactivity opportunities in predictable as well as unforeseen ways. In the mechanism presented herein, this is illustrated in two sequences: i) unsymmetrical addition of chlorine in **1**, which triggers the cascade of Scheme 23.3, and ii) the stepwise addition of chloride and acetate in **4** to afford **11** in Scheme 23.4. Both intermediates are precursors of uncommon skeletal rearrangements of the **8**-oxopurine core presented in this problem.
5. With regards to natural purine and pyrimidine sources, in addition to the large quantity synthesized by our metabolism, we are exposed to external sources in food, beverages, and smokes, under oxidizing conditions that may turn into healthy or damaging results.
6. Check Problem 24 for additional, somewhat challenging, aspects of purine behavior.

REFERENCES

1. Poje N, Poje M. *Org. Lett.* 2003;5:4265. DOI:10.1021/ol035429k.
2. Rosemeyer H. *Chem. Biodiv.* 2004;1:361–401. DOI:10.1002/cbdv.200490033.
3. Biltz H, Krzikalla H. *Liebigs Ann. Chem.* 1927;457:131–189. DOI:10.1002/jlac.199274570106.

PROBLEM 24

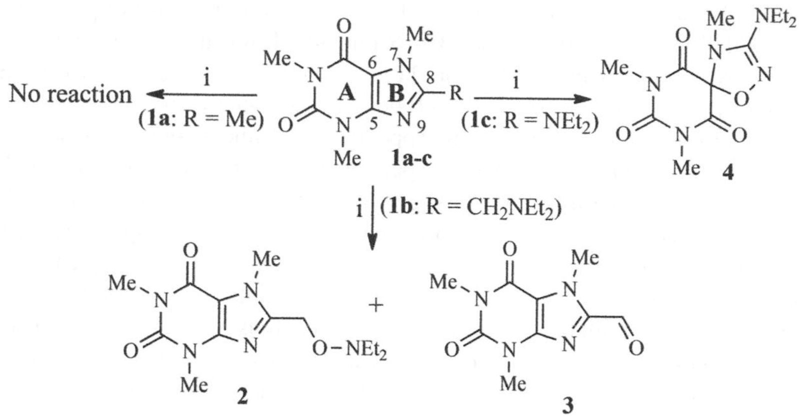

i: *m*CPBA (excess), $CHCl_3$:H_2O 2:1, r.t. 48 h
*m*CPBA: 3-chloroperbenzoic acid; solution added dropwise

SCHEME 24.1 [1] /American Chemical Society; and [2] / John Wiley & Sons.

PROBLEM 24: DISCUSSION

24.1 Overview

Continuing with the preceding problem about purine oxidative rearrangements, we jump back in time to a smart and instructive research performed around 1990–1999 by late professor Hans Zimmer[2] and his coworkers at the University of Cincinnati, United States [1, 2]. Scheme 24.1 portrays the disparate response of three C^8-substituted caffeine derivatives to peroxidation in a two-phase chloroform-water medium. The **1c** → **4** was unknown in purines (xanthines) at that time.

Caffeine is an unobtrusive component of many foodstuffs, isolated for the first time in 1820. It has been a strongly attractive subject of research owing to its psychostimulant properties (the most consumed such substance in the world) and occurrence in many edible products: coffee (globally in 2013, people drank some 26 000 cups of coffee per second, around 608 L per capita and year; in Finland alone, the equivalent of 9.6 kg of pure caffeine per capita and year), chocolate, tea, South American *mate* herb, many industrial energy beverages, appetite suppressants, and several other consumer products. After ingestion, caffeine is partially degraded by the metabolism; among other reactions, it is oxidized by gut wall and liver enzymes (cytochrome P450 oxidase family), thus oxidative degradation is a critical subject. In addition, unchanged caffeine and its metabolites are transported through blood and excreted. These compounds eventually appear in sewage processing plants. If not treated properly, they pass on to exhaust waters as an environmental issue (solubility in water 21.6 g L^{-1}; EC_{50} = 182 mg L^{-1} [3]). Oxidants, radical promoters, and ultraviolet radiation bombard these compounds in water treatment facilities pursuing its decomposition and mineralization [4]. Such interest feeds a flow of current discoveries about caffeine and purines in general.

24.2 Problem Analysis

There are three topics in Scheme 24.1: i) lack of *m*CPBA-induced reaction in **1a**, which indicates the stability of the purine core to reaction conditions; ii) oxidative rearrangement of the CH_2NEt_2 appendage in **1b**; and iii) skeletal rearrangement of the purine nucleus **1c**, the most interesting of the three.

[2] 1921–2001

24.2.1 Exploring the 1b → 2 + 3 Conversion *m*CPBA peroxidations are frequently rationalized in terms of induction of radical intermediates. Secondary and tertiary amines as **1b** yield primarily N-oxides (**5** in our case), which undergo further transformation owing to the labile quaternary ammonium ion thus produced. In the case of product **2**, the oxygen atom of the $R_3N^{(+)} \rightarrow O^{(-)}$ from *m*CPBA oxidation gets inserted in the CH_2–N junction. If it is true that the ammonium group is an excellent leaving unit while the oxide may operate as nucleophile on the allyl methylene, a coordinated intramolecular substitution is hardly feasible. However, homolytic scission of the CH_2–N(O) bond to separate the two molecular entities provides a route for the desired stepwise substitution to construe the CH_2–ONR_2 appendage by radical coupling. The separate components **6** and **7** would be locked in the polar solvent cage preventing their escape. Valence electrons of radical anion **7a** can be handled to create resonance form **7b** depositing the radical character in the oxygen atom, in preparation for the radical coupling to give target **2**. Once O-insertion is completed, oxidative cleavage of the protonated amine **8** gives easy access to final aldehyde **3** (Scheme 24.2, top section). This reaction is known as the *Cope elimination*. The **5 → 6 → 3** sequence is also a named reaction: *Meisenheimer rearrangement*, discovered by Jakob Mesenheimer just after the first World War in allyl aniline submitted to hydrogen peroxide oxidation [5]. In time, both processes became valuable tools in organic synthesis in a variety of forms. Some pose interesting accessible mechanisms like that shown in the lower section of Scheme 24.2, a set of reactions studied by researchers in Portugal and Spain [6] (Scheme 24.2, lower section; feel welcome to solve this bonus problem; a solution is given in Scheme 24.4 at the end of this discussion).

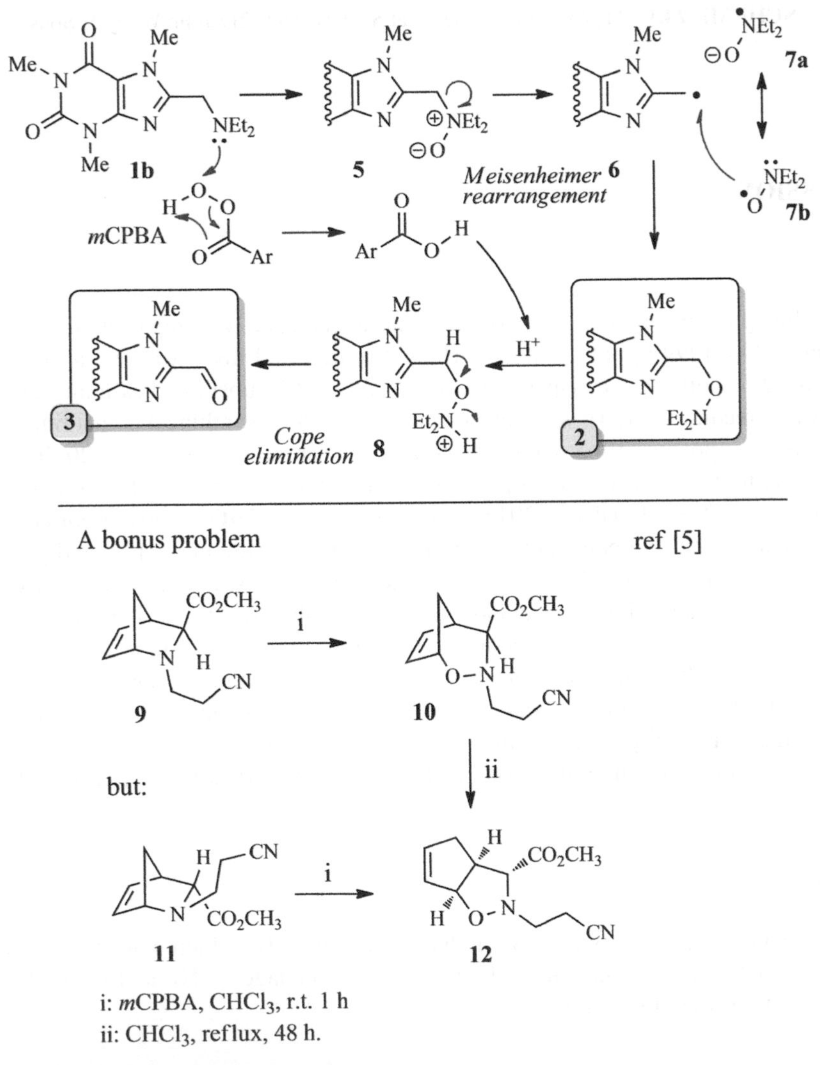

SCHEME 24.2 Top section: rationale explaining targets **2** and **3** from **1b.** The **9 → 10** and **11 → 12** reactions, presented here as a supplementary problem for readers, were adapted from Reference [5]. © 2016, the Royal Chemical Society, used with permission. A likely mechanism is shown in Scheme 24.4.

24.2.2 Exploring the 1c → 4 Rearrangement The major contribution of Zimmer and coworkers, this core rearrangement seems to take place only under certain constraints, which offer insights into the potential mechanism:

- ✓ Ring B rearrangement is observed only when there is a non-bonding pair of electrons (NBP) in C^8 substituents in conjugation with the heterocyclic unsaturations. Thus, amines, ethers and sulfides undergo this transformation [1, 2].
- ✓ Isolation of the NBP-bearing atom by one or more methylenes as in **1b** blocks rearrangement and Meisenheimer and Cope reactions take precedence once oxidation of the heteroatom in the side arm is verified. Similarly, alkyl substituents in C^8 are unreactive (e.g. **1a**).
- ✓ The lack of reactivity of **1a** ($R=CH_3$) toward *m*CPBA (no N-oxidation in the core's nitrogens) suggests that the ED effect of external NBP of compounds like **1c** may contribute to N-oxidation by *m*CPBA of conjugated N atoms in the core. N^9 appears the best candidate (se atom number nomenclature in **1c**, Scheme 24.3).
- ✓ The $N^{9(+)}$–$O^{(-)}$ thus formed would then weaken the N^9–C^5 bond, a requisite for the B ring cleavage preceding the desired spirocyclized target **4**.

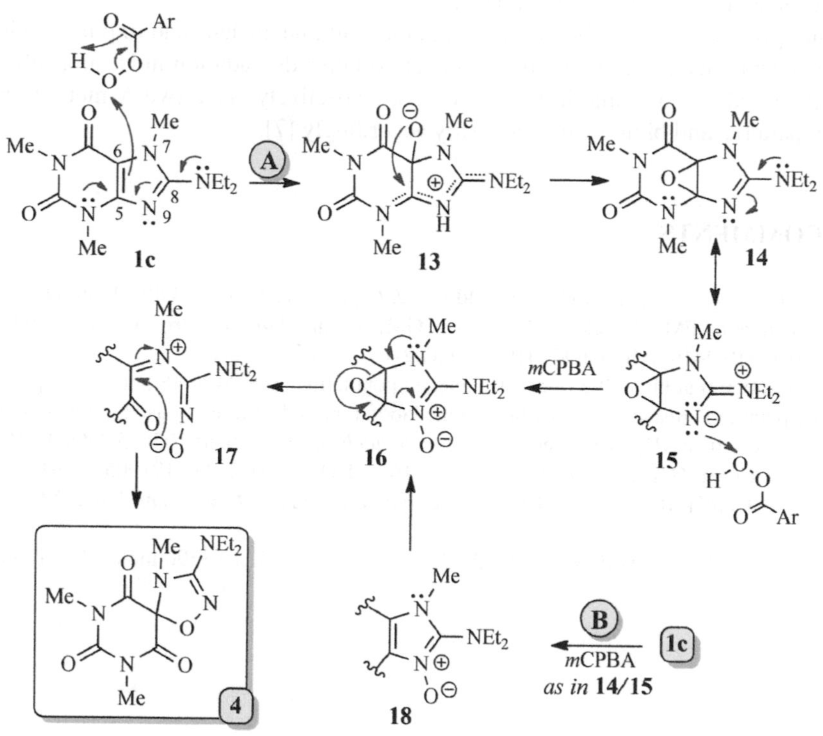

SCHEME 24.3 Oxidative core rearrangement that explains target **4** though routes **A** and **B**.

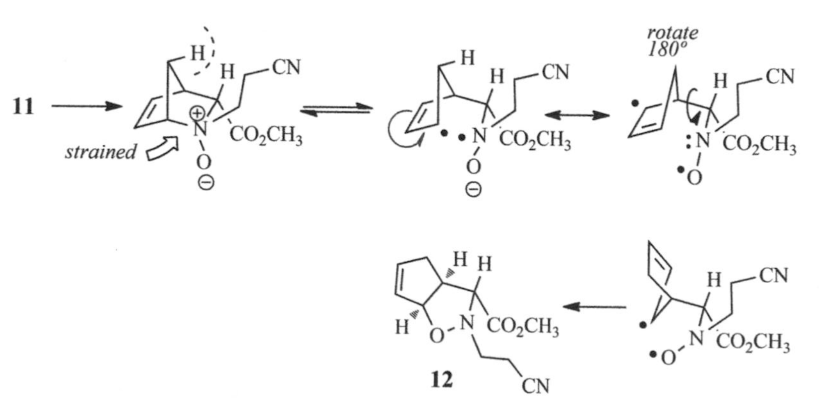

SCHEME 24.4 A feasible mechanism of the reaction in Scheme 24.2 bottom section. Adapted from Reference [6]. © 2016, the Royal Chemical Society, used with permission. The **9** → **10** reaction follows the guidelines of the Meisenheimer rearrangement sufficiently described herein for your perusal in putting it together.

✓ The oxidation of C^5 to C=O in **4** requires intervention of an oxygen atom provider (*m*CPBA). The C^5=C^6 seems a suitable substrate for epoxidation as polarized vinylogous amide reinforced at C^6 by the external nitrogen's NBP, and might as well be the starting reaction (pathway **A**, my choice) although authors invert the order of factors and prefer N^9 oxidation first induced by the C^8 diethylamino group (pathway **B**) [1, 2].

With these hypotheses in hand, a mechanism plan can be contrived (Scheme 24.3).

24.3 Learning Issues and Takeaways

In addition to the points raised in Problem 23, it is worth noting that:

1. Chemical properties of the xanthine nucleus of purines in general, and caffeine in particular, are strongly influenced by non-natural substituents.
2. Although the purine core appears as a robust scaffold, it is open not only to derivatization, oxidation, and reduction but also to skeletal rearrangements, but these are rather rare.
3. Electrons from donor groups in C^8 may be transferred to the xanthine nucleus and impinge on its reactivity.
4. The caffeine core concentrates the reactivity power in peroxidative degradation in N^9 and substituents in C^8. Other siblings like theophylline and theobromine in tea and cacao, respectively, have two *N*-methyl groups instead of three in caffeine. Reactivity patterns and biological effects vary accordingly [7].

REFERENCES AND COMMENTS

1. Zimmer H, Amer A, Ho D, Koch K, Schumacher C, Wingfield RC. *J. Org. Chem.* 1990;55:4988–4989. DOI:10.1021/jo00304a005.
2. Zimmer H, Amer A, Baumann FM, Haecker M, Hess CGM, et al. *Eur. J. Org. Chem.* 1999;2419–2428. DOI:10.1002/(SICI)1099-0690(199909)1999:9<2419::AID-EJOC2419>3.0.CO;2-1.
3. EC_{50} is the effective concentration that kills 50% of a test organism population after 24 or 48 hours of exposure; the EC_{50} value above is for *Daphnia magna*, a tiny plankton crustacean, easily bred in the laboratory and commonly used to test first encounter toxicity.
4. See, for example: Ziylan-Yavas A, Ince NH, Ozon E, et al. *Ultrason. Sonochem.* 2021;76:art.105635. DOI:10.1016/j.ultsonch.2021.105635.
5. Meisenheimer J. *Ber. Deutsch Chem. Gesellsschaft* 1919;52:1667–1677. DOI:10.1002/cber.19190520830.
6. Sousa CAD, Sampaio-Dias IE, García-Mera X, Lima CFRAC, Rodríguez-Borges JE. *Org. Chem. Front.* 2016;3:1624–1634. DOI:10.1039/C6QO00330C.
7. Monteiro JP, Alves MG, Oliveira PF, Silva BM. *Molecules* 2016;21:art Nº 974. DOI:10.3390/molecules21080974.

PROBLEM 25

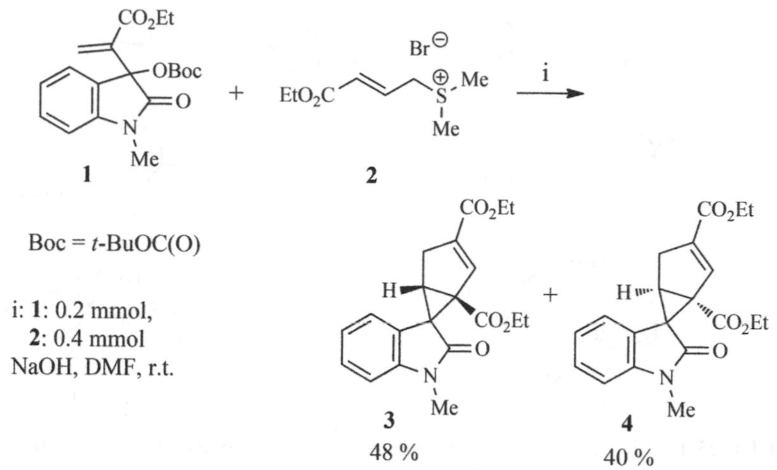

SCHEME 25.1 [1] / American Chemical Society.

PROBLEM 25: DISCUSSION

25.1 Overview

Oxindols equipped with three- and five-membered spirocycles and additional decorations further up like diasteromers **3** and **4** constitute a numerous family of bioactive natural alkaloids. Bioactivity and medical applications are highly attractive poles to synthesis chemists who seek to build these twisted scaffolds in the shortest possible sequence of reactions; better yet if these constructions are performed in one single pot under mild conditions. The present problem, selected from the work of researchers from Nangchang, China, is one interesting example.

The flat renderings of Scheme 25.1 do not convey the stereochemical complexity of these compounds, deemed as a simple set of diasteromers, until one appreciates them in the form of 3D molecular models (Figure 25.1). The models of **3** and **4** are shown at somewhat different angles for viewers to observe the entire scaffolds without eclipsing bond sticks. Because of the spirocyclopropane configuration relative to the oxindole plane, the diasteromers display the northern cyclopentenyl unit in entirely different directions, either away from or facing straight and perpendicularly at the phenyl ring edge. Curiously, the closest interatomic distance between these two sections, rigidly held by the polycyclic core, involves the northernmost alkene-carbethoxy carbon and C^5-H of the oxindol at 2.68 Å (268 pm) only [2]. Steric stress is expected. Meanwhile, in diasteromer **3** it is the ethyl ester moiety bonded to cyclopropane that hovers over the aromatic ring, steering away from it, thanks to σ–C–C bond rotation. Yet, both diasteromers are produced in a nearly 1 : 1 ratio; this suggests kinetic control at a still undefined slow step.

25.2 Problem Analysis

In addition to the overview comments, the reaction is sufficiently complex to demand careful analysis including element balance, HEDZ/LEDZ possibilities, and target deconstruction.

25.2.1 Element Balance *Element balance* to determine what is lost and gained in the 1 + 2 operation. This time we will dismiss H counting (too lengthy and not so informative in this case) and carry out an abridged molecular formula. Selected functional groups appear in [brackets]. The reader will understand this technique easily after looking at the following table. It is quite easy and fast to build and gain insight of atom/functional groups acquisitions and losses.

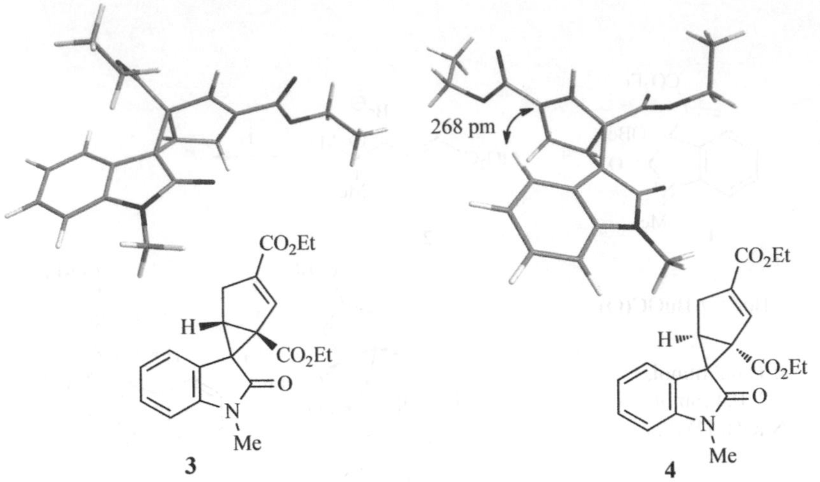

FIGURE 25.1 Minimum energy molecular models of targets **3** and **4** Adapted from [2].

Compound	Abridged molecular formula
1	$C_{11}H_xNO_2$; [COOEt]; [Boc]
2	C_3H_x;[$S(CH_3)_2$]; [COOEt]
1+2	$C_{14}H_xNO_2$; [$S(CH_3)_2$]; 2x[COOEt] [Boc]
3	$C_{14}H_xNO$; 2x[COOEt]
3− (1+2)	C_0–N_0–O_1–[$S(CH_3)_2$]; $[COOEt]_0$; –[Boc]

Target **3** (as well as **4**) includes all C and N atoms as well as all the CO_2Et units present in the starting materials **1** and **2**, while one atom and functions [$S(CH_3)_2$] and [Boc] get removed, the latter probably as OBoc (*O-tert*-butoxycarbonyl). This might be treated as a leaving group upon attack by a carbanion given that it occupies the future spiro-quaternary position (C^3 in oxindole atom numbering).

In a nutshell, initial coupling of **1** and **2** is followed by a cascade of one or more intramolecular rapports to build two carbocycles, three and five membered.

25.2.2 HEDZ/LEDZ Possibilities In the form depicted in Scheme 25.1, both **1** and **2** appear as LEDZs dominated by EWGs; hence, there is no polar interaction as such. Radicals do not seem likely nor do concerted cycloadditions. One of the two starting compounds must have the potential to turn into a HEDZ by base intervention (NaOH [sodium hydroxide]). In the absence of acidic protons in **1**, protons on the α methylene of the sulfide cation should be sufficiently acidic to be removed by NaOH and create a powerful nucleophile there.

25.2.3 Target Deconstruction The 2-oxindole moiety remains unchanged (and kept aside in the renderings); all the action takes place around the spirobicyclic block. In accordance with the minimum group displacement principle, the H_2C=C–CO_2Et section in **1** appears as a convenient beacon to identify it in **3**. There are two forms to represent this fragment in **3** depicted in red, in models **A** and **B** of Scheme 25.2. Note that the three-carbon block remains in one piece all along.

Model **A** identifies a foreign fragment (green) in **3**, which should correspond to sulfonium compound **2**. As one splits red and green sections, the latter piece does not correspond to **2** unless rearrangement has occurred. This is energy costly.

Model **B** maintains the red block in its original shape in **1** and provides a section higlighted in green that clearly embodies the non-rearranged four-carbon chain of **2**. By this procedure we have been able to determine the origin of each CO_2Et unit. The mechanism should therefore be ruled by the following guidelines:

- ✓ Assign carbons labeled C^3, C^4, and C^5 to oxindole **1**.
- ✓ Assign carbons labeled C^6, C^7, and C^8 to the sulfonium cation.
- ✓ Extrude OBoc and $S(CH_3)_2$ in two separate steps. Both are suitable leaving groups in internal substitutions, which create the following bonds:

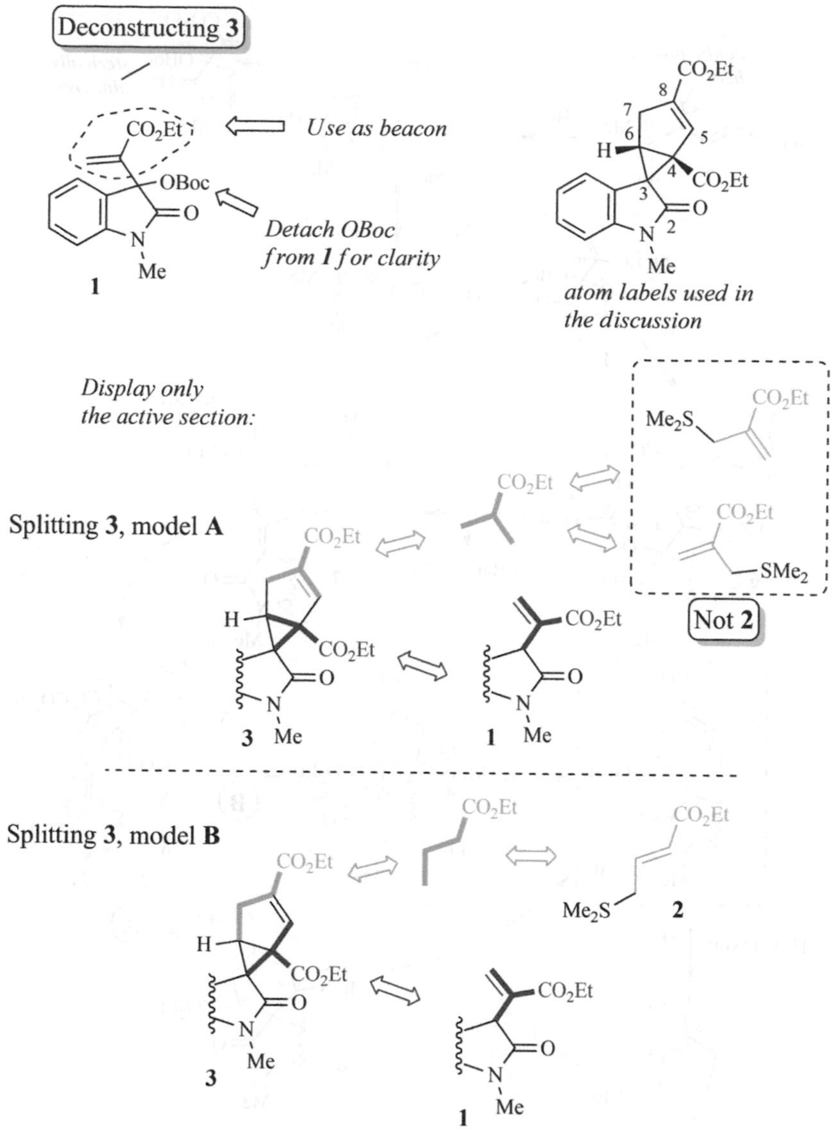

SCHEME 25.2 Fragmentation analysis (FA) of target **3** as a function of starting compounds **1** and **2,** using the $H_2C{=}C{-}CO_2Et$ section as a beacon (green). Two FA models, **A** and **B**, can be devised.

- C^3–C^6 causing the departure of OBoc.
- C^6–C^4.
- C^4–C^5.
- C^5–C^8 by way of a potential 1,4 addition of a carbanionic C^8 onto the red α,β unsaturated ester.

25.3 A Feasible Mechanism

Scheme 25.3 conveys the active electron redeployment from the guidelines just discussed. Briefly:

1. Sulfonium salt **2** gives rise to sulfur ylide in two resonance forms, **5a** and **5b**.
2. The latter performs a Michael addition on starting α,β unsaturated ester **1**.
3. Decarboxylation in **6** ensues, and a second proton is removed from **7**.
4. Key intermediate **8** results with the expression of negative charges in carbons C^6 and C^8. From **8** two options emerge.

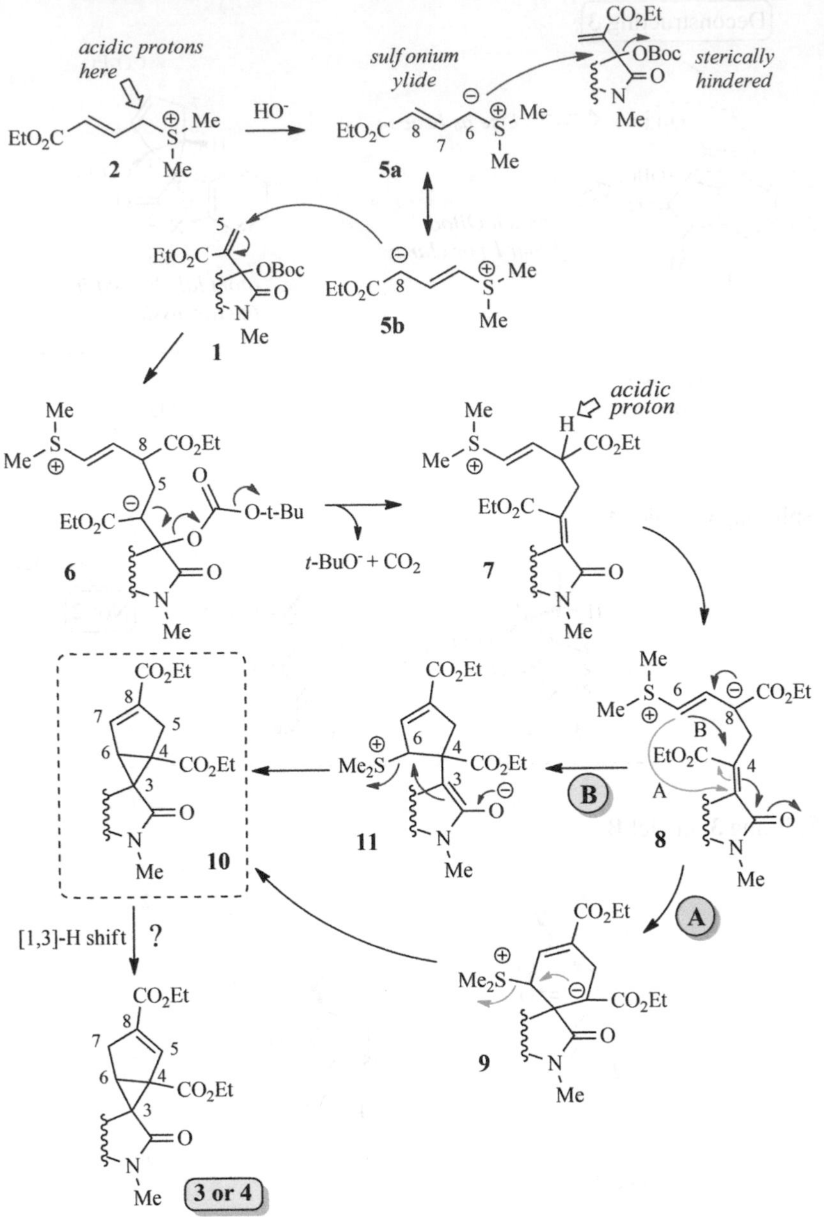

SCHEME 25.3 Feasible mechanisms explaining the reaction of Scheme 25.1. The sequence leading to intermediate **8** and the green option **A** have been adapted from [1] / American Chemical Society, whereas pathway **B** is this M. Alonso's alternative [3]. See also Reference Adapted from [4].

5. Option **A** proposed by the authors [1] (green arrows) assumes a second Michael addition of the anion on C^3 with redeployment of the negative charge *toward the ester*, as shown in **9**, which furnishes a cyclohexenyl body, en route to **10** by displacement of dimethyl sulfide and cyclopropanation.
6. Option **B** (my own; red arrows) conducts the Michael addition so that redeployment of negative charge occurs *toward the more polarized ketone*, yielding enolate **11**. This enolate engages in intramolecular extrusion of dimethyl sulfide ending in cyclopropanation (**10**).

The main difference between pathways **A** and **B** is the order of the cyclization cascade. Because **A** and **B** converge on the same product shortly after anion **8** is formed, there is no experimental evidence to establish which of the two reflects a more accurate mechanism.

Most importantly, **neither A nor B delivers the correct target** (**3** or **4**) since the carbon unsaturation in **10** is clearly defined in C^7=C^8, during the **8**→**9** and **8**→**11** transitions, as opposed to the reported C^5=C^8 configuration [1]. Unless a later migration of this C=C occurs (**10**→**3** or **4**), there is a mistake in the assignment of structures **3** and **4** and the several others reported in the original paper [1].

25.4 Learning Issues and Takeaways

1. Protons in methyls and methylenes next to sulfonium salts are acidic and susceptible of removal by moderate bases. The anion thus produced is known as a sulfur ylide. These ylides have been exploited extensively in organic synthesis, in parallel with imine and carbonyl ylides.
2. When the methylene in question is located sandwich between the sulfonium and a C=C, bond charge delocalization activates the carbanion three carbons removed. Other combinations allowing charge drift are possible.
3. The sulfonium cation moiety is employed herein in two electronically opposed directions. In the first of these, the ylide carbanion operates twice in intramolecular Michael additions to form as many C–C bonds, while in the second direction the sulfur cation operates as leaving group. These properties make sulfur ylides exceedingly useful.
4. All in all, the sequence can be understood as a cascade of unstoppable reactions including two Michael additions and two intramolecular cyclizations, which create substantial molecular complexity.
5. The concurrent formation of adducts **3** and **4** in nearly equal proportion indicates absence of stereoselectivity. This is defined at the **8**→**9** cyclization in option **A** by attack of C^6 at the tethered chain from either side of the molecular plane defined by the oxindole allowed by the free rotation of the two methylene units. Likewise, in pathway **B** the **8**→**11** cyclization can occur on C^4 from both sides but not affecting the oxindole. More stereochemical control might be expected from **11** because of the rigid five-membered ring with two substituents facing the oxindole. DFT or other quantum mechanical studies will be necessary to determine favorable routes and stereochemical restrictions.

REFERENCES AND NOTES

1. Meng Z, Wang Q, Lu D, Yue T, Ai P, Liu H, Yang W, Zheng J. *J. Org. Chem.* 2020;85:15026–15037. DOI:10.1021/acs.joc.0c01919.
2. Energy minimization performed with MMFF94G by Alonso-Amelot ME, this book.
3. Alonso-Amelot ME, this book.
4. Galliford CV, Scheidt KA. *Angew. Chem. Int. Ed.* 2007;46:8748. DOI:10.1002/anie.200701342.

PROBLEM 26

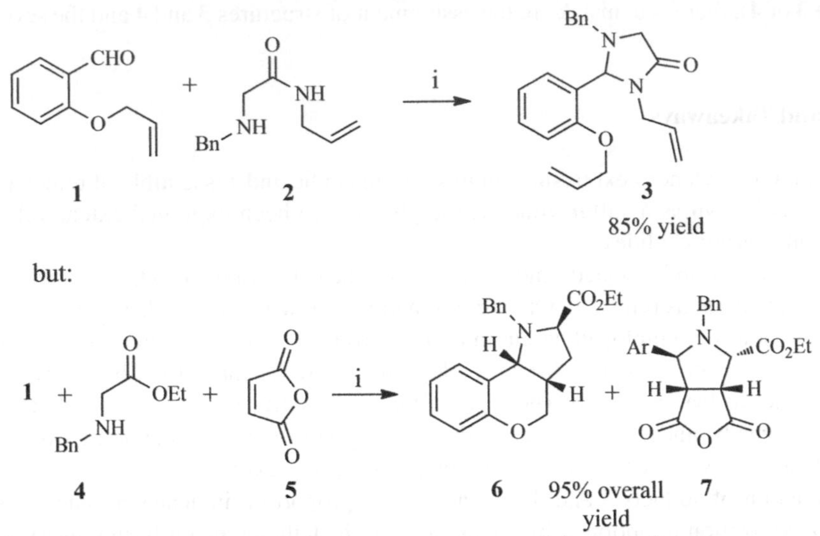

i: no solvent, 200 °C (micro-wave), 15-60 min
Bn = Ph-CH_2-
Ar in **7** = o-allyloxybenzene

SCHEME 26.1 [1] / Elsevier.

PROBLEM 26: DISCUSSION

26.1 Overview

The construction of molecular complexity and diversity in heterocyclic compounds continues to be a major goal of organic synthesis aimed at reproducing Nature's ways to build intricate natural products, confirm structures, and provide the pharmaceutical industry with compound libraries for biological testing; better yet if syntheses are performed in eco-friendly conditions and outcomes: high chemo and stereoselectivity furnishing high yields to minimize side products, usage of neat mixtures of reactants (no solvents to get rid of) and short reaction times.

Since heating is still often necessary for reactions to proceed, the question arises regarding the stability of targets formed in the medium that thermally decompose during classical mantle heating. The desired compounds become reactive intermediates in the way to uninteresting final products.

Ultraviolet radiation (in the electron photoexcitation radiation zone) and microwave heating (MWH; electromagnetic radiation in the bond vibrational zone) offer solutions to this sort of problem in susceptible substrates for reasons discussed in Problem 22 of this collection.

Worth noting, microwaves interact with molecules that bear dipoles through a direct transfer of radiation-molecule energy [2]. As polar molecules rapidly align with the alternating microwave electrical field, friction and collision occur as the field changes orientation to result in heat, which is transferred by molecular clashes to the non-polar fraction of the medium up to unexpected levels. Overheating of organic solvents 13–26°C above their boiling point is on record [2], bringing new heat energy scenarios to kinetics. It is fair to say that the detailed impact of MWH on organic molecules is still not well understood and continues to be a current subject of research [3].

The work of researchers from Brno, Czech Republic [1], part of which constitute your present mechanism problem, took careful and thorough consideration of these effects to stabilize heterocyclic (and hence, polar) reaction intermediates that would otherwise be destroyed by prolonged conventional heating. MWH energy enabled them to determine chemically sensible reaction mechanisms for the systems of Scheme 26.1.

The disparate course of **1**+**2** versus **1**+**4** despite the (apparent) similarity of amide and ester synthons forces us to analyze them as two separate reactions while keeping in mind the possibility of common intermediates.

26.2 Understanding 1+2→3

Adding a bit of strategic design to what seems a visually solvable mechanism, let us compare the abridged element balance leaving aside the unchanged moieties in [**1**+**2**] and **3** (two allyls=R, and the Bn):

Compound	Abridged molecular formula
1+**2**	$C_9H_9N_2O_2R_2Bn$
3	$C_9H_7N_2O\ R_2Bn$
3− (**1**+**2**)	H_2O

As predicted by a bird's eye view, **3** is built from the **1**+**2** addition and water exclusion. Amines and aldehydes do just that to form imines in acidic medium. However, do not lose sight of the absence of solvents or proton sources so one must make do with what is available in this mixture. With this caveat in mind, the sequence of Scheme 26.2 unravels uneventfully. It is worth noting that after the [1,3]-H shift from the iminium ion in **8**, the expected loss of water occurs by the [1,6]-H transfer from the amide, which is further stabilized by dispersal of the developing negative charge by the vicinal C=O.

All in all, for some readers this first reaction may be on the borderline between an exercise and a problem. They may change their minds after a close view of the second reaction of Scheme 26.1.

26.3 Solving 1+4→6

This conversion comprises two separate processes since target **6** does not contain the elements of maleic anhydride **5**. Yet, both take place in the same pot (therefore, there is competition between them, which is part of the problem here). Adduct **7** may provide useful information about the active species evolving from aminoester **4** as we shall see later.

26.3.1 Possible Origin of 6 FA of **6** reveals three new bonds (highlighted in red, Scheme 26.3, top section) between synthons **1** and **4**. This coupling likely begins with the familiar amine-aldehyde addition followed by involvement of the oxyallyl arm. Two conceptually diverging sequences may be conceived from this general vision. In the first of these−**pathway A**−tempted by the high temperature employed (200 °C) radical or ionic intermediacy seems adequate for our comfort zone (Scheme 26.3). This reaction model fails in one aspect, however. If on one hand the anti-Markovnikov diradical *disrotatory* ring closure **12**→**13** justifies the cis-ring junction observed in target **6** and the sortie of water, the ensuing radical ring closure **14**→**6** would not be sterically constrained; a mixture of ester stereoisomers should be produced, a result opposed to actual experimental record in which the all cis structure was obtained [1]. Besides, maleic anhydride adduct **7** is telling a nonradical story (see below).

Turning to ionic characters may be more appropriate to invigorate bonding sites at former carbonyl and α-methylene ester carbons in sufficiently active form to involve the terminal C=C. This idea is developed in pathway **B** and was originally proposed by the Czech authors [1]. A series of zwitterions crop up from the first amine-aldehyde adduct **15** until water is extruded from **17** in a classically predictable imine chemistry manner. For the proficient electron redeployer, three resonance canonical structures **18-I** to **18-III** can be generated. All of them are accessible dipoles combining σ+σ and σ+π expressions, formally constituting an *azomethine ylide* (AMY). Inter and intramolecular cycloadditions of AMYs and alkenes or alkynes of various substitution patterns with or without metal catalysis have been reported many times for the successful construction of molecular complexity. Therefore, AMY **18** may well be trapped by the terminal allyloxy moiety located at a suitable distance and HOMO/LUMO compatibility to craft a [3+2] cycloaddition TS **19TS** (see comment [4]). Approach of the ene moiety from the underside of the molecular plane gives rise to the cis configuration observed in target **6**.

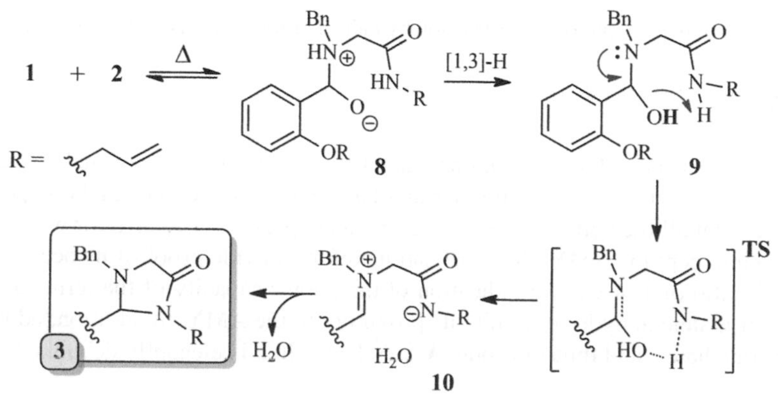

SCHEME 26.2 A feasible mechanism to explain adduct **3** without a solvent-derived proton source.

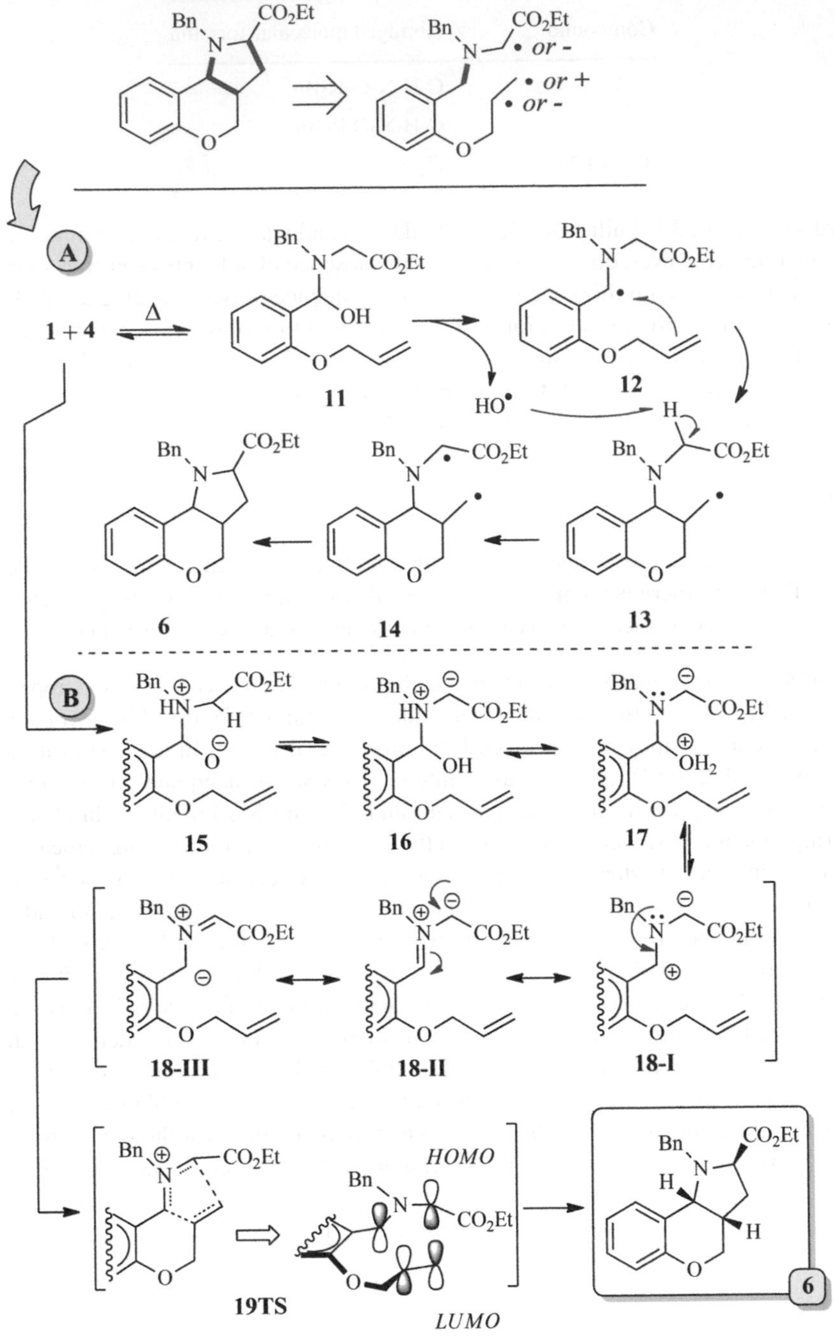

SCHEME 26.3 Possibilities of pyrrolidine ring construction by radical or ionic means and the associated mechanisms, **A** and **B**.

26.4 Solving 1+4+5→7

Once the final step in the construction of **6** is solved, one can easily surmise the role of maleic anhydride, a classical potent dienophile extensively used to trap active dienes and other dipolar species, to intervene and compete against the allyl unit for the AMY dipole in **18**. Notice that all the components of **1**, **2**, and **5** are present in **7**, save for water. Authors [1] added **5** to the mixture seeking precisely to intercept the AMY dipole **18** and therefore provide proof of its occurrence and critical role in the **1**+**4**→**6** process. The **6** to **7** ratio of 1:4 was an indication of the poorer capacity of the terminal alkene compared with the electron deficient π system of maleic anhydride despite its proximity to the AMY. In addition, adduct **7** would not be formed from short-lived radical intermediates, and thus our route **A** of Scheme 26.3 is elegantly debunked by this smart experiment.

Stereochemical considerations: As in any intermolecular electrocyclization without significant stereochemical constraints, there are two possible approaches of the reactants universally known as endo and exo that furnish the corresponding stereoisomers. Because the four stereogenic carbons in *exo* **7**–the only such product detected–are well defined, it is worth examining the reasons behind the observed preference. By drawing a 3D rendering of both TSs in the **18**+**5** combination the stereoelectronic effects (see Chapter 3 regarding the DACA) favor the exo over the endo adduct. It is your job to draw these renderings on paper.

26.5 Learning Issues and Takeaways

1. Azomethyne ylides (AMYs) are obtained from various sources and react with a gamut of intra- and intermolecular dienophiles beyond the reaction of this problem. Extensive reviews are available in the literature [6].
2. To the three AMY resonance forms, a fourth can be added having $C^{(+)}$ at the α-carbonyl methylene. It was not considered for this problem since the ethoxycarbonyl (and other EWGs) discourage the contribution of this canonical form. However, EDGs do favor this structure.
3. According to the frontier molecular orbital theory (FMO), the AMY + dienophile cycloaddition occurs by combining HOMO and LUMO orbital configurations. The AMY section is the electron-rich component (mind the anion character) and contributes valence electrons through its HOMO, whereas the diene (usually an electron poor unit) contributes its vacant LUMO. The specific orbital rendering of **19TS** (Scheme 26.3) reflects this configuration.
4. Successful electron transfer between AMY and alkene is a function of the energy gap between the respective HOMO and LUMO, the closer they are the faster the reaction rate. When more powerful EWGs than carboxyethyl are located at the aminomethylene carbon, one would expect a greater contribution of the **18-II** zwitterion and a faster reaction. The authors [1], however, recorded a much lower cycloadduct yield: EWG=CN, 17% yield of cycloadduct versus 84% for ethoxycarbonyl. The greater EW effect of CN lowers the energy level of the HOMO, increasing the HOMO (AMY)/LUMO (alkene) energy gap, causing a longer half-life of the dipole, and hence, a greater chance to undergo decomposition to other untraceable outcomes.
5. Increasing the steric bulkiness of the ester (Et, *i*-Pro, *t*-Bu) markedly decreased the yield of target **4**. Under the MW radiation, which strongly affects the AMY dipole, the steric effect naturally hampers the formation of **19TS.** The AMY dipole lifetime is thus extended and exposed to uncontrolled thermal decomposition, which lowers product yield.
6. By the same token, substituents on the terminal alkene also affect yield and type-**6** product composition, contingent upon the cis or trans configuration, due to the combined steric interference with the carboethoxy group and stereoelectronic effects on alkene dienophile reactivity.

REFERENCES

1. Pospísil J, Potácek M. *Tetrahedron* 2007;63:337–346. DOI:10.1016/j.tet.2006.10.074.
2. See, for example: Laurent R, Laporterie A, Dubac J, Berlan J, et al. "Specific activation by microwaves, myth or reality?" *J. Org. Chem.* 1992;57:7099–7102. DOI:10.1021/jo00052a022.
3. Churyumov GI (ed.) *Microwave Heating, Electromagnetic Fields Causing Thermal and Non-thermal Effects*. Thieme Publishers, 2021, 202 pp. ISBN 978-1-83968-227-8.
4. In the simplest array of unsubstituted AMY and ethylene, the [3+2] cycloaddition occurs via synchronous concerted TS, according to quantum calculations [5]. Electron localization function (ELF) analysis underlines the pseudo-diradical characteristics of AMY, which explains its high reactivity. Only four electrons, two non-bonding at the AMY carbons and two on the alkene, take part in the calculated TS. The AMY-alkene cyclization is formally a $[2n+2\pi]$ process. Therefore, the ELF assessment supports in part the diradical approach I proposed in pathway **A**, after inserting the diradical shown below upon attack of a hydroxyl radical to R-CH_2-COOEt in **12** before the intramolecular cyclization takes place. One must concede though, that the calculated model is devoid of electronic influences of substituents present in the much more complex case of intermediate **18**, which favors a highly polarized AMY (Scheme 26.2). Additionally, AMYs are well established in the chemical literature.

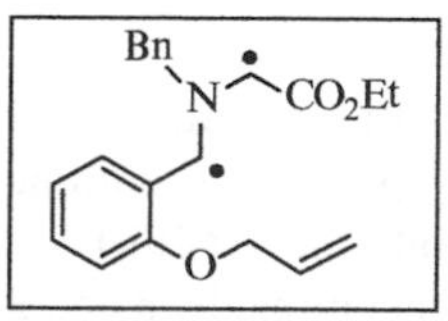

5. Domingo LR, Chamorro E, Pérez P. *Lett. Org. Chem.* 2010;7:432–439. DOI:10.2174/157017810791824900.
6. Selected reviews: Coldham I, Hufton R. *Chem. Rev.* 2005;105:2765–2810. DOI:10.1021/cr040004c. Adrio J, Carretero JC. *Chem. Commun.* 2014;50:12434–12446. DOI:10.1039/C4CC04381B.

PROBLEM 27

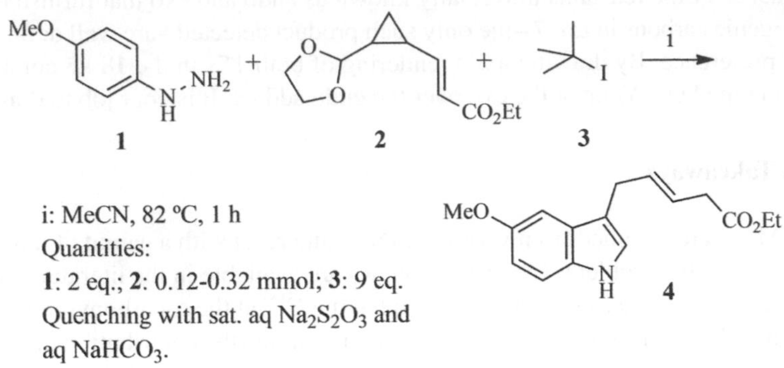

SCHEME 27.1 [1] / American Chemical Society.

PROBLEM 27: DISCUSSION

27.1 Overview

Although apparently accessible to some readers, there are some less obvious issues in this problem that need to be addressed. For one, the elements of **3** do not appear in the isolated target **4** irrespective of the large molar excess used. The logical deduction is that **3** is a reagent added to the reaction mixture by researchers from Kobe, Japan [1] with a clear purpose since no reaction occurs without **3**. Also worth noting, anhydrous conditions were necessary to avoid hydrolytic side reactions, yet the acetal function gets removed, which is classically performed in aqueous acid. The authors achieved this challenge with an indirect strategy based on reagent formed *in situ* from **3** (read on).

27.2 Preliminary Analysis: Mechanism Keys in the Target

Target **4** can easily be dislodged in the perfectly discernible carbon scaffolds of **1** and **2** (Scheme 27.2) save for two orphan groups, which should thus be dropped at given points, momentarily unknown, along the reaction coordinate. Five additional features may be furthered from this scheme:

- ✓ Two new bonds link fragments **A** and **B**.
- ✓ One cyclopropane bond gets cleaved (C^3–C^5), probably in connection with the formation of a C^3-aryl linkage.
- ✓ The rest of the ester sidearm is conserved but the double bond migrates from C^6=C^7 to C^5=C^6. The mechanism must justify the loss of conjugation with the ester.
- ✓ Hydrazine **1** can easily be identified as an *HEDZ*, owing to the nucleophilically active end N. The *LEDZ* counterpart in **2** is only represented by the α,β-unsaturated ester, but this moiety does not play any part in the **1**+**2** rapport. This must involve C^2.
- ✓ C^2 is responsible for the early linking of **A** and **B** sections of **4**. This would be feasible if we had a Brønsted or Lewis acid on hand to activate the acetal group, but there is none in sight; or is there? If only one had a water-free mineral acid, the mechanism would be smoothly scheduled.

27.3 What is *t*-Bu-I Used For?

Because we seem to be at a standstill, let us address this question. *t*-Bu-I is a polar, sterically stressed compound owing to methyl repulsion on a large iodine atom. Heat may promote β-elimination of HI (hydrogen iodide) irreversibly since the by-product is isobutene, which bubbles away. As a point of fact, *t*-Bu-I is used as a convenient *in situ* source of HI in the absence of water [2]. HI is also commercially available as a concentrated water solution (45–67%). In this medium it is completely ionized to $I^{(-)}+H_3O^{(+)}$ giving rise to undesired hydrolytic reactions of intermediates and products (the ethyl ester, or cyclopropane [cp] cleavage, for example). By contrast, anhydrous HI from *t*-Bu-I provides protons for the initial activation of C^3 as a strong *LEDZ*, which enables the mechanism of the lower section of Scheme 27.2.

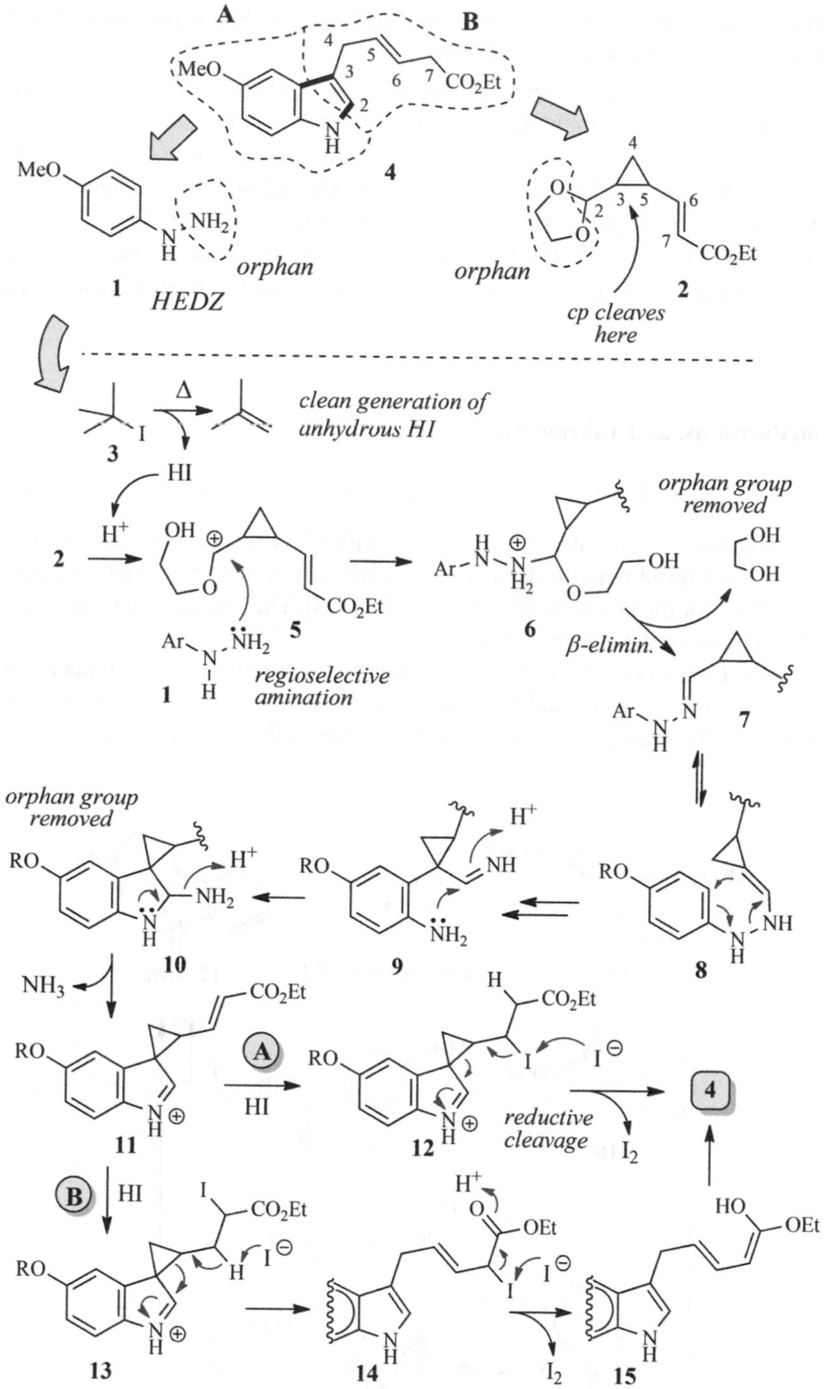

SCHEME 27.2 Fragmentation analysis of target **4** and alternative mechanisms.

27.4 Highlights of Scheme 27.2

✓ The first priority is the removal of the acetal function (*our first orphan moiety*) that enables the first amination by aryl hydrazine. Just one proton from HI is instrumental for this to occur (**2**→**5**→**7**).

✓ Once **1** and **2** are coupled through amination in **7**, the former hydrazine's N end must be removed by first finding a way to break the N–N bond and proceed from there.

✓ This is achieved by way of enamine **8**, which locates the C=C bond in the right position for a [2n+4π] electrocyclic diaza-Cope rearrangement, without alteration of the cyclopropane ring.

✓ Imine **9** invites the aniline intervention on the freed protonated imine to close the five-membered ring of target product **4** and create conditions for the extrusion of ammonia, *our second orphan group*.

✓ It is the turn of the cp to cleave in order to form the five-carbon tether. Two pathways open for the progress of iminium ion **11:** Markovnikov (route **A**), and anti-Markovnikov (route **B**) HI addition. While nucleophilic attack of $I^{(-)}$ on the organic iodide **12** indirectly promotes cp breakage aided by EW from the iminium ion, intermediate **13** may undergo previous elimination-cp cleavage to iodide **14** followed by a similar Nu attack on the organic iodide.

✓ Note that: i) The C=C bond in the two routes ends up in the β,γ position, and ii) **both $I^{(-)}$ interventions are reductive**; I_2 is the oxidized species.

27.5 Learning Points, Clarifications, and Takeaways

In addition to the highlights of Scheme 27.2 just described, the following are instructive issues of substance:

1. The conversion of aryl hydrazines to indoles after coupling with C=O or several other electrophilic receptors (as in **7**→→**10**) has a name: **Fischer indole synthesis.** This is an early development of organic chemistry but nonetheless a powerful tool used many times for the synthesis of indole alkaloids [3]. The mechanism took some work to be solved, and the steps of Scheme 27.2 reproduce the current consensus.
2. During the discovery of novel reactions, researchers often want to learn more about the mechanism involved. Professor Ueda and coworkers [1] were of this opinion and designed two experiments to bring light to the later part (the iodination–deiodination) of the sequence (Scheme 27.3). To this end they performed two experiments:

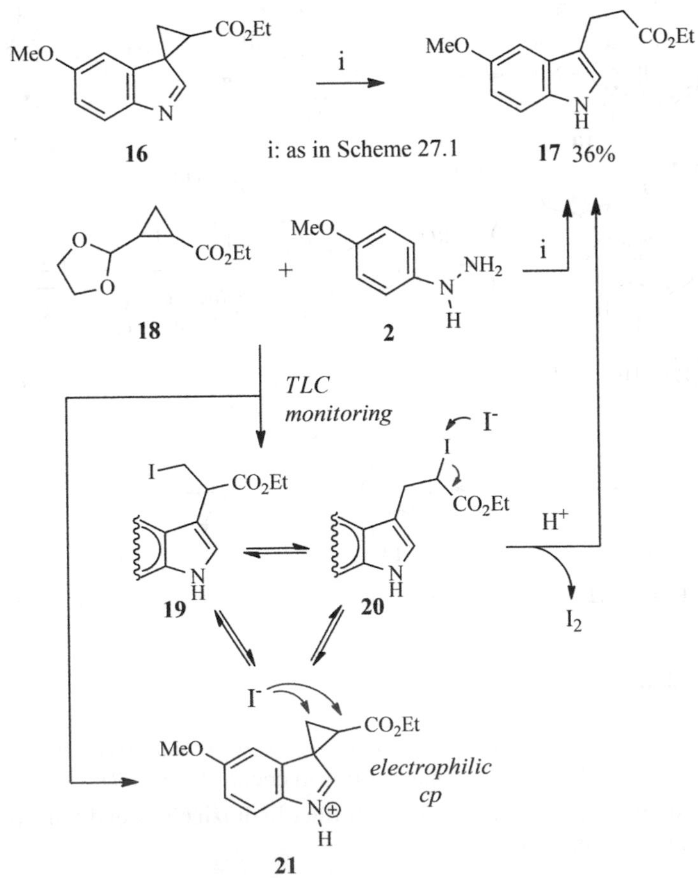

SCHEME 27.3 [1] / American Chemical Society.

a. Synthesize cp-indolenine **16** and submit it to *t*-Bu-I in boiling CH_3CN. That indol ester **17** was the detected product showed that a cp-indoline was a likely intermediate in the sequence.
b. They repeated the aryl-hydrazine cp-acetal ester reaction with model ester **18** not endowed with a C=C bond that might have served for the iodine addition, as annotated in Scheme 27.2. They also cut the reaction short after only 15 minutes and checked up the composition status by TLC. In addition to spots corresponding to methoxyaniline, still unreacted **16**, and ester product **17**, a less polar spot clearly cropped up to which iodide adducts **19** and **20** were assigned in a 5 : 1 ratio. These iodides were not present after the reaction time (60 minutes) was completed. Hence, both iodides underwent conversion to target **17**.

3. Worth noticing, while iodide reduction enables **20** to furnish target **17**, a similar reduction of isomer **19** is not granted. And yet authors are convinced of its existence. Is there any solution to this dilemma in your mind? Here is one; accept the **19** ⇆ **20** equilibrium through cp (**21**) recomposition as shown in Scheme 27.3 (bottom). The **19** : **20** ratio of 5 : 1 can be explained by the conversion of **20** into target **17**, hence escaping from the equilibrium, whereas **19** would return to cp **21** to reconstitute **20** thereafter.
4. Authors further observed that EWGs in the aryl hydrazine gave poorer yields or no reaction at all. This can be interpreted as insufficient reactivity of aniline **9** (Scheme 27.2) to complete the Fischer indole scheme. Exposure to heat for longer times may have destroyed the cp imine or preceding enamine **8** in unknown directions. Additionally, hot *t*-Bu-I is known to reduce a few EWGs such as nitroarenes. The sequence is thus blocked from the start.

REFERENCES

1. Yasui M, Fujioka H, Takeda N, Ueda M. *Org. Lett.* 2022;24:43–47. DOI:10.1021/acs.orglett.1c03607.
2. Ito Y, Ueda M, Takeda N, Miyata O. *Chem. Eur. J.* 2016;22:2616–2619. DOI: 10.1002/chem.201504010. Accessed: September 28, 2021.
3. Heravi MM, Rohani S, Zadsirjan Y, Zahedi N. *RSC Adv.* 2017;7:52852–52887. DOI:10.1039/C7RA10716A.

PROBLEM 28

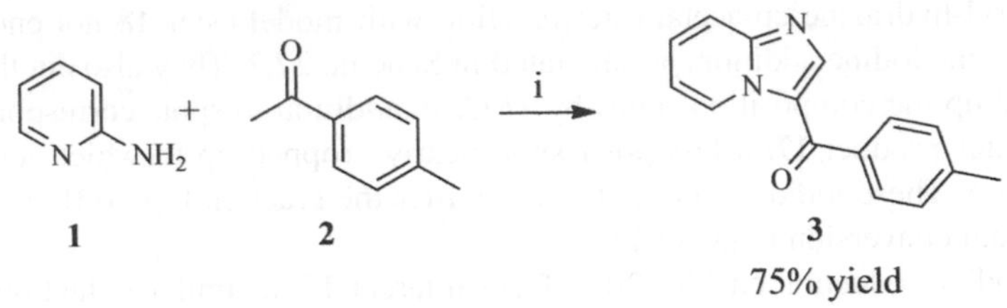

i: $K_2S_2O_8$ (3 mmol), NaOAc (2 mmol),
I_2 (0.1 mmol), DMSO, 110 °C, 16 h
No air exclusion

SCHEME 28.1 [1] / American Chemical Society.

PROBLEM 28: DISCUSSION

28.1 Overview

The imidazopyridine nucleus (e.g. **3**) appears in several well-established and experimental pharmaceutical drugs. A rich chemical literature exists about the numerous ways to build this scaffold adorned with *ad hoc* substituents to enhance their potency. The approach shown herein, presented by researchers from Taiszhou, China [1], however, is still being explored and promises to give access to an expanded chemical space through a peculiar and somewhat hidden mechanism for the neophyte.

28.2 Mechanism Analysis

After examining target **3** as a function of starting compounds by deconstruction (Scheme 28.2), one can seamlessly identify 2-aminopyridine (**1**) and *p*-methylacetophenone (**2**) in a click. Soon enough, though, an unidentified C^6 stands out, playing the role of a bridge between **1** and **2**. Upon assigning HEDZ/LEDZ characters to starting compounds, one realizes that newcomer C^6 is likely a temporary LED carbon.

28.3 Possible Sources of C6

Reagents added to **1** and **2** that put them in motion or transform intermediates:

- ✓ $K_2S_2O_8$ (potassium persulfate): Not a C contributor, but endowed with oxidative properties through radical chemistry [2].
- ✓ Molecular iodine: idem, but through iodide addition followed by S_N2 removal as described in Problem 27.
- ✓ NaOAc (sodium acetate), a mild base and weak nucleophile. Forcing acetate incorporation to the scaffold hoping to insert C^6 would require three reductive stages (for you to figure out, get help from the FN concept, Chapter 4), but the medium is highly oxidative.
- ✓ DMSO: Our only candidate for C^6. However, DMSO is better known for its performance in strong base (NaH for example, not NaOAc) to afford a sulfonyl carbanion, an HED site, whereas we need an LED species for the mechanism to proceed and therefore must find a way to convert the traditional solvent into a one-carbon LED contributor.

Can this conceptually unpolung turn occur in DMSO? The answer is yes (do not miss Section 28.5, "Learning Issues and Takeaways"). Recent developments achieved this goal by exposing DMSO to specific oxidants that enable radical type reactions. Potassium (or sodium) persulfate ($K_2S_2O_8$) is one of these as per a possible mechanism portrayed in Scheme 28.3 (top section). In this compound, two sulfate units are linked by an O–O bond and hence, a peroxide. The substance is stable at room temperature, whereas heat splits it into two radicals. Propagation substracts a hydrogen atom from DMSO, forming one of three possible methylene radicals (only one is depicted for the sake of simplicity). After delocalization of the unpaired electron and additional radical footage, one ends up with methyl sulfenium cation **4**, the sulfur equivalent of an alkyl iminium ion whose electrophilic properties are well known.

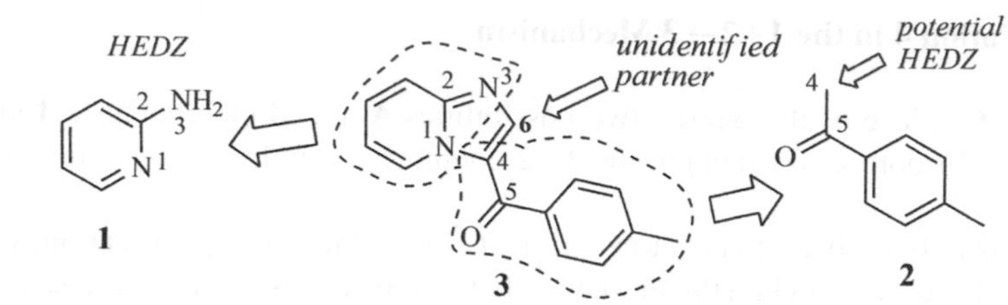

SCHEME 28.2 A straightforward splitting of target **3** into the starting materials that reveals an extra carbon atom not accounted for.

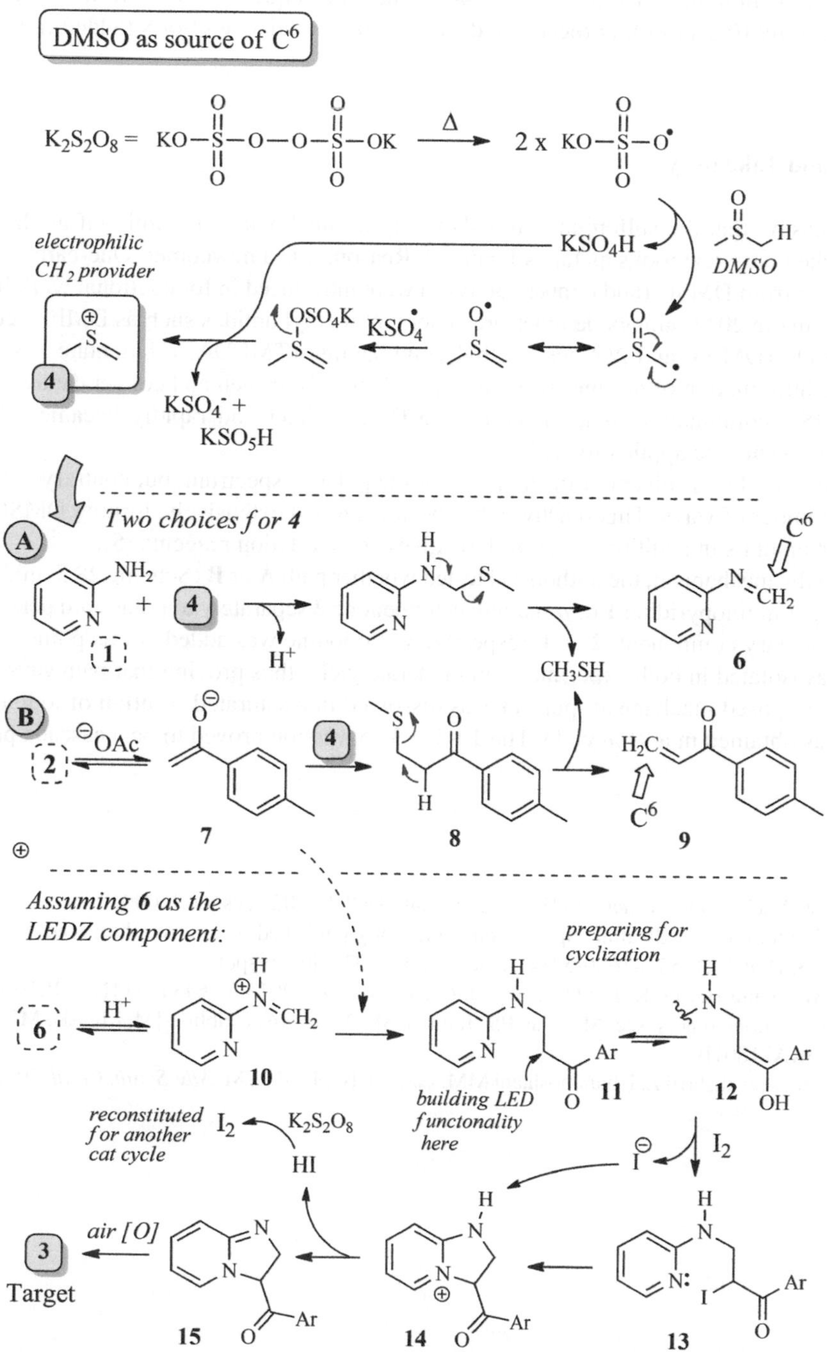

SCHEME 28.3 Dimethylsulfoxide can be a source of electrophilic methylation by a free-radical initated conversion to sulfenium cation **4,** which opens novel routes of alkylation.

28.4 Inserting Sulfenium Cation 4 in the 1+2→3 Mechanism

For this moiety to become the C^6 piece of the puzzle, two possibilities **A** and **B** emerge since **1** and **2** are both nucleophilic (Scheme 28.3, middle section). In both cases though, the **1**+**2** coupling through C^6 eventually converges on the same first adduct.

However, the C^6 alkylation is only halfway to the target, as the N^1–C^4 bond needs justification. While the pyridine N is prepared for a nucleophilic attack, the α-carbonyl methylene can only be activated as an enolate, also a nucleophile. Again, umpolung is required at C^4. How?

Enter molecular iodine for the well-known halogenation of enolates in **12**, which prepares the **13**→**14** cyclization (Scheme 28.3, bottom section). Elimination and air oxidation in **14–15** ends the sequence. Meanwhile, molecular iodine, which was added in catalytic amount only (0.2 mmol), is regenerated by persulfate oxidative efforts (added in molar excess) for another catalytic cycle.

28.5 Learning Issues and Takeaways

1. This is the first occasion that the sulfenium ion **4** shows up in this book. It is rarely–if at all–mentioned in currently available organic-chemistry textbooks as far as I can tell. Reason: **4** is a newcomer. One-carbon building blocks for heterocycle construction from DMSO (and copper catalysis) were introduced in foundational work by professor Qian Zhang from Changchu, China in 2013, alongside other precursors: *N*-methyl amides such as DMF (a commonly used solvent), *N*-dimethyl acetamide (DMA), and *N,N*-tetramethyl ethyldiamine (TMEDA; a bidentate base) [3]. The concept was developed after the introduction of nitrenes (carbene equivalents of nitrogen and copper catalysis).
2. The persulfate-DMSO combination as a one-carbon LED came later and rapidly became a hot topic that led to an increasing number of synthetic applications [4].
3. DMSO is universally used as a solvent at the high end of the polarity spectrum, but contrary to most solvents, DMSO is also reactive in a number of ways. This reactivity has been exploited extensively, turning DMSO into a practical source of one and two carbon units in addition to sulfur derivatives and oxidation reagents [5].
4. To gain insight into the mechanism, the authors [1] tested whether path **A** or **B** (Scheme 28.3, middle section) was feasible or not by submitting 2-aminopyridine **1** or *p*-methylacetophenone **2** separately to potassium persulfate in DMSO and then adding the supplementary component, **2** or **1**, respectively. No iodine was added at this point. Adduct **11** (Scheme 28.3, bottom section) was isolated in both experiments in moderate yield, thus proving that pathways **A** and **B** are possible.
5. To test further the proposed mechanism, pure **11** was dissolved in a saturated solution of iodine in DMSO, from which target product **3** was obtained in a 91% yield. The **1**+**2**→**3** conversion proved to be a cascade process.

REFERENCES

1. Zhang Y, Chen R, Wang Z, Ma Y. *J. Org. Chem.* 2021;86:6239–6246. DOI:10.1021/acs.joc.1c00023.
2. For a compact review of sodium peroxodisulfate applications in recently published organic synthesis see: Evangelista TCS, Ferreira SB. *SynOpen*. 2021;5:291–293. DOI:10.1055/a-1656-5714; Article ID: so-2021-d0046-spot.
3. Lv Y, Li Y, Xiong T, Pu W, Zhang H, Sun K, Liu Q, Zhang Q. *Chem. Commun.* 2013;69:6439–6441. DOI:10.1039/C3CC43129K.
4. For an interesting example, among others, see: Mahajan PS, Tanpure SD, More NA, Gajbhije JM, Mhaske MB. *RSC Adv.* 2015;5:101641–101646. DOI:10.1039/C5RA21801B.
5. For a comprehensive review, see: Tashrifi Z, Khanaposhtani MM, Larijani B, Mahdavi M. *Adv. Synth. Catal.* 2020;362:65–86. DOI:10.1002/adsc.201901021.

PROBLEM 29

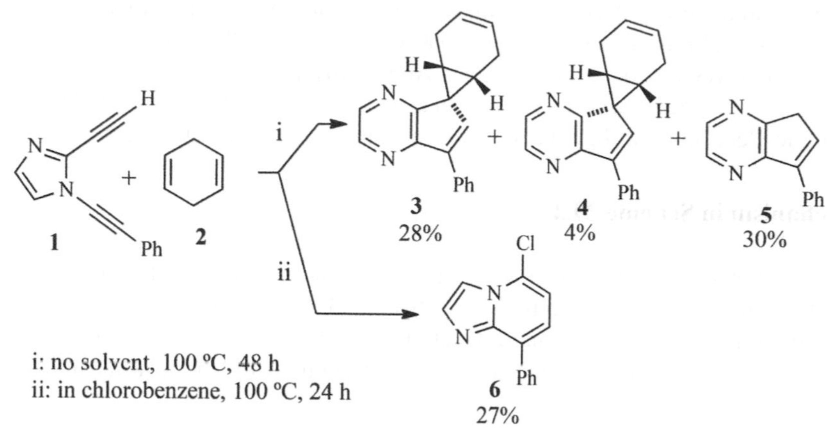

SCHEME 29.1 [1] / American Chemical Society.

PROBLEM 29. DISCUSSION

29.1 Overview

There is a flavor of a past problem in this collection regarding some of the structures of Scheme 29.1. However, this is a false impression since the fused spirocyclopropanes you see here have an entirely different origin, as authors from Texas, United States have discovered [1]. Some rare natural products contain the ene-diyne motif embedded in large and complex structures. Their cytotoxic activity focused on DNA cleavage responds to the curious ways of this molecular block. The mechanism supporting the present problem has a lot to do with the natural behavior of these compounds framed in the *Bergman rearrangement* background. The basic [3,3] electrocyclic reaction is shown in Scheme 29.4, bottom section (please check it up *after* solving this problem).

The unprecedented reactions of Scheme 29.1 constitute a charade of sorts: Under two very similar conditions, compound **1** and 1,4-cyclohexyl diene (**2**) without solvent afford the spirocyclopropyl targets **3** and **4** as a 7 : 1 mixture of enantiomers and **5** as the major product, on one hand, but on the other, the addition of chlorobenzene throws the reaction into a very different outcome. Your mechanism proposal should explain not only the divergent molecular construction but also the partial stereoselectivity. It is also of interest that when the terminal acetylene includes a second benzene ring the reaction gives untraceable mixtures.

29.2 Analysis for Solving 1 + 2 → 3 and 4:

A first visual examination reveals that **3** includes the traits of **1** and **2**, but the ring expansion of the imidazole moiety suggests some sort of rearrangement. The element balance yields the following:

Compound	Abridged molecular formula
1	$C_7H_3N_2(Ph)$
2	C_6H_8
1 + 2	$C_{13}H_{11}N_2(Ph)$
3	$C_{13}H_{11}N_2(Ph)$
3 − (1 + 2)	$C_0H_0N_0(Ph)_0$

There is no net loss or gain of elements, thus **3** is the stoichiometric coupling of **1** and **2.**

Orienting features can be extracted by pulling apart the target structure and analyzing the fragments according to Scheme 29.2 (top section). Adding pertinent labels also helps keep track of events in reverse. A schematic plan arises but we still need a decision regarding the type of intermediates in bond-breaking and forming steps: stepwise (ionic, radical) or concerted (sigmatropic). Having that many π bonds and NBPs packed in such a small space, there is substantial latitude for ideas. Consider, however, that there are no foreign reagents, metal species to build π complexes, Lewis acids or bases, no water to take and give protons conveniently, and a low polarity medium, so chances for ionic species are scarce. We seem to be painted into a corner unless we resort to **radical chemistry** driven by heat. Our familiar yellow arrow brings a manageable, albeit unorthodox mechanistic entry in the lower section of Scheme 29.2. Admittedly, a rare, exceedingly strained cyclic cumulene **8** crops up but, try as I might, there was no other viable alternative. Indeed, this is the authors' choice [1]. Please mind the use of half-headed arrows to move single electrons; for the sake of clarity, two head-on half arrows create a σ bond in these renderings.

29.3 Highlights of Mechanism in Scheme 29.2

✓ Interaction of the end carbons of the tethered alkynes can only occur by π–π orbital overlap of the *sp* carbons. The C^1–C^6 σ bond arises, even though the distance from them is a strong 4.326–4.576 Å as shown by energy-minimized molecular models of the rigid structure of **1** after allowing relaxation dynamics at 100 °C [2]. Furthermore, in the model, the two alkynes are not parallel to each other forming a coplanar structure with the aza-pyrrole ring but stick out at a 64°

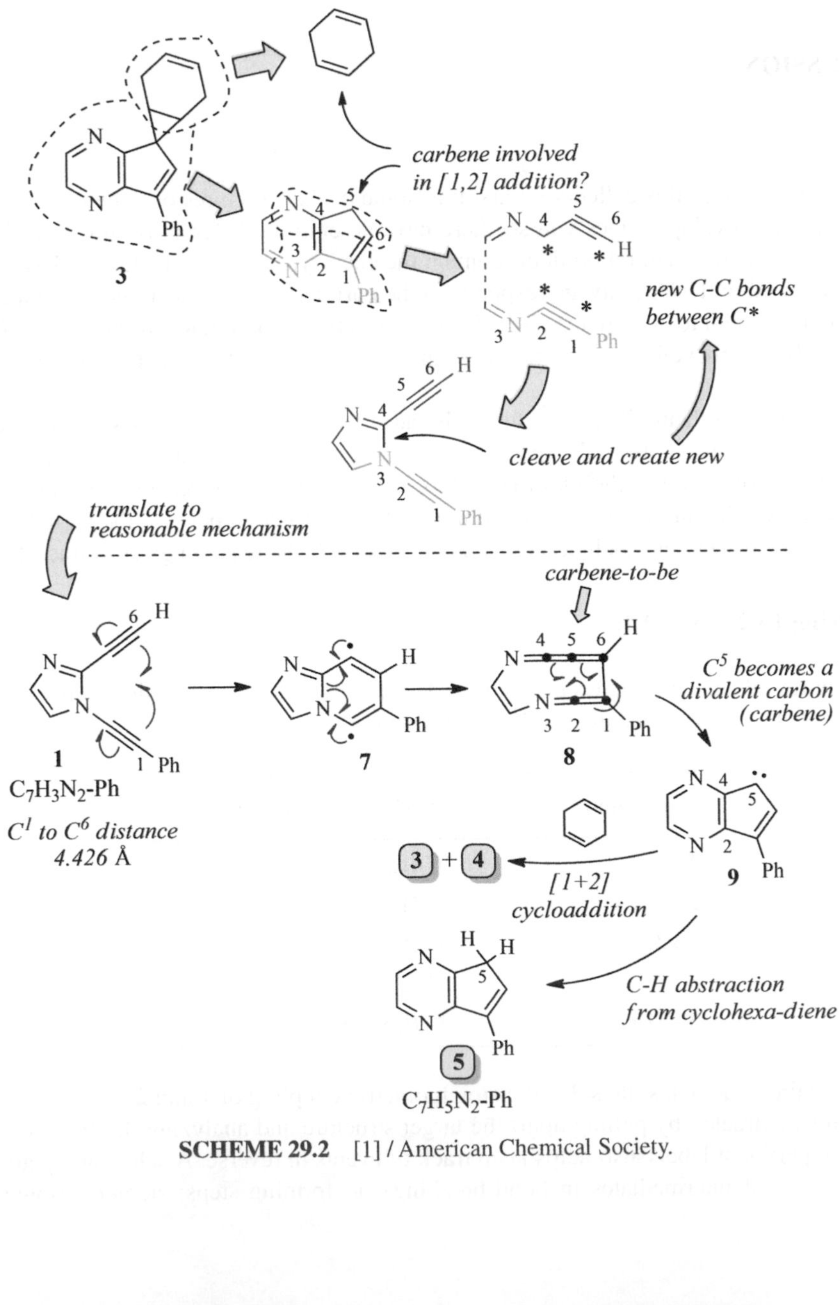

SCHEME 29.2 [1] / American Chemical Society.

C^2–N^3–C^4–C^5 dihedral angle. You can reason this by considering that C^1–C^2 alkyne is bonded to an sp^3 N, whereas the C^5–C^6 terminal acetylene is linked to a trigonal sp^2 carbon in the ring, and hence, is coplanar with it.

✓ Again, one may assume that vibrational degrees of freedom at 363 K allow the approach of the two carbon atoms at a closer distance. Semiempirical energy calculations were not performed by the authors to assess the activation energy required for this seemingly adventurous step. However, the authors trusted that the reaction would follow suit to a known rearrangement in all carbon ene-diynes known as the *Bergman cyclization* (see below).

✓ Emerging diradical **7** is set for the expected aza-pyrrole ring expansion through an astounding electron redeployment that yields cyclic, understandably unstable cumulene **8**.

✓ Before sending our imagination up to the clouds about what to do with **8**, one should pay attention to target **5** (Scheme 29.1), which has remained in the wings until now. Not only does the scaffold register the ring expansion (the desired C^2=C^4 bond is there), but it is strongly suggestive of a *reductive* rearrangement (pay attention to the molecular formulas of **1** and **5** in Scheme 29.2), from a previously formed *carbene* at C^5.

✓ It takes some imagination to redeploy electrons in the parallel cumulenes to create the C^5 carbene, but this task is made easier by remembering that carbenes have six valence electrons instead of eight. Take away two electrons from neutral carbon C^5 one at a time, since this is a radical-led reaction, and you have a *singlet* carbene (one electron per non-bonded AO). Use these traveling electrons to build the desired C^2–C^4 bond. The average C^2–C^4 distance is now 2.761 Å within reach of interacting AOs and the rest is valence bond electron balance.

✓ Once carbene **9** is established, divergent cyclopropanations to **3** and **4** and H trapping from cyclohexadiene to afford **5** complete the sequence smoothly.

✓ Major product **5** supports the carbene participation, and the cumulene route by retro-extension.

29.4 Explaining the 7 : 1 Ratio of 3 and 4

To better appreciate this one should watch computer-generated molecular models of energy-minimized structures as the ones depicted in Scheme 29.3. Adduct **3** displays an exo configuration whereas **4** is endo relative to the heterocycle. While the latter undergoes stereoelectronic repellency between the NBP of N and the double bonds of the cyclohex-1,4-diene during its approach to constitute the transition state of the [1 + 2] cycloaddition, the exo TS is free from this interference and consequently leads to the major epimer.

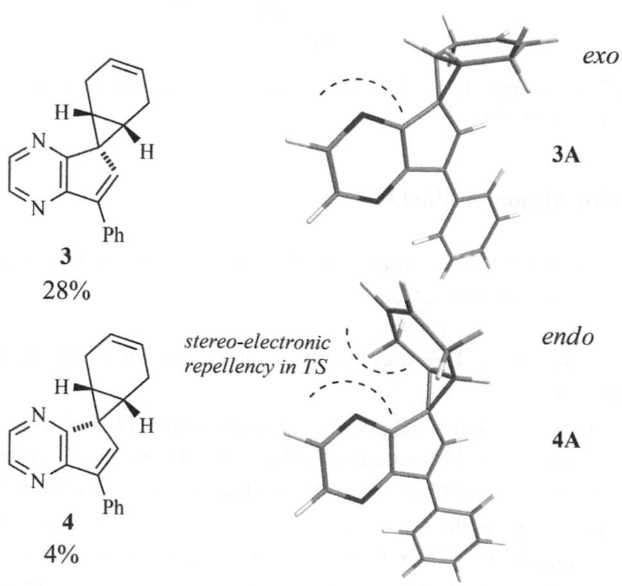

SCHEME 29.3 3-D renderings of targets **3** and **4** that explain the yield differences in consonance with the relative stereoelectronic repulsion of the cyclohexenyl moiety.

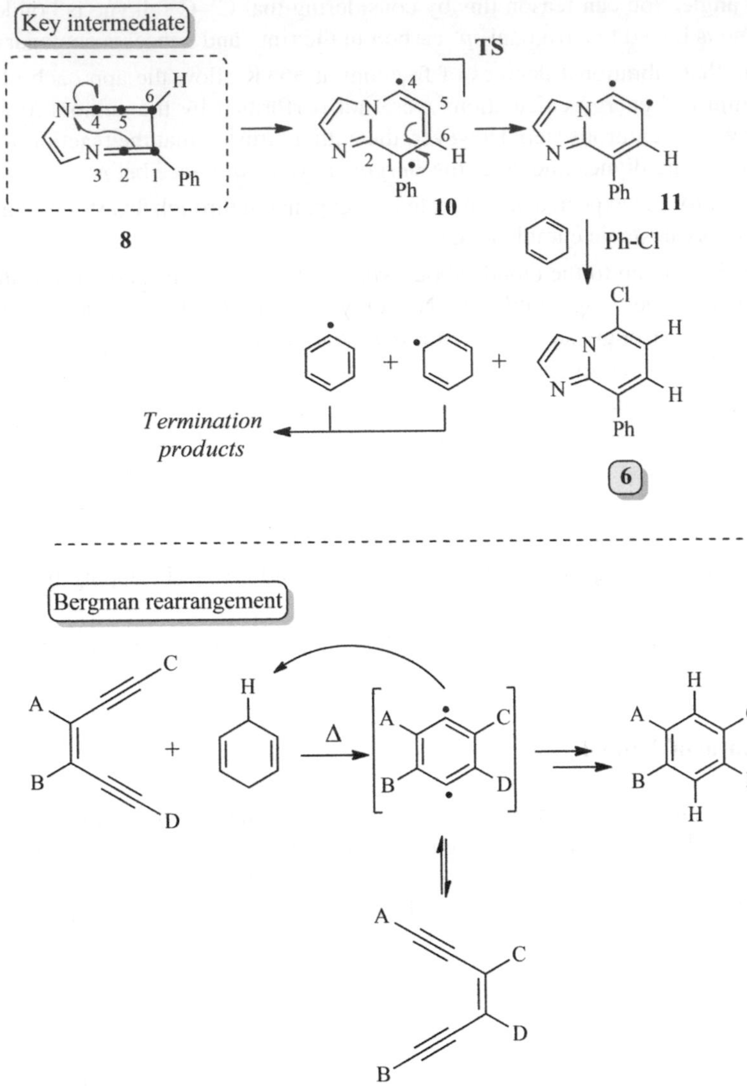

SCHEME 29.4 Alternative electron flow in cummulene **8** that allows the capture of a chloride atom from chlorobenzene and explains target **6**. This is reminiscent of the Bergman rearrangement.

29.5 A Reasonable Mechanism for Major Product 6:

Chlorobenzene changes completely the course of the reaction. Why is that? Before addressing this crux, reassessing the fragments of **6** and the new bonding differences is convenient:

- ✓ Conservation of the imidazole ring in the **1** → **6** conversion carries the idea that this motif remains unfazed during the construction of target **6**. Or does it?
- ✓ Doubts creep in when considering the position of the phenyl ring at the south of **6**, only one carbon away from the angular imine carbon instead of two carbons as in **3** and **4**. This speaks of either rearrangement ([1,2] phenyl shift radical-wise to be allowed suprafacially), or another bonding arrangement linking C^2 in the green section of Scheme 29.2 tops and the red N to the north.
- ✓ For this bonding to occur, one must unravel the imidazole ring as before, then work on cumulene **8** seeking to reconstitute the five-membered heterocycle including the C^2–N bond and a concomitant fused six-membered ring to the east of the molecule.
- ✓ In so doing, a carbon radical should persist in C^4 to capture a chloride atom from chlorobenzene.

In a nutshell, odd cumulene **8** holds the key. All it takes is moving electrons (one at a time) in a different manner to accommodate this plan. Scheme 29.4 portrays this sketch. The sequence forcedly passes through strained cumulene **10**, which is better taken as a TS of partially bonded C=C linkages. The postulation of TSs drives chemical soundness into the gray zone of swiftly passing ephemeral species. Please be reminded that there is only one H atom (at C^6) in the cumulene area of **8** besides the phenyl ring.

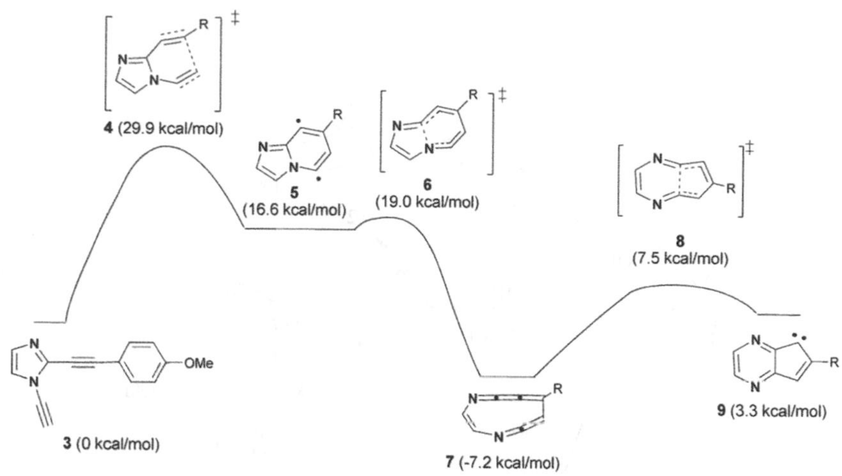

FIGURE 29.1 Relative electronic energies of structures postulated in Scheme 29.2 (p-CH_3O–C_6H_4 instead of phenyl) by unrestricted functional density theory (DFT). Reproduced from Reference [3] with permission from the Royal Society of Chemistry.

It is fair to say that the authors [1] also postulated diradical **11**, although they did not explain how it comes to be from **8**. The TS I depicted in **10** is conceivable but involves undue ring strain. Partial π bonds in the ring may accommodate this strain. Interestingly, the radical electrons in **11** occupy vicinal sp^2 AOs in the plane of the ring, which is a resonance form one finds in benzyne.

29.6 Learning Issues and Takeaways

1. In attempting the N-homolog of the Bergman cyclization (aza-Bergman rearrangement), aimed at enhancing the anti-cancer properties of natural compounds containing the ene-diyne moiety, Kerwin and coworkers [1] discovered a novel transformation passing through most amazing intermediates. Discovery is often serendipitous but takes open-minded scientists with a good handling of electron redeployment to take advantage of these unexpected results.
2. When one is forced to propose off-the-mark intermediates, it may be appropriate to perform kinetic studies and semiempirical calculations to take reaction mechanism hypothesis into the realm of quantitative assessment.
3. This is exactly what Professor Kerwin and his collaborators in Texas did [3] using two approaches:
   a. Running the kinetics of the thermolysis of a sibling compound of **1** (p-CH_3O–C_6H_4 instead of phenyl) to determine the activation energy of the reaction ($\Delta H^{\ddagger} = 30.0\,kcal\,mol^{-1}$). As you know, this value reflects the energy required to overcome the highest barrier of the entire sequence, although a multi-step sequence such as that of Scheme 29.2 may have multiple barriers. In principle, cumulene **8** appeared to be the best candidate, but kinetics alone would not disclose this.
   b. Perform quantum theoretical calculations (DFT) to estimate the electronic energies of intermediates and TSs according to the reaction mechanism model of Scheme 29.2 up to key carbene **9** [3].
4. Their findings are compiled in Figure 29.1. This chart reveals that:
   a. $\Delta H^{\ddagger}$ is in complete agreement with DFT calculations.
   b. The assigned structure is the TS of the first cyclization step (**1** → **7** in Scheme 29.2) and not **8**. We discussed earlier the barrier of this step due to the long distance between the alkyne ends (4.326–4.576 Å).
   c. The electronic energy of cumulene **8** (in our numbering system) is in fact *lower* than that of **1** by $-7.2\,kcal\,mol^{-1}$. Its cyclization to carbene **9** goes through a TS, peaking at $14.7\,kcal\,mol^{-1}$ of activation energy, about 50% of $\Delta H^{\ddagger}$ for the most energy demanding step.
   d. Although the equilibrium between cumulene and carbene is unfavorable for the latter, its further reaction to products leads to effective displacement to the carbene.
5. Further developments in this area are gaining strength in the biochemistry and bioactivity of this sort of compounds to destroy cancer-related DNA through radical species.

REFERENCES AND NOTES

1. Nadipuram AK, David WM, Kumar D, Kerwin SM. *Rg. Lett.* 2002;4:4543–4546. DOI:10.1021/ol027100p.
2. Energy minimum calculated by MMFF94G, M. E. Alonso-Amelot, this book.
3. Laroche C, Li J, Gonzales C, David WM, Kerwin SM. *Org. Biomolec Chem.* 2010;8:1535–1539. DOI:10.1039/b925261d.
4. M. E. Alonso-Amelot is author of RSC articles, and hence, is not required to request copyright permission by RSC.

PROBLEM 30

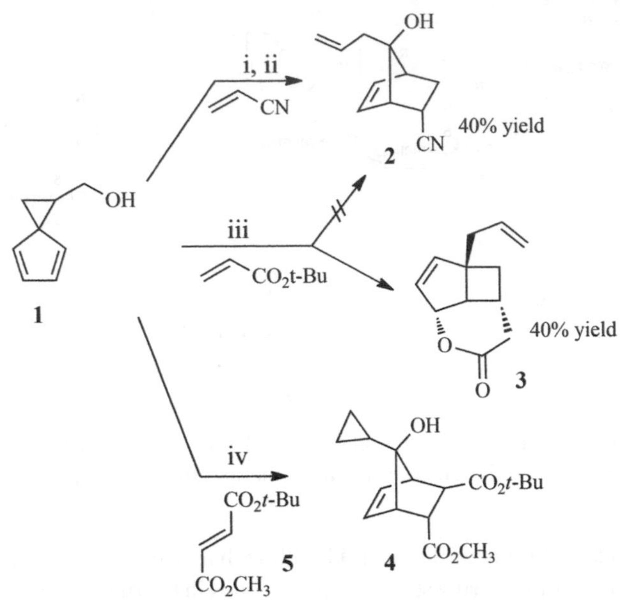

i: acrylonitrile, r.t., overnight
ii: MsCl, DCM, r.t., then silica gel, DCM

iii: same as i and ii but *t*-Bu-acrylate instead
iv: same as i and ii but *fumarate* **5** instead

SCHEME 30.1 [1] / Elsevier.

PROBLEM 30: DISCUSSION

30.1 Overview

This accessible problem, adapted from research work of these two-stage, one-pot reactions from Reading, United Kingdom [1] conceals nevertheless some not-so-easy and instructive questions, if one explores it under the surface (as one should in every problem). It deals with the contrast of outcomes stemming from apparently minor changes in the substituent pattern of involved reagents that operate on the same substrate **1**. Even more so when considering that the authors reported the expected production of **2** in comparable yield when $R=CO_2CH_3$ under similar reaction conditions. The idea of steric hindrance by the *t*-Bu group is inescapable, but from the data I selected in Scheme 30.1, one cannot be certain at which stage this encumbrance effect is more critical, step **iii** or afterwards, to the point of sending the sequence elsewhere to an unexpected side reaction. Indeed, the unanticipated occurred in the isolation of tricyclic lactone **3** and adduct **4** in possession of an odd-located cyclopropane.

Therefore, Scheme 30.1 involves three mechanism problems in a single rendering: explaining outcomes **2, 3,** and **4,** and the failed synthesis of **3**.

30.2 The 1 → 2 Conversion

A bird's eye view of target **2** shows that:

- ✓ A DACA takes care of building the norbornene scaffold.
- ✓ Then, spirocyclopropane is cleaved through a typical cyclopropyl-carbinyl SR elicited by a vicinal $C^{(+)}$. This carbocation emerges by carbinol departure since there is no other candidate.

✓ The role of mesyl chloride (MsCl) in step **ii** is to improve the nucleofuge capacity of the carbinol.

✓ The plan seems simple enough. However, the hydroxide comes back at the unanticipated carbon. You will realize this after labeling the carbon atoms of the partial spiro core (not shown herein). How can this take place?

Scheme 30.2 (top section) shows a feasible route by way of two endo TSs oriented differently relative to the carbinol. In fact, four endo stereo isomers (only **6a** and **6b** are shown) would be anticipated. Although the authors did not dwell on the stereochemistry issue, one might advance **TS6a** as that undergoing the least steric hindrance between CN and methylenehydroxy moieties. Yet, adduct **6a** was not observed. Additionally, *endo-trans* cyclopropyl (cp)-carbinol **6b** appears better suited regarding steric hindrance than the cis alternative (not shown). In the absence of further reagents, **6b** is the most likely primary product of **i**.

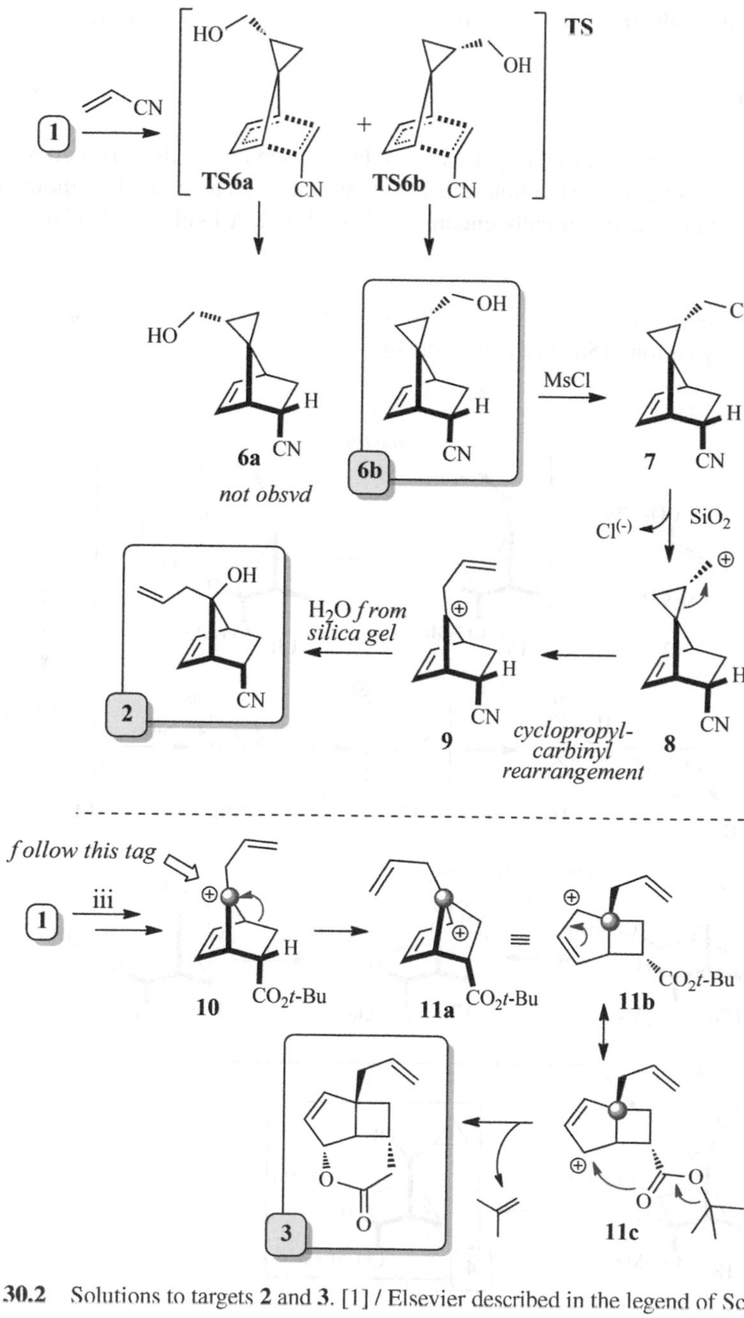

SCHEME 30.2 Solutions to targets **2** and **3**. [1] / Elsevier described in the legend of Scheme 30.1.

The sequence resumes with chloride **7** as a precursor of a cp-stabilized $C^{(+)}$ **8** elicited by surface-active silica gel. After cp cleavage, a tertiary $C^{(+)}$ **9** is produced and trapped by water residing in the silica gel matrix. Target **2** is thus accessed in 40% yield.

30.3 The 1→3 Conversion

Let us jump to the bottom reaction of Scheme 30.1, since a bridge tertiary cation like **9** may very well give us access to target **3**. This lactone is visibly enough the result of a deep-seated skeletal rearrangement of the strained norbornane nucleus usually brought forth by a $C^{(+)}$. An additional clue is the angular allyl appendage in **3**, a most likely relic of cp already set in **9**. Scheme 30.2, lower section portrays this skeletal rearrangement in just two stages portrayed in **11a–c** until the $C^{(+)}$ migrates to a proper distance, five atoms away in **11c**. The ester oxygen atom traps the carbocation and engender the observed lactone **3**, aided by the *t*-Bu group. It operates as electron donor in parallel to the ease of decarboxylation of other *t*-Bu esters. This sequence equally passes through a **6b**-like ester, the primary product (no need to show it, is there?).

30.4 The 1→4 Conversion

This process can be reasoned after realizing that target **4** cannot be reached unless there is a complete reorganization of the cp moiety, possibly cleavage and ensuing reconstruction rather than migration as a cp block without altering the southern molecular sector, irrespective of its strained and sterically encumbered condition. A likely mechanism (Scheme 30.3) can be articulated as follows:

- ✓ For reasons momentarily short of logics, more sterically congested adduct, a synthesis problem in itself (see Section 30.5), generates an **8**-like primary cation (**15a**) from chloride **14**.

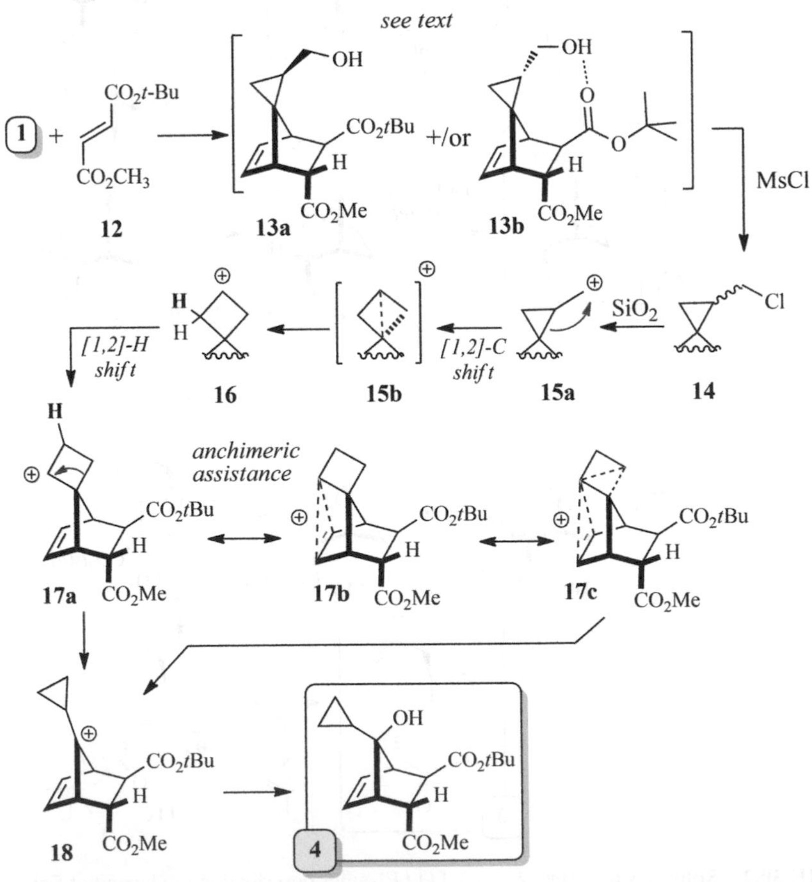

SCHEME 30.3 [1] / Elsevier, described in the legend of Scheme 30.1.

✓ This cp-stabilized $C^{(+)}$ evolves differently to the other cases described in Scheme 30.2. Cp seems to have migrated one carbon whereas the carbinol occupies the bridge tertiary position.

✓ The empty *2p* AO of trigonal cation **15a** interacts with the vicinal cp bonds dispersing the positive charge among three-carbon atoms in bicyclo[1.1.0]butane **15b**. This structure is one of three equivalent semi (partial bonds) bicyclobutane cations, which can also be rendered as cation **16** for the sake of understanding what comes next.

✓ Because we need the $C^{(+)}$ unit next to the bridge quaternary carbon to move forward, a *suprafacial* [1,2]-H shift in **16** (allowed by symmetry rules in $R_2CH–C^{(+)}$) solves this roadblock by rendering cation **17a**.

✓ Anchimeric assistance from the norbornene C=C provides the stability conceptual frame of a *non-classical carbonium ion* **17b**, a heavily discussed topic [2]. This cation is similarly subjected to a stabilizing bicyclo[1.1.0]butane cation structure **17c** closely related to **15b**.

✓ Either **17a** or **17c** isomerize to carbocation **18**, which precedes the desired target **4** due to water provided by silica gel.

30.5 Learning Issues and Takeaways

1. Among other noteworthy issues, this problem illustrates the diversity of molecular interactions and outcomes from a single, small molecular frame (**1**) exposed to just three dienophiles followed by similar treatment with standard reagents (mesyl chloride, silica gel).
2. The detailed exploration of feasible mechanisms gives rise to situations, intermediates and targets not easily justified.
3. When considering the stereochemical implications, the DACA of **1** and acrylonitrile (Scheme 30.2) should furnish four *endo*-CN adducts in compliance of the Alder endo rule (see Chapter 3). The most favorable adduct for stereoelectronic reasons is probably **6a**, yet this product was not observed. Authors favor structure **6b** albeit not defining the R/S configuration of the cp-methylenehydroxy moiety. I did for exclusively scholarly reasons but must concede that authors used a 1 : 1 racemic mixture of **1**. The ensuing steps (hydroxy-chloride exchange, ionization, and rearrangements) are independent of the said R/S configuration anyway.
4. Worth noting, previous work by researchers at the California Institute of Technology in California, United States described a highly exothermic DACA of **1** (as the *t*-butyldimethyl silyl derivative) and N-phenyl maleimide, obtained a 1 : 1 mixture of diasteromers **20a** and **20b** [3] (Scheme 30.4), in consonance with my prediction for **6a/6b**. After their separation by fractional crystallization, **20b** was converted to the more stable **20a** after heating at 100°C in chlorobenzene. Could you explain this conversion? (See answer at the end of this section.) Therefore, only one enantiomer was submitted to rearrangement conditions to furnish **21** in >90% yield. Take note of the different structure of **21** relative to our problem target **2**. How do you account for this?
5. More questions. What reason is there for the stark fate difference of cations **9** and **10** leading to targets **2** and **3**? In other words, which force fuels S_N1 substitution in **9** and scaffold rearrangement in **10**? Both cations are anchimerically assisted by the norbornene C=C, thus are equally long-lived. The cause ought to reside in the endo substituent, CN versus CO_2t-Bu ester. One may appeal to stereoelectronic reasons dangerously close to hand-waiving phenomena. Future semiempirical studies will avoid speculative explanations.

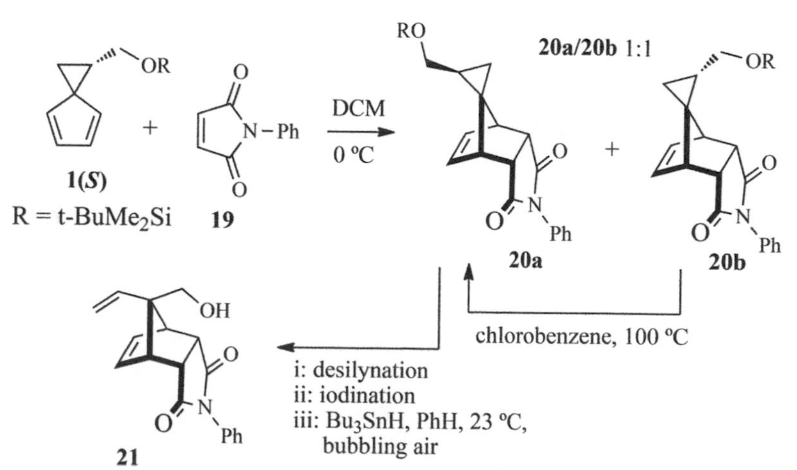

SCHEME 30.4 [3] / American Chemical Society.

Answer to the question in point (4) above: thermal retro-DACA followed by another round of DACA at high temperature favors the kinetic product **20a**.

REFERENCES AND NOTES

1. Nadany AE, Mckendrick JE. *Tetrahedron Lett.* 2007;48:4071–4074. DOI:10.1016/j.tetlet.2007.04.009.
2. For a favorable opinion on the non-classical norbornene cation based on spectroscopic evidence, see: Olah, JA, Prakash GKS. *Acc. Chem. Res.* 1983;16:440–448. DOI:10.1021/ar00096a003. For the emphatically opposing stance published in the same issue, see: Brown HC. *Acc. Chem. Res.* 1983;16:432–440. DOI:10.1021/ar.00096a002. For a highly instructive and thorough computational study of this question by heavyweight scientists, see: Schreiner PR, Schleyer PvR, Schaefer HF. *J. Org. Chem.* 1997;62:4216–4228. DOI:10.1021/jo9613388.
3. Starr JT, Koch G, Carreira EM. *J. Am. Chem. Soc.* 2000;122:8793–8794. DOI:10.1021/ja0019575.

PROBLEM 31

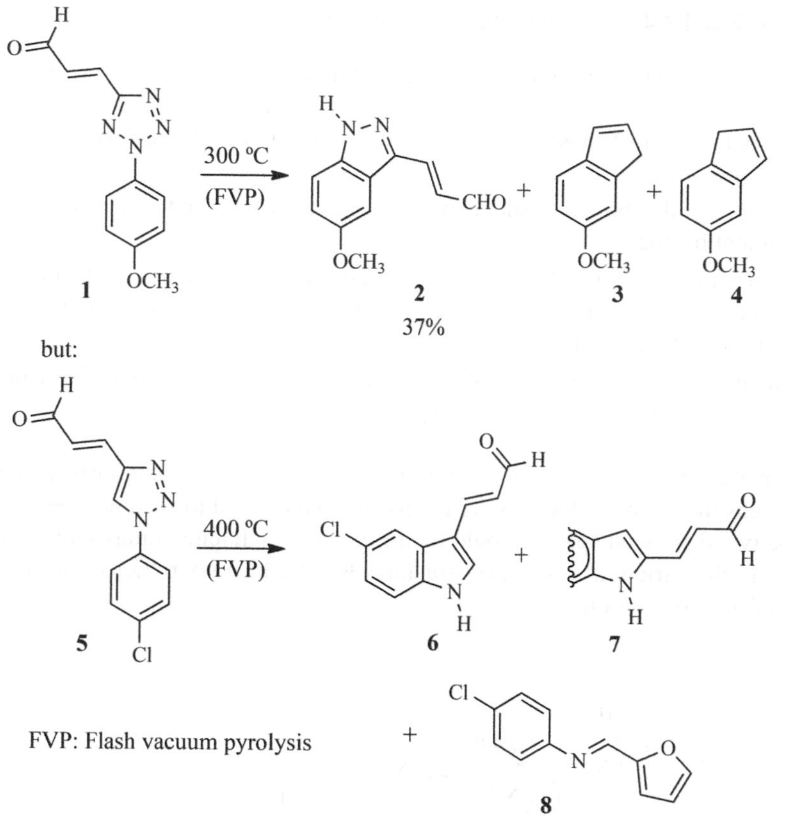

SCHEME 31.1 [1] / Elsevier.

PROBLEM 31: DISCUSSION

31.1 Overview

In the preface, the thermal instability of the great majority of organic compounds at temperatures exceeding 450 °C was emphasized. Organic life and the organic chemical space as we know it reach an overwhelming diversity only under the very narrow range of temperatures that we enjoy on Earth. On the other hand, it is tempting to explore the outcome of subjecting complex compounds to extreme thermal conditions.

Flash vacuum pyrolysis (FVP) is one useful technology to navigate this ocean of surprises, as this work from scientists in Cordoba, Argentina and Budapest, Hungary demonstrate [1]. As its name suggests, in FVP compounds are passed very quickly through a flow chamber at a high temperature then rapidly cooled and analyzed so the exposure time is very short (10^{-2} seconds). Primary products often survive in manageable yields, which provides valuable information of reaction mechanisms involving radicals, carbenes, zwitterions, intramolecular element migration, and concerted cycloadditions. Because neat compounds are tested, there are no solvent effects to complicate the reaction innards. In devising feasible mechanisms, one feels the denuded sensation of being alone with no other help than the molecular structure, its possibilities, and your own logics.

To our good fortunes, however, the somewhat challenging cases presented in Scheme 31.1 include heteroatoms, their polar effects and non-bonding electron pairs (NBP) are always welcome, and an abundance of unsaturated bonds suitable for a variety of electron redeployment pathways, bringing mechanistic diversity and enjoyment.

A peculiar feature of acrylaldehydes **1** and **5** is the disparate course of their pyrolysis due to the occurrence of four (tetrazole) versus three (triazole) nitrogen atoms in the five-membered rings, and the **4**-substituent of the arene ring. Therefore, the analyses of **1** and **5** will be treated individually, even though they may have common intermediates.

31.2 Planning the Pyrolysis of Tetrazole 1

Since two distinct structures, **2** and **3/4**, are produced from **1**, we would rather treat them separately.

31.2.1 1 → 2 By comparing these two structures (Scheme 31.2, top section) one comes to realize that:

- ✓ There is an orphan N=N moiety in **1** (absent in **2**) likely released as molecular nitrogen. Chances are there for an electrocyclic cleavage.
- ✓ The remaining electrons after this breakage should be able to attack the arene to create a C^4–C^8 bond with concomitant H transfer to furnish element balance.
- ✓ The said intramolecular rapport may involve radical or dipolar species, in one direction or another, depending on how one manages electron balance and redistribution after the N_2 exit. The latter can be executed while a triple N–N bond is built using the N=N adjacent σ-boding electrons.
- ✓ In the FVP environment, a neutral compound splits or rearranges into an electronically neutral set considering all fragments, meaning radicals, zwitterionic, or uncharged species.

This conceptual plan is conveyed in the sequence of Scheme 31.2, lower section, which starts with the clockwise electron redeployment shown in **1**, whereas the counter clockwise redistribution would lead to separate species you should be able to construe. If that happened, the two dislodged species would fly away at the high temperature and vacuum to never see each other again. Please pay attention to the variety of resonance structures **9a–d**. Make your choice for the ensuing **9** → **10** cyclization, which ends in a [1,3]-H shift to restore aromaticity.

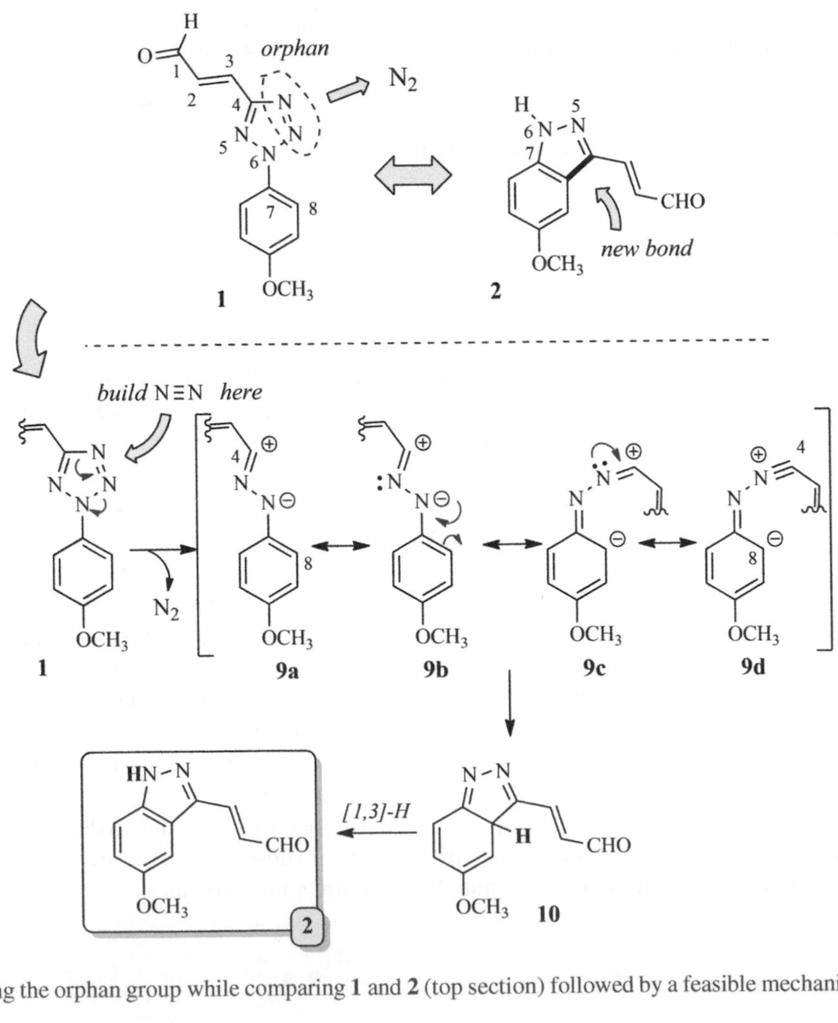

SCHEME 31.2 Spotting the orphan group while comparing **1** and **2** (top section) followed by a feasible mechanism to the first target (lower section).

31.2.2 Explaining 3 and 4: Envisioning a First Plan Putting side by side either **3** or **4** and **1** allows one to see what it takes:

✓ Remove the 1-2 diazole nitrogens althogether.
✓ Create two bonds: $C^8–C^4$ and $C^7–C^2$.
✓ Perform this maneuver without physical separation of the molecular bodies: arene and acrylaldehyde appendage.
✓ Find a way to decarbonylate since the $C^1=O$ moiety has vanished in **3** and **4**.

An impossible task, unless a step-by-step sequence is envisioned that keeps both units properly linked until the desired features are reached. This mission is perhaps feasible by calling up compound **2** to serve as the comparison standard; use it then as the starting species leading to **3** at least, and maybe **4** (Scheme 31.3, top section).

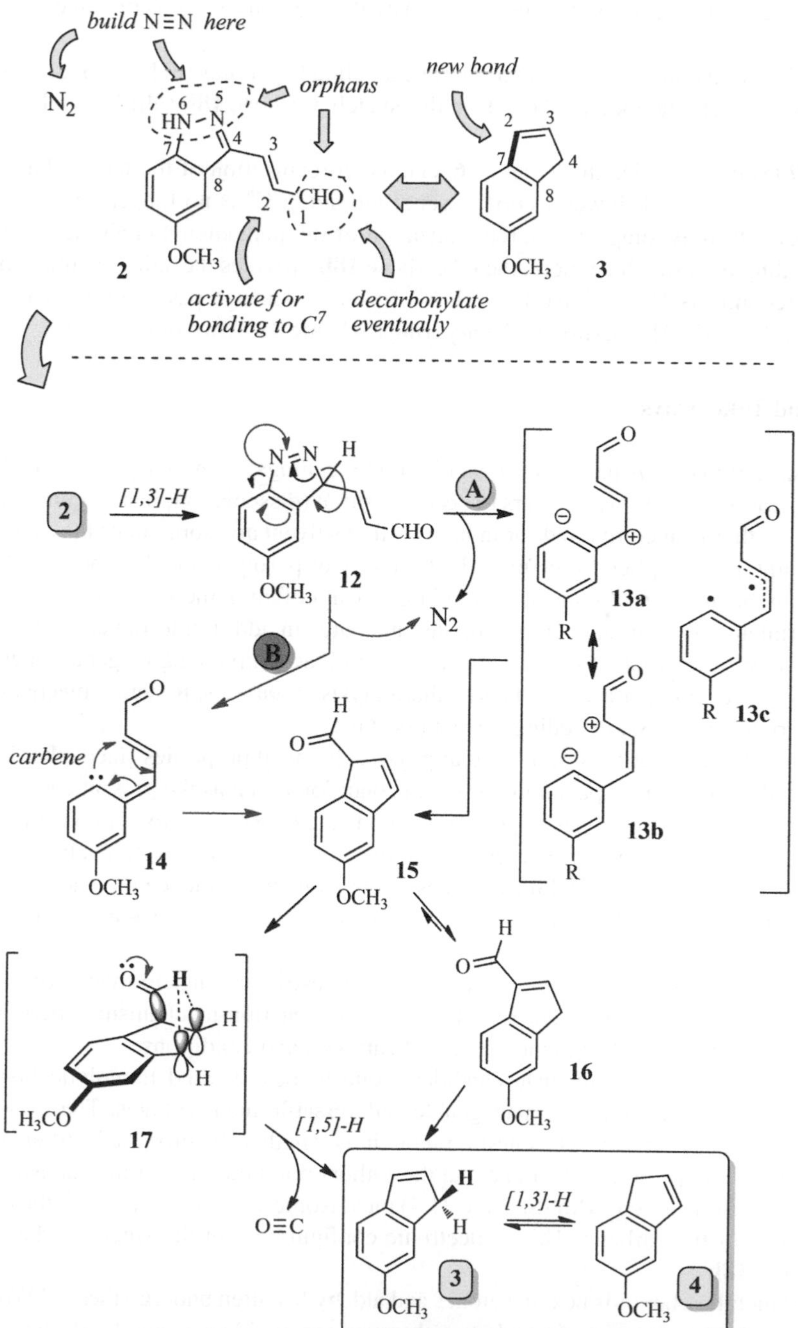

SCHEME 31.3 Strategy to convert the first reaction product **2** to secondary targets **3** and **4**, followed by two alternative reaction mechanisms.

This roadmap turns into a reasonable sequence (Scheme 31.3, lower section) by forming the N=N bond in **12**, which can evolve according to electron distribution to zwitterion or diradical species (pathway **A**, black curly arrows), or more interestingly, to carbene **14** (route **B**, red curly arrows). Both converge on indene aldehyde **15** by intramolecular coupling or $(2n+4\pi)$ electrocyclization (**14**). Authors identified **15** in moderate yield among the products [1].

The pyrolysis could have stopped there were it not for the likely decarbonylation. Authors do not comment about the details, but I wish to add the possibility of anchimeric assistance of the C=C bond to facilitate the [1,5]-H shift portrayed in **17**. Targets **3** and **4** emerge from the thermal turmoil.

31.3 Drawing a Map for the Pyrolysis of 5

The seemingly minor difference between **1** and triazole **5,** just one carbon instead of a nitrogen atom, throws the latter to an entirely divergent set of products. The learning lessons provided by the previous analysis of **1** may speed up our comprehension of what is going on.

By placing compounds **1**, **6**, and **8** in similar orientations, and labeling key atoms (Scheme 31.4, top section), one can rapidly realize the need to enable C^4 to form the indicated bonds in this sketch. Carbene **18b** fulfills the electronic requisites to furnish both targets **6** and **8**.

However, end product **7** (Scheme 31.1), an isomer of **6**, entails reorganization of the tethered aldehyde chain. Labels hold the key to this conclusion (Scheme 31.4, lower section, inset at left side): C^8 is no longer bonded to C^4 but to C^5, while the C^4–N^6 is also a new linkage, strongly suggesting a reorganization of the pentadienyl chain. Either C^4 or C^5 evolve to active centers to promote this bonding pattern. Once more, the C^4 carbene **18b** provides the solution, this time through a transient and strained azeridine **20** that reconnects N to C^4 and C^5 with N^6. Ensuing cleavage passes on the carbene character to C^5. This carbon is thus enabled to perform C^8–H insertion in the aryl ring (Scheme 31.4, bottom section).

31.4 Learning Issues and Takeaways

1. The terms *thermolysis* and *pyrolysis* have somewhat different meanings, even though both involve molecular transformation under high temperature. Thermolysis refers to decomposition (dissociation, rearrangement, elimination, decarboxylation, and so forth) of either neat compounds or in high boiling solvent and sometimes in sealed tubes to heat the mixture beyond its boiling point at atmospheric pressure. The formality of pyrolysis involves heating the neat compound at even higher temperatures in the absence of reagents (including air) and often in the vapor phase.
2. In thermolysis experiments one can appeal to ionic intermediates in addition to radicals, zwitterions, and carbenes that can be stabilized by solvent interaction or close association with accompanying reagents. Intermolecular cycloadditions and several electrocyclic reactions are verified under thermolysis conditions. Reaction mechanisms may be complex due to the variety of contributors to TSs, including solvent assistance.
3. Pyrolysis, in turn, provides a closer look at molecular physicochemical properties since other intermolecular phenomena are reduced to minimal expression, especially in vacuum conditions such as the present case.
4. Reaction mechanisms in organic pyrolysis of pure compounds are constrained to unimolecular radical, zwitterionic, and carbene intermediates but also involve several types of more or less concerted electrocyclic and elimination reactions that take place in the molecular enclosure. Stabilization/destabilization and influence on reaction rates can only be rendered by components of the individual molecular structure. Anchimeric (NGP) assistance may be explored successfully in such conditions.
5. Pyrolysis is also carried out with catalysts in laboratory and industrial settings to treat waste and biomass, a chemically complex and necessary technology to reduce waste pollution. Reaction-mechanism experts are needed to understand what happens in these reactors and design more efficient catalysts and conditions.
6. Although radicals are expected intermediate candidates due to heat-elicited homolytic bond dissociation in several organic structural types, carbenes are also manageable and versatile intermediates. The present study is an instructive lesson in carbene formation-reactivity. The question though, is whether the proposed carbenes are singlet or triplet (see Chapter 2). There is no direct proof of either one, but the authors point out in their mechanism description of furan target **6** that *trans* C^2=C^3 in C^4 allyl carbene **18b** (Scheme 31.4) must isomerize to cis to procure the overlap of the oxygen NBP and the empty *p* orbital of the carbene. This is electronic configuration of the *singlet* carbene and determines electron circulation as shown in **18b**.
7. Singlet carbenes and metal carbenoids next to ketones and aldehydes often undergo thermal Wolff rearrangement, a peculiar and interesting vinylogous version of which was observed nearly 50 years ago by Professor Amos B. Smith III from

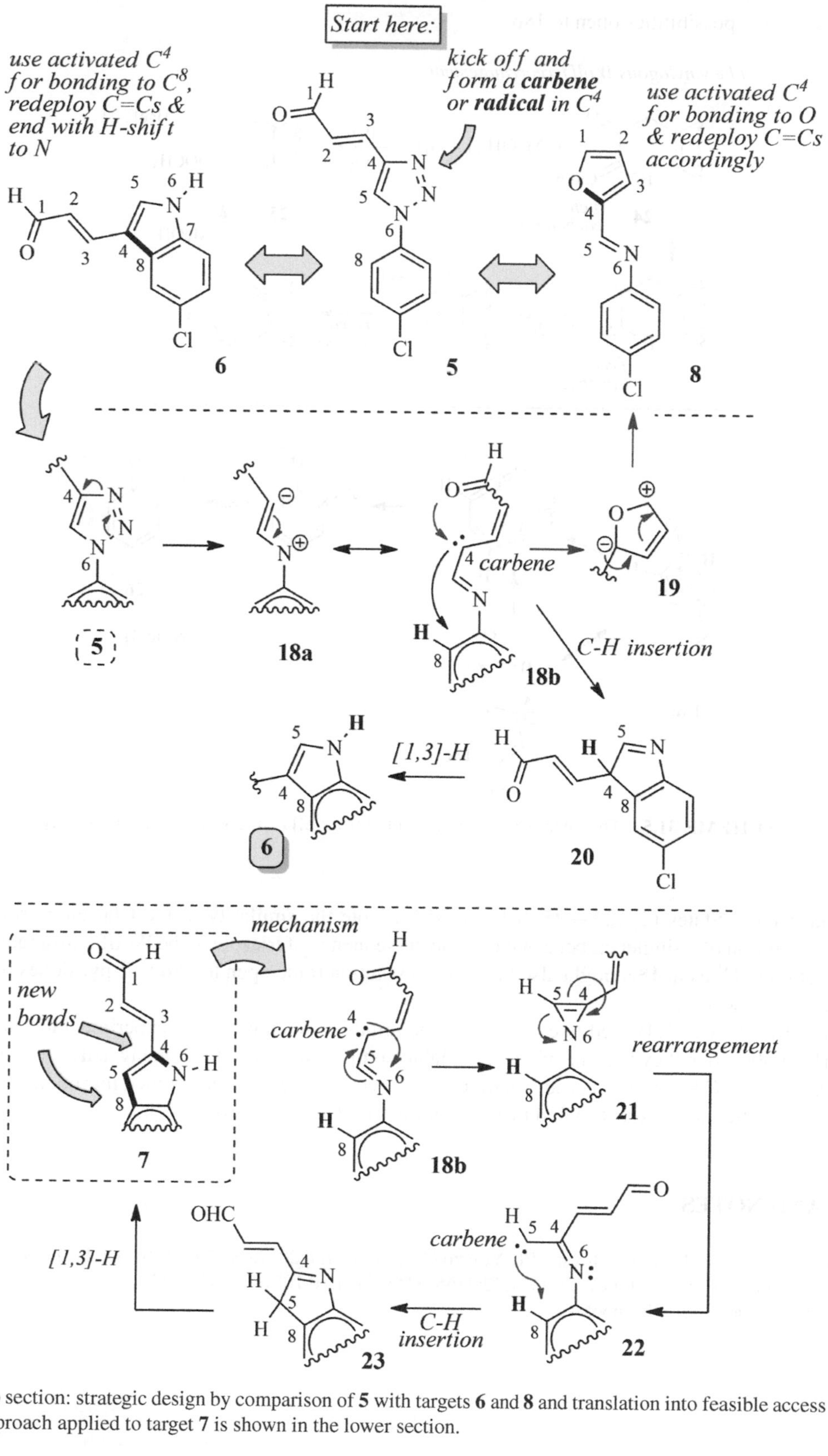

SCHEME 31.4 Top section: strategic design by comparison of **5** with targets **6** and **8** and translation into feasible access sequences (middle section). A similar approach applied to target **7** is shown in the lower section.

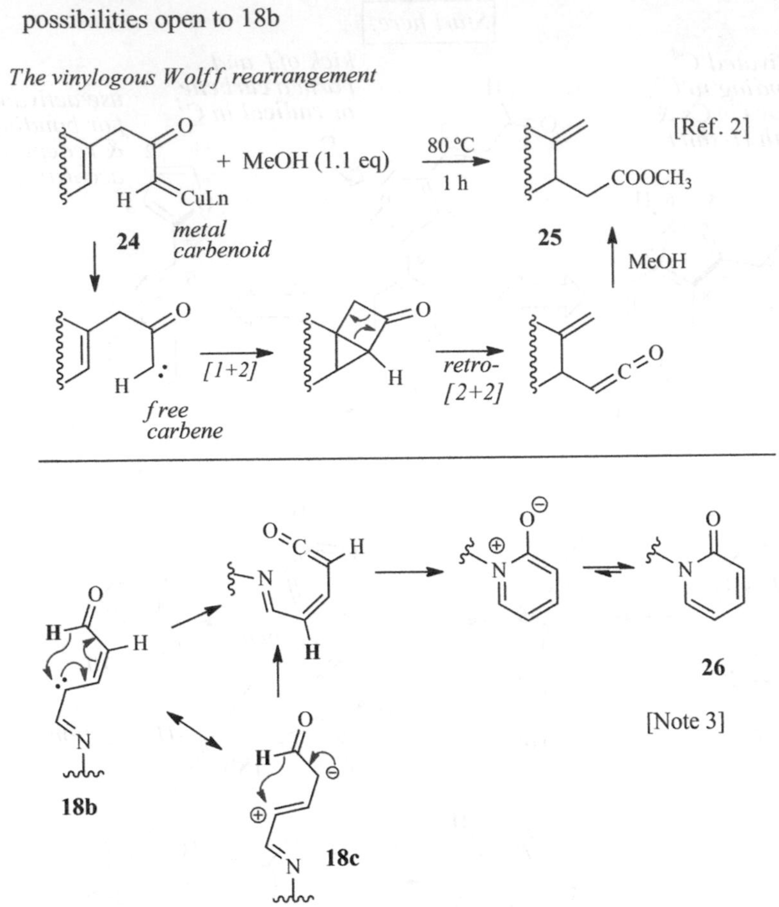

SCHEME 31.5 The upper section was adapted from [2] / American Chemical Society.

Pennsylvania, United States [2] (**24**→**25**, Scheme 31.5). Note the similarity to the Claisen rearrangement. Having a vinylogous expression of a singlet carbene in the pyrolysis sequence of triazole **5**, the Wolff rearrangement, either directly or after migration to C^1 as in **18c**, might also be a novel evolution route open to **18b** to 2-pyridones (e.g. **26**) as shown in Scheme 31.5, lower section [3].

8. Azirines like **21** (Scheme 31.4) used to exist only in the minds of scientists as an experimentally inaccessible curiosity until the early 1960s when they became of age as isolable compounds. Since then, many azirines in their two types (having either a C=N or C=C bonds, the 1- and 2-azirines) have been synthesized and studied thoroughly as stable compounds and valuable synthetic tools; expect to find these in future problems of this workbook.

REFERENCES AND NOTES

1. Lucero PL, Peláez WJ, Riedl Z, Hajós G, Moyano EL, Yranzo GI. *Tetrahedron* 2012;68:1299–1305. DOI:10.1016/j.tet.2011.11.034.
2. Branca S, Lock RL, Smith III Ab-. *J. Org Chem.* 1977;42:3165–3168. DOI:10.1021/jo00439a012.
3. Hypothesis by Alonso-Amelot ME, this book.

PROBLEM 32

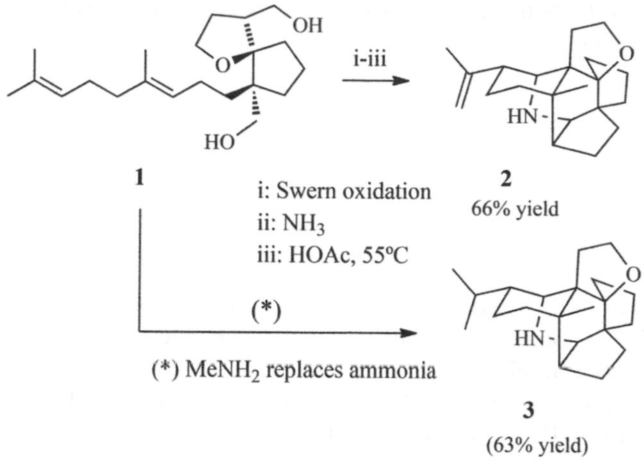

SCHEME 32.1 [1] / American Chemical Society.

PROBLEM 32: DISCUSSION

32.1 Overview

This reaction is a key step in the synthesis sequence of isoprenic alkaloids isolated from *Daphnyphyllum* plants native to Pacific islands (e.g. Japan, New Guinea). The work was published at the turn of the twentieth century and forms part of a collection of brilliant advances of synthetic chemistry of that time. Enormously interesting and challenging chemistry was being carried out in mainstay research groups including Professor Clayton Heathcock's group at Berkeley.

Reasons for selecting this reaction for "The Problem Chest", Part II, were:

- ✓ Although the reagents are trivial, the carefully designed starting material **1** contains the functional and skeletal components necessary to build the substantial amount of complexity found in product **2**.
- ✓ Stereochemistry is not easily grasped in polycyclics like this one until one gets sufficient practice by understanding the components. Any organic molecular structure can be dismounted into convenient fragments no matter how complex. Other problems in this collection offer similar opportunities.
- ✓ The problem can be studied both ways: by *retromechanistic analysis* as frequently suggested in this workbook as well as *rational forward interpretation of reaction possibilities.*

Although some of you may have hinted the cascade sequence, let us examine it anyway for the sake of illustrating the analytical weaponry at our disposal from "The Toolbox", Part I, and learn how to use in more forthcoming complex transformations.

32.2 Turning 2 into Pieces Resembling 1

We will do this in two steps:

1. Because **2** is too complex to discern after a cursory inspection, let us pull apart **2** in at least two scaffold sections using black and gray tones (Scheme 30.2, upper section). I chose Block **A**, owing to its resemblance with the spirotetrahydrofuran moiety of **1**, which is therefore conserved. Block **B** stands out at the front of the 3D rendering. Meanwhile, the orphan NH must stem from an external source since **1** does not bear nitrogen. Ammonia is the only possibility.
2. Start the backbone reconstruction. Put the two blocks **A** and **B** back together at the wiggled C–C bond and tag carbon atoms with numbers (Scheme 30.2 lower section). There is no need to follow IUPAC rules, just build reasonable number sequences that you can use later. Then translate key carbon-atom labels to target **2**.

32.3 Bonding Balance

Bonding balance: multiple C–C and C–N bonds in the **1**→**2** process materialize easily thanks to the atom labels. The new bonds (highlighted in green) in Scheme 32.2 are:

a) C^{10}–C^{18}
b) C^{9}–C^{13}
c) C^{1}–C^{14}
d) C^{9}–N
e) C^{10}–N
f) C^{19}=C^{20} from C^{19}–C^{20}

Note that C^1 and C^{10} of block **A** are C–OH functions in **1**, whereas C^{13}, C^{14}, and C^{18} are sp^2 carbons amenable to C–C bonding by addition reactions through $C^{(+)}$. The acidic medium of *step iii* would be ideally suited to this end; hence, the C–C bonding of the isoprenyl chain should take place at this stage.

This step should be preceded by the incorporation of N to a functionalized form of **1**, since alcohols do not react with NH_3; aldehydes do to furnish reactive imines. The initial oxidative step (Swern oxidation) serves this purpose.

With these conclusions on hand, the mechanism sequence of Scheme 30.3 can be proposed. It amounts to a four intramolecular alkylation cascade with exquisite stereochemical control; an amazing achievement.

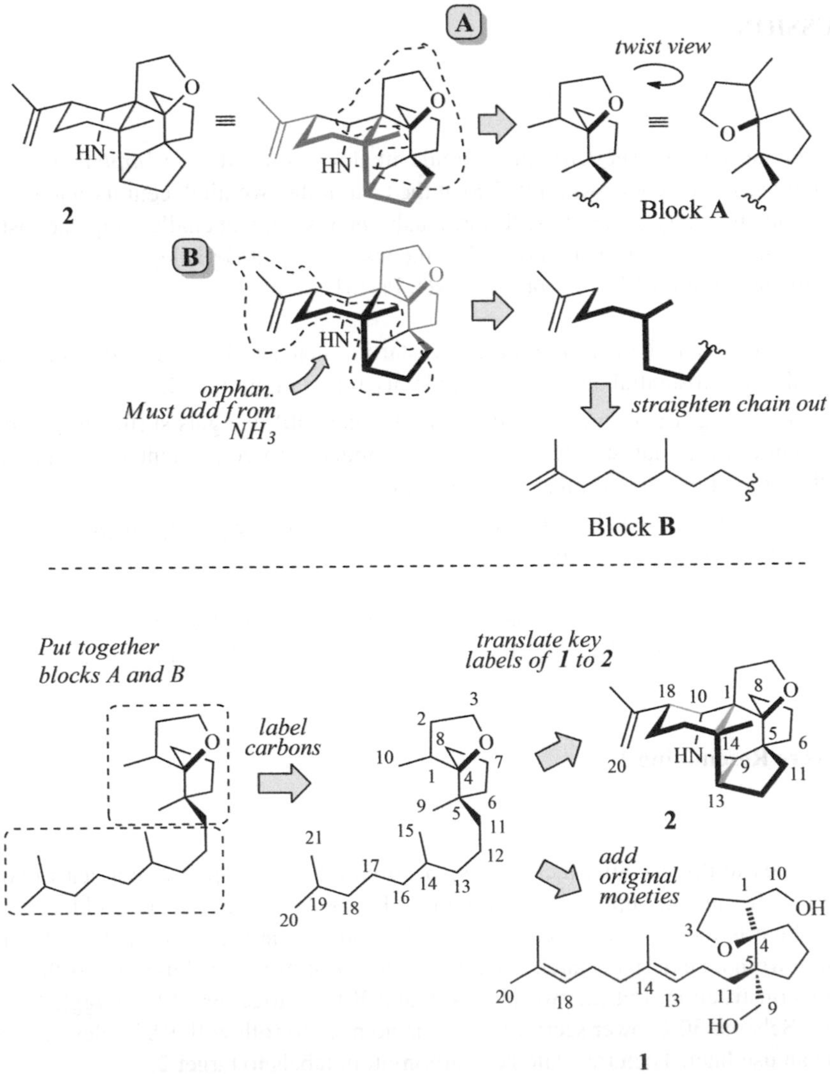

SCHEME 32.2 3D molecular renderings for the spatial appraisal of target **2** to identify molecular sections of starting material **1.** The identification of bond breaking and forming steps is eased by labeling atoms.

32.4 Solving 1 → 3

Authors of this work [1] observed that when ammonia is replaced with methylamine in step **ii**, saturated product **3** is produced. Can you explain this?

The sequence is essentially the same up to the carbenium ion that preceds the final product **2** with the added *N*-methyl group. Before the terminal olefin is formed, the short distance between the hydrogens of the N–CH_3 unit and $C^{(+)}$ in **11** conducts the remote hydrogen transfer depicted in Scheme 30.4. The resulting iminium ion in **12** is hydrolyzed to the secondary amine during workup.

32.5 Learning Lessons and Takeaways

1. The reactions of Scheme 32.1 illustrate the ease of interaction between active moieties in complex molecules. Stimulus triggering domino or cascade sequences may be as simple as a moderately protic medium and mild warming, as it occurs here from enamine **6** (Scheme 32.3) onward.
2. Sequences like this one require careful synthetic planning and a great deal of mechanistic foresight since there is more than one option (the one herein depicted) that would furnish several end products, intermediates stopped in their tracks by proton elimination, H shift, or rearrangement to unknown scaffolds. By exploiting what you have learned herein, you will be able to conceive some such secondary processes based on Scheme 32.3 intermediates.
3. The remote hydride transfer (Scheme 32.4) is an indication of the strong electrophilic power of the tertiary carbenium ion in combination with the HED effect of the amine NBP.

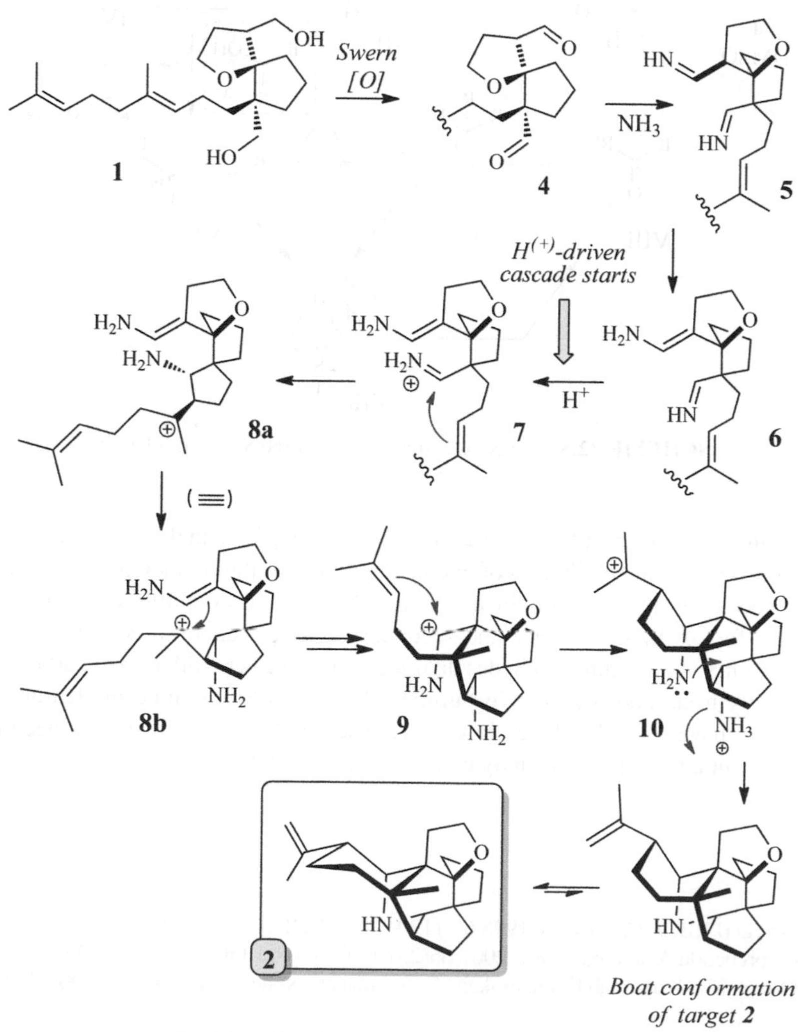

SCHEME 32.3 Reaction cascade from **1** triggered by Swern oxidation-amination and multiple ring closings. The previous stereochemical analysis (Scheme 32.2) greatly facilitates the design of this sequence of events.

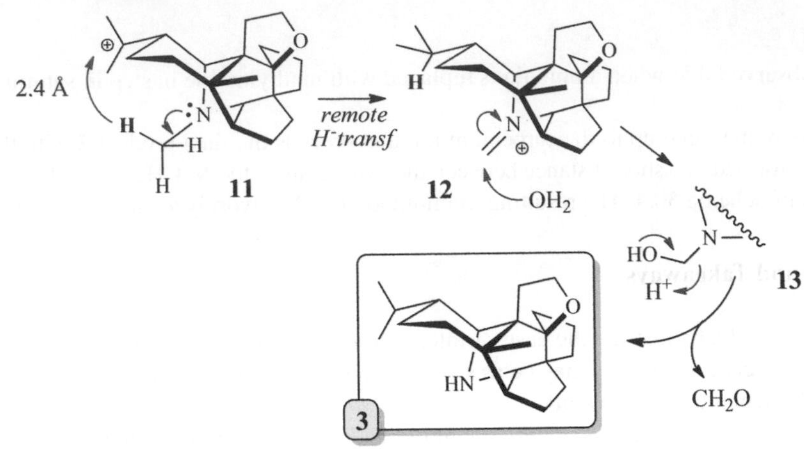

SCHEME 32.4 3D illustration of the remote H-transfer involved in the saturation of the *i*-propyl appendage of target **3**.

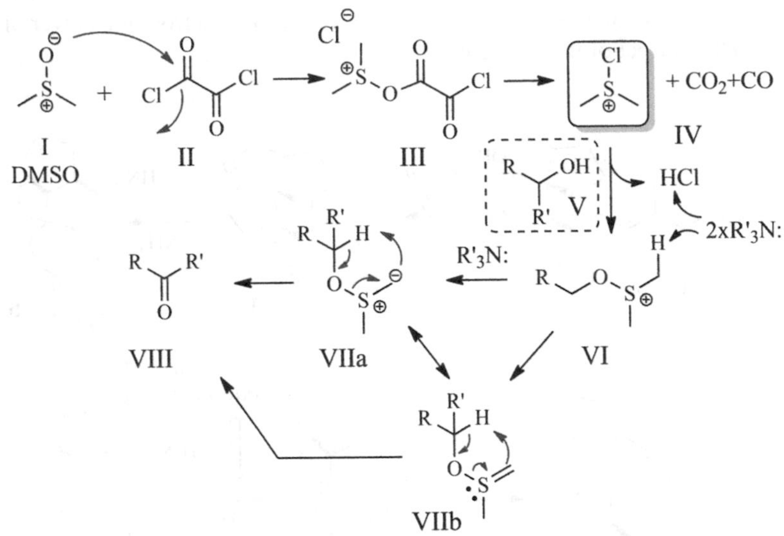

SCHEME 32.5 Consensus mechanism of the Swern oxidation.

4. On the other hand, a significant 63% yield of target **2** suggests a driving force in the desired direction established by the adequate proximity and conformational possibilities of interacting synthons, thus reducing the chance of secondary compounds.
5. The actual oxidative species in the Swern oxidation is the dimethylchloro-sulfonium ion **IV** (Scheme 30.5). This is a widely used procedure for oxidizing primary and secondary alcohols under very mild conditions (−78 °C). Alcohols operate as nucleophiles on S. As a result, aldehydes do not undergo further oxidation to carboxylates. Because co-product oxallyl chloride is highly toxic and the dimethylsulfide odor of DMSO is quite unpleasant, several variations of the original Swern oxidation have been developed [e.g. 2–4]. The reaction mechanism has been re-examined in depth not long ago and includes a tunneling effect that may be of interest to you [5].

REFERENCES

1. Heathcock CH, Kath JC, Ruggeri RB. *J. Org. Chem.* 1995;60:1120–1130. DOI:10.1021/jo00110a013.
2. De Luca L, Giacomelli G, Porcheddu A. *J. Org. Chem.* 2001;66(23):7907–7909. DOI:10.1021/jo015935s.
3. Ohsugia SI, Nishidea K., Oonob K., Okuyamab K, Fudesakaa M, Kodamaa S, Node M. *Tetrahedron* 2003;598393–598398. DOI:10.1016/j.tet.2003.08.055.
4. Tsuchiya D, Moriyama K, Togo H. *Synlett* 2011;2701–2704. DOI:10.1055/s-0031-1289552.
5. Giagou T, Meyer MP. *J. Org. Chem.* 2010;75:8088–8089. DOI:10.1021/jo101636w.

PROBLEM 33

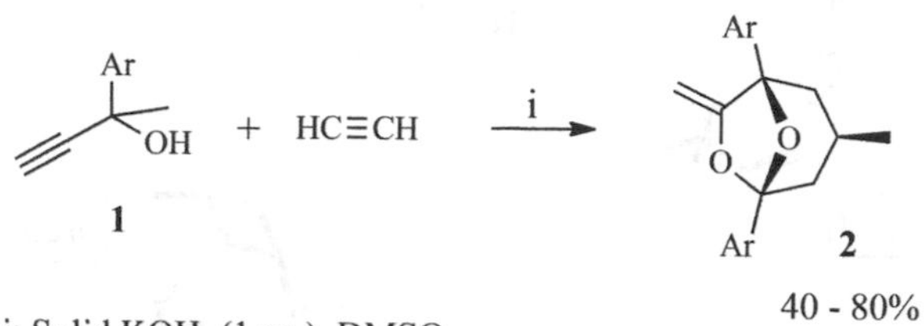

i: Solid KOH. (1 eq.), DMSO
acetylene gas under pressure (14 atm)
70 °C, 15 min

SCHEME 33.1 [1] / Elsevier.

PROBLEM 33: DISCUSSION

33.1 Overview

The chemical universe of propargyl alcohols like **1** has expanded to what appears to be its limits at the present time, so there are few surprises left. From time to time the flash of a previously undetected nova star appears, as this reaction illustrates, discovered by Professor Boris Trofimov and coworkers in Irkutsk, Russian Federation [1]. It gives access to a bicyclooctane nucleus found in several bioactive natural products; for example, behavior-modulating secretions from glands located behind the eye that mature male Asian elephants smear with their trump nostrils. Also, certain bark beetles like *Dendroctonus tenebrans* secrete the same compound, frontalin, in minute amounts as part of their pheromone load; an incredible chemical convergence in vastly separate animals that use similar compounds for social/mating behaviors [2].

The first indication of an unusual course, in fact hitherto unreported, is the question of proton removal by the potent base–solid KOH and DMSO–from three potential sites in both reaction components, which lead to apparently contradicting situations as will momentarily be discussed. Another interesting aspect is the number of chiral atoms formed in **2** in a diastereoselective manner, which suggests predominant TS structuring according to stereoelectronic forces.

Authors tested several tertiary alkyl-aryl propargyl alcohols with similar results and moderate-to-high isolated yields, proof that they were not dealing with a serendipitous curiosity of limited synthetic value [1]. In fact, earlier synthetic efforts by other groups pursuant of the bicyclic scaffold involved several steps and lower overall yields. However, 1,1-dialkyl propargyl alcohols did not perform similarly and gave vinyl ethers as will be shown below.

33.2 Problem Analysis

Comparison of abridged molecular formulas (**1:** C_4H_5OAr; **2:** $C_8H_{10}O_2Ar$) clearly indicate that the latter is the exact stoichiometric replicate of 2 mol of **1**; hence, **2** is a dimer of **1**: a self-assembly case. If dimerization takes place through HED + LED coupling, the monomers must have both types of moieties in a head-to-tail arrangement, although this is not apparent in **1**.

It also seems that acetylene does not participate in the target carbon setting, yet the reaction would not succeed without acetylene. For this reason, authors pumped this gas in to the reactor chamber at 14 atm pressure, a momentary mystery.

Let us appeal to FA (Scheme 33.2). Using the O–C–aryl section for orientation since it remains unmodified in **1** and **2**, one can deconstruct the latter in two breakup forms **A** and **B**. Breakup form **A** gives a partial structure easily associated with the starting propargyl alcohol **I** from the northern section and an *n*-butyl-1-carbinol arene **II** from the southern red portion, which for the moment does not make sense. As one tries the alternative breakup form **B**, it not only leads to the same non-sense **II** but also to fragment **III** in which arene and OH substituents are not on the same carbon. Two non-sense pieces make a worse case than one non-sense and a recognizable section, therefore we will only take fragmentation **A** for guidance.

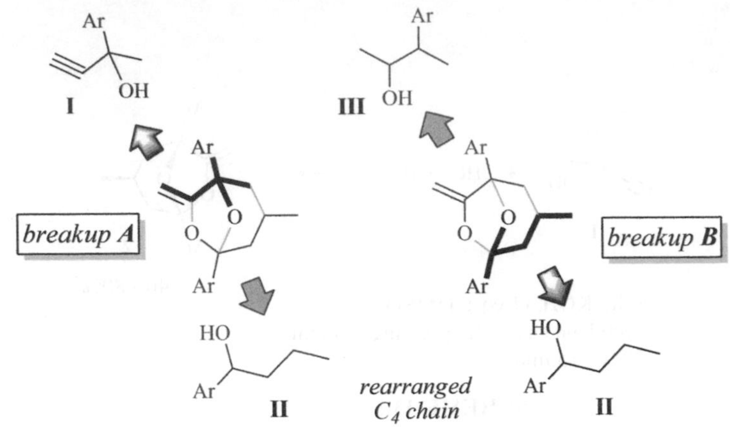

SCHEME 33.2 Fragmentation analysis of target **2**. New bonds to be construed are marked in green.

33.3 Devising a Reasonable Mechanism

Our next task is twofold: Understand how to produce **II** and connect **I** and **II** through three (green) bonds.

Propargyl alcohols like **1** are synthesized from an aryl ketone and acetylene in hot base (KOH). This process, known as the *Favorskii reaction* or condensation, is reversible (Scheme 33.3, top section). Adding excess acetylene drives the equilibrium toward the propargyl alcohol. Since the reverse Favorskii reaction must be occurring in the conditions of the present problem to maintain a manageable concentration of **1**, delivering excess acetylene was why the authors pumped the gas into the reactor. Note that an undetermined amount of ketone **4** will nevertheless survive and feed the acid–base equilibrium extending to **4**+**5**⇆**6** whereby enolate and acetylene combine. The latter compound acts as an electrophile under the auspices of potassium cation. Water from the KOH proton abstraction of **1** returns as a proton donor in **7** followed by [1,3]-H shift to the thermodynamically more stable α,β-unsaturated ketone **8**. This conceptually simple sequence is the basis of the ketone vinylation reaction in synthesis developed by Professor Trofimov's group [3].

Of importance to our mechanism problem, ketone **8** possesses the *n*-butyl backbone of fragment **II** from the fragmentation of target **2** (Scheme 33.2) with all the attributes to form the green C–C and C–O bonds shown there. Our task now is to connect **1** and **8** through HED/LED combinations as shown in Scheme 33.3, middle section. The main roadblock, however, is the activation of the methyl group in **1** as carbanion to the effects of a β-condensation on **8**. We can attribute this rather unusual proton abstraction furnishing dianion **11** to the strength of the KOH/DMSO base (because there seems no other alternative in the realm of the ionic mechanism we have selected). After accepting it [4], the rest of the domino sequence falls in place (Scheme 33.3, second section).

33.4 Are There Additional Alternatives?

There always are. The authors [1] ventured a bolder (in my view) sequence in which ketone **8** does not take part, but a 1,3-methyl migration instead accounts for the methyl (green highlighted) grafted to the three-carbon bridge of target **2**. Follow the green arrows and the course of the highlighted methyl from **11** in the third section of Scheme 33.3. Pay attention to the different origin of the green methyl in routes **11**→**13** trapping **8** versus **11**→**16** via methyl rearrangement. Labeling this methyl group with ^{13}C in the propargylic alcohol and taking a ^{13}C NMR would provide decisive insights in this mechanism.

The recorded *diastereoselectivity* [1] must result from cyclization step at **13**→**2**. For this closing, a chair-like conformation is most suitable with the least steric interaction if the large aryl groups and the methyl are equatorial (Scheme 33.3, bottom).

33.5 Learning Issues and Takeaways

1. The solid KOH/DMSO mix constitutes an inexpensive superbase (pKa=30–32). Protons in DMSO are slightly acidic and can be abstracted by more potent bases than KOH such as alkyl lithium, and NaH to form dimsyl–M (M=alkaline metal). But this is not the case with KOH.

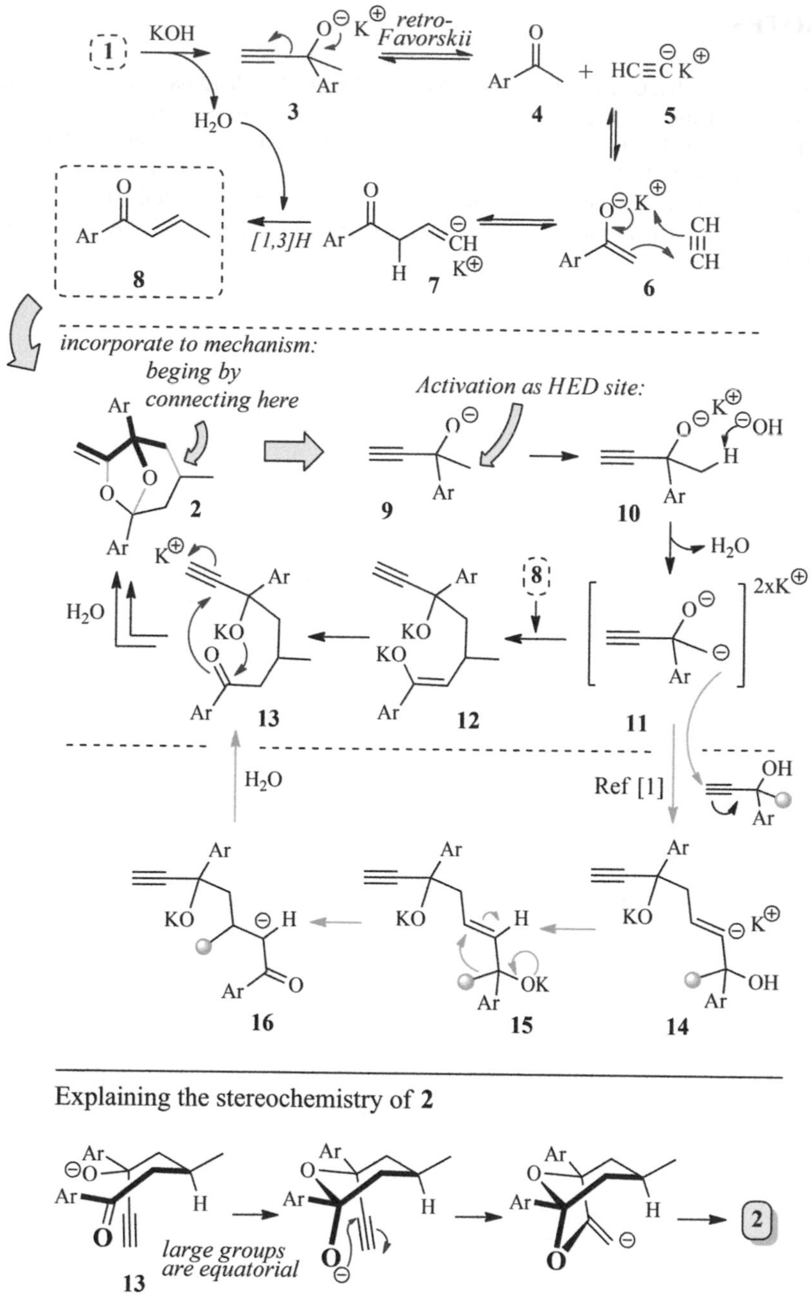

SCHEME 33.3 [1] / Elsevier.

2. Quantum mechanical calculations that explore the vinylation of methanol-acetylene combination in KOH/DMSO [5], that is O-vinylation, a close relative of the present problem, determine that $K^{+}OH^{-}$ ionic pair forms a coordination complex with five molecules of DMSO. KOH, in this complex, coordinates with the carbinol taking up the hydroxyl proton and forming water that remains in the complex. This complexed water molecule serves as proton donor at specific steps of the vinylidation sequence that follows. For this reason, some potassium alkoxide intermediates and even dianions like **11** may survive the water presence as separate ligands.
3. The nucleophilic addition of the alkoxide to acetylene is the rate limiting step [5]. The resulting *sp*^{2} carbanion proposed in various sections of Scheme 33.3 captures a proton from water with very low activation energy. This product is then released from the complex and the initial KOH/DMSO set is reconstituted for another reaction cycle.

REFERENCES AND NOTES

1. Trofimov BA, Schmidt EY, Bidusenko IA, Ushakov IA, Protsuk NI, et al. *Tetrahedron* 2012;68:1241–1246. DOI:10.1016/j.tet.2011.11.050.
2. Rasmussen LEL, Riddle HS, Krishnamurthy V. *Nature* 2002;415:975–976. DOI:10.1038/415975a.
3. Trofimov BA, Schmidt EY, Zorina NV, Ivanova EV, Ushakov IA. *J. Org. Chem.* 2012;77:6880–6886. DOI:10.1021/jo201005p.
4. Organic dianions are no longer the oddballs they used to be. See, for example: Thompson CM. *Dianion Chemistry in Organic Synthesis*. CRC Press, Boca Raton Fl, 1994. Langer P, Freiberg W. *Chem. Rev*. 2004;104:4125–4150. DOI:10.1021/cr010203l.
5. Trofimov BA, Schmidt EY, Ushakov IA, Zorina NV, Skital'tseva EV, Protsuk NI, Mikhaleva AI. *Chem. Eur. J.* 2010;16:8516–8521. DOI:10.1002/chem.201000227.

PROBLEM 34

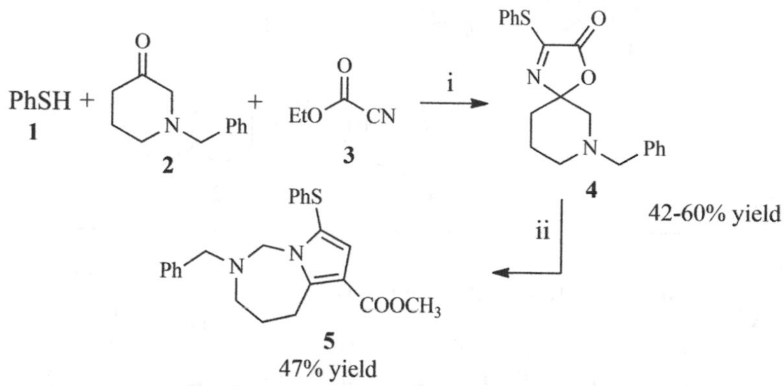

i; Et_2NH, $TiCl_4$, BF_3-Et_2O, DCM, 80 °C, microwave, 15 min
ii: Methyl acetylene carboxylate, chlorobenzene, 150 °C microwave, 10 min

SCHEME 34.1 This and other Schemes in this problem have been adapted in part from [1] / Elsevier.

PROBLEM 34: DISCUSSION

34.1 Overview

This problem puts together two trendy technologies of modern and environmentally sensible organic synthesis: mixing more than two components A + B + C... at a time to get A–B–C... molecules in one shot, and microwave radiation (MWR) as a heat source. The former is known as multi-component reactions (MCR), which belongs in the also stylish one-pot synthesis that include domino reactions, while the latter is an increasingly popular form of focusing heat energy on the reaction medium. The efficiency of MWR is clearly exemplified in the present case: while the reaction took 24–36 hours to reach completion with relatively poor yields when using classical electric mantle heating, MWR expedited the reaction time to only 15 minutes with much improved yields [1]. Professor Wipf and his colleagues from Pittsburg, Pennsylvania, United States devised this MCR aiming to synthesize new derivatives of benzodiazepines [1,2]. If you have sleeping pills at home and bother to check the active component, chances are that it is a benzodiazepine, a blockbuster tranquilizer (anxiolytic) family that became the top drug prescription in the United States by 1977 (sales later declined, but there is a current comeback with novel diazepines being introduced) [2].

Scheme 34.1 portrays two mechanism problems in one shot. Explaining product **4** is relatively easy; a bit pricklier is the **4** → **5** conversion in the second part.

34.2 Problem Analysis, Part 1: Explaining 4

Primary target **4** can easily be broken down into three recognizable sections as the three starting compounds, save for an orphan CH_3O group (Scheme 34.2, upper section). This group can readily leave the premises along the sequence as ethanol, probably during aqueous work up. By assigning electron density character, thiophenol **1** plays the HEDZ role whereas **2** and **3** are LEDZs. It can be inferred that at an advanced stage pyridazone **3** undergoes O-acylation to close the spiroheterocycle. Only one drop of diethyl amine is added to the mixture [3], suggesting a base catalyst, while boron trifluoride (BF_3) etherate and titanium tetrachloride are strong Lewis acids in search of HED sites like C=O for electrophilic activation. This plan is conveyed in a first approach **A** (Scheme 34.2, middle section).

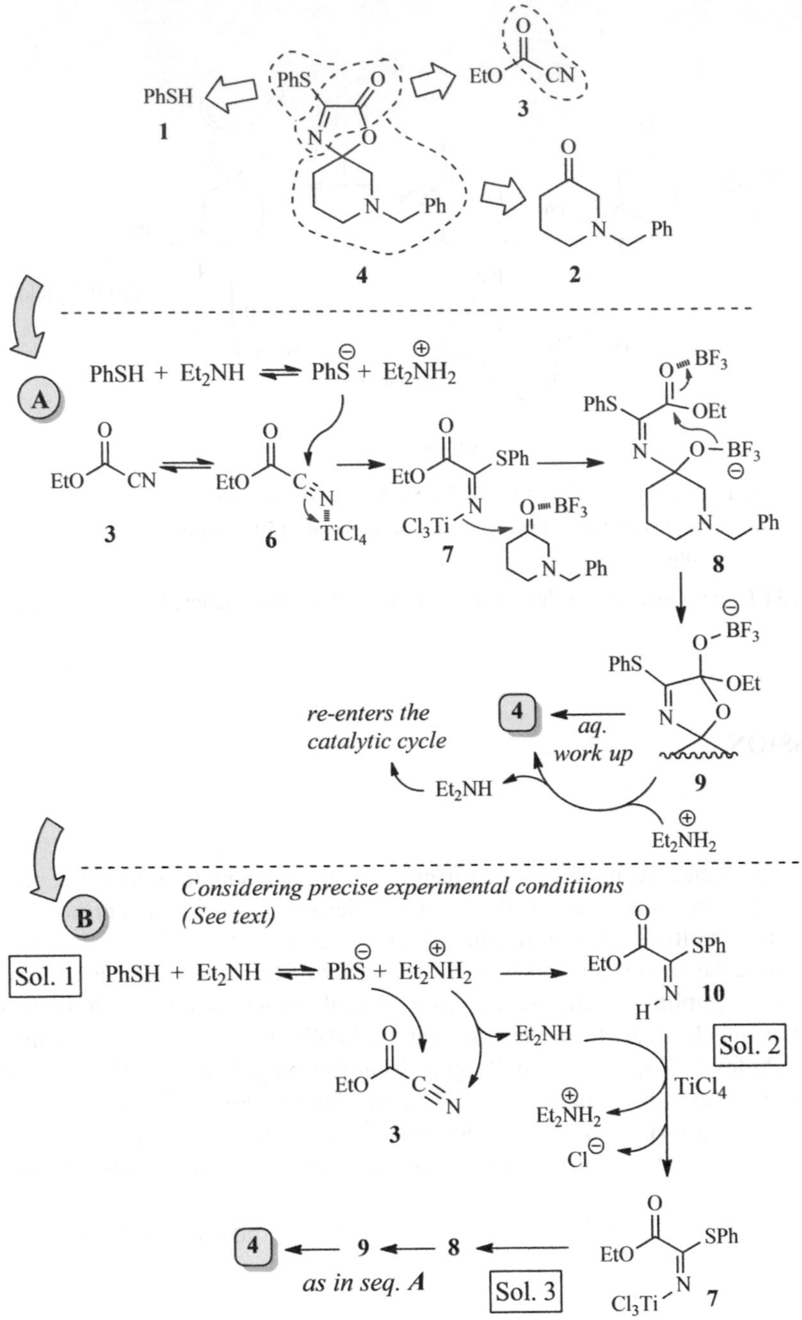

SCHEME 34.2 Top section: A facile fragmentation of the target into the mjultiple starting components, followed by a first general mechanistic proposal **A** (middle section). However, when considering the precise experimental conditions and order of addition of reagents, the mechanism changes sharply to option **B** (lower section).

34.3 Adapting Sequence A to Experimental Conditions

Cautiously speaking, our first approach scheme faces trouble promptly:

Proposal **A** – as any other hypothesis – must be adapted to the specific experimental conditions (below) or be discarded if not, no matter how close it may seem to the rules of chemistry and logics according to the available information. Checking the stoichiometry of all components is always a good idea in mechanism, so please pay attention to the mmol amounts used by authors, even if this is a bit of a bore [1]. Solutions 1–3 are mixed in tandem:

1. *Solution 1*: Compounds **1** (0.625 mmol) and **3** (0.507 mmol) are dissolved in DCM (dichloromethane, 0.98 mL) at 0°C, one drop of ethyl amine is added (**see note 4**), and the solution is stirred at room temperatuere for two hours. Note that piperidone **2** is not added yet.

2. *Solution 2*: A solution of $TiCl_4$ (titanium tetrachloride, 20 drops) and BF_3-etherate (10 drops) is prepared in DCM at room temperature and added dropwise to Solution 1 at 0°C.
3. *Solution 3*: Piperidone **2** (0.516 mmol) is added dropwise to Solution 2 at 0°C and then heated (microwave) to 80°C for 15 minutes. Altogether, we have a one-pot reaction going.
4. Aqueous NaOH is used for quenching.

In a nutshell, compounds **2** and **3** as well as ethylamine amount to approx 0.5 mmol each [4], while there is a small excess of thiophenol (**1**). The mixing order is (**1**+**3**+ethylamine), stir for two hours, then add the Lewis acids followed by piperidone **2**.

Under these conditions, the mechanism furnishing **4** cannot be completed as portrayed in sequence **A**. This is why:

Solution 1 takes ethyl cyanoformate and thiophenol not too far, there is no $TiCl_4$ yet to catalyze the thiolate addition. One can only assume Brønsted catalysis by $Et_2NH_2^{(+)}$ at the beginning of sequence **B** (Scheme 34.2, lower section). Thioimine **10** is produced rather than titanium complex **7** of sequence **A**. And stop here.

Until Solution 2 is added we cannot expect any progress from **10**. Proton-titanium exchange takes place, enhancing the nucleophilic power of the imine, stopping again at **7** until piperidone **2** is added in Solution 3; BF_3-etherate included in Solution 2 traps the C=O moiety, spurring the nucleophilic attack of **7** to furnish **8** and eventually target **4** from there.

Although sequences **A** and **B** run almost in parallel, pathway **A** cannot be accepted as a valid mechanism under the experimental circumstances, and route **B** is an accurate description of the real chemistry [5].

34.4 Analysis of Target 5

To this end, I recommend fragmentation of **5** into recognizable pieces using selected labels, the most successful of our set of strategies (Scheme 34.3). Colors are used instead of dashed fields except for the orphan moiety to emphasize that its departure should perhaps be the starting point of the process. In addition to defining a sequence of events, fragmentation also reveals a deep-seated skeletal rearrangement and the bonds involved. Your job is to put these revelations into a feasible sequence, if only you were courageous enough to keep a blind eye to the lower portion of this scheme.

34.5 Explaining Scheme 34.3 in a Nutshell

✓ Heat-prompted decarboxylation (150°C of MWR); stir the valence electron shell in the thioimine moiety such that five resonance canonical structures (**I** to **V**) can be construed.

✓ These forms include cumulene and dipolar forms of opposite direction (e.g. **II** versus **V**).

✓ The latter is predictably a better contributor to the resonance hybrid than **II** due to the stabilization of positive and negative ends, as well as being best suited to justify the regioselective [3+2] dipolar cycloaddition depicted in **12**.

✓ Ring cleavage (**13**→**14**) responds to a well-established retro-Mannich condensation. The resulting iminium ion loses back onto $N^{(-)}$ to give the desired target **4**.

✓ Atom labels, assigned in the deconstruction analysis, match perfectly.

34.6 Learning Issues and Takeaways

1. The construction of spiro-heterocycle **4** is a relatively simple case of multiple component synthesis, since only three different compounds were put together. Other examples include four, five, and more compounds to be presented herein as more complex problems. No matter how apparently simple, however, underestimating its complexity led us (me, really) to a devious mechanism (pathway **A**) when we skipped the order in which the components and catalysts had been added.
2. At this stage, that single drop of ethylamine carried the (wrong) notion that this was just a speck and played an obvious catalytic role, leading us to find ways to restore it for the next catalytic cycle. In reviewing the author's careful description, one realized the small scale of the reaction (0.5 mmol) in just 0.98 mL of solvent. At that scale, one drop of ethylamine is a stoichiometric amount [4]. Therefore, it is frequently important for your mechanistic musings to check out the details of the experiment.
3. When the order of addition of reagents was considered, pathway **A** was in need of urgent correction. For our good fortunes, brainchild pathway **B** made its way (authors did not publish the mechanism [1,2], but they must have devised one at the time of designing, so careful was the order of reagent addition) by assuming proton activation of cyanoformate.

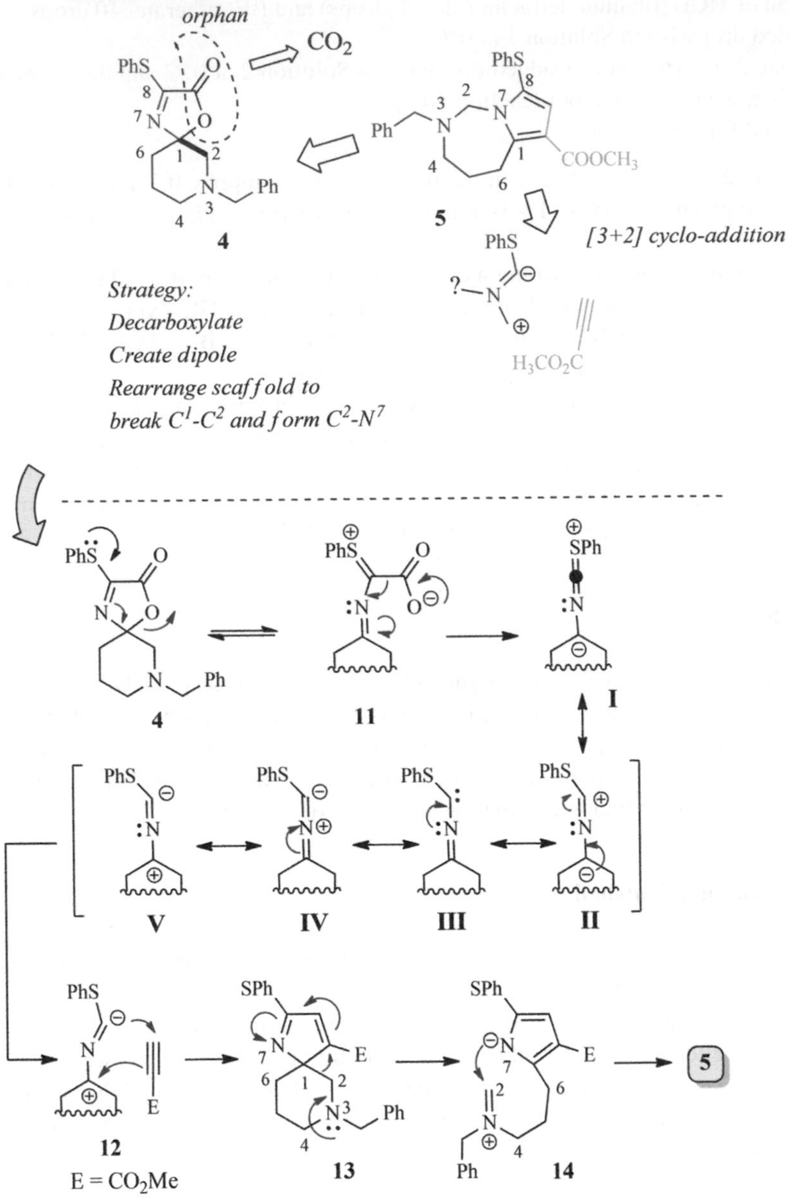

SCHEME 34.3 Mechanistic strategy design from defragmenting target **5** and proposed sequence from primary product **4**.

4. Of note, 2 mol of diethyl ammonium ion were necessary to drive the sequence up to the proton-titanium exchange in **10** → **7** (Scheme 34.2).
5. The **4** → **5** mechanism was far more intriguing, a true problem one might say, as it involved a decarboxylation-cycloaddition-scaffold rearrangement domino not easily perceived at first sight. But our unbeatable FA provided several clues to solve this mechanism. Never underestimate its powers.
6. Although resonance is generally a beginners' subject, it is frequently called upon to provide electron deployment schemes and feasibility to reaction mechanisms, (e.g. Scheme 34.3, resonance structures **I–V**, the only way to build acting dipoles).
7. Our preferred electron distribution here to secure the regiospecific propargyl ester addition was dipole form **V**. The authors' choice [1], however, was cumulene-type **IV**. Their argument was based on testing other substituted acetylenes and observing the loss of regiospecificity (Figure 34.1) albeit there was still some preference for a lower electron density at the tertiary C^1 carbon (which keeps alive our favorite proposal of electron distribution in **V**). In all cases, the reactivity of the propargyl reagent was lower, down to observing no reaction at all, owing to the reduced π electron density caused by two EWGs.

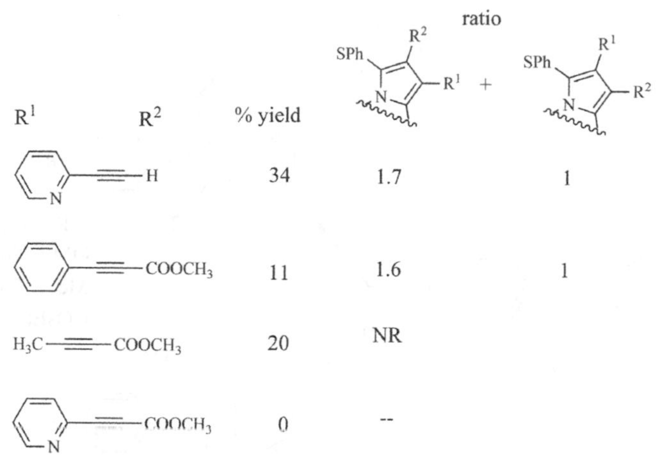

ratio
SPh
R2
R1
+
R1
R2
% yield
H
34
1.7
1
COOCH3
11
1.6
1
H3C
COOCH3
20
NR
COOCH3
0
--

FIGURE 34.1 Additional acetylene dipolarophiles tested by authors of [1] / Elsevier.

8. The final cyclization to the diazaheptane **5** was possible thanks to the inverse course of a Mannich condensation, namely a retro-Mannich cleavage. This is extensively exploited in synthesis design and pose intriguing occult mechanisms (see Problems 3 and 17 of the second edition of this book).

REFERENCES AND NOTES

1. Liang M, Saiz C, Pizzo C, Wipf P. *Tetrahedron Lett.* 2009;50:6810–6813. DOI:10.10116/j.tetlet.2009.09.107.
2. As the future or current organic chemist you are, it may be wise to learn about this group of compounds. For a recent review of new synthesis approaches and biological properties, see: Verma S, Kumar S. *Mini-Rev. Org. Chem.* 2016;14:453–468. DOI:10.2174/15701 93X14666170511121927; for a brief review from the psychiatric stand and an updated set of pertinent references, see: Balon R, Starcevic V, Silberman E, Cosci F, et al. *Braz. J. Psychiatry*. 2020;42:243–244. DOI:10.1590/1516-4446-219-0773.
3. For a preliminary work in which this reaction was first published, see: Wipf P, Prewo R, Jost HB, Heimgartner H, Nastopoulos V, Germain G. *Helv. Chim. Acta* 1987;70:1380–1388. DOI:10.1002/hlca.19870700518.
4. I have no idea how much ethylamine (in mmol) is in one drop of the thing (the authors did not specify this). Let's run a rough estimate: ethylamine density = 0.689 mg mL^{-1}; drops per mL = 20 approx.; therefore one drop = 35 mg; ethylamine MW = 73 g mol^{-1}, thus 35 mg = 0.5 mmol.
5. The authors did not discuss these mechanism details in their paper. Sequences **A** and **B** are of Alonso-Amelot's making and so is solely responsible for these designs.

PROBLEM 35

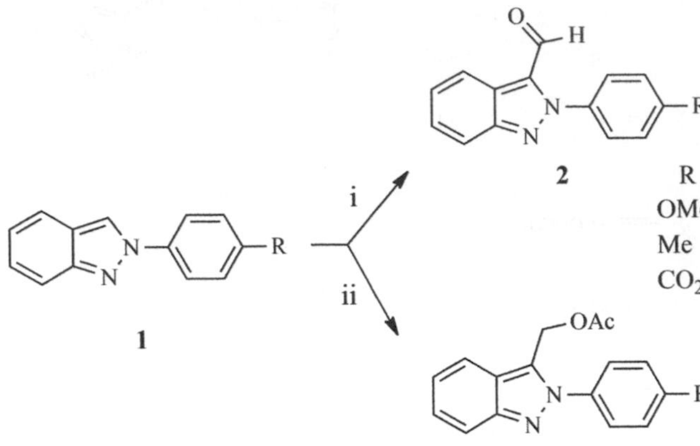

1 → (i) 2; (ii) 3

2: R yields: OMe 80%, Me 78%, CO_2Et 65%

3: 70% yield, R = Me

i: $K_2S_2O_8$ (1 eq), DMSO/HOAc 19:1
100 °C, 24 h, no air exclusion

ii: $K_2S_2O_8$ (3 eq), DMSO/HOAc 1:3
100 °C, 24 h, no air exclusion

SCHEME 35.1 This and other Schemes in this problem have been adapted from [1] / American Chemical Society.

PROBLEM 35: DISCUSSION

35.1 Overview

There is a considerable amount of data in the experimental conditions **i** and **ii** of this chemodivergent set of reactions to pay attention to. Some of it may or may not be necessary to solve the disparity of products, exposed here on purpose to convey the laboratory experimental setting. It must be said though that Professor Alakananda Hajra and coworkers from Santiniketan, India, who authored this work, developed a systematic approach to their discovery by exploring several 2*H*-indazole siblings to gain a broader synthetic and mechanistic insight [1]. The conditions and product yields annotated herein, extracted from their laborious optimization experiments come from my selection of key results to introduce you to the reaction mechanism I wish you to devise.

Some of the reagents here portrayed will be familiar from a past problem, although the course of the reactions deviates from previous results turning this research into an interesting discovery of synthetic impact.

35.2 Problem Analysis and Feasible Mechanisms:

After comparing **2** and **3** with the starting indazole **1**, one sees that the aromatic heterocyclic core is preserved and only C^3 (formal nomenclature is like that of indole) is affected by the substitution of a proton with a CH_2–X unit (X = CHO, OAc). In the absence of skeletal rearrangement this exocyclic carbon must be provided by an external C_1 source. The best and only candidate is DMSO [2], which is activated as methylene donor by potassium persulfate. Authors observe that no reaction occurs when peroxidants or some other oxidizing agents like phenyl diodine diacetate (PIDA) are not included in the mixture.

Authors did not provide kinetic data, only useful enough yields after optimization of reaction conditions. From this information it is worth noting that yields fall within approximately the same range as R varies from OCH_3 and CH_3 to CO_2Et. This suggests little (if any) electronic charge influence of aryl substituents on reaction efficiency. This feature calls for non-ionic intermediates, particularly in the rate-determining step that could limit product yields and enable unforeseen decomposition pathways.

Non-ionic character is associated with concerted couplings as in several – but not all – concerted cycloadditions, for example, or radical sequences. Hot potassium persulfate is an invitation to the latter route (Scheme 35.2), to which DMSO is readily susceptible [3]. As a final issue to account for, authors indicate that N^1-alkyl groups (*n*-Bu, *t*-Bu) in **1** block the desired C^3 alkylation of indazoles. Other points of interest are discussed in the final section below.

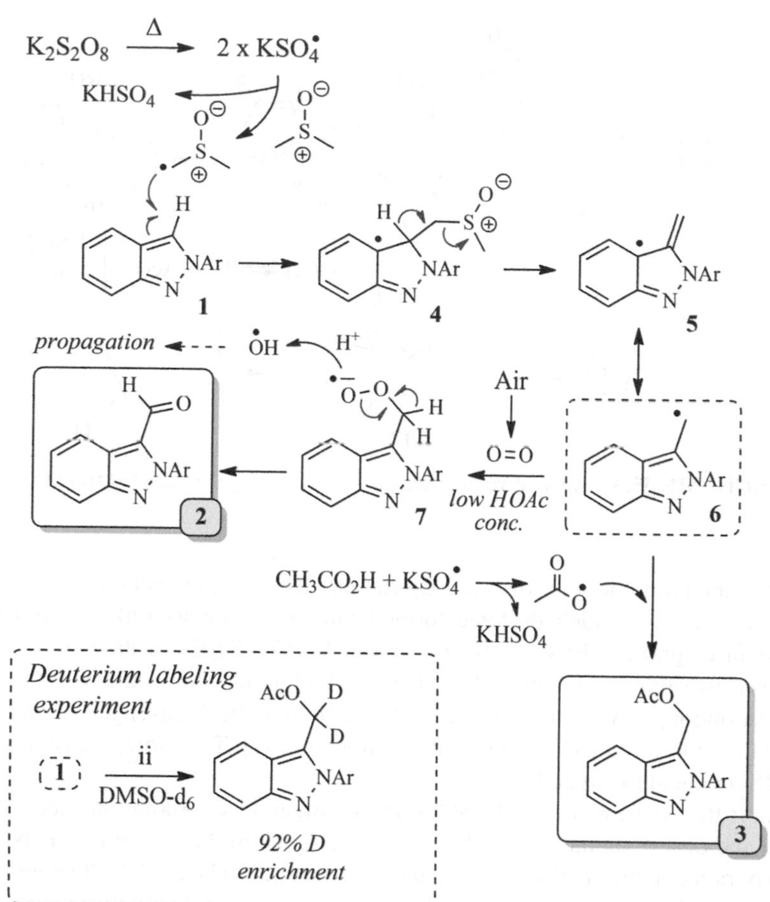

SCHEME 35.2 Radical-based sequence based on the DMSO radical relay explains both targets **2** and **3.** The extra carbon is provided by DMSO as shown by the deuterium-enrichment experiment [1].

35.3 Alternative Mechanism Explaining Acetate 3

Because it is always instructive to propose other possible pathways and then contrast the hypothetical routes with the hard experimental evidence, let us think for a moment of a second choice based on an indirect acetylation after the incorporation of the sulfoxide methylene to C^3. The idea (Scheme 35.3 top section), inspired by previous work from Windsor, Canada [4], is based on a modified Pummerer rearrangement, which gives access to 1-acetoxysulfides from sulfoxides. Acetoxylation at the exocyclic carbon would take place by water elimination from protonated sulfoxide **8** and nucleophilic addition to the emerging sulfenium cation **9** by the freed acetate. Sulfide acetate **10**, however, would be ill suited for accepting a proton at the exocyclic carbon because the C–S bond is polarized toward the sulfur. But reoxidation of **10** would restitute the sulfoxide and as such it would operate as an electron donor enabling the **10** → **11** → **12** sequence. The latter radical that remained stabilized by resonance all the while would terminate by C^3 hydrogen atom exclusion spurred by the potassium sulfate radical, furnishing target **3**.

35.4 Learning Issues and Takeaways: Which Mechanism Is Better Adapted to the Evidence?

1. Professor Hajra and collaborators explored this question by performing control experiments [1]. Scheme 35.1 denotes two divergent reactions, formylation, and methylene acetoxylation. This is explained by a change in solvent composition, while the first one is verified by air oxidation the second results from competition of acetate against air oxygen owing to the much larger concentration of acetic acid in DMSO (5% versus 66% of AcOH [acetic acid], respectively).

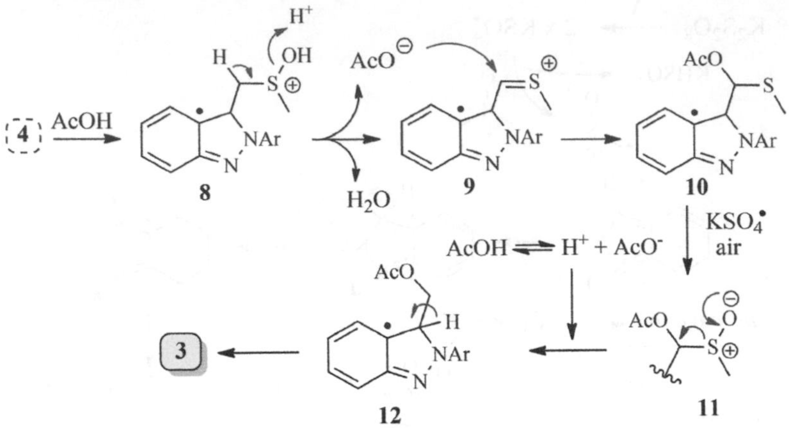

SCHEME 35.3 Alternative mechanism to explain target **3** Adapted from [5].

2. How would you test the intervention of air oxygen? By running the reaction under a N_2 or Ar atmosphere, of course, as authors just did [1]. Only a trace of aldehyde **2** was formed with predominance (70% yield) of acetate **3**.
3. Yields were independent of grafting EDGs (CH_3O, CH_3) or EWG (CO_2Et) in the para carbon of the eastern phenyl; hence, charged species (other than the natural sulfoxide as DMSO or adducts) could be ruled out.
4. If radicals were the best option, how would you test this? By adding radical scavengers, evidently. Indeed, 2,6-di-*t*-butyl-4-methyl phenol (BHT) [6] killed the C^3 methylenation altogether [1]. Therefore, the radical-based sequence could be verified from the DMSO radical to target **2**.
5. How would you support the contribution of DMSO to the C^3 methylene? Adding labeled DMSO, yes. By running the reaction in DMSO-d_6 and observing the size of the methylene signal of **2** and **3** in the ^{1}H NMR authors [1] detected as much as 92% D-incorporation at this methylene (Scheme 35.2, inset). Replacing DMSO with another polar solvent would also provide crucial information. When authors tested other one-carbon synthon-solvents like DMF or DMA, no reaction was recorded [1].
6. Although acetate **3** seems to stem from radical **6** (Scheme 35.2), the combined radical + ionic sequence portrayed in Scheme 35.3 cannot be discarded. Loss of water from alkyl sulfoxides furnishing sulfenium cations (applied here in **8** → **9**) and ensuing addition of acetate in a related system has been reported [4]. Given that the key electronic charges develop far removed from the phenyl EDGs and EWGs, their influence on product yields should be minor, as observed experimentally.
7. In conclusion, the formation of radicals starting from DMSO is indeed the mainstay mechanism explaining the observed targets, but competition from ionic acetate incorporation remains an open contributor.

REFERENCES AND NOTES

1. Bhattacharjee S, Laru S, Ghosh P, Hajra A. *J. Org, Chem.* 2021;86:10866–10873. DOI:10.1021/acs.joc.1c01188.
2. DMSO reactivity makes it an increasingly attractive carbon and sulfide source for organic synthesis. See for example: Wu XF, Natte K. *Adv. Synth. Catal.* 2016;358:336–352. DOI:10.1002/adsc.201501007. See also Problem 28 discussion.
3. Herscu-Kluska R, Masarwa A, Saphier M, Cohen H, Meyerstein D. *Chem. Eur. J.* 2008;14:5880–5889. DOI:10.1002/chem.200800218.
4. McIntosh JM, Leavitt RK. *Can. J. Chem.* 1985;63:3313–3316. DOI:10.1139/v85-548.
5. Alonso-Amelot ME, this book.
6. BHT, also known as butylated hydroxytoluene, is a common antioxidant additive in industrial food preparations. Other compounds find laboratory application as radical scavengers to assess antioxidant properties of many natural products and components of functional foods such as polyphenolics.

PROBLEM 36

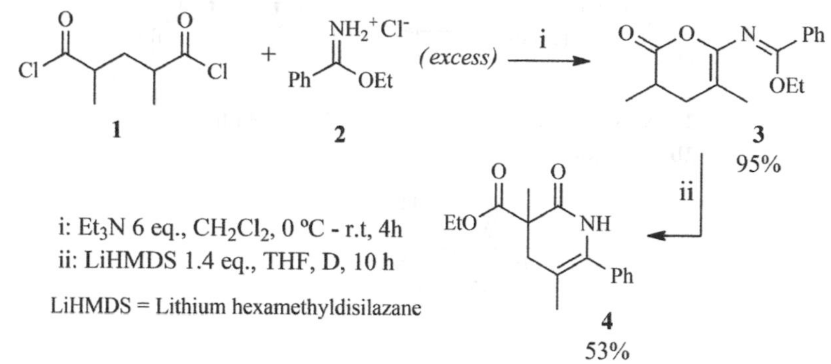

SCHEME 36.1 [1] / Elsevier.

PROBLEM 36: DISCUSSION

36.1 Overview

Advances in organic chemistry – as in many other creative endeavors – often crop up in the realm of the unexpected: serendipity. The reaction presented herein, developed by scientists in Würzburg, Germany [1] is one constructive case, not only for the outcome but also the substantial yields of unexpected product in both reactions **i** and **ii** and the involved mechanism(s). The planned goal was to obtain a series of bis-imides of type **5** with a flexible substitution pattern (only one is shown in Scheme 36.1). These products would be endowed with four active centers of opposed electron density in the same molecule, two LEDZs and two HEDZs (Scheme 36.2, top section), the latter being manageable in basic company as enolates. This arrangement is suitable for intramolecular double cyclization to furnish fused symmetrical diaza-derivatives from which further molecular complexity can be construed. This concept had been applied successfully in previous years in the synthesis of diaza-pentalene **6** and siblings [2, 3]. It is easy to surmise that **5b** is not the precursor of **6** but a close relative one can conceive by means of retro-analysis. Buoyed by these antecedents, Breuning and Täuser [1] procured a homologous ring construction to expand its chemical space. As they failed in accomplishing the original plan, they anyway discovered the novel processes of Scheme 36.1, which spans two mechanisms to solve.

36.2 Analysis and Solution of 1 + 2 → 3

This is the easy part, once structure **3** is analyzed in its composing parts: It amounts to equimolar quantities of **1** and **2**, excess of the latter notwithstanding (Scheme 36.2, middle section). Both sections operate as complementary HEDZ/LEDZ and bonding between imide and carbonyl carbons is thus logical. Keep in mind that the acyl chloride has been regarded as a carbonyl sp^2 cation, although this view has been questioned. At any rate, two acyl chlorides in the same molecule constitute a highly reactive proposition.

After this first analysis, the mechanism unfolds rather smoothly through at least three pathways, two of which are depicted in Scheme 36.2, lower section. Authors', and my own favorite, is the ketene route **1** → **7** → **8**, as it gives sense to the commanding role of ethylamine (added in **6** molar excess) from the start.

36.3 Analysis and Solution of 3 → 4

36.3.1 A First Approach This is where the true problem arises. At first sight, **4** seems a dislocated version of **3:** An ethoxycarbonyl crops up in the absence of ethanol in the solvent that might have justified it from a now familiar ketene. The only ethoxy group available is already present in **3**, meaning that it migrates somehow to a C=O located on the other side of the molecule; hence, the concerted shift is seemingly unlikely. Also, the formed 1,5-dicarbonyl arrangement of **1**, which is conserved in **3**, turns into a 1,3 dicarbonyl moiety in **4**. These features call for a structural rearrangement in more than one step.

Let us apply the atom-label strategy in **3** and **4** keeping in mind that it only *suggests* bond cleavage and formation. The actual chemistry is to be developed from this model but may not necessarily be the most feasible mechanism. Scheme 36.3, top section

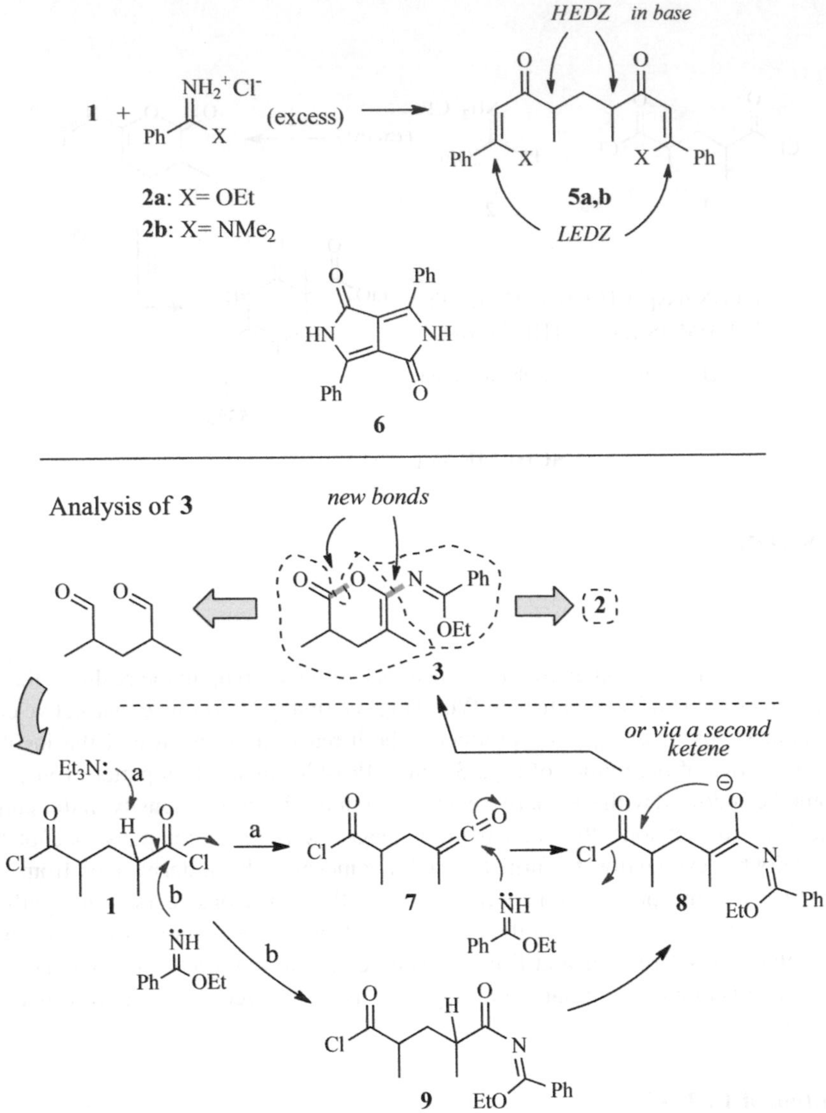

SCHEME 36.2 Top section: Goal originally planned by authors [1]. Middle section: Deconstruction of target **3** and conversion into a feasible mechanism to explain **3**.

compares numbers assigned to **3** that may be interpreted reasonably well (but not with absolute certainty) in **4**. From this model one concludes that:

- ✓ OEt migrates from C^6 to C^5, as said.
- ✓ C^5–O is cleaved, to liberate the carboxyl moiety from the ring.
- ✓ Target **4** shows two new skeletal bonds: C^1–C^4 and C^2–C^6.
- ✓ Given that C^1 in **3** is a pseudo carbonyl (an LEDZ), C^4 should be handled as a $C^{(-)}$, to create C^1–C^4. This is compatible with the strong lithium hexamethyldisilazane (LiHMDS) base in aprotic (anhydrous THF) medium and the vicinity of a C=O group to enhance the acid quality of C^4–H.

With this somewhat nebulous plan the mechanism of Scheme 36.3, second section can be advanced considering that conditions in **3** seem more suitable for forming C^2–C^6 in the first phase.

Highlights of Scheme 36.3, lessons and takeaways:

1. Two consequences emerge from unraveling the lactone ring in **3** early in the sequence via carbanion **10**:
   a. C^2 acquires added nucleophilic capacity to attack C^6, which furnishes a transitory four-membered aza cyclobutenone (**11a,b**).

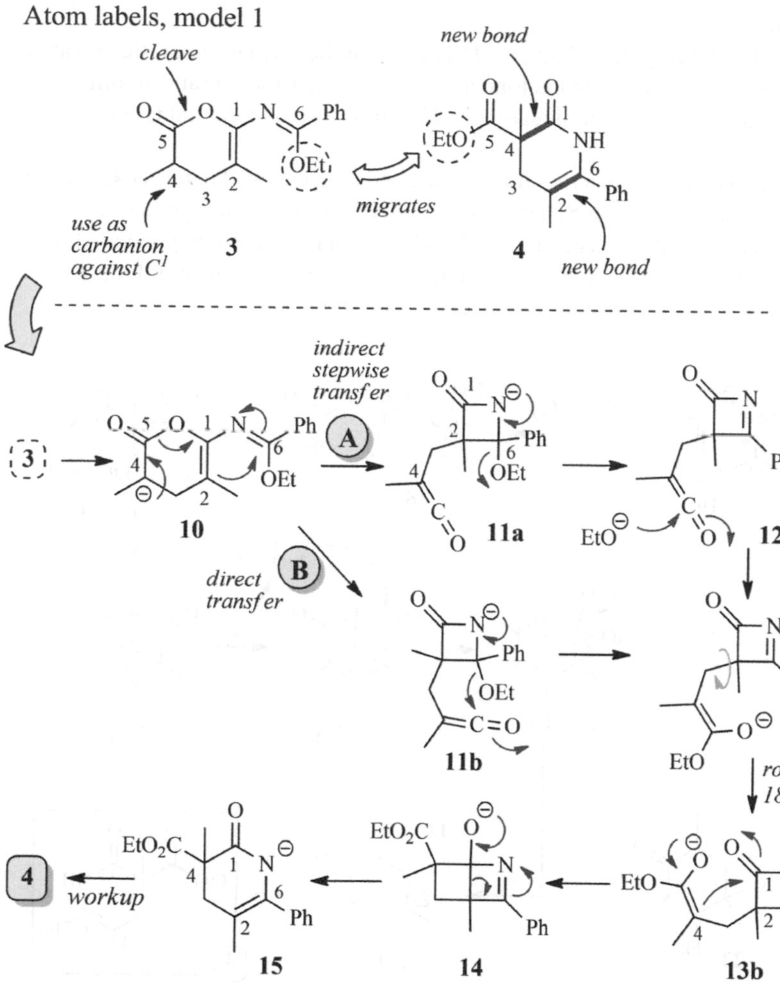

SCHEME 36.3 Analysis by atom labels only and subsequent mechanism emerging from this strategy Adapted from [4].

b. It converts the future ester into a freely rotating moiety, and opens two possible routes, **A** and **B**, which soon converge to **13a**.

2. The basic difference between A and B is the stepwise versus concerted transfer of ethoxide to the ketene carbon. The latter may occur through a six-membered TS from **11b** in a sterically feasible manner and a possibly negative ΔS‡, while it prevents the escape of lithium ethoxide from the THF solvent cage.
3. Rotation of the ester enolate appendage in **13a** places nucleophilic C^4 and electrophilic C^1 at bonding range, thus securing the exocyclic ester.
4. Retro [2+2] cycloaddition in **14** and protonation during workup furnishes all the attributes of target **4**.

36.3.2 A Second Approach If one freezes momentarily the intermediate species of the **10**→**11** step in Scheme 36.2, (ketene enolate **16** Scheme 36.4) another course of evolution (pathway **C**) may ensue from the attack of the C^2 anion to the ketene carbon C^5 placed at accessible distance, rather than C^6 as we did to get **11**. This pathway brings about an entirely new perspective for the major rearrangement in question. The mechanism of Scheme 36.4 was originally proposed by authors of this work [1].

Highlights of Scheme 36.4, *lessons and takeaways*:

1. In **16** the HEDZ/LEDZ combination for bonding is different, while it leads to a cyclobutyl intermediate **17**. This structure is necessary to move closer the two C=O moieties.
2. This enolate then attacks electrophilic imine yielding the aza-bicyclo[1.1.3]heptene **18** with expulsion of ethoxide anion (a lousy leaving group, by the way). This sequence is contingent on the retention of $EtO^{(-)}$ (as EtOLi) in the solvent cage.
3. This anion comes back on the strained ketone followed by ring cleavage in **20**. This rupture may take two separate ways marked in black and green. While the black route furnishes target **4**, the still feasible green pathway leads to an unseen product **23**.

4. Why is this? Please think.
5. Likely reason: Compare the stability of carbanions **21** and **22**. In the former anion, the negative charge is placed at the γ carbon of a hetero-form of a α,β-conjugated ketone and therefore distributed (and stabilized) across the entire MO covering five atoms. By contrast, carbanion conjugation in **22** is limited to the vicinal C=O, as part of an MO spanning three atoms only.
6. When the fate of labels in **10** is followed across this scheme and then checked in target **4**, one ends up with a very different label pattern from the one we designed in Scheme 36.3 (upper section), which guided us to design our first **3**→**4** mechanism. We might have selected route **C** instead if our bonding plan had been designed as shown in Scheme 36.4, (bottom section), but at least for me it would have been a second option. Why? Because linking C^4 to C^6 seemed too farfetched than just looking at structure **3**. Only several intermediates later it became feasible.

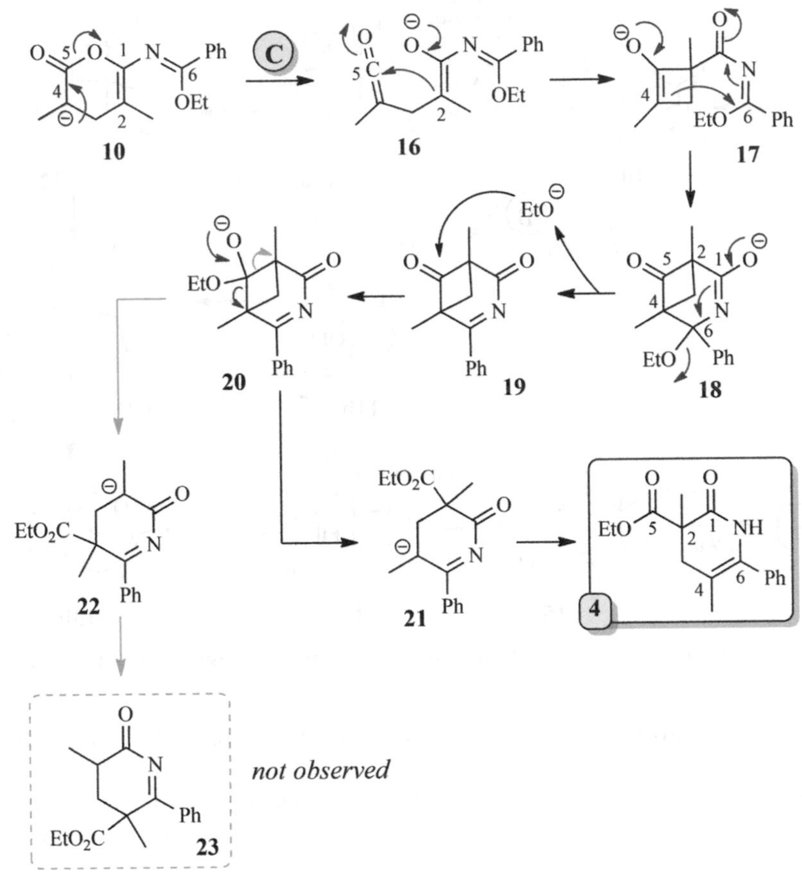

Atom labels and bonding reviewed in retrospective

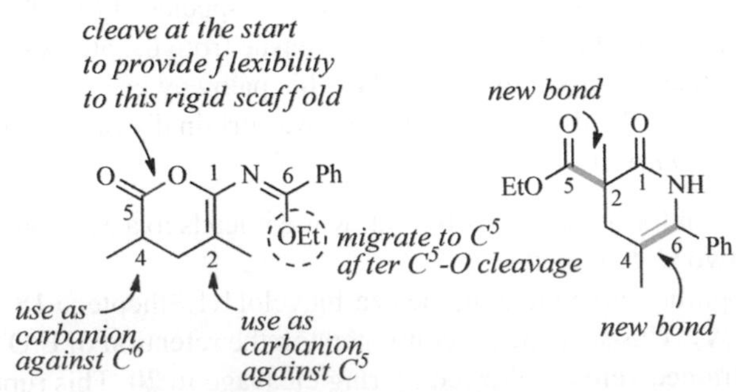

SCHEME 36.4 [1] / Elsevier.

7. In addition, Scheme 36.3 ran smoothly, providing molecular flexibility for the planned bonds, scaffold rearrangement, concerted ethoxide transfer in route **B**, and only one up the hill structure: aza-bicyclobutene **14**.
8. The moral of the story: never believe that atom labels give unique answers to a mechanism, before considering all HED/LED combinations, the accompanying stereochemical constraints, and anticipate intermediates endowed with sufficient degrees of freedom to allow bonding between initially too distant functionalities.

REFERENCES

1. Breuning M, Täuser T. *Tetrahedron* 2007;63:934–940. DOI:10.1016/j.tet.2006.11.033.
2. Closs F, Gompper R. *Angew. Chem. Int. Ed. Engl* 1987;26:552–554. DOI:10.1002/anie.198705521.
3. Closs F, Gompper R, Nöth H, Wagner HU. *Angew. Chem. Int. Ed. Engl* 1988;27:842–845. DOI:10.1002/anie.198808421, and previous work.
4. Proposal by Alonso-Amelot ME, this book.

PROBLEM 37

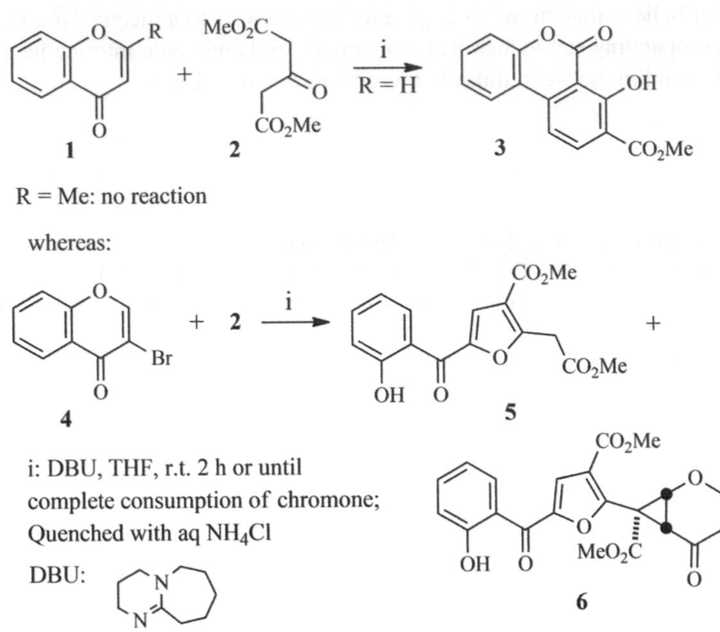

SCHEME 37.1 [1] / Elsevier.

PROBLEM 37: DISCUSSION

37.1 Overview

Despite its molecular simplicity the chromone moiety we see in **1** is a versatile compound with a large curriculum of chemical studies, as this work from scientists in Tesalonike, Greece demonstrate [1]. Nature incorporates **1** in very many compounds; you probably eat a substantial quantity of them every day if vegetables are part of your diet. Chromones are the core of flavones and a long list of other natural secondary metabolites, some of which possess bioactivity and medical applications [2]. Therefore, there is a wide interest in enlarging the chemical space of chromone derivatives.

1,8-Diazabicyclo [5.4.0]undec-7-ene (DBU) needs a word or two. This is a neutral and medium-strong base, no anions or metallic counterions, clean, soluble in solvents and water with poor nucleophilic capacity that allows exposing electrophilic carbons. However, DBU shows selective nucleophilic capacity in certain applications such as the Baylis–Hillman reaction, of interest to your organic reaction mechanism endowment since it activates vinyl carbons as anionic species equivalent in α,β-unsaturated ketones, esters, amides, nitriles, and so forth [3]. The mechanism is quite straightforward (Scheme 37.2 top section). Because a C=C–C=O functionality is present in chromone **1**, a DBU-dependent Baylis–Hillman type process cannot be ruled out. A preliminary analysis will tell us whether DBU operates in this manner or only as a N base.

37.2 Analysis of 1 + 2 → 3 and a Feasible Mechanism

As one dismembers target **3** (Scheme 37.2 middle section), one recognizes the following:

- ✓ Target **3** contains 14 carbon atoms.
- ✓ Compounds **1** and **2** add up to 14 carbon atoms as well, after discounting the orphan methoxy group. Therefore, **3** is the result of the **1** + **2** (–OCH_3) coupling.
- ✓ The upper left section of **3** reproduces part of **1**, leaving the C^2–C^3–C^4 section aside (black). This section is easily detected in **3** but in a distorted configuration.
- ✓ One bond cleavage is required for this disruption: C^2–O followed by C^2 bonding with the ketoester **2**. The two new bonds (red) affect **2** at perfectly identifiable HED carbons, anions in all probability, involving 1,4 and 1,2 additions on the chromene.

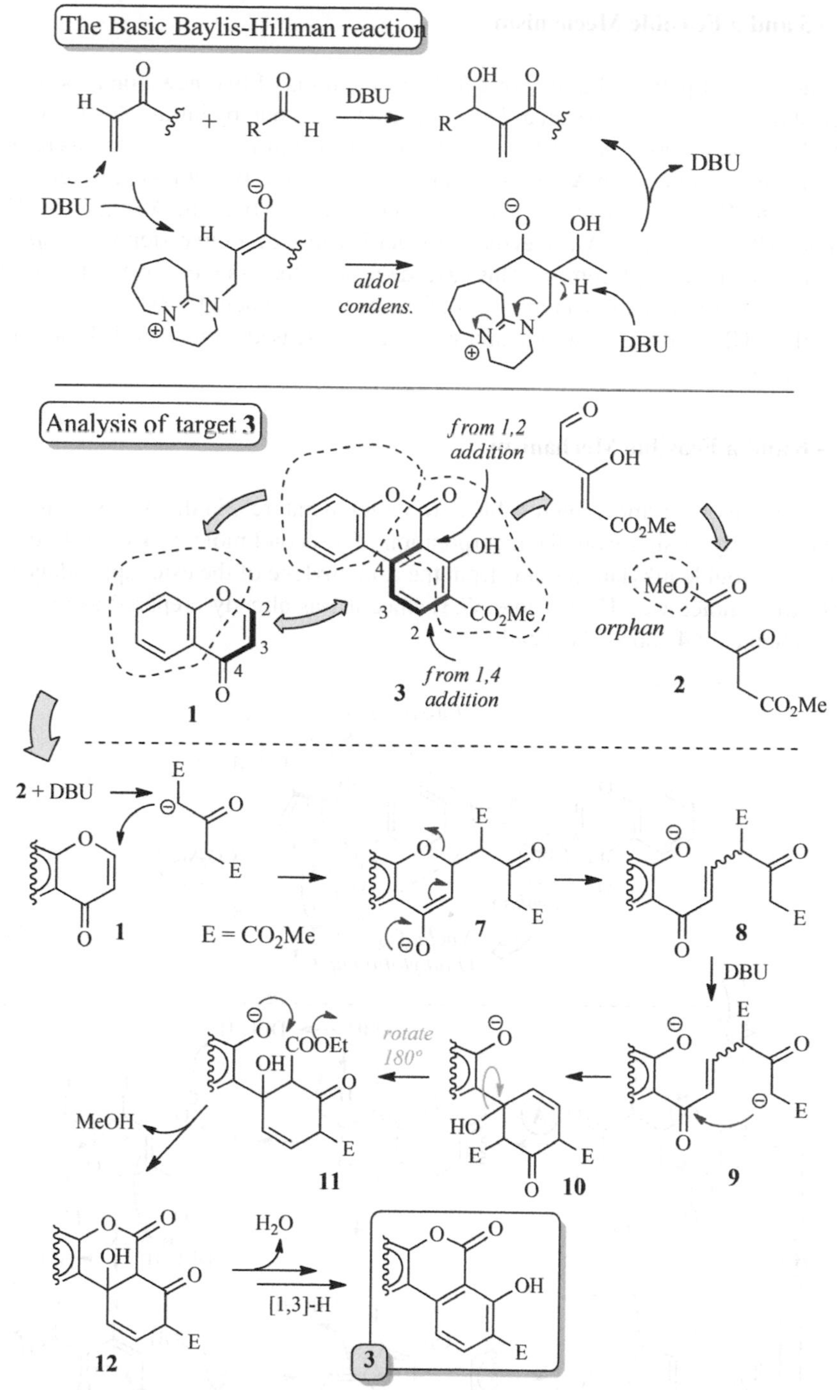

SCHEME 37.2 First section: the basic Baylis–Hillman reaction and mechanism. Second section: Analysis of the **1** + **2** → **3** reaction. Bottom section: a feasible mechanism adapted from the reported sequence of [1] / Elsevier. Copyright permissions as in Scheme 37.1.

- ✓ Two active methylenes in **2** susceptible to proton abstraction by DBU on one side and a well-defined HEDZ in **1** supplement each other neatly. This plan furnishes more than one pathway depending on the order of C–C and C–O bond-forming reactions.
- ✓ A first attack on C^2 rather than C^4 in the starting chromone makes sense since in the opposite case the C^2 δ^+ quality would be lost. In addition, the fact that 2-methyl or alkyl chromones are recovered unchanged under the reaction conditions strongly suggests that the 1,4 addition is the first step of the reaction sequence, composed primarily of aldol condensations afterwards (Scheme 37.2, bottom section).
- ✓ The role of DBU is restricted to substract protons from active methylenes of **2**.

37.3 Analysis of 4 + 2 → 5 and a Feasible Mechanism

The C^3 bromide changes the product pattern. Furan target **5** shows the traits of two new bonds between **1** and **2**, arising from 1,4- addition on C^2 (which should be built at the start as before) and a C^3–O bond by intramolecular O-alkylation with displacement of bromide (Scheme 37.3, top section). A variety of mechanistic possibilities are open to this basic plan, two of which are portrayed in Scheme 37.3, lower section: route **A** (my own) and a shorter route **B** (authors preference [1]).

The difference between **A** and **B** grows from the ring cleavage occurring early in the **A** sequence (**13** → **14**) while in route **B** this is postponed till the end (**19** → **5**). Option **A** is inevitably longer because of a three-step sp^2 to sp^3 conversion of the C^2–Br carbon to enable the intramolecular S_N displacement. This cascade is nevertheless more instructive in that it illustrates the role of DBU not only as a base but as proton carrier in the form of the conjugated acid $[DBU–H]^+$ in intermediates **15** and **16** for example, considering that the THF medium is aprotic and no water is released from **1** or **2**. Proton balance deserves careful attention in reaction mechanism.

37.4 Analysis of 4 + 2 → 6 and a Feasible Mechanism

The unusual–and unexpected outcome ("most improbable at room temperature", in the words of authors [1])–in this instance was the cyclopropyl moiety, visibly arising from the incorporation of a second mole of **4** via 1,4-addition as before, followed by the release of bromide. This double alkylation stems from the α-methylene of the ester appendage on C^2 of furan **5** or some previous intermediate. My first choice was **17** (Scheme 37.3) since it was already prepared as the desired nucleophile in the form of enol. Option **1** of Scheme 37.4 shows the details.

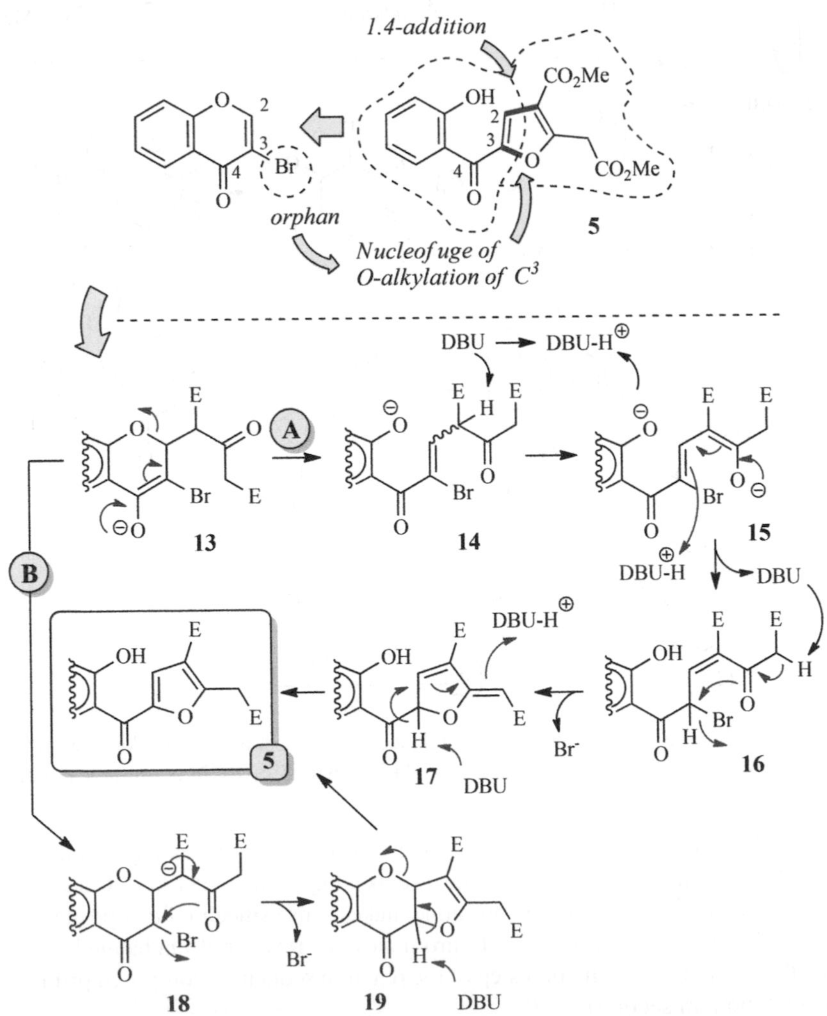

SCHEME 37.3 Analysis of the **4** + **2** → **5** reaction and two feasible mechanisms.

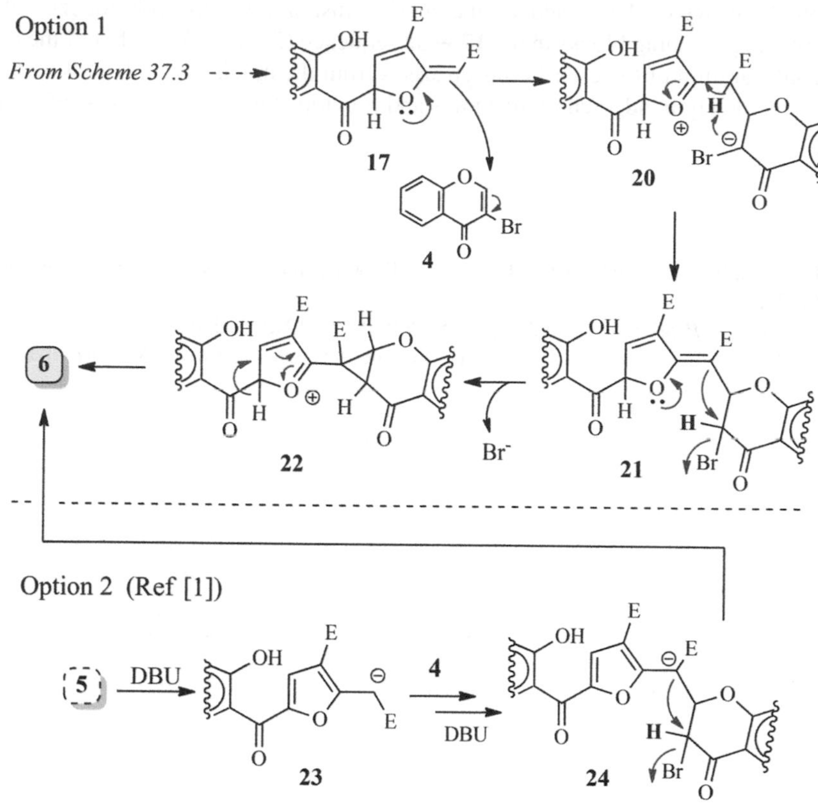

SCHEME 37.4 Reasonable mechanisms for the **4**+**2**→**6** reaction.

Authors offered a different approach (Option **2**, Scheme 37.4) based on the experimental conversion of isolated furan **5** to cyclopropane **6** by exposing the former to a second mole of chromone bromide and DBU in THF at room temperature, a good proof that **5** is the precursor of **6**. Note that **5** contains only one active methylene sandwiched between furan and ester moieties susceptible to double DBU proton abstraction required for cyclopropanation over the LEDZ chromone **4**, as option **1** offered through a preceding intermediate of **5**.

37.5 Lessons and Takeaways

1. In this problem, two highly active reagents, **1**, a typical LEDZ devoid of any HED character (initially) and **2** open to electrophilic and nucleophilic behavior depending on acid/base presence, but in the latter, **2** is typically a HEDZ. Reaction between them is therefore granted.
2. Given the strong response of two expected anions in **2** on electrophilic **1,** authors took care of adding DBU dropwise to a stirred THF solution of **1**+**2** at room temperature. However, the variety of possible HEDZ/LEDZ combinations furnished only low-to-moderate yields, which offered interesting structures responding to a substantial growth of molecular complexity. These results may serve yield optimization exploration using other bases, solvents, temperature, and chromone substituents.
3. All things considered, the reaction types all along the number of schemes portrayed herein were basically: 1,2- and 1,4-additions, and intramolecular substitution and elimination. To connect these reactions through suitable intermediates it was necessary to deploy electrons to strategic atoms, move protons by way of either intramolecular H shifts, or with the aid of the DBU acting as base and proton carrier.
4. Chromones proved to be resilient to this trend of reactions when an alkyl group was grafted at C^2, thus blocking the approach of at least the bulky nucleophile **2**.

5. In regard to the stereochemistry of the cp adduct, the cis-ring fusion is not the consequence of any form of concerted cycloaddition of the two approaching blocks in the **17** → **22** sequence (Scheme 37.4), but of the rigid chromone structure. In addition, the cis configuration of the ester in this cp arises from the much sterically bulkier furan section, which controls the distancing of the newly bonded chromone units and previous furan east and west of cp, respectively.

REFERENCES

1. Terzidis MA, Tsoleridis CA, Stephanidou-Stephanatou J, Terzis A, Raptopoulou CP, Psycharis V. *Tetrahedron* 2008;64:11611–11617. DOI:10.1016/j.tet.2008.10.023.
2. Saengchantara ST, Wallace TW. *Nat. Prod. Rep.* 1986;3:465–475. DOI:10.1039/NP9860300465.
3. Several reviews are available; see for example: Basavaia D, Beeraraghavaiah G. *Chem. Soc. Rev.* 2012;41:68–78. DOI:10.1039/C1CS15174F.

PROBLEM 38

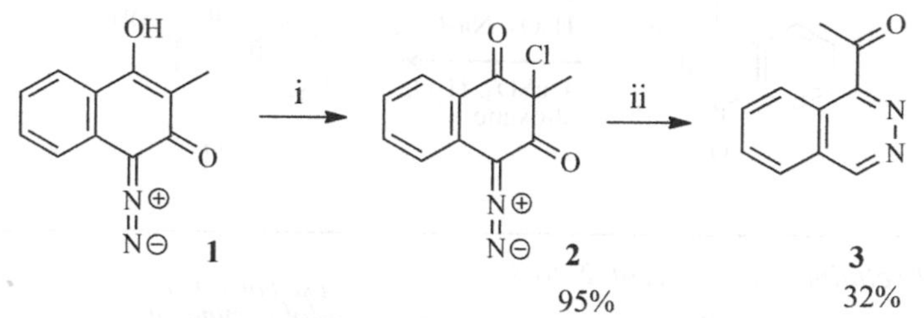

i: NCS, THF, 0 °C
ii: NaOH, DMF-H_2O, 0 °C - r.t.

SCHEME 38.1 [1] / American Chemical Society.

PROBLEM 38: DISCUSSION

38.1 Overview

Up to now we have come across several reactions in which the diazo group is removed. The remaining functionality then evolves in various directions through carbene, anion, and cation species. In the present case, however, the N_2 group remains all along as an active component of the molecular scaffold, calling for a different type of transformation. In addition, there are certain traits in compound **3** that are indicative of a molecular rearrangement of some magnitude.

The sequence of Scheme 38.1 presented by an association of scientists from Australia, China, and the United States [1] is actually a development of a process known as the *Hooker reaction* introduced back in 1936 [2] (Scheme 38.2, top section) and often used in organic synthesis. As one can see, one alkyl group is removed from the core while the R substituent gets moved to the ring. The present diazo version appears to have little to do with Hooker's original discoveries, but it does not.

There are two reactions in Scheme 38.1, the first one deals with a standard oxidation by *N*-chlorosuccinimide (NCS), which follows well-established chemistry. Assessment by functionality number (see Chapter 4) shows this by ΔFNt=+1, a one step oxidation of the substrate. It was included herein to show the diazo-Hooker oxidation in its experimental entirety. The real fun and actual problem for many readers starts in the **2**→**3** conversion, the centerpiece of this reaction.

38.2 Analysis of 2→3

Considering the molecular formulas of compounds **2** and **3**:

Compound	Molecular formula
2	$C_{11}H_7ClN_2O_2$
3	$C_{10}H_8N_2O$
3–2	$C_{-1}H_{+1}O_{-1}$–Cl

Broadly speaking, the loss of CO might suggest decarbonylation, or previous addition of OH followed by *decarboxylation*. The chloride anion can be estranged during the rearrangement since C^2 could serve as anchoring site for the end N atom of the diazo group at some intermediary stage. The N_2 moiety should be retained all along but transformed in a bent configuration suitable for cyclization.

Scheme 38.2 integrates additional advances in mechanism planning for this case:

Vis-à-vis comparisons of structures **2** and **3** would be easier if one replaces Cl for OH as it is likely to occur early in the mechanism. This change may give the chance to a retro-aldol condensation leading to ring cleavage, an operation we need to perform in the way **4**→**3**.

The suggestions highlighted in Scheme 38.2, second section speak for themselves. These need to be orchestrated in a sensitive manner to:

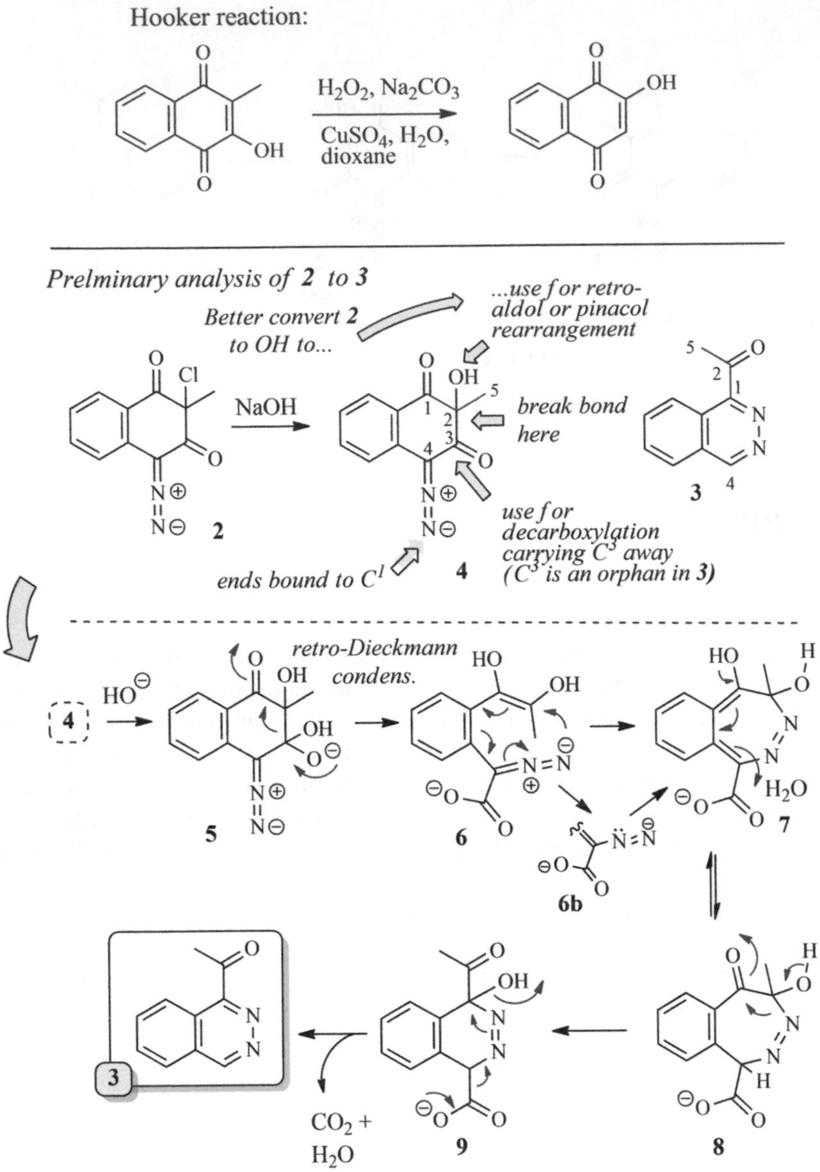

SCHEME 38.2 Top section: The original Hooker reaction. Middle section: Fragmentation analysis is better performed by comparison of a secondary product (**4**) and target (**3**)–see text. Lower section: Translation into a feasible mechanism.

- Split the C^2–C^3 bond by OH addition to C^3=O.
- Break the C^3–C^4 bond to complete the decarboxylation step at a convenient stage not perturbing the nucleophilic activity diazo moiety.
- Build the diazo-C^1 bond once the methyl ketone (in green) has been freed possibly by a reto-aldol condensation.
- With this plan in mind, a likely mechanism unfolds (Scheme 38.2, bottom section), stating from keto-carbinol **4**.

38.3 Highlights and Learning Takeaways in Scheme 38.2

Since element balance predicted decarbonylation or decarboxylation at a higher oxidation level, the first step should prepare **4** for this. The C=O at C^3 appears more sensitive to hydroxide anion attack because of the strong polar effect of the vicinal $C=N^{(+)}$ bond. Therefore, **5** is poised for the desired C^2–C^3 cleavage.

The ensuing electrocyclization may not be concerted because of the orthogonal disposition of the diazo function in **6**, yet a semiconcerted TS or stepwise course through **6b** is also feasible.

After reconstruction of the C^2=O, aromaticity and protonation of the γ benzylic carbon in **8** the semi-pinacol type rearrangement, which is the essence of this sequence, brings about the desired ring contraction to **9**. Then, it is time to estrange the carboxylic acid by decarboxylative elimination. If this step had occurred before the sequence reached this stage it would have stopped it, leaving **8** as a likely primary product.

38.4 Comparison with the Mechanism of the Hooker Reaction, Learning Lessons

Analyzing the relatively close resemblance of the Hooker reaction and its diazo version (diazo-Hooker) is worth the while, after you try the Hooker sequence yourself and match it with that of Scheme 38.3. There are similarities and stark differences, which provide an instructive set of takeaways.

A few dissimilarities: Hooker (H) diazo-Hooker (DH):

1. Electron redeployment in H is more intense than that of DH. For putting together Scheme 38.3 in unabridged form, I required 13 steps to pull electrons at the right places to perform feasible reactions and build credible intermediates.
2. While DH involves one oxidative step to form chloride **4** and proceed from there with the adequate oxidation level, H required *three oxidations*: the first one on account of peroxide, the other two due to electron transfer to copper(II), comprising the crucial oxidative cleavage, which was not necessary in DH.
3. The electron flow direction for the cyclization in H was the opposite of DH: In H bonding electrons were provided by the northern diketone moiety in two occasions whereas the southern diazo group in the latter supplied bonding electrons.

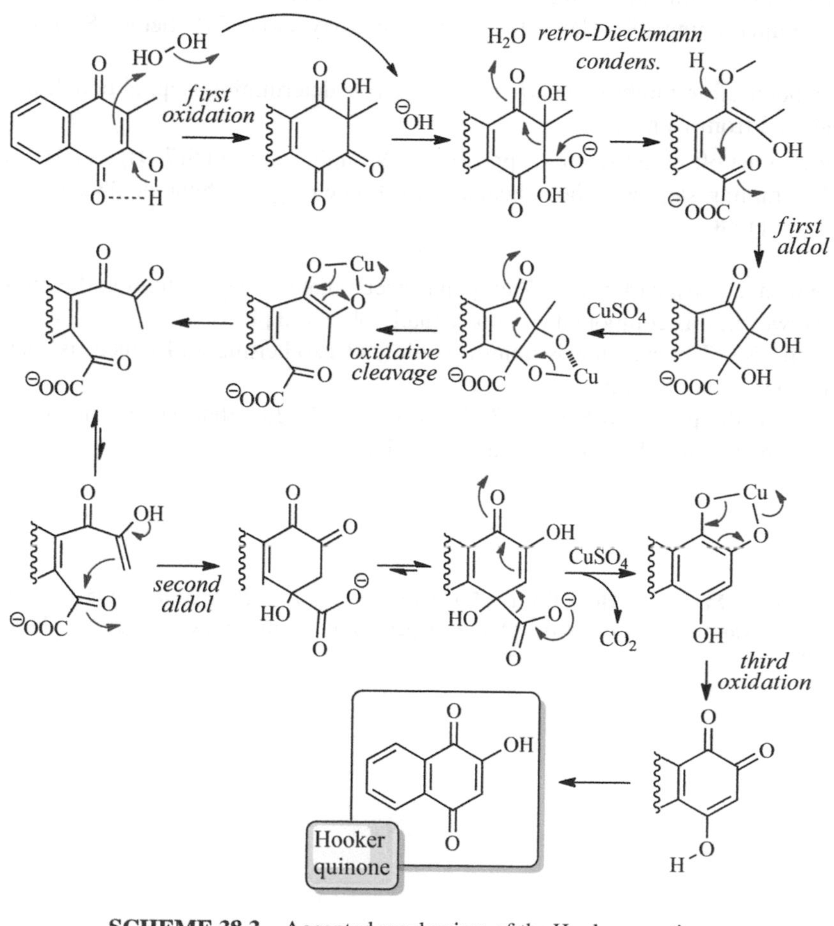

SCHEME 38.3 Accepted mechanism of the Hooker reaction.

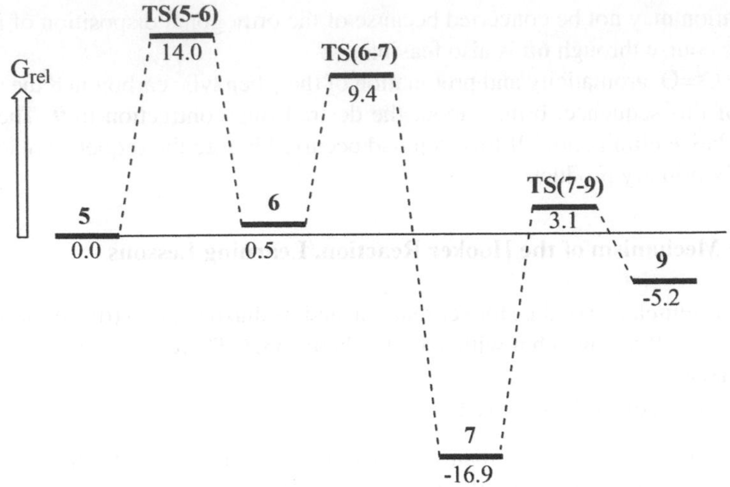

SCHEME 38.4 Gibbs free energy diagram of the diazo-Hooker reaction of Scheme 38.2 calculated by authors [1] using DFT computational resources. Redrawn from Figure 4 published in [1] / American Chemical Society.

38.5 Gibbs Energy Studies and Takeaways

Authors [1] performed DFT calculations to establish energy levels of key intermediates and intervening TSs in the DH sequence. Water solvent was considered in the simulation since water actively participates in H and AH schemes. Gibbs (G) energy levels of key intermediates and the intervening TSs relative to the diazo hydroxyketone **5** (Scheme 38.4) show that:

- The retro-Dieckmann condensation might be interpreted as the rate determining step, although it is also contingent on the continuation of Scheme 38.2 from **7** onwards.
- The ring contraction in **7** to the observed scaffold appearing in **9** implies a ΔG[TS(**7–9**)] = +20.0 kcal mol^{-1}. One concludes that **7**→**9** is the rate determining step with the indicated activation energy. As Scheme 38.2 illustrates, ring contraction is preceded by tautomerization to **8**.

As far as the calculated sequence goes, carboxylate **7** may thus accumulate in the medium at the moderately low temperature of the experiment (0 °C). However, this compound was not actually observed [1].

Along the same lines, the **5**→**7** conversion is nonreversible and exothermic and hence, is thermodynamically favored whereas the continuation **7**→**9** is endothermic.

Calculations give support to the proposed Scheme 38.2, as well as the fact that having the Hooker rearrangement in the background, there appears to be no other feasible alternatives for DH.

REFERENCES

1. Yahiaoui O, Murray LAM, Zhao F, Moore BS, Houk KN, Liu F, George JH. *Org. Lett.* 2022;24:490–495. DOI:10.1021/acs.orglett.1c03810.
2. The Hooker reaction discovery was presented in a series of three papers starting with: Hooker SC. *J. Am. Chem. Soc.* 1936;58:1174–1179. DOI:10.1021/ja01298a030.

PROBLEM 39

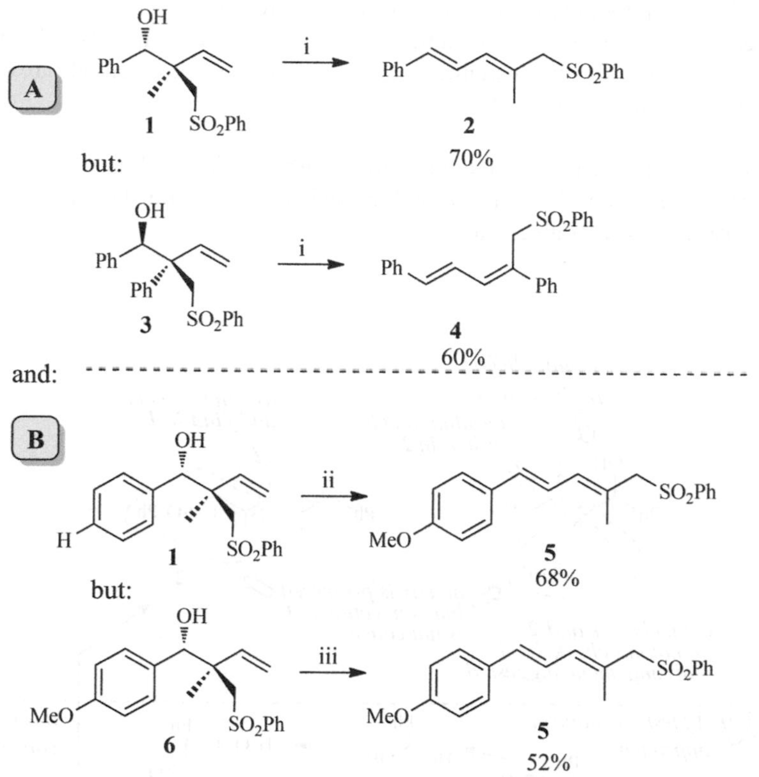

i: benzaldehyde, H^+ (CSA), benzene, reflux, 3.5 h
ii: *p*-MeO-benzaldehyde, H^+ (CSA), benzene, reflux, 3 h
iii: *p*-CN-benzaldehyde, H^+ (CSA), benzene, reflux, 1 h

CSA: 10-camphorsulfonic acid

SCHEME 39.1 [1] / American Chemical Society.

PROBLEM 39: DISCUSSION

39.1 Overview

Scheme 39.1 portrays selected results of a laborious experimental survey carried out by scientists from South Korea [1]. Before confusion grows to scary dimensions, let us organize the presented information as follows:

There are two sections, **A** and **B** in this problem. It was necessary to bring them both to complete the interesting chemistry underpinnings.

Targets in part **A** look alike at first glance, but they are obviously not: while **2** is an all-*E*- (trans) diene, **4** is Z-E diene counting C^1 as the sulfone-substituted end. This result calls for at least one key step in the mechanism that controls the stereoselectivity. Therefore, a common mechanistic sequence must be found first, and then choose at which point this stereoselectivity can be exerted.

The convergent results of part **B** must suggest a different aspect of the mechanistic reasoning since the same product (**5**) is obtained from two different starting materials **1** and **6.** The crux of this difference is in the reaction conditions and thus deserves particular attention.

Annotated yields are only informative, but differences are not significant as regards to mechanism, only marginal. More solid guidance may be found in the type and stereochemistry of products.

This discussion is thus organized in parts **A** and **B** separately. Let me suggest following this process.

39.2 Discussing Part A

39.2.1 Problem Analysis Hint: Without benzaldehyde, **1** would be unreactive.

Before dwelling into stereochemical puzzles let us examine **1** devoid of chiral adornments (which are all lost in its conversion to **2** anyway). Only structural isomerism may be of significance momentarily. Some essentials are depicted in Scheme 39.2, top section. In addition to the notes therein, one concludes that:

- ✓ The branched hydrocarbon chain of **1** turns into a linear hydrocarbon in **2**, which is a sign of skeletal rearrangement.
- ✓ After labeling the linear carbon chain of **2** from the sulfone and attempting translate these tags in **1**, there is a new bond $C^4–C^5$ that should take part in this rearrangement.

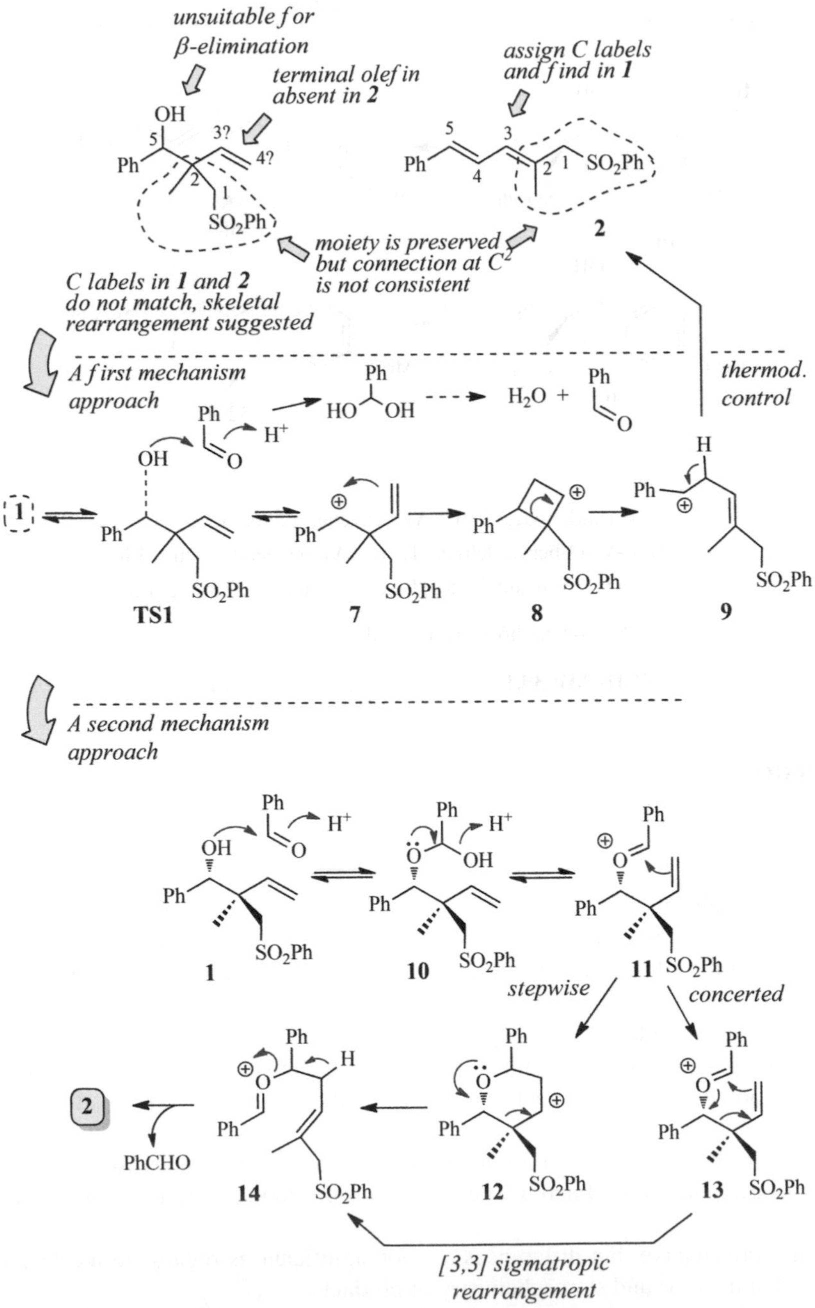

SCHEME 39.2 Issues stemming from comparison of **1** and **2**. Two divergent mechanisms can be devised to address these issues.

- ✓ Because the medium is strongly acidic due to camphorsulfonic acid (CSA), C^5 is activated by the departure of the carbinol as water. Unable to β-eliminate as there are no β protons or receive a nucleophile from the medium, the best and seemingly only choice of $C^{5(+)}$ would be rearrangement.
- ✓ The terminal olefin in **1** *in principle* may turn around and bind $C^{5(+)}$ to create a cyclobutyl cation. Retro [2+2] would furnish the characteristic straight C chain of diene product **2**.
- ✓ Justification of equimolar benzaldehyde might be conceived as an aid to carry away the estranged hydroxyl as acetal, and drive the **1**–$OH^{(-)} \leftrightarrows C^{5(+)}$ equilibrium to the right.

39.2.2 First Mechanism Proposal This first analysis allows me to throw in the putative mechanism portrayed in Scheme 39.2, middle section. After you check it out and compare it with your own proposal, come back to the next line for a critical review.

Pluses:

- ✓ It explains the atom distribution of target **2** from **1** by way of the anticipated rearrangement.
- ✓ The role of benzaldehyde is accounted for through a sustainable **TS1**. Benzaldehyde would be acting as a relay of protic acid transfer. For this reason, a full 1 eq. of this compound is necessary for the reaction success.
- ✓ The structural stereochemistry of all-*E*-(**2**) is easily explained by final proton elimination under thermodynamic control.
- ✓ The E/E configuration of both C=C are explained since the absolute configuration of C^2in **1** is maintained until the retro [2+2] cycloaddition whereby the first C=C is fixed.

Minuses:

- ✓ In principle, cisoid carbinol **3** should experience the same sequence; while it furnishes the all-(*E*)-**4**, it does not explain the Z/E configuration of **4** or the diverging stereoselectivity of **1** and **3** somehow transferred to targets **2** and **4**, respectively.

Given that this liability is a big one, conceiving an alternative mechanism is advisable, although it does not invalidate it altogether. This will be shown in the study of the reactions in section **B** and Scheme 39.4.

39.2.3 A Second Mechanism Proposal

- ✓ Benzaldehyde is also susceptible to addition by oxygen nucleophiles including alcohols. A commencing step, conceptually similar to **1**→**TS1**→**7** may be conceived but the outcome is different. In this case, benzaldehyde would be incorporated to **1** by attack of the carbinol unto benzaldehyde (Scheme 39.2, bottom section). In this analysis, the absolute configuration of **1** is included to facilitate the reasoning of the stereochemical outcome.
- ✓ Water elimination of the formed hemiacetal **10** would give rise to key structure **11**, so constituted to afford the desired C^4–C^5 bond except that the latter carbon *is not C^5 of our original analysis at all but C=O from benzaldehyde*.
- ✓ Intermediate **11** can then evolve toward target **2** through two pathways of diverging characteristics (Scheme 39.2, bottom):
  - *Stepwise* (**11**→**12**→**14**)
  - *Concerted* (**13**→**14**); a [3,3] SR taking advantage of the favorable six-membered conformation of the TS (not depicted).
- ✓ Notably, the **1**→**14** model cascade causes the separation of the original Ph–C^5–OH moiety in **1** as benzaldehyde, in contrast with our first mechanism in which this function is retained in the target.

39.2.4 Stereochemical Considerations In this section, the question of the stereoselectivity of **1** and **3** is addressed. There are two differences worth noting between these two structures: i) The R versus S configuration of carbinols **1** and **3** and ii) the replacement of a methyl by a phenyl group at the quaternary carbon. There ought to be a strategic intermediate in the sequence to define the final stereoselectivity, we should first select a sufficiently rigid structure (a cyclic intermediate for example) to postulate its evolution toward targets in well-defined steric relations between its components.

Given that our first mechanism would not provide answers in this respect, our best choice is examining intermediates **11** or **13** adapted to methyl (**1**) and phenyl (**3**) derivatives. Six-membered pre-cyclic chair form transition structures in 3D are depicted in Scheme 39.3. In a nutshell:

After addition of **1** to benzaldehyde, the adduct can be rendered as a pseudo-chair conformation **15** (left column of Scheme 39.3). Considering the steric repulsion of large substituents, phenyl sulfone>phenyl>methyl, this conformation places the

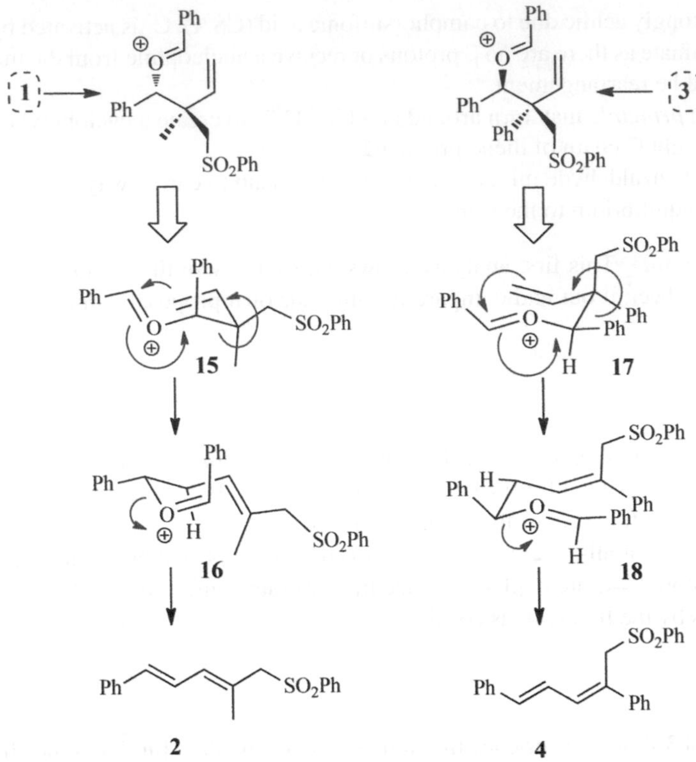

SCHEME 39.3 Stereochemical analysis pursuant on explaining the observed outcomes **2** and **4**.

largest preferentially in the equatorial position whereby the π MOs of terminal alkene and C=O are at close distance. This interaction is relayed through the ring as a [3.3]-sigmatropic shift reminiscent of the Cope rearrangement with an intervening –C=O$^{(+)}$–R unit. It thus adopts the name Oxonia–Cope. The formed primary product **16** is set to undergo β-elimination with loss of benzaldehyde to furnish target **2**.

The corresponding adduct from **3** (right column) undergoes a similar treatment to chair **17**. However, of the two chair conformers possible, a first one displays two axial phenyls and equatorial phenyl sulfone, and a second one, three equatorial phenyls and an axial phenyl sulfone. Steric interactions between substituent are less pronounced in the latter (portrayed in **17**) and the Oxonia–Cope rearrangement proceeds similarly to afford structural isomer **4**. The authors [1] offer this explanation to account for observed products.

39.3 Discussing Part B

39.3.1 Problem Analysis There appears to occur a phenyl → *p*-methoxyphenyl exchange in the **1** → **5** sequence. Note, however, that the second mechanism of Scheme 39.2 precisely accounts for this, since the incoming benzaldehyde is incorporated to the carbon framework while the original benzyl unit is expelled as benzaldehyde.

The real problem is to conceive a sequence that *preserves the p-methoxyphenyl* of the starting carbinol as requested by the **6** → **5** reaction (Scheme 39.2, bottom section). However, this is not allowed by this mechanism. How, then?

There are not one but three possible and chemically admissible ways to achieve this goal.

1. By application of our first mechanism attempt (Scheme 39.2, middle section, sequence **1** → **9** → **2**) whereby the aryl group on C^5 is never affected and remains in place.
2. Having a strongly acidic medium, a Markovnikov proton addition to the terminal alkene or heat (refluxing benzene) may promote SR with liberation of PMB early in the reaction (see green arrows, Scheme 39.4). This PMB then forms an adduct with still unreacted **6** and proceeds through the mechanism proposed by authors [1] (Scheme 39.2 bottom section).

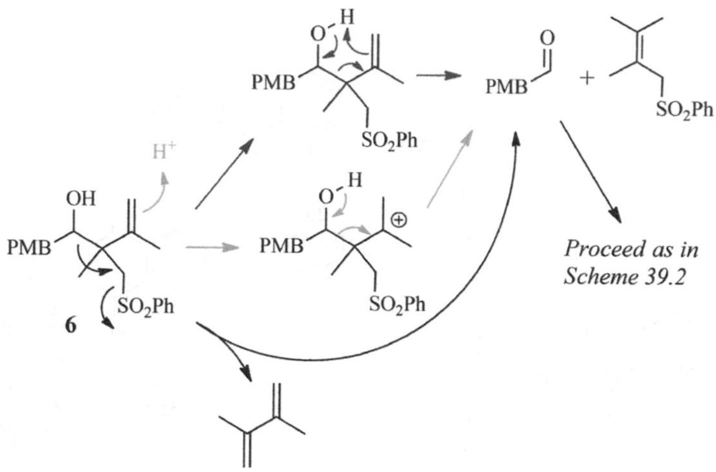

SCHEME 39.4 Three different ways to convert **6** into key reagent PMB-C=O.

3. PMB may also be fragmented from **1** by δ-elimination with loss of the sulfone group, as authors contend [1] (black arrows in this scheme).

The latter two options are shown in Scheme 39.4.

39.4 Learning Issues and Takeaways

1. The stereoselective synthesis of all-(*E*)-polyene stereoisomers with active functional groups at the ends is a useful method to obtain carotenes and other related natural bioactive natural products, in addition to other ends related to cycloadditions of dienes with dienophiles.
2. The indirect formation of these conjugated dienes from apparently unrelated carbon skeletons implies some difficulty at the time of devising mechanisms. Whenever carbon scaffolds with or without heteroatoms are converted to a different skeleton the occurrence of at least one molecular rearrangement must be suspected; in this case, a specific case of the Cope rearrangement.
3. These carbon translocations can be detected easily by atom numbering, as pointed out in "The Toolbox", Part I, hence orienting the identification of bond rupture and formation. Association of these atoms with neighboring substituents greatly facilitates the job of conceiving a mechanism.
4. Depending on reaction conditions, the involved molecular reorganization may occur in steps (more than one rearrangement in tandem) or a concerted process. Frequently, both ways can be depicted but it is contingent on the stabilization of ionic species by suitable substituents.
5. Selecting the triggering step to commence a sequential chain is relatively easy in most compounds, but multifunctional systems may be more difficult to ascertain. For example, compounds **1**, **3**, and **6** include three functional sites susceptible to activation by the acidic medium, C–O, carbinol oxygen atom, and terminal alkene. Each one drives the ensuing process in different directions, which may branch out with progress toward target(s). While this procedure enriches the chemical reasoning space, keeping targets in sight at all times is recommended.

REFERENCE

1. Jung S-Y, Min J-H, Oh, JT, Sangho Koo. *J. Org. Chem.* 2006;71:4823–4828. DOI:10.1021/jo060517e.

PROBLEM 40

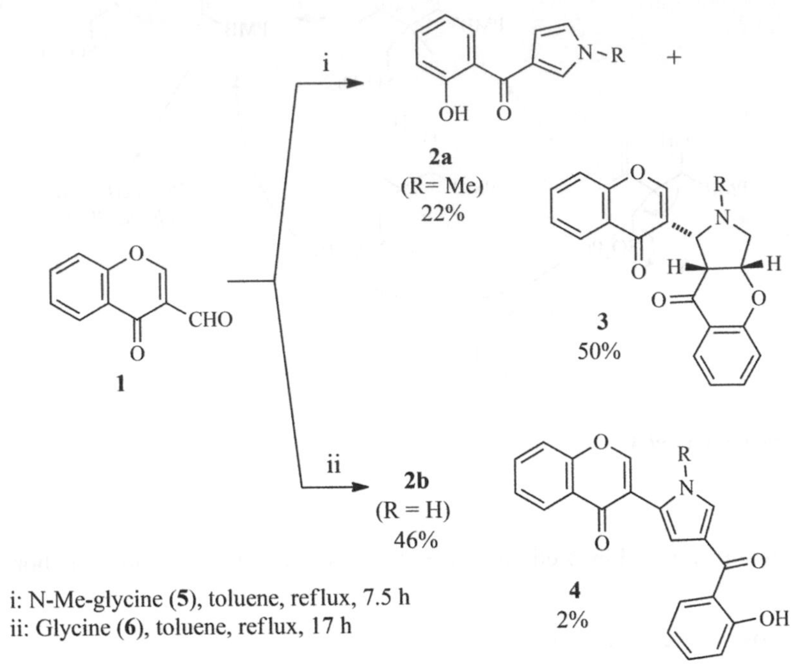

SCHEME 40.1 [1] / Elsevier.

PROBLEM 40: DISCUSSION

40.1 Overview

This is our second encounter with chromone derivatives in this workbook. The apparently accessible problem now presented involves conceptually different issues to electron deployment of the previous problem. The chemistry of chromone carbaldehydes like **1** has been known for quite some time, owing to its electronic versatility as dienophiles and Michael acceptors particularly. Applications go as far as furnishing derivatization of C_{60} fullerenes, the familiar spherical molecules known as buckyballs; yet, these chromones still offer a growing gamut of reactions with several reagents and surprises as this research by Professor Silva and coworkers in Aveiro, Portugal shows [1]. To begin with, notice the stark difference in the outcome of the two reactions of Scheme 40.1 because of using glycine or *N*-methyl glycine, the two smallest amino acids in proteins. Another issue to explain is the stereochemistry of compound **3**.

40.2 Analysis of 1 → 2a and Feasible Mechanisms

This time over we will start with a visual inspection of **1** and **2** only, side by side, throw in N–CH_3–glycine (**5**) as an obvious participant, recognize HEDZ/LEDZ sites, and then assume their natural interaction. A first set of suggestions is indicated in Scheme 40.2, top section, which leads to the mechanism under the dashed line.

40.3 Highlights and Takeaways from Scheme 40.2

Briefly:

- ✓ In accordance with the proposed plan, the expected and most likely HEZ/LEDZ interaction is ruled by the construction of the aldimine **7** involving the aldehyde.
- ✓ Early decarboxylation ensues in **7** in view of the sterically and electronically comfortable TS involving six valence electrons.

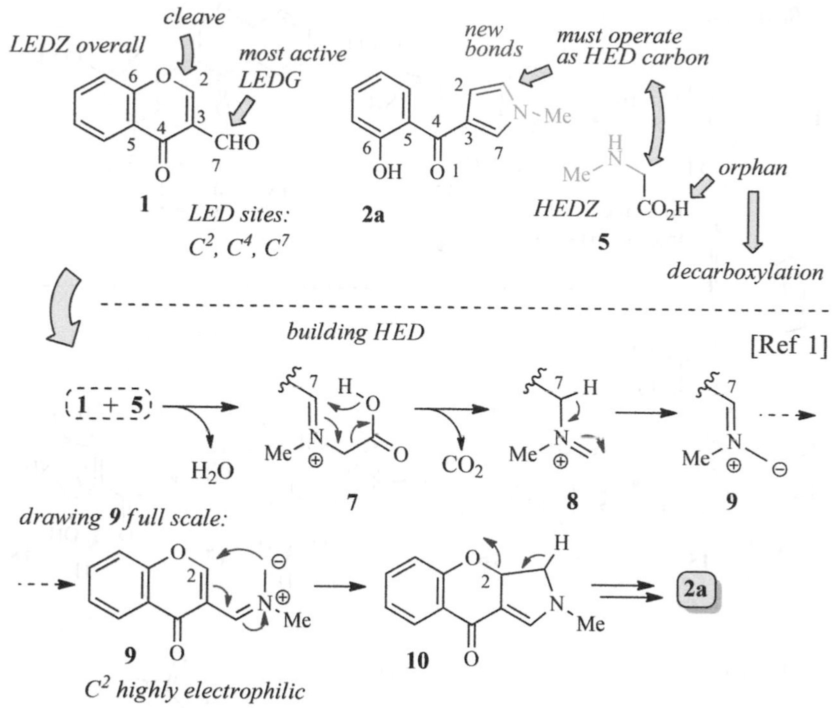

SCHEME 40.2 [1] / Elsevier, same license no. as Scheme 40.1.

✓ Proton elimination turns the strongly electrophilic terminal methylene in **8** into an $N^{(+)}$–$CH_2^{(-)}$ unit in **9**. This intermediate is an *N*-ylide. Mind that there are oxygen, sulfur, and phosphorous ylides widely used in synthesis and comporting interesting in reaction mechanism.

✓ When partial structure **9** is fully drawn, the proximity of the anionic methylene and C^7, a strongly electrophilic carbon by virtue of the double conjugation, including the $C{=}N^{(+)}$ unit, is well suited for the ensuing C–C bonding furnishing **10**.

✓ β-Elimination cleaves the red C–O bond to afford the desired pyrrole.

✓ The sequence is thus: the addition of amine to aldehyde → decarboxylation → formation of *N*-ylide → ring closure → elimination. One mole of water is produced, the only proton source available, which is not needed as such in the entire sequence and may interfere with some of the intermediate species [2].

40.4 An Alternative Mechanism for 1 → 2a

This reaction was examined years earlier (1985) by Professor Suschitzky's research group in the United Kingdom whose mechanism proposal was entirely different [3]. Based on the reaction of glycine ethyl ester and **1**, which furnished pyrrole and pyridine ester derivatives (Scheme 40.3), the first glycine-chromone aldehyde interaction was conceived at the core carbon C^2 as a Michael-type addition rather than the more exposed aldehyde. Although authors portray a simplified version of events that appear in greater detail herein for instructive purposes, the inversion of the order of events is quite apparent:

1,4-addition → ring opening in **16** → cyclization to dihydropyrrole (**17** → **18**) → δ-elimination of water (**18** → **12**).

For the first Michael addition to take place, the glycine ester needs to be poised as the ester enol form **14** to confer nucleophilic capacity to the glycine methylene. This proposition is a bit uphill considering the very small proportion of **14** in the ester ⇆ enol equilibrium, yet without it the first 1,4-addition would not go.

Translation of this cascade to *N*-ethyl-glycine (**5**) (Scheme 40.3) and chromone aldehyde **15** leads to dihydropyrrole carboxylate **20**, in which electron distribution and stereochemical configuration are perfectly suited for decarboxylative elimination (my own proposal). It is worth noting that in this case a catalytic amount of acid (pTsOH, *p*-toluenesulfonic acid) was required to boil toluene. Indeed, the two sequences of Scheme 40.3 require protic acid in three steps. While Schemes 40.2 and 40.3 offer admissible explanations, experimental evidence or computer calculations have not been performed to establish which mechanism is closer to reality or even a third cascade you are welcome to design.

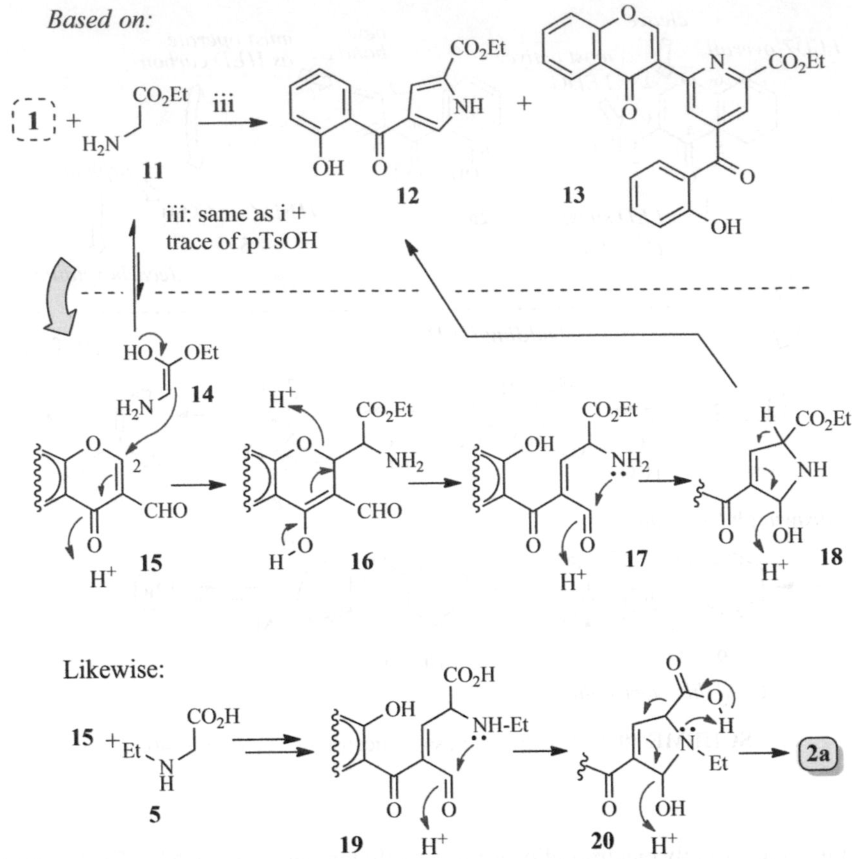

SCHEME 40.3 Alternative mechanism proposed in 1985. Adapted from Reference [3] with my own additions. © 1985, the Royal Society of Chemistry, used with permission.

40.5 Analysis of 1→3 and Feasible Mechanisms

Having Schemes 40.2 and 40.3 at hand, reasoning target **3** is much easier. The molecular structure contains 1 mol of *N*-methyl glycine, already decarboxylated, and 2 mol of chromone aldehyde. Recommendations are suggested in Scheme 40.4, top part, and the sequence flowing from this preliminary plan.

40.6 Highlights and Takeaways of Scheme 40.4

- ✓ The green section of **3** arises from our previous formation of *N*-ylide expressed as **9a** ⟷ **9b** 1,3-dipolar resonance hybrid.
- ✓ Approximation of chromone aldehyde **1** responds to this charge distribution with its own natural dipole. Chromone **1** and other similar compounds have been submitted to dipolar DACAs in the past to build much larger fused benzenoid adducts. As expected from the dipolar attraction, the [2+3] cycloaddition is regiospecific. Stereospecificity is also predicted if the two σ bonds are formed concertedly.
- ✓ Compound **20** is still adorned with an extra carbaldehyde not observed in target **3.** There are two ways to detach it in the absence of nucleophilic base:
  - Homolytic cleavage (not shown) to the angular radical and H capture termination given the high temperature of boiling toluene (111 °C).
  - Attack on **21** by a second molecule of *N*-methyl-glycine (**5**) to promote a peculiar version of retro-aldol condensation shown in **22**. Enol-chromone **23** is the immediate precursor of target **3**.

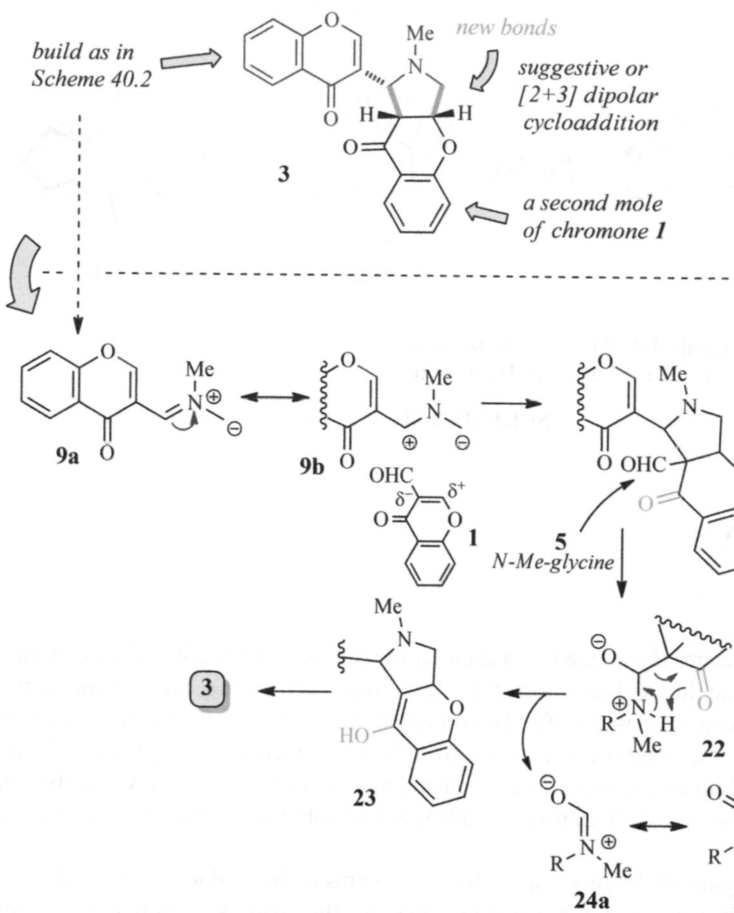

SCHEME 40.4 Solving 1→3.

✓ Meanwhile, the glycine moiety comes off as the oxime. Observe that the overall **21→→24** minisequence is a redox operation.

✓ The decarbonylation mechanism erases the cis-ring fusion owing to the enol C=C bond in 23. However, the cis configuration is thermodynamically more stable but not unique. Authors mention the isolation of two other isomers of **3** in minor amount that may be the two trans-diastereomers.

40.7 Solving 1 → 4

Is it really necessary? With the previous schemes in mind it becomes an exercise rather than a problem, doesn't it?

REFERENCES AND NOTES

1. Figueiredo AGPR, Tomé AC, Silva AMS, Cavaleiro JAS. *Tetrahedron* 2007;63:910–917. DOI:10.1016/j.tet.2006.11.034.
2. Water removal as a reaction progresses can be performed by adsorbents like molecular sieves or the classical method of collecting the refluxing solvent–water mixture in a Dean–Stark trap apparatus, provided that boiling temperatures are required, and water and solvent are not miscible.
3. Clarke PD, Fitton AO, Kosmirak M, Suschitzky H. *J. Chem. Soc. Perkin Trans. I.* 1985;1747–1756. DOI:10.1039/P19850001747.

PROBLEM 41

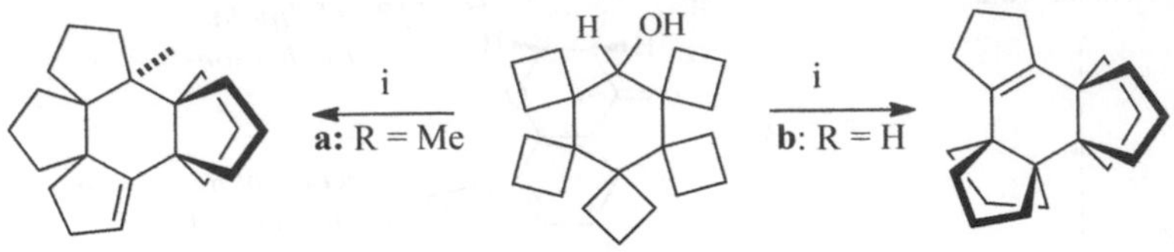

2 1a,b 3

i: anh. TsOH (1.1 eq), benzene, 70 °C
For **1a**: 15 min; for **1b:** 40 min

SCHEME 41.1 [1] / Elsevier.

PROBLEM 41: DISCUSSION

41.1 Overview

This fascinating molecular demeanor may sound a bit familiar to readers of the second edition of this series [2]. Closely related polycyclic carbinols evolved in acidic medium through tandem structural rearrangements initiated by a carbenium ion, whereas a ketone sibling underwent carbanion-driven scaffold recombination to convoluted structures of the propellane family. This line of investigation has continued in the hands of researchers from Göttingen, Germany [1] to provide us with further awe based on bench chemistry, theoretical calculations, and stereochemical analyses of the very many possible intermediates and products. Given the multitude of mechanisms and alternatives, I have selected only the most accessible ones, which all the same conceal challenging issues.

The first question is the outcome difference caused by what seems to be a minor detail: replacing a H with a methyl group at the carbinol carbon. The gross difference in reaction rate given by the fastness of **1a** rearrangement (only 15 minutes) versus the 40 minutes of **1b** and the consequences of the expected secondary versus tertiary carbenium ion must impact the course of the ensuing chain of events. However, the angular methyl must exert further influence, probably conformational, to redirect the cascade in a different direction than the one followed by H-derivative **1b.**

41.2 Solving 1a → 2

41.2.1 Preliminary Analysis

- ✓ There are five spirocyclobutyl units (*cbu*) in **1a** no longer present in **2**. Therefore, these rings have succumbed to ring strain and expanded.
- ✓ The driving force for ring expansion is most probably a $C^{(+)}$, although carbenium ions find little stabilization in benzene. All of it must be provided by the carbon scaffold.
- ✓ In target **2**, there are five fused cyclopentyl units, visibly from the *cbu* ring expansion via [1,2]-C shifts.
- ✓ However, two five-membered rings have overlapped at the eastern portion forming a true propellane partial scaffold. This move follows a peculiar shift pattern to be devised in our mechanism.
- ✓ Meanwhile the other three rings remain in place fused with the cyclohexyl matrix.
- ✓ There are no H shifts.
- ✓ A carbenium ion survives until proton elimination creates the double bond in **2**. This feature indicates the site of this last $C^{(+)}$.
- ✓ Therefore, the initially formed $C^{(+)}$ at the carbinol carbon must "migrate" four or more times before reaching the alkene precursor. Mind the use of the term $C^{(+)}$ *migration*. $C^{(+)}$actually does not. C–C bonds and one of the bonded elements are the actual migration unit.

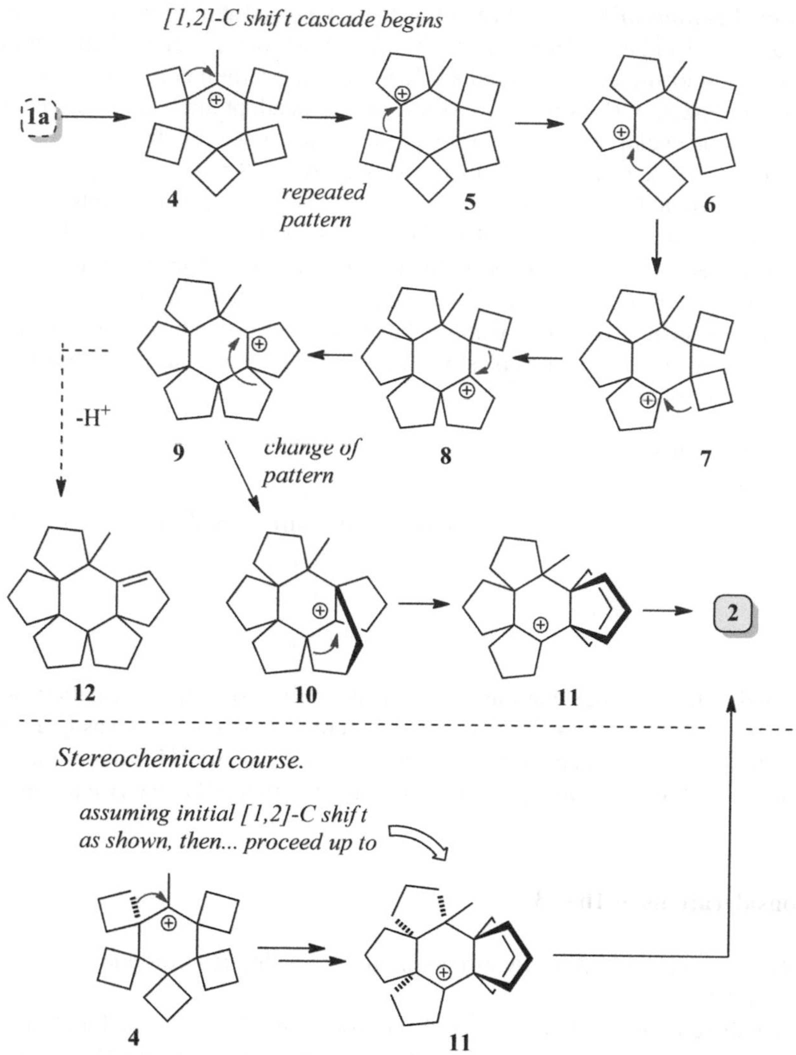

SCHEME 41.2 A repetitive cascade of nine [1,2]-C shifts explains the conversion of **1** to **2,** allowed by Woodward and Hoffman symmetry if C is the migrating carbon.

With these elements of judgment one can devise a reasonable sequence (Scheme 41.2).

41.2.2 The Basic Mechanism in a Nutshell

- The *cbu* rings expand in succession from cation **4** to **8** via similar *antarafacial* [1,2] C-migration. As discussed in "The Toolbox", Part I, while this is not feasible in H migration it is allowed in alkyl shifts owing to the transient trigonal topology of the migrating carbon and the involvement of the *2p* AO. The C-transfer occurs *suprafacially* with inversion of configuration in the case of unsymmetrical sp^3 carbons. This inversion is of no consequence in the present case.
- Once cation **9** is formed it can either stop the rearrangement progress by elimination to afford **12**, a compound observed by authors as the major product [1].
- Alternatively, a change in the rearrangement pattern opens pathway **10** → **11** target **2** by way of a kind of walking ring rearrangement furnishing the unique setup of the three fused tricyclo[0.3.3.4]C_{12} ring section – the actual propellane.
- Taken together, the construction of **2** from cation **4** is a seven step W–M rearrangement cascade.

*41.2.3 Stereochemical Considerations of **1a** → **2*** In addition to the repeated and strained *cbu* motif, compound **1** is sterically encumbered rather heavily; a 3D molecular rendering of energy-minimized conformation of **1a** in stick and space-filling modes (Figure 41.1) illustrate the position of the cyclobutyl units (*cbu*) in the six-carbon core, which assumes a chair conformation after conformational energy minimization (Figure 41.1). As a result, *cbus* adopt alternate pseudo-equatorial and pseudo-axial conformations of essentially planar cyclobutyl rings, forming a compact spherical structure of 331.7 $Å^3$ Connolly's solvent exclusion molecular volume (SEMV; see Chapter 3) and a torsion energy of 88.5 kcal mol^{-1} [3].

Distances between H atoms of vicinal *cbus* in **1a** and **1b** are as close as 2.18 Å, which leads to frequent overlaps of van der Waals radii (vdWR) and a corresponding contribution of 14.37 kcal mol^{-1} to the total energy [3].

Target **2** six carbon core assumes a semi-boat conformation in the energy-minimized structure. Interactions between neighboring methylenes bonded to the quaternary carbons are also intense with distances as close as 2.06 Å, even more compact than **1a** (315.6 $Å^3$ Connolly's solvent exclusion molecular volume) but a substantially reduced torsion energy (36.35 kcal mol^{-1}).

The authors [1, 4] compared the calculated enthalpies of formation ΔH° of all intermediate cations and concluded that the **1a** → **2** reaction is markedly exothermic:

ΔΔH°(**11–4**) = −103.0 kcal mol^{-1} and
ΔΔH°(**2–4**) = −113.0 kcal mol^{-1}.

Most of this energy should stem from the release of torsion, ring stress, and other steric effects of the deep-skeletal rearrangement.

41.3 Solving 1b → 3

Much of what we have learned in the previous analysis is applicable to this novel transformation, which arrives to a synthetically challenging di-propellane scaffold **3**. The second propellane section should be more easily accessible from the preceding cation that gave birth to the first one. This is cation **12**, the equivalent of structure **11** without the angular methyl group. The walking ring rearrangement **12** → **14** and the closing unimolecular elimination (E1) give rise to the convoluted target structure **3** (Scheme 41.3).

41.4 Stereochemical Considerations of 1b → 3

Given the structural complexity achieved in **3** it is worth noting the following features and examine them comparatively with the **1a** → **2** case (see Figure 41.1):

All carbons of the six-membered core of **3** are now coplanar, forced by the C=C bond and the similar repulsion between methylene hydrogens in the two propellane sections above and below the molecular plane. H–H distances are as short as 1.81 Å, clearly overlapping the vdWR of H. As a result, the C–C bond of the central core between the two propellane substructures is stretched to 1.604 Å (for comparison, the C^4–C^5 bond length in cyclohexene is 1.535 Å) [3].

The C^2 axis of symmetry of **3** cutting across the cyclopentane and cyclohexyl rings on the plane of the molecule in addition to the coalescence of four of the original carbocycles into the two propellane substructures engender a more compact molecule relative to target **2** (305.6 $Å^3$ Connolly's SEMV) (space-filling model **VI**, Figure 41.1).

Energy-wise, authors also calculated the enthalpies of formation and the overall change, and arrived to:

ΔΔH°(**14–4***) = −114.5 kcal mol^{-1} and
ΔΔH°(**3–4***) = −122.3 kcal mol^{-1}

(where **4*** is the non-methyl cation **4** in the **1b** → **3** cascade).

The process is also exothermic. In both cases, the highest activation energy ΔH‡ corresponds to the initial formation of the first cation from the carbinol, since the rest of the steps show either negative ΔΔH° or slightly positive values [1, 4]. By comparing Gibbs free energy values of starting alcohols **1ab** and the first carbocation **4**, assuming no other higher energy TS, I estimated a *ΔG‡* = 35.1 kcal mol^{-1} barrier for **1a** → **4** and 34.5 kcal mol^{-1} barrier for **1b** → **4***. The marginally lower value of the latter may result from the increase in SS of the methyl group against the two neighboring cyclobutyl rings.

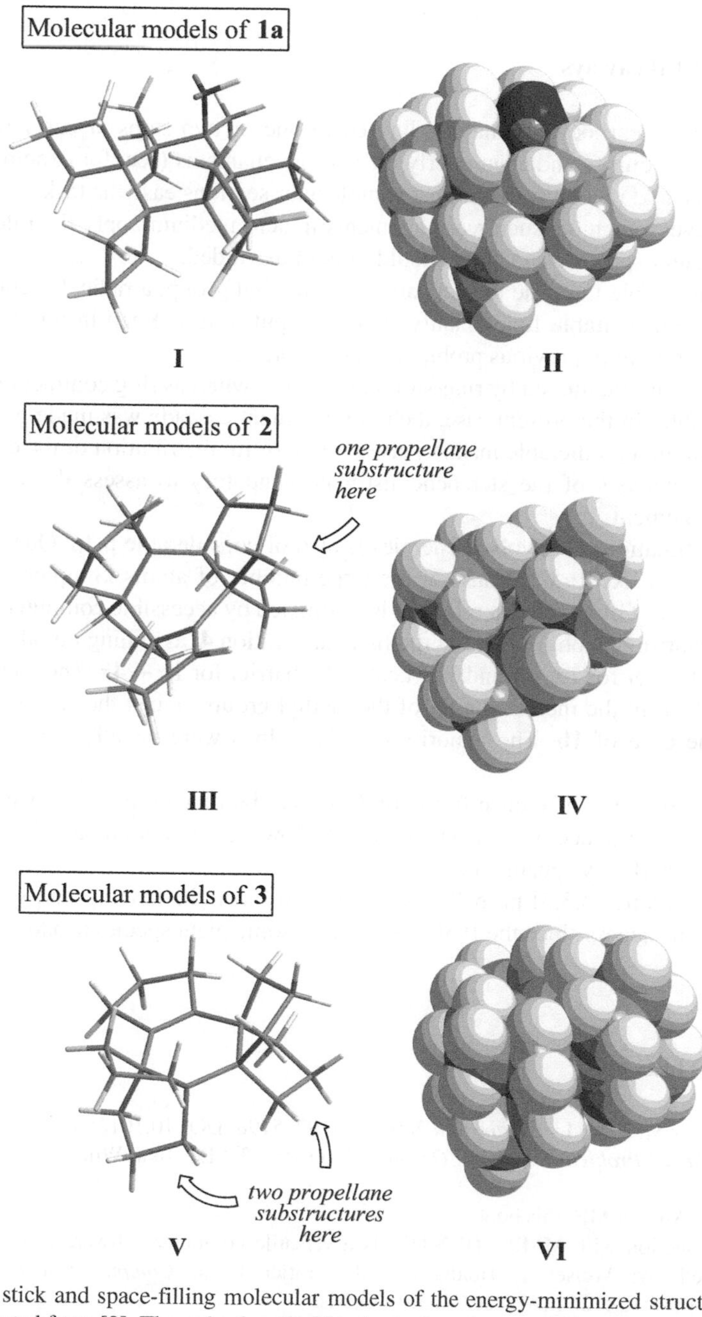

FIGURE 41.1 Three-dimensional stick and space-filling molecular models of the energy-minimized structure of **1a** and **2** drawn at the same scale and perspective view Adapted from [3]. The red sphere in **II** is the hydroxyl group in **1a**.

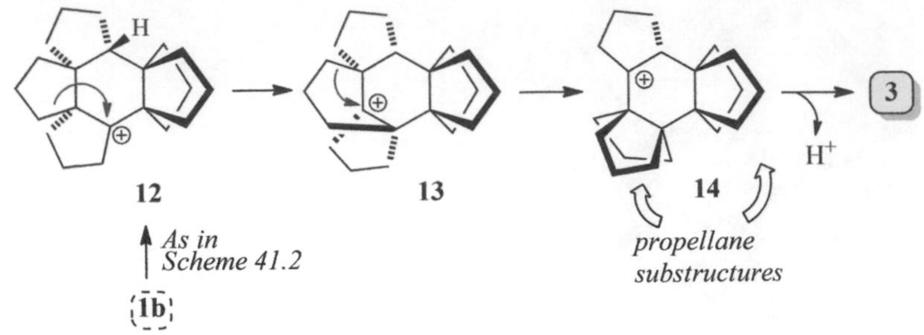

SCHEME 41.3 Origin of the bis-propellane target **3**.

41.5 Learning Points and Takeaways

1. Skeletal rearrangement of carbocycles may be limited to one or two steps or carry forward to complex scaffolds of increasingly untraceable courses and even utterly difficult mechanisms (see, for example, Problem 50). However, good planning, carbon labels, and recognition of unfazed molecular sections ease the task.
2. During mechanism development of such rearrangements in acid medium apply the rule *tert*-$C^{(+)}$ > *sec*-$C^{(+)}$ ⋙ *primary*-$C^{(+)}$ for proposed transient species. The latter should rather be avoided.
3. In case it becomes impossible to evade the activation of a methyl group, a radical if conditions are adequate or carbene stemming from a previous suitable functionality should be put to test. Keep in mind that remote [1.x]-H transfers do occur, as we have seen herein in previous problems and Chapter 2.
4. Skeletal rearrangements are facilitated by ring-strain relaxation, whereas ring contraction to four and three rings is not as common but still feasible. In the present case, the rearrangement cascade was made possible because of the continuous relaxation of ring strain in a predictable manner, given the uniform distribution of the cyclobutyl rings.
5. Frequently a detailed analysis of the stereochemistry is mandatory to assess the feasibility of carbon and H shifts composing the rearrangement.

✓ Access to computer estimations of involved energies is also of considerable help. Quantum mechanical calculations are limited in large molecules like **1a,b** because of the large number of atoms. Simpler representative models are usually studied. In the present case, Gibbs free energies can be estimated by accessible computer algorithms. By comparing Gibbs free energy values of starting alcohols **1ab** and the first carbocation **4**, assuming no other higher energy TS, I estimated a $\Delta G\ddagger = 35.1\,\text{kcal mol}^{-1}$ barrier for **1a** → **4** and $34.5\,\text{kcal mol}^{-1}$ barrier for **1b** → **4*** (*no methyl). The marginally lower value of the latter may result from the increase in SS of the methyl group against the two neighboring cyclobutyl rings in **1a**, which is absent in the case of **1b**. The majority of $\Delta G\ddagger$ values were negative favoring the ensuing cation and final product.

6. With regard to the **2/3** product difference from **1a/1b**, it may be related to the course followed by monopropellane **11** (Scheme 41.2) under the influence of the angular methyl. However, without detailed semiempirical calculations at hand, any explanation becomes flimsy speculation.
7. The synthesis of carbocyclic [3.3.3] propellanes has proved most challenging and several interesting and imaginative strategies have met with success since the 1980s. As well as, some plant species produce from simple to highly substituted propellanes [5].

REFERENCES AND NOTES

1. Justus K, Beck T, Noltemeyer M, Fitjer L. *Tetrahedron* 2009;65:5192–5198. DOI:10.1016/j.tet.2009.05.005.
2. Alonso-Amelot ME. *The Art of Problem Solving in Organic Chemistry*, 2nd Ed. John Wiley & Sons, NY, 2014, Chapter II, pp. 38–40, schemes 11-13.
3. Method: MMFF94, Alonso-Amelot ME, this book.
4. Method: HUNTER in connection with MMP2. HUNTER is a versatile computer software designed for conformational analysis and stereochemistry. Introduced by: Weiser J, Holthausen MC, Fitjer L. *J. Comput. Chem.* 1997;18:1264–1281. DOI:10.1002/(SICI)1096-987X(19970730)18:10<1264::AID-JCC2>3.0.CO;2-L.
5. For a recent review, see: Dilmaç AM, Wezeman T, Bär RM, Bräse S. *Nat. Prod. Rep.* 2020;37:224–245. DOI:10.1039/c8np00086g.

PROBLEM 42

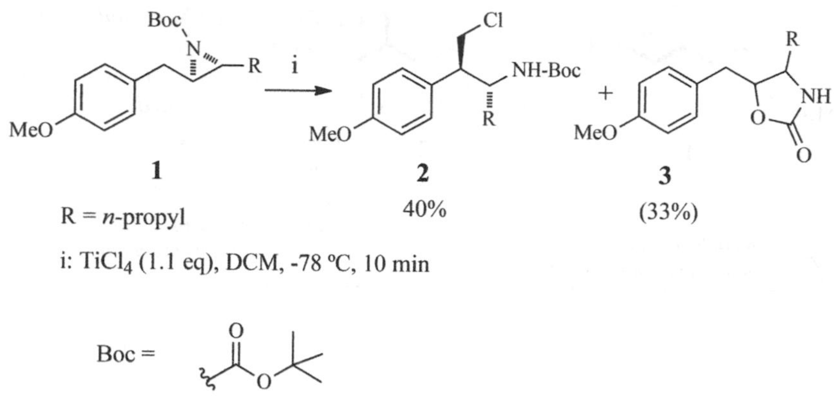

SCHEME 42.1 [1] / American Chemical Society.

PROBLEM 42: DISCUSSION

42.1 Overview

This is our first encounter with a detailed view of organometallic performance in an organic reaction for you to get training in this fundamental area of organic chemistry. The reaction course strictly depends on the type of complex, associated organic functionalities to the metallic element and reaction conditions. As authors from the University of Minnesota, United States demonstrate [1], a variety of related by-products appear with subtle changes in these factors. I selected the reaction in Scheme 42.1 not as the most efficient or best product-focused outcome of the several reaction tested by these researchers but one that illustrates the mechanistic variety open to this particular process, hence a better training court.

This work is not only mechanistically attractive since you will be driven to imagine an unusual intermediate, but it also focuses on the construction of chiral β-phenethylamines having active substituents. These functionalities are suitable handles to expand the chemical space of this medicinally important scaffold, present in several treatment drugs. Authors explored the scope of this reaction in several methoxyphenyl derivatives and found it to furnish the same types of rearranged outcomes in addition to a few side products, depending on the N-protective group. Yields were as high as 74% when *N*-acetate replaced *N*-Boc.

42.2 Analysis of 1 → 2 and a Feasible Mechanism

Conditions are mild in all respects, low temperature and a non-intervening solvent other than organizing a solvent cage. No organic reagents were added. In addition, comparison of **1** and **2** reveals that:

- ✓ Starting and ending compounds are chiral, therefore a dominant stereochemical intermediate must be engendered.
- ✓ Both compounds possess the same number of C, N, and O atoms; hence, there is no fragmentation into loose pieces caused by $TiCl_4$.
- ✓ Other details are scribbled in Scheme 42.2, top section.

With these elements of judgment, a strategy can be devised, beginning with the complexation of $TiCl_4$ (a Lewis acid) with the only Lewis base in the alkyl arm. Activation of the aziridine as an LEDZ is thus provided. To better understand the valence electron redeployment to explain **1** → **2** the stereochemical elements are bypassed momentarily.

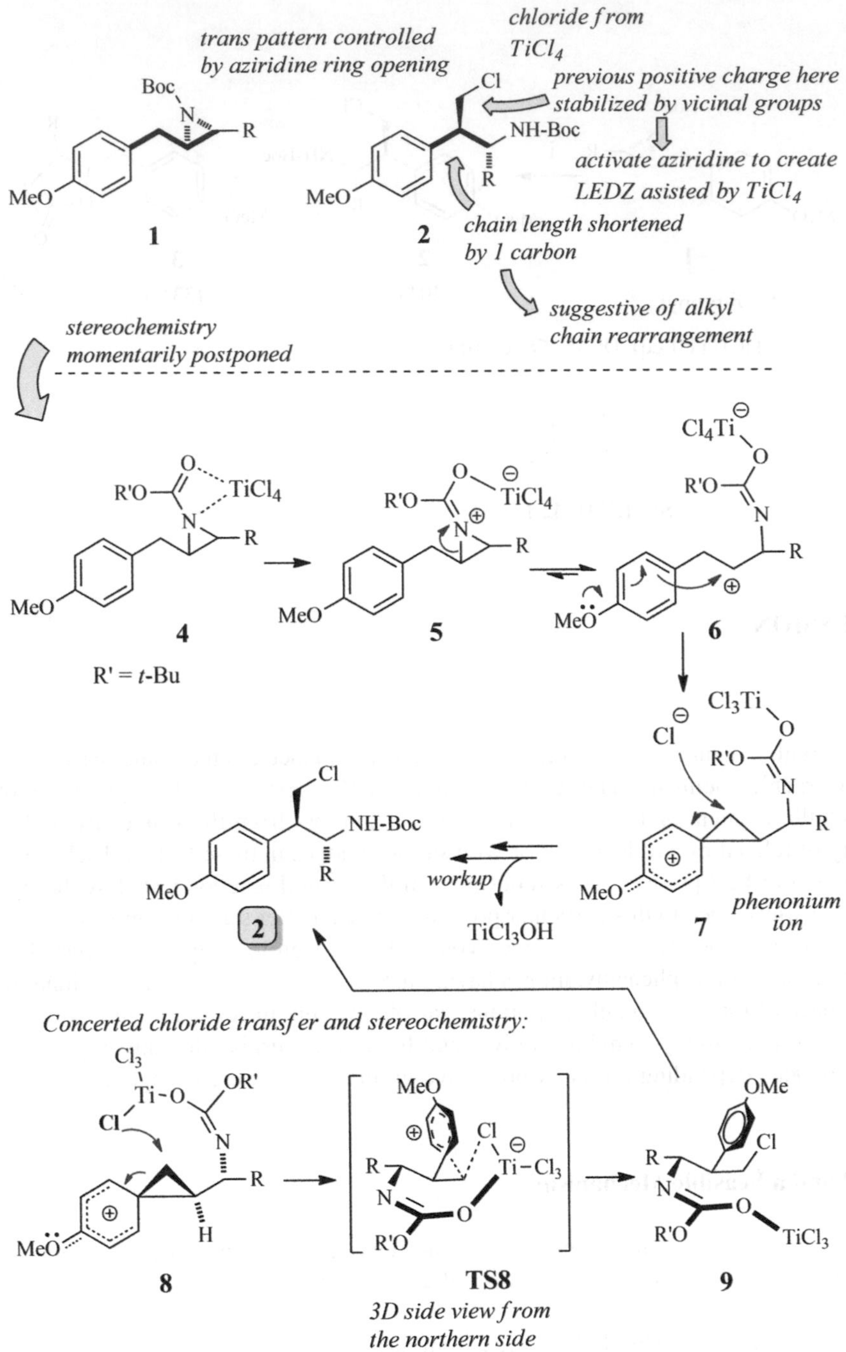

SCHEME 42.2 Analysis of **1** → **2** and a feasible mechanism.

42.3 Learning Features and Takeaways of Scheme 42.2

1. This aziridine involves two chain carbons, which facilitate the apparent rearrangement without the need of dismounting the amide group from the scaffold.
2. Activation for C–C rearrangement is elicited particularly well by a $C^{(+)}$, which is produced by aziridine ring opening (**6**) after its complexation with $TiCl_4$.
3. There is more than one possible complexation constructions, for example either as bidentate as in **5** or monodentate as in **6**, and a few more spanning the *N*-Boc unit (others have not been drawn in Scheme 42.2 to avoid blurring the sequence).

4. No EDGs next to secondary $C^{(+)}$ in **6** are available to bring stabilization. However, the *p*-methoxyphenyl π MO can intervene by virtue of is increased electron density provided by the *p*-methoxy group and supply bonding electrons for cyclopropanation, in preparation for the ensuing alkyl chain shortening.
5. The spirocyclopropane **7** is a curious *phenonium ion*, a phenyl anchimeric assistance; in so doing it losses aromatic character and $\Delta G\ddagger$ is conceivably high but not impossible to surpass (see below). The phenonium ion has been around for quite some time since Professor Donald J. Cram proposed it in 1964 as intermediate in certain W–M rearrangements [2]. A number of experimental and theoretical works have been published since then, including direct observation in an NMR tube in super acid medium [3].
6. The phenonium ion in **7** withdraws electron density from the distal C–C cyclopropane bond, enhancing the electrophilicity of the methylene. This feature enables the entrance of $Cl^{(-)}$ with simultaneous ring cleavage to restitute charge balance and furnish the observed target skeleton.
7. *Stereochemistry of target 2*: The chloride transfer in complex **8** can take place by the molecule adopting a boat conformation of a cyclic TS comprising eight atoms (**TS8**) (Scheme 42.2, bottom section). Careful observation of the carbon chain substituents in this side view of **TS8→9** provides the correct configuration of target **2**.

42.4 Explaining 1→3

This reaction is more easily accessed by redrawing cation **6** with a reoriented titanium-ligated eastern section 180° (**6-b** in Scheme 42.3). LED and HED units are at close range and C–O bonding ensues in **10** securing target **3** in 33% isolated yield. Exclusion of the protective Boc is achieved by [1.5]-H shift and the release of isopropylene.

Authors of this work [1] report changes in product composition when other N-protective groups are used. For example, replacing Boc with tosylate in **1** gives a mixture of the *N*-tosyl derivative of **2** and unrearranged chloride **11** (Scheme 42.3) in 63 and 24% yield, respectively. Given the HED/LED combination at close range there is a chance that target **3** (*N*-Boc) may also arise from the intramolecular S_N displacement of chloride, an event that was not witnessed in the tosylate series.

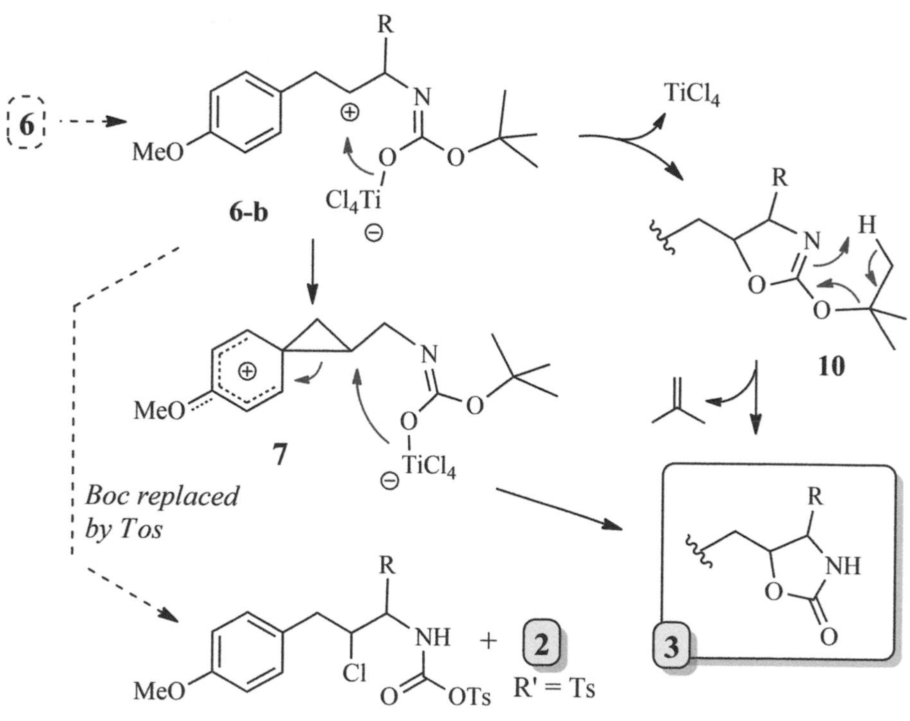

SCHEME 42.3 Analysis of **1→3** and feasible mechanisms.

42.5 Energy Profiles of Phenonium Ions

To develop further insights in the participation of phenonium ions in the **8**→**9** transition (Scheme 42.2, bottom) authors embarked in a DFT calculation of the Gibbs energy profile of a simplified model of phenonium ions possibly involved in the TS. The results of their calculations can be briefly compiled like this:

1. $TiCl_4$ can form mono and bidentate complexes with **1,** which exist in equilibrium with a small ΔG (3.7 kcal mol^{-1}).
2. The bidentate complex evolves to a phenonium ion TS ($\Delta G‡$=9.9 kcal mol^{-1}), which is more favorable than the corresponding monodentate phenonium ion ($\Delta G‡$ =9.9 kcal mol^{-1}).
3. Once the phenonium ion is constructed in the bidentate complex, a chlorine atom from $TiCl_4$ is transferred to the cyclopropane methylene passing through an eight-membered TS (as I construed intuitively in **TS8**–Scheme 42.2–before checking out the G data, as you might have surmised also) of $\Delta G‡$=9.1 kcal mol^{-1} energy barrier that furnishes the rearranged target **2**.
4. Whereas these energy barriers are substantial, they are anyway surmountable even though reaction temperatures employed were low (–78 °C).
5. Phenonium ions have been subjected to detailed DFT scrutiny in the past, showing substantial reductions in Gibbs energy levels relative to the preceding cations [4].

REFERENCES

1. Holst HM, Floreancig JT, Ritts CB, Race NJ. *OrgLett* 2022;24:501–505. DOI:10.1021/acs.orglett.1c03857.
2. Cram DJ. *J. Am. Chem. Soc.* 1964;86:3767–3772. DOI:10.1021/ja01072a032.
3. Olah GA, Head NJ, Rasul G, Prakash KS. *J. Am. Chem. Soc.* 1995;117:875–882.
4. Del Rio E, Menéndez MI, López R, Sordo TL. *J. Am. Chem. Soc.* 2001;123:5064–5068. DOI:10.1021/ja0039132.

PROBLEM 43

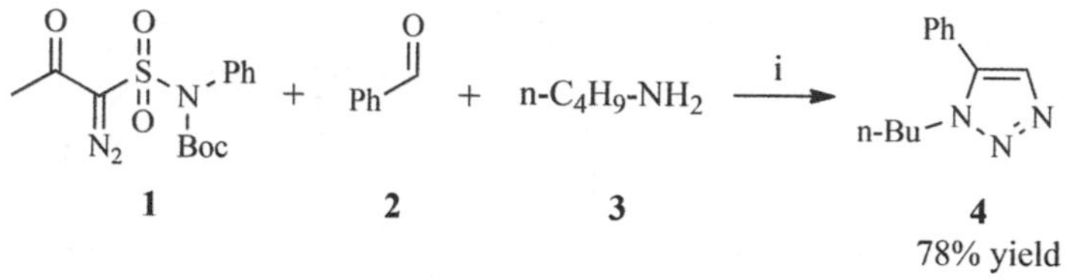

i: MeOH, molecular sieves, r.t. 18 h

SCHEME 43.1 [1] / American Chemical Society.

PROBLEM 43: DISCUSSION

43.1 Overview

The synthesis of heterocycles continues to offer challenges from every possible angle. This work from St. Petersburg, Russia [1] exploits a densely functionalized three-carbon ketone in a MCR approach to a product without much resemblance to the starting materials.

43.2 Fragmentation Analysis of Product and Starting Compounds

The top section of Scheme 43.2 portrays one reasonable way to dislodge target **4** as a function of the MCR participants. From this analysis the following considerations can be advanced.

- ✓ There are two main sections in **4**, here defined as eastern and western.
- ✓ The western part is a reduced form of the **2**+**3** adduct, built by standard aldehyde-imine addition. Because this is an equilibrium process, molecular sieves operate here as a water trap, a classical role for this ceramic material.
- ✓ The origin of the eastern section is more difficult to determine since most of its components associable to **1** are orphan groups relative to **4**. Only the diazo function survives, bearing just one carbon.
- ✓ This means that the reaction mechanism must account for the exclusion of these two orphans either individually or in a single step.

43.3 A Feasible Mechanism for This MCR

Assuming that imine **5**, an LED moiety, is produced initially, the eastern portion must proceed as an HEDZ to create the desired and only C–C bond of the ring structure. There is more than one route to achieve this, either before or after deacylation (option **A** in Scheme 43.2, lower section).

Route A is based on the carbanion character of the diazo moiety owing to the two strong EW groups, ketone and sulfone, directly substituted on this carbon favoring resonance structure **1b** (**1a** ⟷ **1b**). However, addition of **1b** to imine **5** is probably precluded by steric congestion. If this demanding step is overcome, the exclusion of the diazonium ion, an excellent leaving group as long as the remaining carbocation is stabilized, is prevented by the same EWGs, which discourages any further decrease in electron density. The ensuing N–N cyclization in **6**, deacylation by methanol, and Brønsted catalysis provided by molecular sieves [2], followed by sulfonamide β-elimination affords target **4**.

Sequence **A** may be improved by early deacylation of **1a** (methanol and molecular sieves catalysis) followed by nucleophilic addition of the much less sterically congested nucleophile diazo carbon to imine **5**. The rest of the stepwise cyclization to the 1,2,3-triazoline sequence would be homologous to **6** → **9** → **4** without the acyl group.

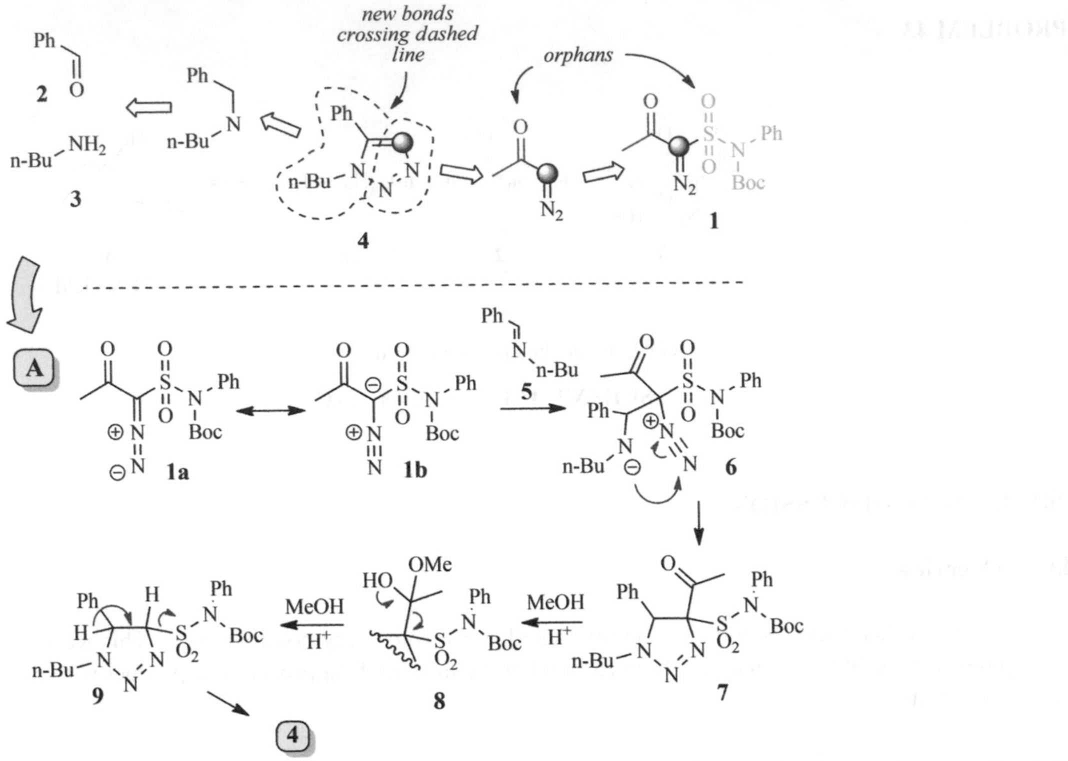

SCHEME 43.2 Sequential fragmentation analysis of target **4** down to the starting compounds of the MCR synthesis (top section) and a first feasible mechanism based on the preliminary analysis (bottom section).

43.4 An Alternative Route to 4

Route B:
We have discussed previously (Chapter 2) the flexibility of electron delocalization in the diazo moiety. Canonical structures place negative or positive charges in all three atoms of the C=N=N functional group. From the various possibilities, the dipolar canonical structure **1b** allows it to operate in [3 + x] dipolar cycloadditions if reaction conditions do not force first the release of molecular nitrogen and carbene formation. The dipolar counterpart in this case is imine **5**. Thus, 1,2,3-triazoline sulfonamide **11** can be formed in a single step after deacylation (Scheme 43.3).

43.5 Which Mechanism is Better Adapted to the Known Chemistry? Important Takeaways

In an earlier report the authors isolated sulfone triazoline **11** (Scheme 43.3) from a three-component reaction – **1** (Boc replaced with alkyl groups) + **2** + **3** in methanol and molecular sieves at room temperature and a shorter reaction time (six to eight hours) [3]. The *trans* configuration in **11** suggested a *concerted type [3 + 2] cycloaddition*, albeit the synchronous or asynchronous formation of the C–C and N–N bonds could not be ascertained from the experimental evidence. Furthermore, when **11** was heated in ethanol at 125 °C in a sealed tube, the elimination of sulfone occurred and target **4** was produced. The scope of the reaction was tested successfully in 11 different triazolines. Therefore, sequence **B** is better adapted to the available evidence. Other transient species are also amenable to [3 + 2] cycloadditions with the diazomethane function as has been recorded for arynes [4].

In another example of diazomethane equivalent congeners, an acyl diazophosphonate series (**12**) was tested recently in their capacity for cycloadditions with aldimines in methanol at room temperature in a three-component arrangement. Initial deacylation followed by cyclization to the triazine phosphonate (**13**) [5]. It is worth noting that these authors propose a stepwise mechanism similar to our sequence **A**.

The diazo-sulfone case in this problem constitutes a novel form of C=N=N moieties amenable to [3 + 2] cycloadditions with polar C=X units (X = N momentarily) without carbene intervention.

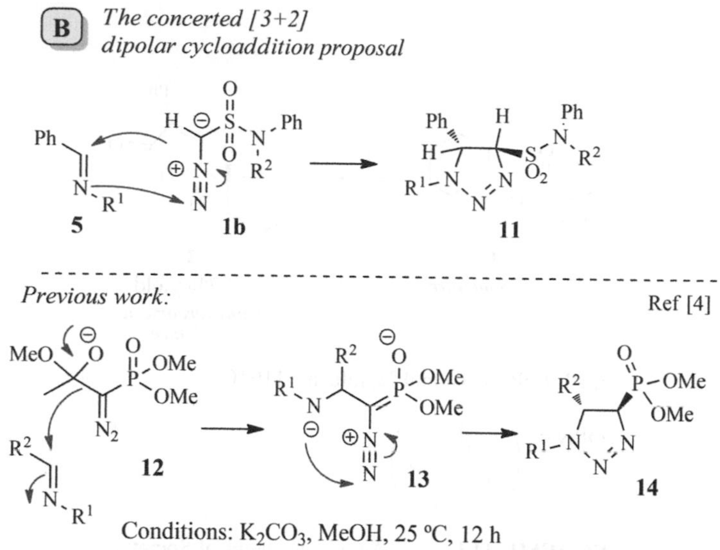

SCHEME 43.3 A different mechanism conception (top section). The lower section of this scheme was adapted from [4] / American Chemical Society.

REFERENCES

1. Bubyrev A, Malkova K, Kantin G, Dar'in D, Krasavin M. *J. Org. Chem.* 2021;86:1756–17522. DOI:10.1021/acs.joc.1c02309.
2. Ding Z, Kloprogge JT, Frost LR. *J. Porous Mat.* 2001;8:273–293. DOI:10.1023/A1013113030912.
3. Bubyrev A, Adamchik M, Dar'in D, Krasavin M. *J. Org. Chem.* 2021;86:13454–13464. DOI:10.1021/acs.joc.1c01552.
4. Cheng B, Li Y, Zu B, Wang T, et al. *Tetrahedron* 2019;75:art 130775. DOI:10.1016/j.tet.2019.130775.
5. Ahamad S, Kant R, Mohanan K. *Org. Lett.* 2016;18:280–283. DOI:10.1021/acs.orglett.5b03437.

PROBLEM 44

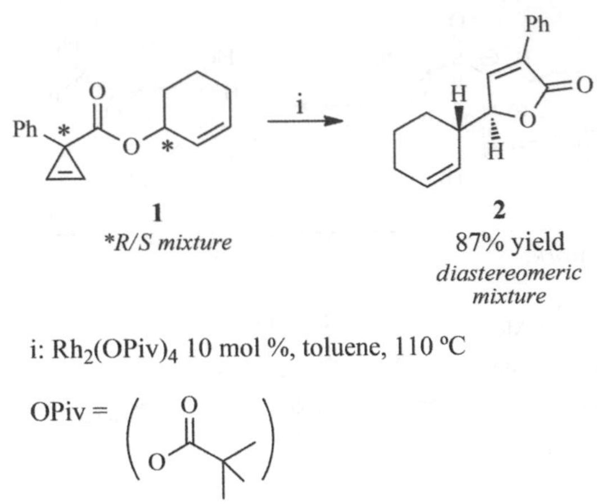

SCHEME 44.1 [1] / American Chemical Society.

PROBLEM 44: DISCUSSION

44.1 Overview

The rearrangement of cyclopropane and cyclopropene ketones/esters to furans by heat, silica gel, or metal catalysis has been on the reaction spectrum for quite some time. In the hands of researchers from Delaware, United States [1] the end product is a butenolide **2** likely arising from a closely related furan. One notable feature is the establishment of two neighboring chiral carbons with a well-defined enantioselective configuration despite the absence of chiral reagents. Butenolide was obtained as a mixture of diastereomers. Additionally, the bond connections between the reaction components seem somewhat difficult to justify, as the first-round problem analysis will show.

44.2 Preliminary Analysis

There are two clearly discernible blocks in **2** that arise from the corresponding sections in **1**. Given that all the elements of **1** are present in **2**, this comports isomerization whose number in organic compounds is far from infinite and most of them can be reasoned quite well. Difficulties may appear in the details, though.

44.3 A Feasible Mechanism: Part 1

The cyclopropene ester section is evidently the unstable part from which all the rest of the mechanism will depart. In Scheme 44.2, the operation of a cyclopropenyl-carbinyl rearrangement prompted by Rh(II) (other transition metal catalysts would be active as well) is suggested by virtue of a well-known σ–π interaction of carbonyl and cyclopropenyl sections and $Rh^{(II)}$–C=C complexation as in **3** (Scheme 44.2, middle section) eventually affording Rh-π complex **6**. After removal of Rh during workup, compound **6** was isolated and characterized alongside a scope of other furan alkenyl ethers [1] in consonance with earlier investigations. Thus, this furan is a likely precursor of our target **2**.

44.4 Performing the 6 → 2 Conversion

This tandem transformation requires additional comparative analysis between **6** and **2** as a separate problem (Scheme 44.3, top section). Two operations are required: i) the transfer of the cyclohexenyl moiety from the ether sector to C^5 and ii) cleavage of the ether with concomitant formation of C=O. Meanwhile, the π MO of furan in **6** must undergo electron redeployment to the observed butenolide at the sacrifice of aromaticity. The reaction must be such that two stereogenic carbons are formed enantiospecifically.

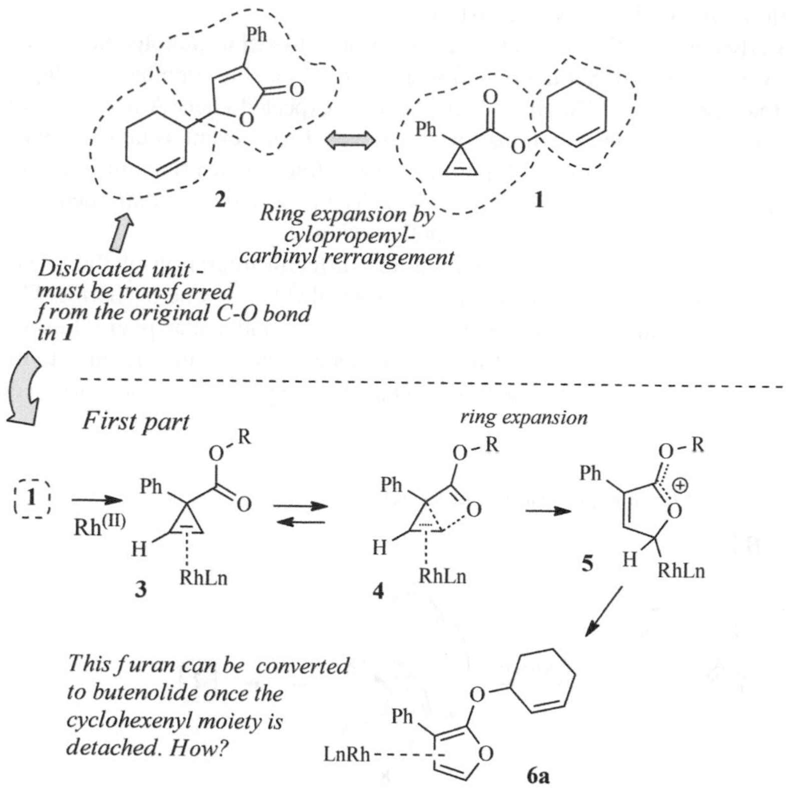

SCHEME 44.2 The 1 → 2 conversion is conceived in two separate steps, starting with the most active western section of **1**.

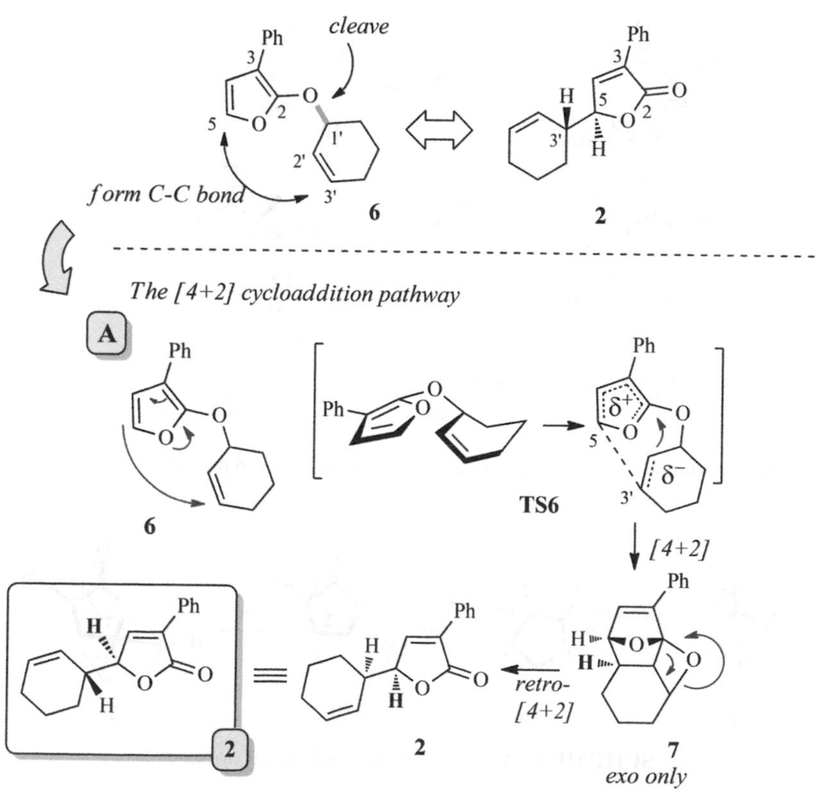

SCHEME 44.3 Likely association of unsaturated carbons in **6** and a first possible mechanism to explain target **2** via a non-concerted intramolecular [4+2] cycloaddition.

There are two conceivable ways to achieve these constraints.

Route A (my own) takes advantage of the stereochemical control of [4+2] intramolecular cycloadditions between furan, a classical diene, and cyclohexene, provided that the two moieties can be suitably oriented to bring $C^{1'}$ and C^5 as close as possible. This is attempted in **TS6** with some difficulty, which turns the expected Diels-Alder type cycloaddition into a non-synchronous process resulting from the assymmetry of the forming bonds. In any event, rotation of the combining moieties would be hampered. Worth noting is the exo-only approach possible that defines the α orientation of the angular proton at the ring fusion ending as the chiral carbon of the butenolide moiety in **2**. The expected, albeit strained, adduct **7** would then undergo retro [4+2] cycloaddition to give target **2** in the recorded configuration.

Route B, preferred by the authors [1], appeals to an entirely different migration of the cyclohexenyl portion along the opposite side of the furan ring in **6** (Scheme 44.4). Homoallyl vinyl ethers are prone to thermal Claisen rearrangement, **6**→**8** in this case, through migration of the allyl unit over the same side of the molecular plane. In the absence of chiral inducers planar compounds such as **6** generally yield **8** as a 1 : 1 mixture of R and S enantiomers relative to the quaternary carbon, while four diastereomers are expected. Further migration to **2** would take place by way of Cope rearrangement as shown in **8**.

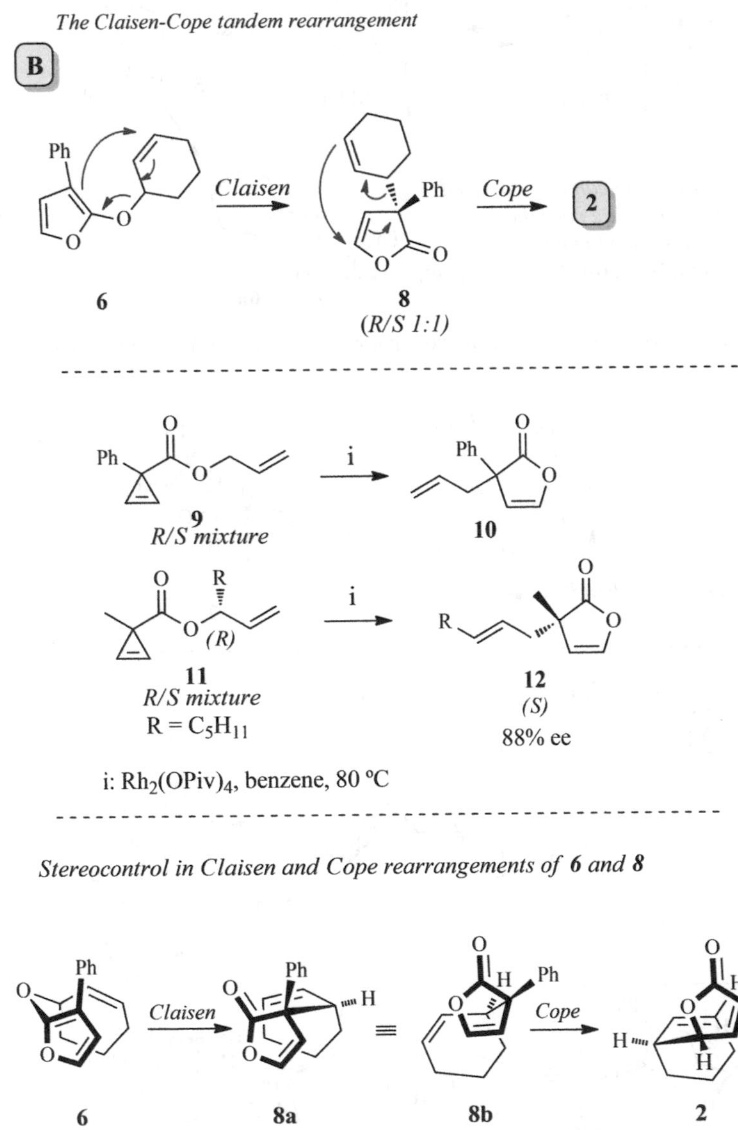

SCHEME 44.4 [1] / American Chemical Society.

44.5 Experimental Evidence, Learning Issues, and Takeaways

1. The reaction of Scheme 44.1 is the result of a double-faceted study [1]. In the first phase, authors studied the conditions and scope of the ring expansion of cyclopropenyl allyl esters of type **1** induced by Rh(II) catalysis at moderate temperature (boiling benzene). High yields of butenolides of type **8** (Scheme 44.4) were recorded. Thus, a two-step tandem ring expansion-Claisen rearrangement of the **1**-type occurred under these conditions serving as proof that target **2** was formed from the butenolides of type **8**.
2. Two selected examples of this work [1] appear in Scheme 44.4, second section. While **9** → **10** is a generic easily understandable ring expansion-Claisen rearrangement, **11** → **12** has the added elements of one chiral carbon bound to the R group in the ester chain. *This chirality does not remain at this atom but is transferred to the butenolide*.
3. The first step of this sequence (ring expansion of cyclopropene carbonyls to furan by rhodium catalysis) has been studied in mechanistic detail including DFT calculations [2]. These authors found that cyclopropene diketone enolates (fixed as trimethyl silyl ethers) form metal carbene precursors and operate as dipolarophiles in [3 + 2] cycloadditions. In the absence of rhodium catalysis, diazocarbonyl precursors only produced cyclopropenes and no furans. Critical parts of this sequence inspired our proposal of Scheme 44.2.
4. In the second phase, Xie, Li, and Fox [1] induced the second rearrangement (Cope) whereby stereocontrol of the rearranged product (of type **2**) was observed. I proposed a Diels–Alder intramolecular cycloaddition assuming that the TS acquired a rigid structure that prevented the free rotation of the cyclohexyl unit. Molecular models suggested that only the exo configuration was possible, although distances between carbons in the process of bonding were different. This led to the notion of a non-synchronous C–C bond formation (sequence A, Scheme 44.3).
5. Authors, however, opted for sequence **B** (Scheme 44.4, upper section) since they were able to isolate **8**-type products after ring expansion-Claisen rearrangement in boiling benzene (80 °C). When the same reaction was performed at higher temperature (toluene, 110 °C) compounds of type **8** formed after ring expansion underwent Cope rearrangement in the same vessel from **1**.
6. The stereochemical control was then justified by a suitable conformation in the Claisen and Cope processes, as shown for the cyclohexenyl case in perspective (Scheme 44.4, lower section: **6** → **8a/b** → **2**) [1].

REFERENCES

1. Xie X, Li Y, Fox JM. *Org. Lett.* 2013;15:1500–1503. DOI:10.1021/ol400264a.
2. Marichev KO, Wang Y, Carranco AM, García EC, Yu ZX, Doyle MP. *RSC Chem. Commun* 2018;54:9513–9516. DOI:10.1039/C8CC05623D.

PROBLEM 45

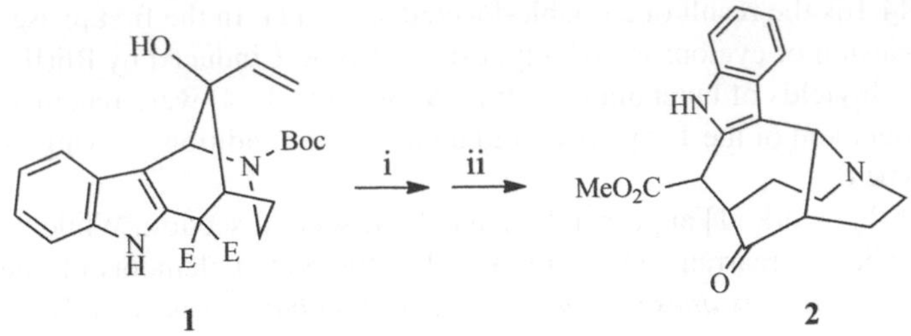

i: 1) TFA, CH_2Cl_2, 0 °C
 2) $(CH_2O)_n$, CSA, benzene, 70 °C

ii: 1) HCl, MeOH, 50 °C
 2) aq. Na_2CO_3

CAS: Camphor sulfonic acid

E: CO_2t-Bu

SCHEME 45.1 [1] / American Chemical Society.

PROBLEM 45: DISCUSSION

45.1 Overview

The convoluted structure **2** is a synthesis intermediate with most of the attributes of a unique alkaloid type (actinophyllic acid) discovered in a plant native to Tropical North Queensland, Australia. No other plant species known to man synthesizes anything like it. It attracted the attention of researchers from Irvine, California, United States [1] owing to the bioactivity of this alkaloid to control blood clotting, a serious killer and common consequence of heart disease.

Our problem herein starts with the understanding of the components of **1** and envision a feasible route connecting it to target **2**. I have reproduced as close as I could the structures as they appear published. Perhaps some cleaning might help to figure out the mechanism. In studying this case we will employ most of our problem-solving strategies.

45.2 Rendering 1 and 2 in the Same Visual Language

At first glance, this seems an insurmountable problem, worse yet since one does not know where to start the analysis of bond forming and cleaving operations. A bit of cleaning up of molecular renderings is necessary. We will attempt this in three phases.

First: Using the indole section for guidance, the original depiction bears the idea of its migration in one piece from the western to the northern section of the tricyclic conglomerate. This would be accompanied by a massive rearrangement of the tricyclic section. However, indole migration is unlikely when one checks the connection of C^2 and C^3 of this indole with the ester and amine functions. It would be to our advantage to fix indole's position on the western side and adapt the eastern section to this orientation. This should place the two parts in a similar visual appearance or language, thus facilitating the analysis.

Second: Yet, we have a lot of atoms to distract our attention. The next issue is determining the molecular area where the **1** → **2** action takes place. Clearly, all of it occurs at the bicyclo [3.3.1] aminononane section at the east side, while the indole portion plays no part.

Third: While stereochemistry is fundamental for mechanism analysis, it may be overwhelming in complex systems like **1** for a first analysis. Let us momentarily use 2D renderings.

The end result of the three phases furnishes simplified structures **1S** and **2S** (Scheme 45.2).

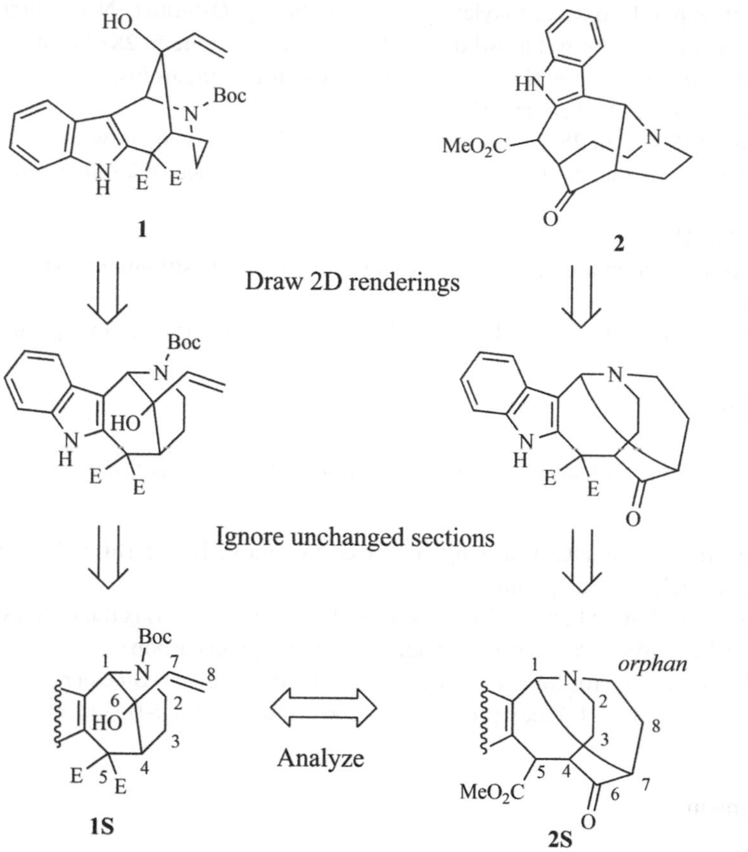

SCHEME 45.2 Visual simplification of the structures in Scheme 45.1.

45.3 Comparative Analysis of 1S and 2S

Completing this paper/pencil exercise (computer graphics in such complex structures are of little value here), brings a series of conclusions. Try your own from Scheme 45.2 and then compare it with the list below:

i) Abridged molecular formulas: **1S:** $C_{10}NO(CO_2t\text{-}Bu)_2$-Boc
   2S: $C_{11}NO(CO_2CH_3)$.
   a. One C is added to form **2S**.
   b. One CO_2*t*-Bu is removed from **1S**.

ii) The other CO_2*t*-Bu is *trans*-esterified to CO_2CH_3 by specific reagents employed.

iii) Amine: passes from secondary to tertiary. Being that N is an HED alkylamino group with an NBP, the new N–C bond should arise by interaction with a LED carbon.

iv) Before this addition takes place, the Boc protecting group is removed.

v) Oxidation level as revealed by functionality numbers (FN_t) (Chapter 4).
   Putting aside indole and Boc, FN_i accounting gives:

$FNt_{(1S)}$ =	11
$FNt_{(2S)}$ =	8
$\Delta FNt_{(2S\text{-}1S)}$ =	-3

a. Reduction? No, it responds to decarboxylation of one of the E [CO_2t–Bu, ($FN_i = 3$, precisely)] functional groups.
b. Why isn't there overall oxidation if a carbinol in **1S** becomes a ketone in **2S**? Because the carbinol ketone step must be accompanied by the loss of one C–C bond next to the O function, an $FN_i = -1$ compensation.

vi) Atom labels (shown in **1S** and **2S**, Scheme 45.2):
a. One identifies an *orphan* C in **2S** (absent in 1S) bonded to N, in agreement with conclusion number 3.
b. The eight-membered ring that contains the amine remains unmodified in **2S**. Rearrangement – if any – does not affect these atoms.
c. C^1–C^6 is cleaved in **1S**.
d. Bond C^7–C^1 is formed (no matter how apparently distant), probably simultaneously. This calls for skeletal rearrangement of the OH-bearing bridge.
e. C^7=C^8 in **1S** interacts with the external *orphan* C (not tagged), possibly as an α,β-unsaturated ketone at C^6.

45.4 Reagents Check Up

With these preliminary conclusions on hand, revision of the employed reagents is timely (please keep an eye on experimental conditions shown in Scheme 45.1):

Step i (1): TFA (trifluoroacetic acid) in DCM: usually employed to remove Boc (*t*-Bu-carbamate). De-protected N can use its NBP as a HED center for nucleophilic operations.
Step i (2): $(CH_2O)_n$, a polymer of H_2C=O under CSA (a strong alkyl sulfonic acid) is the only external carbon source available, ideally suited for a one LED carbon transfer to N. Addition of the orphan carbon.
Step ii (1): HCl in CH_3OH, conditions for trans-esterification of one of the *gem-t*-Bu esters.
Step ii (2): mild alkali (aqueous carbonate) may promote decarboxylation of the *t*-Bu ester.

45.5 A Feasible Mechanism

There are two options:

Pathway A in Scheme 45.3 follows the recommendations extracted from our analysis of points (3) and (4) above. Worth commenting on is the change of 2D to 3D language in the **5**→**6** step. We had earlier concluded that the formation of the C^7–C^1 bond was required to furnish **2**. Yet, these carbons appear too far apart in **5**. However, when turning to the 3D view as in **5-3D**, the axial conformation of the α,β-unsaturated ketone unit is positioned in close proximity to the iminium ylide function; hence, the ensuing Michael addition is perfectly acceptable. Intramolecular Mannich condensation finishes the job.
Pathway B takes advantage of the aza-dienyl substructure in **4** (emphasized in green). The desired C^7–C^1 bond can be formed through a [3,3] SR (Scheme 45.3, inset) of the Cope type: an aza-Cope rearrangement, whereby not only the C^7–C^1 bond is created in the same shot but also the bridge cleavage we had figured in the **4**→**5** transition of route **A**. From enol **8** an intramolecular Mannich condensation completes the construction of the tricyclic eastern section of the alkaloid structure.

The monocarboxylic ester is derived thereafter by regioselective transesterification followed by the tyical decarboxylation of the *t*-Bu ester by mild aqueous base.

45.6 Which Route Better Represents This Reaction?

Given that both routes **A** and **B** are well adapted to the rules of electron deployment, bonding, and stereochemistry, it is not an easy choice. However, the clever retrosynthetic plan of the authors of this remarkable reaction was conceived based on the aza-Cope rearrangement and Mannich reaction, which gives credit to their belief that route **B** would be *the* pathway (in addition to being more elegant). Furthermore, the disposition of the aza-diene function in **4** appears ideally suited for the said rearrangement-Mannich process, which has been used in a variety of reaction models and syntheses, posing interesting reaction mechanisms [3–5].

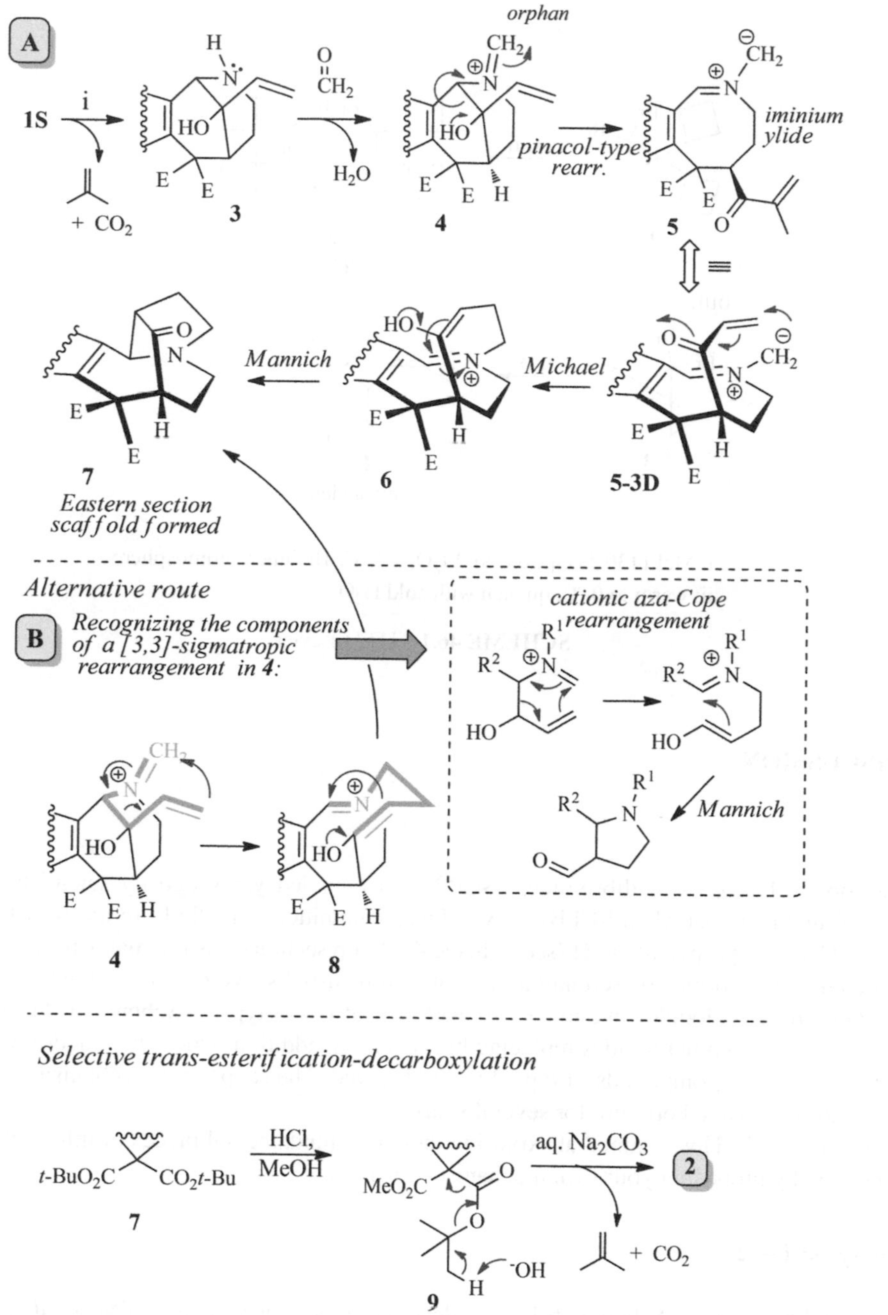

SCHEME 45.3 Two alternative pathways to build the eastern scaffold of the desired product.

REFERENCES

1. Martin CL, Overman LE, Rohde JM. *J. Am. Chem. Soc.* 2008;130:7568–7569. DOI:10.1021/ja803158y. https://pubs.acs.org/doi/10.1021/ja803158y.
2. Notice to readers: further permission related to the material excerpted herein from Reference [1] should be directed to the ACS.
3. Nájera C, Sansano JM. *Angew. Chem. Int. Ed.* 2005;44:6272–6276. DOI:10.1002/anie.200501074.
4. McCormac MP, Shalumova T, Tanski JM, Waters SP. *Org. Lett.* 2010;12:3906–3909. DOI:10.1021/ol101606v.
5. Coldham I, Hufton R. *Chem. Rev.* 2005;105:2765–2810. DOI:10.1021/cr040004c.

PROBLEM 46

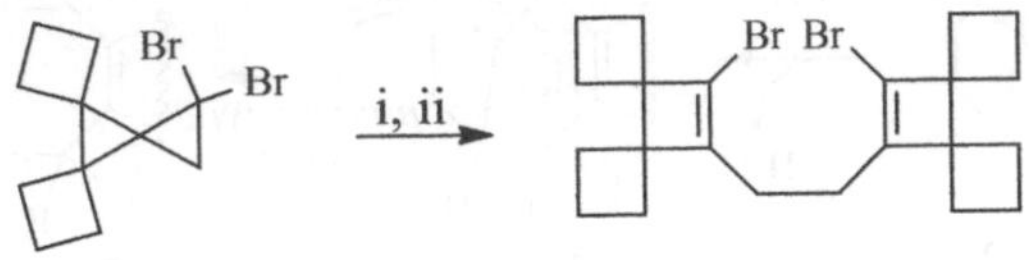

1 2

76% yield

but:

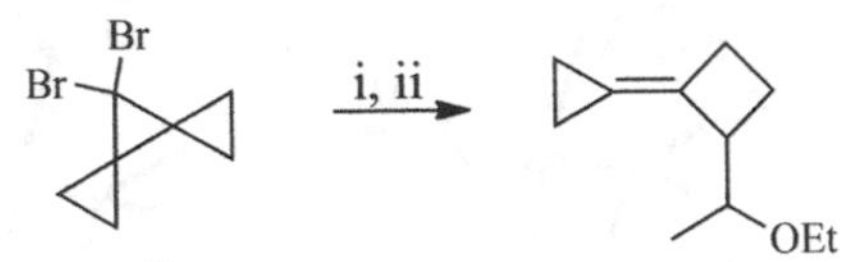

3 4

89% yield

i: MeLi (30 % eq excess), Et_2O, -55 °C, 1h, innert atomosphere
ii: warm to 0 °C, quench with cold H_2O

SCHEME 46.1 [1] / Elsevier.

PROBLEM 46: DISCUSSION

46.1 Overview

The reactions of spirobycyclopropyl-1,1-dibromides (SCPBs) was studied years ago by scientists from Moscow, Russian Federation [2]. It was found then that when SCPBs of type I were submitted to methyl lithium (CH_3Li) in diethyl ether under argon atmosphere at 0°C a cyclopropyl allene **II** (see Scheme 46.2, top section) was the major isolated product.

Other rearranged compounds formed by serendipitous routes were also discovered. For instance, when CH_3Li was prepared from LiI (lithium iodide, meaning iodide contamination) another scaffold of type **2** (Scheme 46.1) appeared (no substitutents other than bromides in the spiro-cp but iodides replacing bromides). In addition, when the reaction was carried out in diethyl ether at low temperature (−55°C) compounds of type **2** were observed. The scope and mechanism of these reactions was not discussed at the time and/or remained obscure for several years.

The research of Scheme 46.1 [1] was aimed at exploring these two fundamental pieces of information, and I hope you may contribute to this question by proposing your own mechanistic ideas.

46.2 Problem Analysis: 1 → 2

✓ While depiction of **2** as shown in Scheme 46.1 is not devoid of topological beauty, MMFF94G computer energy minimization of this structure [3] places cyclobutyl bromide sections as far apart as possible to reduce steric repulsion of the bulky groups (3D rendering inset in Scheme 46.2). This steric compression in each cyclobutyl unit causes geometrical distortion away from planarity in the four cyclobutyl substituents as well as substantial differences in bond lengths of the bromocyclobutyl substructures.

✓ At first glance, target **2** seems a dimer of **1** since it contains twice the number of carbon atoms in **1**, suggesting additionally that carbon from methyl lithium is not incorporated to the original scaffolds.

✓ Scheme 46.2 (top section) shows the comparative analysis of **1** and **2**. The two symmetrical sections may be constructed through very similar processes, but the CH_2–CH_2 bridge between the two cyclobutenyl sections opens the question of how the coupling of these blocks takes place, considering that reaction conditions do not favor radical species (which would have explained this coupling). This leads us to the hypothesis that while the eastern section evolves as a nucleophile (or

Ref [2]

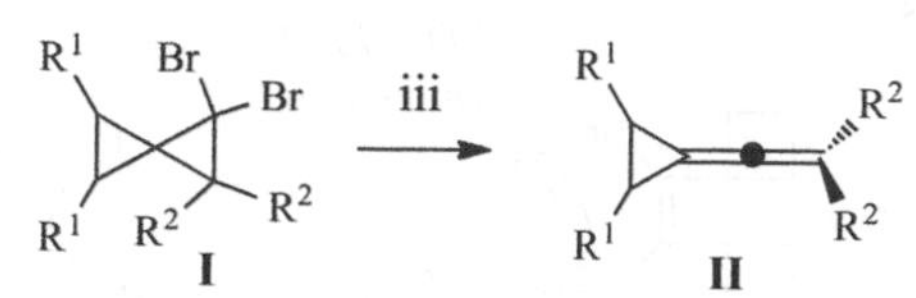

iii: same as i, at 0 °C

energy-minimized structure of 2

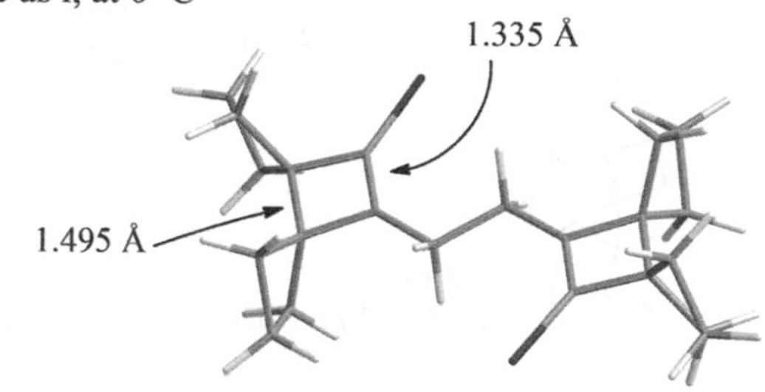

2-3D

SCHEME 46.2 Top section: Reported conversion of spirobicyclopropyl-1,1-dibromides to allenes. Adapted from Reference [2]. © 1992, Elsevier Ltd., used with permission, license no. 5525420196338. Lower section: energy-minimized 3D structure of target **2**.

HEDZ) the western section produces an electrophile (or LEDZ), even though this is seemingly out of chemical order in the same reaction medium.

- ✓ Think for a moment, though, that we had alkyl bromides and methyl lithium. Bromide and lithium can be exchanged to form an alkyl lithium, a powerful LED group and nucleophile as RCH_2–Li in the bridge-to-be end, whereas the other block may retain a bromide atom at its RCH_2–Br bridge end. The methylene bridge would be formed seamlessly.
- ✓ The mechanism would be solved once the CP ring expansion at the base of the carbon skeletal rearrangement of dibromocyclopropanes of **1** to the bromocyclobutenyl fragments in **2** is understood. Scheme 46.3 shows how.

46.3 Solving 1→2

The first hint is offered by the exocyclic CH_2, so activated to allow the coupling of the two cyclobutenyl units. Since this CH_2 emanates from the spirobicyclo-CP, cleavage of this section is the priority. Intermediates should emerge thereafter to operate further toward **2**. Ring rupture is possible by polarization of the C–Br bond of the germinal dibromide under the auspices of methyl lithium (Scheme 46.3, mechanism section). The spiro-quaternary carbon is underscored in red for you to follow its final location in **6**. After lithium and bromide exchange in **6** the exocyclic methylene is formed by cleavage of cation **7**, yielding the necessary structure **8**. Its evolution toward the HEDZ and LEDZ units required for coupling follows standard chemistry, except for the [1,3]-Li migration in the allylic moiety. Such lithium atom migrations and even [1.5] shifts are known [4].

46.4 Problem Analysis and Solution of 3→4

- ✓ According to the FA shown in Scheme 46.4, top part, the solvent (diethyl ether) interferes with the normal course of the reaction by C–H insertion.
- ✓ The only species derived from **3** capable of such insertion ought to be a *carbene*. Thus, we should manage to build one just prior to the C–H insertion that must be the last step of this sequence.
- ✓ Structure **8** (Scheme 46.3) has the flavor of a carbene once the lithium cation is removed. Six electrons remain in this carbon atom of the cyclobutyl lithium cation, a formal singlet carbene (given the very low temperature and absence of metal catalysts), except that in the reaction route **1**→**2** this does not materialize in any product. Yet, the *gem*-dibromide and methyl lithium in **3** are readily prepared for ring expansion to a CB adorned with exocyclic methylene and carbene units (Scheme 46.4, lower section).

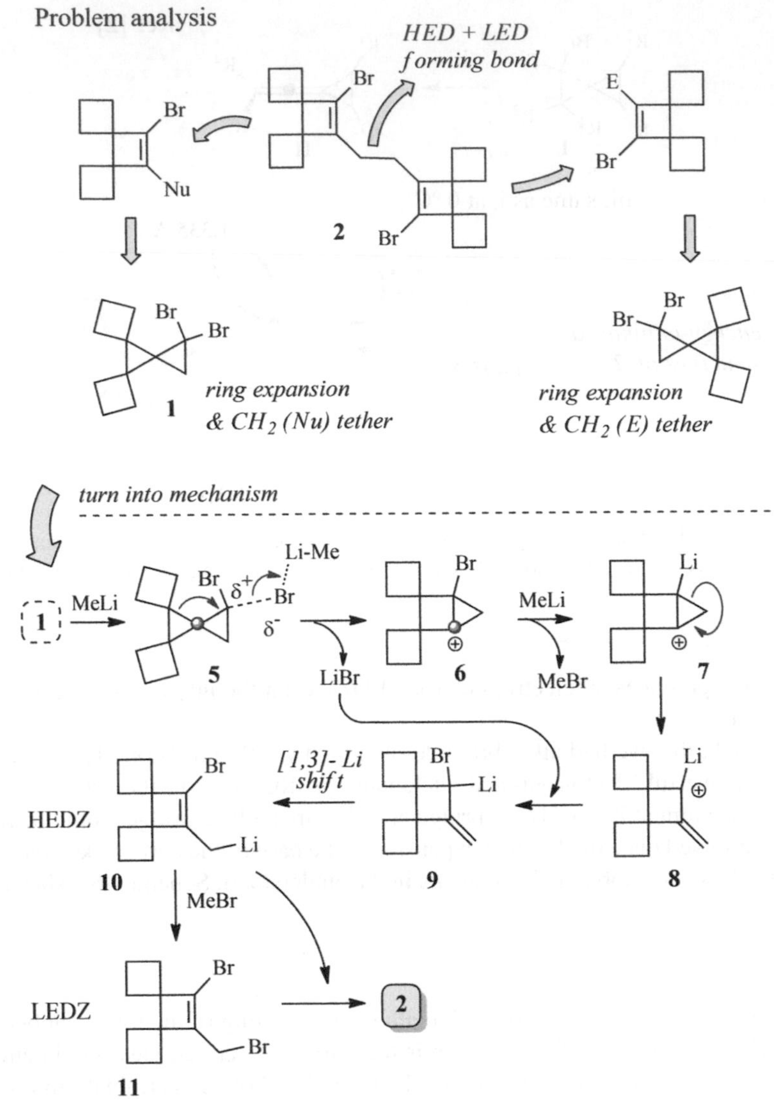

SCHEME 46.3 Analysis and solution for **1** → **2**.

46.5 Learning Issues and Takeaways

1. The reaction of *gem*-dibromo cyclopropanes with alkyl lithium reagents in ether at low temperature has been known for over half a century and developed as a general synthesis method for allenes [5]. CP=CP couplings suggested the intermediacy of carbenes, which was tested by adding cyclohexene to the solvent to trap the carbene. Indeed, the product of the corresponding [1 + 2] cycloaddition was obtained in small yield (10%) whereas this compound was not produced in the absence of cyclohexene [6].
2. In the reactions of Scheme 46.1, carbenes were also suspected [1] and became detectable by the C–H insertion of a carbenic species in high yield (Scheme 46.4). However, this was not tested by the standard method of adding a manageable olefin like cyclohexene.
3. In the Russian investigation, however, the carbenic species were formed not from the *gem*-dibromo CP and methyl lithium but after ring enlargement to CB [1]. The priority of ring cleavage over carbene generation was likely induced by SS of substituents.
4. *gem*-Dibromo CPs continue to be a source of interesting discoveries to this day [7].

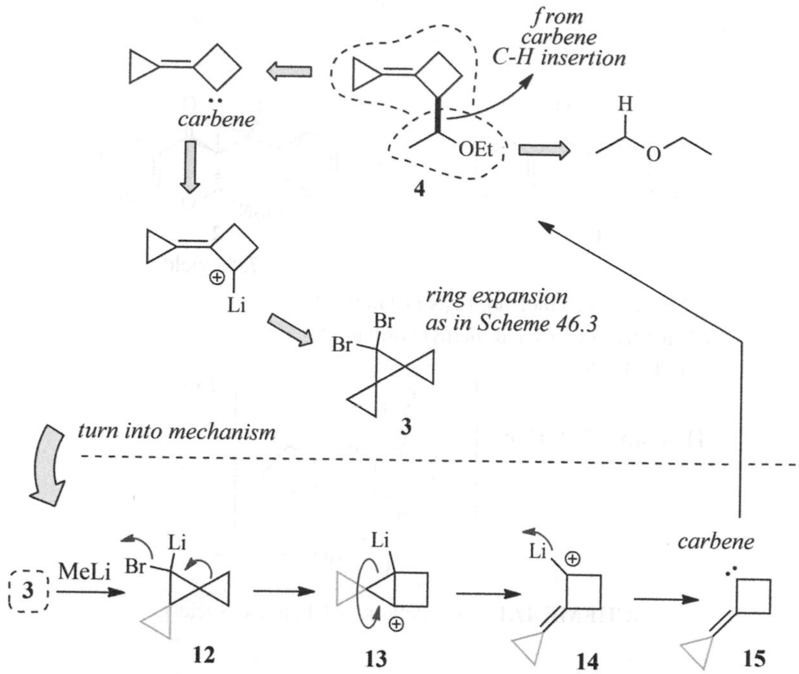

SCHEME 46.4 Analysis and solution for **3**→**4**.

REFERENCES

1. Averina EB, Karimov RR, Sedenkova KN, Grishin YK, Kuznetzova TS, Zefirov NS. *Tetrahedron* 2006;62:8814–8821. DOI:10.1016/j.tet.2006.06.086.
2. Lukin KA, Zefirov NS, Yufit DS, Struchkov YT. *Tetrahedron* 1992;48:9977–9984. DOI:10.1016/S0040-4020(01)92287-2.
3. Alonso-Amelot ME, this book.
4. Cheng D, Knox KR, Cohen T. *J. Am. Chem. Soc.* 2000;122:412–413. DOI:10.1021/ja993325s. See extensive discussion in Problem 25 in M. E. Alonso-Amelot, *The Art of Problem Solving in Organic Chemistry*, 2nd edition. Wiley, NY, p. 202.
5. Von E. Doering W, LaFlamme PM. *Tetrahedron* 1958;2:75-.
6. Moore WM, Ward HR. *J. Org. Chem.* 1960;25:2073-2073. DOI:10.1021/jo01081a054.
7. Lepage RJ, Moore PW, Hewitt RJ, Teesdale-Spittle PH, Krenske EH, Harvey JE. *J. Org. Chem.* 2022;87:301–3015. DOI:10.1021/acs.joc.1c02366.

PROBLEM 47

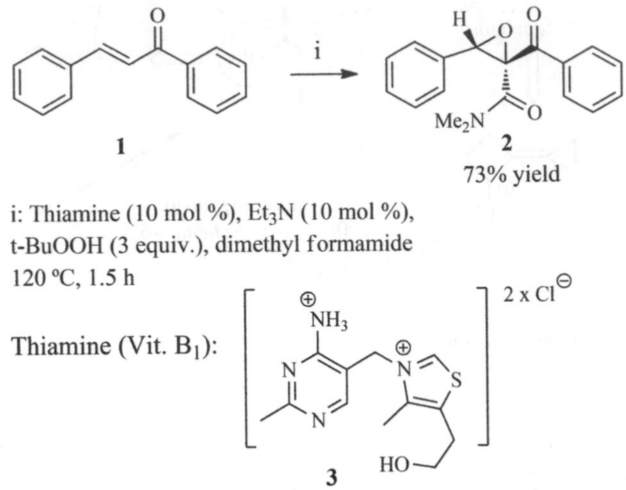

SCHEME 47.1 [1] / American Chemical Society.

PROBLEM 47: DISCUSSION

47.1 Overview

Epoxides are exceedingly useful intermediates (and end products) in the synthesis of many natural products and pharmaceuticals, better yet if they can be constructed diastereoselectively as this case by researchers from India reports. In addition, product **2** includes a trisubstituted, hence sterically congested, oxirane obtained without the use of metallic species and thus is environmentally friendly.

The reaction is apparently mechanistically too simple to occupy a high echelon of "The Problem Chest" collection (a clean chalcone epoxidation by *t*-butyl hydroperoxide and little else). Wrong. However, as you might expect, it does have sufficient complication to spend some time on it and get satisfactory learning issues and takeaways.

47.2 Problem Analysis

A first plan is based on the well-known epoxidation of α,β-unsaturated ketones by alkyl hydroperoxides, first introduced with enantioselective control 1980 [2]. This reaction has progressed from the original discovery to an exquisite diastereomeric control owing to a series of metal catalysts possessing chiral ligands [3] or even natural *Cinchona* alkaloids as hybrid-phase transfer catalysts [4].

A first problem analysis is portrayed in Scheme 47.2 (top section). The main course is defined by the standard oxidation with *tert*-butyl hydroperoxide (TBHP), which should be accompanied by the addition of DMF to C^{α}. This end can be achieved by an initial 1,4 addition of TBHP anion formed by proton removal of R–O–O–H by Et_3N (Scheme 47.2, middle section). The enolate **4** thus formed would proceed to the 1,2-addition on DMF by postponement of *t*-BuOH exclusion. In this manner, the three components of the reaction are bonded in the same molecular entity (**5**). The rest of the sequence follows standard chemistry in reference to the conclusion of epoxidation **6** and radical oxidation of the *southern* unit by thermally produced peroxide radicals (inset).

47.3 Incongruences in Scheme 47.2

While the process is in accordance with classical lines, it does not account for the observed reported diasterospecficity [1]. The two molecular planes of chalcone **1** do not differ in steric or electronic, thus permitting the approach of t–BuO–$O^{(-)}$ from either side. Target **2** would be produced as a 1 : 1 mixture of enantiomers.

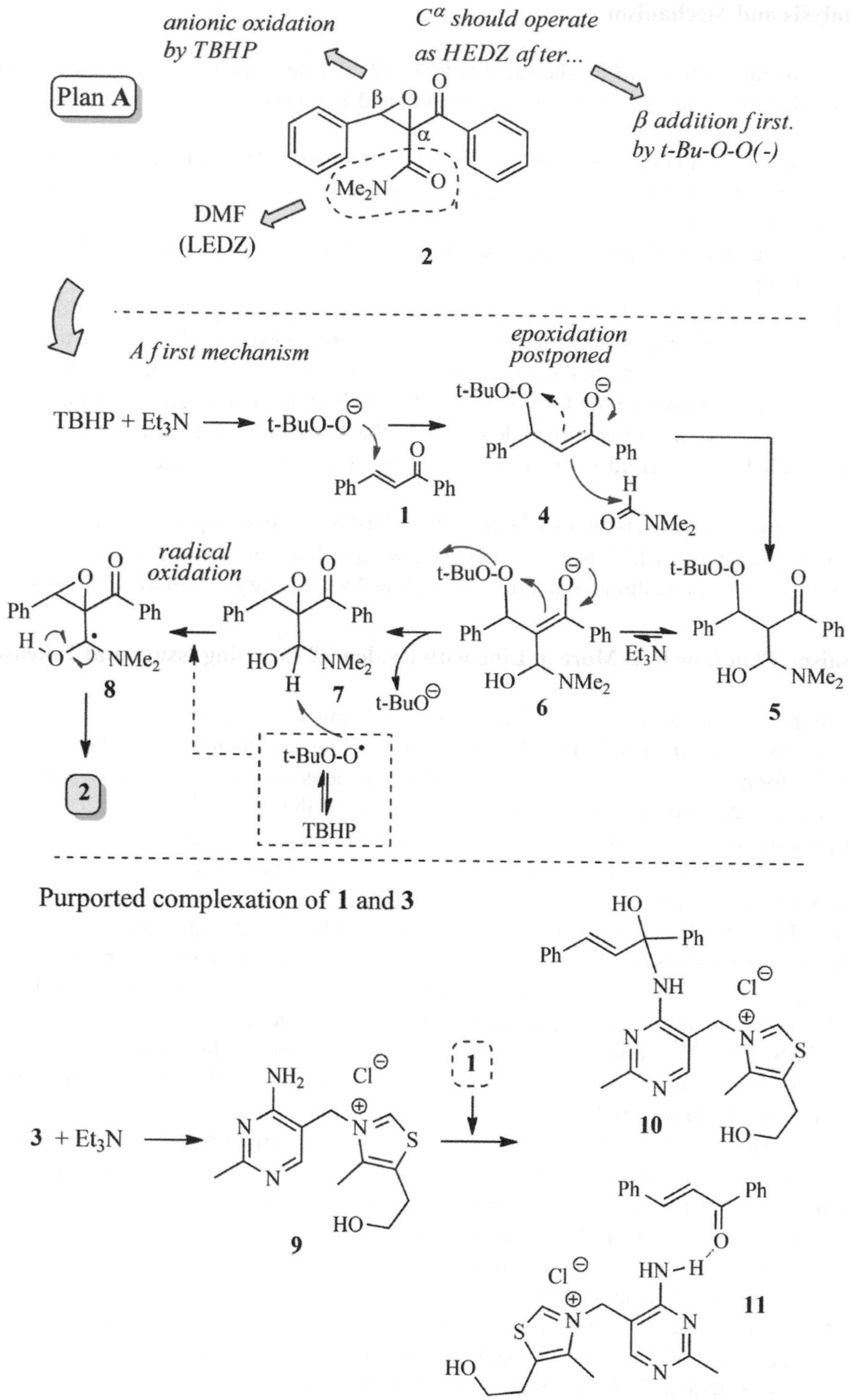

SCHEME 47.2 A first analysis and mechanism plan to explain **1** → **2**.

The symmetric arrangement of **1** may nevertheless be altered by intervention of thiamine in C=O (Scheme 47.2 lower part) after removal of an amine proton by Et_3N. The monochloride salt **9** would introduce conformational differences as **10** or **11** for the ensuing DMF addition-epoxidation. However, this is nothing more than a speculative thought and another mechanistic option using thiamine as the reaction starter–which should make the difference–is mandatory.

47.4 Alternative Analysis and Mechanism

The order of addition of reagents is essential for the success of **1**→**2**. The information about reagents usage given in Scheme 47.1 leads the impression that all components were mixed, heated, and **2** isolated thereafter. The experiment was performed as follows:

A mixture of thiamine and Et_3N in DMF was stirred at 25 °C for 30 minutes. Then, chalcone was added, followed by TBHP in water with stirring and heated to 120 °C while the reaction progress was monitored by TLC; aqueous workup and product isolation/purification finished the process.

This order suggests that thiamine and base gave a secondary product, active toward chalcone, that would grant the entrance of DMF and TBHP oxidation in tandem.

By looking at the structure of thiamine **3**, one realizes the presence of an acidic proton in the thiazole moiety by virtue of the EW power of the proximal $C{=}N^{(+)}R_2$ function. The sp^2 anion can also be expressed as resonance structures **12a** ⟷ **12b**. The latter displays six electrons on this carbon, namely a carbene. Contrary to many congeners, this carbene is nucleophilic, an excellent candidate for a 1,4-addition to the C=C–C=O unit of **1**. Enolate adduct **13** opens the way to DMF trapping at this stage, which can approach only from the underside owing to the steric bulk of the thiamine. Therefore, the early introduction of thiamine in the chalcone scaffold with a given configuration rules the entrance of DMF and TBHP radical in **14** and **16**, respectively.

The oxidation of C^{α} transpires thanks to *t*-Bu–$OO^{(\cdot)}$ radical from TBHP as a first step in the construction of the oxirane under steric restrictions imposed by thiamine, which works as a leaving group. Thiamine is thus recovered for another catalytic cycle. Excess TBHP (3 eq.) completes the semiaminal oxidation via radical **18** to the target with the observed diastereospecificity.

47.5 Which Mechanism, Plan A or B, is More in Line with Evidence? Learning Issues and Takeaways

1. Mechanism **A** suffers from a few serious drawbacks, the first being a dubious participation of thiamine in the way it is bonded to chalcone as depicted in Scheme 47.2, lower part. Thiamine chemistry has been studied independently of enzyme contribution for many years. One of the discarded interactions with active carbonyls such as pyruvic acid – during its thiamine-prompted decarboxylation – is the lack of reactivity of the exocyclic amine to form Schiff bases with carbonyls, as other primary amines do, much less so in 1,4 additions of α,β-unsaturated ketones. Schiff bases had been suggested earlier [5] but proved to be experimentally ineffective toward benzaldehyde. Accordingly, adduct **10** (Scheme 47.2, lower portion) could be discarded.
2. The H-bond model **11** cannot be ruled out, but its role as diastereohemical comptroller of the tandem additions required by target **2** would be dubious unless further bonding or electrostatic sites between DMF and TBHT addition intermediates or TSs and thiamine H-bonded to C=O could be built to produce a more rigid, sterically defined complex.
3. Other active HED sites in thiamine have been proposed, such as an ylide formed by proton abstraction of the methylene bridge in **3**→**19a/b** (Scheme 47.3, lower inset) [6]. This anion or the carbene therefrom could then be used for temporary bonding to the β-carbon of chalcone and exert stereocontrol over the rest of the tandem additions. However, this hypothesis could not be proved experimentally [6].
4. In a quite instructive paper, Professor Ronald Breslow in New York, United States later established unequivocally that thiamine and other simpler thiazoles produced a ring anion at C^2 capable of condensing with aldehydes [7]. Furthermore, by letting thiamine stand in neutral D_2O at room temperature for a few hours, not only C^2 showed H/D exchange as predicted, but also four other protons as well (you are invited to pinpoint which). This foundational observation gives credit to the manner in which thiamine acts as catalyst in option **B** (Scheme 47.3), which is, as a matter of fact, the design proposed by Maheswari et al. of this reaction [1].
5. The primary adduct **13** (Scheme 47.3), which is obtained as an enantiomeric 1 : 1 mixture, offers differential steric hindrance to the ensuing addition of the enolate to DMF. Yet, this is of little consequence considering the following oxidation that creates a nearly trigonal radical **15**. The bonding of *t*-BuOO$^{(\cdot)}$ and displacement of thiamine in **16** by the opposite side creates the diastereomeric configuration observed in target **2**, since the second oxidation of the semiaminal to amide does not involve the vicinal chiral carbon.
6. A hefty body of literature has proved this model correct ever since, which developed the realm of heterocyclic carbenes (NHC).

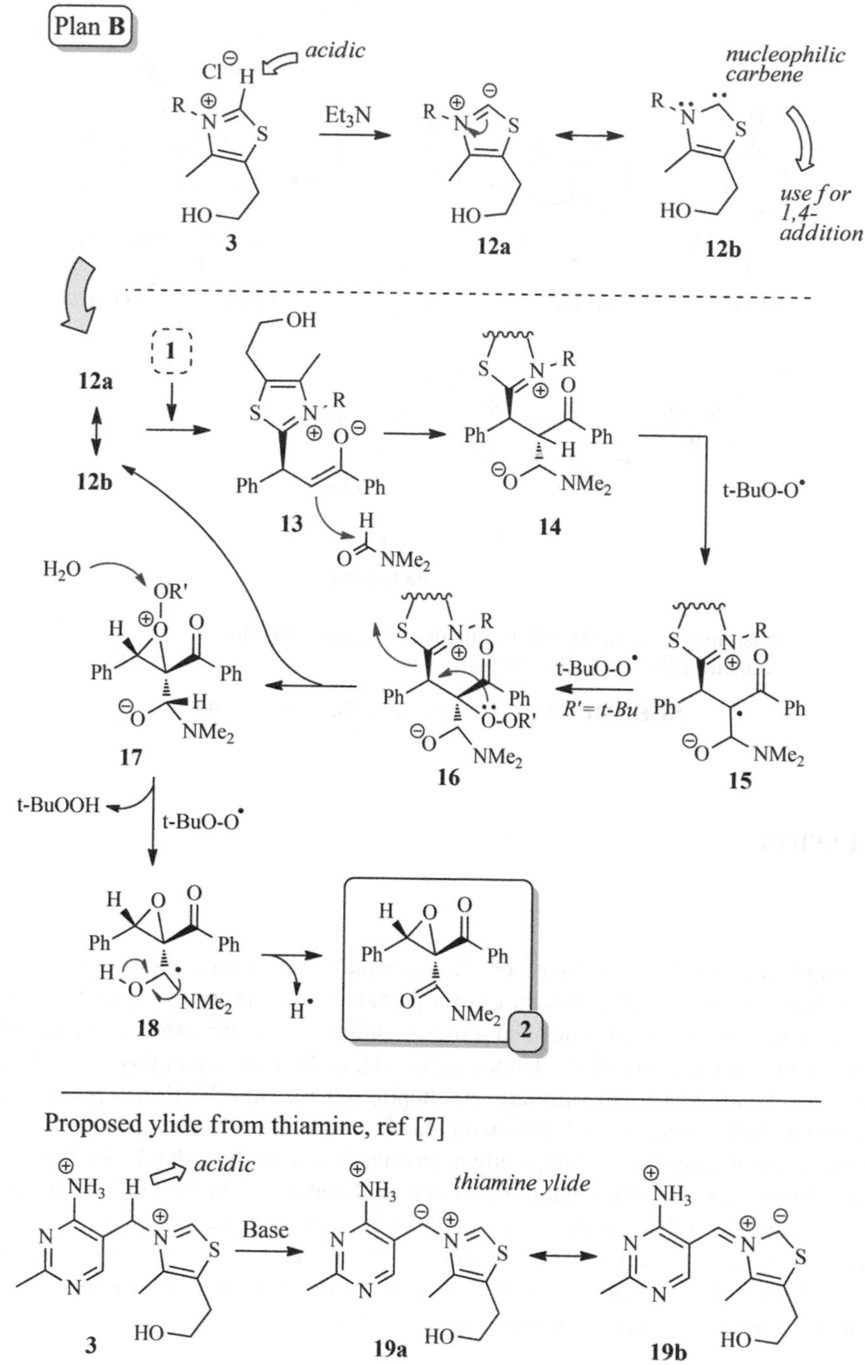

SCHEME 47.3 A second mechanism option.

REFERENCES

1. Devi ES, Pavithra T, Tamilselvi A, Nagarajan S, Sridharan V, Maheswari CU. *Org. Lett.* 2020;22:3576–3580. DOI:10.1021/acs.orglett.0c01017.
2. Katsuki T, Sharpless KB. *J. Am. Chem. Soc.* 1980;102:5974–5976. DOI:10.1021/ja00538a077.
3. Shan H, Lu C, Zhao B, Yao Y. *New J. Chem.* 2021;45:1043–1053. DOI:10.1039/D0NJ05228K.
4. Majdecki M, Tyszka-Gumkowska A, Jurczak J. *Org. Lett.* 2020;22:8687–8691. DOI:10.1021/acs.orglett.0c03272.
5. Stern K, Melnick J. *J. Biol. Chem.* 1939;131:597–613. DOI:10.1016/S0021-9258(18)73456-X.
6. Ugai T, Tanaka S, Dokawa S. *J. Pharm. Soc. Japan* 1943;63:269. Cited in Reference [8].
7. Breslow R. *J. Am. Chem. Soc.* 1958;80:3719–3726. DOI:10.1021/ja01547a064.

PROBLEM 48

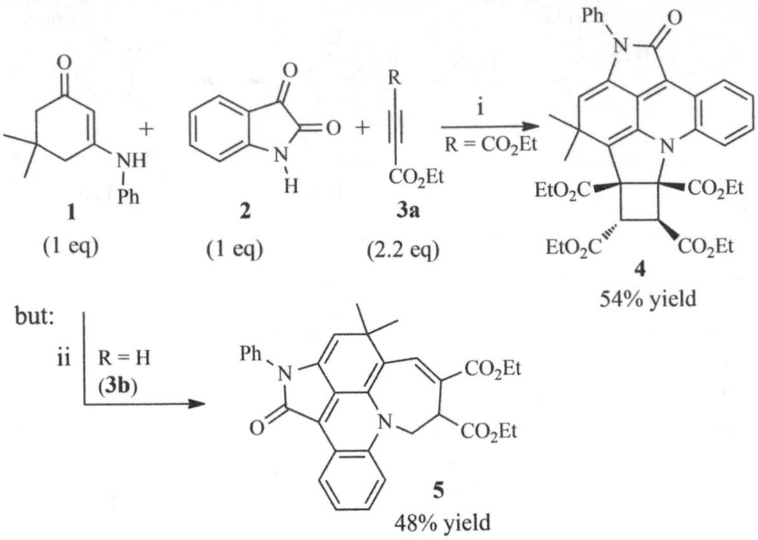

i: iso-butyric acid, 90-100 °C, microwave heating, 30 min
ii: idem, 130-140 °C, 20 - 30 min

SCHEME 48.1 [1] / American Chemical Society.

PROBLEM 48: DISCUSSION

48.1 Overview

MCRs are increasingly employed to build molecular complexity, rapidly, efficiently, and in a fraction of the time and effort required by stepwise synthesis. To be feasible though, careful planning and mixing of components in the reaction vessel is mandatory to avoid reaction runaways and uncontrolled polymerization. There are cases, however, when scientists, like the authors of this problem's work from China and the United States [1], embark in apparently risky designs such as mixing all components in an adequate solvent, heating the mixture, and hoping for the best. Well, this practice is the result of detailed experimentation exploring reaction conditions, solvents, temperature, and time.

In this instance, ingredients were dissolved subsequently in an organic acid in the order **2** + **1** + **3** at room temperature, stirred, and then irradiated in a microwave oven while monitoring indoline **2** consumption by TLC. After several trials, yields were maximized when isobutyric acid, 90–100 °C, and a reaction time of 30 minutes were employed.

These conditions imply the selective reaction of components two by two without interference from the third component until a later stage. A feasible mechanism design will emerge after careful consideration of reaction sites, electronic compatibility, and stereochemical permission. As usual, the mechanism keys reside in the target.

48.2 Analysis of the Conversion to 4

MCR mechanisms are ideal for FA as a first step: recognizing the pieces by approaching the target as a jigsaw puzzle of sorts. After you try your own, feel welcome to visit Scheme 48.2, which probably coincides with your own plan. In our case, the *gem*-dimethyl group serves as landmark to begin identifying the compounds mixed at the start. Save for the acetylene diesters, visible at the southern section, compounds **1** and **2** appear intimately fused and possibly rearranged in part. Thanks to this deconstruction scheme one sees that:

- There are six rings A–F in **4**. Of these, only two, A and D, are defined in the starting materials; hence, the rest are built along the reaction coordinate.

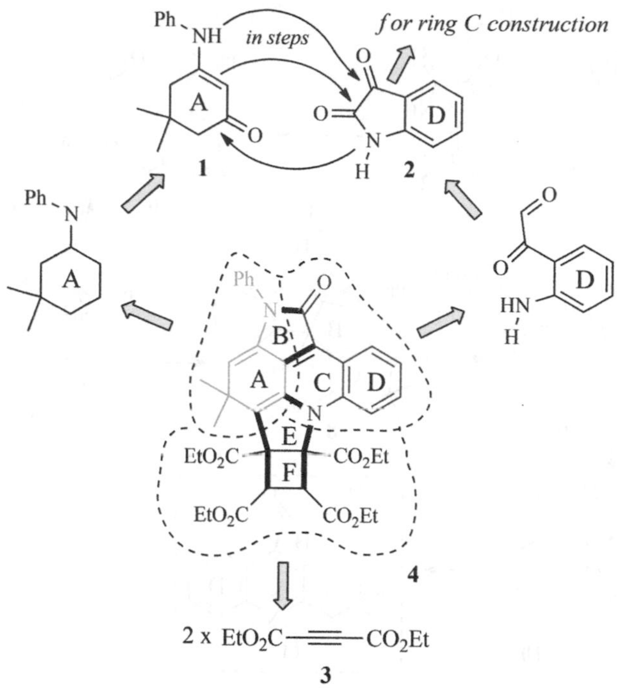

SCHEME 48.2 Deconstruction of target **4** into sections closely resembling the starting components of the MCR process.

- Compounds **1** and **2** are fused such that B and C result from cyclizations. This feature requires the cleavage of the diketoheterocycle of **2.**
- In accordance with the concept of bonds crossed by dotted fields to identify the starting materials in the target, new linkages (shown as thick black lines in Scheme 48.2) define the general construction lines. These bonds are probably formed in tandem to maintain control of the events. The southern tetraester obviously stems from 2 mol of the propargyl diesters and might be postponed for later stage. If so, one should focus on sections holding rings A–D in the early events.

48.3 A Feasible Mechanism for 1→4

To avoid confusion of too many elements in a single scheme, we divide the sequence in two parts.

Part A:

Because the binding of the two northern blocks defines most of the target's scaffold from **1** and **2**, and these compounds have both nucleophilic and electrophilic centers shown at the top part of Scheme 48.2, all the ingredients and electronic capacity to put together rings A, B, and C while dragging aromatic ring D along are contained therein. One must keep in mind that the medium is acidic (isobutyric acid, no other solvent); hence, protonation and cations should be dominant intermediates.

The first decision point is selecting the C=O unit in indoline **2** for its bonding to the vinylogous amide **1**; or the opposite attack of the indoline's nitrogen on the C=O of **1**. The latter is quickly discarded after considering the low nucleophilicity of this N atom, and even forcing a first 1,2 addition here, one is quickly driven to a dead end (not shown, try this yourself). Likewise, addition of **1** to C^2=O is also a bad choice given its pertainance to an amide moiety of low electrophilicity.

The sequence moves on without difficulty up to **8** (Scheme 48.3), whereby ring B is formed. However, the non-C ring is a five-membered cyclic structure whereas the desired C contains six atoms. Dismounting this temporary dihydroindole is feasible as shown in **8**→**9** to rebuild the actual ring C in the Schiff base **10** formation.

It is worth noting that the acetylene diester has not interfered with the nucleophilic atoms developed in the **6**→→**11** pathway. One likely reason is that the nucleophile centers always find a neighboring intramolecular electrophilic unit for binding.

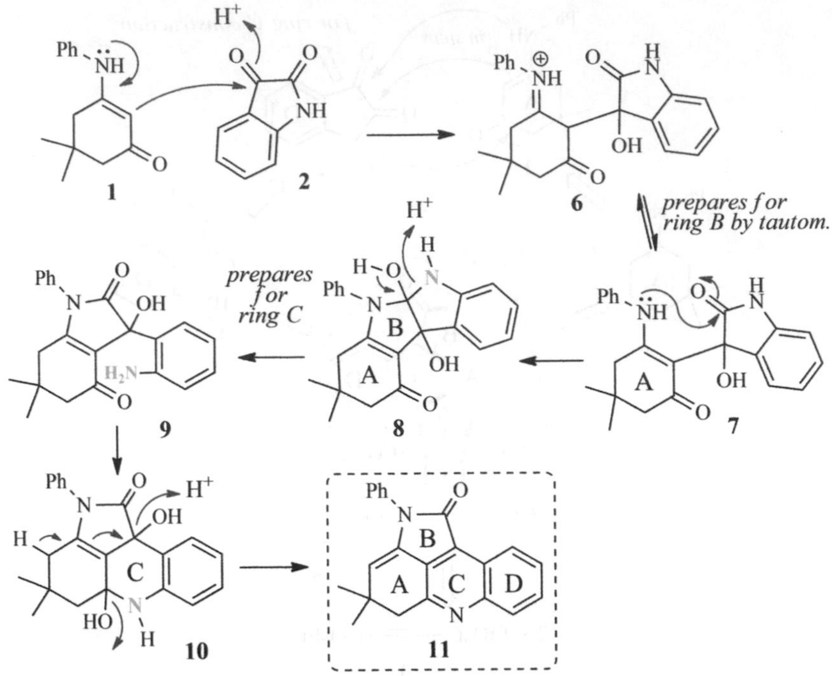

SCHEME 48.3 A feasible construction of rings A–C before intervention of diethyl acetylene dicarboxylates.

Part B:

Comparing the end product **11** of Part A with the final target **4** reveals that the tetracarbethoxycyclobutane moiety is to be built on N and C*, which momentarily lacks activation. Providing this activation is an easy task considering the allylic character of this carbon, right after the first equivalent of acetylene diester is trapped in **12** (Scheme 48.4). The rest of this pathway is equivalent to a [2+2] *stepwise* cycloaddition of the acetylene units. Experimental conditions did not allow a concerted mechanism but redeployment of eight *p* electrons including the NBP of N shown in **14**.

48.4 Analysis of 5 and a Reasonable Mechanism

Scheme 48.5 (top section) shows the division of target **5** in its obvious composing parts. Their coupling is easily conceived as the bonding of an HEDZ in red and two LEDZs in purple. Intermediate **11** should thus be the departure structure to tether two acetylene monocarboxylate units at the most active carbons used for the mechanism justifying target **4** (Scheme 48.4). Two questions come to mind: i) Why is there such a difference in targets **4** and **5** when employing mono and diester acetylenes? ii) How do we manage to bind two LED α carbons of the esters, shall we need an umpolung?

By turning on your chemically practicable imagination, more than one mechanism is possible to achieve goal **5** without subverting chemical principles (the authors [1] offer one solution, others are my own). We will give them names for readers to follow them in Scheme 48.5.

Route I: The diradical pathway.

The main difference between reaction conditions **i** and **ii** in Scheme 48.1 is temperature (90–100 °C versus 130–140 °C), sufficiently large to open opportunities for radical chemistry, especially in highly substituted cyclobutyl moieties. After the addition of 2 mol of carboxyethyl acetylene **3b** at the now familiar active sites of block **11**, the two nearly parallel C–CH=CH–CO_2Et in **16** (Scheme 48.5) undergoes a [2+2] cycloaddition forcibly placing the two carboxylates trans to each other as in **17**, after accepting the *conrotatory* cyclobutane ring closure. A diradical **18** would arise from cleaving the strained carbocycle, evolving to target **5** after [1,3] hydrogen atom transfer through **19**.

Route II: The proton pathway.

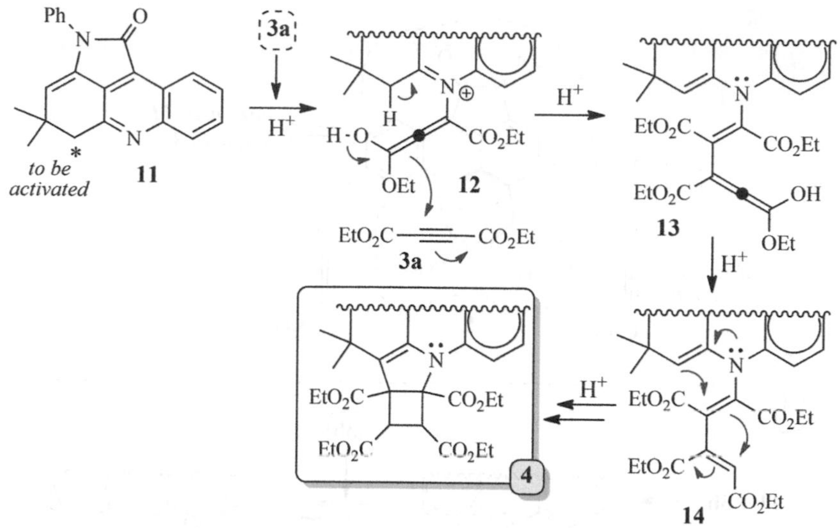

SCHEME 48.4 **Part B** to explain target **4** containing six fused carbocycles.

Key adduct **16** contains a polar enamine-ester susceptible to protonation by the acidic medium. The resulting carbocation would be trapped by the neighboring C=C bond furnishing in two phases (protonation-cyclization) aza-cycloheptene **20** following the electron redeployment of the purple arrows shown in **16** (Scheme 48.5).

Route III: The cycloaddition pathway.

The encounter of acetylene monoester **3b** and **15** (Scheme 48.5, bottom) enables an aza- [5+2] cycloaddition, which would proceed regiospecifically for obvious steric reasons. Proton expulsion in **21** finishes the sequence to target **5**.

48.5 Which Route is More Feasible? Learning Issues and Takeaways

1. Authors [1] report the discovery of an awesome combination of four molecules of three different in organic acid medium, in a typical one-pot multiple component reaction compounds–actually a pseudo four-component reaction–furnishing unique compounds in reasonable yield. All components were mixed at the beginning at room temperature and then heated for half an hour only.
2. Because no previous binary mixing (for example **1**+**2** and isolation/characterization of key adduct **11**, likely stable under reaction conditions and then exposure to acetylenic esters) was attempted or at least reported to support mechanism proposals, one cannot be 100% sure of the mechanism involved. Nor is it possible to buttress any of the route variants with any degree of certainty, as all are chemically logical.
3. Authors did not observe side products of importance, so no clues to other routes are available. Nor was there any attempt to cut short the reaction time, use lower temperature or only one equivalent of acetylene mono and esters to catch any intermediate. Good candidates might have been **9**, **10**, **11**, and especially, **14**, **15**, or **16**.
4. Isolated compounds of the thirty-strong tested behaved similarly and were obtained in comparable yields, irrespective of EW or ED substituents in the aryl rings.
5. Structures were unambiguously obtained from X-ray diffraction studies and confirmed by NMR spectroscopy. Errors in structure assignment are nearly inexistent.
6. Mechanism variants to the synthesis of key intermediate **11** (Scheme 48.3) are marginal so there is a main road sequence involved. Variations in **11**→**4** (not shown) are only in the details but do not affect the concept.
7. Authors thus claim on a good experimental support a tandem sequence involving a surprising gamut of cyclizations: [3+2] – [2+2+1]– [2+2] and [3+2]–[4+2]–[2+2+2+1], with exquisite regiospecific control and tremendous construction of structural complexity in a single pot reaction.
8. Five and up to six (only rarely) component synthesis are on record. Generally, they involve well-known reactions and mechanisms, save a lot of time and reagents and have a lesser environmental impact but require careful planning [2].

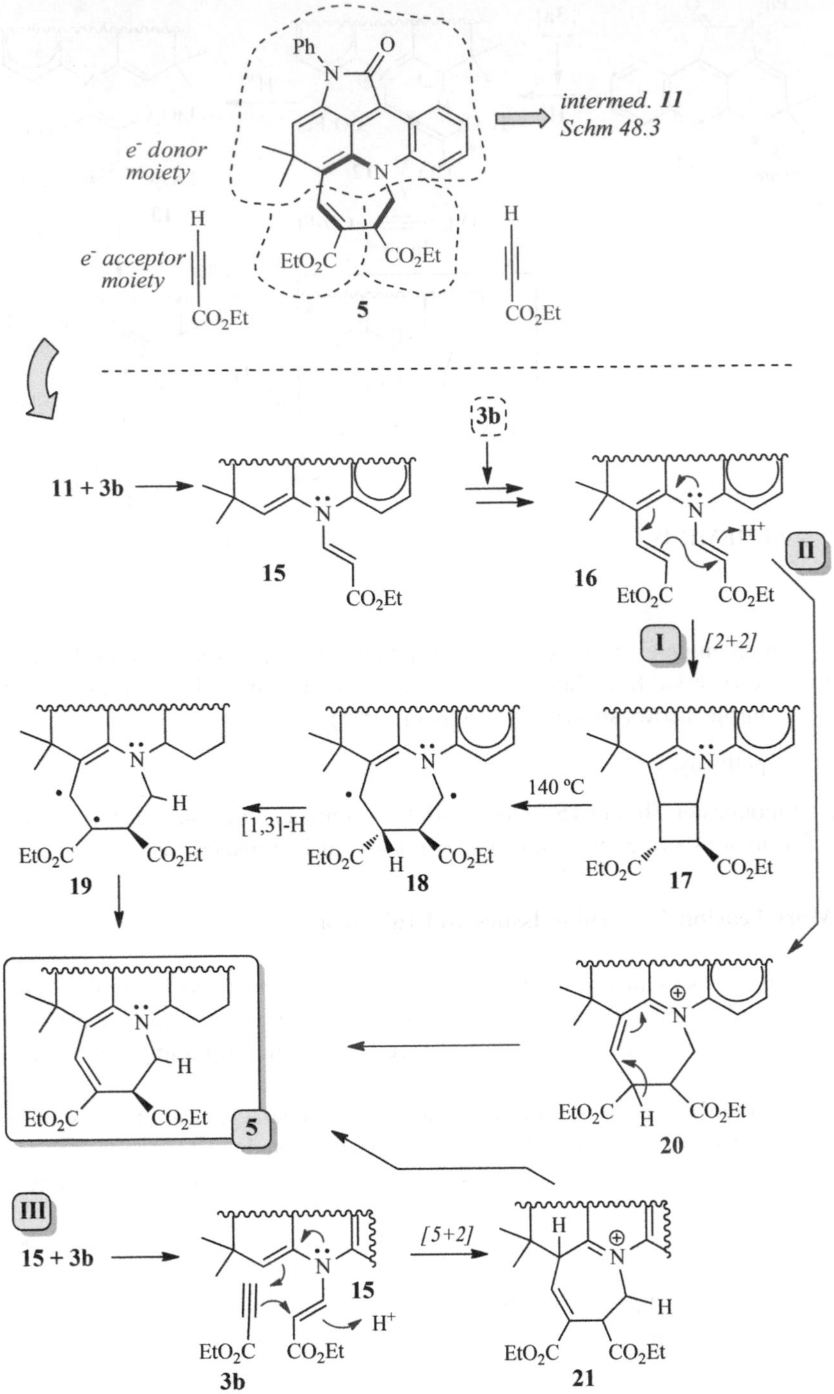

SCHEME 48.5 Three likely sequences to explain the MCR to final product **5**. Routes **I** and **II** are conceptions of this book's author. Route **III** was proposed by authors of this research. [1] / American Chemical Society.

REFERENCES AND NOTES

1. Jiang B, Wang X, Xu HW, Tu MS, Tu SJ, Li G. *Org. Lett.* 2013;15:1540–1543. DOI:10.1021/ol400322v.
2. A veritable flood of MCR research has been published in the last few years. From 2015 to mid 2022, Web of Science registers over 3000 articles after filtering out biochemistry and molecular biology, engineering, science technology, and other topics.

PROBLEM 49

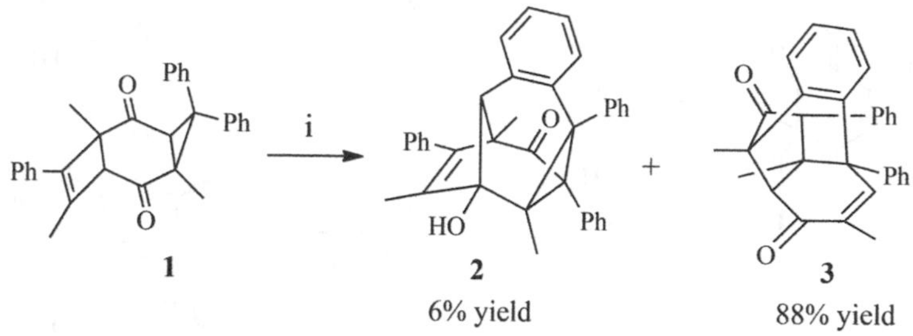

i: $TiCl_4$, $CHCl_3$, r.t., 2 h

SCHEME 49.1 [1] / Elsevier.

PROBLEM 49: DISCUSSION

49.1 Overview

This problem [1] is a development of an earlier photosynthesis study of densely substituted small carbocycles around a quinone heart by researchers in Osaka, Japan [2]. The unexpected performance of alkyl versus bromine substituted quinone cyclopropanes is a short side problem you are invited to solve (Scheme 49.2) and read the universally accessible original article.

Scheme 49.1 depicts a deep-rooted transformation of some complexity, particularly in regard to target **3**, due to the visual overwhelming effect of the 3D molecular scaffold rendering. Explaining compound **2** first will be a worthy training ground for devising a reasonable mechanism for product **3**.

49.2 Reasoning 1 → 2

Based on the accumulated experience acquired in the previous 48 problems, it should be easy to interpret the retro-mechanism analysis of Scheme 49.3, as it provides a short connection between compounds **1** and **2**. We are dealing here with an isomerization, thus bond reorganization. No need for adding tags to scaffold atoms or anything else in this first case, just break the aryl-*sec*-carbon in the bridge, flatten out the rendering for better visualization, and predict the cyclobutene ring expansion of **1** affecting the southern C = O. Is it still necessary to portray the mechanism details? Probably not, as the sequence can be drafted in the **1** → **2** direction seamlessly.

49.3 Reasoning 1 → 3

Comparison of **1** and **3** (Scheme 49.4) suggests a far more complex transformation than the **1** → **2** case. Since the structure of **3** cannot be drawn clearly but in 3D, the use of the familiar dotted areas to recognize similar moieties or sections in starting and final products is too confusing, particularly in the present case. Colors do a much better job. Attentive observation of arguments in roman numerals order of the color road map in Scheme 49.4 is one way to associate **1** and **3**. The bonds in black, namely those that could not be assigned to structures **1** and **3**, pinpoint the C–C bonds that underwent change by cleavage or formation. Bonds **a**, **b**, and **c** in **1** must be broken, whereas **d** and **e** in **3** must be formed, which suggests a deep-seated scaffold rearrangement.

Of central importance to solve the reaction pathway of this novel mechanism is the quaternary benzyl carbon at the bridge of **3**, in contrast with the apparently same bridge carbon in our first target **2**, albeit of secondary substitution pattern. Because the preceding carbocation in **2** arose from the ring expansion of the cyclobutene moiety involving the red southern C=O as the first step, a similar ring expansion is conceivable in **1** → **3** but now involves the green northern C=O and cleavage of bond "**a**" instead.

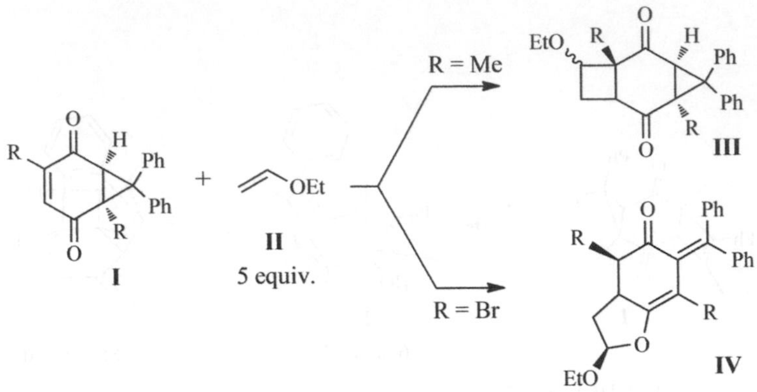

SCHEME 49.2 Earlier work along the same line. [1] / Elsevier.

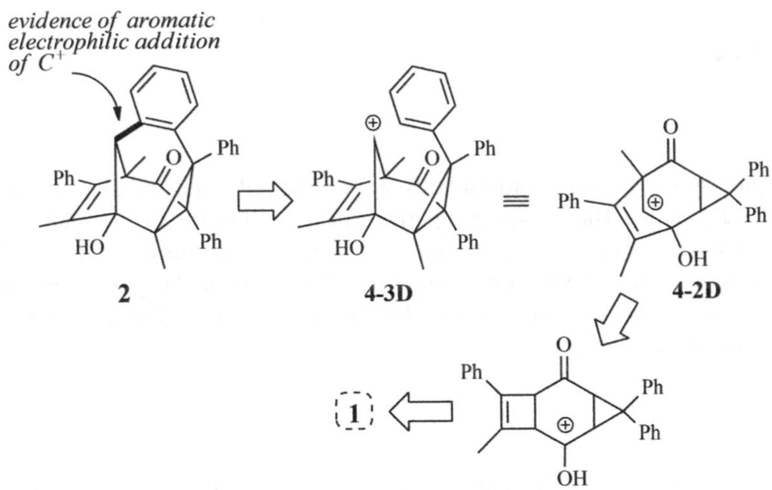

SCHEME 49.3 Retro-mechanism analysis of **1** → **2**.

49.4 A Feasible Mechanism for 1 → 3

With these hints on hand, you should be able to put together a reasonable **1** → **3** mechanism once you overcome the visual impact of stereochemical renderings. An adapted version of the mechanism proposed by the authors [1] is shown in Scheme 49.5. Notice the role assigned to titanium tetrachloride as a Lewis acid.

49.5 Learning Points and Takeaways

1. Small sterically strained compounds offer opportunities for skeletal rearrangements. Compound **1** is not particularly dense in functional polar groups. Yet, its strained scaffold on account of three- and four-membered rings near C=O functions opens it to domino reactions. Large parts of the original molecular structure become hard to recognize during such sequential processes. However, the strategy of sectioning starting material and product(s) in recognizable parts common to both is not only still applicable in complex transformations but constitutes a unique tool to approach bonding changes.
2. Authors tested a variety of Lewis acids: BF_3, Et_2O, $AlCl_3$, and $TiCl_4$, which obtained higher yields of rearranged products and conversion of starting materials with the latter catalyst for unexplained reasons [1]. Results were so diverse that testing several Lewis-acid catalysts for any given susceptible reaction to improve results and explore novel reactions is always a good idea.
3. Among other variations, authors found that **1** subjected to $AlCl_3$ catalysis afforded not only **2** and **3** but also fused tricycle **9** (Scheme 49.5).

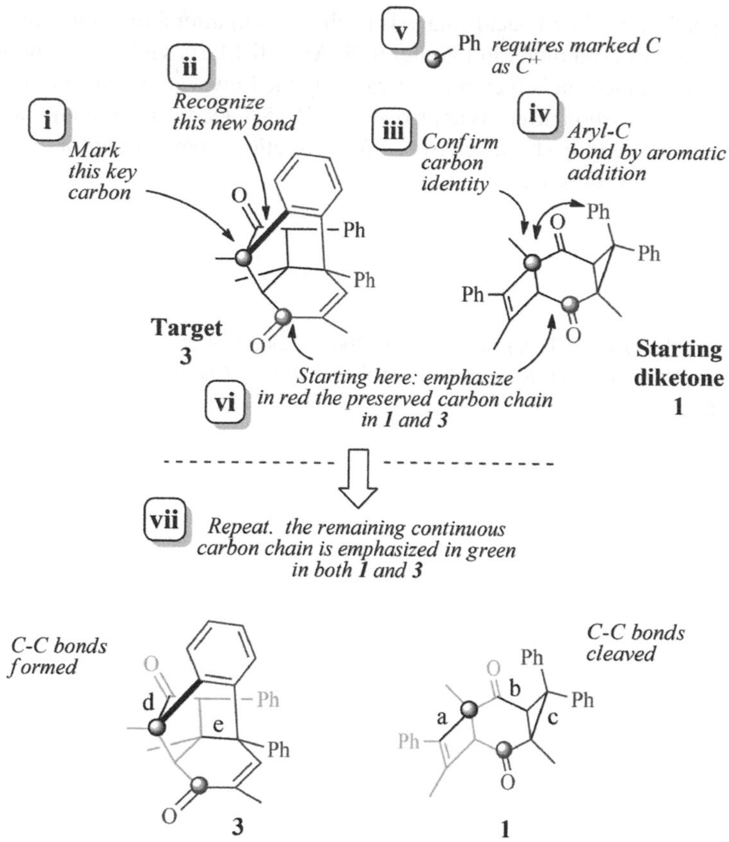

SCHEME 49.4 Reasoning sequence of $1 \rightarrow 3$.

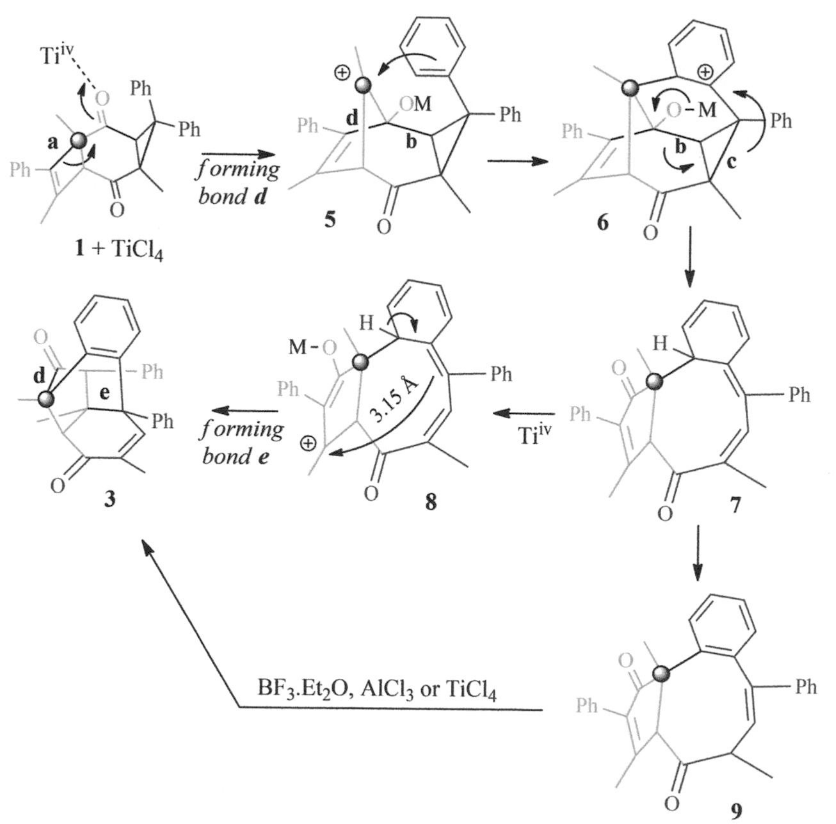

SCHEME 49.5 A reasonable $1 \rightarrow 3$ sequence. Because of the need for colors in structures to illustrate the evolution of the parts, curly arrows of the bond electron course are shown in black.

4. Isolated **9** cyclized to target **3** under Lewis-acid catalysis in chloroform after a few hours at room temperature, thus showing it to be a relatively stable intermediate, in parallel to **8**. An anti-Markovnikov addition of the distant eastern styryl alkene onto the western cyclopentenone's β carbon activated by the Lewis acid assisted the construction of the "**e**" bond. The distance between these two carbon atoms is approximately 3.15 Å according to computer-generated models [3], a still accessible distance for $sp^2 - sp^2$ overlap shortened by scaffold vibrations. Were it not for the reported results, the formation of this bond would not have been foreseen.

REFERENCES

1. Kokubo K, Koizumi T, Yamaguchi H, Oshima T. *Tetrahedron Lett.* 2001;42:5025–5028. DOI:10.1016/S0040-4039(01)00908.
2. Kokubo K, Nakajima Y, Iijima K, Yamaguchi H, Kawamoto T, Oshima T. *J. Org. Chem.* 2000;65:3371–3378. DOI:10.1021/jo991749z.
3. Alonso-Amelot, ME. This book.

PROBLEM 50

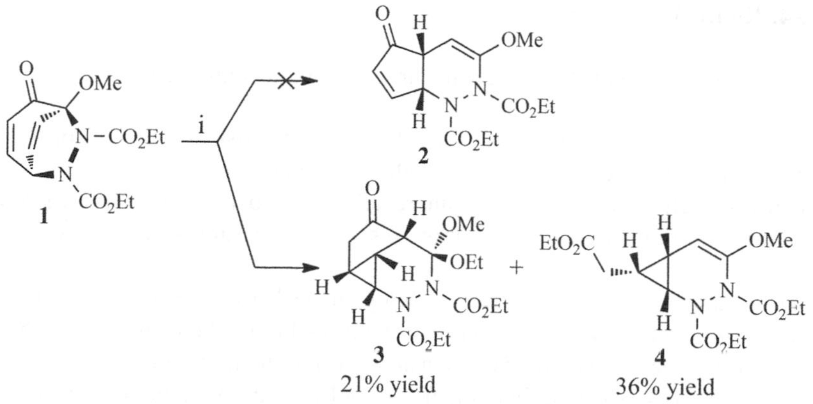

i: hυ, Pyrex glass vessel (>280 nm)
EtOH, 4 h

SCHEME 50.1 [1] / Elsevier.

PROBLEM 50: DISCUSSION

50.1 Overview

Natural or synthetic products with bioactivity against disease usually serve as beacons to pursue structural variation to create compound libraries of enhanced activity. After discovering activity against the parasites that cause two major tropical diseases (of limited interest to large pharmaceutical companies) in certain aza-benzotropolones and aza-tropolones such as **1**, scientists from Uruguay and the United States used these compounds to expand the chemical space and chemotherapeutical impact [1].

Such compounds can be obtained by hetero DACAs with suitable dienes. Adducts possess α,β and β,γ unsaturated ketones, chromophores with potential photoactivity, and therefore, are adequate for photochemical transformation. From their previous experience with aza-tropolones, authors expected the reaction of **1** exposed to UV light in benzene to yield cyclopentanone **2** by photoisomerization via ring contraction. Unfortunately, **1** only furnished decomposition products. After introducing two modifications of the experiment, namely using alcohols as solvent and a Pyrex glass jacket around the UV lamp (>280 nm), **1** yielded novel cyclopropane dicarbamates **3** and **4** in moderate yield [1]. An interesting mechanism between **1** and its products awaits you.

50.2 Mechanism Analysis by Uncomplicated Comparison of Targets and Substrate

By uncomplicated I mean using just one or two of the proposed strategies in this book, not because this problem is particularly easy to solve but to test your practice in spotting structural similarities and differences of significance to the underlying mechanism. Five elements of judgment stir under our eyes in this case:

1. The diazacarbethoxy unit in **1** remains unmodified, inscribed in a six-membered ring all along; therefore, it does not participate in bond cleavage or formation, except perhaps in through-bond electronic effects.
2. That this ring's unsaturation serves as bridge in the bicyclic structure of **1** must be involved in cyclopropanation leading to **3** as well as **4** by interaction of the neighboring allyl moieties.
3. While both **3** and **4** constitute scaffold isomerizations of **1**, the *orphan* ethoxy unit of **3** must proceed from ethanol used as solvent, suggesting a carbocation at the ketal-to-be site.
4. In addition, just one C–C bond is cleaved, and another is formed.
5. Considering the photoexcitation sequence, C=O → C=O* → $^{(\cdot)}$C-O$^{(\cdot)}$, radical intermediates should participate in the bonding process.

Are graphical deliberations really necessary? Based on these criteria, not one but two reaction pathways can be envisioned, routes **A** [1] and **B** [2] (Schemes 50.2 and 50.4).

50.3 Explaining 1→3+4: Route A

Although much of Scheme 50.2 is self explanatory, a few features deserve special attention.

✓ Photoinduced homolysis of the carbonyl alfa C–C bond, known as Norrish type I reaction, furnishes a diradical, which may lead to different outcomes depending on how they split into separate units.

✓ Within the same structure though, several intramolecular reactions may occur, especially when conjugated multiple bonds are present that may carry the radical character to more distant carbon atoms or by way of accessible hydrogen atom transfer.

✓ The authors [1] proposed a Norrish-I reaction since cyclopentane ketones of type **2** (Scheme 50.1) were previously recorded [3] as the major products from irradiation of structural siblings: benzotropolones **8** (similar to **1** with a fused western benzene ring) through the pivotal type **5b** resonance form (Scheme 50.3).

✓ Electron distribution can also exist as ketene **5c** in the free tropolone but not in the benzotropolone case (i.e. **9**) due to the higher Gibbs energy associated with the loss of aromaticity.

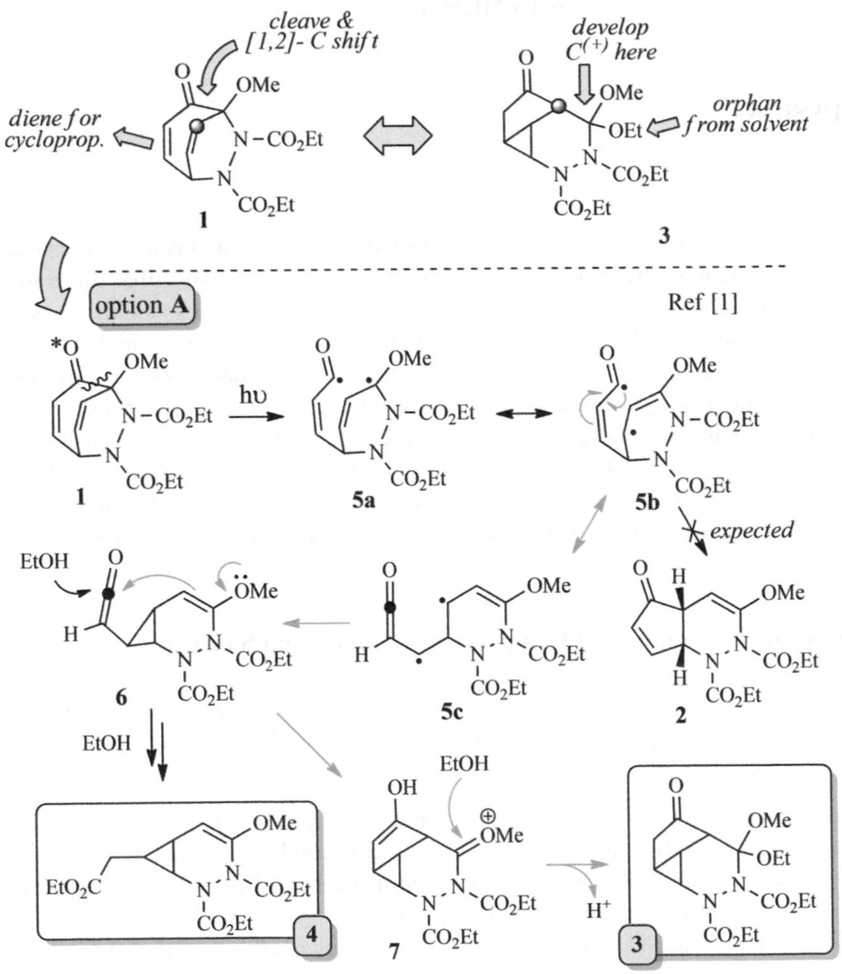

SCHEME 50.2 Brief problem analysis and bond connection/disconnections proposed by the authors of this discovery. [1] / Elsevier. Stereochemical issues are examined in Scheme 50.5 and associated text discussion.

50.4 Explaining 1 → 3 + 4: Route B

As said throughout this book, more than one reasonable mechanism and route branches acceptable to good chemistry are conceivable. Route **B** (Scheme 50.4) is one such alternative [2]. The fundamental difference with route **A** is the very first step after photoexcitation of the carbonyl as described in paragraph five of the problem analysis section above: formation of a C=O* → $^{(\cdot)}$C–O$^{(\cdot)}$ species instead of the Norrish-I cleavage. The resulting diradical is the resonance hybride of **1b** ⟷ **1c**. The homoallyl radical **1c** is sterically set for cyclopropanation that affords cp-stabilized **12** familiar ketene **6** on its way to recorded products **3** and **4**.

From the published data, the Norrish-I approach (option **A**) seems the more acceptable mechanistic course, yet there is nothing solid against option **B** noticeable from stereoelectronic grounds. Pending DFT or other quantum mechanical studies may discriminate the more energy favorable course.

50.5 Stereochemical Appraisal

Authors did not comment on the stereochemical details of the mechanism but established the configuration outcome of targets **3** and **4** by spectroscopic means. That **1** was a mixture of two diastereomers was the result of the diastereomeric control of the DACA employed to synthesize **1**. This is of no consequence to the 1 : 1 diastereomeric mixture of products. For this reason, the 3D aspects of this conversion were not considered in Schemes 50.2 and 50.4 to avoid confusion while determining bond connections and disconnections of the process. Now it is time to have a look at the steric results given the number of chiral carbons in both targets **3** and **4**. And provide further assurance of sequences **A** and **B**.

There are three crucial steps controlling the observed products configuration (Scheme 50.5).

1. Cyclopropanation of **5c-3D** can take place by radical coupling on the β side of the molecular plane defined by the six-membered ring. The three C–H bonds of the ensuing disrotatory closure are oriented to the same western side, ending cis in the cp.
2. As a result, the ketene moiety approaches the enol ether in **6-3D** and enables the construction of the cyclopentanone section. The distance between the ketene central carbon and the enol ether nucleophilic carbon is 3.36 Å in the energy-minimized (MMFF94G [2]) structure of **6**, which may be shortened further by molecular vibration.
3. The cage scaffold **7-3D** thus formed shows two distinct zones for the entrance of ethanol, the underside is clearly more open to incoming nucleophile molecules. Thus, the configuration of the chiral ketal carbon is secured.

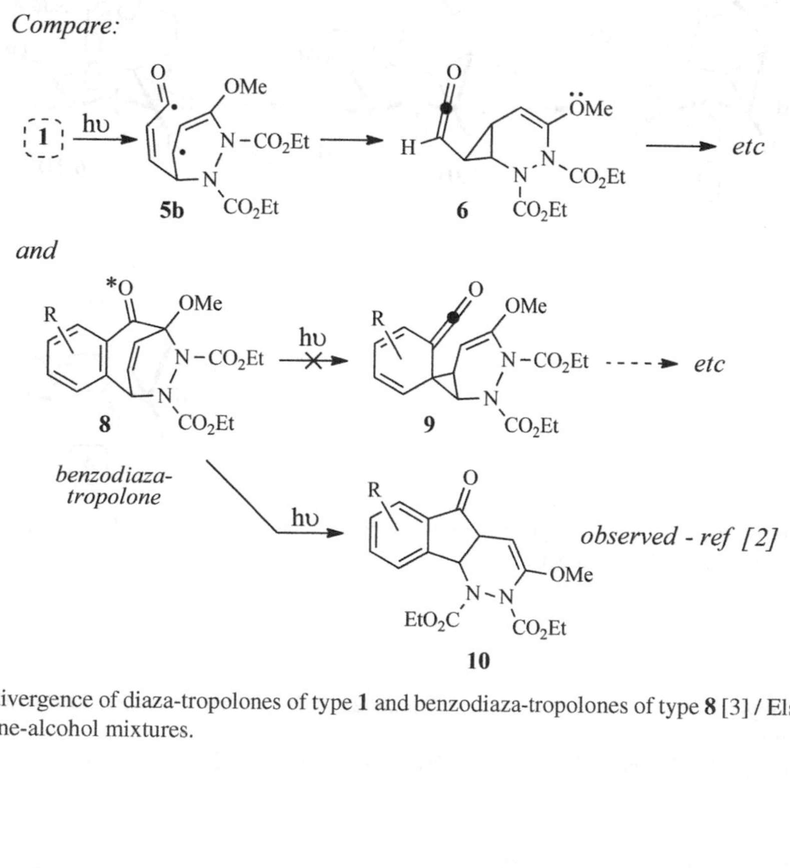

SCHEME 50.3 Outcome divergence of diaza-tropolones of type **1** and benzodiaza-tropolones of type **8** [3] / Elsevier in their UV-photoconversions in benzene or benzene-alcohol mixtures.

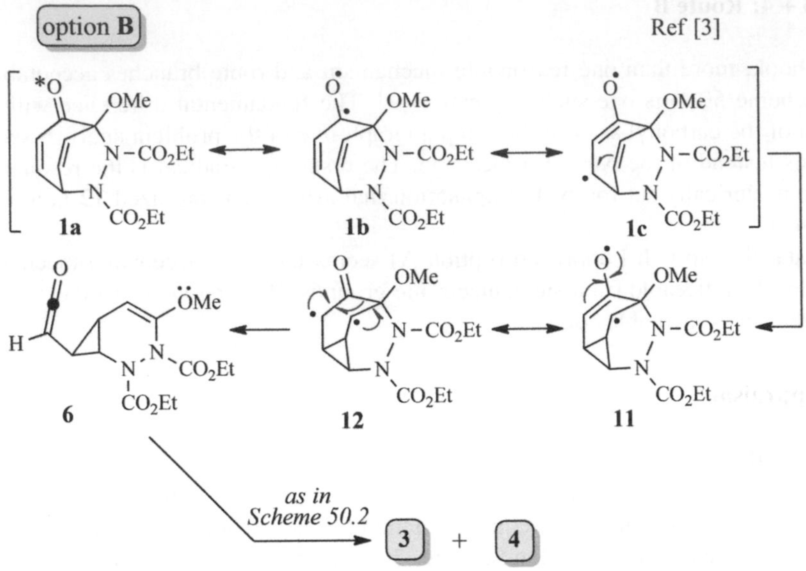

SCHEME 50.4 Alternative route **B** to targets **3** and **4** Adapted from [2].

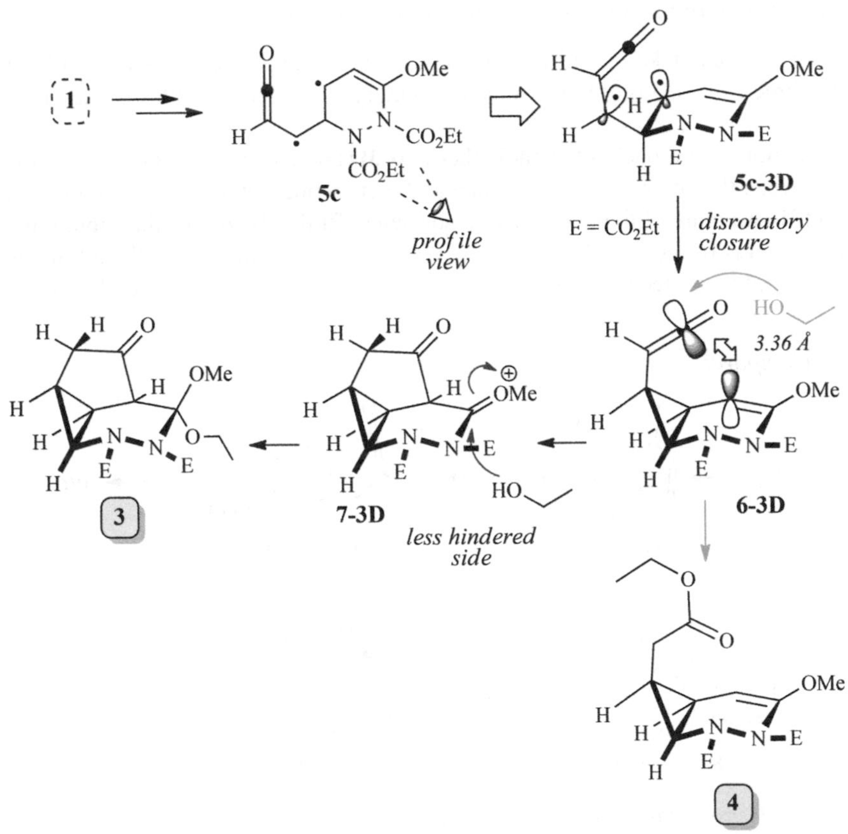

SCHEME 50.5 Stereochemical aspects based on option **A** (Scheme 50.2). Option **B** can be analyzed in the same terms (not shown) and also explains the observed configuration of both targets.

REFERENCES

1. Tabarez C, Khrizman A, Moyna P, Moyna G. *Tetrahedron* 2012;68:8622–8629. DOI:10.1016/j.tet.2012.07.076.
2. Alonso-Amelot ME, this book.
3. Tabarez C, Waterman C, Rapp AL, Moyna P, Moyna G. *Tetrahedron Lett* 2009;50:7128–7131. DOI:10.1016/j.tetlet.2009.09.176.

SUBJECT INDEX

The Art of Problem Solving in Organic Chemistry, Third Edition. Miguel E. Alonso-Amelot.

GRAPHICAL PROBLEMS INDEX

The Art of Problem Solving in Organic Chemistry, Third Edition. Miguel E. Alonso-Amelot.

1

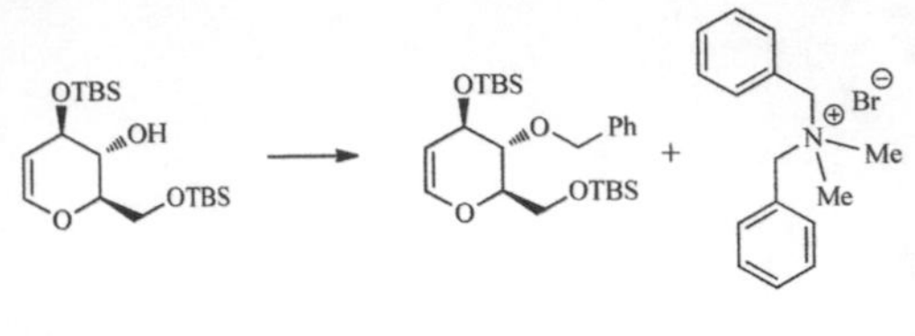

2

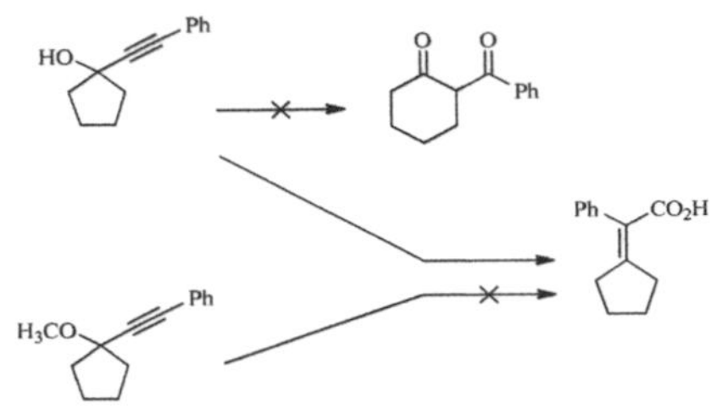

3

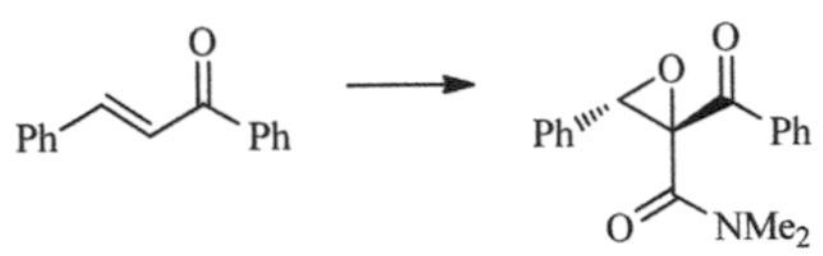

4

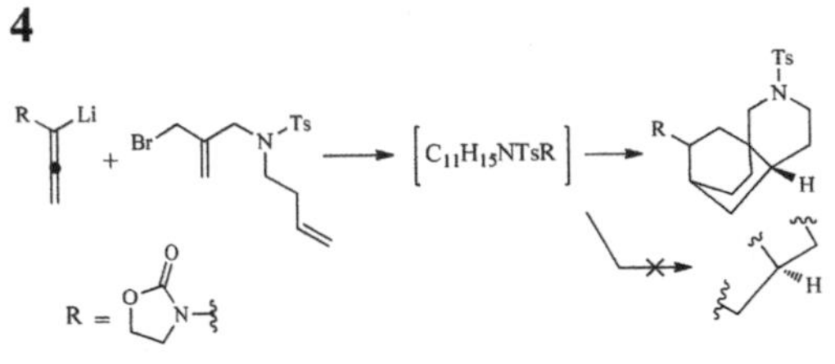

5

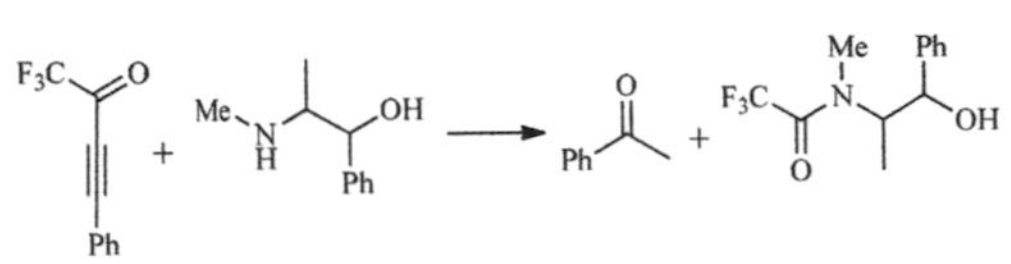

6

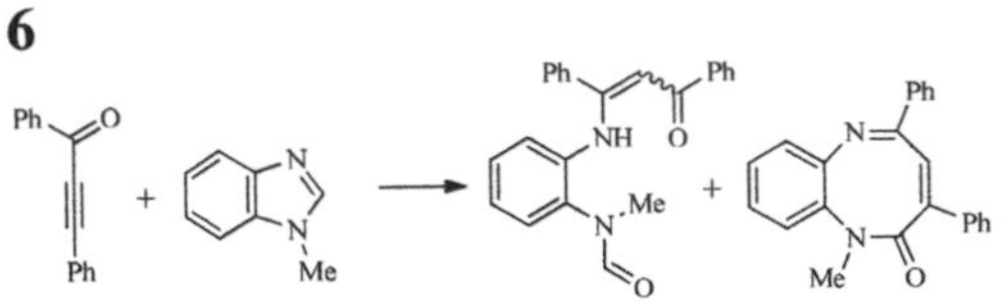

7

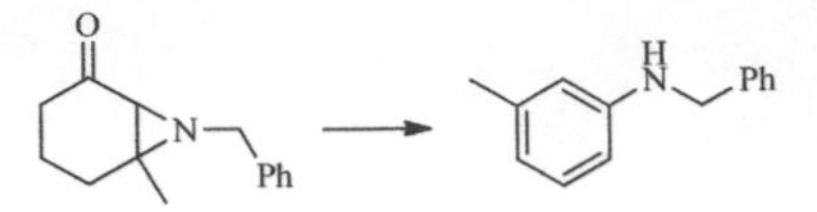

8

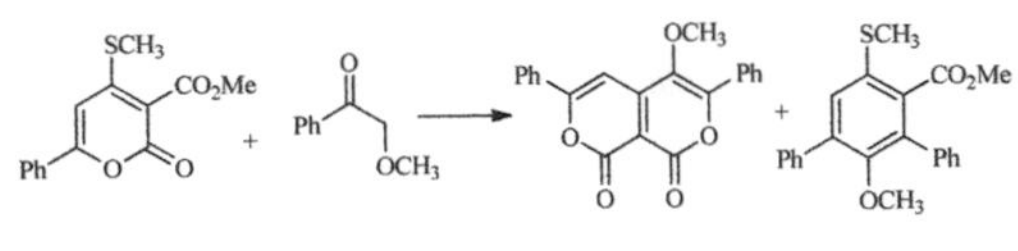

9

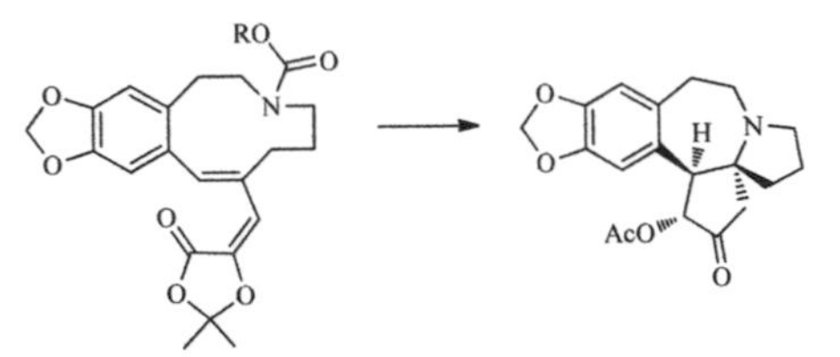

10

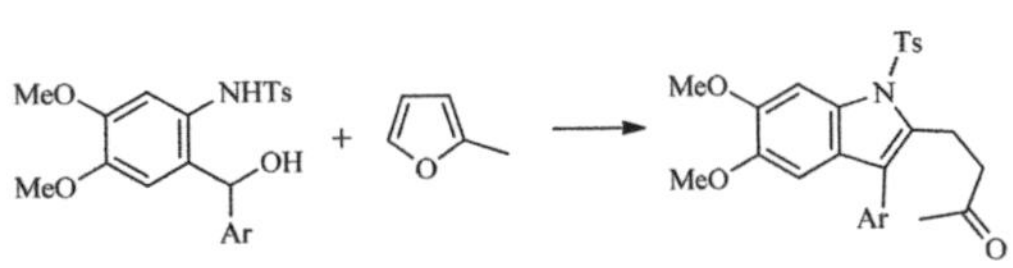

11

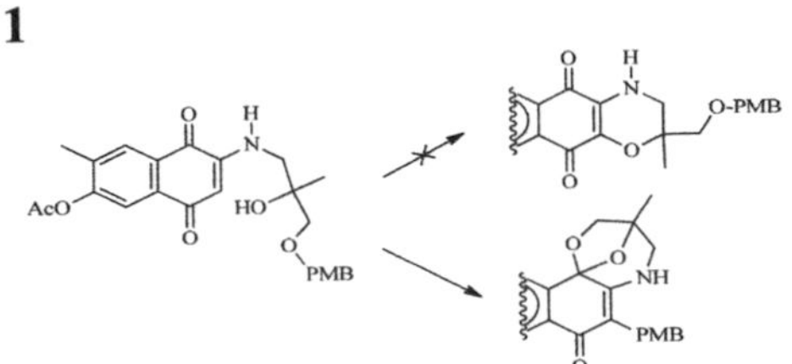

12

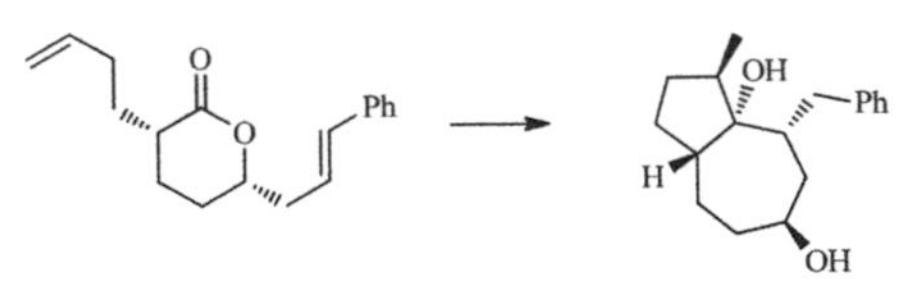

13

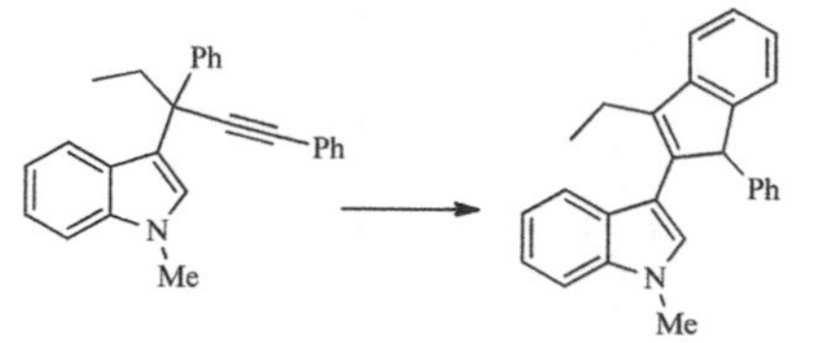

19

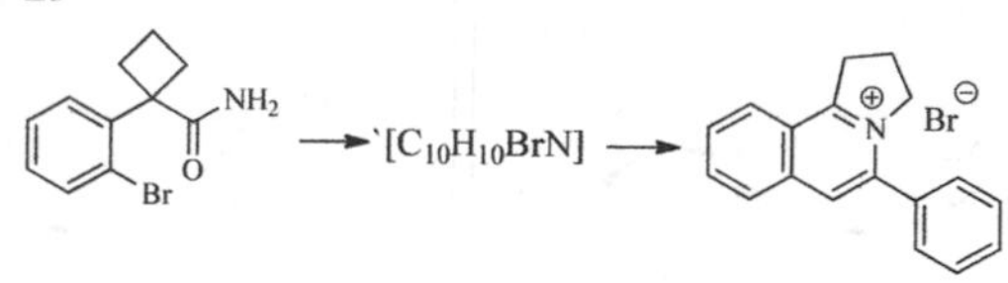

14

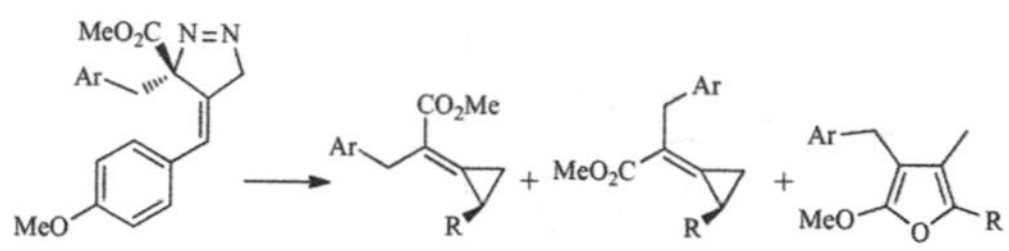

20

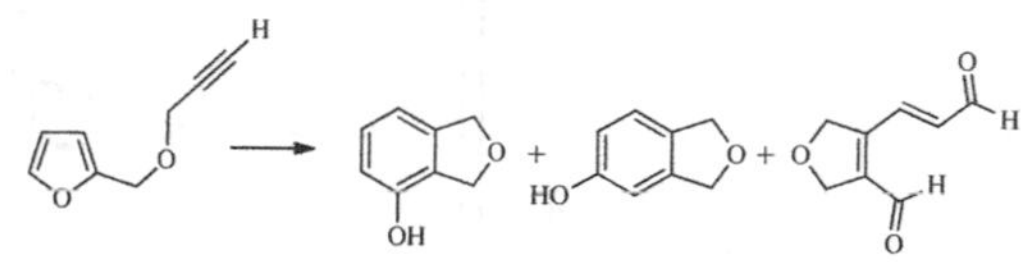

15

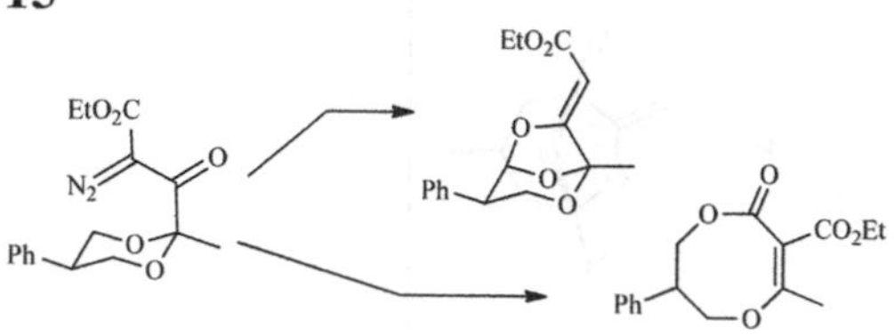

21

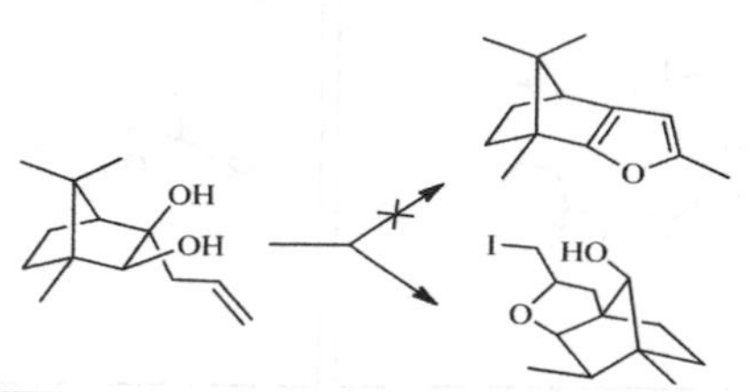

16

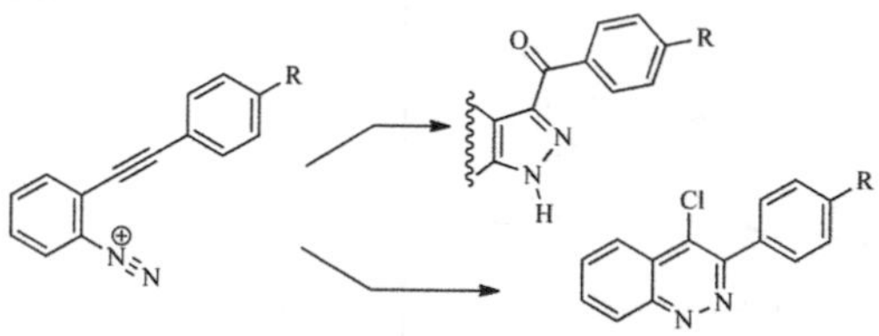

22

17

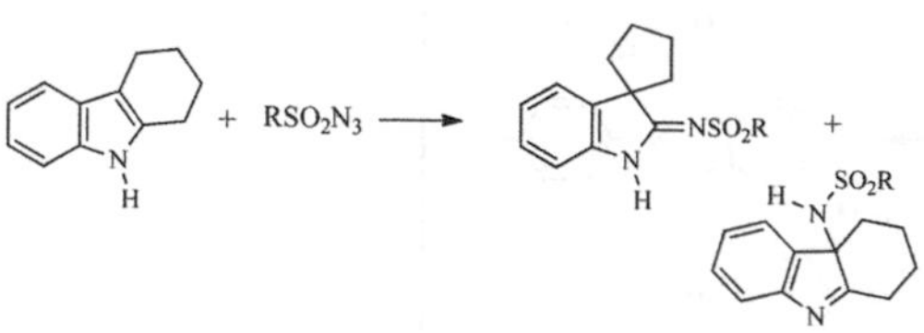

23

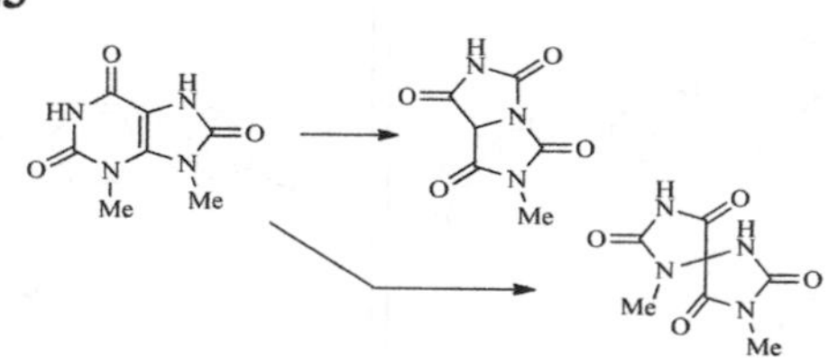

18

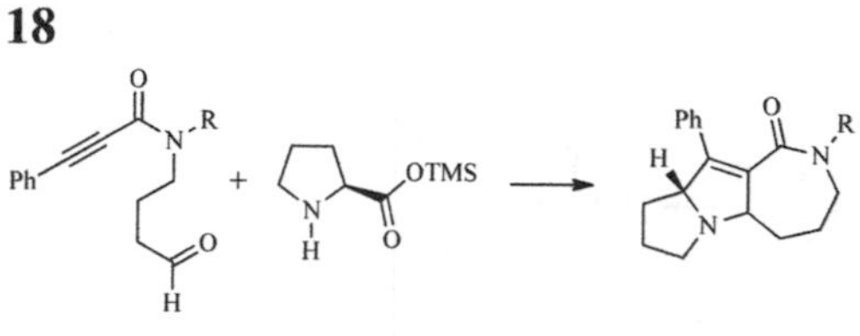

24

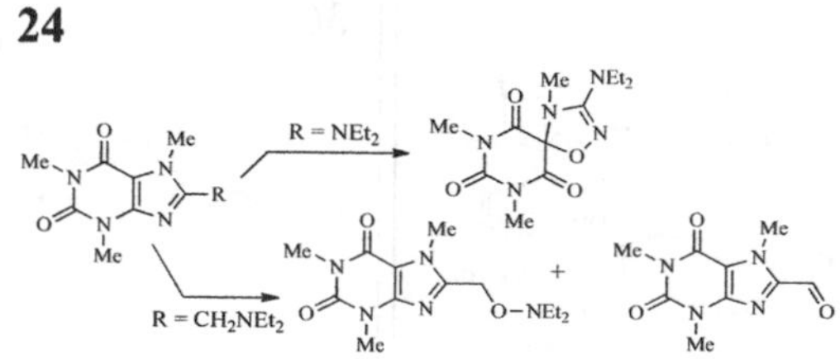

25

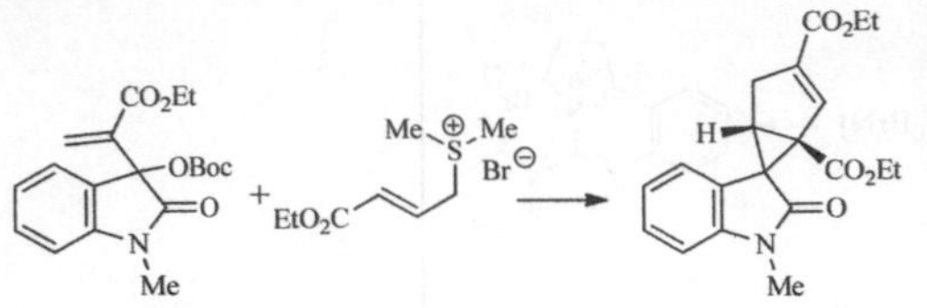

31

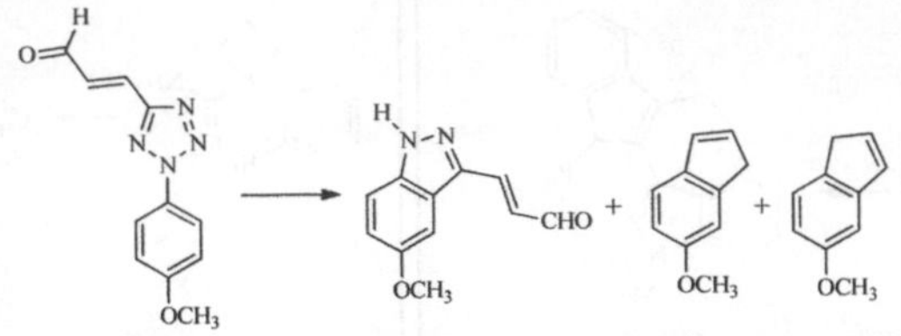

26

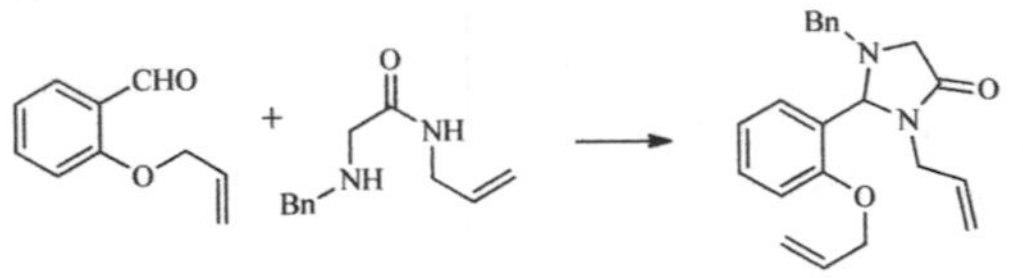

32

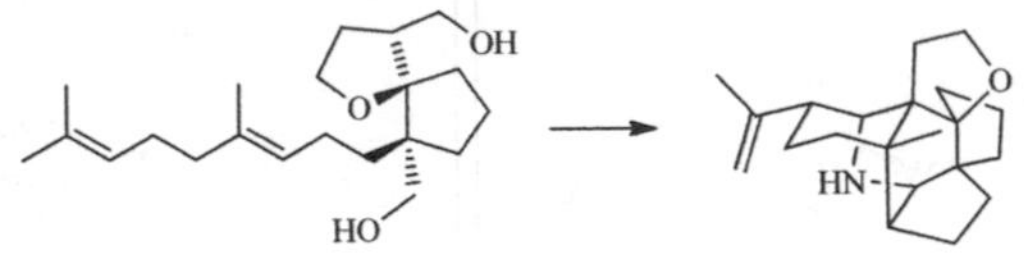

27

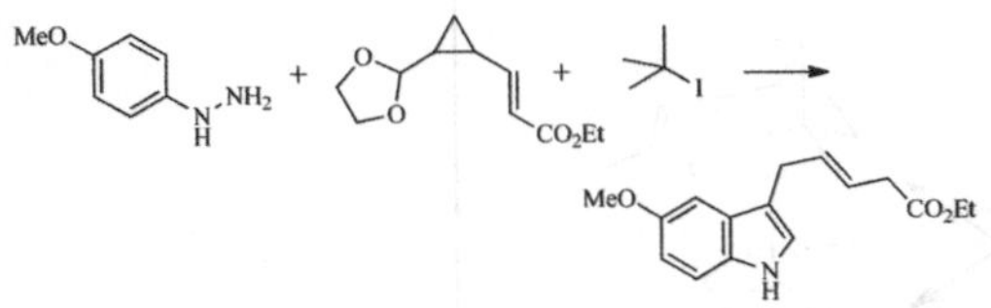

33

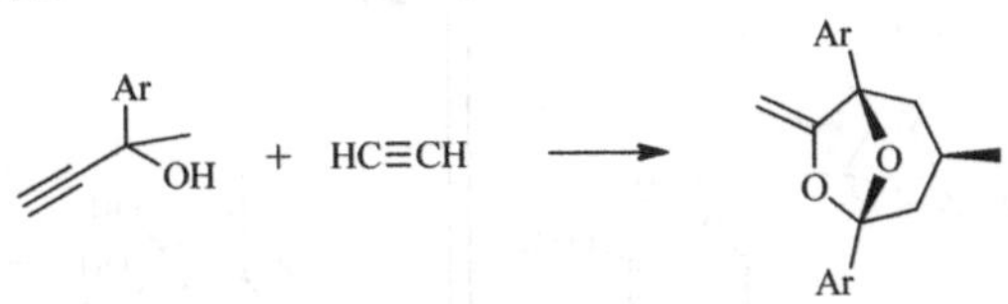

28

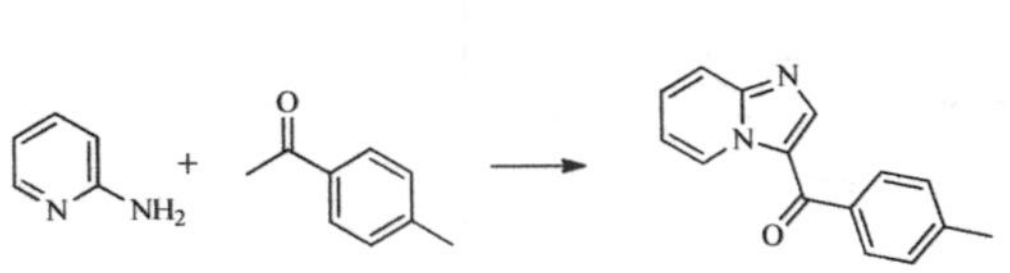

34

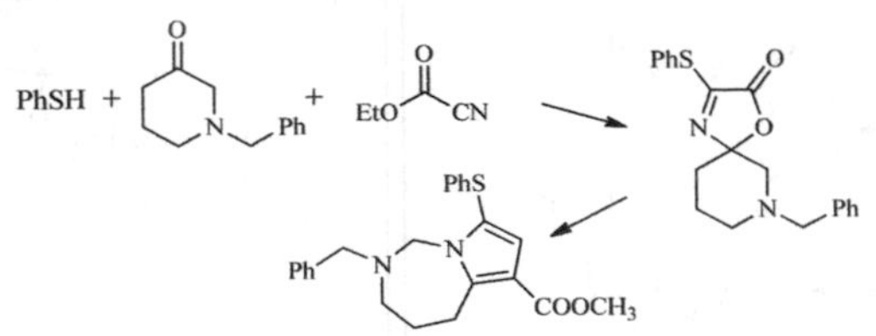

29

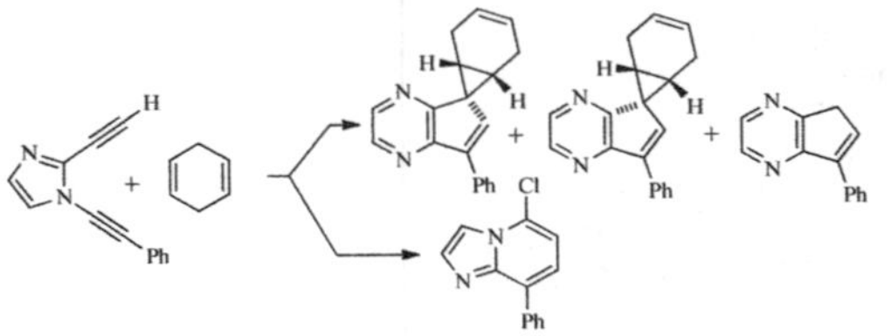

35

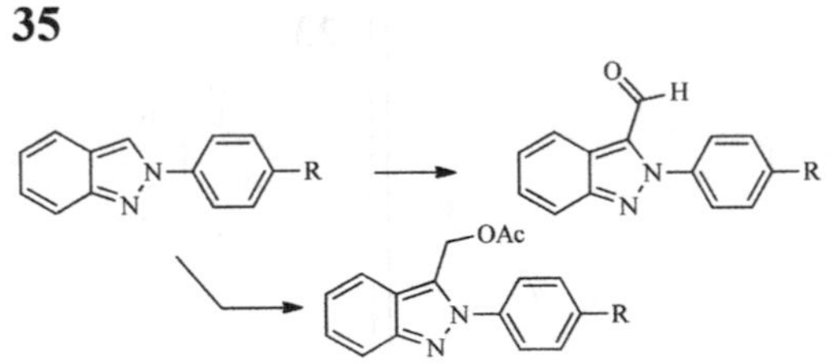

30

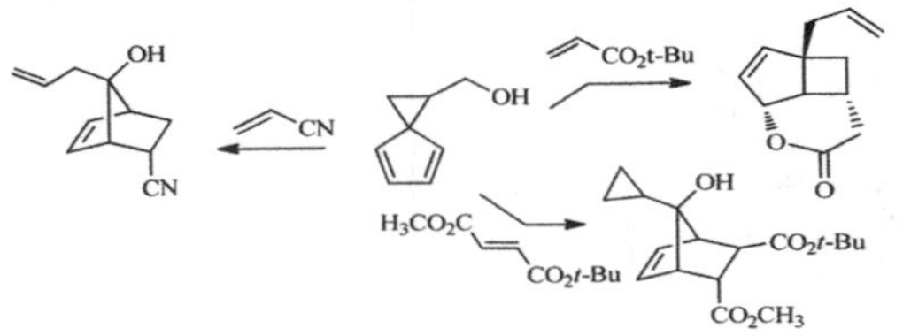

36

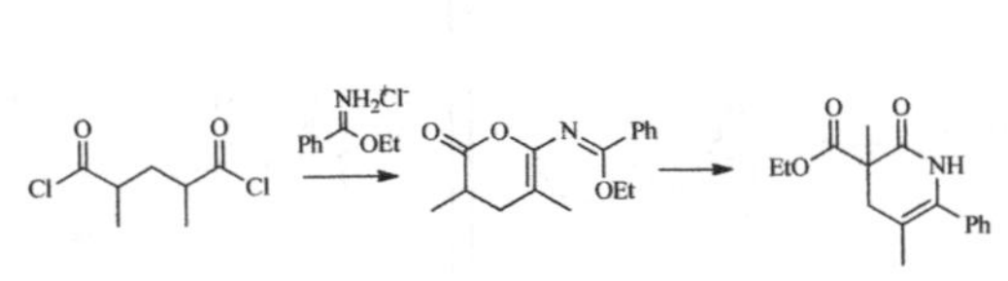

37

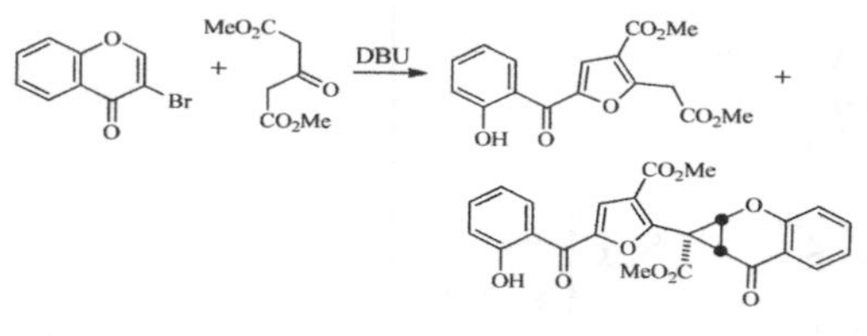

43

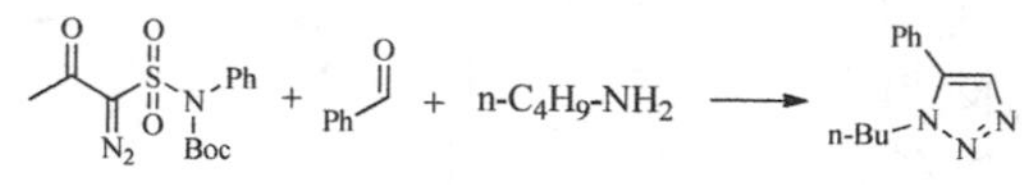

38

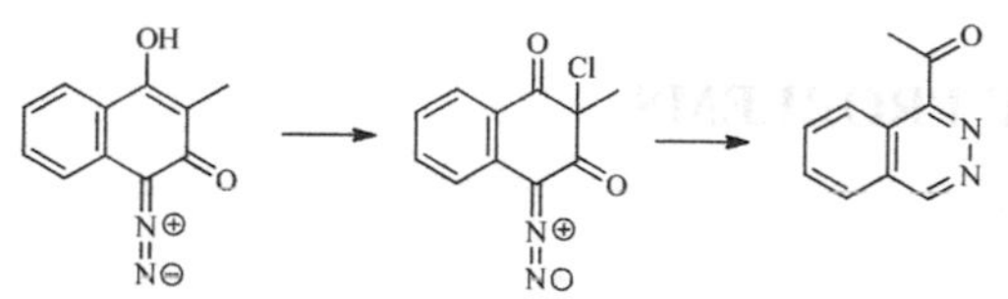

44

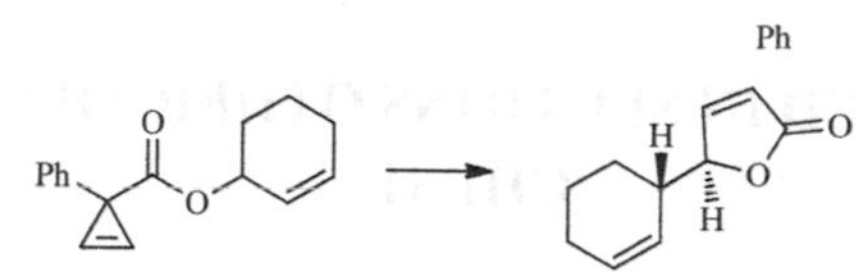

39

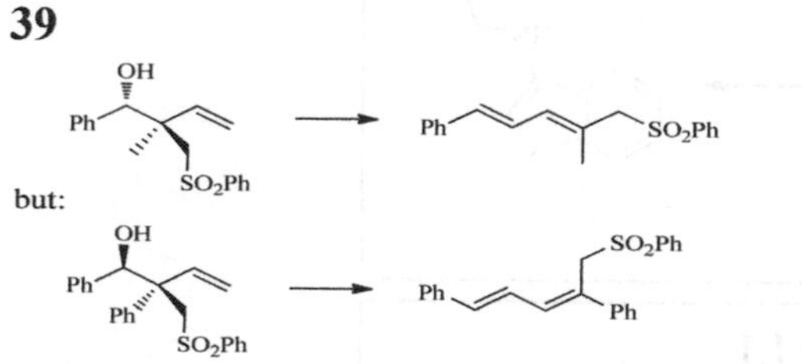

45

40

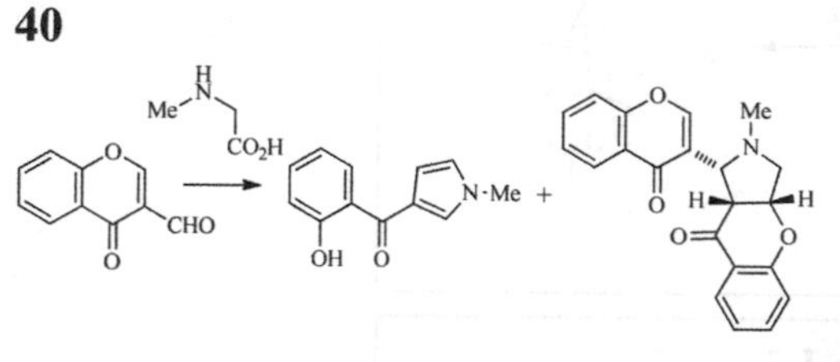

46

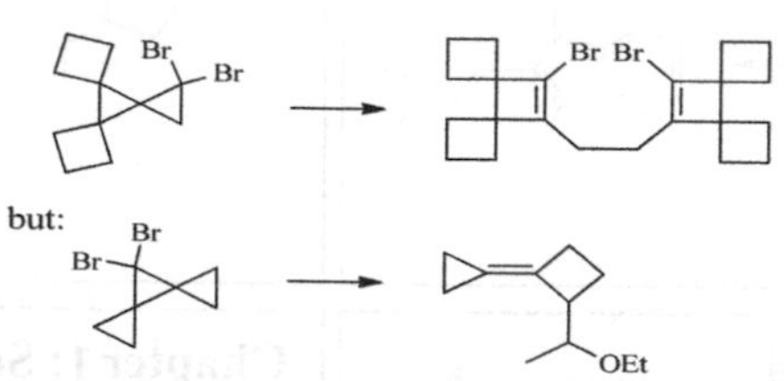

41

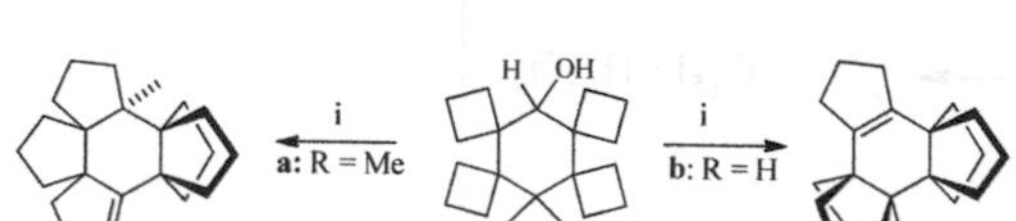

47

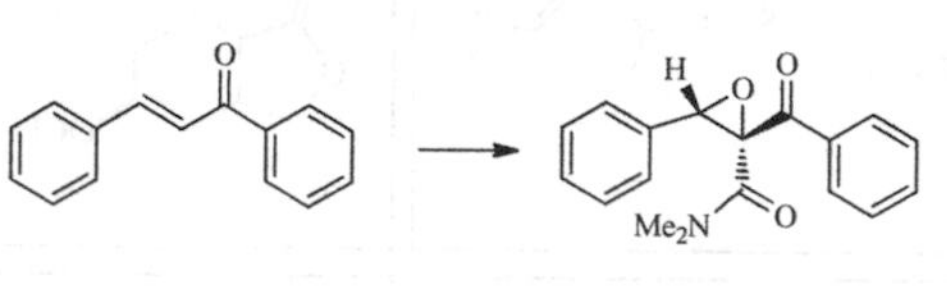

42

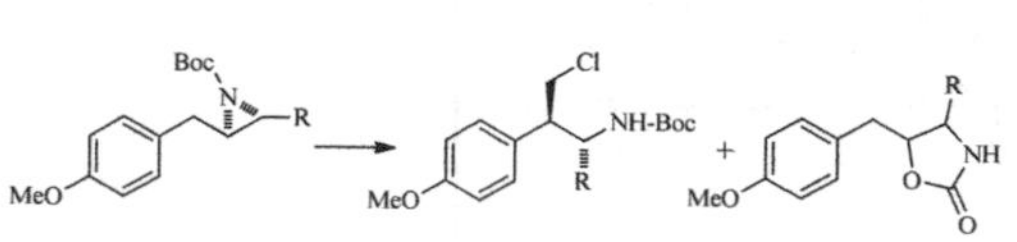

48

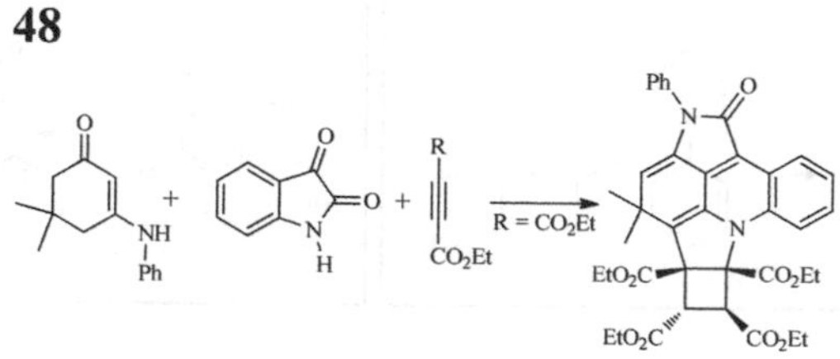

49

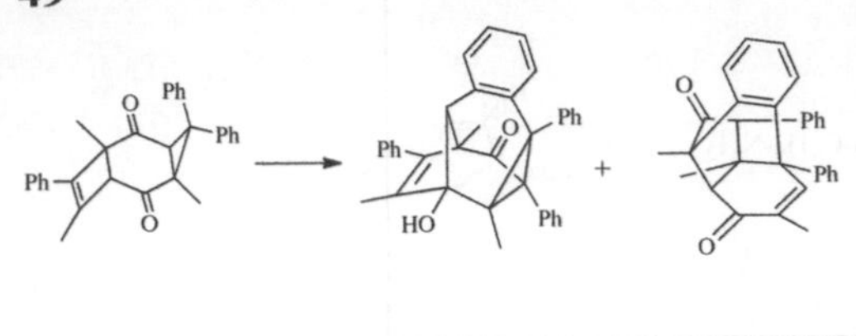

50

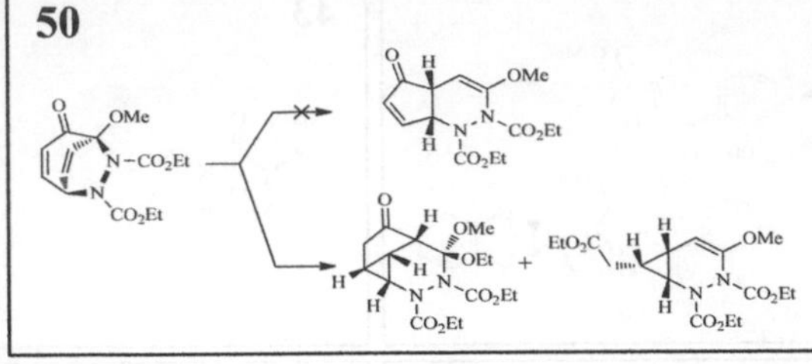

PROBLEMS EMBEDDED IN SECTIONS OTHER THAN THE PROBLEMS CHEST

Preface

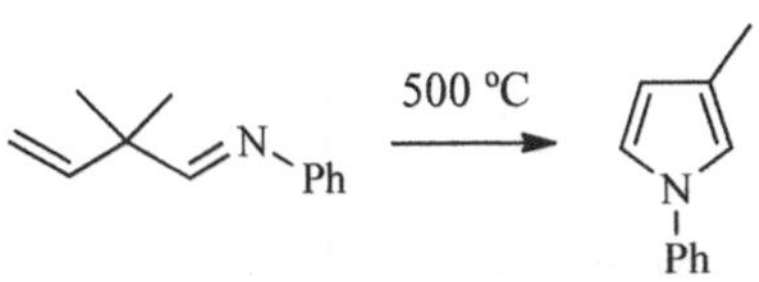

Overview: Scheme 0.7

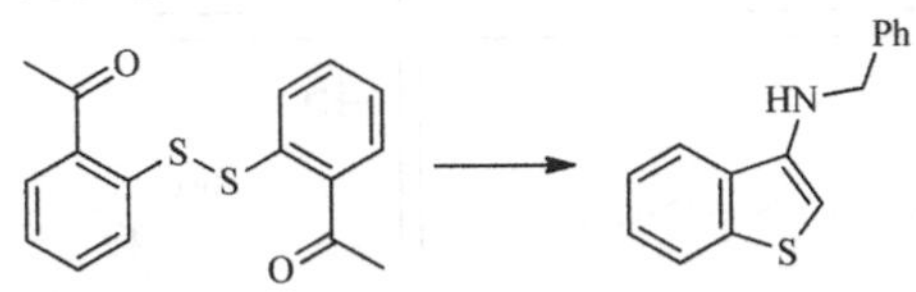

Overview: Scheme 0.1

Chapter 1: Scheme I.1

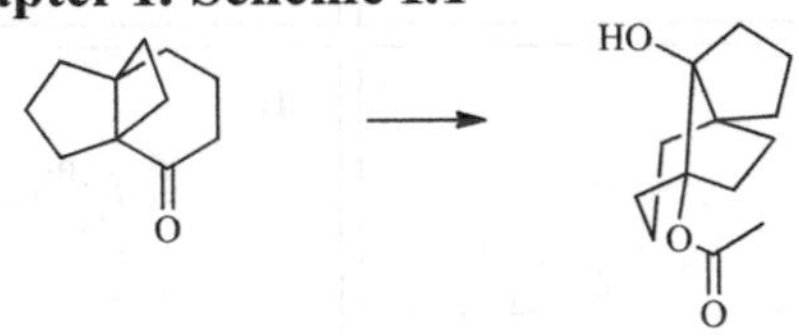

Overview: Scheme 0.4

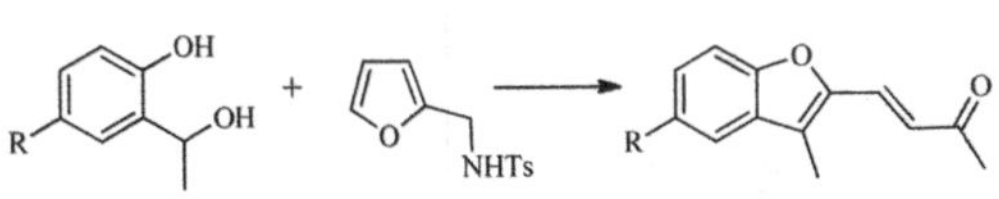

Chapter 1: Scheme I.3

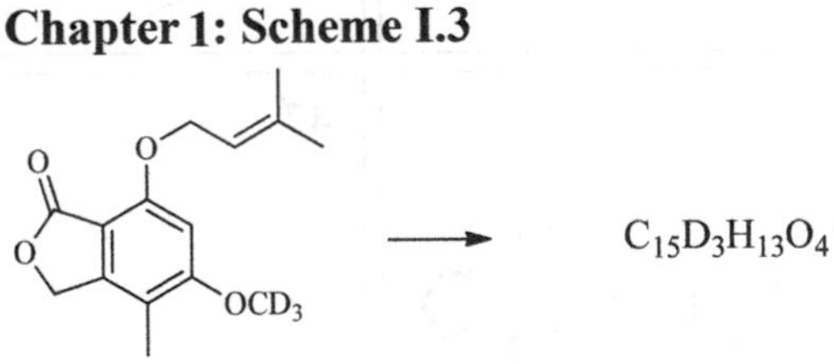

Overview: Scheme 0.6

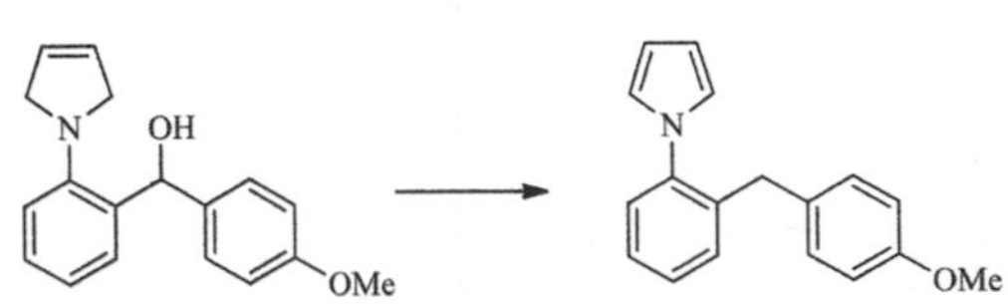

Chapter 1: Scheme I.5

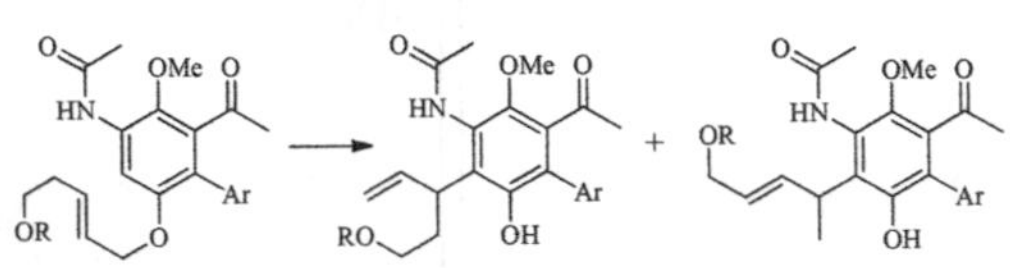

Chapter 2: Scheme II.13-A

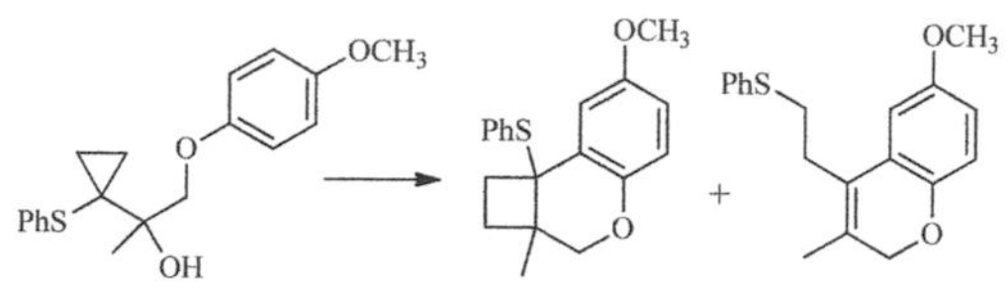

Chapter 2: Scheme II.13-B

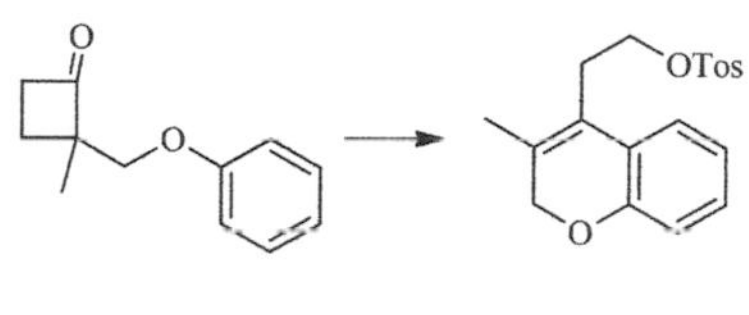

Chapter 2: Scheme II.13-C

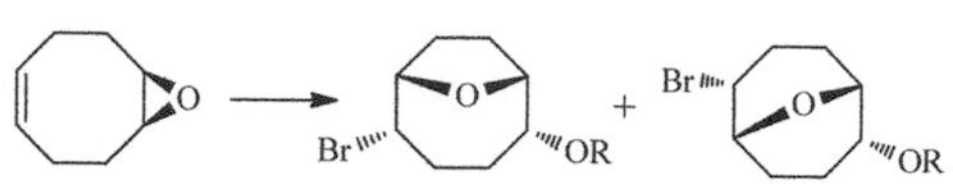

Chapter 2: Scheme II.13-D

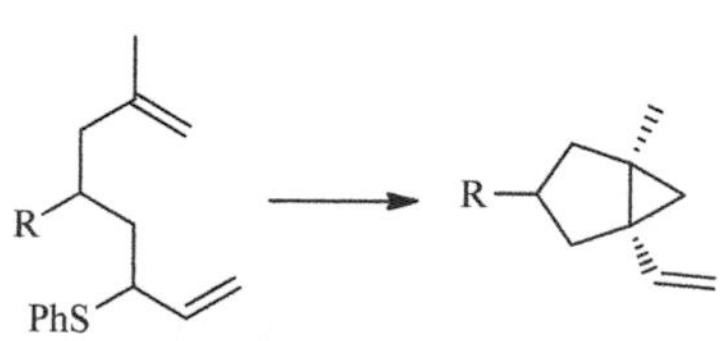

Chapter 2: Scheme II.18

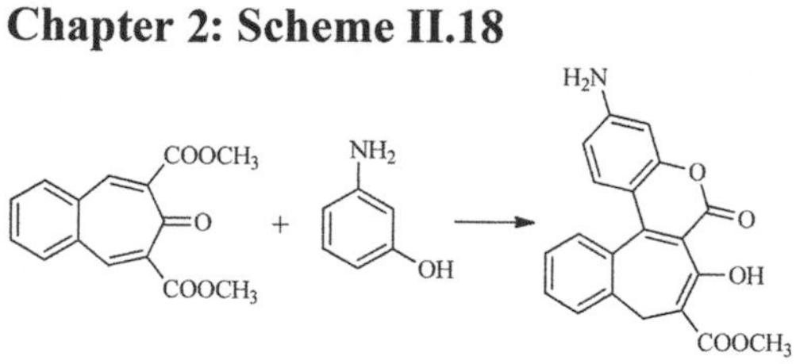

Chapter 2: Figure II.9, Section C

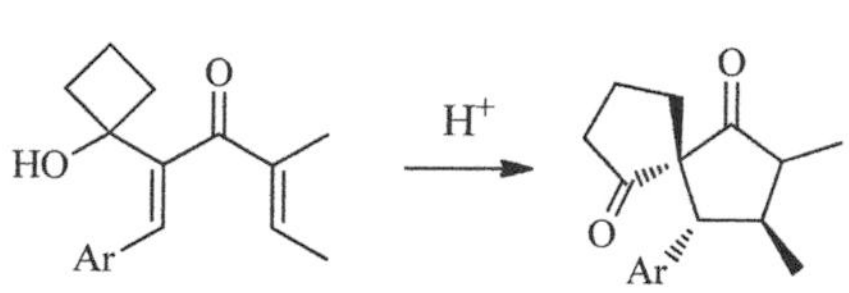

Chapter 2: Scheme II.24

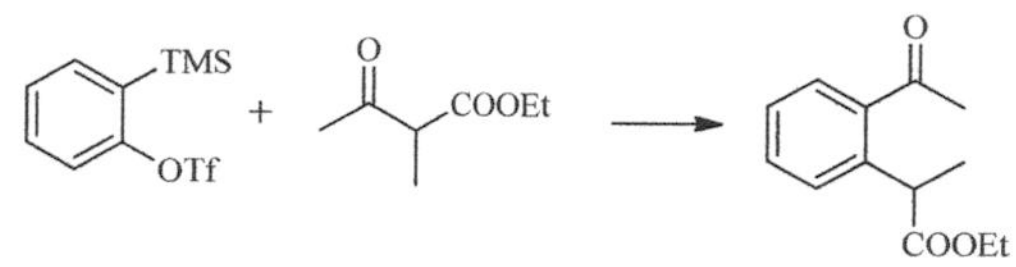

Chapter 2: Scheme II.31, Section A

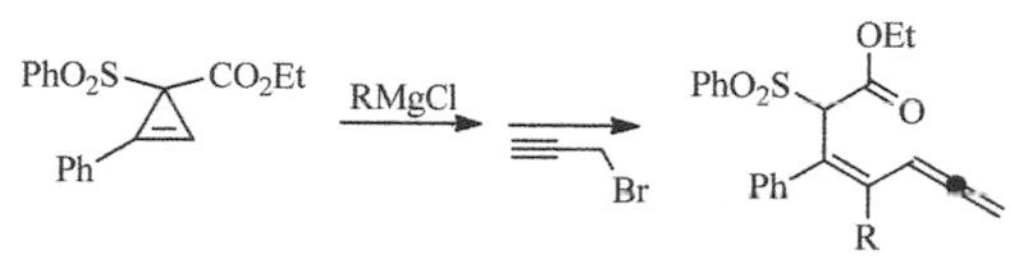

Chapter 3: Scheme III.11

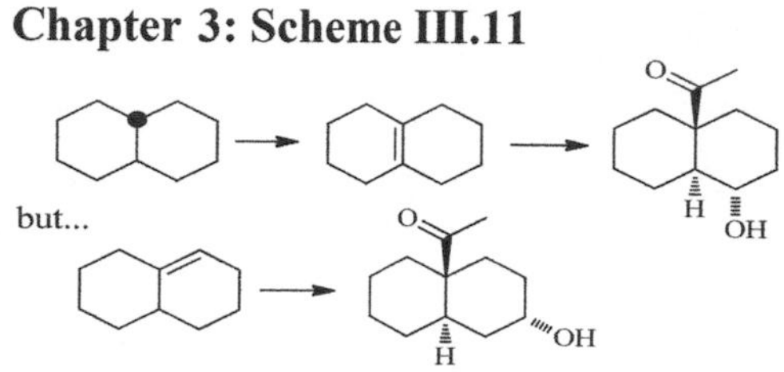

Chapter 4: Scheme IV.9

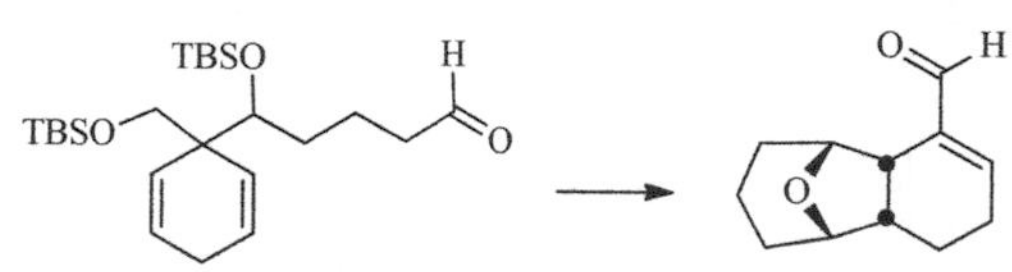

Chapter 4: Scheme IV.12

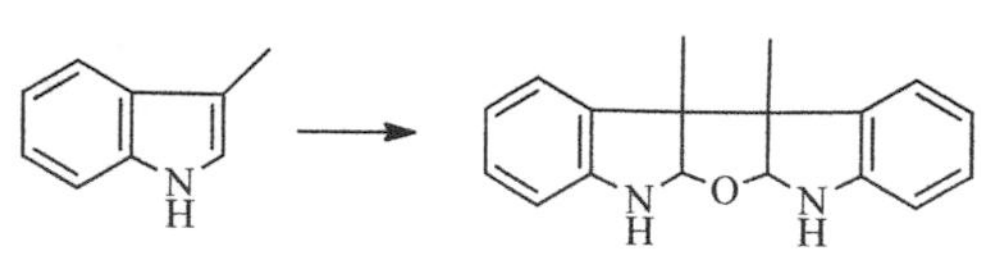

Chapter 4: Scheme IV.13

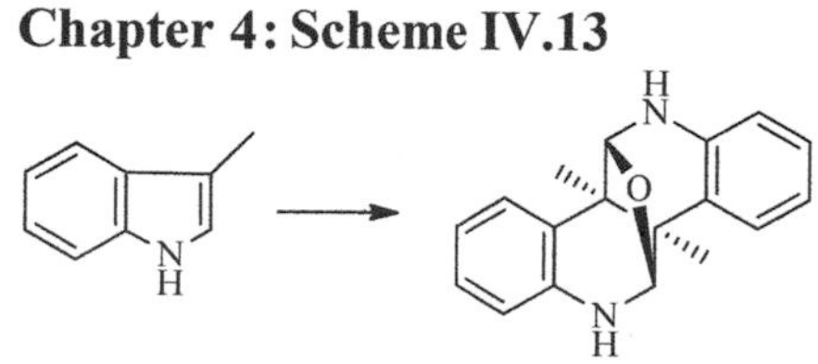